50th American Association of Swine Veterinarians Annual Meeting 2019

Built to Last: Celebrating 50 Years of Progress

Orlando, Florida, USA
9-12 March 2019

ISBN: 978-1-5108-8324-6

Printed from e-media with permission by:

Curran Associates, Inc.
57 Morehouse Lane
Red Hook, NY 12571

Some format issues inherent in the e-media version may also appear in this print version.

Printed by Curran Associates, Inc. (2019)

American Association of Swine Veterinarians
830 26th Street
Perry, IA 50220-2328
USA

Phone: 515-465-5255
Fax: 515-465-3832

aasv@aasv.org

Additional copies of this publication are available from:

Curran Associates, Inc.
57 Morehouse Lane
Red Hook, NY 12571 USA
Phone: 845-758-0400
Fax: 845-758-2633
Email: curran@proceedings.com
Web: www.proceedings.com

BUILT TO LAST

Celebrating 50 Years of Progress

AASV

American Association of Swine Veterinarians

50th Annual Meeting Proceedings

March 9-12, 2019

Lake Buena Vista, Florida

50 ANNIVERSARY

Table of Contents

Howard Dunne Memorial Lecture

Alex Hogg Memorial Lecture

Sunday: Research Topics

Sunday Concurrent Session #1: Student Seminar

Sunday Concurrent Session #2: Industrial Partners

Sunday Concurrent Session #3: Industrial Partners

Sunday Concurrent Session #4: Industrial Partners

Poster Session: Veterinary Students

Poster Session: Research Topics

Poster Session: Industrial Partners

Monday General Session: Built to Last: Celebrating 50 Years of Progress

Monday Concurrent Session #1: Disease Control and Elimination

Monday Concurrent Session #2: Practical Vaccinology and Immunology for the Swine Veterinarian

Monday Concurrent Session #3: Production Innovations

Tuesday General Session: Transboundary Disease Threats and Outbreak Preparedness

Howard Dunne Memorial Lecture

Built to last: 50 years of AASV

John T. Waddell, DVM, MBA

In remembrance and honor of
Howard W. Dunne

Dr. Howard W. Dunne received his veterinary degree from Iowa State University in 1941. He was in engaged in private practice for one year and industry for four years before entering graduate school at Michigan State University. After receiving his PhD in 1951, he spent two years as deputy chief of veterinary microbiology at the Chemical Corps Biological Laboratories in Camp Detrick, MD. In 1953, he joined the faculty of Pennsylvania State University, where he headed the veterinary science research program and the large animal diagnostic laboratory. Dr. Dunne was internationally known for his work on swine diseases, particularly hog cholera and the swine reproductive viruses then referred to as the SMEDI syndrome. Dr. Dunne authored more than 90 scientific publications, and is perhaps best known for initiating and editing the textbook *Diseases of Swine*, which is still widely considered the "bible" of swine disease.

In 1971, the American Veterinary Medical Association presented Dr. Dunne with the XII International Veterinary Congress Prize to recognize his contributions to the international understanding of veterinary medicine. At the 1974 Congress of the International Pig Veterinary Society (IPVS), Dr. Dunne was elected president of the 1976 IPVS Congress, to be held in Ames, Iowa. His sudden death at the age of 61 in September of 1974 saddened the swine veterinary community, and the 1976 IPVS Congress was dedicated to his memory.

At the suggestion of Dr. Allen Leman, a financial gift made by the 1976 IPVS Congress organizers to the fledgling American Association of Swine Practitioners was combined with additional association funds to endow this prestigious annual lecture in remembrance and honor of Dr. Howard W. Dunne.

HOWARD DUNNE MEMORIAL LECTURERS

1978	Michael Muirhead	1989	Steven Henry	1999	Larry Rueff	2009	John U. Thomson
1979	W. J. (Bill) Smith	1990	Stanley Curtis	2000	David Reeves	2010	James F. Lowe
1980	N.C. Nielsen	1991	Roy Schultz	2001	Max Rodibaugh	2011	Locke A. Karriker
1981	Allen Leman	1992	Joseph Connor	2002	Tim Blackwell	2012	R. Dean Boyd
1982	Michael Wilson	1993	Camille Moore	2003	Lonnie King	2013	Craig Rowles
1983	Edward Bohl	1994	Robert Morrison	2004	Robert Desrosiers	2014	Daryl Olsen
1984	Richard Ross	1995	Rodney Johnson	2005	Scott Dee	2015	Gregory Stevenson
1985	Kent Kislingbury	1996	Paul Yeske	2006	Gregg BeVier	2016	John Harding
1986	Ralph Vinson	1997	Barbara Straw	2007	Tom Burkgren	2017	Jeff Zimmerman
1987	Gary Dial	1998	Mike Daniel	2008	Timothy J. Loula	2018	William R. DuBois
1988	Thomas Alexander						

Built to last: 50 years of AASV

John T. Waddell, DVM, MBA

Introduction

It is truly an honor and privilege to present the Howard Dunne Memorial Lecture at the 50th annual meeting of the AASV. I want to thank Dr. Nathan Winkelman for entrusting me to talk to and about an organization that has made such an impact on my and so many swine veterinarians' lives. It is not merely a professional association that makes the AASV special. It is no doubt the people ... both its members and the myriad of producers they serve, as well as all the allied industries, affiliates, universities, and global partners that created, nurtured, cajoled and built the AASV as it is today and to last the next 50 years! Dr. Winkelman charged me with compiling the history of our association as well as a look into the future. I fully embrace the chance to reminisce and honor our past, but as Yogi Berra once quipped: "Predictions are difficult ... especially about the future!" It is implausible that the following will be "a complete history" of the AASV as I will leave most of the details to those better suited and equipped. Instead, what is before you is light on science and relies heavily on (potentially) faded and perhaps biased (to the good) memories of you, the swine practitioner, and our organization! While disease, events, politics, technology and cultural changes are always constant, the AASV was formed by and for the individual swine practitioner. The best reference material on the history of the AASV is our membership directory. The directory contains nearly everything you will ever need to know about the AASV: who the people who built it, those who run it, and those who will make it last! As a bonus, the directory also has all of their contact information! It is truly veterinarians, AASV's staff, and its members (all of you) who built the AASV to last!

A brief biography and history of the Howard Dunne Memorial Lecture precedes this proceedings paper. It includes a list of the individuals who have presented the Dunne Lecture and I urge everyone wishing to be inspired to read or re-read some of the great Dunne Lectures. It is truly a humbling experience to be following such an esteemed group and it is difficult to feel worthy of being placed in their midst. These are but a few of those who built AASV to last!

The Dunne Memorial Lecture was initiated at the 1978 annual meeting of the American Association of Swine Practitioners in honor of the late Dr. Howard Dunne. Throughout the years of the Dunne Lecture series, there have been a wide array of topics addressed. Some of the lecturers have approached this address purely from the hard science standpoint while others of our colleagues have tackled the challenge of looking into the future. Dr. Allan Leman presented the Dunne in 1981 (my first AASV) when his title was: "Preparing for swine practice in the 1980's."[1] Dr. Leman ended with his now famous challenge which impacted me profoundly and continues to be relevant today: *"Would you hire you?"* At the time, I was a senior in veterinary college at the University of Illinois where Dr. Leman had matriculated just a few years prior. Dr. Leman had been my fraternity advisor at the beginning of my undergraduate studies at Illinois in 1973, providing mentoring, guidance and leadership to a group of Illinois farm kids much like himself. While Al's charismatic personality was evident (even to a very naive college freshman), little did I know at the time what an impact he would make on our profession, discipline and industry as well as how many future swine veterinarians he would mentor or how many lives and careers he would influence.

Dr. Michael Muirhead of the United Kingdom gave the inaugural Dunne Lecture in 1978 on the "Effects of the Swann report on veterinary practice and the swine industry."[2] Coincidently, the first Swann report on the use of Antibiotics in Animal Husbandry and Veterinary Medicine was issued in 1969. One of its recommendations was that the only antimicrobials that should be permitted as growth promoters in animals were those that were not depended on for therapy in humans or whose use was not likely to lead to resistance to antimicrobials that were important for treating humans. Either Dr. Muirhead or the program planners for the 1978 annual meeting were clairvoyant! Who would have thought that fifty years later, antimicrobial resistance related to use in livestock production would remain an unsettled issue as our organization continues to provide knowledge and leadership surrounding this important issue.

The Dunne Memorial Lecture has played an important part in the history of AASV and has served well as a bellwether of changes occurring in our association, profession and the swine industry. The presenters and their assigned topics serve as the foundation for tracking and documenting what were considered important or pressing issues at the time of each annual meeting. While the Dunne Memorial Lecture delineated the most important or pressing topics of the times throughout the history of last half century, there were also many important issues addressed within the program of the AASV annual meeting, and both substitute for a written history of AASV.

In the beginning

1969 was a monumental year. One of mankind's crowning achievements occurred in July 1969 when Neil Armstrong became the first human to step onto the surface of the moon after Apollo XI successfully landed in a rocky crater within the Sea of Tranquility. The New York Mets won the 1969 World Series after never having a winning record in prior seasons or finishing higher than 9th place since the team was founded in 1962. The Beatles released their last album together (Abbey Road) and gave their final public performance in London in 1969. An estimated

500,000 gathered on a farm in upstate New York near the village of Woodstock for a rock concert. The Viet Nam War was featured nightly on the network news and anti-war protests were prevalent on college campuses across the nation. The Boeing 747 jumbo jet made its debut and the Concorde supersonic passenger jet made its maiden flight in 1969. The first message was sent over a computer network called ARPANET which was the precursor to the Internet. Richard Nixon was sworn in as President of the United States. Sam Walton incorporated his discount retail business and called it Wal-Mart. Sesame Street, the children's television program, saw its debut on the newly formed Public Broadcasting System. Also of note in 1969, the first women were admitted to membership of the Future Farmers of America.[3] Hog prices received by producers averaged $22.97 per CWT in 1969.[4]

On July 14th, 1969, thirty veterinarians met during the American Veterinary Medical Association (AVMA) Convention in Minneapolis to form a national organization for veterinarians interested in swine practice.[5] Dr. John Herrick was the current president-elect of the AVMA, and having seen both the bovine and equine specialties recently form associations, saw a similar need for those veterinarians interested in and serving the swine industry. Dr. Herrick, along with AVMA staff, drafted a constitution and by-laws for the American Association of Swine Practitioners to make the group an affiliate organization of the AVMA. At the time, to be officially recognized by the AVMA and be afforded membership in the House of Delegates, the requirement for affiliate organizations was to have 200 or more members. The AVMA recognized the synergies that existed for affiliate groups and were fully supportive since membership in the affiliate organizations required membership in the AVMA. The AVMA provided all the necessary administrative assistance and the American Association of Swine Practitioners (AASP) was incorporated in the home state of the AVMA as an Illinois non-profit organization. Officers were named from among those attending that meeting:

- President: Dr. Vaylord Ladwig
- Secretary-Treasurer: Dr. James Bailey
- Board of Directors: Drs. George Bergman, Lewis Runnels, LeRoy Gallantine, Harold Hedlund, David Bechtol and Tom Keefe

Among those present at the inaugural meeting was a young swine practitioner from Avoca, IA, Dr. Roy Schultz, as well as Dr. Richard "Dick" Taylor from Fayette, MO, both of whom would distinguish themselves as renowned swine practitioners. Also of note, Dr. David Bechtol of Canyon, TX attended and was elected to the initial Board of Directors. Interestingly, Dr. Bechtol went on to become President of AASP in 1973. Dr. Bechtol later became known best for his expertise in cattle feedlot consulting in the high plains. Interning with him in 1980, I noted a plaque on his office wall recognizing him as a past president of AASP. When I inquired about how a cattle vet from northwest Texas came to lead the AASP, he told me that he had one large client with pigs so he purchased Dr. Dunne's text book *Diseases of Swine*. This led him to join the AASP. I believe Dr. Bechtol to be the only veterinarian to be president of not only the AASP but also the American Association of Bovine Practitioners (AABP) as well as the Academy of Veterinary Consultants (AAVC).

District representation was not to be implemented until 1972 so the initial officers and board members were all chosen at large from those interested veterinarians that had also attended the AVMA Convention.

Discussion at the initial meeting led to the decision to hold the initial annual meeting with the Iowa Veterinary Medical Association in Des Moines in 1970, but the executive committee met in Sioux City, Iowa in the fall and those plans changed. Prior to the fall "board meeting," the AASP had received an invitation from the American Pork Congress to meet in conjunction with the 1970 Pork Congress and the first annual meeting of the National Pork Producers Council in March of 1970. Seizing on the obvious potential synergies and mutual interests, the AASP officers realized there would be wider exposure and the opportunity to grow their membership and be part of a trade show devoted exclusively to swine. The American Pork Congress rotated between Des Moines, Kansas City and Indianapolis, which geographically provided greater opportunity for swine practitioners from across the Midwest. On March 5th, 1970 Vaylord Ladwig conducted the organizational meeting in Des Moines and 76 attendees signed in. The AASP was off and running with approximately 100 paid members!

The first scientific meeting of AASP was held in Las Vegas in June of 1970 in conjunction with (and included with the $15 registration for) the annual AVMA Convention. This "first annual" meeting was attended by over 100 members as membership swelled to about 200. Dr. Thomas Keefe was the program chair and helped organize the day-long program which included a "Swine Enteritis Symposium." The program included topics on *E. coli* by Dr. E. M. Kohler; transmissible gastroenteritis (TGE) by Dr. E. H. Bohl; *Clostridium perfringens* type C by Dr. M. E. Bergeland; and swine dysentery by a young graduate student at Iowa State, Dr. D. L. "Hank" Harris. Incidentally, in the letter inviting Dr. Harris to speak on "Vibrionic Dysentery," Dr. Keefe noted "we do not have sufficient funds to pay your expenses...I understand though the University will take care of your expenses."

In addition to the Swine Enteritis Symposium, Dr. Keefe also organized an afternoon session for AVMA with topics and speakers attuned to modern swine production:

- *Health of the Pig Reared in Confinement* – John R. Andersen, Ames, IA
- *Nutrition of the Baby Pig* – Ray D. Washam, St. Louis, MO
- *Mycoplasmal Arthritis of Swine* – Richard Ross, Ames, IA and J. R. Duncan, Athens, GA
- *Status of the MMA Complex* – Charles Martin, Columbia, MO
- *Pregnancy Diagnosis of Swine Utilizing the Vaginal Biopsy Technique* – Edward Mather, et. al., Columbia, MO

- *Swine Management and Herd Health Program* – Joseph Hayden, Liberty, IN

This first meeting set the tone and high standards for future scientific meetings.

The AASP continued meeting in conjunction with the American Pork Congress (later changed to the National Pork Congress) for several years. AASP held its annual meeting starting on the day prior to National Pork Congress and overlapped on the first day of the congress. Membership in both organizations grew rapidly and by 1980 the two organizations mutually agreed to split as both membership rolls had reached a critical mass to support separate meetings.

The first decade

AASP growth was brisk the first year and the 1970 membership roll boasted well over the 200 members required by the AVMA for affiliate membership.[6] This news was not received well by all potential new members as some of the initial roster were not AVMA members. Some practitioners refused to join the AVMA and thus elected to not become AASP members. Another good example of "the more things change, the more they stay the same!" Also in 1970, Paul Gorham of ELANCO met with the board to express interest in providing financial assistance to AASP which was the beginning of many corporate sponsorships to come.

Dr. Ken Meyer, now an extension veterinarian at Purdue University was elected Secretary-Treasurer in 1971. Dr. Meyer utilized the resources and staff of the university and extension service to provide valuable information such as the Swine Day Reports from several universities, Midwest Plan Service booklets, swine nutrition and university extension research reports. This provided a large cost savings advantage for AASP as these mailings were sent utilizing the extension services franking privileges for sending educational materials through the US Postal Service. Several extension veterinarians in the swine-producing states were helpful in identifying materials that were helpful to the swine practitioners. Members appreciated the access to current information and word soon spread about the value of the AASP. The AASP Newsletter was launched in 1971 and Al Leman was named the first editor.

In 1972, membership swelled to 475 including many international members. The board committed to sending all materials to international members (even if it cost more than the $15 annual membership dues) and they granted $300 to Al Leman to aid in his attendance at the 1972 IPVS in Hanover, Germany as the AASP representative. Dr. Leman evidently liked what he saw in Germany and reported to the AASP board his interest in bidding on hosting a future IPVS.

The board raised AASP membership dues to $25 in 1973 (effective January 1, 1974) and set the first student membership dues at $5. The board also approved $500 to send Howard Dunne to the 1974 IPVS where he was charged with extending a bid for the US to host the 1976 IPVS in Ames, Iowa (only if enough financial commitment could be secured before the 1974 IPVS). Sadly, Dr. Dunne passed away suddenly in September of 1974 but not before leaving a legacy and planting the seed for the US to host the 1976 IPVS meeting.

The IPVS bid was accepted and Drs. Bob Glock, Norm Hutton and Alex Hogg were appointed to the IPVS Financial Committee to oversee and approve all IPVS expenditures. Earline Hogg, who had been doing hourly secretarial work for AASP, was named the first (part time) Executive Secretary to manage membership files, mailing lists, meeting registrations, collect dues, handle correspondence and provide financial statements for the officers. Late in 1974, over 11,000 recruitment letters were sent to any AVMA members whose practice specialty indicated potential to be AASP members. This recruitment letter listed off the following benefits of AASP membership:

- Co-sponsorship of the 1976 IPVS in Ames
- AASP newsletter
- National Hog Farmer and Hog Farm Management magazine subscriptions
- Other educational materials including Proceedings of the Specific Pathogen Free (SPF) Annual Conference
- Annual scientific meeting held in conjunction with the American Pork Congress.

Al Leman served as AASP president in 1975. Student membership dues were raised to $10 per year and $9000 financial support was pledged for the 1976 IPVS to provide a cushion should it be needed. It was stipulated that those monies were to be refunded to AASP should the IPVS generate a profit. Al Leman began to appoint student representatives for AASP at each college of veterinary medicine. The student representative had their dues waived in exchange for distributing AASP materials, recruiting new members and collecting dues.

Later in 1975 the board, recognizing the need, formed various committees to administer the growing needs of the association. Membership continued to grow and along with the demands of co-sponsoring the IPVS in Ames, the decision was made to hire a full time Executive Secretary. Fred Wertman, the Iowa Veterinary Medical Association Executive Director was hired as the AASP Executive Secretary starting January 1, 1976.

Dr. Bob Glock became president in 1976. The IPVS in Ames was successful and excess proceeds were granted to the AASP. At the suggestion of Dr. Al Leman, the IPVS gift was combined with other funds to establish the Howard Dunne Memorial Lecture, perhaps an early indication that AASP leadership was committed to securing the future while honoring the past.

In 1977, AASP began dipping its toes in the regulatory waters as representatives met with Dr. C. D. Van Houweling, the head of what became the FDA's Center for Veterinary Medicine (FDA-CVM), to discuss proposed rules regulating the use of tetracyclines and penicillin in feed. Concerns regarding increased regulatory oversight from the FDA-CVM continued into 1978 when Dr. Van Houweling announced his resignation. AASP leadership appreciated that Dr. Van Houweling, being a veterinarian, understood veterinary medicine and feared the position might be filled with a non-veterinarian.

Two AASP awards were inaugurated in 1977 to recognize outstanding service: the Howard Dunne Memorial Award and the Swine Practitioner of the Year. Dr. Arthur Starkey, a practitioner and Specific Pathogen Free (SPF) pioneer from Illinois, was the Dunne Memorial Awardee and Dr. Kent Kislingbury of Fairmont, Minnesota was named the first AASP Swine Practitioner of the Year.

Classical swine fever (hog cholera) was officially eradicated from the US in September 1978, an effort that began in the early 1960's and cost about $140 million. Over 800,000 pigs from about 5700 herds had been destroyed in the successful eradication effort.

Canadian veterinarians were offered reduced membership dues in AASP until 1979 when they were incorporated into existing districts and full membership.

Dr. Jim McKean, the ISU Swine Extension Veterinarian, was named "Continuing Education Director," a job that was formerly handled by the treasurer. Jim had already been managing these duties and began instituting some of the policies that remain in place such as "no proceedings paper...no stipend!"

Drs. Ralph Vinson and Alex Hogg served as AASP president in 1978 and 1979 respectively, closing out the first decade with solid leadership of our growing association that then numbered over 1200 members!

The 80's

In the second decade of AASP, change and improvements continued along with solid membership growth. In 1980, AASP leadership and its members, working with the Livestock Conservation Institute (LCI), began to formulate, design and influence the approaching pseudorabies control and eradication program and in May of 1981, after much debate, the Board supported "immediate implementation of a pseudorabies eradication program."

The 1982 Annual Meeting was the first to offer an "industrial partners session" and a new "mini-topic" session where practitioners were encouraged to share brief "practice tips, tricks and short-cuts" in 10- to 15-minute presentations.

Dr. Michael Wilson of Canada presented the 1982 Dunne Lecture, simply entitled "Pigmanship" in a response to Dr. Leman's 1981 Dunne.[5] Dr. Wilson defined swine herd health as "That level of production which meets a producer's objective and is consistent with humane practices."

At the time, producers in the United Kingdom appeared to be far ahead of the rest of the world in swine production. New-sounding metrics such as "pigs per sow per year" (PSY) were being gleaned from detailed production records, and producers were making great strides with improved genetics and management techniques. US and Canadian producers and veterinarians were sojourning to Britain to learn how they too could reach 20 PSY. Dr. Wilson evidently felt compelled to address this in his 1982 Dunne Lecture on "Pigmanship": "I have come to the conclusion that the much-vaunted production of British pigs has little or nothing to do with breeding stock, nutrition, disease control, building design, animal science or veterinary medicine. It comes from centuries of tradition in pig rearing and the empathy that comes from a continuous line of role models."

The AASP bylaws were amended in 1984 to form District 11 to include members in provinces of Canada giving Canadian practitioners representation on the Board of Directors.

IBM launched its first personal computer, the "PC," in late 1981. While Apple and Commodore, among others, were ahead of IBM, IBM used the operating system called DOS which led to an explosion of software. PigCHAMP was first developed in the early 1980's at the University of Minnesota College of Veterinary Medicine, in large part by one of Al Leman's graduate students, Dr. Tom Stein. The original intent of the product was to serve as a platform for data collection for research purposes, but the swine group at University of Minnesota quickly recognized the potential to create a computerized production records system that could lead to benchmarking. Other swine practitioners, Drs. Greg BeVier and Ralph Vinson, had already began developing computerized record systems on other platforms, but PigCHAMP was offered as a beta test version to veterinarians to try in a wide variety of production settings. While PigCHAMP operated as a mail-in service at first, the version that would run on a PC soon followed and the industry had made the leap from hand-kept records to computer-based. As PigCHAMP bureaus sprang up across the swine industry, critical mass was reached, and the power of benchmarking was unleashed.

Dr. Fred Wertman, who served as the Iowa Veterinary Medical Association (IVMA) Executive Director and the AASP Executive Secretary since 1976, retired in 1984. Dr. Tom Neuzil was named as Dr. Wertman's replacement and assumed the role in 1985. Dr. Wertman had served the IVMA and AASP for eight years. It is also important to note that Dr. Wertman's wife, Jo, served as his secretary/administrative assistant for his entire term.

In the mid 80's, Dr. Steve Henry proposed and spearheaded founding of the Food Animal Unity Group (FAUG) with representation from the AASP, American Association of Bovine Practitioners (AABP), American Association of Small Ruminant Practitioners (AASRP) and the Academy of Veterinary Consultants (AVC) to address current issues common to food animal practitioners and the industries they serve. Issues surrounding FDA-CVM regulations, extra-label drug use (ELDU), bulk drugs for compounding, health certificates, AVMA communications, public relations, accreditations, and pharmaceutical distribution methods were among the first to be tackled by FAUG. Spurred by FAUG, the AVMA hosted a meeting of Food Animal Veterinarians Organized for Results (FAVOR) to offer a platform to discuss the issues concerning the food animal sector of the AVMA.

The AASP by-laws were revised in 1986 to allow non-veterinarians to apply for Affiliate Membership. Interestingly, that same year, the suggestion was made to hold the 1989 annual meeting in Orlando, Florida in conjunction with the Eastern States Veterinary Conference. The program planning committee favored staying in the Midwest, voting instead to hold the 1989 annual meeting in Des Moines!

That same year, Dr. Jim Sawatsky was the first to suggest establishing a foundation for AASP. The Executive Committee was charged with exploring the idea. Two years later (1988) the board approved establishing a charitable foundation to promote and develop swine veterinary medicine. $10,000 in start-up funding was appropriated and Drs. Chuck Christensen, Gary Dial, Tim Loula and Tom Neuzil were appointed as conveners of the AASP Foundation.

Late in 1986, the AASP halted the distribution of the pamphlet "Swine Dysentery: Practitioner's Planning Guide for Herd Elimination Programs" due to FDA-CVM's policy on dimetridazole. The CVM announced the plan to withdraw the compound from the market. It was the first of several compounds that would be banned for use in food animals and a prelude to a stricter regulatory environment on veterinary pharmaceuticals.

In 1988 a committee was formed to establish the Board Certification process for swine practitioners. Drs. Wally Brandt, Al Leman, Roy Schultz, Mike Wilson and Ken Meyer were charged to begin formal exploration of board certification for swine veterinarians in what eventually became the American Board of Veterinary Practitioners (ABVP) board certification in Swine Health Management. Drs. Steve Henry, Roy Schultz, Ken Meyer, Al Scheidt and others worked diligently with the AVMA to gain approval for a swine specialty certification under the umbrella of ABVP, which the AVMA had approved in 1978.

In March 1987, AASP President Dr. Bernie Curran recommended discussion on developing a trade show at the annual meeting and in 1988, the first "Tech Tables" were introduced on a limited basis. The committee specified that the tables were to be attended by technical service veterinarians, not sales representatives! The Membership Services Committee led by Dr. Gary Dial, also proposed development of an AASP newsletter, a membership survey, an audio-visual library and the first AASP membership directory with sponsorship of A.L. Labs.

Dr. Wally Brandt, who served as the third AASP president, was tasked with chairing the Historical Committee and presenting on the first 20 years of AASP at the 1989 annual meeting. Dr. David Reeves chaired the development of the first practice tip session at the 1989 annual meeting.

The 20th Anniversary meeting in Des Moines in March of 1989 pre-registered over 500...a new record! President Joe Connor presided over the meeting and handed the gavel to Dr. Kirk Clark for 1989. Veterinary Learning Systems was selected to publish the AASP Newsletter for one year (six issues). Advertising was sold on a one-company-per-issue rate of $7500. Little did they know that this was the first step toward the development of the *Journal of Swine Health and Production*. The AASP Nutrition Committee began to publish reference materials in the proceedings to provide swine practitioners with basic rules of thumb and nutrition information via handy charts and tables, and a "pre-conference" Sunday morning Nutrition Workshop was being planned for the 1990 annual meeting.

The first AASP Newsletter was published as the July-August 1989 issue. Syntrovet was the first sponsor and they featured the PRV/Marker MLV Pseudorabies vaccine in their ad in that issue. The AASP had just formed the Pseudorabies Committee to coordinate information on the national eradication effort.

Students were becoming a focal point for AASP as leadership recognized the importance of cultivating new generations of swine practitioners. The AASV Student Recruitment Committee was formed and immediately set out to identify a faculty advisor at each college, publish a list of benefits, establish student chapters and sponsor a student breakfast at the annual meeting.

A devastating new syndrome emerged in the late 80's and was recognized first in Europe and North America. Our European colleagues first referred to it as "Blue Ear" due to the associated hypoxia of the respiratory symptoms. A severe reproductive form of the syndrome was being described by practitioners in the US but for lack of a definite etiology, the name "Mystery Pig Disease" was coined. The disease syndrome was first recognized in the United States in North Carolina in 1987. An agent was not yet identified and there seemed to be a lull in cases in 1988. But it came back with a vengeance in 1989. After multiple iterations, the veterinary community settled on the name porcine reproductive and respiratory syndrome (PRRS). The virus responsible was nearly simultaneously isolated in the US and the Netherlands in 1992. Testing of stored serum banks indicated some pig herds had likely been infected with the virus as early as 1985. PRRS virus profoundly changed the way pigs were raised.

The Roaring 90's

The 90's was the decade of industry consolidation, PRRS, and AASP change and growth. Terms such as multi-site production, isowean, biosecurity, antimicrobial resistance, PMWS and shackle space made the way into the lexicon of the industry. In his 1986 Dunne Lecture, Dr. Ralph Vinson asked the question, "The 1990's, the last decade of private practice?" Ralph speculated that due to consolidation, the US swine industry might need only 300 veterinarians to service its needs. In 1994, Successful Farming magazine's Betsy Freese compiled the first "Pork Powerhouse" list of the nation's largest producers. That first list reported the largest 25 producers controlled 1,069,123 sows, with the five largest production companies tallying 559,000 sows. Rapid growth and consolidation of our industry were underway. Late in 1998, the number of market hogs exceeded the slaughter capacity and live prices fell into single digits and profitability collapsed. Market forces drove further consolidation and by the end of the 1999, the top five companies controlled 1,213,000 sows and the top 25 producers were responsible for over 2.2 million sows. It was a clear signal that further consolidation and vertical integration was on the way and Ralph's prediction was perhaps closer than most could have imagined.

At the 1990 annual meeting, Mexico was welcomed into the AASP as District 12. Dr. Laura Batista began to organize a membership drive and the election of the first District 12 director. Sulfa drugs were receiving scrutiny and stricter autogenous vaccine regulations were areas of concern. The Pork Quality Assurance Phase III was rolled out and AASP mailed a copy to each member as the National Pork Producer Council had requested AASP participation in the program.

In 1990, Dr. Leroy Biehl, the Swine Extension Veterinarian at the University of Illinois came to the AASP asking for support as he created the Executive Veterinary Program (EVP) for swine veterinarians and enrolled the first class in 1991. Since that time, over 200 swine veterinarians have participated in this in-depth, two-year program to bolster their skills as practitioners and consultants. Dr. Larry Firkins of the University of Illinois assumed the reins of EVP when Dr. Biehl retired.

In 1991, David Reeves coordinated the publication of a book on the care and management of pot-bellied pigs. The proceeds from the sale of this publication were donated to the AASP Foundation.

Eighteen AASP members sat for the initial ABVP Swine Health Management board certification entrance exam in 1991. In 1993, the American Board of Veterinary Specialties approved the Swine Health Management specialty certification process and 13 AASV members were recognized as Diplomates of ABVP in Swine Health Management.

The AASP began to embrace pre-conference meetings in 1992. The Swine Nutrition committee organized a seminar on "Lean Gain." A day-long "Depopulation-Repopulation" pre-conference session was also held that year, with attendance limited to 300. Drs. Al Leman, Steve Henry, Jack Anderson and Walter Heuser were the featured presenters. Dr. Leman referred to repopulation as "the best kept secret in swine production." Proceedings for this seminar were published and made available to all members. In conjunction with Iowa State University, the AASP Foundation opted to produce a video of the seminar and then produced a satellite videoconference for producers and practitioners. It was "high tech" in 1992!

Dr. Al Scheidt proposed a "Student Seminar" for the 1993 AASV Annual Meeting in Kansas City. AASP provided $250 to each of six students selected to make an oral presentation of a scientific paper on Sunday afternoon. While students had been encouraged to attend and present at prior meetings, this move formalized what has become one of the most popular sessions of the conference. Many of us will likely recognize these presenters (if not their topics) from that first Student Seminar:[7]

Sulfamethazine residue detection on-farm – Mark Thompson

Practical bacteriology for the swine veterinarian – Sandy Amass

The use of real-time ultrasonography as a guide for prediction of gestational age in swine – Janet Alsop

Financial evaluation of artificial insemination in commercial swine herds – Martin Mohr

Diarrhea and rectal strictures on a commercial swine farm with enzootic salmonellosis – Luis Fernandez

The pathogenicity of porcine paramyxo-like virus in specific pathogen free pigs – Mary Battrell

The Journal is born!

That same year, the Executive Committee moved forward with evaluation of self-publishing the AASP Newsletter. Dr. Robert Morrison was appointed production editor and his vision soon expanded the newsletter to encompass *Swine Health and Production*. Dr. Morrison proposed establishing a 12-person review board of academic and practicing veterinarians. At the March 1992 Board of Directors meeting, Dr. Morrison proposed (and the board approved) renaming the AASP Newsletter to the *Swine Health and Production* (SHAP). The first issue: **Volume 1, Number 1, January and February 1993** was a success! It contained a variety of original articles all of interest and useful to the practitioner:

Production tools

Adjusting production targets and interference levels: How to help your clients avoid being tricked - William E. Marsh, MS, PhD; and Alvaro J. Soler, MS

Original Research

The survival of Streptococcus suis on farm and veterinary equipment - Scott A. Dee, DVM, MS; and Michelle M. Corey

Preweaning morbidity and mortality in the United States swine herd - Roderick C. Tubbs, DVM, MS; H. Scott Hurd, DVM, PhD; David Dargatz, DVM, MS; and George Hill, MS

Case Report

Streptococcus suis meningitis in finishing pigs of a repopulated herd - Lisa M. Tokach, DVM

Diagnostic Notes

Chromosome damage: A cause of small litters in swine - George R. Ruth, DVM, PhD; Lance C. Buoen; Ting Quing Zhang, DVM, MS and Al F. Weber, DVM, PhD

As the editor, Dr. Morrison made the following plea in the first issue:

"The goal of this publication is to meet your needs. If you have articles, practice tips, case reports or commentary that would benefit the membership, please contact the editor,

Robert Morrison. We will make every effort to assist you in communicating your material. Please consider sharing any new or interesting information with your peers by publishing it in your Journal.[8]"

Dr. Morrison's words solidified the mission of the Journal and laid the groundwork for the future of the *Swine Health and Production*.

Loss of a legend

AASP, our profession, and the swine industry lost one of its greats in 1992. Dr. Al Leman became a legend. Despite his untimely death at the age 48, his impact on the swine industry, our profession, and the AASP cannot be overestimated. Al was born and raised on a farm near Eureka, Illinois and was gripped with a passion for pigs, people and veterinary medicine. He attended undergrad, vet school and garnered his PhD at the University of Illinois, and then accepted a faculty position at the University of Minnesota in 1974. Al served as the AASP president in 1975 and was the first editor of the AASP Newsletter. His charisma and passion were infectious. Al came to Minnesota as the lone faculty member devoted to swine but when he left to join Swine Graphics in 1986, his "Swine Group" had grown to six full-time faculty and multiple graduate students devoted to swine. He impacted untold numbers of veterinarians and producers, first in Illinois and then the across the nation and globally. He also mentored and molded a generation of veterinary students, many of whom went on to become legends in their own right. While at Minnesota, he edited *Diseases of Swine*, initiated the Minnesota Swine Herd Health Programming Conference in 1974, founded the PigLetter newsletter and conducted related PigLetter educational seminars throughout the Midwest for practitioners and their producers. The Allen D. Leman Chair in Swine Health was established at the University of Minnesota after his passing and was funded by all those who Al had influenced and impacted so profoundly around the world.[9]

Enter: Tom Burkgren

Late in 1993, discussion began to take place regarding the Executive Secretary position for the AASP. Additional staff was needed and at the March board meeting in 1994, AASP created the new position of AASP Executive Liaison. On June 1, 1994, Dr. Tom Burkgren filled this new position.[10] While the new position was considered part-time, Tom's skillset quickly became indispensable to the AASP. In late 1996, Dr. Tom Neuzil announced his resignation as the Executive Director of IVMA and the Executive Secretary of the AASP. After a brief search, the AASP Executive Committee (Howard Hill, Larry Rueff, Rick Tubbs, Al Scheidt and Max Rodibaugh), with unanimous support of the Board of Directors, named Dr. Tom Burkgren the new Executive Director of AASP. Dr. Burkgren assumed the position on June 1, 1997, exactly three years after joining AASP as the Executive Liaison. At the same time, the need for a stand-alone office was recognized. Dr. Burkgren lived in Perry, Iowa and offered to move the AASP office to his home until office space could be secured. Shortly thereafter, Dr. Sue Schulteis (also from Perry) was named the Administrative Assistant. Sue's organizational and communication skills elevated the professionalism of the AASP to a new level.

The AASP entered the information age for good in 1997 when `www.aasp.org` went live.[11] Members were issued a logon name and password and could begin accessing the online version of SHAP. David Brown's role as SHAP publications manager expanded to include a new role as AASP webmaster.

The 90's saw a number of milestones including new regulation of the industry. Increased scrutiny of antimicrobial and pharmaceuticals use gave way to the Animal Medicinal Drug Use Clarification Act of 1994 (AMDUCA) that introduced stricter regulations and more oversight responsibilities placed on veterinarians. Terms such as Extra-Label Drug Use (ELDU) and Veterinary Client Patient Relationship (VCPR) became common place among members of AASP as AMDUCA was codified. Prescription (Rx) versus "over the counter" (OTC) was hotly debated and AASP provided strong leadership and input all along the way.

The Veterinary Feed Directive (VFD) was created in 1996, for use in food production veterinary medicine to prevent all feed grade antibiotics from becoming under the control of prescriptions and potentially state pharmacy boards. At first the VFD was created for "new" feed grade antibiotics (specifically tilmicosin). AASP was at the table with other commodity groups and the American Feed Industry Association to hammer out and negotiate a plan that would be workable for all parties while satisfying regulatory needs of FDA-CVM.

Antimicrobial resistance was also an issue coming to the forefront. Multiple AASP members were instrumental in working with FDA-CVM, AVMA, CDC, NPPC, AFIA, AHI (to name a few) and provided leadership in this arena. AASP formed the Pharmaceutical Issues Committee to serve as a clearing house for all issues surrounding these debates. General Judicious Use Guidelines were developed by the AVMA Steering Committee on Antimicrobial Resistance and these guidelines served as a template for development of species-specific guidelines. The AASP was on the field and in the game because Dr. Jim McKean was adamant that "Advocacy is not a spectator sport!"[12]

The National Pseudorabies Eradication Program efforts were in full stride with the goal of elimination by the year 2000. Of course, AASP and its members played a critical role and provided valuable leadership in these efforts.

The AASP held its first "international" meeting in Quebec City, Quebec in 1997. Dr. Larry Rueff was the program chairman. It was very well attended and became the most profitable meeting to that point in AASP history.

The AASP Foundation was growing in strides and bounds as well. In 1996, in honor and memory of Al Leman, the Foundation launched a major capital campaign which allowed AASP members and friends to become donors to the Leman Fellowship. The Foundation established a goal of 300 Leman Fellows by the year 2000.

Bob Morrison handed over the job of Executive Editor of SHAP to the capable hands of Cate Dewey in 1998. Bob dedicated 7 years to SHAP and shepherded it from a newsletter to a respected peer-reviewed scientific journal. Ruth Cronje, the Associate Editor and Dave Brown, the Publications Editor, remained on staff. Dr. Dewey began to usher in more changes for SHAP, including a redesign and other updates in 2000, followed by the name change to *Journal of Swine Health and Production* (*JSHAP*) in 2001.

As the millennium ended, change was accelerating in the AASP and the swine industry. Our industry was experiencing increasing public scrutiny over issues such as environmental impact, antibiotic use, animal well-being, consolidation, integration, and "industrialization."

New millennium/new name

In 2000, by a unanimous vote of the Board of Directors, the American Association of Swine Practitioners (AASP) became the American Association of Swine Veterinarians (AASV). The idea was conceived during the strategic planning process and had been discussed at length.[13]

There were many reasons for changing our association's name. The word "practitioners" was often construed to include only those in private, clinical practice, which excludes many veterinarians in the swine industry. The lines between practice, academia, technical services, research, and production were becoming blurred as the pork industry continued to change and consolidate. The new name acknowledged that we are all veterinarians engaged in the swine industry, regardless of specific discipline.

Another reason for changing the association name was for clarification. The general public did not generally understand the term "practitioner" as applied to veterinarians. The new name was merely easier to convey to the general public. It also was helpful for those working on advocacy and dealing with representatives of regulatory agencies, consumer groups, and elected officials. The less time you spent on describing who you are, the more time you had to make your point.

The name change was not merely a cosmetic modification. It was part of a strategic plan aimed at improving the image and advancing the interests of our members. The mission remained the same: to increase the knowledge of swine veterinarians. This was our core value and it was not altered.

The mission statement was tweaked however, as another outcome of the strategic planning process. The revised mission was made more concise, more proactive and added a focus on mentoring students:

The AASV mission is to increase the knowledge of swine veterinarians by:

- Promoting the development and availability of the resources which enhance the effectiveness of professional activities
- Creating opportunities which inspire personal and professional growth
- Advocating science-based approaches to industry issues
- Encouraging personal and professional interaction
- Mentoring students, encouraging life-long careers as swine veterinarians

The strategic planning process gave rise to other initiatives that focused on providing incremental improvements to the value received by our membership. The web site was improved to get information out to members faster, as well as providing a portal of entry to other resources found on the Internet. Dave Brown was instrumental in the design of these tools and executing the ideas that came from our membership.

In 2000, Dr. Judi Bell was recruited to JSHAP by Cate Dewey as Associate Editor. Judi raised the bar (again) on the Journal! Judi's background in veterinary medicine and science helped elevate JSHAP to world class.

The membership had stabilized in the early years of the new millennium and in some years, domestic enrollment had declined slightly. The effect of the late nineties' markets that drove consolidation also drove a lot of smaller producers out of business and along with them, mixed practitioners that were doing their swine work. Many mixed-species practices found themselves with dwindling numbers of swine clients and thus less need to remain a part of the AASV. On the positive side, the organization that had been created to increase the knowledge of swine veterinarians was building the infrastructure to aid swine veterinarians around the world. Membership outside of North America was continuing to grow and by the end of the decade the AASV had members from 46 countries, and membership had reached nearly 1600.

Advocacy, membership services, and connectivity continued to demand more time from both Drs. Burkgren and Schulteis. In the summer of 2005, Dr. Harry Snelson was hired fulltime as the Director of Communications. Harry made an immediate impact and his background with industry and NPPC was a perfect fit. Much of the travel and advocacy work was shifted to Dr. Snelson.

Some of those important issues that impacted AASV members and the swine industry in the 2000's[14] included:

- Welfare issues, including gestating sow housing and euthanasia techniques
- Antimicrobial resistance and Extra-label Drug Use (ELDU) ban of cephalosporins
- Japan issued Maximum Residue Limits (MRL) on antimicrobials, which had the propensity to impact exports of pork
- A report commissioned by the PEW Charitable Trust was decidedly anti-modern farming

- The last case of pseudorabies in the US was detected in a commercial herd in 2003
- The Veterinary Feed Directive (VFD) went digital
- Certificates of Veterinary Inspection (CVI) or "health papers" went digital
- Preservation of Antibiotics for Medical Treatment Act (PAMTA) and other legislation to further restrict pharmaceutical availability and use in animal agriculture
- The Veterinary Patient Client Relationship (VCPR) was addressed as it related to swine practice
- Growth in the ethanol industry
- PRRS eradication efforts
- Porcine circovirus type 2 (PCV2) emerged causing high mortality in growing pigs
- Pandemic H1N1 IAV was diagnosed in humans and caused immediate impact on demand for pork due in large part to misinformation and fear

Of course, the entire list of issues was much longer but these were some of the most impactful and time consuming. The AASV and our membership tackled these important issues head on and in most cases with lighting speed. The shear number of members, resources and time that our support staff, leadership, and members employ daily and weekly should never be underestimated. The hundreds of hours in committee meetings, conferences, calls, emails and travel should never be downplayed or underestimated. Advocacy is not a spectator sport and the AASV team has always been ready to take the field!

Besides those mentioned previously, new projects, programs, services and milestones continued to evolve after the turn of the century:[15]

- AASV members and staff helped organize and host the second IPVS in the US in Ames, Iowa in 2002
- The first salary survey of AASV membership was conducted in 2002
- Pseudorabies was eradicated from commercial swine herds in 2003
- The AASV e-Letter was introduced as a weekly communication tool under the direction of Drs. Morgan Morrow and Sandra Amass of the AASV Communications Committee, relying on programming expertise provided by Webmaster David Brown. Dr. Harry Snelson assumed editorship of the e-Letter upon his hiring as AASV Director of Communications in 2005.
- A student poster session was added to the Annual Meeting to allow for more students to compete and participate
- Student representatives were added to the Board of Directors
- The Production Animal Disease Risk Assessment Program (PADRAP) risk assessment tool was gifted by Boehringer Ingelheim to AASV for use by practitioners to assess internal and external risks to individual sites and herds for PRRS virus. It was made available as a web-based tool in 2008.
- AASV was instrumental in helping Pork Board execute the Pork Quality Assurance (PQA Plus) program
- An AASV task force was formed to create a case definition for porcine circovirus associated disease (PCVAD) and worked with industry to rush effective USDA licensed vaccines to the market
- A Producer's Guide to Managing PCVAD was co-published with the National Pork Board
- AASV created a task force to explore the goal and methods to eliminate PRRS virus from North America
- Drs. Jeff Zimmerman and John Prickett developed PRRS PCR testing from oral fluid samples
- Efforts began to identify every swine production site and assign a Federal Premises ID number
- The Classical Swine Fever Surveillance Project was initiated with the aid of AASV in 2006. The name of the brochure promoting the effort was: "Got Tonsil?"
- JSHAP was made "open access" via the website in 2007
- New awards were created to recognize additional outstanding members: "Young Swine Veterinarian of the Year" and "Technical Services Veterinarian of the Year"
- The 4th edition of the *Swine Disease Manual,* edited by E.J. Neumann, A. Ramirez, and K.J. Schwartz, was published and made available on the AASV website
- The AASV office was relocated in 2009 into a newly constructed building, providing more room, parking and storage while remaining in Perry, Iowa.

During this first decade of the new millenium, the AASV held its second meeting in Canada (Toronto, Ontario 2005) and its first meeting on the west coast (San Diego, California 2008). It also (likely) held the last annual meetings in Des Moines (2004), Kansas City (2006) and Omaha (2010). A shift in preferences of the membership occurred after having sampled the family-friendly and temperate climates of San Diego and Orlando. Despite the lower cost and convenience of the Midwest meeting venues, the aversion to cold weather and the risk of late winter snowstorms made it an easy decision for the Board of Directors as they considered future annual meeting sites.

A landmark study entitled *Effects of weaning age on growing-pig costs and revenue in a multi-site production system*[16] was published by Rodger Main, Steve Dritz, Mike Tokach and others in JSHAP in 2005. The industry had been gradually reducing wean age to optimize sow productivity to the point that some systems pushed weaning age into the mid-teens. Main's work clearly demonstrated a linear improvement in post-weaning performance of pigs in wean-to-finish facilities by advancing weaning age from 12 to 15 days to 21 to 22 days. The pendulum of weaning age began to swing in the opposite direction.

A novel H1N1 type A influenza virus (later referred to as "pandemic H1N1 2009) emerged as a human pathogen in 2009. The outbreak was first thought to have originated from a swine farm in Mexico and was immediately misnamed "swine flu" since it was genetically similar to strains that commonly circulated in swine.

The popular press began mistakenly calling it "swine flu" which had an immediate and lasting effect on the consumption of pork due to perceived fears by the general public. Subsequently, it depressed the demand for pork which affected the market price of hogs. The AASV Position Statement on Pandemic (H1N1) 2009 Influenza[17] was published on July 29th, 2009.

2010 and Beyond

In this most recent decade of AASV, new disease challenges emerged, further regulations on welfare and antimicrobial use were enacted, consolidation and integration continued. Export demand for pork continued to climb and the swine industry was becoming more globalized. In 2013, Smithfield Foods, the world's largest pork producer was purchased by Shuanghui International Holdings. Ltd. a Chinese firm. At the time, this was the largest purchase of a US agribusiness in history.[18]

JSHAP

Dr. Cate Dewey stepped down as the editor of the *Journal of Swine Health and Production* in 2012 after 13 years of service.[19] Dr. Dewey guided JSHAP toward a world class peer-reviewed scientific journal. In 2011, JSHAP was recognized by Elsevier, the leader in global information analytics for health care, to be included in their list of cited journals. Cate had assembled and managed a world-class team with her staff: Dr. Judi Bell as associate editor, Tina Smith as graphics designer, Karen Richardson as publications manager, and Dave Brown as the IT expert who posts JSHAP on the AASV website to make it available to the world. Cate and Judi also assembled and worked with a long list of reviewers (made up predominately from AASV membership) who, along with the editorial board, provided the authors with invaluable feedback on the submissions in a timely manner. JSHAP has become a cornerstone of AASV, truly fulfilling the mission of "increasing the knowledge of swine veterinarians" with each issue.[20]

Dr. Terri O'Sullivan became the Executive Editor of JSHAP in December of 2012 following Cate Dewey's retirement and she has continued to drive the journal on the path to excellence, raising the bar once again.[21] Dr. Judi Bell retired in 2018 following 17 years of invaluable service to JSHAP as the Associate Editor. Dr. Sherrie Webb accepted the Associate Editor position and was able to work with Judi to assure a smooth transition. The AASV is truly blessed to have had such great leadership and staff from the very beginning that continues still today![22]

Antimicrobial use

Pressure continued to mount on the use of antimicrobials in production with the introduction of Guidance Document 209 and 213 from the FDA in 2011 which outlined new guidelines to broaden the scope of the Veterinary Feed Directive and increase veterinary oversight. These guidance documents direct the use of medically important antibiotics (deemed to be important for human medicine) in livestock for therapeutic purposes only, thereby eliminating medically important antibiotics for growth-promotion purposes. Medically important antibiotics can continue to be used for therapeutic purposes by producers, but only under the guidance of a veterinarian with a valid veterinary-client-patient relationship (VCPR). The regulations were implemented in 2015, essentially ending over the counter access by producers.[23]

Welfare

While welfare of the pig has always been within the primary purview of swine veterinarians, a gradual shift was underway in the social acceptance of pig rearing, fueled in part by the growth of modern pork production. Anti-animal-agriculture activist groups, motivated in part by an anti-meat agenda, began to turn public opinion against what they referred to as "factory farms" with the use of undercover videos and individual state ballot initiatives such as California's Proposition 2, which in part, limited producers' choices of sow housing. AASV members had always worked closely with the National Pork Board as the established programs such as Pork Quality Assurance (PQA) and Swine Welfare Assessment Program (SWAP). Packers and their downstream customers began to request on farm assessments and audits and AASV was at the forefront in establishing guidelines and providing expertise. The AASV Pig Welfare committee was established in 2006 and soon began to create supportive material such as policies and guidelines to educate and aid members with these issues. On-farm euthanasia guidelines, piglet processing and pregnant sow housing policies are among the many issues they have tackled. Recognizing a need for more expertise in the area of swine-specific welfare, the AASV Foundation also began funding scholarships to those members desiring to become board certified via the American College of Animal Welfare (ACAW) process. The first scholarship recipient to become ACAW board certified was Dr. Monique Pairis-Garcia in 2018.[24]

Mission statement update

In 2014-15, the Board of Directors approved a change to our mission statement a third time to update and better reflect our commitment to the health and well-being of the pig which had always part and parcel of the veterinarian's oath.[25] With increased emphasis and focus on welfare, the changes were warranted and were well received by the Board of Directors and membership. The new mission statement:

It is the mission of the American Association of Swine Veterinarians to:

- Increase the knowledge of swine veterinarians,
- Protect and promote the health and well-being of pigs,
- Advocate science-based approaches to veterinary, industry, and public health issues,
- Promote the development and availability of resources that enhance the effectiveness of professional activities,
- Create opportunities that inspire personal and professional growth and interaction and,
- Mentor students, encouraging lifelong careers as swine veterinarians.

The Foundation grows

Our AASV Foundation continued to grow with the addition of new levels of commitment and very successful annual meeting auctions. In 2012, Mary Lou Hogg of MVP Laboratories and Bill Marks of MJ Biologics pledged to match the proceeds of the 2013 auction if it reached $100,000 total between the silent and live auctions. Dr. Daryl Olsen, the auction committee chair for 2013, spearheaded the efforts to assure success! With the matching funds, the total raised for the 2013 auction in San Diego was nearly $240,000! The Foundation surpassed the 1 million dollar level of endowed funds! Not to rest on this success, the AASV Foundation soon set "a big hairy audacious goal" to grow the value of the foundation to $2 million by the 2019 celebration of AASV's 50th anniversary. At the same time, it was important to maintain the Foundation's ongoing commitment to fund research, scholarships, externships, tuition grants, and other programs and activities that benefit the profession of swine veterinary medicine.

A new level of giving named the Legacy Fund was established in 2014 which was marked by $50,000 or more of lifetime giving (inclusive of the Leman and Heritage levels) to recognize a principle donor or named honoree. Dr. Nate Winkelman was the first to establish a Legacy Fund in 2016.

The foundation was created to ensure our future and leave a legacy. It is reaching levels that will ensure that its mission will be fulfilled and become self-sustaining.

The mission of the American Association of Swine Veterinarians Foundation is to empower swine veterinarians to achieve a higher level of personal and professional effectiveness by:

- enhancing the image of the swine veterinary profession,
- supporting the development and scholarship of students and veterinarians interested in the swine industry,
- addressing long-range issues of the profession,
- supporting faculty and promoting excellence in the teaching of swine health and production, and funding research with direct application to the profession.[26]

Disease concerns of the decade

Influenza – After suffering the effects of the pandemic H1N1 in 2009, influenza was at the top of many veterinarians' minds early in the 2010's. Veterinarians were eager for new solutions to what just a decade or so earlier had fairly easy answers. Now they were dealing not only with classic H1N1 and the triple reassortment H3N2 and H1N2 with the still "new" pandemic H1N1. New diagnostic tools and monitoring projects were becoming available and the learning curve was bending upward.

Porcine epidemic diarrhea (PED) – In the spring of 2013, a new epidemic hit the US swine industry with a vengeance: porcine epidemic diarrhea. It first appeared to practitioners as transmissible gastroenteritis (TGE) but the state labs returned negative results for TGE virus. It was soon discovered that while PED was a coronavirus like TGE, it was different enough to avoid detection with the current tests available. It wasn't long until the state veterinary diagnostic labs along with the National Veterinary Services Lab (APHIS NVSL) had created a PCR test specific for PED virus and the trajectory of the epidemic was felt. Within a couple of months, the disease had spread across a dozen states and hundreds of herds. AASV staff and members were instrumental in laying the groundwork for establishing communication with regulatory officials, establishing epidemiological information, diagnostics and control measures. The disease was devastating to naïve herds, impacting neonatal pigs especially hard with reports of 100% mortality for multiple weeks. PED was deemed a transboundary disease instead of a foreign animal disease (FAD), which made it less impactful from a trade, animal movement and regulatory standpoint. It became widespread enough to substantially move the market price. If a bright spot can be gleaned from this outbreak, it would be that it provided an opportunity to clearly see potential weak spots in the nation's ability to respond to an FAD such as classical swine fever (CFS) or foot-and-mouth disease (FMD).

Senecavirus A (SVA) – In the summer of 2015, Senecavirus A (also known as Seneca Valley virus or SVV) was diagnosed in a handful of cases in the upper Midwest. SVA produces vesicles on the snout, oral mucosa and coronary bands of pigs similar to those produced by FMD. Since the lesions are similar to FMD, each case must be handled as a potential FAD, and FMD must be ruled out. SVA is not a severe production disease, but the risk of complacency toward FMD is real. Again, as with PED, this offered another opportunity to hone skills related to dealing with a true FAD event.

African swine fever (ASF) is a foreign animal disease not currently found in North America. It was first described in Africa in the 1920's but has since spread to various parts of the globe. Since 2007, it has gradually spread across southwest Asia and into eastern Europe. In the summer of 2018, the first cases were reported in mainland China, which aroused serious concerns in the US due to the volume of trade between China and the US especially with feed and feed ingredient imports. Dr. Scott Dee began conducting research into the survivability of the ASF virus in various feedstuffs in simulated trans-pacific shipments. While the outbreak in China continues unabated into the winter of 2018, risk mitigation strategies are being developed to safeguard the swine industry in North America. The consequence of an introduction of ASF would be devastating.

Porcine circovirus 2 (PCV2) was the disease of the first decade of the new millennium. In 2015, a North Carolina sow herd experiencing symptoms resembling PCVAD had samples subjected to deep sequencing. From that sequencing, a novel circovirus was identified and named porcine circovirus type 3 (PCV3), since it differed greatly from PCV2. Diagnostic tests were developed to identify PCV3. While PCV3 has been found in tissues from clinical cases it has also been found routinely in samples from non-clinical pigs. At this point, it has been deemed "an association" but more research and information are necessary to be certain it is a new pathogen in swine.

Porcine reproductive and respiratory syndrome (PRRS) – 2008 ushered in the third decade of PRRS. We have come a long way, learned many things, but in the end, PRRS still resides at the top of the list as the most frustrating and costly disease that our industry has faced. The idea of a national eradication program came out of the AASV PRRS Committee in 2006 and the AASV Board approved the formation of the North American PRRS Eradication Task Force. The mission of the task force was to bring all stakeholders to the table to create a National PRRS Elimination Program Framework[27] document which would be a road map for the elimination of PRRS virus nationally. The task force, chaired by Dr. Montserrat Torremorell, created the document to provide a framework by which PRRS could be eliminated from the national swine herd if certain resource and knowledge gaps are filled and producers determine that an elimination effort is warranted. The AASV Board of Directors approved the document in 2016 and it has been shared with the NPB Swine Health Committee. While producers have not yet embraced the goal of a national elimination project, the framework may serve as guidelines for local or regional attempts at elimination should the need arise. The framework also identifies knowledge gaps that will direct research and future tool development in the fight against PRRS.

Disease monitoring

Having just come through the pandemic H1N1 issue that had such a profound effect on the US swine industry, the USDA initiated the National Swine Influenza Surveillance Plan (NSISP) as a pilot in 2009. In October 2010 the NSISP was launched with the cooperation of the National Pork Board (NPB), National Pork Producers Council (NPPC) and the American Association of Swine Veterinarians (AASV). The goal of the plan was to gather a wide range of isolates from the industry at large and make these available to animal and public health analysis as well as to create and improve tools for diagnostics and potential vaccine development. Veterinarians were urged to alter their lexicon by replacing the term Swine Influenza Virus (SIV) with Influenza A Virus of Swine (IAV-S).

In 2010, the PRRS herd classification system was developed by a joint committee of AASV and funded by the USDA PRRS-CAP. Dr. Bob Morrison chaired the committee that was composed of veterinarians representing private practice, academia, industry and the NPB. The purpose of creating this standardized classification system was to create terminology for PRRSV status of swine herds to facilitate communication between stakeholders and aid in development of efforts towards PRRSV monitoring, control and potentially elimination.[28]

In the 2011, Dr. Bob Morrison compiled and released the first Swine Health Monitoring Project (SHMP) report. Bob's idea sprang from a conversation he had with a veterinarian at an AASV PRRS Task Force meeting. Bob had the foresight to see and fill a need for a tool that with wide participation could give the industry an idea of national PRRS incidence and prevalence. It started with reports from three veterinarians and soon expanded to over 25 participating systems comprising close to 50% of the national herd. With the introduction of PED into the national herd, the SHMP was a natural tool for tracking the incidence and prevalence of that disease as well. Then as a natural progression, in 2015 when SVA emerged, it was also added to the SHMP.[29]

The need for improved surveillance and rapid response to emerging or re-emerging and foreign animal disease was recognized by Drs. Tom Burkgren and Harry Snelson following the 2013 introduction of the transboundary disease porcine epidemic diarrhea (PED). PED exposed the vulnerability to new disease introductions and lack of resources and adequate infrastructure to respond, prompting Drs. Burkgren and Snelson to propose the creation of an information center. In the summer of 2015, the Swine Health Information Center (SHIC) was created through a joint effort by AASV, the National Pork Board (NPB), and the National Pork Producers Council (NPPC), with funding supplied by NPB. SHIC's mission: "to protect and enhance the health of the United States swine herd through coordinated global disease monitoring, targeted research investments that minimize the impact of future disease threats, and analysis of swine health data." Dr. Paul Sundberg was named its first Executive Director after spending 21 years with the NPB ending in the role of Vice-president of Science and Technology. Immediately upon its creation, the outbreak of Senecavirus A provided SHIC with an early test of its ability to respond to an emerging threat to swine health.[30]

AASV was truly "built to last"

Jim Collins and Jerry Porras turned a six-year long research project in which they asked: "What makes the truly exceptional companies different from other companies?" into the best-selling tome of the business book: *Built to Last: Successful Habits of Visionary Companies*.[31] They chose eighteen companies that they identified as truly exceptional and visionary and matched them with eighteen "comparison" companies that were founded in the same era under similar circumstances. Collins and Porras had some preconceived ideas that they eventually began to refer to as "myths" and they had to "unlearn" much of what they believed about long-term successful companies. These myths were:

1. A great idea is needed to start companies
2. Visionary organizations need charismatic leaders
3. Maximizing profits is the dominate goal with visionary companies
4. Visionary companies focus on beating competitors
5. Hiring outsiders as CEO's is the best way to spark an organization

Instead, Collins and Porras were able to uncover six fundamentals that allowed great companies to endure and thrive:

1. Make the organization itself the ultimate product: become a clock builder, not a time teller
2. Build your organization around a core ideology
3. Build a cult-like culture
4. Home grow your management and leadership

5. Stimulate progress through big hairy audacious goals, experimentation, and continuous improvement
6. Embrace "the genius of the ***and***": preserve the core ***and*** stimulate progress

It is obvious that the founders of the American Association of Swine Practitioners have done something right in building the organization. In fact, they were checking all the boxes!

Allow me to take the liberty of demonstrating just how well our organization fits these six fundamentals.

The ultimate product

The original mission of the AASP was simple: "to increase the knowledge of swine veterinarians." The organization was built by swine veterinarians, for swine veterinarians. It has always been about the members. It was of no use to have a few great "time tellers" if you could not create a great clock. The first scientific meeting of the AASP included experts in the fledgling specialty field and the cast was struck. A core group of veterinarians with a passion for pigs wanting to create an organization of veterinarians with the goal of making themselves and anyone else interested in pigs more informed on everything swine.[32] Those original organizers knew there was a need and set out to create a "clock" in the form of the American Association of Swine Practitioners to fill that need. The original mission grew in the 1990's as the organization found its footing and leadership reflected on the many issues that the association was tackling:

It is the mission of the AASP to:

- increase the knowledge of veterinarians in the field of swine medicine
- elevate the standards of swine practice
- promote the relationship between swine practice, the swine industry, and the public interest
- promote the interests of swine veterinarians
- improve the public stature of swine veterinarians
- cooperate with veterinary and agricultural organizations and regulatory agencies
- promote goodwill among AASP members

The core ideology

The Veterinarian's Oath is at the heart of that core ideology. Interestingly, the AVMA House of Delegates first adopted the Veterinarian's Oath in 1969 the year that AASP was founded. Members embraced the ideals of the AASP from the very beginning. Veterinarians involved with swine had a passion and yearning for learning everything swine. The AASP satisfied that yearning with newsletters, scientific meetings and any kind of publication related to pigs. The National Hog Farmer and Hog Farm Management magazine subscriptions went to all members. Early meetings were held in conjunction with the National Pork Congress in three Midwest cities. The AVMA also aided the AASP in the early years by including a swine section within their annual meeting. There had been a vacuum for up-to-date information on swine medicine and here was a chance for practitioners to interact with other swine practitioners and fuel that passion for pigs and pig farming. It was a chance to catch the wave of "the new" swine industry just as modern methods were being accepted by their pig producing clients. Swine practitioners were asking themselves: "What do we stand for and why do we exist?" Timeless core values and an organization's enduring purpose should never change. However, operating practices and strategies are more likely to change constantly in response to a changing business climate.

Build a cult-like culture

Cults are defined by great devotion to an ideal or cause. Cult-like cultures do not just happen. They are driven by those ideals that bond a group together. Coming together for a common cause or to achieve a goal much larger than would be achievable as individuals. I am certain that the founders were not driven by a need to create this culture, but by opening the doors to networking and building relationships built on common causes, gave practitioners involved in swine practice and industry the opportunity to improve their skills and increase their knowledge of swine medicine and production. It created a fraternity of sorts with their own set of values and language. Few of the early practitioners to embrace the AASP were involved in predominately swine practices. The AASP, however, gave them the chance to rub shoulders and interact with those who were experts in the field and other practitioners that were being drawn into more and more swine practice. The university extension system was strong and relatively well funded. All of the major Land Grant colleges had or were hiring practitioners with swine experience to create or bolster their presence in swine medicine. Many of these veterinarians formed the basis of early AASP leadership.

Home grow your management and leaders

From the beginning of AASP, home-grown management and leadership has been in place. The first management and leadership positions were filled from among those first members. The early executive secretaries operated the AASP out of their homes or university offices. The only "outside employees" were spouses or administrative assistants from within their institutions. When the membership gradually outgrew those arrangements, it was a natural progression to utilize the offices and executive secretaries of the sister veterinarian organization of the Iowa Veterinary Medical Association. The demands of membership and program growth demonstrated the need for an executive director dedicated to AASP. Dr. Tom Burkgren was home-grown talent to fill the need. Leadership was home grown as well. The bylaws called for election of a vice-president from the membership ranks each year, who after one year of service on the executive committee, automatically filled the position of president-elect and then ascended into the president's role in the third year. The president then remained a part of the executive committee for one year as immediate past president.

Members of the Board of Directors are elected representatives from each of the districts. The Board provides input and consent in all matters and often are leaders within committees and task-forces. All are home grown.

Leadership and management roles do not end with the executive committee. Former leaders and members who serve on and chair committees and task forces are vital to the health of the organization and they are of course, all home grown!

Lastly, the AASV not only utilizes home-grown leadership and management, it mentors, hires and promotes from within!

Stimulate progress through big hairy audacious goals (BHAGs), experimentation, and continuous improvement

Our organization has never been shy when it comes to this fundamental. As if starting a new professional association from scratch wasn't enough of a big hairy audacious goal, within 5 years of forming AASP, leadership launched head first into the bidding process to host the IPVS in the US. I imagine there were some anxious moments and a lot of wondering "what have we gotten ourselves into?" thoughts when they found out that they had won the bid for the 1976 IPVS in Ames! Organizing a charity auction to raise funds for the Foundation and setting a goal of $100,000, now that is a BHAG! So is setting a goal of having $2,000,000 in the Foundation by 2019 or eradicating PRRS from the US by 2025! Having just the "right" BHAG is not as important as having clear, compelling and imaginative goals that drive progress. Imaginative goals literally seep out of the minds of AASV members!

Experimentation and continuous improvement describe our organization. An environment that encourages members to experiment and take risks is a learning environment. From the beginning, the organization has taken risks. Starting a peer-reviewed scientific journal devoted to swine health and production, forming a charitable foundation, organizing a second IPVS meeting, holding an annual meeting in Quebec (and then Toronto) and tackling issues that impact our profession and industry head on such as antibiotic resistance, PRRS, PRV, PCVAD, PED, ASF. Has everything attempted worked? Nope, some things that were attempted didn't persist. Needs change, issues evolve, and some ideas may not be quite ready for prime time, but even failures are opportunities to learn.

Embrace "the genius of the *and*": preserve the core *and* stimulate progress

Collins and Porras[31] refer to this fundamental of built to last as the yin yang from Chinese philosophy for not giving into the "tyranny of the or." In terms that fit our association the "tyranny of the or" may sound like this: "We can invest more in students for the future *or* provide more member services, but not both." Embracing "the genius of the and" sounds more like this: "We will protect and promote the health and well-being of pigs *and* advocate science-based approaches to veterinary, industry, and public health issues." The concept of preserving the core while embracing change stimulates progress in everything except the core ideology. The core ideology is the true north of any organization and must not be compromised. However, progress comes by embracing the "and" to push beyond the comfort zone. Any idea is fair game if the core ideals are conserved.

While *Built to Last*[31] was created for business about business, what readers discovered was that the concepts of this book were applicable to any type of organization including non-profits such as AASV. Fittingly, this year's theme: "50 Years of AASV: Built to Last" evolved from the concepts in this book. Built to Last also spawned a sequel: *Good to Great: why some companies make the leap and others don't.*[33] "Good to Great" was the theme of our 2007 annual meeting and Dr. Tom Burkgren presented the Dunne Lecture that year: "AASV: Good to great."[33]

The forces that shaped AASV

The AASV grew, improved and changed dramatically in 50 years. It has been shaped by many forces, internal and external, major and minor, good and bad, but mostly it has been shaped by our members, past and present, who had a passion for all things pigs and people. From that original idea of creating an association specific to swine with the simple mission of increasing the level of expertise and knowledge of pig veterinarians, a handful of forward-thinking veterinarians grew that idea with a passion into a world-class organization. This organization and its members through a common, uniting cause have changed the trajectory of the global swine industry. Our organization will always be impacted by the same forces that affect the swine industry. A viable, profitable swine industry is essential to viability of those who serve it.

At the time of the founding, pigs were still considered the "mortgage lifters" as the hog checks paid the mortgage on the farm. Economics dictated that hogs were raised where the corn was grown. Advances in nutrition fueled by research at Land Grant institutions soon led to balanced diets utilizing corn and soybean meal to address the limiting amino acids and drastically improved growth rates and efficiency. Genetic improvements provided more productive sows and faster growing and more efficient pigs. With each incremental improvement, another limiting factor to improving the bottom line would reveal itself. From farrowing outside in "A-frames" to controlling the environment with a farrowing barn to save more pigs to bringing the growing pigs out of dirt lots onto concrete and then to slatted floors, each innovation created another opportunity to improve production or lower cost to in turn improve profit. The era of "confinement" hog rearing had been born.

Swine health is also a limiting factor to productivity and profit. Just as the profit motive drove innovation in other aspects of production, it also provided motivation to improve the health of the pigs. As conditions changed, so did the list of diseases and related syndromes that affected the health of the pigs. Through the years, the list of diseases and pathogens continued to grow and change as solutions to one disease created opportunities for the next one to arise. Better vaccines for prevention, better antimicrobials for therapy, improved management and health technologies for reducing the risk of transmission or reducing stressors all play a part in making life better for the pig. From lice and mange to hog cholera to pseudorabies to PRRS to PCVAD to IAV to PED, swine veterinarians have been instrumental in finding solutions for control or elimination. The tools and techniques continue to improve, driven by the desire to better swine health and improve productivity. Tools and methods to eliminate or prevent entry of disease that could not be controlled easily were developed as well. Specific pathogen free (SPF) stock, depopulation-repopulation, partial depopulation, all in-all out, early weaning strategies, multi-site production, air filtration, load-close-expose, wean down, MCREBEL and improved live animal transport and overall biosecurity all had their roots with swine veterinarians.

Porcine reproductive and respiratory syndrome (PRRS) changed the way pigs were raised. Initially called Mystery Pig Disease (because a cause had not yet been identified) the PRRS virus began to wreak havoc in the late 80's. No other disease has had such a profound impact on the swine industry and consumed as many resources. PRRS may appear to some as a failure of our profession, especially if compared to other pathogens that have emerged, been tackled and essentially resolved in a fraction of the time we have been dealing with PRRS. However, we have made and continue to make progress with PRRS and the exogenous benefits and discoveries that have been made as a result of attempting to solve the PRRS issue could (in a small way) be compared to all the spin-off benefits of the space program. PRRS has been the impetus behind so much of what we have learned about biosecurity, diagnostics and epidemiology, virology and immunology. Google "PRRS" today and you will get over 2,120,000 hits!

Advances in production tools and protocols have made major impacts on our industry. Computer-based record-keeping systems were developed in the early 1980's with the advent of personal computers and moved pork production into the information age. This technology opened a whole new world of data analysis that made benchmarking and uses of production records analysis possible. These record-keeping systems allowed swine veterinarians the opportunity to offer another valuable service to producer clients and impact their bottom lines. It is now easy to take for granted the sheer volume of valuable information and advancements that took place due to improved access to information generated from records.

In the 80's producers were making the transition from "pen mating" (where a number of boars were turned in a pen with a larger number of sows) to "hand mating" (where the producer would actually observe sows in estrus and select an individual boar to mate them). As late as the 1980's it was thought that the adoption of artificial insemination (AI) was impractical and would negatively impact productivity by lowering farrowing rate and pigs born per litter as early results with AI were often disappointing compared to natural mating. Gradually, the technology improved with the introduction of disposable supplies, better semen extenders and automated semen analysis tools. The lure of capturing production gains and efficiencies drove the move toward overcoming hurdles to adopt AI. AI has had a nearly undeterminable impact on the velocity of genetic improvements and cost savings.

Capital flows to the highest return, and in the 80's and 90's entrepreneurs could see that a systems approach similar to that applied to the poultry industry offered the potential for very lucrative returns. A new breed of pig farmers evolved, armed with a lot of capital and new ideas. Expansion, consolidation and integration were fueled by efficiencies of scale and high returns. This was an inflection point for the swine industry that continues still today. This trend gave way to new terms such as "contract production" and "coordinated production." Export demand was increasing every year, and with total confinement in the upper Midwest we were seeing the cyclical nature of pork production gradually meld into a steady state. Forward contracting and packer contracts based on futures markets began to reduce what were considered normal marketing channels for small producers, and the number of pigs that were marketed under the normal negotiated price discovery methods gradually subsided. Expansion continued unabated with even many smaller diversified farmers investing in confinement buildings to capture some of the economies and technologies afforded the specialized producers. In the fall of 1998, the expansion curve finally exceeded the slaughter capacity in the US and prices for market hogs crashed to as low as $8.00 per hundred weight. While prices have always been cyclical, and producers moved in and out of production in response to market signals, this dramatic drop in prices produced a tectonic shift. The US breeding herd inventory had been hovering north of 7 million but by the early 2000's it had downsized by nearly a million sows in the US. Those producers who were aligned with packers or had established a solid safety net shrugged off the low prices and continued to expand. Those who had little or no price protection exited the business, especially those who were highly leveraged financially, often from recent expansion. Many small and mid-sized producers shuttered their operations, but those producers with modern facilities that could not weather the storm merged or were purchased by integrators or larger, more financially sound operations. Successful Farming's Pork Powerhouses in 1998 listed the 50 largest producers in the US: 35 of those 50 are no longer on the list in 2018.[34] Most of those facilities and sows were kept in production, many of those by operations that remained on the list. The number of sows owned or controlled by the Pork Powerhouse list has been on an ever upward slope even though the names have changed through mergers, acquisitions and consolidations. In 1998, the list comprised about 2.1 million sows (up from just over 1 million since 1994). In 2018, the list of the top 40 largest producers had topped 4.2 million or about two thirds of the US inventory. The same factors and market pressures driving mergers, acquisitions and consolidation still exist and will likely continue to push the industry to fewer and larger entities that are controlling the pork sector, perhaps at an accelerating pace.

Through all the turmoil of the late 90's and early 2000's, while sow numbers declined, production improved. Productivity in terms of pigs per sow per year continued to climb despite disease issues. Productivity and efficiency continued to improve on the growing pig as well, and producers responded to consumer demand for less fat. Slaughter weights also increased steadily from 200 pounds in the 70's to averaging over 280 pounds today. Where has all that pork gone? The US became a net exporter of pork in 1995 and in 2017 over 26% of the US production was exported.[35] Exports have been the fuel that has spurred demand and maintained profitability and drove production. Becoming dependent on exports has a downside in that it confers a dire consequence of any event such as an introduction of a foreign animal disease or major trade disputes which would interrupt or substantially slow that flow of pork.

The swine industry and hence swine veterinarians have also been affected by government regulations, programs and legislation. While pork producers have not directly benefitted from government subsidies, it is easy to argue that one of the larger factors to affect the cost of pork production in the last 20 years has been the increased use of corn to produce alternative fuel in the form of ethanol. Subsidies aimed at increasing renewable fuels has artificially increased the demand for corn for ethanol production and soybeans for biofuel. The livestock sector must compete for the same bushels that are going into biofuels, which directly raises the cost of feed. Demand for corn by the ethanol sector now exceeds the use of corn for feed. Ethanol subsidies are just one example of government interventions into pork production that directly impact the cost of production. Environmental regulations, trade restrictions, pharmaceutical and biologic use, welfare, transportation, tax laws, labor practices, immigration policies, and nuisance laws are just some of these that directly or indirectly impact the cost of production.

"Once a swine vet, always a swine vet!"

Dr. David Reeves attributed this quote to Dr. Amy Vincent.[36] Amy was talking about her father, Dr. Rodney "Butch" Baker. In my opinion, her simple statement encapsulates the AASV. We are all part of a family connected by a common idea, a common cause and a reason for existing...the pig! It is the one thing we have in common and the only reason we are tied together within the same chain. We are connected by a common core of "doing what's right for the pig."

We are connected by shared leadership. In the case of AASV, leadership comes from within...those who share the same great ideal and core purpose. Leadership is essential to the group and leadership is not always singular. At times one person steps up but mostly within the AASV, leadership is plural. At times, one person steps up, but with AASV, it is usually several. To survive and thrive, any group, whether it be a family, a business or an association, requires leadership. In the case of AASV, past, present or in the future, that strong leadership has (and will continue to) come from within.

We are connected by communication. Members of the AASV have a common language and have developed tremendous tools to communicate. Communication is our primary tool to accomplish our founding mission: "to increase the knowledge of swine veterinarians." There must be a connection between and among the leaders and our members. We have a great narrative: who we are and what we stand for. Depth of knowledge by itself is not enough, it must be shared, passed along and disseminated among the members.

We are connected by a common cause: What we do, what we stand for and why we exist. Our founders aspired to create something bigger and longer lasting than themselves. AASV was created on core values to stand the test of time by virtue of ability to continually renew itself from within. The gifts our founders gave us (and to the future) were their time and efforts to pass it on and pay it forward. We must do the same! We all once were high school students with a dream and a desire to become veterinarians. We were all vet students and interns. We were all new grads. We were all new associates and new members. Every step of the way, we were encouraged, mentored, taught, shown and corrected by someone who had been there and done that! "See one, do one, teach one!" We do not exist or develop in a vacuum. Many have helped us along the way. Many have blazed a trail and smoothed the path. Generous and sincere mentorship will always defeat the selfish efforts of individual accomplishments going it alone just because you can. Believing in what you do, every day, and helping others do the same is a brilliant long-term strategy. Caring is the essential emotion that binds our clan together. Members of the group care what happens to the group and to one another. The Veterinarian's Oath outlines what is important to us as veterinarians and the caring principles we solemnly swear to uphold.

Coincidentally, the Veterinarian's Oath was first adopted in 1954 but revised by the American Veterinary Medical Association's House of Delegates in July 1969 at the same gathering where the AASP was founded. Nearly every newly minted veterinarian since has taken the oath. It has been amended by the AVMA twice since with the latest iteration being approved in December 2010.[37] The current version is:

Being admitted to the profession of veterinary medicine, I solemnly swear to use my scientific knowledge and skills for the benefit of society through the protection of animal health and welfare, the prevention and relief of animal suffering, the conservation of animal resources, the promotion of public health, and the advancement of medical knowledge.

I will practice my profession conscientiously, with dignity, and in keeping with the principles of veterinary medical ethics. I accept as a lifelong obligation the continual improvement of my professional knowledge and competence.

I am known for offering this bit of wisdom to students: "If it were just pigs our jobs would be easy." But we all know that people raise pigs because pigs don't raise themselves. People are who make the difference whether it's raising pigs or improving or growing the AASV. Success resides on a continuum, but it is never a straight line. There are always going to be setbacks and plans that do not come to fruition but that is how we learn. If we require success before commitment, we will get neither! We must be willing to make mistakes and be wrong. The only critical mistake is to stop looking and quit trying.

In May of 2017, our AASV family suffered a great loss. Dr. Bob Morrison passed away in a tragic accident at the age of 64. Bob's contributions to the AASV, our profession and the swine industry are inestimable. His absence from our midst will leave an unfillable hole. Bob often talked about how important it was to do and produce "work that matters."[38] Bob essentially founded the *Journal of Swine Health and Production* and then served as its editor for seven years. He served on the AASV leadership team and as our president in 2000. Bob presented the Dunne Memorial Lecture in 1994 titled *"Opportunity: A matter of perspective."* Bob was a brilliant and practical researcher, a lifelong learner and a gifted teacher. Bob's legacy is a portrait of work that mattered. We are all better for having known Bob and have benefitted from his "work that mattered."

I can think of no better challenge to current and future members of AASV than to follow Dr. Morrison's mantra and philosophy of work and life: "Do things that matter!" The AASV was truly built to last with a firm foundation from which future generations will build it stronger and better! Our destiny is our own, so instead of looking back and just tallying the score, let's look ahead to fifty more!

Acknowledgements

We all, at some point, decided to pursue this noble profession and in some small way, every day, commit ours lives to something greater than bringing home a paycheck. We help feed the world while caring for, respecting, and doing what's right for the pig. It is not just a job, it is a commitment.

I am proud to be a swine veterinarian and to be a part of this great organization. It is not the organization itself that makes it so. It is the people ... colleagues, co-workers, pork producers, friends, mentors, students (past, present and future), as well as all our spouses, children, parents and grandparents. I want to thank all those from every category just listed who have been so impactful to my life and career. There are truly too many to name and many of whom likely have no idea to what degree they have influenced my life. It is to all of you that I dedicate this lecture, with special thanks to Drs. Burkgren, Schulteis and Snelson for all their help and encouragement, as well as their many years of dedicated service to AASV, making it truly an organization that was "built to last" for future generations.

References

1. Leman, Al. Preparing for Swine Practice in the 1980's. American Association of Swine Practitioners Proceedings, Annual Meeting, 1981.

2. Muirhead, Michael. Effects of the Swann Report on Veterinary Practice and the Swine Industry. American Association of Swine Practitioners Proceedings, Annual Meeting, 1978.

3. Major events of 1969: www.wikipedia.com. Accessed 29 September 2018.

4. US calendar year average hog prices received for 1969-2014: www.farmdoc.illinois.edu /manage/uspricehistory. Accessed 10 October 2018.

5. Brandt, W., Meyer, K. History of the American Association of Swine Practitioners 1969-1994 AASP Annual Meeting 1994.

6. Anonymous. Notes from AASP board and executive committee meeting minutes 1969-1994.

7. Schulties, Sue. Personal communication and notes of brief history of the student seminar. 2018.

8. Morrison, R. Note from the editor. J Swine Health Prod. 1993;1(1).

9. Allen D. Leman tribute. https://ccaps.umn.edu/allen-d-leman-swine-conference Accessed 15 October 2018.

10. Burkgren, Tom. Personal communication. 2018.

11. Anonymous. Notes from AASP board and executive committee meeting minutes 1995-2000.

12. Personal recollections 2018.

13. From the Executive Director: https://www.aasv.org/shap/issues/v8n6/v8n6ex.html Accessed 4 November 2018.

14. President's messages: *Journal of Swine Health and Production* online editions 2000-2010.

15. Anonymous. AASV board of director's meeting minutes 2000-2010. https://www.aasv.org/aasv/board.html Accessed 9 November 2018

16. Main, R. et.al. Effects of weaning age on growing-pig costs and revenue in a multi-site production system. *J. Swine Health* 2005 https://www.aasv.org/shap/issues/v13n4/v13n4p189.pdf Accessed 8 November 2018.

17. AASV Staff. American Association of Swine Veterinarians Position Statement on Pandemic (H1N1) 2009 Influenza *JSHAP news*. https://www.aasv.org/news/story.php?id=3700 Accessed 8 November 2018.

18. Freese, B. Successful Farming Pork Powerhouses 2013. 30 September 2013 https://www.agriculture.com/livestock/hogs/pk-powerhouses-2013-disease-hits-growth_283-ar34203 Accessed 8 November 2018.

19. Dewey C. From the Executive Editor. 2012 https://www.aasv.org/shap/issues/v20n1/v20n1editor.html.

20. Burkgren T. From the Executive Director.2012 https://www.aasv.org/shap/issues/v20n2/v20n2exdir.html.

21. O'Sullivan, T. From the Executive Editor. 2012 https://www.aasv.org/shap/issues/v20n4/v20n4editor.html.

22. O'Sullivan, T. From the Executive Editor. 2018 https://www.aasv.org/shap/issues/v26n3/v26n3editor.html.

23. Schulz LL, Rademacher CJ. Food and Drug Administration Guidance 209 and 213 and Veterinary Feed Directive regulations regarding antibiotic use in livestock: A survey of preparation and anticipated impacts in the swine industry. J Swine Health Prod. 2017;25(5):247–255.

24. AASV staff. AASV News. 1 August 2018. https://www.aasv.org/news/story.php?id=10953.

25. AASV Board meeting minutes. Fall 2014. https://www.aasv.org/members/only/board/board_f14.php.

26. AASV staff. AASV Foundation News 2006-2018 https://www.aasv.org/news.

27. AASV PRRS Task Force. Nation PRRS Elimination Framework. 2016. Document located at https://www.aasv.org/documents/PRRSEliminationFramework-Final.pdf.

28. Holtkamp, D., et.al. Terminology for classifying swine herds by porcine reproductive and respiratory syndrome virus status. *J. Swine Health* 2011 https://www.aasv.org/shap/issues/v19n1/v19n1p44.pdf.

29. Morrison, R., et.al. PRRS epidemiology: Data from Swine Health Monitoring Project. AASV 2016 proceedings https://www.aasv.org/library/swineinfo/Content/AASV/2016/367_Morrison.pdf.

30. AASV staff. AASV news. 2015 https://www.aasv.org/news/story.php?id=8118.

31. Collins JC, Porras JI. *Built to Last: Successful Habits of Visionary Companies.* New York, New York: HarperCollins Publishers Inc; 1994.

32. Meyer, Ken. Personal communication. June 2018

33.Collins JC. Good to Great: Why Some Companies Make the Leap...and Others Don't. New York, New York: HarperCollins Publishers Inc; 2001.

34. Freese, B. Successful Farming Pork Powerhouses 2018. https://www.agriculture.com/livestock/hogs/pork-powerhouses/pork-powerhouses-2018-ramping-up Accessed 15 December 2018.

35. https://www.pork.org/facts/stats/u-s-pork-exports/ Accessed 31 December 2018.

36. Reeves, David. Personal communication. October 2018.

37. AVMA News 2011. https://www.avma.org/News/JAVMANews/Pages/110615d.aspx Accessed 8 November 2018.

Alex Hogg
Memorial Lecture

Today's swine veterinarian: Challenges and opportunities for the future

Deborah Murray, DVM

In remembrance and honor of Alexander Hogg

Alexander Hogg
1920-2006

A native of Scotland, Alex (pronounced "Alec") Hogg immigrated to the United States with his family when he was six years old. After reaching adulthood, he served in the US Navy during World War II before entering veterinary school at Kansas State University. He received his DVM degree in 1950, and was a mixed animal practitioner in Iowa for twenty years. Dr. Hogg returned to college – this time, Iowa State University – to complete a Masters Degree in 1972. He spent the following years serving and educating swine veterinarians and producers in his roles as Nebraska Swine Extension veterinarian and professor at the University of Lincoln-Nebraska. Upon "retirement" from the university in 1990, Alex worked as a technical services consultant for MVP Laboratories. Ever an enthusiastic learner, he graduated from the University of Illinois Executive Veterinary Program in 1995.

An active and longstanding member of the AASV, Dr. Hogg joined what was then the American Association of Swine Practitioners shortly after it was formed. He served as secretary/treasurer of the association from 1974 to 1976, and was elected president in 1979.

The many awards and honors bestowed upon him over the years included several presented by the AASV: The Howard Dunne Memorial Award in 1985, the Meritorious Service Award in 1996, and in 2005, the first-ever AASV Foundation Heritage Award for a lifetime of achievement in veterinary medicine.

ALEX HOGG MEMORIAL LECTURERS

2011	Joseph F. Connor	2015	C. Scanlon Daniels
2012	Lisa Tokach	2016	Peggy Anne Hawkins
2013	John T. Waddell	2017	Matthew Turner
2014	Mark J. Engle	2018	Rodger Main

Today's swine veterinarian: Challenges and opportunities for the future

Deborah Murray, DVM

Time

Todays swine veterinarians are great multitaskers by necessity. Many are working moms and dads with more distance to travel between sites (and many times between states) which takes them away from family on a frequent basis. As farms have consolidated and systems have grown many swine veterinarians oversee more sows and downstream production than what they historically have. With this distance and number of farms to cover, technology has become an essential piece and tool of practice. The good news is that young swine veterinarians tend to embrace technology and are generally excited to try a new app or to test out a new device. The challenge with technology is to remember that it is a tool but not a substitute for being in the barn and walking the pens yourself. Telemedicine, which many of us have unofficially practiced for years, consists of taking histories and gathering information as well as looking at necropsy pictures or video taken from a group of pigs, through a phone/tablet or computer. This has been a great tool however, telemedicine still legally requires a VCPR and knowledge of the pigs and or premise. One could argue whether this it is really necessary to be at the site when we can gather this vast amount of information, but ultimately the small observations you make or conversations you have while at the site can many times lead you to a clue to solve the root cause of the problem. Without rigorous training many times a field manager/grower or producer may not describe what they are seeing the same way that you would, leading you to a misdiagnosis or at least a delayed diagnosis. A good experiment to conduct with growers or field staff, courtesy of Dr. Mike Eisenmenger, involves asking these individuals to perform a necropsy (after some training to do so) and describe to you what they see, but be there in person to observe the animal yourself. The responses one receives range from fairly close to what you observe, to nothing remotely close with normal organs being considered a potential main cause of clinical signs (when they truly had no pathology at all). With our busy schedules and limited time, continuing to get into barns on a regular basis will be one of our biggest challenges.

Feeding the world

Experts agree that as the population grows agriculture will need to become more efficient, more sustainable, and higher producing than ever before. Veterinarians will be challenged to reduce/eliminate pathogens, to increase efficiencies of food producing animals, and likely with increased scrutiny on antibiotic use and pressure to reduce usage. In other words, do more with less, which is certainly easier said than done. We will definitely need the help of all experts in this endeavor, including universities, researchers, industrial partners, the Pork Board, and the USDA/FDA to name a few.

Communication

As our animals have increased their efficiencies a continued challenge is to simply meet and understand their requirements. Do we need to design our barns differently? Has feeder space and water allowance changed? Have we not only met their nutritional requirements but optimized them? Do we have the research capability of measuring gain and feed conversions while also having studies large enough to take into account small differences in mortality at a statistically significant level? We have valuable resources to help us find answers to these questions, such as geneticists and nutritionists, but have we been able to work together and build off each other's knowledge in the most productive way? Communication and team work will be a huge opportunity to take advantage of.

Public perception

Public perception, although not a challenge that we alone face, will continue to be critical to the economic survival of our producers. There has not been a trend to move back to the farm and with each generation away from the farm, connecting with consumers to help them understand where their food comes from will be more and more difficult. We need to be able to explain what we do and how pigs are raised, or we will lose consumers to protein sources like fake meat, or continue to have activists groups drive legislation and increase the cost to raise pigs to the point of loss of business and loss of consumers looking for a more affordable option.

Young talent

Ultimately our most valuable resource and opportunity for future progress is our young talent. Armed with new ideas and the ability to see farms/records/procedures without any preconceived ideas or historic baggage, their thoughts and comments can be invaluable at improving or even recreating procedures. Frequently they trigger new ideas, even with seasoned practitioners, as they assume very little and ask fundamental questions. These questions can be quite important as practices/procedures that may no longer be relevant have been glossed over (and may no longer practical or the most efficient ways of doing things). I had the privilege of judging the student poster competition at AASV several years ago. While having a conversation with the student at his poster on McREBEL he mentioned to me that he had gotten nearly all negative results when testing common equipment and fomites in the farrowing house that were traditionally thought to spread PRRSV, but that he had noticed the employees disseminating tails and testicles through out the room while processing. He then decided to grab some of them (not part of the project but he took it on his own to do this) and submit them for PRRSV PCR testing. Low and behold some of these tails and testicles were PRRSV positive. With this student's observations, it triggered a system change within our system as we began to collect tails and testicles while piglets were being processed to reduce the risk of PRRSV transmission within the farrowing house. This became a standard practice on all of our farms. Eventually it led to the observation that there was fluid that had accumulated in the bucket which could potentially be used for PRRSV testing to test more of the population with less labor. Today many farms and producers use processing fluids for testing of multiple pathogens, which originally came from a student idea/observation!

So how does the AASV retain and cultivate talent? This is truly both an ongoing challenge as well as an opportunity.

How does AASV cultivate and retain talent?

To better understand how we are doing today and how we can continue to improve this process as an organization, AASV members who graduated from veterinary school 2005 to present were surveyed and asked these questions. How are we doing as an organization to help prepare new graduates for practice? 39% of respondents strongly agreed that the AASV has helped prepare them for practice, an additional 45% agreed, 13% as neutral and only 3% that disagreed. A big part of a new graduates first months/years of practice is mentorship, which our members strongly agreed with. 90% of respondents say that they had a mentor who helped them their first few years in practice. Who are these mentors? They are many of our members. When asked who mentored them their first years in practice, the list was quite extensive with well over 100 members having served as mentors to these respondents, and several individuals were considered a primary mentor to many new graduates. Did these mentors need to practice for 20+ years to have this influence on new graduates? No, they did not. Many mentors listed had only been out of school for 10 years or less themselves. Was our experience at New Fashion Pork unique in getting an idea or procedure from a student presentation or poster at AASV that was then applied in their practice? No, it was not. 87% of respondents agreed that they too were able to apply in their practice what a student worked on and presented at AASV as either a poster or an oral presentation. With companies and practices benefiting from these student projects, once these students graduate do these new or recent graduates pay it forward? Yes, they are with 91% of respondents saying they host students for preceptorships or summer internships. The breakdown was as follows; 18% of individuals surveyed saying they host students a great deal, 26% saying they host students a lot, 19% host students a moderate amount, 28% saying they host students a little, with only 9% saying they don't host students at all (and may not have been out of school long enough to do so). One of the most common responses which I have heard from students in regard to AASV reflects just this, "AASV was the most welcoming organization for me as a student", and that they were included and welcomed by nearly all of the members that they have interacted with. With all of these relatively new graduates hosting students which worked on projects with their clinic/company, did the student project pay off and were they able to apply any of this work? 66% of respondents say that their practice or company changed or modified a protocol or procedure as a result of work done on a summer internship or project that their company or practice hosted. Many times change is not easy and good science behind the change is critical, this speaks volumes about the quality of the student projects that are accomplished by our up and coming members. What were some of the things that were changed? The responses ranged from production related changes such as gilt breeding timing, sow feeding, troubleshooting positive pressure filtration, employee training program development for new employees, change of light bulb brands for farrowing, to many responses involving biosecurity practices which were changed or modified due to new information they received. Diagnostics achieved the highest number of responses, but disease control and antibiotic/analgesic use/applications received quite a few mentions as well. With all of the positive comments on student projects, did AASV influence the career choice of recent graduates during or even after vet school? Examples of this may have been students that would have chosen small animal practice, or students that may have practiced mixed and decided to practice on just swine? Over 50% of respondents agreed that the organization had influenced their career choice. As a group the AASV has done a great job in mentoring students and has gained some valuable information in return from these student interactions. Can we improve upon this? What do our students and recent graduates feel would be valuable going forward? Many are very happy with what is currently being done by the AASV. Would you be surprised to learn that most respondents would like MORE mentorship and networking? Never underestimate the value that mentorship brings to these young graduates and if you are not already a mentor this is certainly a call to become one. In addition to mentoring and cultivating up-and-coming talent, when asked what technologies do they feel are needed to maintain or improve their level of practice over the next 15-20 years here were some of the responses: "pen side diagnostics", "better (and easier) data integrations and analysis", "autopilot vehicles", "telemedicine", "diagnostic capabilities outside of infectious disease (metabolic, nutrition, auto-immune, degenerative)", "technology which allows data consolidation and analysis", "precision agriculture technologies", "robotics", "better vaccines and therapies", "digital records and standardization of medical records", "artificial intelligence technologies to identify issues in a barn with real time data reporting", as well as many more responses on data management, interpretation, and integration. As swine veterinarians we may also need to learn additional languages to better communicate with our employees and staff. Several responses to the survey felt very strongly about this as our farm staff has culturally changed over the years and will likely continue to do so. A few responses discussed people and communication. As we move forward with more and more data and information, we will also need to find a way to continue to communicate this information with clients and farm staff, "which is something they don't teach in vet school" as many have mentioned. One response takes us back to the future and I could not agree more with this individual's view on the future challenge of a swine veterinarian, "We need to focus on basics because animal welfare will become more important going forward and how the public views our industry. Attracting and maintain a good work force and compensating them fairly to maintain a good welfare standard, or we need technology to do their job better. How do you explain 10% + mortality to the public and the volume of water meds used at times, we need to figure out how to get ahead of major swine diseases."

Doing what is right for the pig

As stated in the Veterinarians Oath we have sworn to, "...the protection of animal health and welfare, the prevention and relief of animal suffering, the conservation of animal resources...", and in the quest to maximize profit and minimize losses as well as to improve health, sometimes the cheapest option is not the best for the pig. As swine veterinarians it is our duty to find a way to accomplish both and yet continue to keep our clients and employers in business successfully as we have done for many years. With all of these opportunities and challenges our work is cut out for us but I am confident that as an organization the AASV will help to guide us in this path and continue to challenge us to improve and progress like many of the producers have that we work with on a day to day basis.

Comparison of different cell lines for improving PRRSV virus isolation from clinical samples

Wannarat Yim-Im[1], MS; Haiyan Huang[1], BS; Jie Park[1], PhD; Chong Wang[1], PhD; Gabriela Calzada[2], PhD; Phillip Gauger[1], DVM, PhD; Karen Harmon[1], PhD; Rodger Main[1], DVM, PhD; Jianqiang Zhang[1], MD, PhD
[1]Veterinary Diagnostic and Production Animal Medicine, Iowa State University, Ames, Iowa;
[2]Aptimmune Biologics, St Louis, Missouri

Introduction

Porcine reproductive and respiratory syndrome (PRRS) causes tremendous economic loss to global swine industry. The causative agent, PRRS virus (PRRSV), includes PRRSV-1 (type 1) and PRRSV-2 (type 2). Due to the high rate of genetic and antigenic diversity of PRRSV, the commercial vaccines are not always effective against field isolates. Virus isolation (VI) is often conducted to obtain PRRSV isolates for producing autogenous vaccines. The MARC-145 cell line is commonly used for PRRSV VI but the success rate is frustratingly low. Recently, a ZMAC cell line, derived from porcine alveolar macrophages, has become available. The objectives of this study were to compare PRRSV VI efficiency in ZMAC and MARC-145 cells, compare the infectious titers of PRRSV isolates obtained in two cell lines, and investigate if ZMAC-derived PRRSV isolates grow in MARC-145 cells and vice versa.

Materials and methods

Clinical samples

264 PRRSV-2 PCR-positive clinical samples including 109 serum samples with C_T 13.2-36.2, 96 lung samples with C_T 13-30.2, and 59 oral fluid samples with C_T 22.0-34.0 and 70 PRRSV-1 PCR-positive clinical samples including 31 serum samples with C_T 18.4-36.8, 8 lung samples with C_T 21.6-35.4, and 31 oral fluid samples with C_T 28.0-36.8) were selected from diagnostic cases submitted to the Iowa State University Veterinary Diagnostic Laboratory (ISU VDL) and used for virus isolation attempts in MARC-145 and ZMAC cells in this study.

Cells

MARC-145 cell line is a clone of the African monkey kidney cell line MA-104.[1] ZMAC cell line was derived from the lung lavage of porcine fetuses.[2]

Virus isolation

Clinical samples were filtered through 0.45 μM membranes to sterilize samples. For inoculation, 200 μl of samples were inoculated per well of 24-well plates containing monolayers of MARC-145 cells and ZMAC cells, respectively. After incubation for 1 h at 37°C incubator with 5% CO_2, inoculum was removed and 1 mL of the culture medium was added per well. Inoculated cells were checked daily for development of cytopathic effects (CPE) for 3-5 days. Cell culture supernatants were harvested and the cell plates were fixed with 80% cold acetone followed by immunofluorescence staining using PRRSV nucleocapsid protein-specific monoclonal antibodies conjugated to fluorescein isothiocyanate. Supernatants were tested by PRRSV real-time RT-PCR to confirm VI results in case of unclear immunofluorescence staining.

Determination of PRRSV-2 genetic lineages

PRRSV-2 ORF5 sequencing was performed using the Sanger method following the previously described procedures.[3] The PRRSV-2 ORF5 sequences were aligned with the reference sequences representing the nine lineages of PRRSV-2[4,5] using the MEGA6 MUSCLE program and the maximum likelihood phylogenetic tree was generated using the MEGA6 software.

Virus titration

Virus isolates from 20 PRRSV-2 PCR-positive samples with VI positive in both MARC-145 and ZMAC cells were selected for virus titer comparisons. Virus isolates obtained in MARC-145 cells were titrated in MARC-145 cells and virus isolates obtained in ZMAC cells were titrated in ZMAC cells. Each virus isolate was serially 10-fold diluted and inoculated into MARC-145 cells or ZMAC cells grown in 96-well plates, 100 μl per well, five replicate wells per dilution. The plates were incubated at 37°C with 5% CO_2 for 3-5 days. Viral CPE was recorded daily. The cell plates were fixed with 80% cold acetone and followed by immunofluorescence staining. Supernatants were tested by PRRSV real-time RT-PCR to confirm VI results in case of unclear immunofluorescence. The virus titers were calculated according to the method described by Reed and Muench and expressed as $TCID_{50}$/mL.[6]

Virus growth cross-checking

Seventy-eight PRRSV-2 isolates obtained in ZMAC cells were inoculated into MARC-145 cells to evaluate their growth. Similarly, 42 PRRSV-2 isolates obtained in MARC-145 cells were inoculated into ZMAC cells to evaluate their growth.

Statistical analyses

McNemar's test was performed to investigate the effects of cell lines or lineage classification on PRRSV VI outcomes. The effects of C_T values and specimen types on PRRSV VI were evaluated by Penalized Maximum Likelihood Estimates. Fisher's exact test was performed to examine the results of virus growth cross-checking.

Results

Comparison of PRRSV-1 and PRRSV-2 VI from clinical samples in ZMAC and MARC-145 cells

None of the 31 PRRSV-1 and 59 PRRSV-2 oral fluid samples was VI positive in either ZMAC or MARC-145 cells (Table 1). Among the 39 PRRSV-1 serum and lung samples (31 serum and 8 lung samples), 25.8% (8/31) and 0% (0/31) of serum samples, 62.5% (5/8) and 12.5% (1/8) of lung samples, and 33.3% (13/39) and 2.6% (1/39) of serum plus lung samples, were VI positive in ZMAC and MARC-145 cells, respectively (Table 1). Among the 205 PRRSV-2 samples (109 serum and 96 lung samples), 47.7% (52/109) and 9.2% (10/109) of serum samples, 68.8% (66/96) and 45.8% (44/96) of lung samples, and 57.6% (118/205) and 26.3% (54/205) of serum plus lung samples, were VI positive in ZMAC and MARC-145 cells, respectively (Table 1). Statistical analysis indicated that PRRSV-1 and PRRSV-2 VI success rates were significantly higher in ZMAC than in MARC-145 for serum and lung samples. In addition, the PRRSV-2 VI success rates were significantly higher in lung than in serum samples regardless of in ZMAC or MARC-145 cells.

Impact of C_T values on PRRSV-2 VI outcomes

The success rates of PRRSV-2 VI from serum and lung samples with different C_T values were 90.6% and 49.1% ($C_T < 20$), 71.2% and 32.9% (C_T 20-25), 31.5% and 7.4% (C_T 25-30), 5.6% and 0% (C_T 30-33) in ZMAC and MARC-145 cells, respectively (Table 2). The success rates of PRRSV-2 VI decreased in serum and lung samples with increasing C_T values. Evidently, PRRSV-2 VI success rate was very low for samples with C_T 30-33 and there was no VI success from samples with $C_T > 33$ either in ZMAC or MARC-145 cells. Statistical analysis indicated that C_T values significantly affected the success rates of PRRSV-2 VI in ZMAC and MARC-145 cells.

Impact of genetic lineages on PRRSV-2 VI outcomes

Based on phylogenetic analysis, ORF5 sequences of 220 PRRSV-2 samples from 91 serum, 85 lung, and 44 oral fluid samples were determined to belong to four lineages. These included lineage 1 (102 samples including 48 serum, 33 lung and 21 oral fluid samples), lineage 5 (69 samples including 27 serum, 30 lung and 12 oral fluid samples), lineage 8 (44 samples including 16 serum, 17 lung and 11 oral fluid samples), and lineage 9 (5 lung samples). The lineage information and the VI outcomes from 176 serum and lung samples are presented in Table 3. For lineage 1 and lineage 8 samples (serum plus lung), the VI success rates were significantly higher in ZMAC cells than in MARC-145 cells. In contrast, for the lineage 5 serum plus lung samples, the VI success rate between ZMAC and MARC-145 cells was not significantly different. Due to the small number of samples in lineage 9, they were excluded from statistical analysis.

Comparison of virus titers of PRRSV-2 isolates obtained in ZMAC and MARC-145 cells

As shown in Figure 1, most of the isolates (14/20) obtained in ZMAC and MARC-145 cells had similar infectious titers whereas 5 ZMAC isolates had higher titers than MARC-145 isolates and one ZMAC isolate had lower titer than MARC-145 isolate.

Cross-checking of PRRSV isolates growth in ZMAC and MARC-145 cells

Among the 78 PRRSV-2 isolates obtained in ZMAC cells, 45 (57.7%) grew and 33 (42.3%) did not grow in MARC-145 cells while all of the 42 PRRSV-2 isolates obtained in MARC-145 cells grew in ZMAC cells.

Discussion

Under the conditions of this study, PRRSV-1 and PRRSV-2 was not isolated from any oral fluid samples regardless of using ZMAC or MARC-145 cells, indicating that PRRSV VI from oral fluid is still a big challenge. For clinical serum and/or lung samples, the success rates of PRRSV-2 and PRRSV-1 VI were significantly higher in ZMAC cells than in MARC-145 cells. When PRRSV VI is negative in MARC-145 cells, PRRSV VI attempts in ZMAC cells should be explored. The PRRSV-2 VI success rates were significantly higher in lung than in serum samples in either MARC-145 or ZMAC cell line. The preferred

Table 1: Outcomes of PRRSV virus isolation from clinical samples in MARC-145 and ZMAC cells

Type	Specimen	Number	ZMAC VI+	MARC-145 VI+
PRRSV-2	Serum	N = 109	47.7% (52/109)	9.2% (10/109)
	Lung	N = 96	68.8% (66/96)	45.8% (44/96)
	Oral fluid	N = 59	0% (0/59)	0% (0/59)
	Total	**N = 264**		
PRRSV-1	Serum	N = 31	25.8% (8/31)	0% (0/31)
	Lung	N = 8	62.5% (5/8)	12.5% (1/8)
	Oral fluid	N = 31	0% (0/31)	0% (0/31)
	Total	**N = 70**		

Table 2: Distribution of PRRSV-2 PCR C_T values and virus isolation outcomes in ZMAC and MARC-145 cells based on specimen types

	Serum & lung		
C_T value	Total	ZMAC VI+	MARC-145 VI+
< 20	53	48 (90.6%)	26 (49.1%)
20-25	73	52 (71.2%)	24 (32.9%)
25-30	54	17 (31.5%)	4 (7.4%)
30-33	18	1 (5.6%)	0
33-37	7	0	0
Total	205	118 (57.6%)	54 (26.3%)

Table 3: Virus isolation outcomes of different PRRSV-2 lineages in MARC-145 and ZMAC cell lines

	Serum & lung		
Lineage	Total	ZMAC VI+	MARC-145 VI+
1	81	53 (65.4%)	13 (16.0%)
5	57	19 (33.3%)	22 (38.6%)
8	33	19 (57.6%)	9 (27.3%)
9	5	3 (60%)	2 (40%)
Total	176	94 (53.4%)	46 (26.1%)

specimen types for PRRSV VI are lung > serum > oral fluid. The success rates of PRRSV VI decreased in samples with increasing C_T, hence PRRSV VI had low success in samples with $C_T > 30$. Besides, genetic lineages also affected the PRRSV VI outcomes. If there is no information about genetic lineages, PRRSV VI in ZMAC will overall give a better outcome than in MARC-145. It is noteworthy that not all of the PRRSV-2 isolates obtained in ZMAC cells grew in MARC-145 cells. Currently, the majority of autogenous vaccine manufacturing companies/labs use MARC-145 cells for PRRSV propagation; they may need to consider introducing ZMAC cells to their labs if they want to use PRRSV isolates obtained in ZMAC cells by veterinary diagnostic laboratories. In summary, this study provides important guidelines to the swine veterinarians, producers, and researchers to better serve their needs of obtaining PRRSV isolates for further characterization or for producing autogenous vaccines.

References

1. Kim HS et al. Enhanced replication of porcine reproductive and respiratory syndrome (PRRS) virus in a homogeneous subpopulation of MA-104 cell line. *Arch Virol.* 1993; 133:477.

2. Calzada-Nova G, Husmann RJ, Schnitzlein WM, Zuckermann FA. Effect of the host cell line on the vaccine efficacy of an attenuated porcine reproductive and respiratory syndrome virus. *Vet Immunol Immunopathol.* 2012; 148:116.

3. Zhang J et al. High-throughput whole genome sequencing of Porcine reproductive and respiratory syndrome virus from cell culture materials and clinical specimens using next-generation sequencing technology. *J Vet Diagn Invest.* 2017; 29:41.

4. Shi M et al. Phylogeny-Based Evolutionary, Demographical, and Geographical Dissection of North American Type 2 Porcine Reproductive and Respiratory Syndrome Viruses. *J Virol.* 2010; 84:8700.

5. Shi M et al. The spread of Type 2 Porcine Reproductive and Respiratory Syndrome Virus (PRRSV) in North America: A phylogeographic approach. *Virology.* 2013; 447:146.

6. Reed and Muench. A Simple Method of Estimating Fifty Percent Endpoints. *Am J Hyg.* 1938;27:493.

Figure 1: Comparison of infectious titers ($TCID_{50}$/mL) of 20 PRRSV-2 isolates obtained in ZMAC and MARC-145 cells

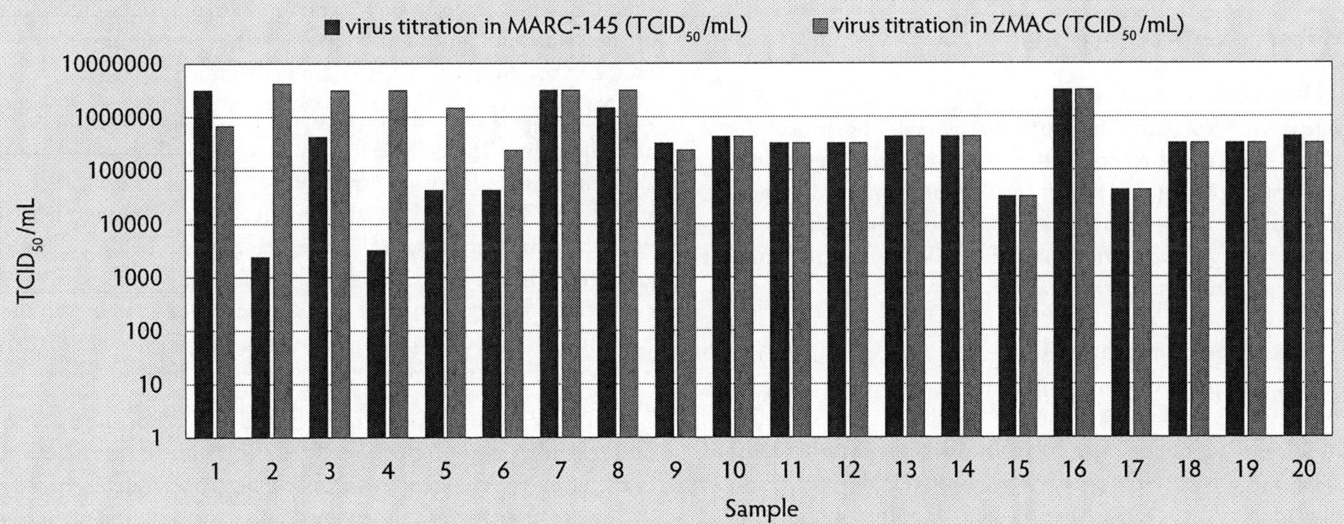

Investigating biosecurity aspects related to PRRSV outbreaks

G. S. Silva[1], DVM, MS, PhD; G. Machado[2], DVM, MS, PhD; K. Baker[1]; D. Holtkamp[1], DVM, MS; D. C. L. Linhares[1], DVM, MBA, PhD

[1]Veterinary Diagnostic and Production Animal Medicine Department, Iowa State University, Ames, Iowa; [2]College of Veterinary Medicine, North Carolina State University, Raleigh, North Carolina

Introduction

Biosecurity is defined as the set of practices to prevent introduction of infectious agents such as porcine reproductive and respiratory syndrome virus (PRRSV) in a herd. In other words, biosecurity can be viewed as a way to keep pathogens out of the farm and control further spread of the pathogens that are already present.[1] Even after years of research over the last three decades, PRRSV is still causing productivity losses also considered a major health problem affecting the global swine industry.[2]

The assessment of biosecurity practices in breeding herds is usually done through surveys. However, biosecurity surveys may be long, time consuming and thus to obtain such information from producers may be exhausting. Therefore, the identification of a few biosecurity aspects typically associated with the occurrence of PRRSV outbreaks would facilitate the process to obtain biosecurity information from producers, allowing to compare biosecurity vulnerabilities among herds, and within herds over time. The objective of this study was to identify the minimum number of practices that when combined have a strong association with the likelihood of PRRSV introduction into swine breeding herds. This was done by comparing the biosecurity aspects between farms with and without PRRSV outbreaks through a set of machine-learning algorithms.

Materials and methods

1. Data and survey

Data from a cross-sectional study that assessed the biosecurity aspects in 84 breeding herds from 14 US production systems in 2016 was used.[3] The biosecurity aspects of each breeding herd were assessed using a biosecurity survey with questions about herd demographics, swine density, PRRSV history, frequency of risk events, and biosecurity practices.

2. Classification method

In the 1st step, selected machine learning algorithms (Random Forest, Gradient Boosting Machine and Support Vector Machine) were trained to classify (yes/no) whether a farm reported a PRRS outbreak in the past 5 years, given the biosecurity aspects collected in the surveys.[4] After the best algorithm was chosen, a second step reduced the number of biosecurity aspects based in a set of parameters including measures of accuracy.[5]

3. Predictive positive value as a risk index

To evaluate the feasibility of using the positive predictive value (PPV) obtained by the model as a risk index of PRRSV outbreak, the PPV of each farm was obtained, and the association with the frequency of PRRSV outbreaks reported by each farm in the last 5 years were assessed. Pearson correlation coefficient was used to test this association.

Results and discussion

The mean herd size was 3,437 sows (range: 543-7,200) in the herds that did not have an outbreak (n = 34) and 4,006 sows (range: 1,000-10,852) in the herds that had at least one outbreak (n = 50) during the investigated period.

After the first step, Random Forest shown to outperform other algorithms with an accuracy: 76.3%; sensitivity: 86.3%; specificity: 62.5%. Only 20 shown to be relevant and these variables were related to were related to farm characteristics and requirements (e.g. being a commercial herd or downtime required for visitors), monthly frequency of risk events (e.g. frequency of weans per month or frequency of rendering per month), animal density around the farm (e.g. number of swine sites within 3 miles or number of finishing pigs within 3 miles) and operational connections to other sites (e.g. number of sites that share the same breeding replacement trailer). The variables in order of relevance are described in Figure 1 and represent the most common risk factors associated with PRRSV transmission.[6]

After 1st step, these variables were offered to the reduction variable algorithm. The best performance model was obtained using Random Forest algorithm (accuracy: 80%; sensitivity: 86%; specificity: 72%) with 6 variables retained in the final model. The variables with increasing scale of relevance were: being a commercial herd, number of finishing pigs within 3 miles radius, frequency of monthly dead animal removal, type of dead disposal in the farm, if PRRSV positive animals were hauled on a trailer that also hauls culled sows and frequency of weans per month.

In summary, the probability to be classified as reporting a PRRS outbreak increased proportionally to the frequency of risk events in the farm (e.g. frequency of weaning pigs, or frequency of rendering per month, or to the swine density surrounding the farm (e.g. number of swine sites or pigs within 3 miles). Likewise, the risk of reporting PRRS outbreak increased when the farm did not have onsite disposal method for dead animal, or when the farm shared trailers with other farms to transport breeding replacement animals or culled sows. However, the risk decreased with the increase in the frequency of weaning pigs per month, with the number of employee working daily in the farm, and with the increase in the downtime period requirement for visitors.

The scoring system (PPV) obtained by the first step showed a strong positive correlation with the frequency of PRRSV outbreaks reported by the farms in the 5 past years (Pearson r = 0.71, $P < 0.01$). Using the second approach, the association was higher, with a Pearson correlation coefficient of 0.77 ($P < 0.01$).
Figure 2 describes this relationship for step 1 and 2, respectively.

Conclusion and implication

The results prove the concept that there is correlation in biosecurity aspects of swine breeding herds. In other words, it is possible to measure and compare the biosecurity vulnerabilities of swine sites by 'asking' fewer questions. In this dataset, both methods had a good accuracy (80%) to identify farms that reported a PRRS outbreak(s) in the previous 5 years. This can help producers and veterinarians to measure and benchmark biosecurity vulnerability more frequent, identify sites at relatively higher (or lower) risk of PRRSV introduction, and set priorities to manage the risk of virus introduction over time. In addition, this method can be used to screen farms before the adoption of a comprehensive biosecurity assessment.

References

1. Food and Agriculture Organization. Good Practices for Biosecurity in the Pig Sector - Issues and Options in Developing and Transition Countries. Food and Agriculture Organization of the United Nations/World Organization for Animal Health/ World Bank. 2010. http://www.fao.org/docrep/012/i1435e/i1435e00.pdf.

2. Holtkamp DJ, Kliebenstein JB, Neumann EJ, et al. Assessment of the economic impact of porcine reproductive and respiratory syndrome virus on United States pork producers. *J Swine Heal Prod.* 2013;21(2):72-84. http://www.aasv.org/shap/issues/v21n2/v21n2p72.html.

3. Baker KL, Silva G, Morrison R, Vilalta C, Holtkamp D, Linhares D. Comparison of biosecurity aspects between swine breeding herds with low and high incidence of porcine reproductive and respiratory syndrome virus outbreaks. In: *49th Annual Meeting of the American Association of Swine Veterinarians.* 2018:80.

4. Breiman L. Random Forests. Mach Learn. 2001;45(1):5-32. doi:10.1023/A:1010933404324

5. Kursa MB, Rudnicki WR. Feature Selection with the Boruta Package. J Stat Softw. 2010. doi:10.18637/jss.v036.i11

6. Zimmerman JJ, Benfield D, Dee SA, et al. Porcine Reproductive and Respiratory Syndrome Virus (Porcine Arterivirus). In: Zimmerman JJ, Karriker LA, Ramirez A, Schwartz KJ, Stevenson GW, eds. Diseases of Swine. 10th ed. Wiley-Blackwell; 2012:461-486.

Figure 1: Variable relevance plot. Top 20 most relevant biosecurity aspects.

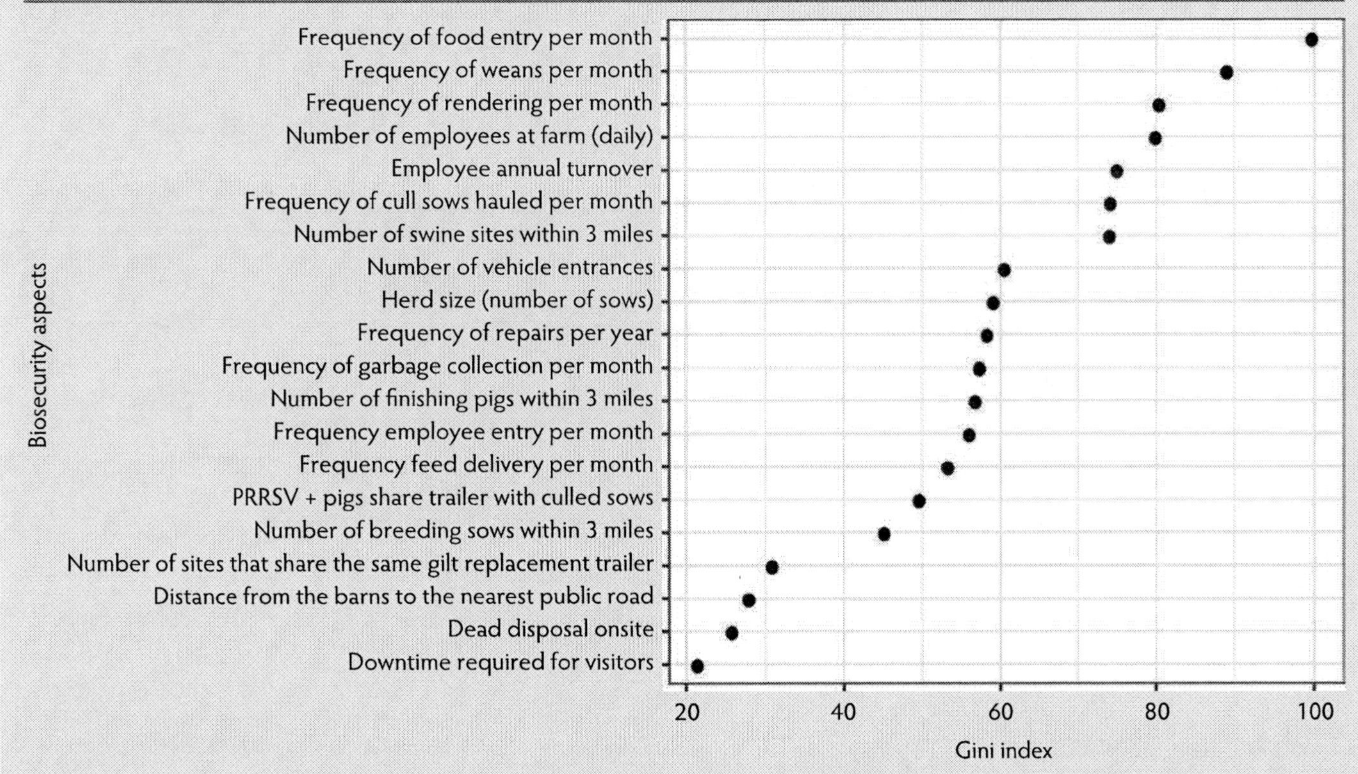

Figure 2: Scatter plot of the observed number of outbreaks in the x-axis and the positive predicted values (PPV) in the y-axis. In "black" the PPV from farms that did not report PRRSV outbreak during the study period, and in "brown" the PPV from farms that reported PRRSV outbreak(s) during the study period. Left side A – First step and right side B – second step.

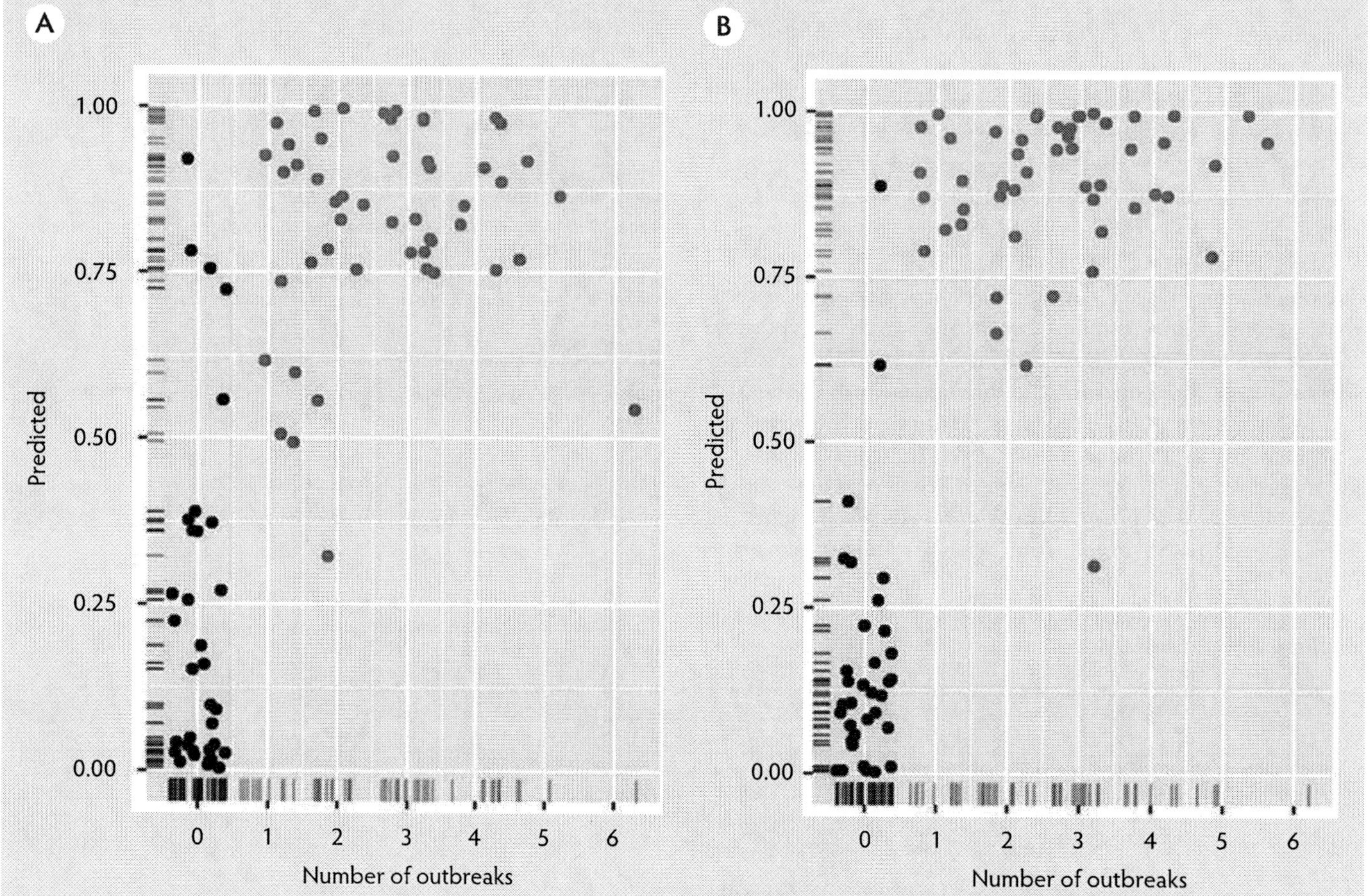

Comparison of production impact in breeding herds following PRRS clinical outbreaks adopting killed or attenuated PRRS virus vaccination protocols

Gaurav Rawal; Christopher Rademacher; Daniel Linhares
Veterinary Diagnostic and Production Animal Medicine, College of Veterinary Medicine, Iowa State University

Introduction

Porcine reproductive and respiratory syndrome virus (PRRSV) has been endemic for more than 30 years in most pig producing countries across the globe, and is still one of the most economically important pathogen affecting swine. The use of modified live vaccines (MLV) on breeding herds has been reported by several investigators and is a common practice in the industry for the control of PRRSV. There is limited published information on efficacy of killed vaccines (KV) as part of PRRS control programs in breeding herds.

Objective

The objective of this study was to describe productivity losses from breeding herds using KV (MJPRRS®) protocol compared to MLV protocol for herds following an outbreak with PRRSV.

Materials and methods

A retrospective observational study was conducted, following 78 PRRSV eligible outbreaks in swine breeding herds. More specifically we followed 27 outbreaks from 19 breeding herds using the KV protocol,[1] and 51 outbreaks from 50 breeding herds using the MLV protocol.

A survey was used to record key demographic information including herd size (number of breeding sows in the inventory), location of gilt acclimation site (offsite or onsite), type of gilt development units (dedicated or shared), duration of herd closure (in weeks), prior PRRS status before an outbreak (stable or unstable), frequency of weaning events per week (1 and 2 or 3+), and PRRS vaccination protocol of breeding sows and/or gilts (KV, or MLV).

Moreover, weekly productivity data was gathered. Outbreaks were defined by American Association of Swine Veterinarians (AASV)[2] when there was evidence of PRRS-associated changes in productivity such as significant decrease in weaned pigs, or spike in the number of weekly aborts.

Time to baseline production (TTBP), and total loss per thousand sows (TL/1000 sows) were reported for outbreaks, as previously established.[3] Eligible outbreaks were censored if dropped from study between enrollment and reaching TTBP, or at 52 weeks if not having reached TTBP. 'Time to baseline production' (TTBP), and 'total loss per thousand sows' (TL/1000 sows) attributed to PRRS was compared between treatment groups (eligible outbreaks reporting to use KV, compared to those using MLV). Statistical analyses were performed with SAS version 9.4.

Results

The univariate analysis demonstrated that there was a shorter median TTBP of 6.2 weeks and significantly lower 1,614 pigs per 1000 sows in outbreaks adopting the KV protocol, as compared to those adopting the MLV protocol. In the final model (multivariate analysis), there was no significant difference between groups for TTBP after adjusting for weaning frequency, but outbreaks reporting use of KV still had significant lower ($P = 0.0130$) TL/1000 sows compared to outbreaks reporting use of MLV.

Conclusion

This study provided information about changes in productivity of commercial breeding herds using KV protocol as part of PRRS management strategy when compared to those using MLV protocol. After adjusting for significant covariates, there was no difference on TTBP between treatment groups. However, outbreaks using KV vaccine had significantly lower TL/1000 sows compared to those using MLV.

References

1.Rawal G, Rademacher C, Linhares D. Description of herd-outbreaks Using Killed PRRSV Vaccine with an Attempt to Boost Immune Response. Proceedings Paper 2018 ISU James D. McKean Swine Disease Conference, November 1–2, 2018, Page no. 35-38.

2.Holtkamp, D., Polson, D., Torremorell, M., Morrison, R., Classen, D., Becton, L., Henry, S., Rodibaugh, M.T., Rowland, R.R., Snelson, PRRSV., Straw, B., Yeske, P., Zimmerman, J., 2011. Terminology for classifying swine herds by porcine reproductive and respiratory syndrome virus status. J. Swine Health Prod. 19 (1), 44–56.

3. Linhares, D.C., Cano, J.P., Torremorell, M., Morrison, R.B., 2014. Comparison of time to PRRSV-stability and production losses between two exposure programs to control PRRSV in sow herds. Prev. Vet. Med. 116 (1–2), 111–119.

Economic and production benefit of 2 PRRS MLV doses compared to a single dose vaccination program on growing pigs

C. A. A. Moura[1], DVM; R. Philips[2], DVM; G. S. Silva[1], DVM, MS, PhD; E. Mondaca[2], DVM, PhD; A. Scheidt[2], DVM, MS; Jens Kjaer[2], DVM; D. C. L. Linhares[1], DVM, MBA, PhD
[1]Veterinary Diagnostic and Production Animal Medicine Department, Iowa State University, Ames, Iowa;
[2]Boehringer Ingelheim Animal Health, Duluth, Georgia

Introduction

PRRSV continues to cause significant impact in whole-herd productivity.[1] It has been demonstrated that PRRS modified-live virus (MLV) vaccination, with one full dose, significantly reduces the clinical consequences of PRRSV challenge compared to non-vaccinates.[2-8] Aiming to reduce clinical signs and shedding, some veterinarians have recommended vaccination of growing pigs with two doses of attenuated vaccines. In fact, it has previously been described that the application of two vaccinations was effective in reducing the spread of wild-type virus in nursery and finishing pig populations.[9-13] The objectives of this study were to 1) evaluate the performance of growing pig groups vaccinated with 2 doses of PRRSV MLV and placed in high pig dense areas, as compared to matched groups vaccinated with single dose, and 2) to assess the economic benefit of using a second dose of PRRSV MLV on a vaccination program of growing pig groups placed at high pig dense area.

Methods

A randomized field study was conducted from fall of 2017 to summer of 2018 to compare the production impact and economic benefit of two Ingelvac PRRS®MLV vaccination doses (Boehringer Ingelheim Vetmedica Inc., St Joseph, MO) compared to a single dose protocol using the same product.

Fifty-five growing pig groups were conveniently selected from five PRRSV-stable and 2 PRRSV-positive[14] sow farms in the US, randomly allocated, by a coin toss, into two treatment groups and placed at high pig dense areas. All pigs received one full intramuscular dose of Ingelvac PRRS® MLV vaccine around weaning or processing age. Lots from the 2-dose treatment group received the second dose 4 weeks after the initial dose. Six cotton ropes were spatially distributed in the barns for oral fluid sampling,[15] and samples were collected on the week of placement for PRRS rRT-PCR testing and PRRSV ORF-5 sequencing to confirm the PRRSV status of the group (negative, positive for wild-type PRRSV, or positive for vaccine-like PRRSV). Production records from weaning to market were collected from all groups for performance analysis.

Poisson regression models were built to measure the effect of the treatment on mortality. Number of deaths were used as the response variable and number of pigs weaned were used as an offset variable. A linear regression model was built to measure the effect of treatment on ADG. The mean differences and 95% CIs were reported. A production-economic analysis was done, using the mortality, feed conversion and ADG adjusted for 165 days on feed, to assess the economic benefit of using an extra MLV dose. The differences on the revenue and on the variable costs were collected to define the margin over variable costs for each treatmet group. The benefit of the second dose was calculated by the difference on the margin over variable costs. The cost of the second dose was calculated by the vaccine dose price and the cost of vaccination labor per dose. The benefit-cost ratio was reported as a measure of economic benefit of the second vaccination dose. The economic analysis was done assuming a price point of $0.99 per vaccine dose, $0.10 of vaccination labor per dose, feed costs of $0.23 per pound and market pig price of $0.58 per pound. All production and economic analysis were performed by PRRSV status of the respective breeding sow farm sources.

Results

The study included 55,230 growing pigs (Table 1). Lots from the single dose treatment from PRRSV-stable breeding herds had a significantly higher mortality (mean 7.20%) when compared to the 2-dose treatment group (6.51%; $P = 0.0144$) with a mean mortality difference of 0.69% (95% CI: 0.61% - 0.77%). (Figure 1A) Additionally, lots from the single dose treatment from PRRSV-positive breeding herds had a significantly higher mortality (mean 20.31%) when compared to 2 doses treatment (15.39%; $P < 0.0001$) (Figure 1B) with a mean mortality difference was 4.92% (95% CI: 4.99% - 4.85%). There was no statistical difference on ADG (Table 2) between treatment groups from either PRRSV-stable ($P = 0.7563$) and -positive breeding herds ($P = 0.9123$). The benefit-cost ratio of the extra dose of MLV vaccine was 1.11 for groups from PRRSV-stable farms and 4.08 for groups from PRRSV-positive farms, showing an economic benefit of a double dose vaccination program (Table 3).

Table 1: Baseline production for each treatment group.

Treatment	1 dose group	2 dose group
Weight started (lb)	16.02	16.05
Average days on feed	161	160
Groups started	26	29
Pigs started	26370	28860

Table 2: Regression model for ADG comparison between treatment groups.

PRRSV status	Adjusted* ADG for 1 dose group	Adjusted* ADG for 2 dose group	Adjusted* ADG difference	Number of groups	*P*-value
Overall	1.756	1.766	0.01	55	0.7546
Stable	1.747	1.758	0.011	40	0.7563
Positive	1.776	1.770	-0.006	15	0.9123

* ADG was adjusted for 165 days.

Figure 1: A) Poisson regression model for mortality comparison between treatment groups from PRRSV-stable breeding herds. **B)** Poisson regression model for mortality comparison between treatment groups from PRRSV-positive breeding herds.

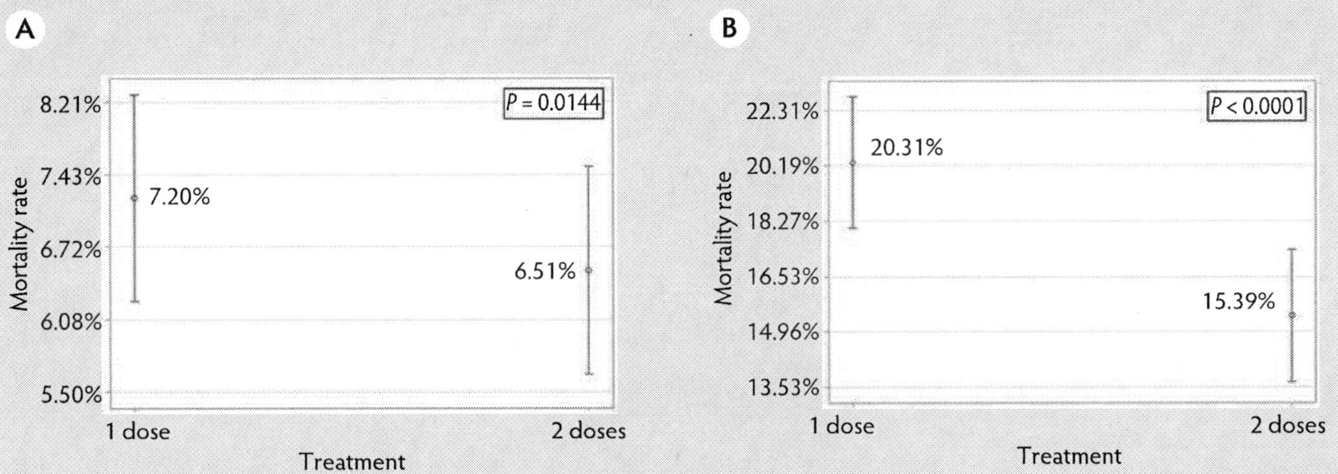

Table 3: Production-economic benefit of using a second dose of MLV 4 weeks after the first one (simulation based on 10,000 heads started per group).

PRRSV status	Economics	1 dose group	2 dose group	Difference	Benefit of intervention*	Cost of intervention	Benefit-cost ratio
Stable	Revenue	$1,569,006	$1,601,666	$32,660.78	$12,093	$10,900	1.11
	Variable costs	$1,755,656	$1,766,224	$20,568.14			
Positive	Revenue	$1,410,007	1,504,848	$94,841.45	$44,496	$10,900	4.08
	Variable costs	$1,617,308	$1,667,654	$50,345.75			

* Margin over variable costs

Conclusion

Under the conditions of this study, it was economically beneficial to apply 2 doses of Ingelvac PRRS® MLV vaccine on growing pig groups from either PRRSV-stable and -positive sow farms, raised in high pig dense areas, when compared to matched groups that received one dose. Double dose groups had significantly lower mortality rate than single dose groups on PRRSV stable and positive flows.

References

1. Holtkamp D, Kliebenstein J, Neumann E, et al. Assessment of the economic impact of porcine reproductive and respiratory syndrome virus on United States pork producers. *Journal of Swine Health and Production.* 2013;21(2):72-84.

2. Pileri E, Mateu E. Review on the transmission porcine reproductive and respiratory syndrome virus between pigs and farms and impact on vaccination. *Vet Res.* 2016;47(1):108.

3. Linhares DC, Cano JP, Wetzell T, Nerem J, Torremorell M, Dee SA. Effect of modified-live porcine reproductive and respiratory syndrome virus (PRRSv) vaccine on the shedding of wild-type virus from an infected population of growing pigs. *Vaccine.* 2012;30(2):407-413.

4. Opriessnig T, Pallares FJ, Nilubol D, et al. Genomic homology of ORF 5 gene sequence between modified live vaccine virus and porcine reproductive and respiratory syndrome virus challenge isolates is not predictive of vaccine efficacy. *Journal of Swine Health and Production.* 2005;13(5):246-253.

5. Zuckermann FA, Garcia EA, Luque ID, et al. Assessment of the efficacy of commercial porcine reproductive and respiratory syndrome virus (PRRSV) vaccines based on measurement of serologic response, frequency of gamma-IFN-producing cells and virological parameters of protection upon challenge. *Vet Microbiol.* 2007;123(1-3):69-85.

6. Cano G, Cavalcanti MO, Orveillon FX, et al. Production results from piglets vaccinated in a field study in Spain with a Type 1 Porcine Respiratory and Reproductive virus modified live vaccine. *Porcine Health Manag.* 2016;2:22.

7. Lyoo KS, Choi JY, Hahn TW, Park KT, Kim HK. Effect of vaccination with a modified live porcine reproductive and respiratory syndrome virus vaccine on growth performance in fattening pigs under field conditions. *J Vet Med Sci.* 2016;78(9):1533-1536.

8. Kristensen CS, Kvisgaard LK, Pawlowski M, et al. Efficacy and safety of simultaneous vaccination with two modified live virus vaccines against porcine reproductive and respiratory syndrome virus types 1 and 2 in pigs. *Vaccine.* 2018;36(2):227-236.

9. Dee SA, Philips R. Using vaccination and unidirectional pig flow to control PRRSV porcine reproductive and respiratory syndrome virus transmission. *Swine Health and Production.* 1998;6(1):21-25.

10. Dee S, Philips R. Using vaccination and unidirectional pig flow to control PRRSV transmission. *Swine Health and Production.* 1998;6(1):21-25.

11. Philips RC, Dee SA, Aasv. Evaluation of mass vaccination and unidirectional flow for the elimination of PRRS virus from grow/finish populations. *American Association of Swine Veterinarians, Proceedings.* 2002:165-166.

12. Philips R Dee S. Elimination of the PRRS virus from a farrow to finish single site farm using mass vaccination and unidirectional pig flow strategy. Paper presented at: American Association of Swine Practitioners1999; St. Louis.

13. Philips RC Dee S. Evaluation of mass vaccination and unidirectional flow for elimination of PRRS. Paper presented at: 4th International Symposium on Emerging and Re-emerging Pig Diseases 2003; Rome.

14. Holtkamp D, Polson D, Torremorell M. Terminology for classifying swine herds by porcine reproductive and respiratory syndrome virus status. *Journal of Swine Health and Production.* 2011;19(1):44-56.

15. Rotolo ML, Sun Y, Wang C, et al. Sampling guidelines for oral fluid-based surveys of group-housed animals. *Vet Microbiol.* 2017;209:20-29.

Modeling the dilution effect of PRRSV RNA in processing fluid field samples on the probability of virus detection by qRT-PCR

Will A. Lopez[1], DVM; Phil Gauger[1], DVM, MS, PhD; Karen Harmon[1], PhD; Laura Bradner[1], MS; Jean Paul Cano[2], DVM, PhD; Gustavo Silva[1], DVM, MS, PhD; Nubia Macedo[1], DVM, MS, PhD; Jose Angulo[3], DVM; Daniel C. L. Linhares[1], DVM, MBA, PhD

[1]Veterinary Diagnostic and Production Animal Medicine, Iowa State University, Ames, Iowa; [2]Pig Improvement Company, Hendersonville, Tennessee; [3]Zoetis, Parsippany, New Jersey

Introduction

Understanding diagnostic methods capabilities is of vital importance to design porcine reproductive and respiratory syndrome virus (PRRSV) monitoring and surveillance programs. Using population-based samples such as oral fluids (OF) or processing fluids (PF) generates questions about the effect of pooling samples on the probability of PRRSV RNA detection by qRT-PCR, especially when herd prevalence is low. PF-based sampling schemes have demonstrated the ability to be a reliable method for PRRSV monitoring and surveillance in sow farms undergoing PRRSV control/elimination and it is particularly cost-effective in low prevalence scenarios.[1] Previous data on PF supports that when litters are pooled by farrowing room (all litters processed within a farrowing room), or by whole day of collection (all litters processed in a day) the room sensitivity of the PF method was 100%. The specificity for PF testing was 100% for all sampling schemes (litter sampling, whole room sampling, and whole day sampling). When qRT-PCR was performed individually per each litter, PF-based sampling had lower sensitivity than bleeding all piglets at the litter level, but had 100% sensitivity and 100% specificity at the whole room, or whole day level.[2] Thus, the objective of this study is to model the dilution effect of PRRSV in PF on the probability of PRRSV RNA detection by qRT-PCR.

Materials and methods

PRRSV qRT-PCR conducted on 77 PF and matching blood samples (77 litters and 834 individual pig samples) revealed that among positive litters, those having a single viremic piglet were commonly found, and the median cycle threshold (CT) value of PRRSV qRT-PCR on PF from those litters was 30.21. For the purpose of this study, litters having a single viremic piglet are defined as 'low prevalence litters'. Thus, to represent a conservative scenario, we targeted to obtain a field PF sample with a CT value of 30. PF previously tested and with CT of 28 was used to account for increase in CT value on the sample after a freeze-thaw cycle. The field PF sample was used to represent a litter of 11 pigs having a single viremic piglet in it (low prevalence litter). Using PF from PRRS-naïve herds, 6 replications of 8 two-fold serial dilutions were performed on the PRRSV-positive sample. The serial dilutions were made to model the dilution effect of pooling PRRSV-positive PF with PRRSV-negative PF on the probability of PRRSV RNA detection by qRT-PCR. For each two-fold dilution, the number of PRRSV-negative pigs represented in the PF pool doubled relative to the previous PF sample. For instance, the first dilution would represent a PF sample from 22 pigs where only 1 piglet is viremic. After 8 dilutions, the PF pool would represent 2,816 pigs with 1 viremic pig. All samples were tested individually for PRRSV RNA by qRT-PCR using the RealPCR IDEXX PRRS commercial kit. The analyses of qRT-PCR results were done using simple linear regression model for the analysis of CT values and logistic regression models to calculate the probability of PRRSV RNA detection using RStudio statistical software.

Results

The results of serial dilutions showed an average increment of 1.37 points in CT values for each two-fold dilution across the 6 replications. The mean CT values for each serially diluted group of 6 replications are shown in Table 1. The cut-off value for the qRT-PCR test was set at 37. Thus, most qRT-PCR results were negative after the 6th serial dilution when the CT value were higher than the cut-off point. The raw probabilities of detecting PRRSV by qRT-PCR are presented in Table 2. We can observe that the proportion of positive tests decreased to 33% at the 6th dilution (704 piglets) then to 17% at the 7th dilution (1408 piglets) and 0% at the 8th dilution (2816 piglets). Figure 1 shows the predicted probabilities for PRRSV detection by the number of pigs in the pooled sample, derived from the data.

Conclusions

Regression analysis for the serially diluted PF data indicates that one viremic pig with a CT value of 29 could be detected in a pooled PF sample with up to 780 PRRSV-negative pigs with 61% probability. Similarly, data suggests that pooling PF from about 1 viremic piglet along with 400 non-viremic piglets (about 28 litters) gives 90% probability of detecting PRRS RNA by qRT-PCR. The scenario presented here models the 'worst case scenario' of having just one viremic pig in the PF sample which was a scenario previously described. There is an obvious dilution effect due to the increased pooling of PRRSV-negative PF with

Table 1: Mean CT values for each serially diluted group of 6 replications and the equivalent number of pigs in the pool at each corresponding dilution.

	CT values of serially diluted processing fluids (pooled)								
	Dilution								
	0	1	2	3	4	5	6	7	8
qRT-PCR CT value*	29.05	30.12	31.14	32.26	33.92	35.05	37.04	38.12	39.95
No. of pigs in the pool	11	22	44	88	176	352	704	1408	2816

* Mean CT value of the 6 replications

Table 2: Raw probability for PRRSV RNA detection by qRT-PCR in different levels of pooled processing fluids

	Raw probability for PRRSV RNA detection in pooled processing fluids								
	Dilution								
	0	1	2	3	4	5	6	7	8
Probability detection (%)*	100	100	100	100	100	100	33	17	0
No. of pigs in the pool	11	22	44	88	176	352	704	1408	2816

* Proportion of PRRSV qRT-PCR positive tests from 6 replications

Figure 1: Predicted probabilities for PRRSV RNA detection in pooled processing fluids with one (1) 'low prevalence litter.'

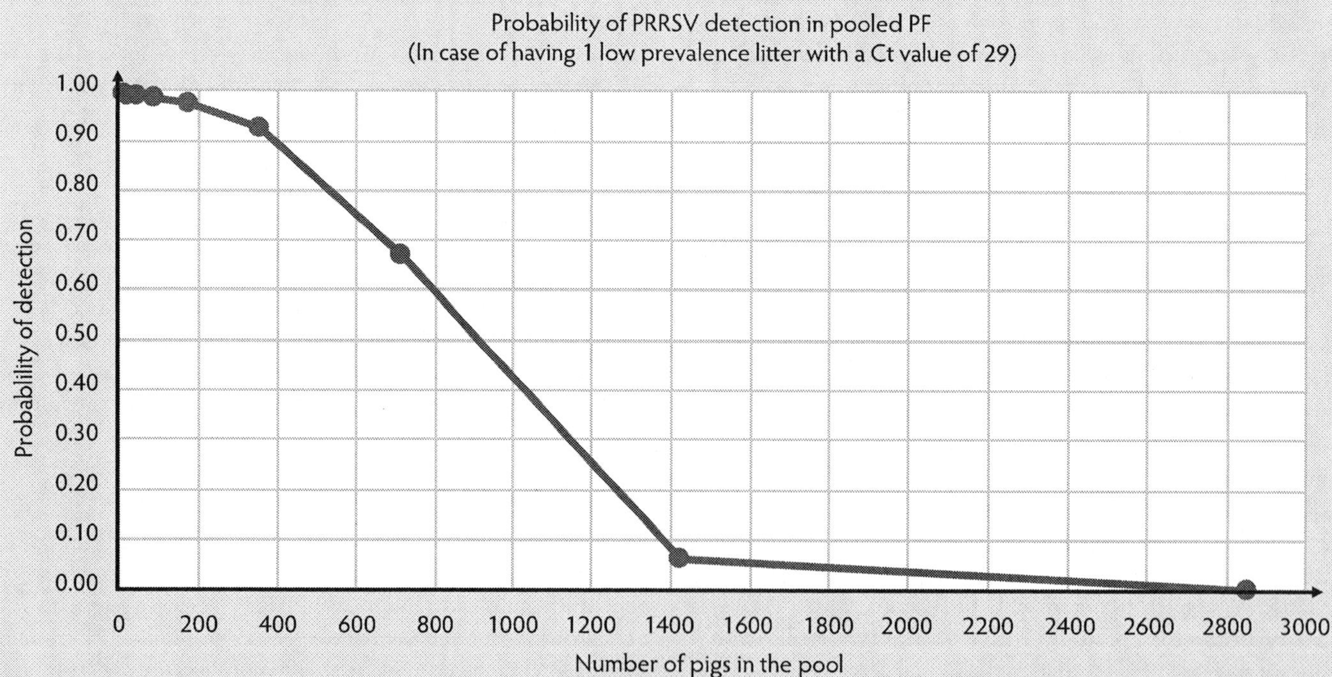

the original PRRSV-positive sample, however, pooling PF (e.g. whole-room as opposed to PF from few litters) may also increase the coverage (number of pigs contributing to the PF sample) which in turn may increase the probability of PRRSV detection by qRT-PCR. PF sampling method is capable of detecting one single PRRSV-positive pig among a high number of negative pigs depending upon the magnitude of viremia of the positive pig. Results suggest that the probability of PRRSV RNA detection decreases the number of PRRSV-free pigs in the pool increase. However, it is important to remember that only one negative qRT-PCR test result at a given point in time is not sufficient to determine the absence of PRRSV in a pig population. Continuous sampling over time is recommended to better characterize virus circulation at piglet processing time. Therefore, the ability to pool large numbers of pigs in a PF sample will allow to test more pigs more frequently. Results of this study can be used as reference for designing PF-based sampling protocols.

References

1. Yeske P. Application of new tools to declare a herd as PRRS- negative. In: *49th Annual Meeting of the American Association of Swine Veterinarians* (March 3-6, 2018).; 2018:14-16.

2. López L. WA, Linhares D. Processing fluids sensitivity and specificity for RT-PCR assays to detect PRRS virus under field conditions. In: *Allen D. Leman Swine Conference*. Minneapolis; 2018.

Put a CLAMP on it! PCR-based strategy to selectively sequence wild-type PRRSV in vaccinated herds

K. Harmon, PhD; L. Bradner, MS; M. Bieber; P. Gauger, DVM, PhD; J. Zhang, PhD
Veterinary Diagnostic and Production Animal Medicine, Iowa State University, Ames, Iowa

Introduction

Porcine reproductive and respiratory syndrome (PRRS) is one of the most impactful diseases of the U.S. swine industry. PRRS control and prevention methods often include herd management, biosecurity, and vaccination with a modified-live virus (MLV) PRRSV product. Swine producers frequently utilize sequences of open reading frame 5 (ORF5) to track the movement and/or introduction of new strains in their herds. However, routine Sanger sequencing methods are often unable to differentiate wild-type from MLV PRRSV in vaccinated herds if both are co-circulating at the time of sample collection. Advancements in Next Generation Sequencing (NGS) hold promise as a tool for such investigations but attempts to sequence a mix of two PRRSV-2 viruses in a clinical sample may be problematic unless the sequence of one PRRSV is known.[1] Additionally, the PCR cycle threshold (Ct) value of samples often needs to be in the early-to-mid 20s in order to obtain whole genome sequences, which can be problematic for some clinical specimens. Cost and time of NGS compared to Sanger sequencing may also be a concern. Therefore, a tool which preferentially amplifies ORF5 of wild-type PRRSV in the presence of MLV vaccine virus would be highly beneficial. "PCR clamping" technology has been described in which a modified oligonucleotide ("clamp") is used to block PCR amplification of specific sequences.[2-4] The clamp sequence binds to complementary nucleic acid and prevents PCR amplification from occurring, increasing the likelihood of amplification of other, non-blocked sequences that may be present. The objective of this study was to design and evaluate a PCR clamping assay to suppress ORF5 amplification of Ingelvac PRRS® MLV (I-MLV) vaccine to promote preferential amplification of wild-type PRRSV if also present in the sample.

Materials and methods

For clamp design, 204 PRRSV sequences from GenBank and 74 sequences generated from clinical cases at the Iowa State University Veterinary Diagnostic Laboratory (ISUVDL) were used to identify an I-MLV "signature sequence" present in sequences highly similar to I-MLV but not in wild-type PRRSV virus sequences. For initial evaluation of this technique, samples were selected based on ORF5 sequence homology to I-MLV, previously determined by Sanger sequencing that had been performed on samples submitted to the ISUVDL. PCR amplification of ORF5 was attempted both with and without a PCR clamp designed to specifically prevent amplification of I-MLV, and resulting PCR products were sequenced. Next, 3 sets of different PRRS virus isolates previously characterized as "I-MLV-like" and "wild-type" were combined in ratios from 10:0 to 0:10 in 10% increments, using a variety of Cts for the mixtures, designed to represent a range of viral loads. The mixtures were subjected to PCR with and without the clamp and resulting PCR products were sequenced.

Results

Samples with a Ct of approximately 29 or greater generally did not amplify regardless of whether or not the clamp was present in the reaction. Of the 52 samples which amplified in the absence of the clamp, 29 had sequence homology ≥ 98% to I-MLV. With addition of the clamp, complete quenching of the PCR reaction occurred in 25 of the 29 samples. Partial quenching, as evidenced by a decrease in intensity of the band rather than complete absence of the band, was observed in the other four samples, which all had low Cts (Ct < 24). No quenching was observed in 22 samples with sequences < 91% similar to I-MLV. The remaining sample, which was 97.2% similar to I-MLV, was fully quenched, indicating that clamping can occur in sequences in the 97-98% similarity range to I-MLV. See Figure 1 for gel depictions of the various representative scenarios that were observed. Mixtures with varying ratios of previously sequenced PRRS virus isolates at various Cts were tested to simulate a dual infection of I-MLV and wild-type virus. Results varied with the viruses used. For the first set, which was constructed to generate a Ct of approximately 24, a clean wild-type sequence devoid of any ambiguous bases was obtained from mixtures without the clamp only when there was no vaccine-like virus in the sample. However, when the clamp was incorporated into the reaction, clean wild-type sequence was obtained when the wild-type virus was present at levels as low as 20%. For the second and third sets, mixtures of a range of Cts (20, 27 and 33) were prepared. For the second set, at Cts of 20 and 27 in the absence of the clamp, the lowest wild-type concentrations that yielded a clean wild-type sequence were 80 and 70%, respectively, while with the clamp, both mixtures generated clean wild-type sequence consistently at levels down to 10% wild-type virus. At a Ct of around 33, clean wild-type sequence was inconsistently obtained between the levels of 70 to 100 % without clamp, and also inconsistently between levels of 20 to 100% with clamp. For the 3rd mixture, without the clamp, clean wild-type sequence was consistently obtained from mixtures of Cts 20 and 27 at wild-type levels as low as 20 and 10%, respectively, and with clamp at levels down to 10%. At a Ct of around 33, clean wild-type sequence was obtained inconsistently at ranges of 20 to 100% wild-type virus without clamp but when the clamp was included in the PCR reaction, wild-type sequence was consistently obtained at

Figure 1: Gel picture depiction from capillary gel electrophoresis for samples of various Cts and percent identities to Ingelvac MLV ORF 5 sequence. For each set, A is the PCR product run in the absence of the vaccine-specific clamp; B includes the clamp in the PCR reaction. Addition of the clamp to sample 1 did not result in any visual decrease in amplification. Sample 2, with a Ct of 14.3, was partially quenched by adding the clamp to the reaction. Sample 3, with a Ct of 26.8 was totally quenched with the addition of the clamp.

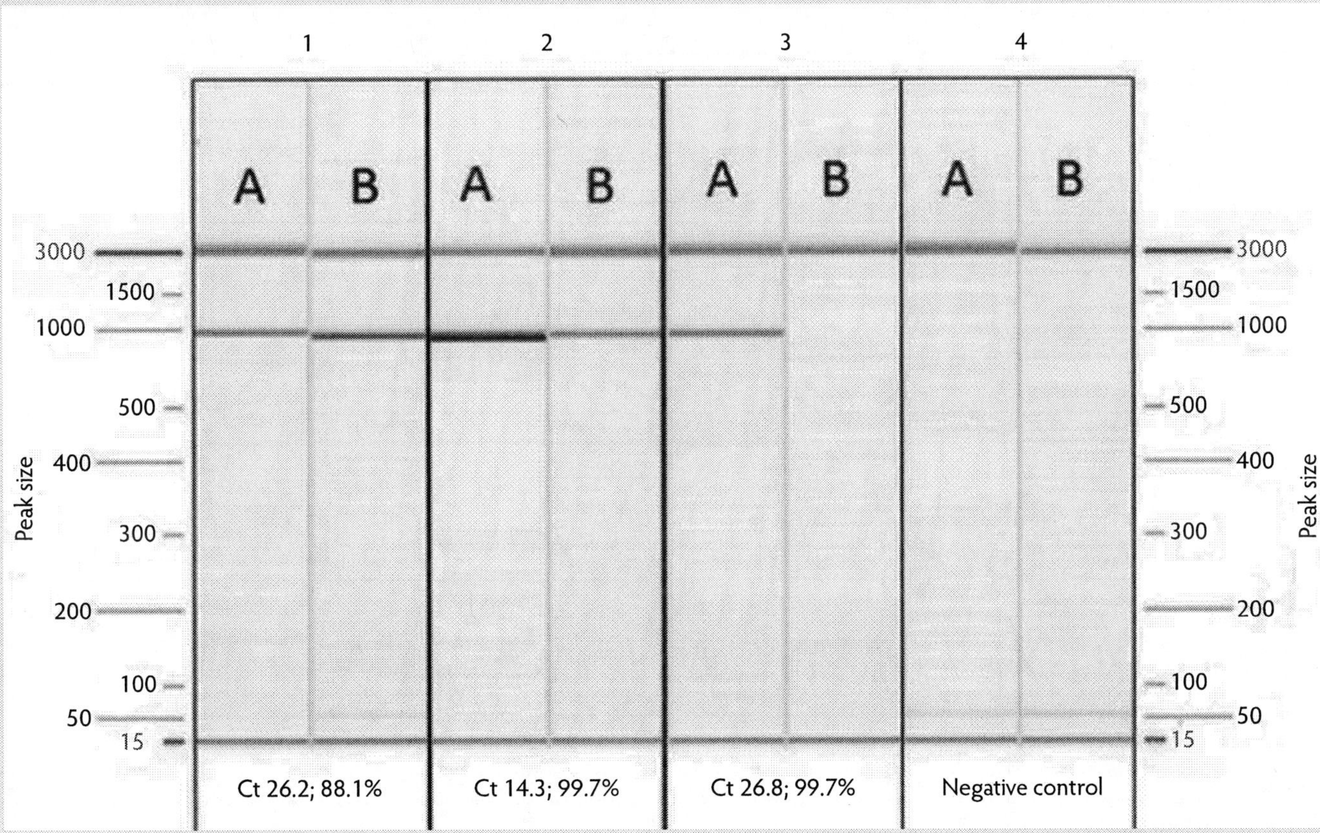

levels from 10 to 100% wild-type virus. Figure 2 is an example of a portion of four sequencing electropherograms from the same portion of the ORF5 sequence generated from an I-MLV-like virus, a wild-type PRRS virus, and a mixture containing both viruses at a Ct of approximately 27, with and without the clamp. These results indicate that the clamp can successfully quench amplification of I-MLV in samples from a mixed infection. Success of sequencing wild-type virus out of a mix of wild-type and vaccine like viruses with and without the clamp varies with Ct and individual virus sequences, but at Cts of 27 and less, sequencing of wild-type virus out of a mix was consistently successful at levels as low as 20% wild-type virus and often when the level was only 10% of the sample. Further optimization may allow for more complete quenching of I-MLV with high vaccine virus titer or on samples with later Cts which reflect a lower viral load. Evaluation is ongoing for a second clamp for I-MLV. To date, we have successfully applied the clamping assay to identify wild-type PRRSV in three clinical cases originally sequenced as I-MLV-like. This tool will be highly beneficial to producers tracking the health of their vaccinated herds by monitoring wild-type PRRSV through ORF5 sequencing. This technology could be applied in other scenarios in which preferential amplification is desired in a mixed population. Future goals include expanding this technology to other PRRSV vaccines and potentially other pathogens.

References

1. Zhang J et al. High-throughput sequencing of porcine reproductive and respiratory syndrome virus from cell culture materials and clinical specimens using next-generation sequencing technology. J Vet Diag Invest. 2017;29:41.

2. Dabritz J et al. Detection of Ki-ras mutations in tissue and plasma samples of patients with pancreatic cancer using PNA-mediated PCR clamping and hybridization probes. British Journal of Cancer 2005;92:405.

3. Kim S et al. Bridged Nucleic Acids (BNAs) as Molecular Tools. J Biochem Mol Biol Res 2015 1:67.

4. Orum H. PCR Clamping. Curr Issues Mol Biol. 2000;2:27.

Figure 2: Sequencing electropherograms of a portion of ORF5 of an Ingelvac PRRS MLV vaccine-like virus (A), 50:50 mix of Ingelvac MLV and wild-type PRRS virus without (B) and with (C) the clamp, and wild-type PRRS virus (D). All four electropheorgrams represent the same portion of ORF5. Note the presence of several ambiguous peaks in the absence of the clamp in the mixed sample that are resolved for the wild-type virus when the clamp is included in the reaction.

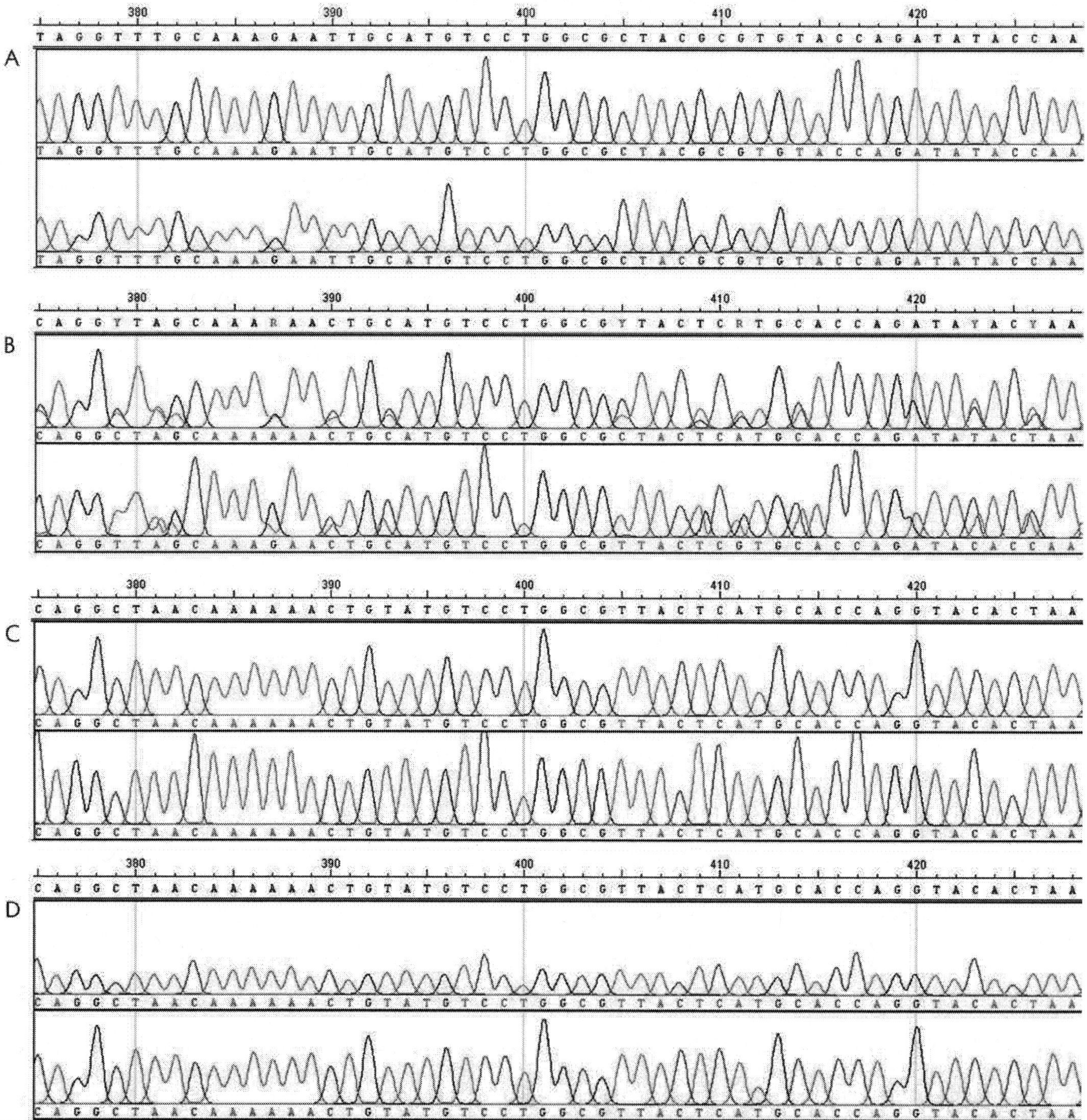

Use of oropharyngeal swabs and udder wipes to monitor porcine reproductive and respiratory syndrome virus in breeding herds

Jorge Garrido-Mantilla[1]; Juan Sanhueza[1]; Carles Vilalta[1]; Nathan Winkelman[2]; Cesar Corzo[1]; Montserrat Torremorell[1]
[1]College of Veterinary Medicine, University of Minnesota, St. Paul, Minnesota
[2]Swine Services Unlimited, Inc. Rice, Minnesota

Introduction

Porcine reproductive and respiratory syndrome (PRRS) is an important economic disease in the swine industry. PRRS virus (PRRSV) monitoring and surveillance is crucial to advance control and elimination programs. rRT-PCR of due-to-wean piglet serum samples is a common procedure for monitoring PRRSV in breeding herds. Recently, the use of processing fluids (PF) has become a common sample to monitor PRRSV. Nevertheless, PF can only be collected at processing which is about 3-8 days of age and do not represent PRRSV status along the suckling period. Oropharyngeal swabs (OP) and udder skin wipes (UW) are sampling strategies that have proven their value for influenza surveillance in suckling piglets. OP and UW are easy to collect and represent minimum stress for pigs or sow. The purpose of this study was to evaluate the use of OP and UW for detection of PRRSV in piglets prior to weaning.

Materials and methods

One sow farm in the Midwestern US that had a recent PRRSV outbreak was conveniently selected for the study. Sampling was performed 5 weeks after the outbreak. Twenty litters were randomly selected at 12 (ten litters) and 21 days old (ten litters). All piglets were bled and an OP was introduced into the mouth to sample the deeper areas of the mouth. One udder skin sample was collected from each sow using a sterile gauze impregnated with 10 mL of transport media by wiping the areas of contact between the pig's nose and udder skin during suckling. Samples were tested by rRT-PCR for detection of positive samples using a Ct value of 40 as cut off. Cohen's kappa statistic and McNemar test were used to assess the agreement between OP, UW and serum rRT-PCR results. Sensitivity (Se) and specificity (Sp) were also calculated.

Results

Seventy seven (159/205) percent of the piglet serum samples were positive by rRT-PCR. Sixty nine (143/205) percent OP samples and 95% (19/20) udder wipes tested positive. Average Ct values (and standard deviation) were 21.23 (5.1), 27.96 (3.5) and 31.91(2.27) for serum, OP and UW, respectively. Se and Sp were 89% (95% CI 83%-93%) and 96% (95% CI 85%-99%) for OP. Se of OP was 98.6%, 92% and 38.5% for those samples from animals with serum Ct values were < 20, between 20-30 and over 30, respectively. Estimated Kappa value between serum and OP was 0.75 (95% CI 0.62-0.89). McNemar test was highly significant ($P < 0.001$). UW Se to detect a serum PRRS positive litter was 94.7%. However, SP could not be calculated due to lack of negative litters.

Discussion

Our results indicate that collection of OP samples is a reliable option to determine PRRSV status in due-to-wean piglets under the study conditions, particularly when Ct values of the paired serum are below 30. Udder skin wipes are also a good sample to detect PRRSV during an acute outbreak. Nevertheless, their Ct values are higher compared with serum and OP. OP swabs are a reliable method to monitor PRRSV status in in breeding herds in a welfare friendly and easier way of sampling.

Experimental transmission of influenza A virus and porcine reproductive and respiratory syndrome virus from nurse sows to adopted pigs during lactation

Jorge Garrido Mantilla; Marie Culhane; Montserrat Torremorell
College of Veterinary Medicine, University of Minnesota, St. Paul, Minnesota

Introduction

Influenza A virus (IAV) and porcine reproductive and respiratory syndrome virus (PRRSV) transmission in pigs occurs mainly by direct contact and indirectly by exposure to infectious aerosols or contaminated materials, and in utero in the case of PRRSV. However, less is known about the sources of exposure in pigs during the pre-weaning period. In a previous study, we isolated viable IAV and PRRSV from the udder skin of lactating sows which suggests that skin can support virus viability at least temporarily. In swine farms, nurse sows are used regularly to adopt younger pigs to maximize productivity. If the skin of nurse sows is contaminated with IAV or PRRSV it may allow the transmission and persistence of IAV and PRRSV in farms. The objective of this study was to evaluate transmission of IAV and PRRSV between litters using a nurse sow transmission model.

Materials and methods

A nurse sow model was established by obtaining 2 IAV and PRRSV-negative pregnant sows that gave birth in separate BSL-2 research rooms. In the case of IAV, Sow 1 and half of her pigs at 6 days of age were intranasally inoculated with $10^{\wedge 5}$ TCID50/mL of an H1N1 IAV isolate. After confirming infection 4 days later, the rest of the litter which had been moved to a separate room, was moved back with the litter of origin and challenged with IAV similarly to the first inoculation. The second inoculation was done to maintain infection at the litter level and to have an immune but not shedding sow at the time of adopting the IAV negative pigs from Sow 2. Prior to that, Sow 1 was moved into a clean room and IAV-negative pigs from Sow 2 were transferred onto her to mimic nurse sow pig adoption.

For PRRSV infection, after IAV study completion, the Sow 1 piglets were intramuscularly inoculated with $3.2 \times 10^{\wedge 6}$ TCID50/mL of PRRSV VR-2332 strain. After confirming PRRSV infection and at 7 days post inoculation, Sow 1 was moved into the room where sow 2 PRRSV negative pigs were held.

In both studies, all pigs and both sows were sampled daily before and post-movement. Blood, nasal swabs (NS) and udder skin wipes (UW) were collected. All samples were tested for IAV or PRRSV rRT-PCR and positive samples cultured for virus isolation using MDCK or MARC 145 cells.

Results

In the case of IAV, UW from Sow 1 tested IAV rRT-PCR and virus isolation positive at time of adoption. NS from Sow 1 were positive by rRT-PCR (Ct value, 28.22) but negative for virus isolation. After adoption, an adopted pig became IAV positive as early as 1 day post-movement (DPM) and the whole litter became positive at 4 DPM.

For PRRSV, the UW from Sow 1 tested PRRSV positive by rRT-PCR and virus isolation at time of adoption. At 3 DPM, 3 out of 11 pigs became PRRSV rRT-PCR positive. The study was terminated at 7 DPM with 7 pigs PRRSV positive.

Conclusions and discussion

In this study we report for the first time the IAV and PRRSV infections of pigs after suckling from sows with virus-contaminated udder skin. This study identified a pathway of mechanical transmission involving nurse sows that need to be considered in disease control and elimination programs. The study highlights how a common management practice can contribute to the persistence and perpetuation of these important pathogens in breeding herds. The impact and role of nurse sow transmission of IAV and PRRSV under field conditions needs further investigation.

Acknowledgements

Funding for this study was provided by the University of Minnesota Swine Disease Eradication Center.

Shedding of a newly commercialized live attenuated influenza vaccine under field conditions

Gustavo Lopez[1], DVM; Jayaveeramuthu Nirmala[1], PhD; Christa Goodell[2], DVM, MS, Dipl ACVPM, PhD; Marie Culhane[1], DVM, PhD; Montserrat Torremorell[1], DVM, PhD

[1]Department of Veterinary Population Medicine, College of Veterinary Medicine, University of Minnesota, St. Paul, Minnesota. [2]Boehringer Ingelheim Vetmedica, Inc. Duluth, Georgia

Introduction

Influenza A virus (IAV) is one of the most important respiratory pathogens affecting swine health. Though H1N1, H1N2 and H3N2 are endemic subtypes in pigs, there is significant genetic diversity within them.[1] Changes in the IAV genome by mutation (antigenic drift) or reassortment (antigenic shift) can result in IAV antigenic changes that limit the efficacy of killed vaccines currently used in pigs. A new live attenuated influenza vaccine (Ingelvac Provenza™) is now available in the U.S market and could be an important tool to control IAV in swine. Because the vaccine is live and can replicate in vaccinated animals, it is important to assess how much and for how long the vaccine virus can be shed and whether it can transmit to other pigs. Therefore, the objectives of this study were to: a) assess the onset and duration of vaccine virus shedding, b) investigate the ability of vaccine virus to transmit from vaccinated to unvaccinated pigs and c) evaluate whether the vaccine virus can be aerosolized and detected in the environment.

Material and methods

A 5,200 sow farm that vaccinated pigs at processing with Ingelvac Provenza™ was selected for the study. Thirty-three litters in three rooms were selected for the study. Within vaccinated litters (n = 21) six pigs per litter were tagged from which, three pigs were left unvaccinated and the rest were vaccinated to assess direct contact transmission to the unvaccinated pigs. There were 12 litters left unvaccinated (4 litters/room) and three pigs within these litters were tagged to assess transmission by indirect contact.

Samples were screened for IAV using a vaccine strain specific rRT-PCR. Samples with cycle threshold (ct) ≤ 35 were considered positive, ct >35-≤ 40 suspect and ct > 40 negative. A subset of positive samples by rRT-PCR were selected for virus isolation on MDCK cells. Blood samples were tested for presence of antibodies against influenza A virus.

Sample collection

Nasal swabs were obtained from all tagged pigs before vaccination and at 1, 2, 3, 4, 6, 8 and 12 days post vaccination (DPV). Udder wipes from lactating sows of tagged pigs and surface wipes of the farrowing crates from those same sows were collected before vaccination and at 1,2,3,4,6,8 and 12 DPV.

Blood samples were collected before vaccination and at 15 DPV for antibody testing.

Air samples were collected immediately after vaccination of the entire room and at 1, 2, 3, 4, 6 and 8 DPV. Air samples were collected for 30 min, using an air cyclonic sampler (Midwest Micro-Tek, Brookings, SD, USA), capable to sample 200 L/min.

Samples from surfaces representing the deposition of airborne particles were also collected immediately after vaccination of the entire room and at 1, 2, 3, 4, 6 and 8 DPV. These samples were collected using 6 pieces of aluminum foil paper, 1 meter long each, located on top of farrowing crates for 60 min.

Results

Ninety percent (19/21) of the vaccinated litters tested vaccine virus positive 1 DPV (Table 1). At 6 DPV, 4% (1/21) of the litters tested positive which was the last time when vaccine virus was detected in vaccinated litters. All direct contact non-vaccinated pigs tested negative throughout the study. Unvaccinated indirect contact litters did not test positive, there were suspect results observed in both 2 DPV (9%) and 12 DPV (25%). Twenty-three percent (5/21) of udder wipes in vaccinated litters tested positive at 1 DPV and positives were detected up to 4 DPV (4% of litters). Using surface wipes, vaccine virus specific rRT-PCR tested positive at 1 DPV (9% of crates), 2 DPV (9% of crates) and 8 DPV (4%) in vaccinated litters. Vaccine virus was detected in 12.5% (1/8) of air samples on the day of vaccination, 16% (1/6) at 4 DPV and

Table 1: Number of influenza A virus RT-PCR positive pigs (Ct ≤ 35) in nasal swabs by group and days post vaccination

	Days post vaccination							
	0	1	2	3	4	6	8	12
Vaccinated	0/21	19/21	10/21	6/21	4/21	1/21	0/21	0/21
Direct contact	0/21	0/21	0/21	0/21	0/21	0/21	0/21	0/21
Indirect contact	0/12	0/12	0/12	0/12	0/12	0/12	0/12	0/12

10% (1/10) on 6 DPV. Samples from surfaces that represented the deposition of airborne particles had a 40% (10/25) positivity on the day of vaccination and 4% at 1 DPV. Suspect rRT-PCR results were obtained up to 8 DPV. Virus isolation was successful on 11% (4/36) of the selected samples. These samples corresponded to nasal swabs of vaccinated pigs from 1 and 3 DPV. The blood samples before vaccination showed that 100% of the pigs had presence of antibodies against influenza A virus.

Conclusions and discussion

Under the conditions of this study, shedding of the live attenuated influenza Ingelvac Provenza vaccine virus from vaccinated pigs was limited in time, resulted in limited shedding into the air and environment and was minimally transmitted to in-contact unvaccinated pigs.

Acknowledgements

This study was funded by Boehringer Ingelheim Vetmedica, Inc. Duluth, GA 30096, United States.

Reference

1.Vincent, A. L., Ma, W., Lager, K. M., Richt, J. A., Janke, B. H., Sandbulte, M. R., Garcia-Sastre, A. (2012). Live Attenuated Influenza Vaccine Provides Superior Protection from Heterologous Infection in Pigs with Maternal Antibodies without Inducing Vaccine-Associated Enhanced Respiratory Disease. Journal of Virology, *86*(19), 10597–10605. https://doi.org/10.1128/JVI.01439-12

Identification of metabolite markers for enzootic pneumonia in pigs

M. Surendran Nair[1]; D. Yao[2]; C. Chen[2]; M. Pieters[1]
[1]Department of Veterinary Population Medicine; [2]Department of Food Science and Nutrition, University of Minnesota, St. Paul, Minnesota

Introduction

Mycoplasma hyopneumoniae is a primary pathogenic bacterium that significantly affects worldwide swine production through the causation of enzootic pneumonia.[1] Mostly *M. hyopneumoniae* infections are marked by chronicity and lack, or delay, of systemic and detectable immune responses in the host.[2] Hence, there is a growing need for practical surveillance tools that complement diagnostics to demonstrate that populations are negative or to detect early infection in swine populations. Therefore, the objective of this investigation was to identify metabolite biomarkers associated with pathophysiological events occurring in early *M. hyopneumoniae* infections in pigs through metabolomic analysis.

Materials and methods

Twelve 8-week old *M. hyopneumoniae*-free pigs were included in this study. Two pigs served as non-infected controls, while 10 pigs were intratracheally inoculated with *M. hyopneumoniae* strain 232. Serum samples, laryngeal swabs (LS) and tracheo-bronchial lavages (TBLF) were collected on day 0, 2, 5, 9, 14, 21 and 28 of the experiment. The metabolites in serum samples were profiled by high-resolution liquid chromatography-mass spectrometry (LC-MS) analysis. The metabolite markers were identified by the multivariate statistical analysis of LC-MS data and structural analysis[3] the uptake of these nutrients and phytochemicals is expected to affect the metabolic system. In this study, 2 groups of young pigs (17.8 \u00b1 2.7 kg initial BW. Genetic material from *M. hyopneumoniae* was detected in LS and TBLF using a species-specific real-time PCR.[4]

Results

In *M. hyopneumoniae* infected pigs, profound changes were observed in host amino acid and fatty acid metabolism, as early as 5 days post-infection (dpi). A decrease in serum branched-chain amino acids (BCAAs) concentrations were observed at 5 and 9 dpi along with reduced levels of other proteinogenic amino acids, including proline, alanine, arginine, histidine, and glycine ($P < 0.05$). A significant increase in the levels of 2-aminobutyric acid (AABA), a non-proteinogenic amino acid was also identified 14 and 21 dpi ($P < 0.05$). Interestingly, serum AABA levels were positively correlated with *M. hyopneumoniae* DNA detected in both LS and TBLF (LS: $r = 0.21$, $P = 0.053$; TBLF: $r = 0.32$, $P = 0.0028$). Furthermore, the levels of free fatty acids in serum samples were dramatically elevated on 14 and 21 dpi in infected pigs ($P < 0.05$).

Conclusions

The identification of the mycoplasma-responsive serum metabolites indicated that *M. hyopneumoniae* infection was associated with systemic metabolic changes in pigs. The changes in serum amino acids and fatty acids could become useful indicators of mycoplasma infection. Further studies are required to determine the biochemical and physiological mechanisms underlying these metabolic changes.

References

1. Thacker, E. L. & Minion, F. C. Mycoplasmosis. in *Diseases of Swine* (eds. Zimmermann, J. J., Ramirez, K. L. A. & S.K. J, S. G. W.) 779–797 (2012).

2. Maes, D. *et al.* Update on *Mycoplasma hyopneumoniae* infections in pigs: Knowledge gaps for improved disease control. *Transbound. Emerg. Dis.* (2017). doi:10.1111/tbed.12677

3. Chen, C. *et al.* Identification of redox imbalance as a prominent metabolic response elicited by rapeseed feeding in swine metabolome1. *J. Anim. Sci.* 96, 1757–1768 (2018).

4. Pieters, M., Daniels, J. & Rovira, A. Comparison of sample types and diagnostic methods for in vivo detection of Mycoplasma hyopneumoniae during early stages of infection. *Vet. Microbiol.* 203, 103–109 (2017).

Mycoplasma hyosynoviae diagnostics: What experimental data can tell us about field strains

N. Macedo[1], DVM, MS, PhD; B. Arruda[1], DVM, PhD; P. Canning[2], DVM, PhD; L. Karriker[2], DVM, MS, DACVPM; J. Mora[3], DVM; M. Clavijo[1], DVM, PhD

[1]Veterinary Diagnostic and Production Animal Medicine, Iowa State University, Ames, Iowa; [2]Swine Medicine Education Center, Iowa State University, Ames, Iowa; [3]Eco Animal Health, Surrey, England

Introduction

The host-specific bacterium *Mycoplasma hyosynoviae* (MHS) is considered one of the most frequent causes of polyarthritis in growing pigs.[1] MHS also colonizes the upper respiratory tract of pigs, especially the tonsil, and pigs can become persistent carriers.[2] MHS-associated disease is usually observed in pigs 3 to 6 months of age. The determinant factors for systemic spread and disease development seems to be associated with the decay of maternal immunity,[3] in addition to environmental factors and genomic diversity among isolates.[4] This pathogen is an emerging concern among producers and veterinarians, particularly due to a rise in recognized cases.[5] An accurate clinical history and detection of the pathogen from lesions is necessary to diagnose MHS-associated arthritis.[5] Antimicrobial treatment can be effective in reducing lameness and improving welfare of affected animals if administered at the right moment.[1] If the disease is not diagnosed early, MHS lameness can become a difficult clinical problem to control. Moreover, there is no effective commercial vaccine for prevention of disease. Therefore, more comprehensive studies are needed to better understand the pathogenesis of MHS infection and develop appropriate control measures for this disease. The objective of this study was to describe the clinical presentation, detection, lesions and immune response after inoculation of two different strains of MHS in pigs.

Methods

Experimental design

Sixteen cesarean-derived colostrum-deprived (CDCD) pigs were obtained when they were 8-weeks-old. Prior to inoculation, all pigs tested negative for MHS and PRRSV by ELISA, and *M. hyorhinis* and MHS by PCR on tonsil swab. All pigs were randomized into one of three groups, and housed individually within each group. Group T1 (n = 4), T2 (n = 6) and T3 (n = 6) were inoculated via three routes (intranasal, tonsil painting and intravenous) with either Friis media only, or 10^8 CFU/mL of two distinct MHS strains clinically characterized as low virulence and high virulence, respectively.

Clinical evaluations and sampling

After inoculation, every other each pig received a lameness score[6] and rectal temperature was recorded. With the objective of evaluating shedding/carriage patterns, tonsil swabs were collected on 0, 3 7, 10 14, and 17 days post inoculation (DPI) and oral fluids were collected on 0, 2, 4, 6, 8, 11, 14 and 16 DPI. Tonsil swabs and oral fluid samples were tested by PCR for MHS. Serum samples were collected at days 0, 3, 7 and 17 DPI, and were tested with an in-house MHS species-specific ELISA. Additionally, the levels of IL-1β, IL-6, IL-8, IL-10, TNF-α and IFN-γ cytokines in the serum were measured by a Luminex Multiplex assay. At 18 DPC, pigs were euthanized and necropsied. Synovial fluid was collected from the humeral-radial, femoro-tibial and tibio-tarsal joints, and tested by PCR for MHS. These joints were also examined grossly for synovial hyperemia/hyperplasia, fibrin, cartilage defects and increased synovial fluid. Synovium from each joint was collected and evaluated histologically. Each evaluated joint received a gross and histologic arthritis/synovitis severity score.

Statistical analysis

The presence of clinical signs and lesions at necropsy, and PCR results were analyzed for significant differences between groups using Fisher's exact test ($P < 0.05$). Serology results and scores of lesions were analyzed for differences between groups using Kruskal-Wallis test followed by Mann-Whitney pairwise comparison ($P < 0.05$).

Results and discussion

Clinical findings

Lameness, pyrexia, and gross lesions were not observed in control pigs. Lameness was evident at 1 DPI in T3 (2/6) with 5/6 pigs affected in T2 and T3 at 15 DPI. Clinically, the severity of the lameness was not significantly different between T2 and T3, with most pigs exhibiting mild to moderate lameness.

MHS detection by PCR

Control pigs remained negative for MHS by PCR. All tonsil swabs from T3 were MHS positive starting at 3 until 17 DPI, being significantly higher than T2 (2/6) at 7 DPI (Figure 1). MHS detection in oral fluids was transient and significantly higher in T3 compared to T2 at 6 and 8 DPI (Figure 2). At 16 DPI all oral fluids tested negative. Oral fluid samples are quick, easy and inexpensive to collect, and they may be a good sample for monitoring the circulation of MHS in swine populations. In this study, samples were collected individually, and lower detection in oral fluids may reflect a potential drop in sensitivity compared to tonsil swabs. There is no evidence that the detection of MHS in the tonsils or oral fluids correlates with clinical outcome.

Figure 1: Percentage of *Mycoplasma hyosynoviae* positive pigs by tonsil swab PCR. Different letters indicate difference between groups ($P < 0.05$).

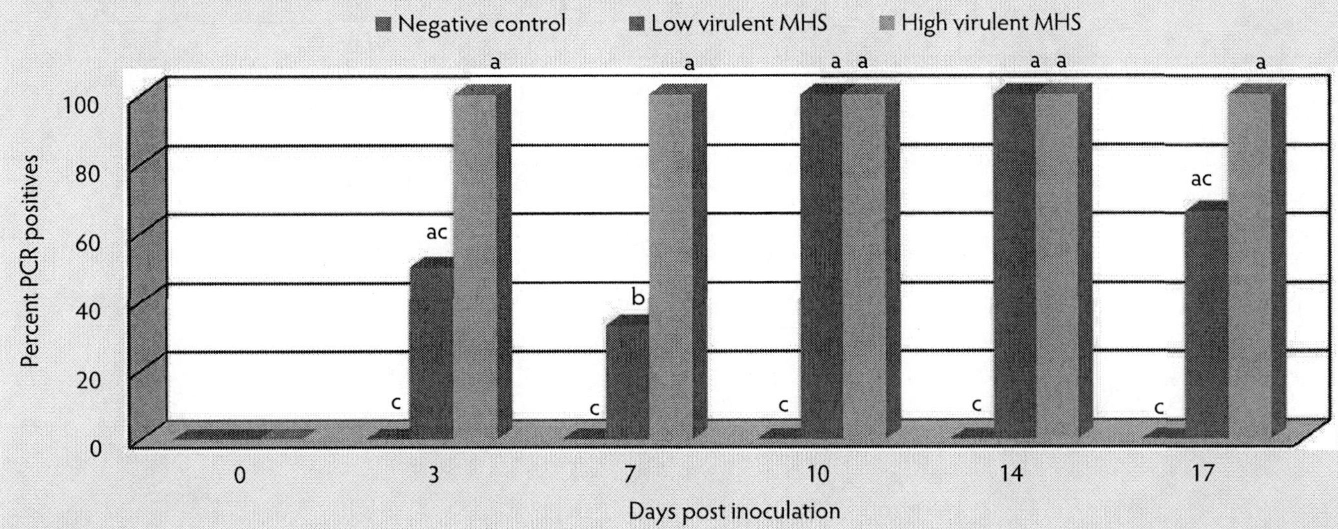

Figure 2: Percentage of *Mycoplasma hyosynoviae* positive pigs by oral fluids PCR. Different letters indicate difference between groups ($P < 0.05$).

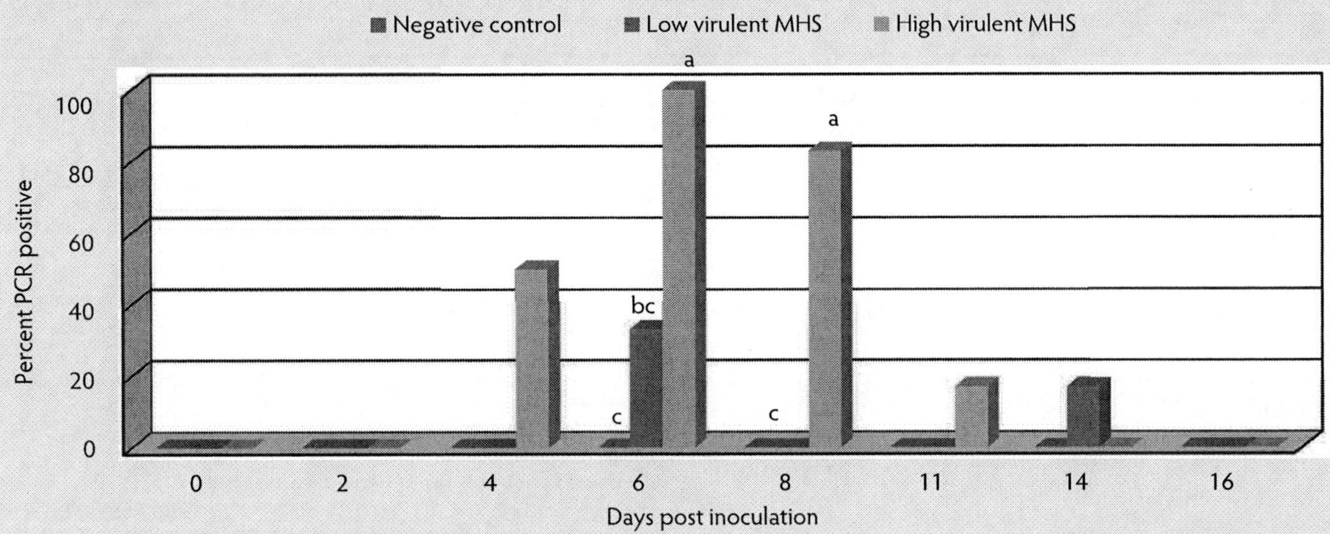

Pathological examination

At necropsy, 4/6 pigs from T3 had gross lesions characteristic of arthritis including synovial hyperemia and hyperplasia and increased synovial fluid, compared to 1/6 in T2 (Table 1). Moreover, all pigs (6/6) in T3 had one or more joints with histologic lesions consistent with MHS including synovial membrane villous hypertrophy and hyperplasia, hypertrophy and hyperplasia of synoviocytes, and mixed inflammation in the synovial subintima. The synovium of pigs in T1 and T2 had similar histologic changes, which included rare perivascular lymphocytes and plasma cells. Interestingly, although lameness and arthritis were observed in T2, MHS was not detected by PCR in synovial fluids at necropsy, while the synovial fluid from 5/6 pigs in T3 were positive (Table 1), suggesting that some MHS strains might be cleared from the joints quicker.

Serology

A serologic IgG response was evident in the MHS inoculated pigs, regardless of the group at 17 DPI. Tonsil colonization by MHS has been previously associated with higher serological responses,[2,7]

Table 1: Gross score evaluation, histopathology and PCR results for joints of all animals sampled.

Groups	Gross/Histopathology/PCR				
	Pig ID	Hock	Stifle	Elbow	Total
Control	130	0* / - / -	0 / - / -	0 / - / -	0 / - / -
	128	0 / - / -	0 / - / -	0 / - / -	0 / - / -
	145	0 / - / -	0 / - / -	0 / - / -	0 / - / -
	139	0 / - / -	0 / - / -	0 / - / -	0 / - / -
Low virulent	130	0 / - / -	0 / - / -	0 / - / -	0 / - / -
	128	0 / - / -	0 / - / -	0 / - / -	0 / - / -
	145	0 / - / -	0 / - / -	0 / - / -	0 / - / -
	139	1 / - / -	0 / - / -	0 / - / -	1 / - / -
	148	0 / - / -	0 / - / -	0 / - / -	0 / - / -
	142	0 / - / -	0 / - / -	0 / - / -	0 / - / -
High virulent	118	0 / + / +	0 / - / +	0 / - / +	0 / + / +
	132	0 / + / +	0 / - / +	0 / + / +	0 / + / +
	137	1 / + / -	0 / + / +	0 / - / +	1 / + / +
	150	2 / + / -	0 / - / +	2 / + / +	4 / + / +
	114	1 / + / +	0 / - / +	2 / + / -	3 / + / +
	125	1 / + / -	0 / - / +	0 / - / -	1 / + / +

* Gross score 1: Slight synovial hyperplasia and hyperemia; score 2: Synovial hyperplasia/ hyperemia and increased joint fluid/decreased viscosity; score 3: Marked synovial hyperplasia/ hyperemia and increased joint fluid/decreased viscosity.

even though there is no evidence that those antibodies would prevent the development of arthritis. Serum cytokines could be detected on all pigs before inoculation, and their detection oscillated throughout the study. Overall, the levels of IL-1β, IL-6, IL-10 and TNF-α cytokines were statistically higher in pigs from both MHS inoculated groups compared to control pigs at 3 and 7 DPC, while no differences were observed in the levels of IL-8 and IFN-γ. *M. hyorhinis* has been shown to induce local inflammatory responses in joint, with release of IL-1β and IL-6.[8] To our knowledge, this is the first study describing the dynamics of serum cytokines in pigs following MHS inoculation. The significance of the higher levels of cytokines at 3 and 7 DPC needs further investigation, but it may be associated with the intensity of the inflammatory response and severity of lesions.

Conclusion

These findings highlight potential differences in the pathophysiology between MHS strains circulating in pig populations that can hinder appropriate diagnosis and timely treatment of arthritis. In pigs inoculated with a low virulent MHS, the pathogen was not detected by PCR in synovial fluid, suggesting that some MHS strains might be cleared from the joints quicker, or may not reach the joint. It seems that systemic infection with MHS induces inflammatory and humoral responses, but their protective role needs more investigation. Studies are needed to identify the underlying genetic differences that give rise to distinct MHS phenotypes.

References

1. Nielsen EO et al. Mycoplasma hyosynoviae arthritis in grower-finisher pigs. *J. Vet. Med. A.* 2001;48, 475-486.

2. Hagedorn-Olsen T et al. 1999. Progression of Mycoplasma hyosynoviae infection in three pig herds. Development of tonsillar carrier state, arthritis and antibodies in serum and synovial fluid in pigs from birth to slaughter. J. *Vet. Med. A.* 1999; 46, 555-564.

3. Lauritsen KT et al. Absence of strictly age-related resistance to *Mycoplasma hyosynoviae* infection in 6-week-old pigs. *Vet Microbiol.* 2008; 130, 385-390.

4. Bumgardner EA et al. Comparative genomic analysis of seven *Mycoplasma hyosynoviae* strains. *Microbiology Open.* 2015; 4:343-359.

5. Gomes Neto JC et al. Mycoplasma-associated arthritis: Critical points for diagnosis. *J Swine Health Prod.* 2012; 20: 82-86.

6. Karriker L et al. Validation of a lameness model in sows using physiological and mechanical measurements. *J Anim Sci.* 20013; 91:130-136.

7. Nielsen EO et al. Use of a novel serum ELISA method and the tonsil-carrier state for evaluation of *Mycoplasma hyosynoviae* distributions in pig herds with or without clinical arthritis. *Vet Microbiol.* 2005; 111:41-50.

8. Reddy NRJ et al. Cytokines in *Mycoplasma hyorhinis*-induced arthritis in pigs bred selectively for high and low immune responses. *Infection and Immunity.* 2000;68:1150-1155.

Oral infectious dose of African swine fever virus consumed naturally in feed or liquid

M. C. Niederwerder[1], DVM, PhD; A. M. M. Stoian[1], MS; R. R. R. Rowland[1], MA, PhD; S. S. Dritz[1], DVM, PhD; V. Petrovan[1], MS; L. A. Constance[1], MS; J. T. Gebhardt[1], BS; M. Olcha[1], DVM; C.K. Jones,[2] MS, PhD; J. C. Woodworth[2], PhD; Y. Fang[1], PhD; J. Liang[3], BS; T. J. Hefley[3], PhD

[1]Department of Diagnostic Medicine/Pathobiology, College of Veterinary Medicine, Kansas State University, Manhattan, Kansas; [2]Department of Animal Sciences and Industry, College of Agriculture, Kansas State University, Manhattan, Kansas; [3]Department of Statistics, College of Arts and Sciences, Kansas State University, Manhattan, Kansas

Introduction

African swine fever virus (ASFV) is a highly contagious foreign animal disease and a significant threat to animal agriculture in the United States. Disease caused by ASFV is characterized by severe disseminated hemorrhage and high mortality rates in swine. There are currently no available vaccines to protect pigs against ASFV infection, creating significant challenges to disease control. Current mitigation strategies during an ASFV outbreak are centered around large-scale culling and elimination of infected animals. Since spreading into Eastern Europe and Russia in 2007, and most recently into China in 2018, the concern for further spread and introduction of ASFV into the US has heightened. In 2013, the introduction of porcine epidemic diarrhea virus (PEDV) into the US unveiled the risk of contaminated feed as a vehicle for swine viral disease introduction and subsequent spread. However, very little is known about the oral infectious dose and transmission risks of other foreign animal diseases, such as ASFV, in contaminated complete feed.

Objectives

The objectives of this study were to determine the minimum infectious dose and median infectious dose (ID_{50}) of ASFV Georgia 2007 through oral exposure via natural drinking and feeding behaviors.

Materials and methods

Individually housed high-health commercial pigs (n = 68) approximately 7 weeks of age were exposed to specific doses of ASFV in media (drinking) or complete feed (feeding). Doses administered ranged between 10^0 $TCID_{50}$ and 10^8 $TCID_{50}$. Negative control pigs (n = 14) were administered sterile media or complete feed without virus. Pigs naturally consumed liquid in gravity-fed drinkers and naturally consumed feed in stainless steel creep feeders after a 3-4 day acclimation period. At 5 days post-exposure, pigs were euthanized and assessed for ASFV infection by PCR and virus isolation.

Results and discussion

The minimum infectious dose of ASFV in liquid was 10^0 $TCID_{50}$ whereas 10^4 $TCID_{50}$ was the minimum infectious dose in a complete feed matrix. The mean ID_{50} was $10^{1.0}$ $TCID_{50}$ (95% CI: 10^0, $10^{2.3}$) for liquid and $10^{6.8}$ $TCID_{50}$ (95% CI: $10^{4.6}$, 10^{8+}) for feed. Taken together, this study demonstrates that ASFV Georgia 2007 can be easily transmitted through the oral route when consumed naturally, with higher doses required for infection in complete feed compared to liquid media.

Acknowledgements

This study was supported by the National Pork Board grant #17-057 and the State of Kansas National Bio and Agro-Defense Facility Fund.

Reference

1. Niederwerder, M.C., A.M.M. Stoian, R.R.R. Rowland, S.S. Dritz, V. Petrovan, L.A. Constance, J.T. Gebhardt, M. Olcha, C.K. Jones, J.C. Woodworth, Y. Fang, J. Liang, T.J. Hefley. 2019. Infectious dose of African swine fever virus when consumed naturally in liquid or feed. *Emerging Infectious Diseases*. Accepted *In Press*.

Efficacy of an inactivated Seneca Valley virus vaccine in nursery-aged pigs

Alexandra Buckley[1]; Abby Patterson[2]; Albert van Geelen[1]; Kelly Lager[3]
[1]Oak Ridge Institute for Science and Education and National Animal Disease Center and US Department of Agriculture, Ames, Iowa; [2]Boehringer Ingelheim Vetmedica, Inc., Ames, Iowa; [3]National Animal Disease Center, US Department of Agriculture, Ames, Iowa

Introduction

Senecavirus A, commonly known as Seneca Valley virus (SVV), causes vesicular disease in swine that is clinically indistinguishable from foot-and-mouth disease (FMD). Therefore, when a vesicle is observed, a foreign animal disease investigation must be performed by trained personnel. These investigations have economic ramifications for federal and state governments investigating the case as well as for producers and packers who cannot move animals with active lesions. A vaccine that provides clinical protection from SVV infection could be a valuable tool for the swine industry to improve swine health and reduce disruptions related to foreign animal disease investigations.

Materials and methods

Weaned pigs were split into 4 groups: Group 1 (no vax + sham, n = 8), Group 2 (no vax + challenge, n = 15), Group 3 (vax + sham, n = 8) and Group 4 (vax + challenge, n = 15). At 4 and 7 weeks-of-age Groups 3 and 4 were vaccinated intramuscularly (IM) with 2 mL of whole-virus inactivated SVV with an oil-in-water emulsion adjuvant. Pigs were intranasally challenged at 9 weeks-of-age (0 dpc) with 5 mL of SVV (10^6 $TCID_{50}$/mL, challenge) or Minimum Essential Medium (sham). All pigs were rectal swabbed daily for the first 10 dpc and checked for vesicular lesions. Pigs were bled at -42, -21, 0, 3, 5, 7, 14, and 21 dpc. Rectal swabs and serum were tested for SVV RNA by PCR and the antibody response was determined by virus neutralization (VN) assay.

Results

In Group 1 (no vax + sham), there were no clinical signs, SVV RNA was not detected in serum or feces by PCR, and none of the pigs developed a neutralizing antibody response. In Group 2 (no vax + challenge), 80% had vesicular lesions, SVV RNA was detected in serum in most animals (from 1 – 7 dpc), and 80% of pigs had detectable SVV RNA in rectal swabs by 1 dpc with shedding detected until 21 dpc in 60% of pigs. No animals had a VN titer greater than 1:16 prior to challenge (0 dpc) and after challenge titers ranged from 1:16 to 1:256. In Group 3 (vax + sham), all animals developed a VN titer of 1:256 following two doses of vaccine (21 dpc). Following sham challenge, no animals had vesicular lesions and SVV RNA was never detected in serum or fecal samples. In Group 4 (vax + challenge), all animals had VN titers between 1:64 and 1:1024 following two doses of vaccine (21 dpc). Following virulent challenge, no animals had vesicular lesions and SVV RNA was never detected in serum. However, SVV RNA was detected in fecal samples in 13/15 pigs on 1 dpc, but all were negative by 2 dpc.

Conclusion

Whole-virus inactivated SVV plus an oil-in-water emulsion adjuvant provided protection against clinical disease and prevented viremia. While viral nucleic acid was detected in rectal swabs from vaccinated animals on 1-2 dpc, it is not known if this material was passed through from the intranasal inoculation or reflective of active replication within the pig. Moreover, it is not known if this material was infectious. An effective vaccine could have a positive impact on welfare in the swine industry and reduce the economic burden of investigating vesicular disease.

Transmammary delivery of firocoxib from sows to piglets reduces stress and improves average daily gain after castration, tail docking and teeth clipping

Johann F. Coetzee[1,2]; Pritam K. Sidhu[3]; Jon Seagen[4]; Teresa Schieber[4]; Katie Kleinhenz[3]; Michael D. Kleinhenz[3]; Larry W. Wulf[2]; Vickie L. Cooper[2]; Kelly Lechtenberg[4]

[1]Department of Veterinary Diagnostic and Production Animal Medicine (VDPAM), College of Veterinary Medicine, Iowa State University, Ames, Iowa; [2]Veterinary Diagnostic Laboratory (VDL), College of Veterinary Medicine, Iowa State University, Ames, Iowa; [3]Department of Anatomy and Physiology, College of Veterinary Medicine, Kansas State University, Manhattan, Kansas; [4]Midwest Veterinary Services, Inc., Oakland, Nebraska

Introduction

Each year, more than 50 million piglets in the United States undergo processing procedures (e.g., castration, tail docking, and teeth clipping) that result in post-procedure pain. Pain-related behaviors have been shown to persist for up to 4 days after castration in 5-day-old piglets.[1] Consumer concern about the animal welfare implications of painful processing procedures has resulted in investigation of methods to provide analgesia and alleviate distress associated with these practices.[2] In the US, no drugs are labeled by the Food and Drug Administration (FDA) to mitigate pain in swine. Therefore, the development of pain mitigation strategies that address consumer concerns about animal welfare while still being safe, effective, practical and economical to administer would be beneficial to the swine industry.

The transmammary delivery of analgesic drugs is an attractive option for providing long lasting pain relief to piglets and reducing labor costs because this reduces the need for individual injections and repeated handling events at the time of processing. A study evaluating a transmammary route of administration of the non-steroidal anti-inflammatory drug (NSAID), meloxicam, reported a significant reduction in expression of biomarkers associated with pain and stress (i.e., prostaglandin E_2 and cortisol) after processing in piglets nursing medicated sows.[3] However, transmammary delivery of meloxicam required a dose of 30 mg/kg bodyweight to be administered orally to sows over several days. This is 75 times higher than the approved dose for swine in the European Union and is therefore not practical, safe or economically feasible for producers to implement in swine production systems.

Firocoxib is a NSAID with FDA-approval in the USA for control of pain and inflammation associated with osteoarthritis in horses and dogs. Firocoxib is 384 times more selective for canine COX-2, the inducible isoform of the enzyme that synthesizes prostaglandins that mediate pain and inflammation, than COX-1, the constitutively induced isoform that produces prostaglandins that maintain normal renal function and gastrointestinal integrity.[4-6] Furthermore, compared to meloxicam, the tendency of firocoxib to move from the blood into the tissues, as reflected by the volume of distribution (Vd), is reported to be 12 times greater in dogs (2.9 ± 0.3 L/kg vs. 0.23 ± 0.03 L/kg).[6,7] Taken together, these properties suggest that, compared to meloxicam, firocoxib may be safer and more likely to distribute into the milk of medicated sows for delivery to nursing piglets prior to processing.

In the present study, we tested the hypothesis that transmammary delivery of firocoxib, from the lactating sow to the nursing piglets achieved safe and effective analgesic drug concentrations in suckling offspring that underwent processing procedures (e.g. castration, tail docking and teeth clipping). The specific aims of this project were to (1) describe the pharmacokinetics and transmammary delivery of firocoxib to piglets after intramuscular (IM) administration to sows, (2) investigate the effects of transmammary-delivered firocoxib on the stress response and performance of piglets after castration, tail docking and teeth clipping, and (3) to assess the safety and firocoxib tissue residue concentrations in both sows and piglets at weaning.

Materials and methods

Pregnant Yorkshire x Landrace sows (mean ± SEM bodyweight, 250.3 ± 7.61 kg) that were approximately 1 week before farrowing were sourced from a commercial swine farm (Table 1). Each sow was examined by a veterinarian to confirm that she was healthy and pregnant. A unique numerical ear tag was placed in the right ear of each sow for identification. The sows were housed in accordance with the recommendations in the Guide for the Care and Use of Agricultural Animals in Agricultural Use and Research and Teaching (2010). Sows were housed in a farrowing crate measuring 0.6 m × 2.1 m. Farrowing crates had a minimum piglet creep area of 6.4 m^2. After farrowing, a heat lamp was provided on one side of the creep area for the piglets in each crate. All sows were fed a diet that met or exceeded National Research Council's 2012 nutrient requirements and water was provided ad libitum.

On Study Day 0, before dosing, 16 sows each nursing nine piglets [aged 5 (± 2) days] per sow were selected to receive clinical examinations that included recording body weights. Each litter provided six male and three female piglets for sampling at time point 0. Body weights were used to randomly assign sows to 1 of 4 treatment

Table 1: Individual doses of firocoxib administered to lactating sows via intramuscular injection at 0.5 mg/kg, 1.0 mg/kg, 1.5 mg/kg, or 2.0 mg/kg.

Sow treatment group	Sow tag	Sow ID	Parity	Body weight, Kg	Dose, mg	Dose volume (mL) administered IM
0.5 mg/kg	1	1427	8	266	133	6.5
	2	1471	7	255	127.5	6.5
	3	1578	5	238	119	6.0
	4	16175	1	179	89.5	4.5
1.0 mg/kg	5	1351	9	211	211	10.5
	6	1380	9	292	292	14.5
	7	1448	8	258	258	13.0
	8	15177	4	241	241	12.0
1.5 mg/kg	9	1441	7	292	438	22.0
	10	1590	4	248	372	18.5
	11	14145	6	260	390	19.5
	12	16149	2	224	336	17.0
2.0 mg/kg	13	304	10	262	524	26.0
	14	1381	8	294	588	29.5
	15	14156	6	253	506	25.5
	16	16148	2	231	462	23.0

groups (n = 4 sows/ group). At the time of study commencement (T0), sows received a single dose of firocoxib (Equioxx Injection, Merial, Duluth, GA) administered at either 0.5 mg/kg, 1.0 mg/kg, 1.5 mg/kg, or 2.0 mg/kg by IM injection into the right lateral neck (Table 1). In cases where the calculated dose volume exceeded 20-mL, the remaining volume was administered in the left lateral neck muscle.

Blood samples for firocoxib determination were collected from the sows and three piglets/litter (two male and one female piglet at each time point) at 0, 2, 4, 6, 8, 12, 24, 48, 72, 96, and 120h post-drug administration to the sow. Blood samples were transferred to 6-mL evacuated tubes that contained lithium heparin (Vacuette plasma tubes, Greiner Bio-One, Monroe, NC).

Castration, teeth clipping, and tail docking were completed for each male piglet between the 6h and 8h blood collection time points. Female piglets did not undergo any of these procedures and thus served as procedural controls as previously described.[8] Castration, teeth clipping and tail docking were done using standard industry practices.

On day 21 post-drug administration, the body weight of each sow and piglet was recorded before weaning to calculate average daily weight gain. Each animal was then humanely euthanized according to American Veterinary Medical Association guidelines. Necropsies were performed on all animals to inspect for macroscopic signs of NSAID toxicity. Samples of kidney, liver, small intestine and stomach were sectioned for histopathology examination. Approximately 200g of muscle, liver, kidney, fat, and injection site tissue were also collected from each sow for firocoxib residue determination. The liver, the two kidneys, and 50g muscle and fat were collected from at least three piglets/ litter (two male and one female) for tissue drug residue determination. The tissue samples were stored at -20 °C until analysis.

Firocoxib analysis was conducted using a previously described method developed in bovine plasma that was adapted and validated for use in swine.[9] A quadratic (1/X) fit was used for the sow plasma samples with a concentration range of 1-5,000 ng/mL. The samples from the piglets were analyzed using a narrower range of calibration spikes of 1-500 ng/mL. A linear (1/X) fit was used for the piglet plasma samples and the narrower 1-500 ng/mL concentration range.

Samples of muscle, injection site, kidney, liver, and fat tissue were analyzed for firocoxib and the descyclopropylmethyl metabolite. Eight calibration spikes were prepared in blank porcine tissue in the concentration range of 0.05-10 µg/g.

Plasma and tissue firocoxib concentrations were determined using high-pressure liquid chromatography (Agilent 1100 Pump, Column Compartment and Autosampler, Agilent Technologies, Santa Clara, CA) with mass spectrometry detection (LC-MS) (LTQ Ion Trap, Thermo Scientific, San Jose, CA). The limit of quantitation of the analysis for both firocoxib and the descyclopropylmethyl metabolite was 0.05 µg/g; the limit of detection was 0.01 µg/g.

Plasma cortisol concentrations were determined using a commercial radioimmunoassay kit (CortiCote I-125, MP Biomedicals, Santa Ana, CA). The samples were incubated at 4°C for 2h to improve assay sensitivity.

Formalin-fixed sections of kidney, liver, small intestine and stomach from sows and 3 piglets/litter were examined by a veterinary diagnostician with experience in the histological examination of swine tissues.

The firocoxib plasma concentration vs. time profile from each sow from the treatment groups injected with firocoxib at doses of 2 mg/kg, 1.5 mg/kg, 1 mg/kg, and 0.5 mg/kg, was evaluated using pharmacokinetic analysis software (Phoenix® Win-Nonlin® 7.0, Certara, Inc. Princeton, NJ). The data were analyzed using non-compartmental methods.

For each treatment group, the plasma firocoxib concentrations vs. time data for the nursing piglets (n = 36 per treatment) were evaluated using non-compartment analysis and the sparse data option available in the software (Phoenix® Win-Nonlin® 7.0, Certara, Inc. Princeton, NJ). The non-compartment analysis (NCA) sparse method calculates pharmacokinetic parameters based on the mean profile for all the subjects in the dataset.

The piglet exposure to firocoxib via milk consumed from treated dams was evaluated by comparing the firocoxib AUC for the piglets with the drug AUC calculated for the corresponding dams from each treatment. Exposure percentages were determined using the equation:

$$\%\text{ exposure} = 100 \times \frac{\text{AUC (piglet)}}{\text{AUC (sow)}}$$

All statistical analyses were performed using statistical software (JMP Pro. 13.0, SAS Institute, Cary, NC). Firocoxib pharmacokinetic parameters in sows were not normally distributed and were thus compared using Wilcoxon Rank Sum tests. Dose linearity was investigated by calculating the square of the Pearson correlation coefficient (R^2) for linear regression. Plasma cortisol responses were analyzed using a nonlinear mixed-effects analysis incorporating both fixed effects and random effects. Fixed effects for time series were assumed to be the same at each time data collected, and random effects were associated with each sample (e.g. piglets) from the population. Average daily weight gain (ADG) after processing was analyzed using ANOVA. Interactions were investigated using the Tukey HSD test for multiple comparisons when $P \leq 0.1$. Statistical significance was set a priori at $P \leq 0.05$.

Results

The mean ± SEM plasma firocoxib concentrations in the sows from the four treatment groups that received a single IM dose of firocoxib at 0.5, 1.0, 1.5, or 2 mg/kg are presented in Figure 1. The pharmacokinetic parameters for firocoxib in sows following IM administration are presented in Table 2. Mean ± SEM firocoxib C_{max} values of 107.90 ± 15.18, 157.50 ± 24.91, 343.68 ± 78.89, and 452.83 ± 90.27 ng/mL were recorded at 3.5, 5.5, 3.0, and 4.5h, respectively, in sows receiving 0.5, 1.0, 1.5, or 2 mg/kg firocoxib. Mean C_{max} was higher in sows that received 2 mg/kg compared to sows that received 1 mg/kg and 0.5 mg/kg ($P = 0.03$). Firocoxib demonstrated a prolonged plasma elimination half-life ($T_{1/2}\lambda z$) ranging from 26.71 ± 5.77h to 31.09 ± 6.73h. Dose linearity was investigated by plotting the C_{max} and $AUC_{0\text{-last}}$ values against the administered doses. The results suggest that both C_{max} (R^2=0.60) and $AUC_{0\text{-last}}$ (R^2 = 0.55) demonstrated dose linearity across the four doses that were investigated.

The results for the plasma firocoxib concentrations (mean ± SEM) in the piglets after the sows were injected with firocoxib (0.5, 1.0, 1.5, or 2 mg/kg) are presented in Figure 2. The firocoxib serum concentrations versus time profiles for the four groups of piglets were similar except that there was a more rapid decline in drug concentrations after C_{max} in the piglets that nursed sows that received a 0.5 mg/kg dose. Mean peak plasma concentrations (C_{max}) of 9.53 ng/mL, 31.04 ng/mL, 53.30 ng/mL, and 44.03 ng/mL were found at 24 h after 0.5, 1.0, 1.5, or 2 mg/kg firocoxib administration to the sows, respectively. The results for the pharmacokinetic parameters following transmammary delivery to the piglets are presented in Table 3. Firocoxib had a prolonged plasma $T_{1/2}\lambda z$ between 30.86h and 48.71h in the piglets after transmammary delivery. The transmammary exposure in the four groups of piglets ranged between 26.7% and 56.5%; exposure was not dose-dependent. The results for C_{max} and AUC indicated that dose linearity was absent.

The plasma cortisol concentrations in the piglets were determined 6h after firocoxib was administered to the sows (Figure 3). This time point occurred immediately before castration, tail docking, and teeth clipping were performed in male piglets. The results of the analysis indicated that there were effects of treatment group ($P = 0.0009$), time ($P < 0.0001$), and a time X treatment interaction ($P < 0.0001$) on plasma cortisol concentrations after castration, tail docking, and teeth clipping (Figure 3).

Average daily gain in body weight (ADG) over the 21 days from processing to weaning was calculated by subtracting the body weight at processing from the body weight at weaning and dividing the result by 21d (Figure 4). The results suggest that there was an effect of treatment on ADG over the 21d from processing to weaning ($P = 0.0257$). Specifically, ADG increased with the increasing doses of firocoxib administered to the lactating sows.

At 21d after IM administration, no detectable concentrations of firocoxib or its descyclopropylmethyl metabolite were found in concentrations above the limit of quantitation (0.05 µg/g) in any of the muscle, liver, kidney, fat or injection site tissue samples in sows. Similarly, there were no detectable concentrations of firocoxib in the tissues harvested from the piglets at weaning.

Table 2: Summary of pharmacokinetic parameters (mean and SEM) for lactating sows administered firocoxib via intramuscular injection at 0.5 mg/kg, 1.0 mg/kg, 1.5 mg/kg, or 2.0 mg/kg (a-c: $P < 0.05$).

Dose (IM)		0.5 mg/kg		1.0 mg/kg		1.5 mg/kg		2.0 mg/kg	
Parameter	Units	Mean	SEM	Mean	SEM	Mean	SEM	Mean	SEM
λz	1/h	0.03	0.006	0.03	0.002	0.02	0.003	0.02	0.002
$T_{1/2\lambda z}$	h	28.87	6.92	26.70	2.88	29.89	4.84	31.09	3.36
T_{max}	h	3.50	0.96	5.50	0.96	3.00	0.58	4.50	1.26
C_{max}	ng/mL	107.90^{b}	15.18	157.50^{b}	24.91	343.68^{ab}	78.89	452.83^{a}	90.27
CL/F	L/h/kg	0.33	0.04	0.21	0.05	0.19	0.03	0.18	0.04
$AUC_{0\text{-}24h}$	h×ng/mL	1030^{c}	62	2673^{b}	0.486	4841^{ab}	961	6639^{a}	1421
$AUC_{0\text{-}last}$	h×ng/mL	1534^{b}	182	5331^{a}	1.357	8323^{a}	1814	12721^{a}	3548
AUC_{INF}	h×ng/mL	1586^{b}	184	5624^{a}	1.534	8656^{a}	1760	13651^{a}	4045
AUC_Exp	%	3.39	0.66	4.15	1.50	4.98	2.92	5.61	1.83
AUMC	h^2×ng/mL	31765^{b}	8530	164892^{a}	58409	215365^{a}	56451	392753^{a}	142492
$MRT_{0\text{-}last}$	h	27.16	7.64	29.38	2.89	26.35	3.66	29.07	2.86
MRT_{INF}	h	29.36	9.93	34.80	4.87	33.48	7.73	36.80	5.44
V_z/F	L/kg	13.82	4.25	7.82	1.44	9.12	3.24	7.75	1.55

λz = slope of the terminal phase; $T_{1/2\lambda z}$ = terminal half-life; Tmax= time to achieve peak concentration; C_{max} = maximum plasma concentration; CL/F = apparent systemic clearance; $AUC_{0\text{-}24h}$ = area under the concentration-time curve for the first 24 h after administration $AUC_{0\text{-}last}$ = area under the curve to the last measured time point; AUC_{INF} = area under the curve extrapolated to infinity; AUC Exp= % of the AUC extrapolated to infinity; AUMC=area under the first moment of the concentration-time curve; $MRT_{0\text{-}last}$ =mean residence time to the last time point; MRT_{INF} =mean residence time extrapolated to infinity; V_z/F= apparent volume of distribution during the terminal phase.

Table 3: Summary of pharmacokinetic parameters (mean ± SEM) for piglets (n=36/treatment) nursing lactating sows administered firocoxib via intramuscular injection at 0.5 mg/kg, 1.0 mg/kg, 1.5 mg/kg, or 2.0 mg/kg bodyweight. The non-compartment analysis sparse method was used to calculate pharmacokinetic parameters based on the mean profile for all the subjects in the dataset. Therefore, in this analysis, standard error of the mean (SEM) was only calculated for Tmax, Cmax and $AU_{C0\text{-}last}$.

Sow dose		0.5 mg/kg		1.0 mg/kg		1.5 mg/kg		2.0 mg/kg	
Parameter	Units	Mean	SEM	Mean	SEM	Mean	SEM	Mean	SEM
λz	1/h	0.02		0.01		0.02		0.02	
$T_{1/2\lambda z}$	h	30.86		48.71		37.91		32.42	
T_{max}	h	24.00	12.56	24.00	14.83	24.00	14.17	24.00	12.21
C_{max}	ng/mL	9.53	1.21	31.04	6.79	53.30	11.10	44.03	7.47
$AUC_{0\text{-}24h}$	h×ng/mL	175.98		406.67		662.37		608.69	
$AUC_{0\text{-}last}$	h×ng/mL	635.36	55.82	2468.00	315.10	3897.55	459.80	3220.90	279.30
AUC_{INF}	h×ng/mL	690.47		3178.00		4615.17		3652.32	
AUC_Exp	%	0.08		0.22		0.16		0.12	
AUMC	h^2×ng/mL	28433.00		136805.00		204246.00		166016.00	
$MRT_{0\text{-}last}$	h	44.76		55.42		52.41		51.54	

λz = slope of the terminal phase; $T_{1/2\lambda z}$ = terminal half-life; T_{max} = time to achieve peak concentration; C_{max} = maximum plasma concentration; $AUC_{0\text{-}24h}$ = area under the concentration-time curve for the first 24 h after administration $AUC_{0\text{-}last}$ = area under the curve to the last measured time point; AUC_{INF} = area under the curve extrapolated to infinity; AUC Exp= % of the AUC extrapolated to infinity; AUMC=area under the first moment of the concentration-time curve; $MRT_{0\text{-}last}$ =mean residence time to the last time point.

Figure 1: Mean (± SEM) plasma firocoxib concentrations (ng/mL) for lactating sows administered firocoxib via the intramuscular route at 0.5 mg/kg, 1.0 mg/kg, 1.5 mg/kg, or 2.0 mg/kg.

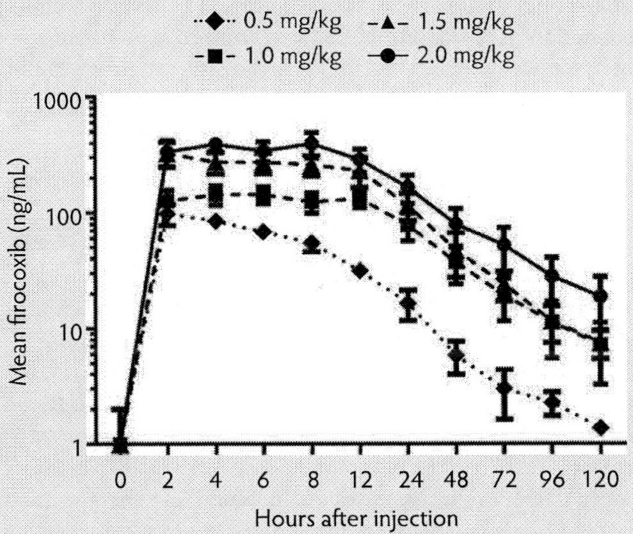

Figure 2: Mean (± SEM) plasma firocoxib concentration (ng/mL) for piglets nursing lactating sows administered firocoxib at 0.5 mg/kg, 1.0 mg/kg, 1.5 mg/kg, or 2.0 mg/kg via intramuscular injection.

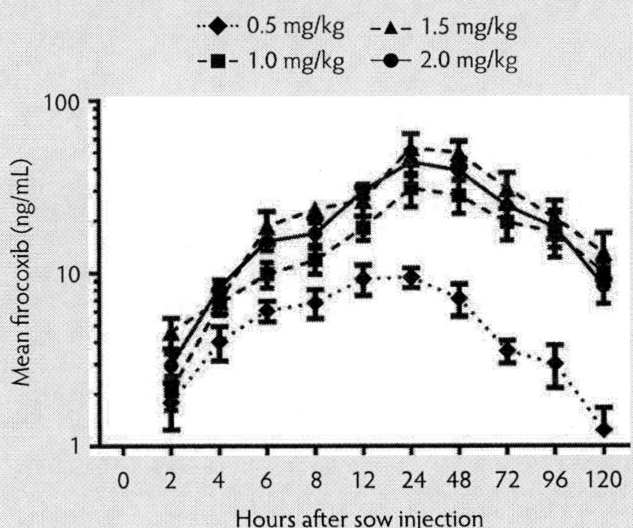

Figure 3: Mean (± SEM) plasma cortisol concentrations (ng/mL) following castration, tail docking and teeth clipping in male piglets nursing lactating sows administered firocoxib at 0.5 mg/kg, 1.0 mg/kg, 1.5 mg/kg, or 2.0 mg/kg via intramuscular injection. Female piglets served as procedure controls and were not subject to processing (teeth clipping or tail docking) (a-b: $P < 0.05$).

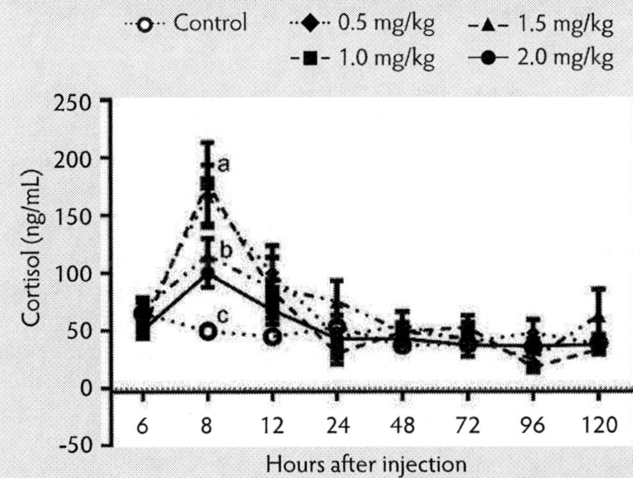

Figure 4: Mean (± SEM) average daily weight gain (ADG) (kg) at 21 days after castration, tail docking and teeth clipping of male piglets nursing lactating sows administered firocoxib at 0.5 mg/kg, 1.0 mg/kg, 1.5 mg/kg, or 2.0 mg/kg via intramuscular injection. Female piglets served as procedure controls and were not subject to processing (teeth clipping or tail docking) (a-b: $P < 0.05$).

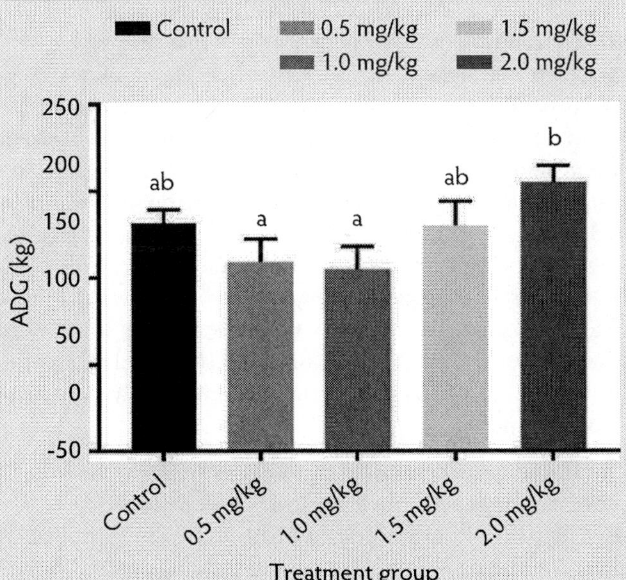

Discussion

The results of the present study suggest that a single injection of firocoxib administered to sows resulted in successful transmammary delivery of analgesia to nursing piglets prior to processing. This discovery could potentially address many of the impediments to routine analgesic drug use in piglets at the time of processing by reducing labor costs and improving piglet welfare through reduced stress. Furthermore, the cost of analgesia may be offset by enhanced production through increases in piglet weaning weights.

Plasma elimination half-life is the pharmacokinetic parameter that describes the time taken for the plasma drug concentrations to decrease by half. The long plasma elimination half-life of firocoxib reported in sows in the present study (26.7-31.1h) was similar to the results obtained in horses (29.6-31.1h) and calves (31.8h).[9-11] A long terminal half-life is desirable from a clinical perspective because this may result in a longer duration of analgesia following a single dose and may also reduce dosing frequency. These data support the hypothesis that firocoxib is a suitable analgesic for single dose administration in swine which reduces labor costs and stress associated with frequent injections.

Piglets nursing sows that received oral meloxicam at 30 mg/kg for three consecutive days demonstrated a significant reduction in plasma cortisol concentrations over 10h after processing compared to piglets nursing unmedicated sows. Furthermore, piglets nursing medicated sows demonstrated higher cranial skin temperatures and lower concentrations of prostaglandin E2 compared to placebo-treated piglets. The results of the present study advance our understanding of transmammary delivery of analgesic compounds to manage pain in the offspring by demonstrating that this can be accomplished with a single injection using a dose volume that is attainable in a swine production environment.

The results of the pharmacokinetic analysis of the plasma firocoxib concentrations in the piglets suggest that the passage of firocoxib from the sow plasma into the milk was not linear, suggesting that this is a saturable process. Therefore, an increase in the sow dose above 1.5 mg/kg may not result in higher firocoxib concentrations in the milk. Based on the AUC values calculated from 0h to the last time point, piglets nursing on sows administered 0.5 mg/kg to 1.5 mg/kg firocoxib as a single IM injection received between 41% and 46% of the total sow firocoxib exposure. In contrast, the piglets nursing on sows that received 2 mg/kg firocoxib were exposed to 25% of the sow exposure. These results suggest that a dose of 1.5 mg/kg firocoxib IM is likely the optimal sow dose for achieving effective transmammary delivery of firocoxib to the piglets.

In most studies that evaluated the use of NSAIDs around the time castration is performed, the drugs were delivered to each test piglet. This study was designed to evaluate transmammary delivery of firocoxib via treated sows to suckling piglets. There was a positive effect of dose on piglet ADG. Some studies of postpartum meloxicam effects on sow behavior and performance found increases in ADG in the lower birth weight piglets, compared with placebo sow piglets.[12,13] A review of pain associated with parturition concluded that more research is needed to better understand costs and benefits of intervention strategies.[12] Results of studies focusing on the use of periparturient NSAID therapy indicate that this drug class can benefit dam behavior and milk production.[14,15] Milk production was not measured in our study. However, the increases in ADG across treatment groups could have resulted from the effects of firocoxib on postpartum physiology in the sows.

In conclusion, the results of this study suggested that maternal firocoxib administration at 8h before performing piglet processing procedures resulted in successful transmammary delivery of firocoxib to the nursing piglets. A 1.5 mg/kg dose provided the maximal percentage of drug transfer from sow to piglet and significantly reduced the cortisol concentrations found in piglets after processing. There were also positive effects on piglet growth rates as maternal growth increased. Drug concentrations found in tissue samples taken 21d post-maternal administration were below the level of detection of the assay. When given via the transmammary route, firocoxib has potential as a therapeutic drug used for analgesia, to reduce processing-induced stress, improve piglet welfare, and enhance production through increases in weaning weights.

References

1. Hay, M., A. Vulin, S. Genin, P. Sales, and A. Prunier. 2003. Assessment of pain induced by castration in piglets: behavioral and physiological responses over the subsequent 5 days. Applied Animal Behaviour Science 82(3):201-208. doi: 10.1016/s0168-1591(03)00059-5.

2. Sutherland, M.A. 2015. Welfare implications of invasive piglet husbandry procedures, methods of alleviation and alternatives: a review. New Zealand Veterinary Journal 63(1):52-57. (Review) doi: 10.1080/00480169.2014.961990.

3. Bates, J.L., L.A. Karriker, M.L. Stock, K.M. Pertzbom, L.G. Baldwin, L.W. Wulf, C.J. Lee, C. Wang, and J.F. Coetzee. 2014. Impact of Transmammary-Delivered Meloxicam on Biomarkers of Pain and Distress in Piglets after Castration and Tail Docking. Plos One 9(12). doi: 10.1371/journal.pone.0113678.

4. Vane, J.R., and R.M. Botting. 1995. A better understanding of antiinflammatory drugs based on iso forms of cyclooxygenase (COX-1 and COX-2). Prostaglandins and Related Compounds: Ninth International Conference, Florence, Italy 23:41-48 (Review).

5. McCann, M.E., D.R. Anderson, C. Brideau, D.H. Black, W.C. Zhang, and G.J. Hickey. 2002. In vitro activity and in vivo efficacy of a novel CQX-2 inhibitor in the horse. Proceedings of the Academy of Veterinary Internal Medicine. 2002. Abstract 114, p. 789.

6. McCann, M.E., D.R. Anderson, D. Zhang, C. Brideau, W.C. Black, P.D. Hanson, and G.J. Hickey. 2004. In vitro effects and in vivo efficacy of a novel cyclooxygenase-2 inhibitor in dogs with experimentally induced synovitis. American Journal of Veterinary Research, 65, 503-512.

7. Montoya, L., L. Ambros, V. Kreil, R. Bona¢ne, G. Albarellos, R. Hallu, and A. Soraci. 2004. A pharmacokinetic comparison of meloxicam and ketoprofen following oral administration to healthy dogs. Veterinary Research Communications, 28(5), 415-428.

8. Kielly, J., C.E. Dewey, and M. Cochran. 1999. Castration at 3 days of age temporarily slows growth of pigs. Swine Health and Production 7(4):151-153.

9. Stock, M.L., R. Gehring, L.A. Barth, L.W. Wulf, and J.F. Coetzee. 2014. Pharmacokinetics of firocoxib in preweaned calves after oral and intravenous administration. Journal of Veterinary Pharmacology and Therapeutics 37(5):457-463.

10. Holland, B., C. Fogle, A.T. Blikslager, A. Curling, B.M. Barlow, J. Schirmer, and J.L. Davis. 2015. Pharmacokinetics and pharmacodynamics of three formulations of firocoxib in healthy horses. Journal of Veterinary Pharmacology and Therapeutics 38:249-256. (Article) doi: 10.1111/jvp.121777.

11. Kvaternick, V., M. Pollmeier, J. Fischer, and P.D. Hanson. 2015. Pharmacokinetics and metabolism of orally administered firocoxib, a novel second generation coxib, in horses. Journal of Veterinary Pharmacology and Therapeutics 30:208-217. (Article) doi: 10.1111/j.1365-2885.2007.00840.x.

12. Mainau, E., and X. Manteca. 2011. Pain and discomfort caused by parturition in cows and sows. Applied Animal Behaviour Science 135(3):241-251. doi: 10.1016/j.applanim.2011.10.020.

13. Tenbergen, R., R. Friendship, G. Cassar, M.R. Amezcua, and D. Haley, 2014. Investigation of the use of meloxicam post farrowing for improving sow performance and reducing pain. Journal of Swine Health and Production 22(1):10-15.

14. Viitasaari, E., M. Raekallio, M. Heinonen, A. Valros, O. Peltoniemi, and L. Hanninen. 2014. The effect of ketoprofen on post-partum behaviour in sows. Applied Animal Behaviour Science 158:16-22. doi: 10.1016/japplanim.2014.06.005

15. Carpenter, A.J., C.M. Ylioja, C.F. Vargas, L.K. Mamedova, L.G. Mendonca, J.F. Coetzee, L.C. Hollis, R. Gehring, and B.J. Bradford. 2016. Hot topic: Early postpartum treatment of commercial dairy cows with nonsteroidal antiinflammatory drugs increases whole-lactation milk yield. Journal of Dairy Science 99(1):672-679. doi: 10.3168/jds.2015-10048.

Potential risk factors for pelvic organ prolapses: Survey of 104 US commercial breeding herds

Chris J. Rademacher; Amanda L. Chipman; Colin D. Johnson; Ken J. Stalder; Anna K. Johnson; Aileen F. Keating; John F. Patience; Nick K. Gabler; Daniel C.L. Linhares; Kent J. Schwartz; Suzanne T. Millman; Jamie M. Studer; Zoe E. Kiefer; Gustavo Silva; Jason W. Ross

Iowa Pork Industry Center, Iowa State University, Ames, Iowa

Introduction

Incidence of pelvic organ prolapses (POP - uterine, rectal or vaginal) has increased at an alarming rate, particularly over the past 5 years. Although several causes have been speculated, the available literature is lacking, and additional research is needed to further clarify the etiology. The inability to execute hypothesis driven research is due to a lack in fundamental understanding of the root cause(s) contributing to the increased incidence POP in the swine industry.[1] In response to this industry issue, the Iowa Pork Industry Center (IPIC) enrolled 104 U.S. swine breeding herds using a retrospective survey and analysis to generate hypotheses in order to identify potential risk factors and eventually test mitigation strategies.

Materials and methods

A comprehensive survey was administered to 104 swine breeding herds across 15 states including farms ranging from breeding herds within large productions systems to smaller, independent producers, totaling approximately 400,000 sows. The survey asked sow farm managers to provide data and information enabling examination of potential risk factors of POP. Questions solicited data on herd dynamics and management approaches, facility types, nutritional strategies, and animal-based measurements. On 62 of these farms, IPIC staff collected individual sow measurements including, but not limited to, perineal score, tail length, and body condition. Each week, farms reported sow death numbers along with putative causes categorized into prolapse and non-prolapse deaths. This information was used to create the weekly POP incidence rate per 1,000 sows and an annualized POP rate. All data was collected from January through July, 2018. A Poisson mixed regression model using PROC GLIMMIX in SAS® (SAS Institute, Inc., Cary, NC) to assess risk factors associated with POP incidence rate was applied to these data. The weekly count of POP per farm was the outcome variable and the log of the inventory as an offset variable. Production system was included as a covariate in the analysis.

Results

The annualized POP mortality across the entire dataset was 2.7% with a range from 0.3% to 10.3% during weeks 6 through 30 of 2018; 22% of all mortality reported was due to POP. When analyzing weekly POP mortality by 20% highest POP incidence, 60% average POP incidence, and 20% lowest POP incidence, larger variation over time was observed in the 20% of farms with the highest incidence rate (Figure 1).

Preliminary analysis demonstrated that potential root causes not statistically associated with POP incidence were herd size, induction protocols, farrowing assistance protocols, and tail length when using $P < 0.05$ as the threshold of significance.

Interestingly, feeding strategy prior to farrowing was identified as being important, as utilization of bump feeding reduced ($P = 0.03$) POP rate in farms bump feeding animals with a lower body condition score compared to farms not bump feeding (Figure 2). The observation that bump feeding was able to influence POP-induced mortality is consistent with our observation that sows with the lowest body condition score had a greater probability of POP compared to sows in optimal body condition or overweight. A total of 4953 body condition scores (BCS) were assigned during late gestation. A total of 884 sows were considered a BCS 1 (thin), 3,378 sows had a BCS 2 (normal body condition), and 691 sows BCS 3 (overweight). Retrospective analysis after the scored sows underwent opportunity for farrowing indicated that 65 of them prolapsed (1.3%). Twenty-one of the BCS 1 sows prolapsed (2.4%), 41 of the BCS 2 sows prolapsed (1.2%), and 3 of the BSC 3 sows prolapsed (0.4%). The average BCS of the farm tended to negatively correlate with prolapse incidence ($r = -0.25$; $P = 0.06$).

Additionally, a perineal scoring system was developed and utilized to assess potential for prolapse during late gestation. Individual sows were scored with a three point perineal score (1, a presumed low risk of prolapse; 2, a presumed moderate risk of prolapse; and 3, a presumed high risk for prolapse). Out of the 2906 animals scored, 14 sows prolapsed that scored a 1 (1.1%), 11 sows prolapsed that scored a 2 (0.8%), and 17 sows prolapsed that scored a 3 (7.2%). The average perineal score of the farm was correlated to herd level prolapse incidence ($r = 0.42$, $P < 0.001$). This scoring system has potential for further investigations to better assess the biological causes and the effectiveness of POP mitigation strategies.

Water treatment also appeared to reduce POP incidence with those farms treating water (using either a chlorine or peroxide based treatment system) having lower POP-induced mortality than those farms that did not (Figure 3). When analyzing water source, farms on this study using water treatment on well or pond water sources had lower POP incidence than those utilizing untreated well water, while rural water was not different in POP incidence from treated or untreated water (Figure 3).

Figure 1: Weekly pelvic organ prolapse caused mortality presented as an annualized rate from February through July 2018. Weekly changes in POP are presented as the farms for which the average POP rate is in the lowest 20% of all farms (orange line), the farms for which the average POP rate is in the highest 20% of all farms average (green line), and the 60% of farms considered to have an average POP rate (blue line). The average annualized POP rate was 5.7%, 2.4%, and 0.9% for the highest incidence herds, average incidence herds, and lowest incidence herds. Separating the farms into these groups revealed that farms most afflicted by higher annualized mortality as the result of the POP experience greater change over time is not observed in the farms experiencing lower incidence rate of POP.

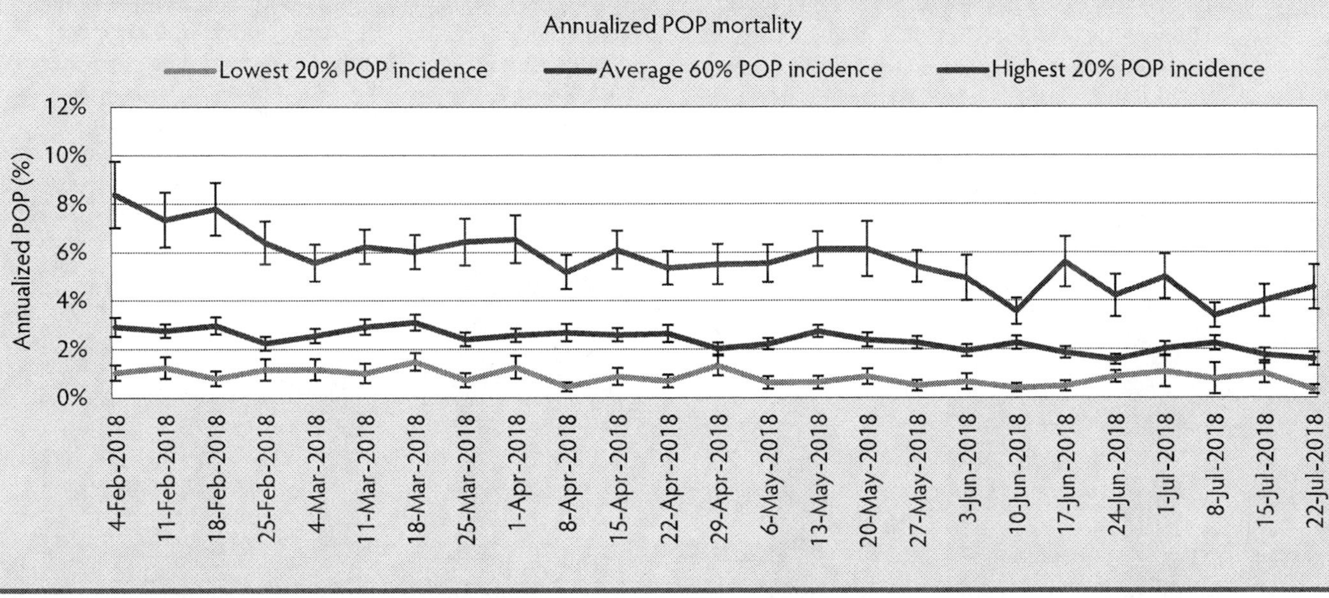

Summary and conclusion

Results from this study have preliminarily identified several different risk factors needing further investigation to verify their causality for increased POP risk and potential mitigation strategies. Additional evaluation will be completed to further define and prioritize risk factors likely contributing to a greater incidence in POP in the farms evaluated in this study for additional testing.

In addition, this project supports the concept of building an ongoing collaboration from commercial farms from multiple production systems across the US swine industry for field research allowing within and between production system comparisons of risk factors associated with relevant outcomes. This project was funded by the National Pork Board project #17-224 and supported by a vast network of industry collaborators representing producers from 15 states managing approximately 400,000 sows.

Reference

1. Supakorn et al., 2017, Open Journal of Veterinary Medicine; 7:85-97.

Figure 2: Impact of bump feeding during late gestation on pelvic organ prolapse (POP) induced mortality incidence rate. Farms provided weekly mortality from calendar year weeks 6 through 30 during 2018. Bump-feeding strategy by sow farms was detected to influence ($P = 0.03$) POP-induced mortality/1000 sows/week, being greatest in those farms that did not utilize bump feeding as a management strategy during late gestation, being numerically greater great than farms that bump feed all animals and markedly different than those farms that bump feed sows in low body condition scores. Bars with different superscripts differ significantly ($P < 0.05$).

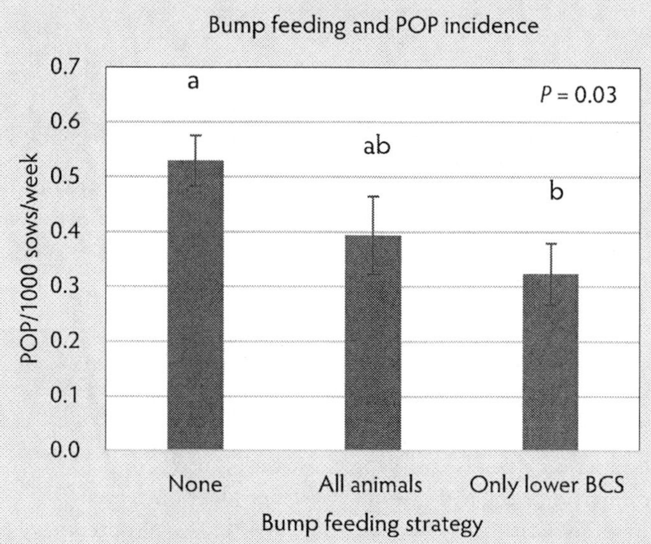

Figure 3: Impact of water source and treatment on pelvic organ prolapse (POP)-induced mortality. The source of water and treatment strategy was associated with differences in POP-induced mortality for the farms included in this study. Specifically considering farms using well water, those using water treatment systems had reduced POP-induced mortality compared to those not using a water treatment system. Bars with different superscripts differ significantly ($P < 0.05$). Treated included well or pond water that has been treated with chlorides or peroxides.

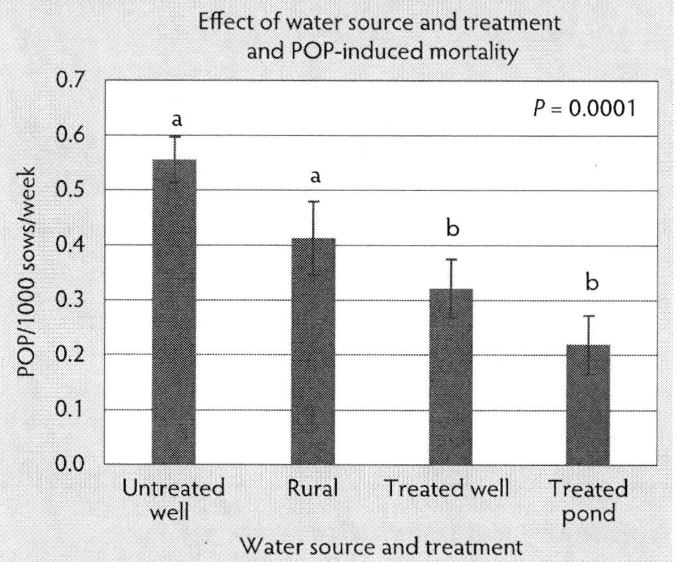

Clinical validation of an infrared thermometer in periparturient sows

E. DeCaluwe-Tulk, HBSc; T. M. Rossi DVM, PhD; A. L. Greer, MSc, PhD; E. Luo; T. L. O'Sullivan, DVM, PhD
Department of Population Medicine, Ontario Veterinary College, University of Guelph, Guelph, Ontario, Canada

Introduction

Individual monitoring and care of sows around farrowing is important for welfare and productivity in a farrowing operation. Body temperature can be a useful indicator of health status in sows.[1] Infrared thermometers (IR) work by collecting and quantifying infrared radiation at the body surface.[2] Infrared thermometers offer a quick, touchless alternative to taking rectal temperatures (RT) and reduces animal handling and labor for the producer. The InstaTemp MD™ thermometer is validated for use in humans, however has not been previously validated in swine.

Objectives

There were three objectives of this study: (1)To determine the agreement between RT and InstaTemp MD™ IR measurements at three anatomical locations (the ear base, caudal udder and vulva), (2)To determine the precision of the IR thermometer when taking repeated measurements in the same location, and (3)To determine if the use of a correction factor would improve agreement between IR and RT measurements.

Materials and methods

Study population and enrollment

Periparturient sows (n = 48) from a commercial swine operation were enrolled in the study, divided into four 1wk cohorts. For the purposes of this study, periparturient sows were defined as those expected to farrow during each cohort week. The producer allocated sows to crates within a farrowing room containing 24 crates according to standard operating procedures for the operation. Four crates (two on each end of the farrowing room) were designated to be enrolled in each cohort to account for any differences in ambient temperature between the two sides of the farrowing room. The remaining crates in the room were assigned a number from 1 to 20. For each cohort, the sows in the 4 designated crates were enrolled plus sows in 8 of the crates numbered from 1-20 selected by a random number generator (**random.org**) for a total of 12 sows per cohort.

IR thermometer validation

Three sets of temperature measurements were taken for each enrolled sow once per day for 5 consecutive days: IR temperatures followed by a RT and an ambient temperature measurement. IR measurements were taken using the InstaTemp MD™ non-touch thermometer (ARC Devices USA, Inc., Boca Raton, FL) at three anatomical locations: the ear base, caudal udder and vulva. Ear measurements were taken directly caudal to the ear at the base of the pinna. Caudal udder measurements were taken on the lateral aspect of the udder approximately 3 inches cranial to the stifle joint. The vulva measurements were taken by pointing the IR thermometer directly at the center of the vulva. If the measurement sites were accessible, the IR measurements were taken in the position the sow was in; however, if the measurement sites were inaccessible (for example, the caudal udder while the sow was in sternal recumbency), researchers would get the sow to stand. The order in which the three IR measurements were taken was randomized for each sow on each day using a random number generator (**random.org**). The IR thermometer was used in accordance with the user manual; it was held 2 inches from the skin surface and held still during the 2.5 second measurement time. If the sow changed position during this time, the measurement was repeated. The RT measurement was completed following the IR measurement using a digital thermometer (Sharptemp®, Cotran Corp, Partsmouth, RI, USA). The RT measurements were done following the IR measurements to reduce the potential for handling to influence body temperature. For the purposes of this study, the RT measurement was considered the gold standard. The ambient temperature was recorded at the time of temperature measurements at each crate, this was recorded by a room thermometer (AccuTemp®, Springfield, USA). All temperatures (IR, rectal and ambient) were recorded in degrees Celsius (°C).

IR thermometer precision

For each day of a cohort (n = 20), one of the enrolled sows was randomly selected using a random number generator (**random.org**) to have precision measurements taken in addition to their IR and RT measurements for the validation aspect of this study. Precision measurements involved 20 consecutive IR temperature measurements at each IR site (ear base, caudal udder and vulva). If the sow changed her body position partway through the precision measurements, the precision measurements were repeated.

Statistical analysis

Data was entered into Microsoft Excel® and cross-validated by a member of the research team. Agreement was calculated using concordance correlation coefficients (ρC) using STATA (Stata Statistics/Data Analysis® v 14.0); precision was determined by calculating coefficients of variation (CV) in R-Studio (RStudio® Desktop 1.1.456). Scatter plots were made to plot IR temperatures against RT in STATA.

Correction factor investigation

To investigate the possibility of using a correction factor, IR measurements were corrected by adding the average difference from RT. Concordance correlation coefficients were then calculated again for the corrected values (Stata Statistics/Data Analysis® v 14.0).

This study was reviewed and approved by the University of Guelph Animal Care Committee. Animal use protocol #3987.

Results

Agreement

Agreement, as determined by the concordance correlation coefficient, was poor for all IR sites investigated (Table 1).[3,4] At the ear base, the concordance correlation coefficient was 0.232 (95% CI [0.176, 0.287]) with an average difference of 1.068°C below the RT. At the caudal udder, the concordance correlation coefficient was 0.369 (95% CI [0.308, 0.430]) with an average difference of 0.921°C below the RT. The vulva measurements performed the poorest of all three sites as it had the lowest concordance correlation coefficient of 0.131 (95% CI [0.088, 0.175]) and the greatest average difference from the RT (1.636°C below the RT).

Precision

The coefficient of variation was close to zero for all three IR sites, indicating the IR thermometer is precise when taking repeated measurements at the same site (Table 1).

Correction factor investigation

When IR was corrected by the average difference from RT, the agreement between IR (corrected) and RT was improved at all three sites compared to the agreement of IR (uncorrected) with RT (Table 1). The ρC for the corrected ear base IR was 0.549 (95% CI [0.047, 0.642]), which was improved from 0.232. The ρC for the corrected caudal udder IR was 0.691 (95% CI [0.624, 0.757]), which was improved from 0.369 (Figures 1a and 1b). The ρC for the corrected vulva IR was 0.373 (95% CI [0.271, 0.475]), which was improved from 0.131. However, even though ρC was improved by the correction factor, the agreement was still considered poor for all sites.[3,4]

Conclusions and applications

Infrared thermometers could offer a convenient alternative to RT for producers to monitor temperature in sows. In order to consider IR thermometry for use in sows, it should be accurate compared to an accepted gold standard i.e. RT, and it should be precise. The InstaTemp MD™ IR thermometer was precise when tested at the ear base caudal udder and vulva. However, the agreement observed between the InstaTemp MD™ IR and RT was poor for all three anatomical sites investigated, even when a correction factor was applied. Of the three sites investigated, the caudal udder was the most promising IR site; as it had the greatest ρC, the lowest average difference from RT and a CV close to zero. The results of this study do not support the incorporation of InstaTemp MD™ IR thermometer into swine herd health protocols.

References

1. Bressers, H. P. M., Te Brake, J. H. A., Jansen, M. B., Nijenhuis, P. J., & Noordhuizen, J. P. T. M. (1994). Monitoring individual sows: radiotelemetrically recorded ear base temperature changes around farrowing. *Livestock Production Science, 37*(3), 353-361.
2. ARC Devices (2015). InstaTemp MD™ Non-Touch Thermometer User Manual.
3. Lawrence, I., & Lin, K. (1989). A concordance correlation coefficient to evaluate reproducibility. *Biometrics*, 255-268.
4. McBride, G. B. (2005). A proposal for strength-of-agreement criteria for Lin's concordance correlation coefficient. *NIWA Client Report: HAM2005-062.*

Table 1: Summary of agreement (concordance correlation coefficient, ρC) and precision (coefficient of variation, CV) analysis of Instatemp MD™ infrared thermometer with rectal temperature.

	Agreement			Precision
IR measurement site	ρC prior to correction	*Correction factor (°C)	Corrected ρC [95% CI]	Coefficient of variation (CV)
1. Ear base	0.232 [0.176, 0.287]	1.068	0.549 [0.047, 0.642]	0.0047
2. Caudal udder	0.369 [0.308, 0.430]	0.921	0.691 [0.624, 0.757]	0.0047
3. Vulva	0.131 [0.088, 0.175]	1.636	0.373 [0.271, 0.475]	0.0058

IR = infrared thermometer
* Correction factor = average difference RT - IR in degrees Celsius (°C)
ρC = concordance correlation coefficient
CI = 95% confidence interval

Figure 1: A) Scatter plot showing the caudal udder IR temperature (udder in °C, y-axis) vs. the rectal temperature (rectal in °C, x-axis). B) Scatter plot showing the corrected caudal udder IR temperature (corrected udder in °C, y-axis) vs. the rectal temperature (rectal in °C, x-axis).

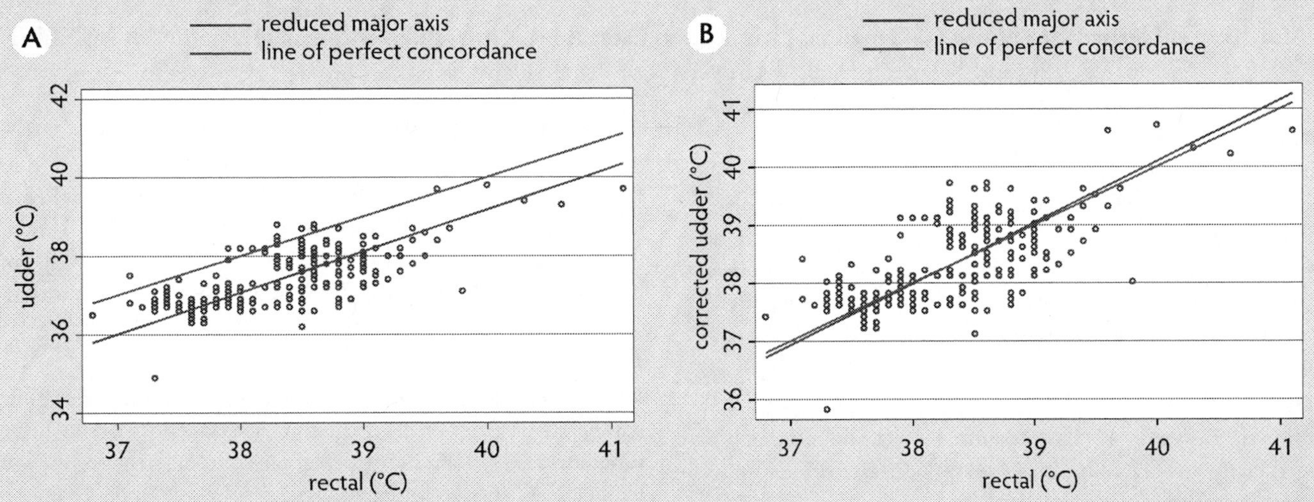

Opening of a hinged farrowing crate increases the variety of behaviors expressed by a lactating sow

Matthew T. Herber, BS; Maria C. Ceballos, PhD; Karen Camille R. Gois, PhD; Thomas D. Parsons, VMD PhD
Swine Teaching and Research Facility, University of Pennsylvania, Kennett Square, Pennsylvania

Introduction

Public discourse about the confinement of mother sows in stalls is being extended to the farrowing house where concerns for pig welfare historically have been prioritized over that of the crated sow. Hinged farrowing crates, as seen in Figure 1, provide sows the opportunity to express more varied and natural behaviors once opened at some point during lactation.

Objective

The aim of this study was to determine how the opening of a hinged farrowing crate on different days post-farrowing impacts sow behavior and piglet mortality.

Materials and methods

A total of 36 sows (DNA Genetics, Line 241) were studied. At approximately 110 days into their gestation, they were moved into a farrowing room with their breeding cohort of 10 individuals. The sows were randomly allocated to 1 of 3 treatment groups: TC- crate remained closed until weaning ($n = 13$), T4- crate was opened on day 4 post farrowing ($n = 12$), and T7- crate was opened on day 7 post farrowing ($n = 11$). On days 3-8 post farrowing, sow behavior was captured by continuous video recording of each individual sow. Sow behavior was quantified from 6 AM to 6 PM for each day of recording by using instantaneous focal animal sampling at a 2-minute fixed interval. An established ethogram was used in conjunction with Observer XT by Nodulus to assist and perform the behavioral data collection. The following behavioral categories were measured: Resting/sleeping-(RES), Investigating piglet-(IP), Nosing piglet-(NP), Biting fixture-(B), Exploring environment-(EE), Drinking-(D), Eating-(E), Nursing-(N), Nose piglet-(NP), Sitting-(SIT), Standing-(ST), Ventral lying-(VL) and Lateral Lying-(LL). Additionally, piglet mortality was tracked daily for each sow until 21 days of age. The behavioral data was then analyzed for the days prior (day 3 and 6) and post (day 4 to 5 and day 7 to 8) to opening the crate for both T4 and T7, respectively, and compared with TC by using the Wilcoxon-Mann-Whitney test.

Results

Before opening the crate, there were no significant differences in sow behavior. After opening the crate at 4 days post-farrowing, differences ($P < 0.05$) in the percentage of total activity were found between TC and T4 for EE (1.2 vs 3.3%, $W = 37$), N (16.0 vs 20.0%, $W = 41$), RES (71.5 vs 64.9 %, $W = 119$) and VL (25.7 vs 16.9%, $W = 121$). When the crate was opened 7 days after farrowing, differences ($P < 0.05$) were found between TC and T7 for EE (1.0 vs 3.4%, $W = 15$) and IP (0.41 vs 0.63%, $W = 30$). Finally, we observed no significant difference in the piglet mortality between treatment groups ($P > 0.05$) albeit given our limited sample size.

Discussion

The hinged farrowing crate allowed sows the freedom to express a greater variety of behaviors as reflected in the increased time spent interacting with their environment and their piglets as well as the decreased time spent resting/sleeping following crate opening. This study provides further evidence to support the use of a hinged farrowing crate as a tool to improve the welfare of lactating sows. However, additional studies with larger sample sizes are needed to more carefully quantify the impact of a hinged crate on piglet mortality and to understand fully the economic implications of this alternative farrowing system.

Figure 1: Closed and open configuration of a hinged farrowing crate.

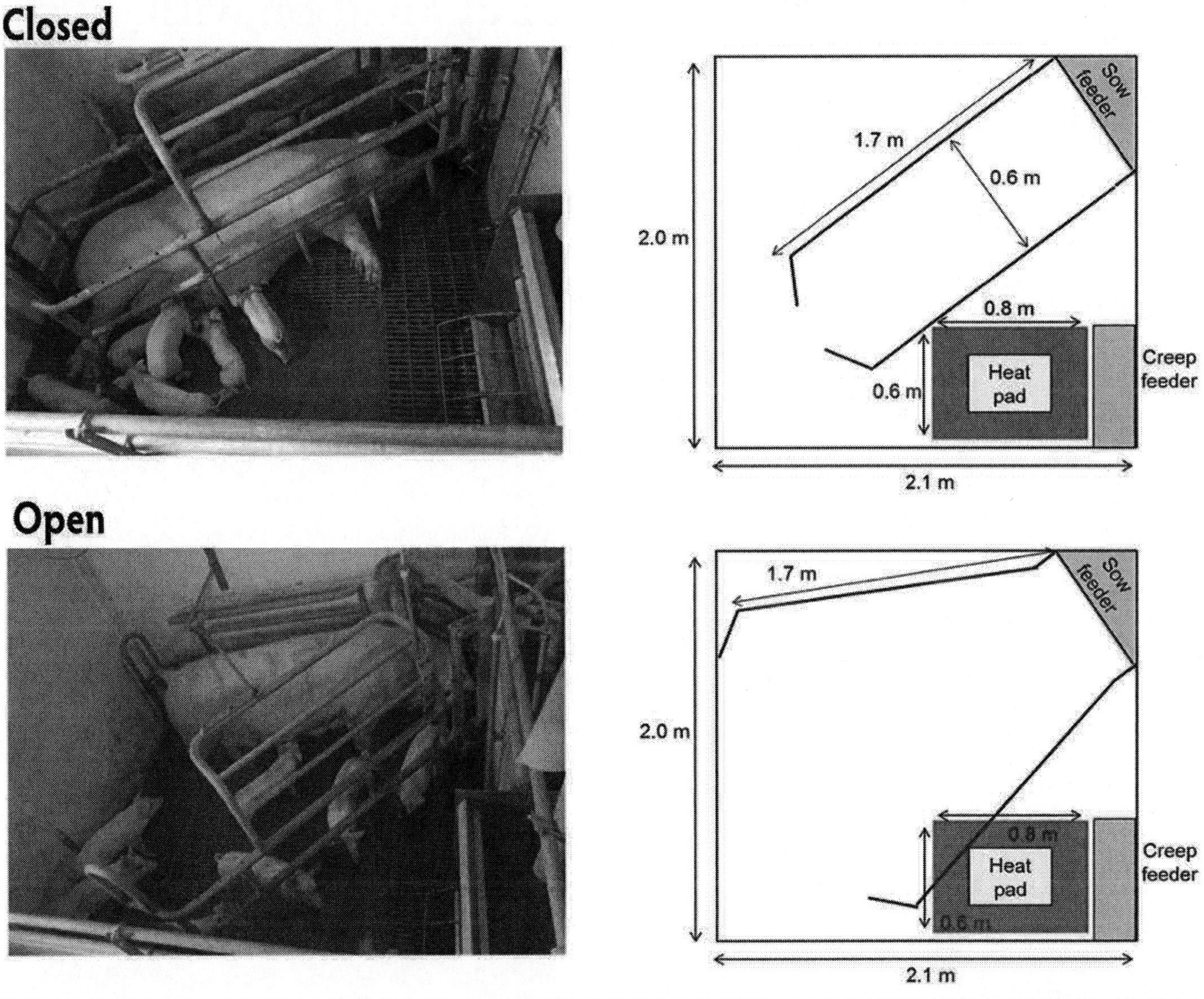

Validation of a modified method of processing fluid collection to mitigate cross-contamination with a porcine reproductive and respiratory syndrome modified live virus vaccine

D. Brown[1]; L. Ochoa[2], PhD; D. Amodie[3], MS; M. Jansen[3], DVM
[1]University of Illinois, Urbana, IL; [2]Carthage Innovative Swine Solutions, Carthage, Illinois;
[3]Zoetis, Parsippany, New Jersey

Introduction

Processing fluids (PF) are a convenient and affordable sample for porcine reproductive and respiratory syndrome virus (PRRSV) surveillance.[1] However, cross-contamination (resulting in false positive samples) with PRRS modified live virus (MLV) vaccine has been observed in systems that administer a PRRS-MLV vaccine at processing. Limited literature exists on cross-contamination of processing fluids with vaccine. The objective of this study was to validate a modified method of collecting processing fluids to mitigate cross-contamination with a PRRS-MLV vaccine.

Materials and methods

A 6,500 head commercial sow farm vaccinating piglets with a PRRS-MLV vaccine at processing was utilized for this study. For several weeks prior to starting the study, all PF samples collected by farm staff yielded positive results for PRRSV by polymerase chain reaction (PCR), and sequencing revealed a 100% homology to the PRRS-MLV vaccine used by the farm. When the study began, litters were processed by teams of technicians according to the following protocols:

- Control protocol (CP): A PRRS-MLV vaccine was not administered to piglets at processing (n = 3 pools).
- Standard protocol (SP): Processing and PRRS-MLV vaccination were performed by same technician, and reflected the farm's typical procedure for collecting PF (n = 31 pools).
- Modified protocol (MP): Technicians either vaccinated with a PRRS-MLV vaccine, or castrated and docked tails. The vaccinating technician was not allowed to touch pigs before they were castrated and docked (n = 32 pools).

Tails and testicles were placed into a plastic bag which was stored in the same portable container as the processing instruments (tattooing equipment, scalpels, etc.). PF pooled from 10 litters was the experimental unit. Observations of the order of processing and instrument disinfection with a chlorhexidine dip were recorded for each sample collected. Every 10 litters, bags from each team were collected, and new bags provided to each technician by a separate individual. Bags were cut in one corner to drain the PF into a 50-mL Falcon tube (Thermo Fisher Scientific, Waltham, MA). PF were harvested from each bag in the same area that technicians used to prepare their processing instruments, including the PRRS-MLV vaccine. After PF from every 10-litter pool was collected, the table used for PF collection was cleaned with a phenol-based disinfectant. PF samples were sent to Iowa State University Veterinary Diagnostic Laboratory. They were pooled by the 10 litters processed by each team, and tested for PRRSV by PCR. In the event of a positive pool, each technician's sample was individually tested.

PRRSV PCR results were considered negative for Ct values greater than 37, and analyzed with a mixed model using the SAS Glimmix procedure with treatment as a fixed effect and the residual error as a random effect (SAS 9.4, Cary, NC). Observations of the order of processing and instruments disinfected, were summarized by the number of samples submitted by individual technicians, but not statistically analyzed.

Results

No positive PCR results were found for CP or MP. SP yielded 12.9% positive pools, and MP yielded zero positive pools (Table 1). There was no significant difference between the SP and MP ($P = 0.97$).

Samples from technicians following the SP were run individually to look for correlation between risk factors and incidence of positive PCR. Vaccinating before castrating and docking tails yielded positive results 12.5% of the time, while technicians who vaccinated after castrating and docking tails yielded positive results only 5% of the time. Technicians that disinfected only their scalpels yielded positive results 23.5% of the time, opposed to technicians who disinfected neither instrument and yielded positive results 10.3% of the time (Table 2). Of the 11 technicians observed utilizing the SP, two yielded 50% of the positive samples.

Table 1: PCR results based on protocol

Treatment group	LSMeans, % positive
Control protocol (n = 3)	0
Standard protocol (n = 31)	12.9
Modified protocol (n = 32)	0

Table 2: Comparison of potential risk factors for cross-contamination of individual PF[1]

	% positive (PCR)
Vaccine order	
Vaccinated before castrating/tail docking	12.5
Vaccinated after castrating/tail docking	5
Instruments disinfected	
Disinfected neither instrument	10.3
Disinfected scalpel only	23.5
Disinfected docker only	0
Disinfected both instruments	4.6

1 Risk factors assessed only for technicians using SP

Discussion

The results suggest that there may be a relationship between processing protocols and cross-contamination of PF with vaccine virus. While the percentage of positive PF samples was not significantly different between the SP and MP, the number of positive samples collected using the SP was lower than previously observed through routine PRRSV-surveillance at the farm, and may have limited the ability to detect significant differences. Although they were not statistically analyzed, the variation in positive and negative results based on order of processing and instruments disinfected could possibly demonstrate potential risk factors for cross-contamination. The greater number of positive results for technicians who disinfected their scalpel blades when compared to technicians who did not disinfect either of their instruments may be attributed to a technician using the same container of disinfectant to dip his instruments in throughout one day. Cross-contamination of the container could occur as the technician processes pigs, and chlorhexidine has been shown to be ineffective at reducing detection of PRRSV with PCR.[2] The greater number of positive results for technicians who vaccinated before castrating or docking tails could potentially be attributed to an increase in the amount of vaccine on a technician's hands if vaccine were to leak from the injection site. Given the evidence of PF sample cross-contamination, it is advisable that production systems which elect to administer a PRRS-MLV vaccine at processing consider risk factors potentially associated with cross-contamination events, and that risk factors be individually studied to assess their potential impact on cross-contamination rates. While further study of the ability for the MP described in this study to control cross-contamination of PF is warranted, assigning processing and vaccinating to separate technicians may help mitigate cross-contamination of PF with a PRRS-MLV vaccine at processing.

References

1. Lopez WA, Angulo J, Zimmerman JJ, et al. Porcine reproductive and respiratory syndrome monitoring in breeding herds using processing fluids. *J Swine Health and Prod.* 2018;26(3):146-150

2. Prickett JR, Cutler S, Kinyon JM, et al. Stability of porcine reproductive and respiratory syndrome virus and antibody in swine oral fluid. *J Swine Health Prod.* 2010;18(4):187–195.

Methods for achieving successful and safe gilt inoculation with *Mycoplasma hyopneumoniae*

M. Schleper[1], B.S.; L. Garza-Moreno[2], DVM; S. Tousignant[3], DVM, PhD; Maria Pieters[1], DVM, PhD
[1] Veterinary Population Medicine Department, College of Veterinary Medicine, University of Minnesota, St. Paul, Minnesota;
[2] IRTA, Centre de Recerca en Sanitat Animal (CRESA, IRTA-UAB), Barcelona, Spain;
[3]Boehringer Ingelheim Vetmedica, Inc., Duluth, Georgia

Background

Mycoplasma hyopneumoniae (M. hyopneumoniae) is the causative agent of enzootic pneumonia (EP), a respiratory disease frequently complicated with other pathogens. EP can be complicated and lead to a more significant condition known as Porcine Respiratory Disease Complex.[1] The prevalence of *M. hyopneumoniae* at weaning has been correlated with disease severity in growing and fattening pigs.[2] Is has been suggested that gilts and young parity sows are significant bacterial shedders.[3] Thus, gilt acclimation has been proposed as a strategy to control *M. hyopneumoniae* in breeding herds.[4] Intentional exposure to *M. hyopneumoniae* is commonly used for gilt acclimation.[5] However, methods for exposure to *M. hyopneumoniae* in the field may lack effectiveness, consistency, or safety.

Objective

The aim of this study was to compare four *M. hyopneumoniae* exposure techniques to induce EP in gilts. The effectiveness and consistency of the techniques were assessed based on *M. hyopneumoniae* colonization, lung lesions, coughing index, and humoral immune response against *M. hyopneumoniae*.

Materials and methods

Forty 5-week old gilts from a *M. hyopneumoniae* negative sow farm were selected and tested to confirm their negative status by ELISA (-13 days post inoculation; dpi) and rt-PCR (0 dpi). At -4dpi, gilts were transferred to an isolation facility and distributed into five groups according to the following inoculation methods: 1) Intra-tracheal catheter (ITc; n = 5); 2) Laryngo-Tracheal Mucosal Atomization (LTMA; n = 10); 3) Intranasal with device (INd; n = 10); 4) Rope chewing (RCh; n = 10); 5) Mock-inoculated (MI; n = 5). Gilts in groups 1-4 were inoculated with 10 mL of M. hyopneumoniae lung homogenate (Strain 232; 1 x 105 CCU/mL) at 0 dpi. The MI group was inoculated with 10mL of Friis medium. Laryngeal swabs and blood samples were collected at - 0, 14 and 28 dpi to detect M. hyopneumoniae by rt-PCR and to assess M. hyopneumoniae antibodies by ELISA. The number of coughs over 15 minutes were recorded for each treatment group from 10-25 dpi and a cough index was calculated. All gilts were euthanized at 28 dpi. At necropsy, lungs were scored on the percentage of affected tissue associated with M. hyopneumoniae infection and bronchial swabs were collected to be evaluated with rt-PCR. M. hyopneumoniae detection by PCR, lung lesions, and coughing index were statistically analyzed using the Chi-square test. P-values <0.05 were considered statistically significant.

Results

All gilts resulted seronegative at -13 and 0 dpi, and negative by rt-PCR at 0 dpi. *M. hyopneumoniae* was detected in gilts in the groups ITc, LTMA and INd at 28 dpi, showing significant differences in ITc and LTMA compared to INd, RCh and MI (Table 1). Gilts from RCh and MI groups remained negative thought the study. All gilts remained seronegative with the exception of gilts from the ITc (2/5, 40%) and LTMA (5/10, 50%) groups, which seroconverted by 28 dpi. *M. hyopneumoniae*-like gross lung lesions were observed in different proportions of gilts in all inoculated groups. The LTMA and ITc groups exhibited higher cough index than the INd and RCh groups, although significant differences were not detected.

Discussion and conclusions

These results showed that LTMA and ITc were the more effective inoculation methods to induce EP in gilts while INd route was less effective. In addition, gilt infection was not achieved in the RCh group. The LTMA group showed the greatest percentage of *M. hyopneumoniae* colonization and seroconversion in this study. A potential hypothesis for the increase in *M. hyopneumoniae* colonization in LTMA over ITc (which appears to deliver the inoculum very close to the same anatomical area as LTMA), may be a difference in particle size of the inoculum. The LTMA device produces 30-100 µm particles while the ITc delivers a stream of homogenate. These smaller particles could result in inoculum reaching areas of the respiratory tract that the stream of fluid from the catheter might not reach. In addition to being highly effective, the LTMA device was more user-friendly than the traditional rubber catheter, resulting in shorter gilt restraint times. These results suggest that inoculation using the LTMA is an efficient and safe route to induce MP in gilts, representing a potential exposure strategy during gilt acclimation for controlling *M. hyopneumoniae* in the field.

Table 1: Detection of *Mycoplasma hyopneumoniae* by rt-PCR (Ct mean ± SD) in laryngeal swabs.

Exposure technique	Sampling points							
	Laryngeal swab PCR						Bronchial swab PCR	
	0 dpi		14 dpi		28 dpi		28 dpi	
	Prop (%)	Ct	Prop (%)	Ct	Prop (%)	Ct	Prop (%)	Ct
ITc	0/5 (0.0)	NA	1/5 (20.0)	38.8 ± 1.3	3/5 (60.0)	34.6 ± 3.4	5/5 (100.0)	29.4 ± 2.5
LTMA	0/10 (0.0)	NA	2/10 (20.0)	38.7 ± 0.6	10/10 (100.0)	34.6 ± 3.3	10/10 (100.0)	27.8 ± 2.8
INd	0/10 (0.0)	NA	1/10 (10.0)	35.7 ± 0.0	4/10 (40.0)	34.9 ± 2.9	4/10 (40.0)	29.6 ± 1.7
RCh	0/10 (0.0)	NA	0/10 (0.0)	NA	0/10 (0.0)	NA	0/10 (0.0)	NA
MI	0/5 (0.0)	NA	0/5 (0.0)	NA	0/5 (0.0)	NA	0/5 (0.0)	NA

ITc: Intratracheal catheter; LTMA: laryngo-tracheal mucosal atomization; INd: Intranasal device; RCh: Rope chewing; MI: Mock inoculation; NA: non-applicable; Ct values (mean±SD) were calculated from positive samples.

References

1. Maes et al. Update on *Mycoplasma hyopneumoniae* infections in pigs: Knowledge gaps for improved disease control. Transbound. Emerg. Dis. 2017; 62, 1-15.

2. Fano et al., Effect of *Mycoplasma hyopneumoniae* colonization at weaning on disease severity in growing pigs. Can. J. Vet. Res. 2007; 71, 195-200.

3. Boonsoongnern et al., The Prevalence of *Mycoplasma hyopneumoniae* in Commercial Suckling Pigs in Thailand. World J. Vaccines. 2012; 161-163.

4. Pieters and Fano, *Mycoplasma hyopneumoniae* management in gilts. Vet. Rec. 2016; 178, 122: 1-123.

5. Garza-Moreno et al., Acclimation strategies in gilts to control *Mycoplasma hyopneumoniae* infections. Vet. Microbiol. 2017; 219, 23-29.

Evaluation of various ante-mortem sentinel pigs sampling techniques for detecting *Mycoplasma hyopneumoniae* following seeder pig exposure

S. Leone[1], BS; A. Maschhoff[2], DVM; J. Lehman[3], DVM, MS

[1]Iowa State University, Ames, Iowa; [2]The Maschhoffs, Carlyle, Illinois; [3]Merck Animal Health, DeSoto, Kansas

Introduction

Mycoplasma hyopneumonia (Mhp) is a widespread respiratory pathogen in the swine industry causing significant economic losses.[1] Due to industry efforts to eliminate Mhp in breeding herds, early detection of Mhp in replacement populations is critical for preventing infection of Mhp-negative sow farms. The gold standard for Mhp diagnosis is PCR from post mortem lung tissue. The objective of this study was to evaluate the sensitivity of various ante mortem diagnostic approaches.

Materials and methods

Mhp-positive lung homogenate was prepared and stored as previously described.[2] Study pigs were 15 weeks old and verified Mhp-negative by PCR on study day -2. Pigs were housed in a 360-head, naturally ventilated barn with six pens. On day 0, thirteen pigs in the westernmost pen were inoculated intratracheally with 10mL of Mhp positive lung homogenate. Thirty non-challenged sentinel pigs distributed in other pens throughout the barn were ear tagged and serially tested at days 14, 24, 35, 42 and 48 post inoculation. Four sample types were collected from each sentinel pigs at each test date: A tracheal swab (TS) using a 14 FR urinary catheter, a tracheobronchial swab (TBS) via a cotton swab attached to a post-cervical artificial insemination (PCAI) catheter, a laryngeal swab (LS) via a spoon attached to a PCAI rod, and serum. The distal 3-4 inches of the TS and TBS were removed, then placed into a 5 mL falcon tube containing 1 ml phosphate buffered saline. For LS samples, a BBL CultureSwab was wiped over the spoon surface then placed in a vial of Starplex Liquid Stuart's Medium. All swabs were tested for Mhp by PCR at the Iowa State University Diagnostic Laboratory (ISU-VDL), pooled in groups of five. The poolings were consistent throughout the study with the same animals used for each pool. All samples were collected by the same investigator except for serum samples on day 68. Investigators recorded collection time required for each sample and scored ease of sampling for each sample obtained.

Results and discussion

At day 14, all inoculated pigs were confirmed Mhp positive by individual PCR (Average ct = 19.4). All sampling methods eventually detected Mhp infection in sentinels at various time (Table 1), however time to detection and average Ct counts were different between sample types. Mhp was first detected in sentinel pigs via PCR in TB. Average collection time was fastest for LS. In comparing sample types, TS produced a unique sound and vibration when placed in the trachea, allowing more confidence in proper sample collection. TBS may be preferable to the TS in older animals (e.g. sows) due to the swab reaching deeper into the trachea. The results of this study imply that the most sensitive sample type to detect

Table 1: Results summary of various ante-morten sampling techniques for *Mycoplasma hyopneumoniae*

	Percentage of pools positive or suspect (PCR or ELISA)					Average Ct count of positive PCR samples			Time required to collect sample (Seconds)			
Days post exposure	Tracheal swab	Tracheo-bronchial swab	Laryngeal swab	Serum		Tracheal swab	Tracheo-bronchial swab	Laryngeal swab	Tracheal swab	Tracheo-bronchial swab	Laryngeal swab	Serum
				Positive	Suspect							
-7	0	0	0	0	0							
14	83	17	0	0	0	31.7	29.5		18.6	30.5	7.3	16.5
24	50	50	17	0	0	22.9	26.3	33.7	15.9	30.8	8.3	12.8
35	83	83	83	0	0	24.1	25.1	31	23.9	28.6	14.2	18.2
42	100	100	83	7	3	21.6	26.8	32	17.3	31.8	17.3	22.8
48	100	100	100	13	7	20.6	19.6	32.2	23.2	54.6	11.2	37.3
68				77	13							

Mhp post infection is a tracheal swab. Further investigations are warranted to determine the best variation of swab in adult animals and other alternative diagnostic approaches for Mhp.

References

1. Sibila, M., Pieters, M., Molitor, T., Maes, D., Haesebrouch, F., Segales, D., 2009. Current perspectives on the diagnosis and epidemiology of *Mycoplasma hyopneumoniae* infection. *The Veterinary Journal* 181, 221-231.

2. Nickel, M., Toohill, E., Lehman, J., 2018. Use of a hurricane fogger for *Mycoplasma hyopneumoniae* inoculation in nursery age gilts. *49th Annual Meeting of the AASV, Proceedings 2018*. 97-98.

Evaluation of efficacy of ultraviolet germicidal chambers in swine farms

K. Rieland; M. Yang; M. Torremorell, DVM
Veterinary Population Medicine Department, College of Veterinary Medicine, University of Minnesota, St. Paul, Minnesota

Introduction

Ultraviolet (UV) germicidal chambers, also known as UV boxes, are used in swine farms to decontaminate objects entering farms. In these chambers, UV type C (UV-C) light is used to inactivate pathogens by destroying or disrupting the pathogen's nucleic acid. Our goal was to develop procedures that producers could use to evaluate the efficacy of UV germicidal chambers in their swine farms, more specifically to assess the efficacy of UV germicidal chambers at inactivating *Escherichia coli*. We were also interested in obtaining information on how UV germicidal chambers are being used since there is limited data on protocols used across farms. Currently, there are no validated methods to evaluate the efficacy of UV germicidal chambers in farms. Additionally, limiting the number of pathogens entering a swine farm is important for animals to remain healthy, as well as maintain animal welfare, production standards, and decrease the use of antibiotics.

Materials and methods

An inoculum of non-hemolytic *Escherichia coli* was grown for 6 hours (exponential phase) to 10^8 CFU/mL in Luria-Bertani (LB) broth, aliquoted in 1 ml tubes, and kept frozen at -80°C until use. The inoculum was used to create test plates which were obtained by placing 0.5 mL of the *E. coli* inoculum in blank petri dishes and allowing them to air dry in the biosafety cabinet for three hours. The *E. coli* concentration post drying was 10^6 CFU/mL. These plates were part of a pre-assembled kit that was sent to 46 farms conveniently selected. Each kit contained six test plates to be UV tested in two distinct treatment cycles, one control (untreated) test plate, the protocol sheet, questionnaire, shipping label, scotch tape, measuring tape, ice packs, and two ziploc bags. The plates were shipped on ice overnight. Farm personnel were asked to treat the six plates as per the on-farm standard UV protocol and answer questions regarding UV chamber operation.

Plates within the UV chamber were placed in different locations to evaluate the homogeneity of UV effectiveness within the chamber. Three plates were tested during the same run and placed in a diagonal line to evaluate the variation of UV exposure within the UV chamber. The other three plates were tested in a second run that consisted of having one plate in the left corner of the UV chamber (similar to run 1), another plate was taped to the side of a 3-D object (the cardboard box that the kits were shipped in) with the plate facing the right side of the UV chamber, and the last plate was taped on top of the cardboard box. After UV treatment, plates were shipped to the laboratory for bacterial quantification on ice. Bacteria were resuspended in LB broth, plated on MacConkey plates in serial dilution and the plates were incubated overnight at 37°C. Colonies were counted within 48 hours to determine the CFU/mL concentration and percentage inactivation was calculated using bacterial counts in the control plate as reference. The questionnaire obtained information about the construction of the UV chamber, disinfection practices on items entering the farm, placement of items in the chamber, time of exposure, and frequency of use.

For purposes of analysis, control plates with $\leq 10^3$ CFU/mL were removed from the analysis. Results were analyzed using ANOVA to determine whether differences among test plates and type of UV box existed. Results were considered significant if $P < 0.05$.

Results and discussion

Of the 41 kits sent to farms, only 36 were received. Out of those, only 26 plates had *E. coli* growth in the control test plates and only 18 of those reached the minimum concentration of 10^3 CFU/mL set as the minimum quality control. Only 1 farm had inactivation rates of 100% in all test plates. The average farm percent inactivation was 93.81% and ranged between 62.6% and 100%. Differences between the test plates were not statistically significant ($P < 0.05$) indicating that there were no significant differences between inactivation rates obtained between cycles, or location within the UV box. However, it is important to note that although there were not statistical differences, average inactivation at the top of the 3-D object was higher (92.93%) in most of the farms compared to the lower (85.33%) inactivation rate obtained at the side of the 3-D object. In addition, the homemade UV chambers had higher inactivation rates compared to commercial chambers with values of 99% vs. 88% for the homemade and commercial chambers respectively ($P = 0.04$). Most commercial chambers were operated for 5 minutes and the homemade chambers were operated for 10-15 minutes.

The questionnaire revealed differences in operation protocols and design among UV chambers in farms. Forty five percent (14/31) were homemade UV chambers, 45% (14/31) were commercial UV chambers, and 9% (3/31) were UV rooms. Forty five percent (14/31) of UV chambers had lights on top and bottom of the chamber only. Seventy one percent (10/14) of commercial chambers run for 5 min while 93% (13/14) of the homemade chambers were run for 10 to 15 minutes. All UV rooms were run for over 15 minutes with the maximum being 30 minutes. Ninety four percent (33/35) of survey participants did not have an empty cycle before placing items in the UV chamber and 56% (14/25) of producers that had changed light bulbs in their UV chamber changed the bulbs only when the bulbs burn out. There was a notable difference in the interior dimensions of the UV germicidal chambers and rooms. The dimensions of the commercial UV chambers ranged from 19.5-112 inches in length, 20.5-96 inches in width, and 16.25-96 inches in height compared to the homemade UV chambers which ranged from 28-60 inches in length, 18-66 inches in width, and 22.75-57 inches in height. The UV rooms ranged from 120 -348 inches in length, 70-174 inches in width, and 94-114 inches in height.

Conclusion

The level of bacterial inactivation in on-farm UV chambers varied among farms. There was variability in inactivation rates depending on location within the chambers with the lowest inactivation rates obtained in the lateral part of the 3-D object although these differences were not statistically significant. There was also significant variability in the protocols, procedures, and UV box dimensions reported by farm employees. In summary, current UV protocols used to decontaminate fomites in swine farms are not homogenous and do not appear to fully inactivate bacteria that may be present on the fomites.

Acknowledgements

Special thanks goes out to the University of Minnesota Veterinary Diagnostic Laboratory for supplying the non-hemolytic *E. coli* isolate, Once Inc. for supplying the UV germicidal chamber, and Gustavo Lopez for conducting the statistical analysis. Additionally, the help and input from Aaron Stephan, PhD and Deb Murray, DVM was greatly appreciated. This evaluation would not be possible without the help and time of the producers and veterinarians involved in the field study. This project was supported by the University of Minnesota Swine Disease Eradication Center and the student Katelyn Rieland was supported by Boehringer Ingelheim through the Summer Scholars program.

References

1. Abedi-Moghaddam, N., Bulic, A., Henderson, L., and Lam, E. (2004, April). Survival of *Escherichia coli* to UV irradiation during exponential and stationary phases of growth. *Journal of Experimental Microbiology and Immunology*, 5, 44-49. Retrieved from https://www.researchgate.net/publication/268415940_Survival_of_Escherichia_coli_to_UV_Irradiation_During_Exponential_and_Stationary_Phases_of_Growth.

2. Cody et al (2008, September). Skim milk enhances the preservation of thawed 80°C bacterial stocks. *Journal of Microbiological Methods*, 75 (1), 135-138. doi:10.1016/j.mimet.2008.05.006.

3. Ribbens, S. et al (2008). A survey on biosecurity and management practices in Belgian pig herds. *Preventative Veterinary Medicine*, 83, 228-241. Retrieved from https://www.sciencedirect.com/science/article/pii/S0167587707001791.

Use of processing fluids to monitor sows for porcine circovirus type 2 and determine parity influence on viral detection

Kayla Castevens[1]; Ethan Spronk[2], DVM; Noel Garbes[3], DVM; Giovani Trevisan[4], DVM
[1]North Carolina State University, Raleigh, North Carolina; [2]Eichelberger Farms, Wayland, Iowa;
[3]Zoetis, Parsippany, New Jersey; [4]Iowa State University, Ames, Iowa

Introduction

Emerging in the 1990s, porcine circovirus (PCV) commonly affects nursery and growing pigs worldwide and is now considered ubiquitous and hardy in the environment. Clinical manifestations are termed porcine circovirus associated disease (PCVAD). Diagnosis is made via clinical signs and confirmation of porcine circovirus type 2 (PCV2) antigen or nucleic acid within a lesion. Vertical transmission of PCV2 continues to be concerning for some veterinarians and producers. Monitoring methods on sow farms currently include sampling placental umbilical cord serum (PUCS), colostrum, and pre-suckling serum;[1-4] these can be difficult and time consuming to collect. The study objective was to determine if processing fluids (PF) could be used as an easier monitoring tool to establish sow vaccination protocols and if parity influenced PCV2 detection.

Case history

Seven sow farms (A,B,C,D,E,F,G) were selected for this case study. Farm A, a start-up farm, observed elevated downstream mortality with clinical signs of PCVAD and confirmatory virus detection within associated lesions. Clinical signs of PCVAD were not observed in established farms (B-G). All farms vaccinated piglets with a commercial, two dose PCV2 vaccine at processing and weaning. In addition, farms B-G's gilts received a full dose PCV2 vaccine at both 10 and 23 weeks of age; whereas, farm A's gilts received a single additional full dose PCV2 vaccine at 20 weeks of age. A full dose of PCV2 vaccine was added to the vaccine protocol at 10 weeks of age after PCVAD was diagnosed in Farm A's downstream flow. Sow farms (A-G) were not routinely vaccinating sows with PCV2 vaccine. At the time of this study, Farm A was Porcine Respiratory and Reproductive Syndrome Virus (PRRSV) and Porcine Epidemic Diarrhea (PED) negative. Farms B-G were historically PRRSV and PED positive but not currently showing clinical signs of either. As a result, biosecurity measures were increased on farms B-G for processing piglets and included changing boots between farrowing rooms, wiping down aprons with disinfectant, and changing gloves between litters. All farms dipped castration tools in disinfectant between litters.

Methods and materials

In April and May 2018, all farms began weekly collection of PF for PCV2 testing via PCR. Farms collected all tails and testicles daily and extracted PF. The daily samples were combined and each farm submitted one pooled sample for the week. In June 2018, Farm A (due to PCVAD) and Farms B and C (due to intermittent weekly PCV2 PCR positives), began PF collection by litter for five days. A total of 655 litter-specific tail and testicle samples were collected daily in resealable plastic bags, frozen overnight, thawed, PF were drained into a 5-mL Falcon tube (Thermo Fisher Scientific, Waltham, MA), and refrozen. All PCV2 PCRs were performed at Iowa State University Veterinary Diagnostic Laboratory, Ames, IA by farm in pools of 20 by parity (P0, P1, P2, P3, P4+) except for Farm A which only had P0 and P1 dams. Individual PF samples were then re-pooled and assayed by farm and day collected.

Results

For Farm A, all the weekly samples tested positive with Ct values 20-30 whereas farms B-G had varying positive and negative results with 15/20 (75%) of the positive results having a Ct value ≥ 30 and 100% having a Ct > 25 (Table 1). Positive results for the weeks collected were 100%, 25%, 30%, 100%, 86%, 25% and 13% for farms A-G, respectively. Mean Ct results for farms A-G weekly pooled samples were 25.5, 35.7, 31.0, 33.3, 29.1, 31.7, and 36.3, respectively. Ct range for weekly positive results on all farms was 20.6-36.8. When pooled by parity, Farm A (P0-1) had Ct values between 18.4-28.1 whereas P0 and P4+ pools (Farm B) and P1 and P3 (Farm C) had Ct values >33. All other parity pools on farms B and C were negative (Table 2). When pooled by day, Farm A tested positive 5/5 days (Ct 19.7-24.2) and positive for the week, Farm B tested positive 4/5 days (Ct 33.3-36.8) and positive for the week, and Farm C tested positive 2/5 days (Ct 33.5-35.0) and negative for the week (Table 3). No statistical analysis was applied.

Discussion

Our findings from the three farms where individual litter PF were collected and tested in pools correlated well with previous studies confirming that parity was not significant when detecting PCV2 in a sow herd as measured by PUCS and colostrum.[1,2,5] However, when looking at results from weekly PF samples and the daily results from farms A-C, a few differences were noted. Considering this, it may be important to look beyond just positive or negative, but to consider the degree of "positivity". For this system, a proposed scale was developed to determine "pos-

Table 1: Weekly pooled processing fluids PCV2 PCR results by farm

Week collected	A Ct	A Result	B Ct	B Result	C Ct	C Result	D Ct	D Result	E Ct	E Result	F Ct	F Result	G Ct	G Result
18	24.8	Pos	>37	Neg	25.8	Pos					>37	Neg	>37	Neg
19			34.6	Pos	35.9	Pos					34.2	Pos		
20	30.2	Pos	>37	Neg	>37	Neg					29.2	Pos		
21	26.7	Pos	>37	Neg	>37	Neg			>37	Neg	>37	Neg		
22	25.4	Pos	>37	Neg	>37	Neg	32.5	Pos	25.9	Pos	>37	Neg	>37	Neg
23			>37	Neg	>37	Neg	33.2	Pos	28.3	Pos			>37	Neg
24			>37	Neg	>37	Neg			33.6	Pos	>37	Neg	>37	Neg
25			36.8	Pos	31.4	Pos	34.6	Pos	24.8	Pos	>37	Neg	>37	Neg
26	20.6	Pos			>37	Neg	36.5	Pos	31.0	Pos			>37	Neg
27					>37	Neg	30.0	Pos	31.1	Pos			36.3	Pos
28							33.0	Pos			>37	Neg	>37	Neg

Neg = negative; Pos = positive

Table 2: Pooled processing fluid PCV2 PCR Ct values by parity at each farm

Farm	Parity 0	Parity 1	Parity 2	Parity 3	Parity 4+
A	**21.6 (16)** **26.4**	**18.4** **18.4** **20.8** **21.1** **22.4** **22.9 (19)** **25.3** **25.8** **28.1 (19)**	N/A	N/A	N/A
B	**33.8** **35.1 (15)** >37	>37 >37 >37 >37 (12)	>37 >37 >37 (27)	>37 >37 (14)	**34.5 (13)** **36.2** >37
C	>37	**36.3** >37 (19)	>37 >37 (7)	**34.0 (11)**	>37 >37 (21)

>37 is considered a negative result and anything less than 37 is considered a positive result
N/A – not applicable, no samples collected in category
All pooled samples were tested in pools of 20 litter samples except the results in which the number within the () indicates the number of samples for that pool

Table 3: Pooled processing fluid PCV2 PCR results by day collected

Farm	Day	Number of samples in pool	Ct	Result
A	1	43	19.7	Pos
	2	45	20.2	Pos
	3	42	24.2	Pos
	4	49	21.2	Pos
	5	36	23.5	Pos
	Pooled	215	20.7	Pos
B	1	86	>37	Neg
	2	73	33.3	Pos
	3	66	36.1	Pos
	4	44	34.4	Pos
	5	32	35.0	Pos
	Pooled	301	36.8	Pos
C	1	24	>37	Neg
	2	24	35.0	Pos
	3	43	33.5	Pos
	4	15	>37	Neg
	5	33	>37	Neg
	Pooled	139	>37	Neg

Day 1 = Monday, Day 2 = Tuesday, Day 3 = Wednesday, Day 4 = Thursday, Day 5 = Friday
Pooled = all samples for the week were combined to get the weekly test result after the daily results were acquired
Neg = negative; Pos = positive

itivity" and a subsequent plan of action: Ct>30–negative→no action, Ct25–30→monitor closely by increasing testing frequency and checking downstream flows for PCVAD, Ct<25→likely active infection/diagnostics needed/potential vaccination. Disagreement amongst daily samples in positivity and disagreement in individual days compared to the weekly samples in farms B and C indicate daily fluctuations in viral detection amongst dams/litters which further supports findings in other studies indicating varying viremia.[1,4,6] Without a known dilution factor of individual positive PF with negative PF in this study, the findings cannot be fully understood. In this system, monitoring for PCV2 via PF will continue as it proved helpful in making an intervention decision. The proposed "positivity" scale is a start but is still being tested and validated by monitoring PF PCV2 PCR as well as clinical PCVAD manifestations and tissue diagnostics. This study supports the use of PF to detect PCV2. However, further studies on relationship to other methods like PUCS, contamination from the environment, co-morbidities, dilution factor from pooling and impact on farrowing house internal biosecurity measures should be completed to further validate use of PF as a good monitoring method for PCV2 in a sow herds.

References

1. Dvorak, C; M. Lilla, et al. Transmission of PCV2 from sow to piglet in the farrowing room. *Allen D. Leman Swine Conference*, 2010; 57-62.

2. Seate, J; P. Thomas, et al. Assessing Sow Herd PCV2 Stability and Cross-Sectional Antibody Status of Dams. *European Symposium of Porcine Health Management.* 2015.

3. Seate, J; T. Fangman, et al. Management characteristics of PCV2 PUCS stable sow herds. *European Symposium of Porcine Health Management.* 2015.

4. Baumert, D. In-field research and PCV2 sow herd stability: *Our process. AASV BIVI Symposium.* 2014

5. Murtaugh, M. Natural infection, immunity, and prevalence of PCV2. AASV BIVI Symposium. 2014.

6. Dvorak, C; B. Payne, et al. Effect of maternal antibody transfer on antibody dynamics and control of porcine circovirus yype 2 infection in offspring. *Viral Immunology*. 2018. 31(1): 40-46.

Evaluation of PCV3/2d prevalence in placental umbilical cord blood and processing fluids and reproductive outcomes following mass vaccination with an RNA-particle platform PCV3/2d vaccine

A. Vennekotter[1], BS; E. Byers[2], DVM; B. Thacker[3], DVM, PhD; J. Cagle[3], DVM
[1]University of Illinois College of Veterinary Medicine, Urbana, IL;
[2]Prestage Farms, Inc., Clinton, North Carolina; [3]Merck Animal Health, De Soto, Kansas

Introduction

Porcine circoviruses (PCV) are single-stranded, non-enveloped, circular DNA viruses. PCV2 has three major genotypes (a, b, and d)[1] and has been recognized as a devastating pathogen in the US. The first cases linking PCV3 to clinical disease in the US were reported in 2016.[2-3] Both viruses have been associated with a number of clinical manifestations. PCV2 is a ubiquitous pathogen that causes a systemic infection that leads to porcine circovirus associated disease (PCVAD) where the virus can be detected in multiple tissues and blood.[4] Clinically affected pigs experience varied manifestations of PCVAD including post-weaning multi-systemic wasting syndrome (PMWS), porcine dermatitis and nephropathy syndrome (PDNS), weight loss, slow growth, and death.[4] Additionally, PCV2 has been associated with reproductive failure.[5-6] PCV3 has been associated with a number of conditions similar to PCV2 infection including PDNS,[2,7] multi-systemic inflammation and myocarditis,[3] congenital tremors,[8-9] and reproductive failure manifested as abortions, increased mummies and non-viable piglets, low litter size and delayed return to estrus.[2,7,9] PCV3 can be detected in multiple tissues and blood, similar to PCV2.[2,7] Both share the same name but are only 37% related genetically.[2] Although the increasing clinical significance of PCV3 is relatively recent, the virus already appears to be widely distributed in the US swine population[10] and has been documented in multiple publications worldwide since 2017.[9]

Placental umbilical cord (PUC) blood and processing fluid have been utilized for monitoring the PRRSV[11-15] and PCV2[16-17] status of breeding herds. However, there is no data supporting one sample type over the other regarding PCV2 testing, and the use of either sample type has been minimally investigated for PCV3.

The Sequivity™ Technology RNA-Particle (RP) platform is an alpha replicon vector system that is used to produce vaccines containing a pathogen's gene of interest that codes for a protective antigen. Following injection of the RP vaccine, the protective antigen is produced in dendritic lineage host cells, which in turn elicits an immune response against the pathogen.[18] This vaccine platform has experimentally been shown to be effective in protecting against PCV2d challenge.[1] No studies have been reported using the RP vaccine platform to control PCV3. Therefore, the objective of this study was to determine the impact of mass vaccination of a Sequivity™ Technology RNA-Particle (RP) platform, PCV3/2d vaccine on the prevalence of PCV3 and PCV2d in PUC blood and processing fluid based on PCR testing along with evaluating reproductive performance at farrowing.

Materials and methods

The study was conducted in a farrow-to-wean complex consisting of six PRRSV-positive, stable herds (approximately 13,000 multiparous sows) that share a common replacement gilt source. All farms had a history of PCV3/2d disease with clinical presentations of reproductive failure and downstream porcine circovirus associated disease (PCVAD). In all six herds, odd numbered sows were vaccinated with Sequivity™ by intramuscular injection (IM) twice at a 4-week interval. Even numbered sows serving as controls received synchronous IM injections of an equivalent volume of sterile water. Weekly service groups (groups 28-34 in 2018) from 5 to 11 weeks prior to farrowing at the time of the first vaccination were used. The vaccine was produced based on open reading frame (ORF) 2 sequences that were at least 98% homologous with the sequences of PCV3 and PCV2d isolates obtained from this complex.

Three herds were chosen for PUC blood and processing fluid collection based on the results of prior routine PCV3 and PCV2d processing fluid PCR monitoring. Placenta umbilical cord blood was collected from a single litter at the time of farrowing, and processing fluid (testicles only) was collected from the same litter at 3-5 days of age. Paired PUC blood and processing fluid samples from each litter were tested individually for PCV3/2 using a bivalent PCR at the Iowa State University Veterinary Diagnostic Laboratory (Ames, Iowa). Farrowing data obtained from the operation's PigKnows bureau included farrowing date, dam parity and total born, born alive, stillborn and mummified fetuses per litter. Statistical analysis was performed using Statistix (Tallahassee, FL). For binary variables, logistic regression and either Chi square or Fishers Exact test were used. For continuous variables, linear regression and analysis of variance were used. P-values <0.05 were considered significant.

Results and discussion

Initial analysis indicated that parity and weekly service group influenced results such that grouping of these independent variables would provide a better representation of the data. Accordingly, sows were categorized as parity 1 and 2, or parity 3 or more. Weekly service groups were combined to represent sows in late gestation (groups 28 to 31) or mid-gestation (groups 32-34) at the time of first vaccination.

Overall, the percent PCV3 positive PUC bloods was 5.4% and the percent positive processing fluids was 6.4% for the 624 litters tested. Table 1 presents a cross tabulation of PCV3 PCR cycle time (CT) values for each sample type. The lowest CTs measured in PUC blood and processing fluid were 16.8 and 16.2, respectively. Comparing PUC blood and processing fluid results, the overall agreement between the two sample types was 92.6%. For vaccinated litters (n = 311), the percent PCV3 positive for PUC blood and processing fluids was 5.1% and 4.2%, respectively, while the percent positives in the controls (n = 313) was 5.8% and 8.6%, respectively. By logistic regression the number of PCV3 positive processing fluids was impacted by vaccine group and parity group but not by farm, gestation stage or any litter parameters (total and live born per litter; percent stillborn or mummies per litter). The percent positive PUC bloods were impacted by parity group, live born per litter and percent stillborn per litter. With regard to parity, the percent PCV3 positive with either sample type was highest in parities 1 and 2 then quiescence followed by a spike at parity 6. PCV3 PCR results by sample type, gestation stage, parity group and vaccination status are presented in Table 2. The percentage of positive samples within gestation stage and between vaccine groups was statistically different for processing fluids but not PUC blood. The rate of positive processing fluids was significantly higher in control litters for parities 1 and 2 and vaccinated litters for parities >3.

Table 1: PCV3 PCR cycle times (CTs) by sample type

Percent PCV3 PCR positive (n = 624)		Processing fluids							
		>37	34 - <37	31 - <34	28 - <31	25 - <28	22 - <25	<22	Total
PUC bloods	>37	90.4	1.4	0.6	1.4	0.3	0.3	0.0	94.6
	34 - <37	1.0	0.0	0.0	0.0	0.0	0.0	0.0	1.0
	31 - <34	0.0	0.0	0.0	0.0	0.0	0.0	0.0	0.0
	28 - <31	0.8	0.2	0.0	0.0	0.0	0.0	0.2	1.1
	25 - <28	1.0	0.0	0.0	0.0	0.0	0.0	0.0	1.0
	22 - <25	0.3	0.0	0.0	0.0	0.2	0.0	0.0	0.5
	<22	0.2	0.0	0.0	0.0	0.2	0.2	1.4	1.9
	Total	93.6	1.6	0.6	1.4	0.6	0.5	1.6	100.0

Table 2: Results of PCV3 PCR testing of PUC bloods and processing fluids

			Breed groups 28-31*		Breed groups 32-34**	
			Treatment groups		Treatment groups	
Data type	Parameter	Parity group	Vaccine	Control	Vaccine	Control
No. of sows	Count	1 & 2	46	53	40	44
		3 >	115	110	110	106
PCV3 PCR	% PUC bloods positive	1 & 2	15.2	9.4	7.5	15.9
		3 >	1.7	1.8	3.6	3.8
	% Processing fluids positive	1 & 2	13.0	13.2	5.0^{a}	22.7^{b}
		3 >	2.6	2.7	1.8^{a}	6.6^{b}
	% PUC bloods or processing fluids positive	1 & 2	21.7	17.0	10.0	27.3
		3 >	4.3	4.5	4.5	9.4

^{a,b} Values with different superscripts within a row were significantly different ($P < 0.05$).
* First vaccination 5-8 weeks prior to farrowing. Second vaccination 1-4 weeks prior to farrowing.
** First vaccination 9-12 weeks prior to farrowing. Second vaccination 5-8 weeks prior to farrowing.

Overall, the percent PCV2 positive PUC blood was 16.0% and the percent positive processing fluids was 9.5% for the 624 litters. Table 3 presents a cross tabulation of PCV2 PCR cycle time (CT) values for each sample type. The lowest CTs measured in PUC blood and processing fluid were 19.5 and 22.0, respectively. Comparing PUC blood and processing fluid results, the overall agreement between the two sample types was 81.6%. For vaccinated litters (n = 311), the percent PCV2 positive for PUC bloods and processing fluids was 15.3% and 7.5%, respectively, while the percent positives in the controls (n = 313) was 16.7% and 11.3%, respectively. By logistic regression, the number of PCV2 positive processing fluids was impacted by parity group only. The number of PCV2 positive PUC bloods was impacted by parity group, gestation stage and total born per litter. PCV2 PCR results by sample type, gestation stage, parity group and vaccination status are presented in Table 4. There were no statistical differences between vaccinated litters and controls within any of the parity groups, gestation stages and sample types.

Reproductive data collected at farrowing are presented in Table 5 by parity group, gestation stage and vaccination status. By linear regression, total born per litter was impacted by parity group but not by vaccine group, farm, parity or gestation stage. Born alive per litter was impacted by parity and parity group; percent stillborn by parity and gestation stage; and percent mummies by parity. Vaccination had no impact on reproductive parameters except for the percent mummies in mid gestation (groups 32-34). For parities 1 and 2, the percent mummies was higher in controls while in parities ≥ 3 the percent mummies was higher in vaccinates.

In summary, the prevalence of PCV3 in placental umbilical cord blood and processing fluids was influenced the most by parity with higher rates in younger parities. Vaccine group influenced the percentage of positive processing fluids. For PCV2, prevalence with either sample type was also influenced the most by parity but the pattern was not as distinct as observed with PCV3.

Table 3: PCV2 PCR cycle times (CTs) by sample type

Percent PCV2 PCR positive (n = 624)		Processing fluids							
		>37	34 - <37	31 - <34	28 - <31	25 - <28	22 - <25	<22	Total
PUC bloods	>37	78.0	5.3	0.5	0.2	0.0	0.0	0.0	84.0
	34 - <37	8.0	1.0	0.8	0.0	0.0	0.0	0.0	9.7
	31 - <34	2.2	0.5	0.2	0.0	0.2	0.0	0.0	3.0
	28 - <31	1.8	0.0	0.2	0.0	0.0	0.0	0.0	1.9
	25 - <28	0.5	0.2	0.0	0.2	0.0	0.0	0.0	0.8
	22 - <25	0.0	0.0	0.0	0.2	0.2	0.0	0.0	0.3
	<22	0.0	0.0	0.2	0.0	0.0	0.2	0.0	0.3
	Total	90.5	6.9	1.8	0.5	0.3	0.2	0.0	100.0

Table 4: Results of PCV2 PCR testing of PUC bloods and processing fluids

			Breed groups 28-31*		Breed groups 32-34**	
			Treatment groups		Treatment groups	
Data type	Parameter	Parity group	Vaccine	Control	Vaccine	Control
No. of sows	Count	1 & 2	46	53	40	44
		3 >	115	110	110	106
PCV3 PCR	% PUC bloods positive	1 & 2	13.0	15.1	7.5	22.7
		3 >	17.4	13.6	17.3	17.9
	% Processing fluids positive	1 & 2	6.5	17.0	2.5	6.8
		3 >	9.6	10.9	8.2	10.4
	% PUC bloods or processing fluids positive	1 & 2	17.4	26.4	10.0	27.3
		3 >	23.5	21.8	21.8	22.6

* First vaccination 5-8 weeks prior to farrowing. Second vaccination 1-4 weeks prior to farrowing.
** First vaccination 9-12 weeks prior to farrowing. Second vaccination 5-8 weeks prior to farrowing.

Table 5: Results of farrowing data analysis

Data type	Parameter	Parity group	Breed groups 28-31* Treatment groups Vaccine	Control	Breed groups 32-34** Treatment groups Vaccine	Control
No. of sows	Count	1 & 2	46	53	40	44
		3 >	115	110	110	106
Reproductive performance	Total born/litter	1 & 2	14.70	14.21	13.97	14.23
		3 >	15.20	15.81	15.56	15.57
	Live born/litter	1 & 2	13.54	13.42	13.25	13.11
		3 >	13.44	13.58	13.55	13.93
	% Stillborn	1 & 2	5.75	3.82	4.16	4.84
		3 >	8.77	10.06	8.48	7.59
	% Mummies	1 & 2	1.63	1.23	0.72[a]	2.56[b]
		3 >	2.25	3.52	4.12[a]	2.18[b]

[a,b] Values with different superscripts within a row were significantly different ($P < 0.05$).
* First vaccination 5-8 weeks prior to farrowing. Second vaccination 1-4 weeks prior to farrowing.
** First vaccination 9-12 weeks prior to farrowing. Second vaccination on 5-8 weeks prior to farrowing.

Vaccine group had no impact on the prevalence of PCV2 in either sample type.

Because a bi-valent PCV3/2d vaccine was used, the impact of vaccination with-regard-to the prevalence of PCV3 or PCV2d in PUC blood or processing fluid could be evaluated separately by testing for each virus by PCR.

Conversely, the impact on reproductive parameters from either the PCV3 or PCV2d components of the vaccine could not be directly determined. However, the finding that vaccination impacted the prevalence of PCV3 but not PCV2d in PUC blood or processing fluid suggests that the impact of vaccination was mainly PCV3 related. The beneficial impact of vaccination appeared to be limited to younger parity sows and within the three breeding groups that were earlier in gestation at vaccination. This finding along with the higher prevalence of PCV3 in younger sows further supports that vaccination mainly impacted PCV3. These findings also suggest that PCV3 and possibly PCV2d vaccination should be applied earlier in gestation or prior to breeding to control these viruses for most, if not all, of gestation.

Using processing fluids rather than placental umbilical cord blood as a monitoring or research tool on an individual litter basis for PCV3 or PCV2 infection in sow herds was supported by the 92.6% and 81.6% agreements, respectively, between the two sample types. Extrapolation of these findings to routine herd monitoring where aggregate processing fluids are collected from a number of litters is limited. These findings support the use of aggregate, herd-level processing fluids for determining if the viruses are present in a herd, however, assessing the level of virus or prevalence of either PCV2 or PCV3 may be misleading due to apparent sensitivity of the method and potential sampling bias, especially parity if younger parity dams are not sampled. Based on the results of this study, it is recommended to collect litter-based PUC or processing fluid samples from parity 1 and 2 sows to evaluate PCV3 prevalence within a herd. This study did not evaluate downstream PCVAD, and the data did not provide insights into this problem.

References

1. Opriessnig T, Gerber P, Ziao C, et al. 2014. A commercial vaccine based on PCV2a and an experimental vaccine based on a variant mPCV2b are both effective in protecting pigs against challenge with a 2013 U.S. variant mPCV2b strain. *J Vaccine* 2014; 32(2):230-237.

2. Palinski R, Pineyro P, Shang P, et al. 2016. A novel porcine circovirus distantly related to know circoviruses is associated with porcine dermatitis and nephropathy syndrome and reproductive failure. *J Virol.* 2016; 91(1):e01879-16. https://doi.org/10.1128/JVI.01879-16

3. Phan TG, Giannitti F, Rossow S, et al. 2016. Detection of a novel circovirus in pigs with cardiac and multi-systemic inflammation. *Virol. J.* 2016; 13:184.

4. Segales J, Allan GM, Domingo M. Porcine circoviruses. In: Zimmerman J, Karriker L, Ramirez A, Schwartz K, Stevenson G, eds. *Diseases of Swine*, 10th edition. Ames, Iowa: Wiley-Blackwell; 2012.

5. Pittman JS. Reproductive failure associated with porcine circovirus type 2 in gilts. *J Swine Health Prod.* 2008; 16(3):144-148.

6. Madson D, Opriessnig T. 2011. PCV2-associated reproductive failure. Proc. 42nd AASV Annual Meeting, 427-432.

7. Byers E. 2018. Porcine circovirus type 3: Practitioner perspective. Proc. 49th AASV Annual Meeting, 445-447.

8. Chen GH, Mai K, Zhou L, et al. 2017. Detection and genome sequencing of porcine circovirus 3 in neonatal pigs with congenital tremors in South China. *Transbound. Emerg. Dis.* 64: 1650-1654.

9. Byers E. 2018. Allen D. Leman Swine Conference, 17 September 2018, Saint Paul, Minnesota.

10. Yang Z, Knutson T, Chen F, et al. 2018 Geographic distribution and genetic diversity of porcine circovirus type 3 from clinical samples in the US swine farms. Proc. 49th AASV Annual Meeting, 96.

11. Lopez W, Linhares D. 2017. Processing fluids, blood serum, and tail blood swabs to detect PRRSV RNA and PCV2 DNA by PCR-based analysis. Proc. ISU James D. McKean Swine Disease Conference, 69.

12. Lopez W, Angulo J, Zimmerman J, Linhares D. 2018. Porcine reproductive and respiratory syndrome monitoring in breeding herds using processing fluids. J Swine Health Prod. 2018; 26(3):146-150.

13. Baker J, Murray D, Sponheim A, et al. 2018. Utilizing piglet-processing fluids to detect PRRSV by PCR in a low-prevalence population. Proc. 49th AASV Annual Meeting, 95.

14. Hood M, Hough SD, Angulo J, et al. 2018. A comparison of the use of tails versus testicles in the production of processing fluids for detection of porcine reproductive and respiratory syndrome virus. Proc. 49th AASV Annual Meeting, 99-100.

15. Mason B, Garbes N, Pollard C, Johnson E. 2015. An investigation of PRRSV positive PCR findings in weanling pigs using a novel sampling method. Proc. 46th AASV Annual Meeting, 89-90.

16. Ruston C, Pittman JS. 2017. Evaluation of methods for determining the porcine circovirus type 2 status of sow herds. Proc. 48th AASV Annual Meeting, 77-78.

17. Fangman T, Baumert D, Rustvold J. 2014. Comparing placenta and presuckle piglet PCV2 status between two breeding sites. Proc. 23rd IPVS Congress, 106.

18. Vander Veen RL, Loynachan AT, Mogler M, et al. 2012. Safety, immunogenicity, and efficacy of an alphavirus replicon-based swine influenza virus hemagglutinin vaccine. Vaccine 2012; 1944-1950. doi:10.1016/j.vaccine.2012.01.030.

Effects of antibiotic and non-antibiotic interventions applied to swine umbilici within the first 24 hours of life on the incidence of umbilical infection, growth, and mortality

A. Szczotka[1], BS; K. Hayman[1], DVM; B. Ratliff[1], BS; C. Rademacher[1], DVM; R. Swalla[2], DVM; M. Hoogland[3], DVM; J. Brown[1], DVM; H. Kittrell[1], DVM; K. Skoland[1], BS; M. Breuer[1], BFA; A. Forseth[1], DVM; C. Wang[4], PhD; L. Karriker[1], DVM, MS, DACVPM

[1]Swine Medicine Education Center, Iowa State University, Ames, Iowa; [2]Zoetis, Parsippany, New Jersey; [3]Smithfield Hog Production Division, Algona, Iowa; [4]Veterinary Diagnostic and Production Animal Medicine, Iowa State University, Ames, Iowa

Introduction

The umbilical cord in neonates can serve as an entry point for establishing infection in the animal by environmental bacteria. Insufficient closure of the abdominal wall due to umbilical infection early in life is one factor that increases the prevalence of abdominal herniation.[1] Infection starting at the umbilical cord in a neonate can increase the potential for septicemia and localized infection such as joint swelling. Any infection early in life including those starting at the umbilicus could potentially lead to poor production performances in individuals. Research on antimicrobial interventions applied within the first 24 hours of a pig's life to impact the incidence of umbilical infections in all phases of production is minimal. The objective of this study was to evaluate four treatment groups of animals utilizing two parenteral antibiotic products, a non-antibiotic topical navel dip, and a control, no treatment applied, within the first 24 hours of life on the incidence of umbilical infection, development of secondary clinical signs, and common production parameters measured in the pre-weaning and weaning phases.

Materials and methods

This study was conducted on a 7,500-head sow operation located in west-central Missouri. Prior to the initiation of the study, 10 freshly deceased or euthanized piglet umbilici were aseptically swabbed and submitted to Iowa State University Veterinary Diagnostic Laboratory for bacterial culture and susceptibility. Individually ear-tagged piglets (n = 2,030) were randomly allocated to one of four treatments applied within the first 24 hours of life: penicillin G procaine (Penicillin,PenOne Pro™, VetOne), ceftiofur crystalline free acid (CCFA, Excede®, Zoetis), accelerated hydrogen peroxide animal shampoo (PeroxiWash™, Virox Animal Health), and negative control. Body weights and navel diameters were obtained before treatment administration. Penicillin and CCFA were delivered intramuscularly at doses of 33,000 IU/kg and 2.27 mg/lb, respectively, using 0.5 cc adjustable Prima Tech syringes. 15 mL of undiluted 3.75% hydrogen peroxide animal shampoo was administered topically using a bovine dip cup (Ambic Non Return DipCup™, Ambic) to the umbilical stump, umbilical cord, and abdomen for 5 seconds. The negative control group received no injectable antibiotic or topical treatment. All study pigs received a 200 mg dose of iron dextran (Uniferon® 200, Pharmacosmos) intramuscular injection at the time of treatment administration. Body weights, navel diameters, navel scores, and incidence of abdominal herniation and joint swelling were measured immediately after weaning. Navel scores were determined using a scoring system established by PIC and Bradley Wolter at The Maschhoffs, Carlyle, Illinois. Statistical analysis was completed utilizing a generalized mixed linear model. Differences between the least squares means obtained from this model were assessed via the Tukey-Kramer method based upon treatment.

Results and discussion

Bacterial culture and susceptibility reports from the veterinary diagnostic lab revealed 4 of 10 swabs to have no significant growth and 6 of 10 swabs to have various pathogenic bacterial growth with CCFA susceptibility and with intermediate to susceptible results against penicillin. Results, summarized in Table 1, indicated pre-wean mortality to be significantly lower in the CCFA treatment group compared to that of the topical treatment group ($P = 0.0049$). Navel score was significantly lower for CCFA treatment group compared to negative control and topical treatment groups ($P = 0.0031$, $P = 0.0465$, respectively). Navel diameter at weaning were significantly lower for the CCFA treatment group compared to the negative control group ($P = 0.0053$). Weaning body weight and average daily gain were significantly higher for the negative control group compared to the topical treatment group ($P = 0.0216$, $P = 0.0059$, respectively). No significant differences between treatments on abdominal herniation or joint swelling at the time of weaning were observed. Under the conditions of this study, topical treatment showed no indication of having significant effects on the measured parameters of interest. Identification of a therapeutic agent that significantly impacts production parameters and the incidence of umbilical infection and subsequent abdominal herniation would lead to more meaningful and judicious treatment decisions by the swine practitioner.

Table 1: Summarization of the LS Means (± standard error) based upon treatment for the measured study variables and the reported significant differences between treatment as determined by the Tukey-Kramer method.

Measure	Negative control	Penicillin G Procaine	CCFA	Topical hydrogen peroxide animal shampoo
Enrollment body weight (kg)	1.57 (0.01)[a]	1.55 (0.01)[a]	1.57 (0.01)[a]	1.56 (0.01)[a]
Weaning body weight (kg)	6.22 (0.08)[a]	6.13 (0.08)[a,b]	6.16 (0.08)[a,b]	6.01 (0.08)[b]
Pre-weaning average daily gain (kg/d)	0.197 (0.003)[a]	0.193 (0.003)[a,b]	0.195 (0.003)[a,b]	0.187 (0.003)[b]
% Preweaning mortality	11.96%[a,b]	12.30%[a,b]	8.28%[a]	15.03%[b]
% Abdominal herniation at weaning	0.00%[a]	0.00%[a]	0.00%[a]	0.00%[a]
% Joint swelling at weaning	1.28%[a]	0.43%[a]	0.63%[a]	1.35%[a]
Navel score at weaning	0.40 (0.03)[a]	0.32 (0.03)[a,b]	0.24 (0.03)[b]	0.36 (0.03)[a]
Navel diameter at weaning (mm)	7.39 (0.11)[a]	7.23 (0.11)[a,b]	6.92 (0.10)[b]	7.22 (0.11)[a,b]

LS Means with the same letter superscript ([a,b]) are not significantly different ($P > 0.05$).

Reference

1. James H, Harding J, Waldner C. Umbilical hernias and abscesses: a result of management issues, not genetics. AASV. 2007.

Effects of perinatal antibiotic administration on fecal microbiota and selection of antimicrobial resistant genes in piglets

B. Burton[1]; M. Zeineldin[1]; A. Megahed[1]; B. Blair[1]; A. Briscoe[2], J. Lowe[1,2]
[1]University of Illinois College of Veterinary Medicine, Urbana, Illinois; [2]Lowe Consulting Ltd., Mahomet, Illinois

Introduction

Antibiotic usage has been and continues to be common in the swine industry. Historically, one of the purposes for using antibiotics was to increase growth rate and improve feed utilization.[1] Although antibiotic usage has been documented to increase overall performance in nursery and finishing pigs,[2,3] recent studies have demonstrated no difference in individual weight gain, litter weight gain, or mortality when compared to piglets not receiving antibiotics.[4]

Additionally, an increasing awareness by the public of antibiotic-resistant bacteria has pressured public health officials to create policy urging a reduction in the use of antimicrobials in food producing animals. Because of this, antibiotic use is limited to the prevention, control, and treatment of various bacterial diseases as stated by the FDA's Guidance for Industry #213. Several lines of evidence indicate that the extensive use of existing antibiotics increases the numbers of antimicrobial resistant gene (ARG) copies present in bacteria and the risk of these ARGs spreading amongst other commensal bacteria.[5,6] There is a poor understanding of the effects dose, frequency, and numerous other factors may have on the formation of new ARG.

To better understand the complexity of this issue, it is important to appreciate how common management practices, such as antimicrobial administration, might affect the complex microbial ecosystem of the gut and ultimately impact animal health. Given the instability of microbiota in the early phase of life, the microbial population has the potential to disseminate and transfer ARG's, which could have significant effects on development of disorders afterward.[7,8] Therefore, not only is there a potential risk for developing ARG's, there could also be detrimental health effects on these pigs throughout the nursery and finishing stages.

The interaction between the gut microbial population, antimicrobials, and the host are not well documented. Understanding the impact of antibiotic administration on host/microbe interactions is crucial in the development of strategies to utilize antibiotics while reducing possible adverse consequences, such as antimicrobial resistance. To investigate this multifaceted question, this study strives to understand the effect of commonly used perinatal antibiotics on the formation of selected ARG's in the fecal microbiota using high-throughput sequencing.

Materials and methods

Study design

The study was conducted over the summer in a commercial sow farm in southern Illinois. A randomized complete block design was used. Forty-eight litters were blocked to one of six treatments by farrowing day, and dam parity group (P1 vs P2+). Litters of less than 11 piglets were excluded from the study. Within litter, all piglets received the same treatment. Piglets (n = 678) were weighed, and treatments administered within 24 hours of birth. All litters consisted of pigs only from the birth sow. Following treatment administration piglets were not added or removed from the litter except for mortality. Treatments were as follows: control (CONT; saline 1cc IM; n = 113), tulathromycin (TUL; 2.5mg/kg IM; n = 108), ceftiofur crystalline free acid (CCFA; 5.0mg/kg IM; n = 115), ceftiofur hydrochloride (CHC; 5.0mg/kg; n = 113), oxytetracycline (OTC; 22mg/kg IM; n = 114) and Procaine Penicillin G (PPG; 33,000 IU/kg IM; n = 115). Two piglets per litter were randomly selected and individually identified with ear notches. Deep fecal swabs were collected from the identified piglets and individual weights were recorded on days 0 (pre-treatment), 5, 10, 15, and 20. Samples were on dry ice immediately following sample collection and then stored at -80°C until analysis. Weight collection was on days 0 and 20 for all pigs. Recording of all mortalities occurred throughout the study period until weaning at day 21 of age.

Extraction of genomic DNA and Illumina sequencing

Four piglets from each group (CONT, TUL, CCFA, CHC, OTC and PPG) at each of the different sampling time points (days 0, 5, 10, 15, and 20) were selected for the microbiota analysis. In a sterile environment, microbial DNA was extracted from each fecal swab along with negative controls (sterile cotton swab and extraction kit reagent) using Power Fecal DNA Isolation kit. The extraction process was completed according to Power® Fecal DNA extraction Kit manufacturer manual. The concentration and integrity of DNA were assessed using a Nanodrop™ spectrophotometer and agarose gel electrophoresis respectively.

The hypervariable region V3-V4 of the16S rRNA gene was amplified and sequenced on the Illumina MiSeq™ sequencer (Illumina, San Diego, CA, USA) at the DNA Services lab at the W. M. Keck Center for Comparative and Functional Genomics (U. of Illinois, Urbana, IL). The extracted DNA was subjected to Fluidigm Access

Table 1: Primers targeting selected antibiotic resistance genes used in this study.

Target gene	Primer	Sequence 5'-3'	Amplicon size (bp)	Reference
tet O	tetO_F	ACGGARAGTTTATTGTATACC	1920	(Ng et al., 2001)[15]
	tetO_R	TGGCGTATCTATAATGTTGAC		
tet W	tetW_F	GAGAGCCTGCTATATGCCAGC	1920	(Ng et al., 2001)[15]
	tetW_R	GGGCGTATCCACAATGTTAAC		
tet C	tetC_F	GCGGGATATCGTCCATTCCG	1191	(Ng et al., 2001)[15]
	tetC_R	GCGTAGAGGATCCACAGGACG		
sul I	sul I_F	CGCACCGGAAACATCGCTGCAC	840	(Pei et al., 2006)[16]
	sul I_R	TGAAGTTCCGCCGCAAGGCTCG		
sul II	sul II_F	TCCGGTGGAGGCCGGTATCTGG	816	(Pei et al., 2006)[16]
	sul II_R	CGGGAATGCCATCTGCCTTGAG		
erm B	erm B_F	GGTTGCTCTTGCACACTCAAG	738	(Martineau et al., 2000)[17]
	ermB_R	CAGTTGACGATATTCTCGATTG		
bla_{CTX-M}	CTX-M_F	CTATGGCACCACCAACGATA	876	(Birkett et al., 2007)[18]
	CTX-M_R	ACGGCTTTCTGCCTTAGGTT		

Array Amplification (Fluidigm Corporation, South San Francisco, CA, USA). All extracted DNA concentration was assessed on a Qubit (Life Technologies, Grand Island, NY, USA) using the High Sensitivity DNA Kit.

A total of seven primer sets targeting seven different ARGs against the most popular antibiotic classes used in the swine industry were used (Table 1). A mastermix for amplification was prepared using the Roche High Fidelity Fast Start Kit and 20x Access Array loading reagent according to Fluidigm protocols. PCR reactions consisted of DNA sample, 20X Access Array Loading Reagent, forward and reverse primer, Fluidigm Illumina linkers with unique barcode. The final fluidigm pools were quantitated by qPCR. All pools were combined in an equimolar ratio and the final genomic libraries were sequenced from both ends following manufacturer's guidelines.

Sequence data processing and microbial community analysis

After preprocessing, the 16S rRNA gene sequences data was analyzed using the open source Quantitative Insights into Microbial Ecology (QIIME) software, version 1.9 (`http://qiime.org/`).[9] Sequence reads were quality filtered using the default parameters.[10] The de novo open-reference operational taxonomic unit (OTU) clustering process was conducted by QIIME with 97% as similarity using UCLUST clustering[11] and was taxonomically assigned against the Silva reference database.[12] One sample from the TUL-treated piglets was not included in the analysis due to unsuccessful sequencing. Two OTUs detected as a contaminant in negative controls (classified as Stenotrophomonas and Xanthomonas) were removed prior to analyses. The alpha diversity (within community) was calculated within QIIME using Chao 1 richness index and the Shannon diversity index. To standardize our analysis for uneven sequencing depth, all samples were randomly rarefied to 1358 sequences per sample. To compare overall microbiota composition among samples, a beta diversity analysis was performed considering the abundance of each detected OTUs in each sample using weighted UniFrac distance and was displayed using principal coordinate analysis (PCoA).

Statistical analysis utilized PAST (version 3.13), JMP 13 (SAS Institute Inc.) and RStudio (version 1.1.383, R Studio, Inc.). Data were logarithmically transformed or ranked when necessary to achieve normality and homogeneity of variance prior to statistical analyses. Significance difference was stated at $P < 0.05$. Statistical comparisons of weighted UniFrac distances between treatment groups at the different sampling days were determined using analysis of similarity (ANOSIM) statistical methods with 9999 permutations and Bonferroni corrected *P* values in PAST. Because of repeated sampling of the same piglets over the course of the study, repeated measures ANOVA with post hoc Tukey's honestly significant difference (HSD) pairwise comparisons was performed to compare each of the differences in microbial relative abundance and alpha diversity indices between the treatment groups. Comparison of microbial communities between the treatment groups utilized discriminant analysis.-

For ARG sequence classification, we developed a customized version of the Antibiotic Resistance Gene Database (ARG-ANNOT). We used the customized ARG-ANNOT database to align the different ARG raw sequences reads obtained from the Illumina sequencing. To avoid the bias, normalization of the ARG reference sequence length by the 16S-rRNA gene sequence length was conducted.[13]

The difference in ARG abundance, between the treatment groups at the different sampling times points were analyzed using repeated measures ANOVA with pairwise post hoc Tukey's HSD comparisons. We used Dunnett's multiple comparisons procedure to compare the mean ARG abundance in different treatment group by sampling day. Comparison of the difference in the abundance of ARGs among treatment groups utilized a fitted PCA.[14]

Results

Summary of sequence data analysis

2,508,268 sequences were obtained from all samples. The number of sequence per sample ranged from 5307 to 48524 (mean 15201, SD 7324). 1296 OTUs were identified among all samples, all of which were taxonomically assigned according to Silva databases classification.

Microbial taxa affected by the early-life antimicrobial intervention

Alpha-diversity was computed using Chao1 richness and Shannon diversity indices (Figure 1). Metagenomics analysis revealed that microbial diversity and richness indices increased with age ($P < 0.001$). Compared to CONT group, only TUL treated piglets exhibited a significant difference in Chao 1 at day 20 ($P = 0.031$) (Figure 1). Beta diversity demonstrated that the fecal microbiota composition at baseline (day 0) did not differ between groups (ANOSIM, $P = 0.17$; Figure 2). Changes induced by early life antimicrobial administration in the microbial community composition were not significant at any time point (ANOSIM, $P > 0.05$; Figure 2). Taxonomic-based analysis revealed that the relative abundance of several microbial taxa differed considerably by age but not by treatment.

Relationships among the overall microbiota composition of the six treatment rroups

A multiple group similarities tree was constructed using Unifrac distance metrics to identify the similarities and differences among the antimicrobial treatments (Figure 3). Collectively, the microbiota composition of the different treatments group showed no significant changes when compared to the control group (ANOSIM, $P > 0.05$, Table 2).

Effect of early life antimicrobials administration on selected ARGs

Across all samples, the highest level of ARGs were *erm B* (33.85%), *tet W* (11.65%), and *Sul* II (9.06%). The effect of the different antimicrobial treatments on ARGs abundance is summarized (Figure 4). Compared to CONT group, TUL resulted in significant increase in the abundance of *tet W* at days 10 and 20 ($P < 0.05$) and *erm B* at day 20 ($P < 0.05$) (Figure 4). PPG-treated piglets exhibited a significant increase in the prevalence of *erm B* and *tet W* at day 20 of life ($P < 0.05$). PCA of ARG prevalence across all treatment days revealed that samples from CONT and TUL treated piglets were clustered into two distinct groups (ANOSIM; $P = 0.015$) with all other groups being similar to the CONT group (ANOSIM; $P > 0.05$, Figure 5).

Discussion

The results indicate that the microbial composition in the neonatal gastrointestinal track is primarily a factor of age with an insignificant influence of perinatal antibiotic administration. This suggests that regardless of class perinatal administration of an antibiotic does not alter that trajectory of microbiota development. There is a limited understanding of the significance of these dynamic changes. However, these findings suggest that factors other than antibiotics, such as their rearing environment, are more important in influencing the development of the piglet's microbiota during the suckling period.

There were significant increases in *tet W* and *erm B* resistant genes for the TUL and PPG treatment groups on days 10 and 20. The biological importance of these increases is unknown, but the magnitude of the changes in both *tet W* and *erm B* at both 10 and 20 days is very small (< 5% change in prevalence, Figure 6) suggesting that the implications of these increases are transient and very limited. These results suggest that a single dose of parenteral antibiotics has only a limited impact on the overall abundance of ARGs throughout the suckling period. Future studies should examine if there are post-weaning implications of perinatal antibiotic administration, to further elucidate the effects of repeated multiple or sustained dosing of antibiotics and the implications of route (oral vs. parenteral) and dose on ARG development and the composition of the fecal microbiota.

References

1. Cromwell GL: Why and how antibiotics are used in swine production. Anim Biotechnol. 2002, 13: 7-27. 10.1081/ABIO-120005767.

2. Tangen, J. et al. A comparison of the effect of ceftiofur crystalline free acid and procaine penicillin-G administered at either birth or processing on suckling piglet health and performance. AASV Annual Meeting. 2012:109-112.

3. Zhou, C et al. A global multilocation study to evaluate the effect of ceftiofur sodium on preweaning mortality and growth performance. Swine Health and Production. 2000; volume 8, number 3:113-117.

4. Burton, B., Zeineldin, M., Megahed A., Harding, A., Lowe, J., et al. (2018). Assessing the impact of perinatal antibiotics on performance and antibiotic resistant bacteria in pre-weaned pigs. *Proceedings of the Annual Meeting of the* AASV, 2018, pg 309.

5. Czaplewski, L., Bax, R., Clokie, M., Dawson, M., Fairhead, H., Fischetti, V. A., et al. (2016). Alternatives to antibiotics-a pipeline portfolio review. *Lancet Infect. Dis.* 16, 239–251. doi:10.1016/S1473-3099(15)00466-1.

6. Roca, I., Akova, M., Baquero, F., Carlet, J., Cavaleri, M., Coenen, S., et al. (2015). The global threat of antimicrobial resistance: Science for intervention. *New Microbes New Infect.* 6, 22–29. doi:10.1016/j.nmni.2015.02.007.

7. Gibson, M. K., Crofts, T. S., and Dantas, G. (2015). Antibiotics and the developing infant gut microbiota and resistome. *Curr. Opin. Microbiol.* 27, 51–56. doi:10.1016/j.mib.2015.07.007

8. Su, J. Q., Cui, L., Chen, Q. L., An, X. L., and Zhu, Y. G. (2017). Application of genomic technologies to measure and monitor antibiotic resistance in animals. *Ann. N. Y. Acad. Sci.* 1388, 121–135. doi:10.1111/nyas.13296.

9. Caporaso, J. G., Kuczynski, J., Stombaugh, J., Bittinger, K., Bushman, F. D., Costello, E. K., et al. (2010). QIIME allows analysis of high-throughput community sequencing data. *Nat. Methods* 7, 335–336. doi:10.1038/nmeth.f.303.

Figure 1: The difference in bacterial diversity indices (Chao 1 and Shannon) between the different treatment groups (CONT, TUL, CCFA, CHC, OTC, and PPG) at different sampling days (Day 0, 5, 10, 15, and 20).

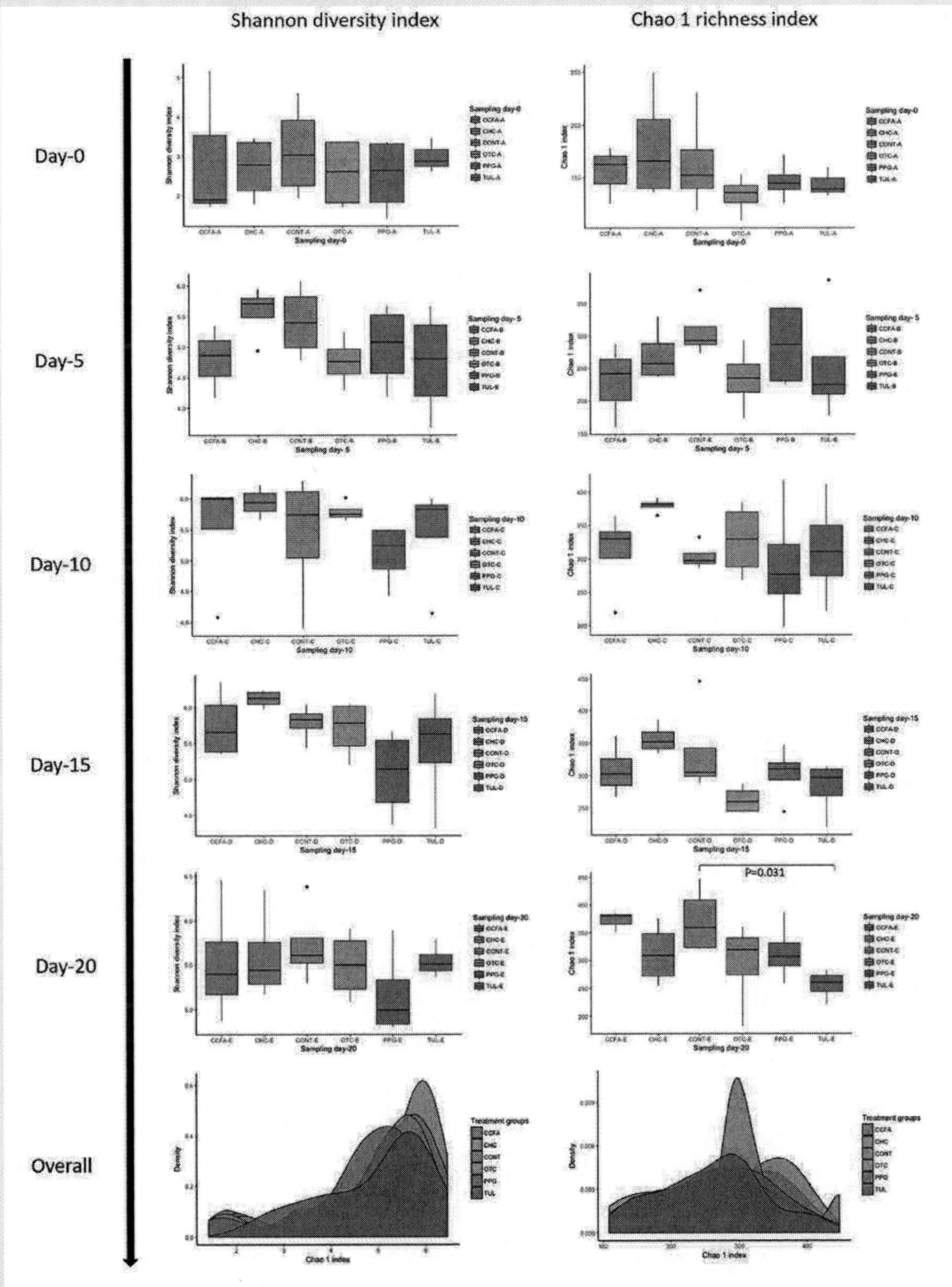

Figure 2: Principal Coordinate analysis (PCoA) for different treatment groups (CONT, TUL, CCFA, CHC, OTC, and PPG) at different sampling days. The percent variation explained by each coordinate is indicated on the axes. The individual data points, which represent total fecal microbiota compositions of each piglet, are depicted. Significance between groups was analyzed using analysis of similarity (ANOISM) with 9999 permutations and Bonferroni corrected *P* values.

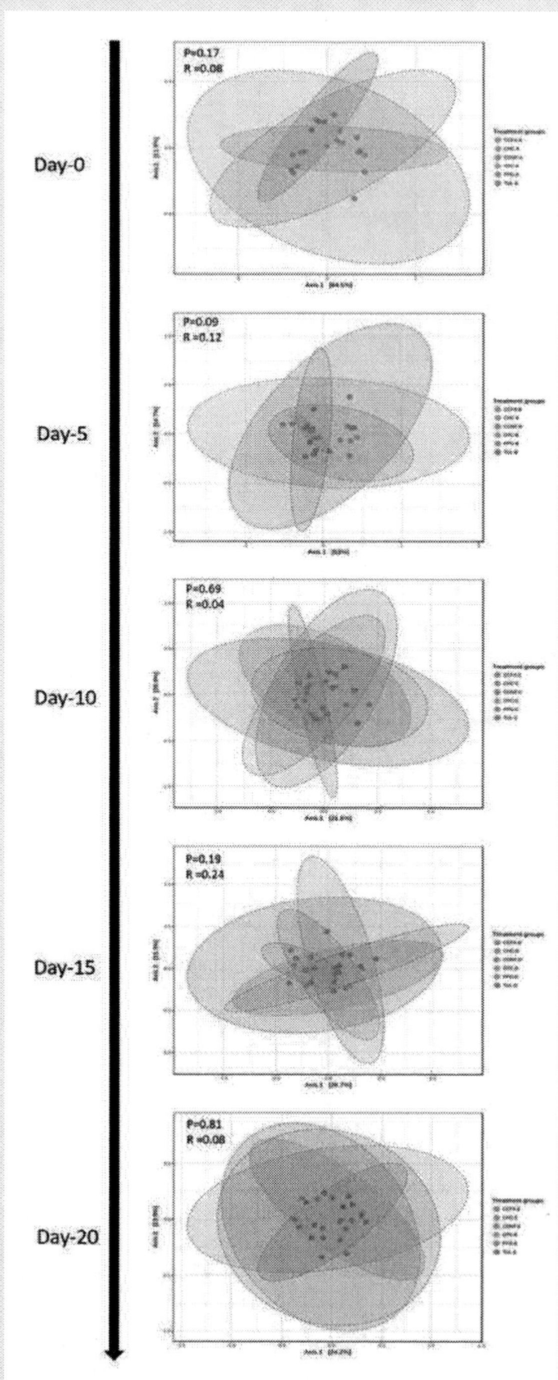

Figure 3: Multiple group similarities tree was constructed using Unifrac distance metrics to identify the similarities and differences among the antimicrobials treatment.

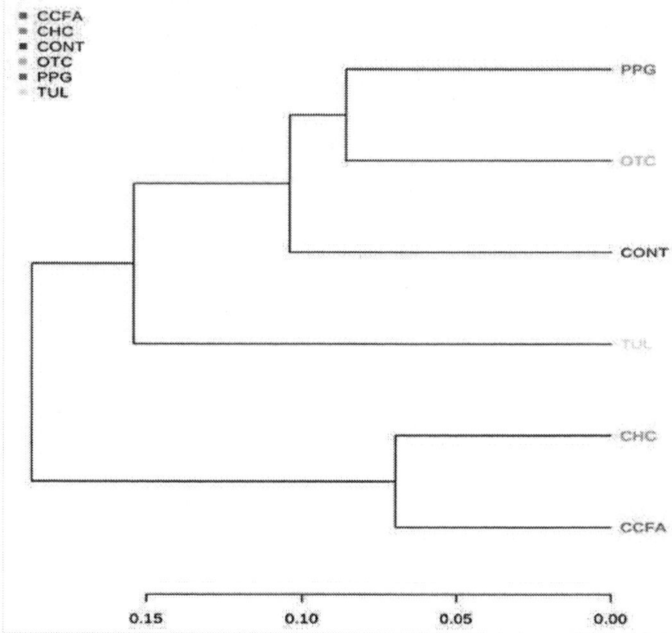

10. Bokulich, N. A., Subramanian, S., Faith, J. J., Gevers, D., Gordon, J. I., Knight, R., et al. (2012). Quality-filtering vastly improves diversity estimates from Illumina amplicon sequencing. *Nat. Methods* 10, 57–59. doi:10.1038/nmeth.2276.

11. Edgar, R. C. (2010). Search and clustering orders of magnitude faster than BLAST. *Bioinformatics* 26, 2460–2461. doi:10.1093/bioinformatics/btq461.

12. Quast, C., Pruesse, E., Yilmaz, P., Gerken, J., Schweer, T., Yarza, P., et al. (2013). The SILVA ribosomal RNA gene database project: improved data processing and web-based tools. *Nucleic Acids Res.* 41, D590-6. doi:10.1093/nar/gks1219.

13. Li, B., Yang, Y., Ma, L., Ju, F., Guo, F., Tiedje, J. M., et al. (2015a). Metagenomic and network analysis reveal wide distribution and co-occurrence of environmental antibiotic resistance genes. *ISME J.* 9, 2490–2502. doi:10.1038/ismej.2015.59.

14. Parks, D. H., Tyson, G. W., Hugenholtz, P., and Beiko, R. G. (2014). STAMP: Statistical analysis of taxonomic and functional profiles. *Bioinformatics* 30, 3123–3124. doi:10.1093/bioinformatics/btu494.

15. Ng, L.K. et al. (2001). Multiplex PCR for the detection of tetracycline resistant genes. Molecular and Cellular Probes. 15, 209-215. doi:10.1006/mcpr.2001.0363.

16. Pei, R., Kim, S.C, Carlson, K.H., Pruden, A., et al. (2006). Effect of River Landscape on the sediment concentrations of antibiotics and corresponding antibiotic resistance genes (ARG). Water Research. 40, 2427-2435. doi:10.1016/j.watres.2006.04.017

17. Martineau, F., et al. (2000). Correlation between the Resistance Genotype Determined by Multiplex PCR Assays and the Antibiotic Susceptibility Patterns of Staphylococcus aureus and Staphylococcus epidermidis. Am. Soc. For Micro. 44, 231-238. doi: 10.1128/AAC.44.2.231-238.2000

18. Birkett, C.I., Ludlam, H.A., Woodfoord, N., Brown, D. F., et al. (2007). Real-time TaqMan PCR for rapid detection and typing of genes encoding CTX-M extended-spectrum Beta-lactamases. J of Med. Micro. 56, 52-55. doi: 10.1099/jmm.0.46909-0

Table 2: The result of the nonparametric ANOSIM test (analysis of similarities) with 9999 Monte Carlo permutations to evaluate the UniFrac distances significance between the treatment groups.

	CCFA	CHC	CONT	TUL	OTC	PPG
CCFA		0.8935	0.4816	0.0243	0.4357	0.3249
CHC			0.3806	0.015	0.3179	0.2128
CONT				0.196	0.6121	0.1837
TUL					0.1124	0.0759
OTC						0.2548
PPG						

Figure 4: Line graphs illustrating the difference in abundance of antibiotics resistance gene (*erm B*, *Sul I*, *Sul II*, *tet C*, *tet O*, and *tet W*) between the CONT group and TUL, CCFA, CHC, OTC, and PPG treated piglets at different sampling days.

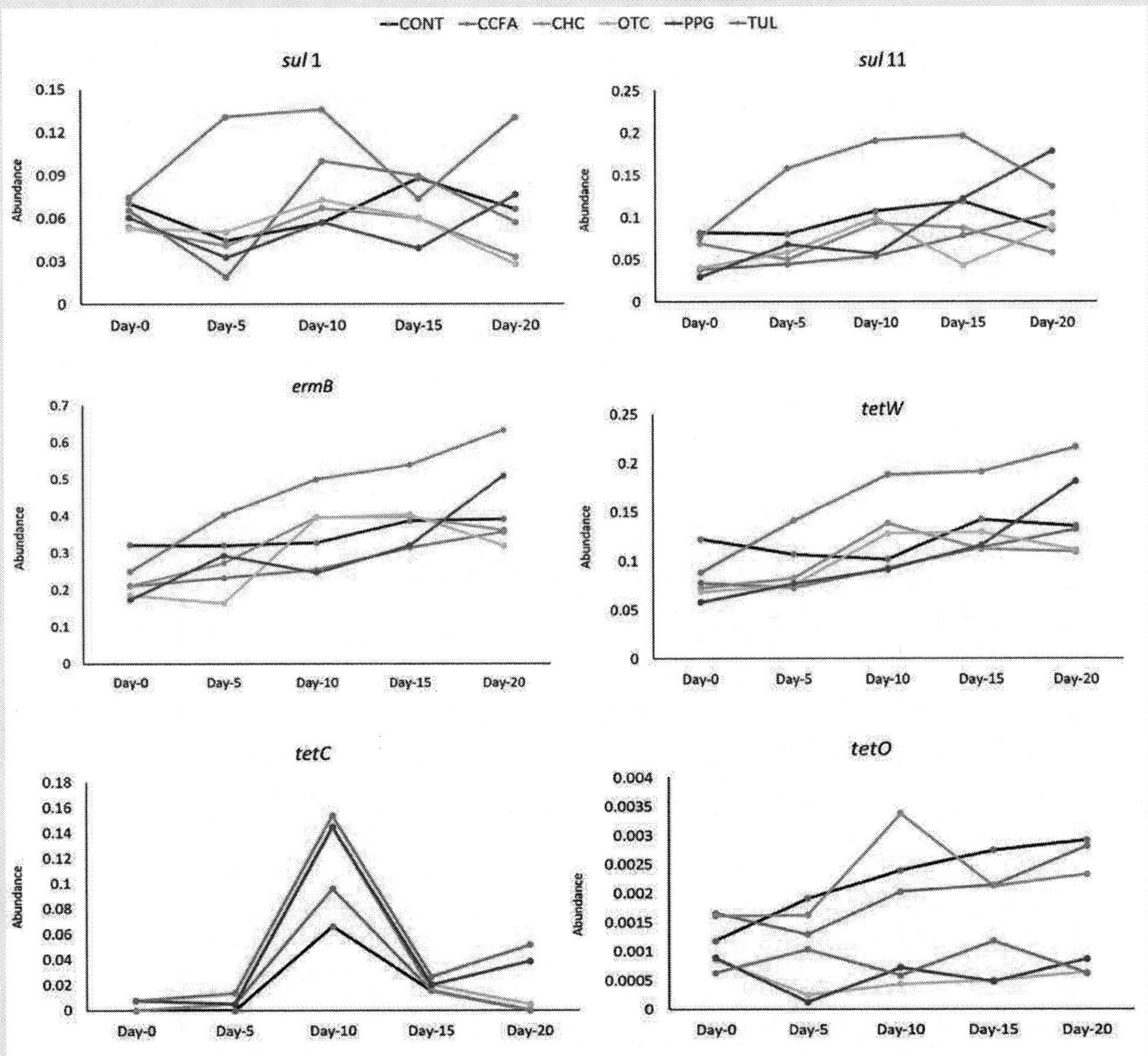

Figure 5: PCA for the overall antibiotics resistance gene (*erm B, Sul I, Sul II, Tet C, Tet O,* and *Tet W*) between the CONT group and CCFA, CHC, OTC, PPG and TUL treated piglets.

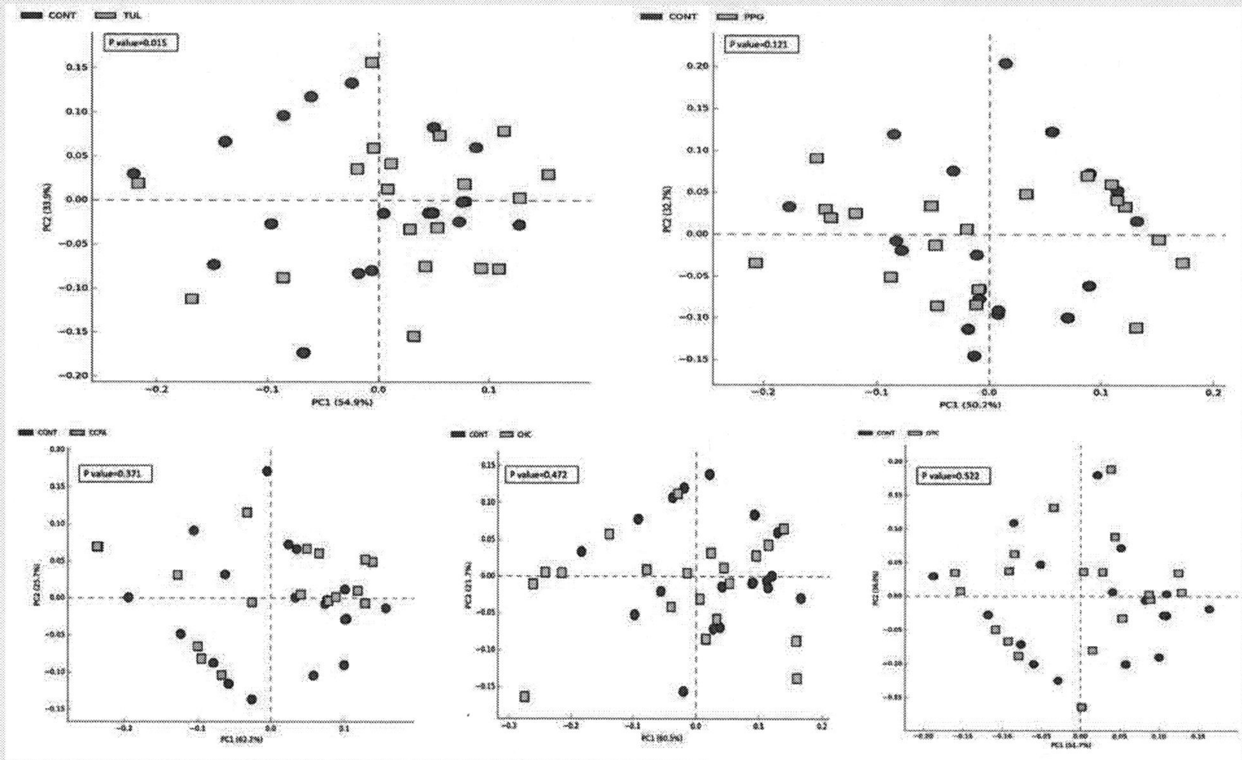

Figure 6: Relative abundance of ARG's per treatment group per sampling day.

Development of a scoring system to determine the prevalence, severity, and associated risk factors of tail-biting and lameness

S. Baker[1]; S. Hough[2], DVM; E. Jablonski[3], DVM, MS; J. Angulo[3], DVM; E. Nemechek[3], DVM
[1]Iowa State University College of Veterinary Medicine, Ames, Iowa; [2]Prestage Farms, Inc., Clinton, North Carolina; [3]Zoetis, Parsippany, New Jersey

Introduction

Lameness and tail-biting affect the welfare and productivity of growing pigs. Both can be sporadic, but there are several risk factors that may promote occurrence, including availability of feed and water, overcrowding, ventilation, and concomitant disease. The objective of this study was to develop a common scoring system for producers to communicate severity of tail-biting and lameness and use that scoring system to determine prevalence and severity in a production system, while also identifying relevant risk factors.

Materials and methods

This was a prospective observational, cross-sectional study across 108 barns, 4,472 pens and 130,437 pigs during the summer of 2018 in one production system. The pigs selected were randomly allocated across several sow sources. All observations were conducted by one trained observer to ensure consistency of results. Based on previously reported incidence by production staff, 14 -18 weeks post wean-to-finish placement was the window of observation. Each barn was surveyed to quantify prevalence and severity while recording risk factors. The survey included entering every pen to measure feed and water availability, making observations on welfare of the pigs, and scoring tail-bitten and lame pigs on a 0-3 scale (Table 1). Blood and minor skin loss on the tail was assigned a score of TB1. A more reddened, bloody tail with swelling received a score of TB2. Loss of tail length, more redness and swelling received a score of TB3. Mild to moderate stiffness when rising/walking received a score of L1. Lameness in one or more limbs, unwillingness to bear full weight on limb, and minor swelling was assigned a score of L2. Inability to get to feed/water or bear any weight on the limb received a score of L3. Each pig in each pen was observed rising and walking, and feed and water availability parameters were documented by pen. Two oral fluids were collected in the same four pens per barn for influenza A virus and *Mycoplasma hyosynoviae* PCR testing based on herd veterinarian's knowledge of health concerns in these pigs. Testing was completed at the Iowa State University Veterinary Diagnostic Laboratory. If at least one oral fluid result from a barn tested positive, the group of pigs were considered positive. Chi-Square test for association and odds ratios (OR) were used to measure the association between risk factors and tail-biting and lameness.

Results

Of the 130,437 pigs observed, the total number and prevalence of tail-bitten and lame pigs is reported in Table 2. There was significant association between *Mycoplasma hyosynoviae* PCR positive oral fluids and the presence of ≥ 1 level of lameness ($P < 0.005$) with an estimated OR of 1.30 (95% CI = 1.108 to 1.534). The association test for feed outage and ≥ 1 level of TB suggested a trend bordering on statistical significance ($P = 0.076$) with a calculated OR of 2.64 (95% CI = 1.855 to 3.754).

Discussion

The scoring system generated creates a common language for producers and veterinarians to better communicate severity and prevalence, which may lead to earlier identification of the issue and improvements in animal welfare and production performance. The association between feed outages and tail-biting is not very strong, but the risk is still significant at pen level. Moreover, it appears that the severity of lameness will increase when *Mycoplasma hyosynoviae* is present. Further work should be done to determine the association of diagnostic evidence of *Mycoplasma hyosynoviae* disease (histopathology vs PCR) to assess causation.

Table 1: Proposed lameness (L) and tail-biting (TB) scoring guide

Lameness	
L0	Not affected.
L1	Shifts weight away from affected limb when standing, shows stiffness walking.
L2	Reluctant to stand and/or walk, obvious limp.
L3	Non-weight bearing on the affected limb when either standing or walking. Swelling commonly seen.
Tail-biting	
TB0	Not affected.
TB1	Visible blood and minor skin loss on the tail.
TB2	More visibly red and chewed, generally seen around the whole tip. Swelling observed.
TB3	Loss of tail length is observed and visibly red/bloody.

Table 2: Prevalence, association, and odds ratio of tail-biting (TB) and lameness (L)

		TB1	TB2	TB3	Cumulative	L1	L2	L3	Cumulative
Count		136	181	259	576	744	119	85	948
Prevalence[1]		0.10%	0.14%	0.20%	0.44%	0.57%	0.09%	0.06%	0.72%
		TB1	**TB2**	**TB3**	**Chi-Square test for association***	**L1**	**L2**	**L3**	**Chi-Square test for association***
Mycoplasma hyosynoviae	+	74	101	176	*P*-value 0.113	293	74	52	***P*-value < 0.005**
	-	48	76	88		316	33	29	
Feed outage	+	15	12	33	***P*-value 0.076**	50	11	8	*P*-value 0.681
	-	96	148	186		583	101	72	
Water pressure low/out	+	3	7	8	*P*-value 0.772	38	4	3	*P*-value 0.452
	-	108	153	211		595	108	77	

*Pearson Chi-Square

	Odds ratio[2]	95% CI
***Mycoplasma hyosynoviae* / Lameness**	1.30	(1.108 , 1.534)
Feed outage / Tail biting	2.64	(1.855 , 3.754)

[1] Prevalence is based on 130,437 total pigs as the denominator.
[2] OR is based on number of pens with TB or L pigs.

The limitation of this analysis is that this was a cross-sectional designed study, looking for a relationship between exposure and disease prevalence in a defined population at a single point in time. The temporality for tail-biting and lameness related to feed outage and *Mycoplasma hyosynoviae* is unknown. Closeout performance of groups relative to tail-biting and lameness will be shared at the AASV 2019 meeting.

Detecting porcine reproductive and respiratory syndrome virus (PRRSV) via polymerase chain reaction (PCR) by pooling pen-based oral fluid samples

K. L. Baker[1], BS; J. Seate[2], DVM; J. J. Zimmerman[1], DVM, PhD, DACVPM; J. Ji[1], MS; K. Dion[3], DVM
[1]Department of Veterinary Diagnostic and Production Animal Medicine, Iowa State University, Ames, Iowa; [2]Aptimmune Biologics Inc., St. Louis, Missouri; [3]The Hanor Company Inc., Enid, Oklahoma

Introduction

After decades of research, porcine reproductive and respiratory syndrome virus (PRRSV) still compromises the health of millions of pigs and costs the industry $664 million annually. Therefore, swine veterinarians need reliable cost-effective methods for routine PRRSV surveillance. Oral fluids provide a straightforward, efficient method to monitor PRRSV in all swine. However, diagnostic costs often prohibit testing a large number of samples from one population. Pooling oral fluid samples might be a more cost effective method to determine a population's PRRSV burden. The objective of this study was to determine the sensitivity of detecting PRRSV via quantitative polymerase chain reaction (PCR) in pooled pen-based oral fluid samples.

Materials and methods

Seventy-six PRRSV positive oral fluid samples were obtained from pens of growing pigs in Iowa and Illinois. One pen-based oral fluid sample was the experimental unit. Each of the 76 oral fluid samples was tested individually and in pools of 3:1, 4:1, 5:1, and 6:1 for the presence of PRRSV via quantitative PCR at the Iowa State University Veterinary Diagnostic Laboratory (ISU VDL). Samples were pooled by adding PRRSV negative oral fluids (1mL for 3:1, 1.5mL for 4:1, 2mL for 5:1, and 2.5mL for 6:1) to a 0.5mL aliquot of a PRRSV positive oral fluid by using a sterile pipet. PRRSV negative oral fluids were collected from a naïve genetic nucleus breeding herd in Wisconsin and were confirmed negative by PCR at the ISU VDL prior to use in this study.

The oral fluid samples were categorized into three groups based on their individual cycle time (Ct) value: A) Ct 26-30 ($n = 12$); B) Ct 30.1-33.9 ($n = 38$); and C) Ct > 34 ($n = 26$). The diagnostic sensitivity for this study was calculated for each group by using a Ct value ≥ 37 as the negative cutoff. The Ct value of ≥ 37 was chosen as the negative cutoff because it is the cutoff value used by the ISU VDL. Sensitivity is defined as the probability of obtaining a positive test result in subjects who have the disease of interest. Sensitivity was calculated for a particular group and pool level by taking the number of oral fluid samples that tested positive (true positive samples) divided by the sum of the number of true positive oral fluid samples and the number of false negative oral fluid samples within that group.

Four linear regression models were fit by using the *Proc GLM* in SAS (Cary, NC) to determine the rate of change of Ct values and genomic copies by pool and to determine if the rate of change was different between groups A, B, and C.

Results

All oral fluid samples in group A were PCR positive for PRRSV across all pool levels; therefore, group A had 100% diagnostic sensitivity across all pool levels. The diagnostic sensitivity for group B was 95% for 3:1 pool, 92% for 4:1 pool, 92% for 5:1 pool, and 84% for 6:1 pool. The diagnostic sensitivity for group C was 38% for 3:1 pool, 27% for 4:1 pool, 23% for 5:1 pool, and 19% for 6:1 pool. Figure 1 shows the average Ct value and Log_{10} of the genomic copies by pool. No significant differences in the rate of change of Ct value or genomic copies were detected between groups A, B, and C. On average, the Ct value of a pooled oral fluid sample is expected to increase by 0.50 for every additional sample in the pool. On average, the Log_{10} of genomic copies is expected to decrease by 0.15 logs for every sample added to the pool.

Discussion

Swine veterinarians may pool oral fluid samples from swine populations with a moderate to high expected viral load, such as growing pigs with respiratory disease or a breeding herd early in a virus elimination project, with minimal loss of diagnostic sensitivity. This may allow veterinarians to decrease diagnostic costs of their current PRRSV monitoring program or test more pigs at their current costs. Pooling oral fluid samples is not recommended for swine populations with a low viral load, such as gilts in a gilt developer, a PRRSV-negative breeding herd, or a breeding herd at the end of a virus elimination project because the diagnostic sensitivity is low and the cost of a false negative result is extremely high.

Figure 1: Average cycle time (Ct) value and genomic copies by pool level and starting Ct value.

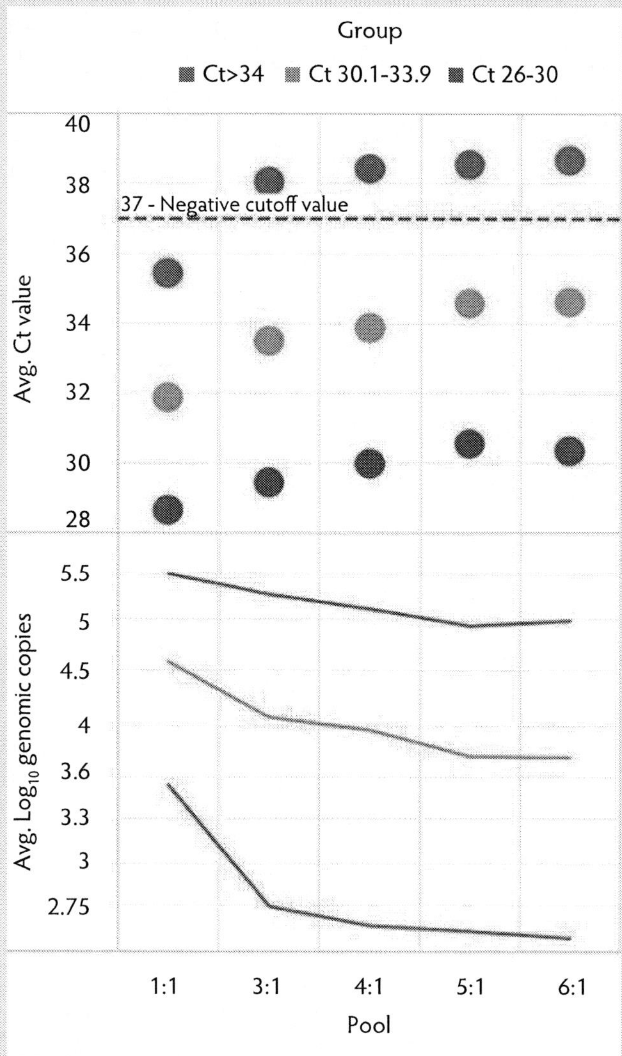

Efficacy of a T cell epitope DNA vaccine, an inactivated vaccine, and use of both in the H1N1 pig challenge model

J. Hewitt[1], BS; A. Karuppannan[1], DVM, PhD; P. Gauger[1], MS, DVM, PhD; P. Halbur[1], MS, DVM, PhD; Anne S. De Groot[2,3], MD, Leonard Moise[2,3], PhD, Swan Tan[2], T. Opriessnig[1,4], Dr med vet, PhD
[1]Iowa State University College of Veterinary Medicine, Ames, Iowa, USA;
[2]Institute for Immunology and Informatics, University of Rhode Island, Providence, RI, USA;
[3]EpiVax Inc., Providence, RI, USA; [4]The Roslin Institute, University of Edinburgh, Midlothian, UK.

Introduction

Influenza A virus (IAV) in pigs is highly infectious, genetically diverse, and difficult to control. Most IAV vaccines available commercially utilize inactivated whole virus; however, they are often not effective in providing heterologous protection. A novel swine IAV T-cell epitope driven DNA vaccine (EPITOPE), PigMatrix EDV,[1] which encodes conserved structural and non-structural T cell epitopes predicted to bind with swine leukocyte antigens 1 and 2, was developed and has previously been shown to induce specific cell-mediated-immunity (CMI) in vaccinated pigs. Cell-mediated immunity has demonstrated protection across heterosubtypic IAV strains in pigs.[2] The objective of this study was to compare the ability of the EPITOPE vaccine, an inactivated commercial whole virus vaccine (INACT), and a heterologous prime boost regimen, utilizing the EPITOPE vaccine followed by the INACT vaccine, to protect pigs from IAV challenge with a contemporary pandemic H1N1 strain.

Materials and methods

Thirty-nine IAV-free, 3-week-old pigs were randomly assigned to one of five groups including a NEG-CONTROL group (unvaccinated, sham-challenged), an INACT-IAV group (vaccinated with FluSure XP® at 4 and 7 weeks, pH1N1 challenged), an EPITOPE-INACT-IAV group (vaccinated with PigMatrix EDV at 4 and with FluSure XP® at 7 weeks, pH1N1 challenged), an EPITOPE-IAV group (vaccinated with PigMatrix EDV at 4 and 7 weeks, pH1N1 challenged), and a POS-CONTROL-IAV group (unvaccinated, pH1N1 challenged). Vaccination was done intramuscularly (INACT) or intradermally (EPITOPE). Intradermal vaccinations were completed using a commercial needle-free high-pressure injection device set at 65 pounds per square inch. The challenge (pH1N1 strain A/swine/Iowa/A01104104/2017) and sham-challenge (saline) were done at 9 weeks of age, and all pigs were necropsied at day post challenge (dpc) 5. Data collected included presence of IAV RNA in nasal swabs (real-time RT PCR), seroconversion to IAV (IDEXX nucleoprotein-based blocking ELISA), clinical assessment of the pigs after challenge (average daily weight gain, rectal temperatures, cough and respiratory distress scores),[3] presence of CMI (ELISpot)[4], and assessment of macroscopic lung lesions,[5] microscopic lung lesions (histopathology),[5] and presence of IAV antigen in lung tissues (immunohistochemistry).[6] Summary statistics were calculated for groups to assess the distributional property. Quantitative RT-PCR data was $\log_{10}$ transformed prior to analysis. Analysis of variance (ANOVA) was used to detect differences among treatment groups followed by Tukey HSD test for significant differences.

Results and discussion

All INACT-IAV pigs were ELISA positive at the time of challenge and the EPITOPE-INACT-IAV pigs seroconverted by dpc 5. Compared to non-vaccinated control pigs, significant increases in the number of IFNγ secreting cells against the pH1N1 challenge virus were measured in all treatment groups, with the EPITOPE-INACT-IAV group being numerically highest. There was significantly less IAV antigen in the lungs of both groups receiving the EPITOPE vaccine compared to INACT-IAV and POS-CONTROL-IAV groups. Viral shedding was significantly reduced by the INACT vaccine compared to the EPITOPE vaccine; however, differences in gross lesions were not observed. Overall, the EPITOPE vaccine induced a strong CMI response while the INACT vaccine was associated with a humoral immune response. Both humoral and CMI were improved in the EPITOPE-INACT-IAV group. Based on immunohistochemistry results, the conserved class 1 and 2 T cell epitopes in the EPITOPE vaccine elicited some protection against a virulent challenge. In addition, the needle-free injection device used here is effective, safer for both pigs and people, faster, and cost efficient to use. The experimental EPITOPE vaccine, once commercially available, in combination with intradermal delivery, will be an effective tool to combat diverse IAV infections in pigs and will compliment current inactivated IAV vaccines.

Acknowledgements

We would like to thank the Iowa Pork Producer Association for the funding of this study and the Iowa Livestock Health Advisory Council for the support of J. Hewitt.

References

1. Gutierrez AH, Loving C, Moise L, Terry FE, Brockmeier SL, Hughes HR, et al. In vivo validation of predicted and conserved T cell epitopes in a swine influenza model. Plos One. 2016;11.

2. Heinen PP, de Boer-Luijtze EA, Bianchi ATJ. Respiratory and systemic humoral and cellular immune responses of pigs to a heterosubtypic influenza A virus infection. Journal of General Virology. 2001;82:2697-707.

3. Halbur PG, Paul PS, Frey ML, Landgraf J, Eernisse K, Meng XJ, et al. Comparison of the pathogenicity of two US porcine reproductive and respiratory syndrome virus isolates with that of the Lelystad virus. Veterinary Pathology. 1995;32:648-60.

4. Olson ZF, Sandbulte MR, Souza CK, Perez DR, Vincent AL, Loving CL. Factors affecting induction of peripheral IFN-gamma recall response to influenza A virus vaccination in pigs. Veterinary Immunology and Immunopathology. 2017;185:57-65.

5. Gauger PC, Loving CL, Khurana S, Lorusso A, Perez DR, Kehrli ME, et al. Live attenuated influenza A virus vaccine protects against A(H1N1) pdm09 heterologous challenge without vaccine associated enhanced respiratory disease. Virology. 2014;471:93-104.

6. Vincent LL, Janke BH, Paul PS, Halbur PG. A monoclonal-antibody-based immunohistochemical method for the detection of swine influenza virus in formalin-fixed, paraffin-embedded tissues. Journal of Veterinary Diagnostic Investigation. 1997;9:191-5.

Evaluation of the efficacy of ceftiofur crystalline free acid versus autogenous vaccination for control of *Streptococcus suis* and *Haemophilus parasuis* in a commercial swine production system

M. Finch[1], BS; P. Thomas[2], DVM; R. Swalla[3], DVM; D. Amodie[3], MS
[1]Iowa State University, Ames, Iowa; [2]Iowa Select Farms, Iowa Falls, Iowa; [3]Zoetis, Parsippany, New Jersey

Introduction

Streptococcus suis (S. suis) and *Haemophilus parasuis (H. parasuis)* are bacterial pathogens that often affect pigs shortly after weaning and inflict costly damage to the health and production of pigs. The objective of this study was to compare the efficacy of a ceftiofur crystalline free acid (CCFA) (Excede, Zoetis, Parsippany, NJ) treatment at weaning versus an autogenous vaccination program and untreated group. This study is important because of the implications it has on responsible antibiotic use, production efficiency, and animal health.

Materials and methods

A PRRS provisional negative 4,000 head sow farm (Farm A) and a PRRS positive unstable 4,300 head sow farm (Farm B) with a confirmed history of *S. suis* and *H. parasuis* associated mortality were used for this study. A week's worth of litters from each farm were assigned to 3 treatments: (1) 1 milliliter (mL) intramuscular (IM) neck injection of autogenous vaccine (Newport Laboratories, Worthington, MN; *S. suis* Types 2, 1, 4, and 27; *H. parasuis* Type 4) at processing (day 2-4) and weaning (day 18), (2) 0.3 mL IM neck injection of CCFA at weaning, and (3) a control receiving no treatment. The 3 most uniform gilts and 3 most uniform barrows per litter were randomly assigned to a treatment group. Farm A had 230 pigs in each of the CCFA and autogenous vaccine groups, and 229 pigs in the control group. Farm B had 322 pigs in each of the CCFA and autogenous vaccine groups, and 320 pigs in the control group. Each treatment group was double ear tagged at processing. At weaning, pigs were co-mingled into a wean-to-finish (W/F) barn. The flows were kept separate and housed at different W/F sites. Individual weights were recorded at processing, weaning, and after 7 weeks in the W/F barn. Mortalities and treatments were recorded daily. The primary variables of interest in the study were mortality and weight from processing to 7 weeks post-weaning. A secondary variable of interest was the number of pigs that received treatment post-weaning. Using the Proc Mixed Procedure (SAS 9.4, Cary, NC), weights were analyzed for each farm separately with a model that considered the fixed effect of treatment and the random effects of room, pen-within-room and the residual error. Treatment Least Squares Means (LSMeans) were calculated for each group and compared by the two-sided Student's t-test at the 5% level of significance. Mortality and post-weaning treatments were defined as a binary variable (yes = 1, no = 0) and analyzed by SAS Glimmix procedure using the model stated above. The analysis utilized a binomial error and logit link. The 5% level of significance was used to assess statistical differences.

Results and discussion

This study showed no significant difference in mortality and final weight between treatment groups in both flows. There was a significant difference in the percent of antibiotic treatments post-weaning on Farm A (Table 1). This difference was not

Table 1: Effects of ceftiofur crystalline free acid (CCFA), autogenous vaccination, or no treatment on weight, mortality, and antibiotic treatments (LSMean) from processing to 7 weeks post-weaning on two farms with a history of *S. suis* and *H. parasuis* infection.

	Farm A			Farm B		
	CCFA	Autogenous	No treatment	CCFA	Autogenous	No treatment
Allocation weight (lbs)	3.97[a]	3.97[a]	4.00[a]	4.02[a]	4.01[a]	4.02[a]
Weaning weight (lbs)	11.10[a]	11.08[a]	11.49[b]	11.21[a]	11.18[a]	11.36[a]
Final weight (lbs)	40.46[a]	39.43[a]	40.04[a]	42.71[a]	42.90[a]	42.61[a]
Mortality (%)	10.49[a]	10.49[a]	6.68[a]	4.69[a]	7.68[a]	4.72[a]
Treatments (%)	8.34[b]	15.12[a]	14.29[a]	15.07[a]	16.93[a]	19.53[a]

[ab] LSMeans within a row with no common superscript are significantly different. ($P \leq 0.05$).

observed on Farm B. The health of both farms was high and that may have contributed to the limited effectiveness of CCFA and no effectiveness of the autogenous vaccine compared with the untreated group. By reducing the number of treatments post-weaning, CCFA improved pig health, decreased the labor associated with individual pig treatments, and supported the responsible use of antibiotics on this farm.

Impact of item orientation on the efficacy of supply decontamination with aerosolized chemical disinfectants

E. Kettelkamp[1]; J. Kordas[1]; B. Blair[1]; J. Lowe[1,2]
[1]College of Veterinary Medicine, University of Illinois, Urbana, Illinois; [2]Lowe Consulting Ltd., Mahomet, Illinois

Introduction

Significant economic losses due to the disruption of production and increased operating costs often result from the introduction of a novel pathogen into a herd. The industry has invested significant effort in understanding the potential routes of pathogen introduction into herds and developing methods to mitigate those risks. One of those routes is the entry of supplies. Supplies may arrive containers with a complex (e.g. cardboard) or smooth (e.g. plastic) surface that can act as fomites for novel pathogens for that herd. To mitigate those risks, chemical decontamination of supplies entering farms is common practice to protect herd health in swine farms. Typical disinfection practices involve foggers (small droplet) or mechanical sprayers (larger droplet) to apply chemicals to supply items prior to entry.

A glutaraldehyde- quaternary ammonium chloride mixture (Synergize, Neogen Corp.; SYN) is a common choice of swine farms for the decontamination of supplies because it has a neutral pH and non-corrosive properties. SYN is a multipurpose cleaner and disinfectant that kills Gram-negative and Gram-positive bacteria, viruses, and fungi. The quaternary ammonium chloride acts to penetrate cell membranes and the glutaraldehyde fixes nucleic acids, killing the micro-organism.[1]

Accelerated hydrogen peroxide (Intervention, Virox Animal Health; IVN) is another commonly employed disinfectant on swine operations. IVN is labeled as a broad-spectrum bactericidal, fungicidal and virucidal disinfectant. The active ingredient is hydrogen peroxide, an environmentally friendly oxidizing agent, that breaks down into water and hydrogen. Food-grade surfactants present in the mixture allow for added microbial activity in the presence of organic matter.[2-3]

Ozone (O_3) does not leave a residue and therefore is attractive as a disinfectant in swine production facilities. O_3 is known as a strong oxidizing agent, effective on bacteria, molds, viruses, and protozoa through enzymatic disruption of cellular membranes, allowing for damage of nucleic acids.[4]

Research has elucidated the efficacy of individual chemicals on swine pathogens, however, the effectiveness of the process and factors that influence process efficiency rates are unknown. Measurement of process effectiveness on farms is impossible because of the risk of disease introduction. Therefore, a model system constructed to allow for the manipulation of factors of interest is required to quantify these factors. Based on commercial availability and applicability, this study utilized SYN, IVN, and O_3 in the model system. The aim of this study is to evaluate the factors that influence the efficiency of chemical decontamination of supplies entering swine farms.

Materials and methods

The study evaluated the impact of three aerosolized chemical disinfectants on two substrates in multiple orientations relative to the source of the aerosol under conditions typical on swine production sites. Salmonella, which is resistant to chemical disinfection, was the indicator pathogen to create the most robust model possible.

An enclosed 1.5 × 1.5 × 3m room fitted with an electric portable aerosol applicator (Hurricane Ultra II Fogger) served as the decontamination chamber. Treatments consisted of SYN and IVN at 1:256 and 1:8.5 dilutions respectfully (manufacturer recommendations), and 8 ppm aqueous O_3 and at 10-15°C. O_3 was produced using OOG1X0 O_3 generator manufactured by Origin, Inc. (Princeton, NJ, USA). For all experiments, the operation of the fogger for 20 minutes followed the manufacturer's instructions.

Substrates were either plastic or cardboard. There were four orientations for the surfaces of the substrates used in the model system: 0° (facing the ceiling), 180° (facing the floor), 90° (vertical and away from the airflow of the fogger) and 270° (vertical and in the direct airflow of the fogger). Orientation for substrates for all experiments is shown in Figure 1. Negative (samples prior to contamination) and positive (contaminated surfaces not exposed to treatment) controls were collected for all experiments.

In experiment 1, the substrate was 9cm^2 cardboard squares of which 10 were suspended in a 90/270 orientation. Each substrate was inoculated with 100µl of 1:4 Enterisol® Salmonella T/C vaccine and peptone water suspension and allowed to dry prior to exposure to the disinfectant. Substrates were exposed to SYN, IVN or O_3 for 20 minutes at room temperature from 20 to 25 degrees Celsius. Following treatment, the entire substrate was suspended in 5 mL of peptone water and mixed until it broke down and then was centrifuged for 5 minutes. One ml of each sample mixture was plated onto 3M Petrifilm Rapid Aerobic Count Plates for 24 hours before automated plate counts were evaluated according to manufacturer interpretation guidelines.

Figure 1: Orientation of substrates and sampling locations relative to room and chemical source. Hurricane fogger was located on the floor; substrates were 0.6m above the floor surface. The chamber was 1.5m × 1.5 × 3m in size.

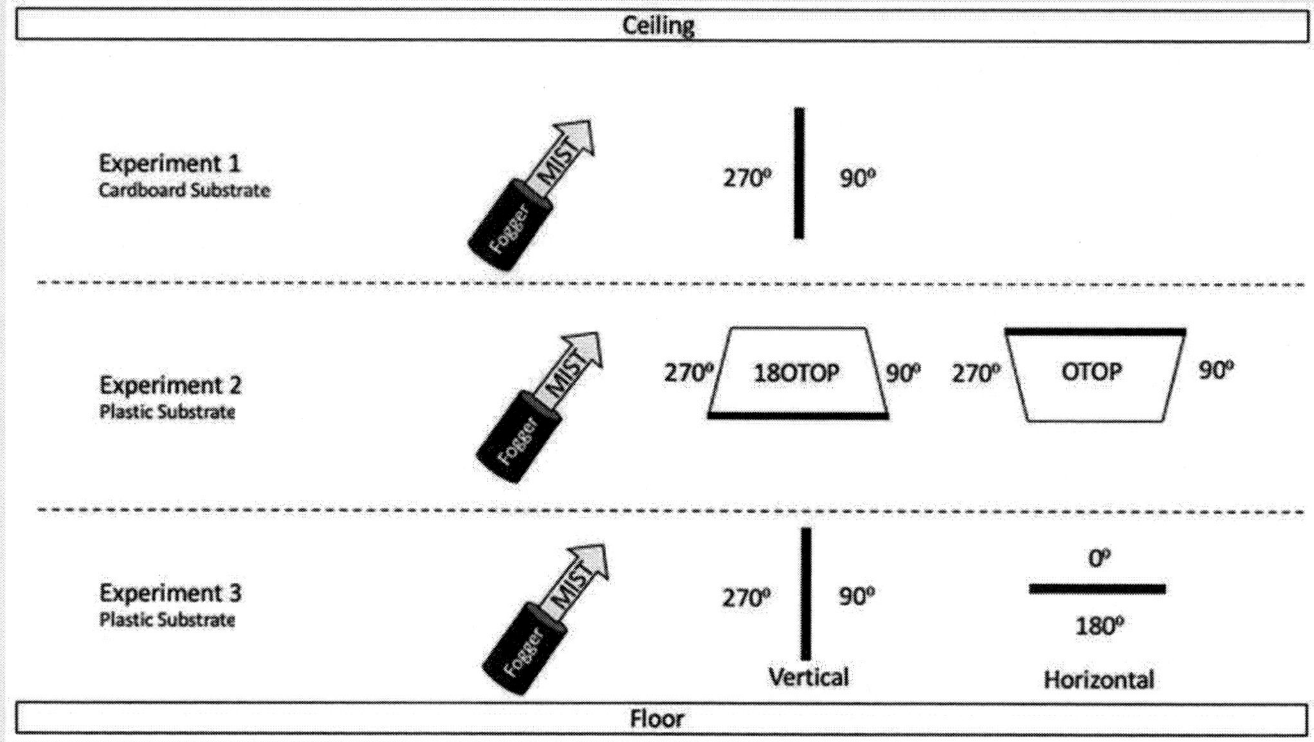

In experiment 2, plastic containers with a sloped side of approximately 10° from vertical were the substrate. 5 replicates were done with the top lid in the 0 position (0TOP) and 5 replicates with the lid in the 180 position (180TOP). This resulted in the vertical sides (90) either being sloped inward (top wider than bottom) for the 0TOP placement or sloped outward (top narrower than bottom) for the 180TOP treatment. Inoculation occurred in a 18cm² area of the surfaces with the 90 and 270 orientations with 40µl of a Salmonella culture (Enterisol® Salmonella T/C, Boehringer Ingelheim Animal Health). Substrates were exposed to SYN, IVN or O_3 for 20 minutes at room temperature from 20 to 25 degrees Celsius. Following treatment, the 18cm² contaminated area on each substrate was swabbed with a single sterile cotton swab (both the 90 and 270 sides), which was then vortexed in 5 mL of peptone water for one minute. One mL of the peptone water was plated onto 3M Petrifilm Rapid Aerobic Count Plates for 24 hours before automated plate counts were evaluated according to manufacturer interpretation guidelines.

Experiment 3 focused on the importance of direct aerosol contact. 10 plastic substrates were placed on wire racks in the center of the chamber with a 0/180 orientation and 10 plastic substrates were suspended in a 90/270 orientation. Inoculation consisted of 9cm² areas on opposing sides of sterilized plastic with 20µl of Enterisol® Salmonella T/C vaccine. All surfaces were allowed to dry before use. Treatments were SYN and IVN with the operation of the fogger for 20 minutes at room temperature from 20 to 25 degrees Celsius. Following treatment, each replicate was swabbed with a sterile cotton swab then vortexed in 5 mL of peptone water for one minute. One ml of each sample mixture was plated onto 3M Petrifilm Rapid Aerobic Count Plates for 24 hours before automated plate counts were evaluated according to manufacturer interpretation guidelines.

Results and discussion

In experiment 1, exposure of cardboard to any of the three treatments regardless of their orientation resulted in only 10 to 40% of the substrates being decontaminated (Table 1). These data suggest that current practices in the industry are inadequate for effective decontamination of items entering farms in cardboard packaging.

In experiment 2, there were differences in the effectiveness of decontamination of plastic surfaces depending on the orientation of the surface to the airflow of the aerosolized chemicals. Vertical surfaces on the 180TOP, which had an increased exposure to aerosol were decontaminated at variable rates (60% O_3, 60% SYN,100% IVN) while surfaces in the OTOP orientation were not decontaminated for with any of the three disinfectants (Table 1).

In experiment 3, vertical plastic surfaces were partially decontaminated. 20% (IVN) to 30% (SYN) of the surfaces 270° position with 20% (IVN), and 0% (SYN) of the surfaces 90° position remained contaminated after treatment. All of the horizontal surfaces in the 0° position were decontaminated with both IVN and SYN while none of the surfaces in the 180° were decontaminated for either product (Table 2).

Taken in total these results indicate that current practices for decontaminating materials prior to entry into swine farms are inadequate for all pathogens. Surfaces that are where gravitational accumulation of the disinfectant droplets is reduced (vertical or sloped sides) demonstrated only partial efficacy on clean smooth surfaces. On surfaces where there was no gravitational accumulation of the disinfectant (bottom) or the surface was complex (cardboard) there was decontamination did not occur. Salmonella is sensitive to H_2O_2, the active ingredient in IVN, with an MIC of 0.002%.[5] In this experiment, delivery of H_2O_2 was at 0.5% exceeding the published MIC by 250 fold. This suggests that the inability to decontaminate surfaces was not a problem of disinfectant efficacy but of disinfectant contact. In addition, in this model, items were placed in an ideal manner with adequate (2× the width of the item) spacing between substrates to ensure adequate air and aerosol movement around the items.

Any surface in this study that did not come direct contact with the chemical remained contaminated, this indicates that aerosolized chemicals do not permeate with the properties of a gas, but rather travel through the air and fall like a mist under the forces of gravity. These data suggest that regardless of disinfectant, there are significant physical barriers to effective decontamination of supplies entering farms. Current practices used on farms are not effective in decontaminating supplies prior to entry. This study reports a previously undescribed weakness of biosecurity systems on swine farms.

In response to these findings, further investigations of design and operation of facilities and equipment used to decontaminate supplies entering farms to determine if modification of current practices is possible to improve decontamination efficacy or it will require the development of alternative designs.

Table 1: Decontamination data of cardboard substrates suspended in a 90/270 orientation and plastic substrates with the lid in the 0 position (0TOP) and in the 180 position (180TOP) within the chamber. Positive recordings represent the number of replicates that remained contaminated following treatments of SYN, IVN, and O_3

Treatment		#Positive	#Total
	Cardboard 90/270 orientation	9	10
SYN	Plastic 0TOP orientation	5	5
	Plastic 180TOP orientation	3	5
	Cardboard 90/270 orientation	4	10
IVN	Plastic 0TOP orientation	5	5
	Plastic 180TOP orientation	0	5
	Cardboard 90/270 orientation	10	10
O_3	Plastic 0TOP orientation	5	5
	Plastic 180TOP orientation	3	5

Table 2: Decontamination data of plastic substrates in 0/180 and 90/270 positions within the chamber. Positive recordings represent the number of replicates that remained contaminated following treatments of SYN and IVN.

Treatment		#Positive	#Total
	0° Orientation	0	10
SYN	180° Orientation	10	10
	270° Orientation	3	10
	90° Orientation	0	10
	0° Orientation	0	10
IVN	180° Orientation	10	10
	270° Orientation	1	10
	90° Orientation	2	10

References

1. Dee S, Deen J, Burns D, Douthit G, Pijoan C. An assessment of sanitation protocols for commercial transport vehicles contaminated with porcine reproductive and respiratory syndrome virus. *Can J Vet Res.* 2004;68(3):208-14.

2. Holtkamp DJ, Myers J, Thomas PR, et al. Efficacy of an accelerated hydrogen peroxide disinfectant to inactivate porcine epidemic diarrhea virus in swine feces on metal surfaces. *Can J Vet Res.* 2017;81(2):100-107.

3. Baker KL, Thomas PR, Karriker LA, et al. Evaluation of an accelerated hydrogen peroxide disinfectant to inactivate porcine epidemic diarrhea virus in swine feces on aluminum surfaces under freezing conditions. *BMC Vet Res.* 2017;13(1):372. Published 2017 Dec 1. doi:10.1186/s12917-017-1300-4

4. Megahed, Ameer & Aldridge, Brian & Lowe, J. (2018). Evaluation of the Decontamination Power of Aqueous and Gaseous Ozone on Various Contaminated Surfaces with Cattle Manure containing Salmonella. Advances in Dairy Research. 06. 10.4172/2329-888X.1000205.

5. Aarestrup F Hasman H, Susceptibility of different bacterial species isolated from food animals to copper sulphate, zinc chloride and antimicrobial substances used for disinfection. Veterinary Microbiology, 2004 vol: 100 (1-2) pp: 83-89. doi.org/10.1016/j.vetmic.2004.01.013.

Dietary sodium diformate and glycerine monolaurate affects fecal pathogen load in lactating sows

C. Lückstädt, PhD
ADDCON GmbH, Bitterfeld-Wolfen, Germany

Introduction

Compound feed is not only an effective way of delivering nutrients to animals, but it has long been utilised as a delivery strategy for non-nutritive additives used to promote health. To this end, the industry relied heavily on the use of antibiotics as growth promoters in the feed for decades, until it became clear that the development of bacterial resistance against these compounds could jeopardise their future in the fight against bacterial disease, not only in animals, but also in human health care. The search for alternatives became critical around 15 years ago, beginning in Europe, where it led to a ban of antibiotic growth promoters in feed in 2006. It is generally agreed that good gut health is effective against intestinal pathogens, a strategy that has only been made possible through the removal of antibiotic growth promoters in feed. Creating and maintaining a healthy intestinal environment has become essential to productivity and food safety programmes alike. Maintaining a healthy gut requires up to 25% of the daily protein and 20% of the dietary energy supplied with the feed. This strategy should be carefully planned into the dietary programme, to avoid wasting resources. Acidifiers have a long history of safe and effective use in diets for pigs, especially in Europe where their application was pioneered. Their mode of action on the growth and health of pigs at all life stages has been well documented, and their magnitude of action has been estimated in different husbandry conditions through statistical analysis. To date, there have been a number of modes of action proposed to explain their effects on feed hygiene, pig health and performance. Although there have been some contradictory reports as to the magnitude of these benefits, research and analysis of the data shows that the variability in the performance results stems from environmental and feed composition variables, rather than an inconsistency in the efficacy of acidifiers *per se*.[1] Although growth performance benefits have been shown in numerous studies over the past half-century;[2] the ban on antimicrobial growth promoters in the European Union in 2006 resulted in an increased scientific focus on organic acids. Meanwhile, experience has shown that acidifiers are the most reliable product group of the non-antibiotic growth promoters available in Europe and can also be used safely and effectively with other additives. Papatsiros and Billinis stated in this regard "...Dietary acidifiers can actually become the most common and efficacious alternative solution to antibiotics, in order to improve health status and performance in pigs".[3] In pig diets, organic acids and their salts take effect in the gastrointestinal tract, mainly in the proximal tract - the stomach and small intestine. Firstly, organic acids lower the pH of the stomach contents, which can be especially beneficial at weaning, when gastric acid secretion capacity of the animal's stomach is often insufficient to take on the dietary stress of weaning and the social stress of mixing. While pH reduction can inhibit pathogen growth and optimise pepsin activity, it cannot alone account for the numerous benefits reported when organic acids are included in diets for pigs. Supplementing diets with organic acids reduces the pH in the stomach, especially in weaning pigs, where it stimulates the activation of pepsinogen to pepsin and thus may improve protein digestibility and decrease the rate of gastric emptying; further improving protein digestion by increasing the rate of proteolysis of large protein molecules. The bacteriostatic and bactericidal effects of organic acid anions also take effect in the proximal gastrointestinal tract. Here, the differential inhibition of Gram-negative pathogens (e.g. *E coli*, *Salmonella*) compared to beneficial bacteria such as lactobacilli and bifidobacteria improve the microbial load (eubiosis) in the tract, preventing post-weaning diarrhea. Several studies have shown that the bactericidal effect of organic acids can still be strong in the absence of a significant decrease in pH. They are therefore also useful in combating bacterial pathogens in grower-finisher pigs and sows.[4] While the antimicrobial impact of organic acids and their salts, including diformates, is mainly directed against Gram-negative bacteria, medium chain fatty acids (C6 to C12) have been shown to have also an antibacterial impact against various Gram-positive bacteria.[5] This is especially true for lauric acid and its monoglyceride ester monolaurate.[6] *Streptococcus suis*, a Gram-positive bacterium, is one of the most common harmful bacteria in pig production. It also proliferates in the digestive tract. Most piglets contract the bacteria from the sow during parturition. Thus, decontamination of the sow feces may help in reducing infection. However, data on the combined impact of these additives on gut health in sows under commercial conditions had not yet been generated. This formed the impetus of this trial.

Material and methods

The study tested the efficacy of a blend of sodium diformate with the monoglyceride of lauric acid – monolaurate, traded as Formi GML (ADDCON, Germany) on decontamination of Gram-negative and Gram-positive bacteria in sow feces. Forty multiparous sows on a research farm in Saxony-Anhalt, Germany, were allocated to two equal groups and fed a commercial lactation diet from one week before farrowing until the piglets were weaned at 26 days. The test diet contained 1.0% Formi GML. Water was available *ad libitum* throughout the trial. The lactation diet was fed according to a feeding curve, from the last week before farrowing to day 13 of lactation and thereafter *ad libitum*. On the 21st day of lactation, freshly excreted fecal matter was collected from all sows and analysed for *E coli*, enterococci and streptococci counts. Data were analysed using the t-test and a significance level of 0.05 was used in all tests.

Results and discussion

Results of the microbial analysis revealed a strong impact of the diformate-monolaurate mixture on the bacterial population in the fecal matter of sows. This holds true for *E coli*, streptococci and enterococci counts. The data are displayed in Table 1. The *E. coli* count in the feces differed significantly ($P = 0.0012$) between the control and treatment groups. Feces of treated sows had a 98% lower *E coli* count (CFU/g). Similar results were measured for enterococci ($P = 0.027$) and streptococci ($P = 0.014$), which were reduced by 99% and 75%, respectively. Addition of Formi GML caused a significant improvement in the health status of sows. The impact against the Gram-positive streptococci is especially noteworthy. This is particularly important since the EU-funded Focus Group is calling for actions to reduce the use of antibiotic treatments on swine farms. The above findings support the use of Formi GML as an effective and sustainable tool for improving gut health in sows by reducing pathogen load at intestinal level and decontaminating the feces of the sow. Therefore, the risk of cross-infection to the piglets is reduced. Furthermore, the use of the additive has a supporting function for the world-wide antibiotic reduction initiatives.

References

1. Theobald P and Lückstädt C. Experience on the usage of organic acids in swine diets in Europe. In: Lückstädt, C.: *Standards for acidifiers – Principles for the use of organic acids in animal nutrition*. Nottingham University Press, Nottingham. 2011:39-48.

2. Cole DJA et al. The effect on performance and bacterial flora of lactic acid, propionic acid, calcium propionate and calcium acrylate in the drinking water of weaned pig. *Veterinary Record*. 1968;83:459–464.

3. Papatsiros V and Billinis C. The prophylactic use of acidifiers as antibacterial agents in swine. In: Bobbarala, V.: *Antimicrobial Agents*. 2012;14:295-310.

4. Kirchgessner M et al. Zum Einfluß von Ameisensäure, Calciumformiat und Natrium-hydrogencarbonat auf die Keimzahlen der Mikroflora und deren Zusammensetzungin verschiedenen Segmenten des Gastrointestinaltraktes. *Journal of Animal Physiology and Animal Nutition*. 1992;67:73-81.

5. Preuss et al. Minimum inhibitory concentrations of herbal oils and monolaurin for gram-positive and gram-negative bacteria. *Molecular and Cellular Biochemistry*. 2005;272(1-2):29-34.

6. Ruzin A and Novick RP. Equivalence of lauric acid and glycerol monolaurate as inhibitors of signal transduction in Staphylococcus aureus. *J Bacteriology*. 2000;182 (9):2668-2671.

Table 1: Bacterial counts (CFU/g) in sow feces (n = 40) fed with or without Formi GML and their respective reduction rates in per cent

	Control (n = 20)	1.0% Formi GML (n = 20)	Reduction rate (%)
E. coli	1.20×104^{a}	$2.00 \times 10^{2*b}$	-98.3
Standard deviation	(1.58×104)	(0)	-
Enterococci	5.85×105^{a}	6.36×103^{b}	-98.9
Standard deviation	(1.27×106)	(5.29×103)	-
Streptococci	1.92×104^{a}	4.78×103^{b}	-75.1
Standard deviation	(2.42×104)	(1.30×104)	-

* Detection limit: [a,b] means with a different superscript differ significantly ($P < 0.05$).

The effect of feeding an alternative protein technology to replace animal plasma on piglet nursery performance

B. Bass[1], PhD; S. Crowder[2], MS; T. Weeden[2], PhD, PAS; B. DeRodas[2], PhD; T. Karnezos[1], PhD
[1]PMI, Arden Hills, Minnesota; [2]Purina Animal Nutrition, Gray Summit, MO

Introduction

Weaning introduces piglets to several stressors including dietary changes, change in environment, establishment of hierarchy, and disease challenge, which may negatively impact performance.[1] Various feeding strategies and additives have been implemented in an attempt to overcome the diminished performance that may be observed following weaning. Several products, including spray-dried plasma, have been used to improve weaned pig performance.[2,3] However, many producers have begun to look for alternative proteins to reduce the use of animal plasma due to cost, entry into animal byproduct-free diet markets, and the concern it may be a carrier for certain viruses. The objective of this project was to evaluate an alternative protein technology (TecPro™, PMI, Arden Hills, MN) to replace animal plasma in nursery diets.

Materials and methods

A series of studies were conducted in Midwest commercial wean-to-finish research facilities. In the first study, 600 pigs were individually weighed and randomly assigned to one of two treatments (20 mixed-sex pigs/pen; 15 pens/treatment). Treatments consisted of 1) Positive control with 2.75% bovine plasma in phase 1 diets and 2) TecPro at 0.625% in phase 1 and 0.3125% in phase 2. A feed budget of 6.7, 13.3, and 40 lbs/pig was provided in phases 1, 2, and 3, respectively. Mecadox (50 g/ton) was provided in all phase 1 diets. Pigs weight and feed intake were recorded on d 0, 7, 14, 21, 28, and 43 to determine ADG ADFI, and F/G. In a second study, 660 pigs were individually weighed and randomly assigned to one of three treatments (20 mixed-sex pig/pen; 11 pens/treatment). Treatments consisted of 1) Positive control with 2.75% bovine plasma in phase 1 diets; 2) TecPro at 0.625% in phase 1 and 0.3125% in phase 2; and 3) TecPro at 1.25% in phase 1 and 0.625% in phase 2. All diets were devoid of in-feed medication. A feed budget of 5, 13, and 40 lbs/pig was provided in phases 1, 2, and 3, respectively. Pigs weight and feed intake were recorded on d 0, 7, 14, 21, 28, and 42 to determine ADG ADFI, and F/G. In the third study, 484 pigs were individually weighed and randomly assigned to one of two treatments (22 mixed-sex pigs/pen; 11 pens/treatment). Treatments consisted of 1) Positive control with 5% bovine plasma in phase 1 diets and 2) TecPro at 1.25% in phase 1 and 0.625% in phase 2. A feed budget of 5, 13, and 40 lbs/pig was provided in phases 1, 2, and 3, respectively. Pigs weight and feed intake were recorded on d 0, 7, 13, 20, 27, and 42 to determine ADG ADFI, and F/G. Pigs in the final study arrived at the wean-to-finish facility infected with PRRS. Treatment through the water, and individual treatments, were provided as needed.

Results

In the first study, TecPro enhanced overall feed conversion (1.42 vs. 1.46; $P = 0.03$). TecPro performed similarly to bovine plasma for final body weight and overall ADG and ADFI (Table 1). However, feeding TecPro provided a cost savings of $0.50/pig compared to the diet with 2.75% bovine plasma.

For the second study, BW was significantly increased when 1.25% TecPro was provided compared to 2.75% bovine plasma on d 14, 21, and 28 ($P \leq 0.05$; Table 2). Additionally, there was a significant increase in ADG and ADFI, and improved feed conversion, during the period between d 0 and 21 in pigs provided 1.25% TecPro compared to 2.75% bovine plasma, with 0.625% TecPro being intermediate. Overall, there were no significant differences in ADG, ADFI, feed conversion, or final BW. There was a numerical improvement in livability by 1.3 and 2.2% for

Table 1: Performance of nursery pigs (Study 1) provided bovine plasma or an alternative protein technology

	Bovine plasma (2.75%)	TecPro (0.625%)	SEM	*P*-value
Initial BW, lb	13.65	13.60	0.35	0.93
Ending BW, lb	58.55	59.61	1.44	0.61
Overall ADG, lb/d	1.01	1.04	0.02	0.19
Overall ADFI, lb/d	1.47	1.48	0.02	0.81
Overall F/G, lb/d	1.46	1.42	0.01	0.03
Feed cost savings/pig	--	$0.50	--	--

Table 2: Performance of nursery pigs (Study 2) provided bovine plasma or two levels of an alternative protein technology[1]

	Bovine plasma (2.75%)	TecPro (0.625%)	TecPro (1.25%)	SEM	*P*-value
Initial BW, lb	13.48	13.51	13.81	0.24	0.56
d 7 BW, lb	15.78	16.12	16.51	0.27	0.18
d 14 BW, lb	19.92	20.37	21.31	0.39	0.05
d 21 BW, lb	24.5	25.31	26.83	0.59	0.03
d 28 BW, lb	32.88	33.61	35.26	0.67	0.05
Ending BW, lb	49.86	51.2	53.09	1.27	0.21
d 0 to 20 ADG, lb/d	0.519	0.559	0.619	0.026	0.03
d 0 to 20 ADFI, lb/d	0.629	0.679	0.716	0.027	0.09
d 0 to 20 F/G, lb/d	1.22	1.21	1.16	0.02	0.05
Overall ADG, lb/d	0.854	0.891	0.934	0.029	0.16
Overall ADFI, lb/d	1.17	1.22	1.27	0.04	0.25
Overall F/G, lb/d	1.37	1.37	1.36	0.01	0.34
Livability, %	97.3	98.6	99.5	0.77	0.13
Feed cost savings/pig	--	$0.28	$0.37	--	--

pigs provided TecPro at 1.25 and 0.625%, respectively, compared to bovine plasma ($P = 0.13$). Additionally, providing TecPro reduced feed costs by $0.28 and $0.37 per pig when provided at 0.625 and 1.25%, respectively.

In the third study, pigs arrived at the research facility infected with PRRS, which impacted performance and mortality. Overall, there were no differences in final body weight, average daily gain, average daily feed intake, or feed conversion between pigs provided 5% bovine plasma or TecPro (Table 3). However, there was a numerical reduction in mortality of 2.4% when pigs were fed TecPro compared to bovine plasma. Additionally, feeding TecPro saved $0.35/pig in feed costs.

Discussion

TecPro was provided to pigs in three different scenarios. In the first, pigs were provided diets with in-feed antibiotics during the first dietary phase and an enhancement of feed conversion was observed when TecPro was added compared to animal plasma. In another, an improvement was observed in early performance when TecPro was provided in diets without antibiotics. Finally, performance of pigs provided TecPro was not different from pigs provided animal plasma when pigs were naturally infected with PRRSV when they entered the wean-to-finish facility. In all scenarios, feeding TecPro provided feed cost savings over the diets formulated with animal plasma. In conclusion, the alternative protein technology, TecPro, provides an option to producers who choose to reduce, or eliminate, animal plasma from nursery diets.

References

1. Pluske et al. Factors influencing the structure and function of the small intestine in the weaned pig: a review. *Livest Prod Sci.* 1997;51:215.
2. van Dijk et al. Growth performance of weanling pigs fed spray-dried animal plasma: A review. *Livest Prod Sci.* 2001;68:263.
3. Torrallardona and Polo. Effect of spray-dried porcine plasma protein and egg antibodies in diets for weaned pigs under environmental challenge conditions. *J Swine Health Prod.* 2016;24:21.

Table 3: Performance of nursery pigs (Study 1) provided bovine plasma or an alternative protein technology

	Bovine plasma (5.00%)	TecPro (1.25%)	SEM	*P*-value
Initial BW, lb	10.6	10.6	0.2	0.94
Ending BW, lb	43.9	42.8	0.7	0.26
Overall ADG, lb/d	0.743	0.726	0.021	0.58
Overall ADFI, lb/d	0.933	0.907	0.027	0.49
Overall F/G, lb/d	1.26	1.25	0.02	0.80
Livability, %	87.6	90.0	2.1	0.41
Feed cost savings/pig	--	$0.35	--	--

Early-life administration of Tonisity Px™ isotonic protein drink to pigs improves farrowing livability and growth to end-nursery

A. Carlson[1], DVM; M. Eisenhart[2], DVM; K. Bretey[1], DVM; S. Buzoianu[3], DVM, PhD;
Ava M. Firth[3], DVM, MVS, DACVECC
[1]Standard Nutrition, Omaha, Nebraska, [2]Tonisity Inc, St. Joseph, Missouri
[3]Tonisity International, Dublin, Ireland

Introduction

Increases in litter size and concurrent decreases in birth weights have been shown to decrease livability from farrowing to weaning. Similarly, various management approaches or on-farm logistics often result in large spreads in weaning ages. In addition, the small intestine of the newborn piglet goes through tremendous development over the first ten days of life. This period has been identified as a "window of opportunity"[1] for nutritional interventions to improve the development of intestinal structure and the maturation of the immune system which could have lifelong effects. These factors create an opportunity to provide supplemental nutrition in the first days of the piglets' lives to increase livability and growth performance. However, the variables outlined above also have the potential to obscure the results of any intervention if the data is analyzed at a population level. Therefore, the aim of this study was twofold, with the first being to test if an isotonic protein solution (Tonisity Px) provided at days 2 through 8 of age would increase livability through weaning and growth performance to the end of nursery. Secondly, the study examined the effect of Tonisity Px when piglets were stratified into light, medium and heavy birthweights and into three weaning-age peer groups.

Materials and methods

Test ingredient

The test ingredient consisted of an isotonic protein solution (Tonisity Px™), which provides easily-absorbable nutrients (glucose, amino acids, and peptides) and electrolytes that can be used directly by the enterocytes.

Study population and feeding plan

Farrowing: 3862 commercial piglets farrowed from 353 gilts (PureTek genetics) in a 3200 head breed-to-wean farm in Indiana undergoing repopulation were enrolled in the study. Litters were allocated to one of 2 groups, control (Control—1969 piglets) or supplementation with Tonisity Px (TPX—1893 piglets). All piglets were identified at day 2 with two individual ear tags.

Control litters received no liquid nutritional supplementation during the suckling or pre-weaning phase. TPX litters were given 250 ml of 3% Tonisity Px solution on day 2 of age, and from days 3-8 of age TPX litters were given 500 mL of 3% Tonisity Px once daily all in an open pan. Both groups had pelleted creep feed available from day 3 of life to weaning. TPX litters also received 500 ml of 3% Tonisity Px solution three, two, and one days before weaning. Weaning was at 15-37 days.

Nursery: Pigs were moved to two separate nursery sites in Indiana and Illinois at weaning. At each nursery site, all pigs received their standard nursery diet free choice in their feeder and free choice water. TPX pigs also received Tonisity Px solution in open feeding pans on the following schedule: day of arrival → 2 gallons per pen of 3% solution (no feed added), day 1 → 2 gallons per pen of 1.5% solution mixed with feed, day 2 → 2 gallons per pen of 0.75% solution mixed with feed.

Measurements

Pigs were individually weighed at birth, weaning and end of nursery. Pre-weaning mortality was calculated as the number of dead pigs pre-weaning divided by the total number born alive. Nursery mortality rates were not available at the time of this writing. Weight gain data was analyzed by litter and also individually. Weight gain data was further stratified in sub-populations by weaning age (15-18, 19-22, and 23-26 days) and birthweight (< 2.2 lb, 2.2 – 3.5 lb, and > 3.5 lb).

Statistical analysis

Data were analyzed using SAS 9.4 (SAS Inst. Inc., Cary, NC). Mortality was tested using Fisher's Exact test. Weight gain data were assessed for normality with the Kolmogorov-Smirnov test. The individual weight data were not normally distributed, and were therefore analyzed using the Kruskal-Wallis test. Data are presented as medians and 5th-95th percentiles. When stratified by weaning age, the litter data was normally distributed and was analyzed using PROC MIXED, with treatment as a fixed effect, and birth weight, weaning weight and weaning age as covariates in the model, which minimized the variability due to these factors. Results were significant at $P \leq 0.05$ and considered a trend at $P > 0.05$ and $P \leq 0.10$.

Results and discussion

Farrow-to-wean data was received for 3,254 piglets (TPX n = 1618, Control n = 1636), representing 353 first-parity litters. Litter-based production data was analyzed. However, the individually-analyzed data was considered to be more relevant to examining the difference in key production metrics for this study which was focused on uncovering the differences that occurred in different age and birthweight sub-populations of pigs, as described in Materials and Methods. This allowed comparison of a much larger number of samples and increased the strength of the analysis. Furthermore, only the individual data could provide the stratification per birth weight. Where relevant, litter-based data is provided for comparison.

Mortality

Pre-weaning mortality in the TPX group was 10.9%, which was significantly lower ($P < 0.001$) than the 14.2% mortality in the Control group. This was a difference of 3.2 percentage points, and represents a 22.8% reduction in preweaning mortality, with the odds ratio for survival at weaning being 1.35 times greater in the TPX group than in the Control group.

This reduction in mortality amounts to the control group losing 63 more pigs compared to the TPX group. Based on the control group of 176 litters, eliminating this loss would amount to a gain of 0.36 pigs weaned/litter, which extrapolates to an extra 0.85 pigs weaned per sow per year (based on 2.36 farrowings per year). The mortality reduction revealed in this study indicates that the micro-enteral nutrition provided during the day 2-8 "window of opportunity" increases piglets' chances of survival to weaning. Furthermore, these results were obtained in healthy pigs, with no disease outbreaks recorded during the trial. It should be noted that all sows were of first parity and thus differences in milking ability were minimized.

Weight gain

Despite the large number of pigs enrolled in the study and the randomization of those pigs into treatment groups, the TPX pigs weighed 0.16 lb more at birth than the control pigs, representing a 4.7% advantage ($P = 0.0001$). The TPX pigs subsequently had higher weaning weights compared to the control group at 13.9 lb vs 13.2 lb ($P = 0.0001$), representing a 5.1% advantage. Piglets in the TPX group gained an average of 0.52 lb/day, which was significantly higher than the control piglets who gained an average of 0.50 lb/day ($P = 0.001$), representing a 2.1% advantage in the TPX group. When analyzed by litter, virtually the same numeric differences were found (0.52 and 0.51 lb), however, they were not statistically significant (data not shown). Similarly, at end of nursery, the TPX piglets outperformed the control pigs, with final individual weights of 55.0 lb vs 53 lb and 55.4 lb vs 54.4 lb as litters; $P < 0.01$). Weight gain in nursery was 42.1 lb vs 40.1 lb based on individual weights and 42.0 lb vs 40.2 lb as litters. Even at the litter analysis level, though, the TPX pigs tended to have a 0.3 lb advantage in weaning weight ($P = 0.07$). The magnitude of this difference increased to 2 lb by the end of nursery, becoming highly significant ($P < 0.01$).

The fact that the P values were different for individual and litter data indicates the existence of subpopulations within the data. Such subpopulations may not be detected when analysis is done at a litter level, as this latter is only an average. Weight data were therefore stratified by birth weight and weaning age categories.

Growth performance by birth weight category (Table 1)

Stratifying the growth performance data by birth weight category (< 2.2 lb, 2.2 – 3.5 lb, and > 3.5 lb) removed the significant birthweight differences between treatment groups in the light- and medium-birthweight pigs. The weaning weights were 0.5 lb higher for TPX pigs in the medium birthweight strata ($P < 0.01$) and at the end of nursery, the light and medium birthweight TPX pigs were heavier by 3.5 lb (8.4%; $P = 0.09$) and 1 lb (1.9%; $P < 0.05$), respectively. Similarly, the medium weight TPX pigs showed a significantly higher weight gain pre-weaning and in the nursery phase than control pigs ($P < 0.05$; Table 1). It should be noted that the medium- and light- birthweight pigs accounted for 58% of the study population.

In the heavy-birthweight pigs, although the median birthweight was not different, the TPX pigs had a higher weight range than control pigs. For the heavy birthweight pigs, this translated into a 1 lb numerical advantage for the TPX group. A post-hoc sample size calculation revealed that the 1 lb difference found at the end of nursery would have been statistically significant if 3000 pigs per group had been enrolled. These results demonstrate that Tonisity Px resulted in heavier pigs at the end of nursery across all birthweight ranges.

Growth performance by weaning age category (Table 2)

Due to labor constraints and pig flow, the weaning age of the pigs varied from 15-37 days, which markedly increased the spread of the weaning weights. Performance data was therefore stratified into 3 age categories, namely, early-weaned pigs (15-18 days of age), medium-age-weaned pigs (19-22 days of age) and late weaned pigs (23-26 days of age). Seven pigs over 26 days of age were removed from this analysis. This allowed assessment of whether specific weaning age subgroups reveal any growth trends in relation to the treatment and to make a more sensible estimate of growth performance. For both litter and individual pig analysis, approximately 50% of the pigs in each group were weaned between 19-22 days, with the remaining 50% allocated almost equally to the 15-18 day and 23-26 day weaning age categories (Table 2).

Piglets weaned at 15-18 days of life

For piglets weaned at 15-18 days of life, in the individual analysis, the TPX group had heavier weights from birth to weaning ($P < 0.05$) (Table 2). However, with the exception of a significantly higher birth weight, the litter analysis revealed only numerical differences. Similarly, based on the individual analysis, the pre-weaning weight gains were significantly higher in the TPX group (9.1 lb vs 8.4 lb; $P < 0.01$) but litter analysis showed only a numerical increase. In this weaning age group, there were no significant differences between groups for end-nursery weight and nursery weight gain.

Piglets weaned at 19-22 days of life

In this weaning age category, no significant differences were observed for birth weight, weaning weight or weight gain in the farrowing house (overall or per day) or end-nursery weight as

Table 1: Growth performance of piglets stratified by birthweight[a]

	Light birthweight piglets (<2.2 lb)				Medium birthweight piglets (2.2-3.5 lb)				Heavy birthweight piglets (>3.5 lb)			
Parameter	**Control**	**Tonisity Px**	**P value**	**Change***	**Control**	**Tonisity Px**	**P value**	**Change***	**Control**	**Tonisity Px**	**P value**	**Change***
Birth weight (lb)	2.0 (1.5-2.2)	2.0 (1.3-2.2)	0.67	-	3.0 (2.4-3.5)	3.0 (2.3-3.5)	0.28	-	3.9 (3.5-4.8)	3.9 (3.5-4.9)	0.03#	-
Weaning weight (lb)	10.0 (7.4-13.1)	10.0 (7.4-13.8)	0.17	-	12.5 (8.6-16.8)	13.0 (9.1-17.4)	0.0001	(0.5 lb) 4.0%	14.7 (10.1-19.2)	15 (10.6-20.1)	0.26	(0.3 lb) 2.0%
Weaning weight gain (lb)	8.2 (5.3-11.1)	8.5 (5.8-11.7)	0.14	0.3 lb (4.0%)	9.6 (5.8-13.7)	9.9 (6.3-14.2)	0.0001	(0.3 lb) 3.4%	10.8 (6.2-15)	11.0 (6.6-15.8)	0.33	(0.2 lb) 1.6%
n/% of row total	108/3%	71/2%			921/28%	794/25%			607/19%	745/23%		
Nursery weight (lb)	41.5 (24.5-57.0)	45.0 (30.0-59.0)	0.09	(3.5 lb) 8.4%	52.0 (34.0-65.0)	53.0 (33.0-69.0)	0.02	(1 lb) 1.9%	58.0 (41.0-73.0)	59.0 (38.0-75.0)	0.16	(1 lb) 1.7%
Nursery weight gain (lb)	31.7 (15.4-46.4)	34 (20.2-50.2)	0.14	(2.3 lb) 7.2%	39 (23.7-50.8)	40.4 (21.5-54.2)	0.04	(1.4 lb) 3.4%	43.8 (27.9-57.1)	44.5 (24.6-58)	0.13	(0.7 lb) 1.8%
n/% of row total	80/3%	59/2%			779/29%	641/24%			511/19%	600/23%		

a All weight data is presented as medians with 5-95 percentiles in parentheses
* For the % change, the Control group value was considered the baseline (100%)
\# Although the median is not different between groups, the ranking of the pigs by body weight during the analysis showed that the TPX pigs had significantly higher ranks than the Control pigs.

Table 2: Growth performance of piglets stratified by weaning age[a]

Parameter	Early weaned pigs (15-18 days)				Medium age weaned pigs (19-22 days)				Late weaned pigs (23-26 days)			
	Control	Tonisity Px	P value	Change *	Control	Tonisity Px	P value	Change *	Control	Tonisity Px	P value	Change *
Birth weight (lb)	3.3 (2.0-4.7)	3.6 (2.3-4.8)	0.0001	(0.3 lb) 8.3%	3.3 (2.1-4.5)	3.3 (2.2-4.4)	0.34	0%	3.2 (2.1-4.3)	3.5 (2.3-4.7)	0.0001	(0.3 lb) 10%
Weaning weight (lb)	11.7 (7.8-15.9)	12.6 (8.8-16.7)	0.0001	(0.9 lb) 7.3%	13.7 (9.2-18.2)	13.6 (9.1-17.9)	0.25	(-0.1lb) - 0.6%	14.8 (9.4-19.3)	15.6 (11.1-20.8)	0.0001	(0.8 lb) 5.5%
Weaning weight gain (lb)	8.4 (5.1-11.8)	9.1 (5.7-12.3)	0.0001	(0.7 lb) 7.6%	10.4 (6.3-14.2)	10.2 (6.3-14.0)	0.15	(-0.2lb) -1.3%	11.6 (6.7-15.1)	12.0 (8.0-16.6)	0.001	(0.4 lb) 3.5%
n/% of row total	492/15%	388/12%			871/27%	794/25%			272/8%	430/13%		
Nursery weight (lb)	50.0 (33.0-66.0)	50.0 (32.0-67.0)	0.51	0%	56.0 (37.0-72.0)	56.0 (35.0-73.0)	0.28	0%	54.0 (34.0-67.0)	58.0 (38.0-75.0)	0.0001	(4 lb) 7.4%
Nursery weight gain (lb)	38.2 (24.3-51.5)	37.8 (21.7-53.1)	0.64	(-0.4 lb) - 1.1%	41.6 (25.4-56.1)	42.7 (22.8-56.5)	0.05	(1.1 lb) 2.6%	38.7 (23.1-50)	43.3 (25-57.7)	0.0001	(4.6 lb) 11.9%
n/% of row total	408/15%	319/12%			726/27%	646/24%			235/9%	327/12%		

a All weight data is presented as medians with 5-95 percentiles in parentheses
* For the % change, the Control group value was considered the baseline (100%)

individuals. However, when analyzed individually, the total nursery weight gain was 1.1 lb higher in the TPX group compared to Control ($P < 0.05$; Table 2). This was mirrored by a numerical difference when the data were analyzed per litter. No other differences were observed for this age category, when analyzed by individual or per litter.

Piglets weaned at 23-26 days of life

For the pigs that were older at weaning, the individual analysis revealed a significant difference in the weights of pigs in the TPX pigs compared to Control for birth weight, weaning weight and pre-weaning gain (both overall and per day) ($P < 0.05$; Table 2). However, only numerical differences were observed when the analysis was performed per litter, except for a tendency for higher birth weights for the pigs in the TPX group.

A significant increase in end-nursery weight and overall nursery weight gain was observed for the pigs in the TPX group compared to controls. This was evident for both individual and litter analyses (Table 2). The analyses show a distinct advantage of 4 lb (7.4%) for nursery weight and 4.6 lb (11.9%) for weight gain. For this weaning age the advantage for the TPX group increased from 0.3 lb for birth weight to 0.8 lb for weaning weight to 4 lb for end-nursery weight for individual analysis. These data show that the TPX group outperformed the Control group for the difference even at the litter analysis level and amounted to a statistically significant outcome.

Conclusions

Pre-weaning mortality was reduced by 3.2 percentage points, from 14.2% in the control group to 10.9% in the TPX group. This represents a 22.8% reduction in preweaning mortality. This reduction can be attributed to Tonisity Px and is consistent with results we have seen in multiple previous studies. This is an important benefit at a time where mortality in pig production adds significantly to the losses under tight profit margin conditions. The difference in pre-weaning mortality amounts to 63 pigs that could have been alive in the Control group if they had received Tonisity Px. Those 63 when added to the 62 extra pigs in the TPX group totals an extra 125 pigs from 353 litters. This translates into 0.35 more piglets weaned per litter, and over 1 year, would amount to 0.85 more piglets weaned/sow/year (assuming 2.36 farrowings per year). Assuming that the subsequent mortality in nursery was 3%, i.e., loss of 4 pigs, then an extra 121 pigs out of 3862 (31 pigs/1000) would have finished nursery. At an average nursery weight of 55 lb, the extra 121 pigs would amount to 6655 lb extra weight at the end of nursery.

Despite the spread in weaning ages and birth weights, overall, TPX pigs showed a statistical tendency for increased weaning and nursery weight, total pre-weaning and nursery weight gain, even when analyzed by litter. Stratification of the data into "light", "medium" and "heavy" birthweight piglets clarified the distinct advantages in the TPX group over the control from weaning through the end of nursery. These differences were seen even in the absence of a difference in birth weight. The 8.4% increase in nursery weight in the TPX group for the light birthweight pigs indicates that Tonisity Px allows the piglets to catch up to their growth potential. The medium birthweight pigs, which accounted for ~ 50% of the population (as is typical in most production systems), showed significant weight gain advantages. Furthermore, if the study had contained 3000 pigs per treatment group, the end-nursery weight advantage found in heavy birthweight pigs would also have been significant.

Overall, for the three birthweight categories, at the end of nursery, TPX pigs had 1447.5 lb more body weight (number of pigs in each birthweight category multiplied by the weight advantage (Table 1), distributed as follows: Light birthweight TPX pigs (n = 59) weighed 206.5 lb more overall than Control; medium birthweight TPX pigs (n = 641) weighed 641 lb more; heavy birthweight TPX pigs (n = 600), weighed 600 lb more. At a median weight at nursery of 55 lb, this 1447.5 lb weight difference equals the body weight of an extra 26 nursery pigs.

The growth data from the current study, linked with data from a previous study (data on file) in which mortality was evaluated up until 42 days of age, showed that mortality improvements were achieved equally across light-, medium- and heavy-birthweight pigs. These results indicate that the benefits of administering Tonisity Px to pigs are evident for all weight categories, not only for light birthweight pigs.

Similarly, stratification of pigs by weaning age also revealed better performance for TPX pigs. Piglets weaned between 23-26 days of age weighed 4 lb more at end of nursery and had 4.6 lb higher weight gain in the nursery period. Similarly, piglets in the 19-22 weaning age category gained 1.1 lb more during nursery, without any differences observed for birth weight. These gains amounted to an overall 1308 lb more body weight for the TPX group compared to Controls. Based on a median weight of 55 pounds per pig, this weight difference equals the body weight of 24 nursery pigs.

The data presented in this study provide evidence that sub-populations present in the normal pig production in relation to birth weight and weaning age can behave differently and may be overlooked when analysis is performed only at an overall population level. These data indicate that supplementation early in life has the most effect when the weaning age is higher, and provides evidence that all pigs responded positively to Tonisity Px administration, which adds up to a significant advantage to the producer.

Overall, administration of Tonisity Px from days 2-8 of life and at weaning provide an economically-important advantage to the pork producer by increasing the number of pigs weaned and their subsequent growth performance.

Reference

1. Torow N. et al. The neonatal window of opportunity: setting the stage for life-long host microbial interaction and immune homeostasis. *Journal of Immunology* 2017;198:557-563

The effects of administering Tonisity Px™ isotonic protein drink to piglets on gut microbiota as assessed through 16S rRNA sequencing

S. G. Buzoianu[1], DVM, PhD; C. Putrino[2]; A. M. Firth[1], DVM, DACVECC
[1]Tonisity International, Dublin, Ireland; [2]California Polytechnic State University, San Luis Obispo, California

Introduction

Tonisity Px is an isotonic protein drink which is administered to piglets from as early as day 2 of life. When administered to piglets between days 2 and 8 of life and peri-weaning, Tonisity Px has been shown to modulate intestinal development by increasing villus height. Similarly, administration of Tonisity Px to piglets has resulted in reduced pre-weaning mortality and improved post-weaning growth performance.[1,2]

Objective

Since the intestinal microbiota in piglets has been shown to be an important factor affecting the life-long production and health of pigs, the effects of Tonisity Px on intestinal microbiota were studied in the current trial.

The aim of this study was to assess whether the gut-modulating effects of Tonisity Px were linked to changes in intestinal microbial populations and whether any effects observed on intestinal microbiota are stable over time.

Materials and methods

The test ingredient was an isotonic protein solution (Tonisity Px™), which provides easily-absorbable nutrients (glucose, amino acids, and peptides) and electrolytes that can be used directly by the enterocytes.

15 gilts (Yorkshire × Landrace × Duroc) and their litters (161 piglets) were enrolled in the study which was performed in a dedicated high-health research facility. Litters were allocated to one of two study groups, Control or supplementation with Tonisity Px (TPX). Control litters received no supplementation during the pre-weaning phase, and were given creep feed from weaning up until two days after weaning. From days 2-8 of age, TPX litters were supplemented with 500 mL of 3% Tonisity Px solution once daily in an open pan. TPX litters also received the 3% solution 3 and 2 days before weaning followed by a gruel mixture of creep feed and the 3% protein solution from 1 day before weaning up until 2 days after weaning. Both groups had ad lib access to normal feed after weaning. Weaning was at ~ 21 days. A total of 56 piglets were randomly selected from each litter for euthanasia at 9 and 17 days of age and a further 20 piglets were randomly selected from each treatment group at 30 days of age.

Two pigs/litter were randomly selected from the Control and TPX groups at days 9 (n = 13 vs 12), 17 (n = 13 vs 14) and 30 (n = 10 vs 8) of the trial. Fecal samples were collected for 16S rRNA gene sequencing. 16S rRNA gene sequencing was performed using the MiSeq (Illumina) platform after PCR amplification of the 16S v4 region. Data were normalised to the lowest number of reads per pig (3157), as the rarefaction curves revealed a plateau after this point.

Statistical analysis

Data were analyzed using a generalized mixed model within the GLIMMIX procedure in SAS 9.4 using a binomial distribution. Treatment, age and their interaction were included as main effects in the model. Body weight at euthanasia was included as a covariate and the litter origin was included as a random effect in the model. Relative abundance of microbiota was calculated as the number of reads for individual taxa divided by the total number of reads per sample. Data are presented as means and standard errors of the mean (SEM). Values of $P \le 0.05$ were considered statistically significant, while $0.05 < P \le 0.10$ were considered a near-significant trend. Since data were non-normally distributed, they were normalised by logit transformation within the model to facilitate analysis and then transformed back to the original data scale.

Six pigs from the same litter in the TPX group received antibiotics during the trial and were therefore removed from the analysis, as antibiotics are known to severely alter the intestinal microbial population and could skew the data.

Initially, a phylum level analysis was performed for the 9 main phyla that accounted for > 90% of the sequences from this study. The analysis also showed that a core of 23 genera with a relative abundance of 0.5% comprised the majority of reads analyzed over days 9, 17 and 30. These amounted to 91.2%, 87.3%, 86.0% of the total abundance of genera uncovered in these samples at days 9, 17 and 30, respectively and represent the main bacterial groups that are present in all pigs. A genus-level analysis was then performed to reveal any changes within bacteria that have established roles in the intestine. Genera such as known potentially-pathogenic bacteria (e.g., *Escherichia*, *Helicobacter*, *Streptococcus*) and known beneficial bacteria (e.g., *Lactobacillus* and *Bifidobacterium*) were also included in the analysis.

Results and discussion

Both treatment groups had a similar richness and diversity of bacterial population. Age had a significant effect on all bacterial populations studied and therefore data were stratified by age to investigate treatment differences at each time point. The microbiome at Day 30 of age (which was post-weaning) showed distinct shifts in population compared to days 9 and 17 of age (which was pre-weaning) population. This was most likely due to the fact that solid feed had been introduced at the time of weaning and this has been shown to severely affect the microbial balance around the weaning transition.[3]

The phylum-level analysis correlated to what was found at the genus level analysis and therefore the results presented here are at the genus level. The results are presented based on the potential role of the bacteria in the gut, as some genera are considered beneficial bacteria (such as *Lactobacillus*), while others are potentially-pathogenic (such as *Escherichia*). Some bacterial genera (such as *Prevotella*) are known to have variable effects, depending on species, sub-species and the intestinal conditions. The bacterial groups analyzed are categorized in Table 1 with a representative species in parentheses.

Tonisity Px effects on beneficial bacteria

Lactobacillus - Tonisity Px administered at days 2-8 of life resulted in a highly significant 3.5-fold increase in *Lactobacillus* at day 9 of life compared to control (4.0% vs 1.2%, respectively; $P = 0.0001$), amounting to a 233% increase (Figure 1). The *Lactobacillus* are considered indicator beneficial gut bacteria and are linked to improved gut maturation and digestion.[4-6] The presence of lactobacilli in the gut has been shown to provide the host with advantages such as fighting off potential pathogens and modulating the immune response.[5-7] Furthermore, lactobacilli have been shown to also improve pig growth when administered as probiotics.[7,8]

In the current trial, the availability of the Tonisity Px as a rapidly-digestible ingredient in the intestine could have stimulated the lactobacilli to proliferate, since these bacteria are specialised in using simple carbohydrates.[5] A higher number of intestinal lactobacilli have been linked to a reduction in *E. coli*[9] and the increase in the *Lactobacillus* genus observed in this study could be linked to the lower frequency of *E. coli* cultured from the ileum of these pigs (Putrino et al, AASV 2019 poster).

Bacteroides are another genus of bacteria considered to be beneficial in the intestine and were significantly more abundant at day 17 in the TPX compared to the control group (8.5% vs 4.3%; $P < 0.05$). The increase in *Bacteroides* observed at day 17 was most likely due to their reciprocal relationship with *Lactobacillus*. While both *Lactobacillus* and *Bacteroides* are known for their beneficial effects on the host, their abundance has been shown to be negatively correlated,[10] likely due to competition for specific nutrients. Therefore, at day 17, the *Bacteroides* may have occupied the niche that was occupied by *Lactobacillus* at day 9. Nevertheless, a higher population of *Bacteroides* in the intestine is considered beneficial, since they have been shown to improve gut function, and to modulate the host immune system.[11,12]

A numerical reduction in *Bacteroides* in the TPX group compared to Control (9.6% vs 14.8%; $P > 0.10$) was observed at day 9, possibly due to the above-mentioned reciprocal relationship with *Lactobacillus* although it did not reach statistical significance (Figure 2). Since no detrimental effects are associated with lower intestinal *Bacteroides*,[11] the numerical decrease at day 9 is not believed to negatively affect the host, especially as this occurred in conjunction with an increase in the abundance of the beneficial *Lactobacillus*.

Ruminococcus, Oscillospira, **and *Veillonella*** are all less well-known species of beneficial bacteria. All of them were significantly more abundant in the TPX group at day 9. *Ruminococcus*

Table 1: Overview of the changes in bacterial populations in response to Tonisity Px administration

	Beneficial bacteria	Variable-role bacteria	Potentially pathogenic bacteria
Increased in TPX group	*Lactobacillus (L. acidophilus)* *Bacteroides (B. fragilis)* *Ruminococcus (R. flavefaciens)* *Oscillospira (O. guillermondii)* *Veillonella (V. parvula)* *Romboutsia*	*Streptococcus (S. salivarius subsp. thermophilus)*	*Actinobacillus (A. pleuropneumoniae)*
Decreased in TPX group		*Prevotellaceae (P. ruminicola)*	*Actinobacillus (A. pleuropneumoniae)* *Clostridiales (C. perfringens)* *Helicobacter (H. pylori)* *Spirochaetaceae (Brachyspira hyodisenteriae)*
No Change in TPX group	*Bifidobacterium (B. longum)*		*Escherichia (E. coli)* *Salmonella (S. typhimurium)*

Figure 1: Relative abundance of *Lactobacillus* in pigs receiving Tonisity Px between days 2 and 8 of life

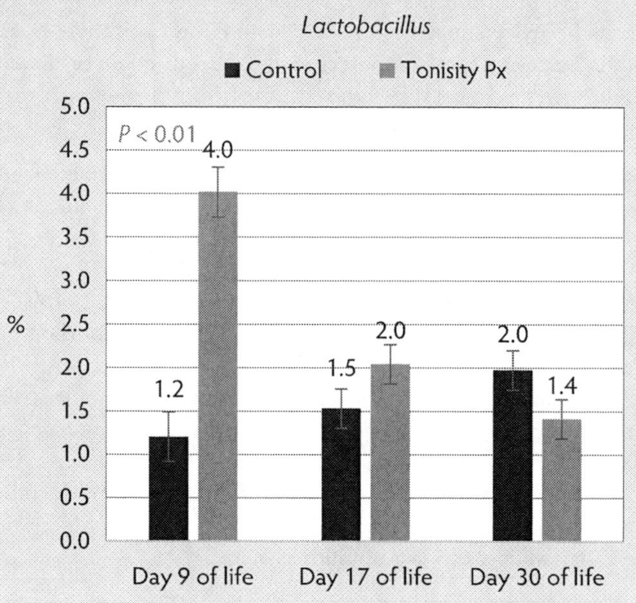

Figure 2: Relative abundance of *Bacteroides* in pigs receiving Tonisity Px between days 2 and 8 of life

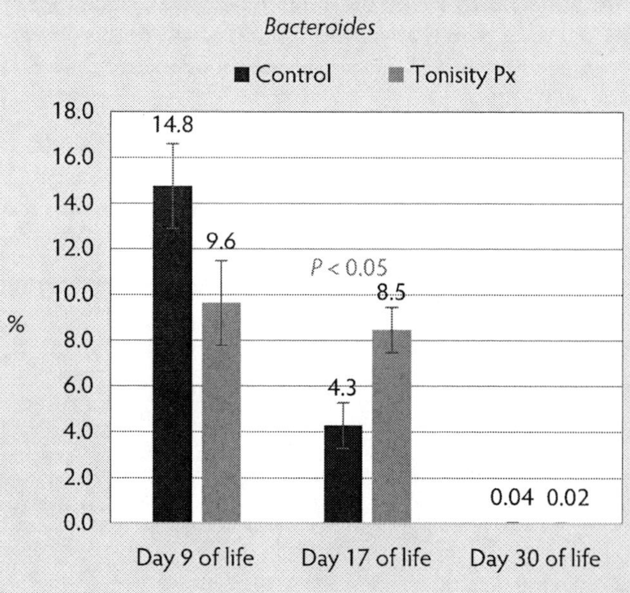

($P = 0.08$) and *Veillonella* ($P < 0.01$) were both increased at day 30. These changes indicate that the effect of Tonisity Px persists post-weaning, in relation to improving fiber fermentation (*Ruminococcus*) and immunomodulation (*Veillonella*).

Tonisity Px effects on bacteria with variable roles

Prevotellaceae are bacteria with variable roles, depending on the strain, but they comprise 30-40% of all bacteria in the intestine. As a family, the *Prevotellaceae* were significantly lower in the TPX group compared to Control at days 9 and 17 (20.5% and 27.6% vs 29.0% and 37.0%, respectively; $P < 0.01$). These differences represent a 29% reduction at day 9 and a 25% reduction at day 17. No differences were observed at day 30 of life (Figure 3).

Members of the *Prevotellaceae* family are generally carbohydrate fermenters in the intestine and can use both ammonia and peptides as nitrogen sources.[13,14] Some *Prevotella* have been associated with intestinal disturbances due to weaning stress in pigs[9] and are known intestinal mucus degraders,[10] which could link them to intestinal pathology. The large decreases in *Prevotellaceae* in the preweaning phase further support the improvement in digestion due to administration of Tonisity Px in pigs. However, the *Prevotellaceae* also have a role as beneficial fiber fermenters post-weaning, when fiber substrate is available through the diet. The lack of a difference at day 30 of life indicates that the fiber degrading capacity of the pigs in the TPX group was not affected. This further indicates that Tonisity Px allows the intestinal microbiota to adapt their abundance in accordance to nutrient availability and indicates that Tonisity Px does not have a detrimental effect long-term on fiber fermenters.

Streptococcus is a bacterial genus which contains species with variable roles. Their ability to cause pathology or act as beneficial bacteria is strain-specific. While some *Streptococcus* are known pathogens, intestinal streptococci have been used as probiotics in human and pig nutrition and are one of the most consumed probiotics worldwide.[7,8,15,16] A higher population of *Streptococcus* was found in the TPX pigs at day 30 (1.5 vs 0.2%; $P < 0.01$) and a tendency for higher *Streptococcus* was observed at day 9 (0.5% vs 0.4%; $P = 0.08$). The streptococci have also been shown to have beneficial immune-stimulating effects and are normal inhabitants of the digestive tract of pigs.[17,18] Resident intestinal streptococci do not seem to be associated with pathology as it is known that the streptococci from the respiratory tract and tonsils are the strains most likely to cause pathology.[19] Since the microbiota sequencing only classified the bacteria to genus level, there was no information as to the identity at a species level. However, the differences observed in this study did not translate to any pathology present in these pigs, either respiratory or intestinal.

Tonisity Px effects on potentially pathogenic bacteria

Actinobacillus tended to be lower in the TPX group compared to Control at day 9 (1.0% vs 2.2%; $P = 0.09$). However, the abundance of *Actinobacillus* was significantly higher at day 30 in the TPX when compared to Control (1.2% vs 0.4%; $P < 0.01$).

Figure 3: Relative abundance of the *Prevotellaceae* family in pigs receiving Tonisity Px between days 2 and 8 of life

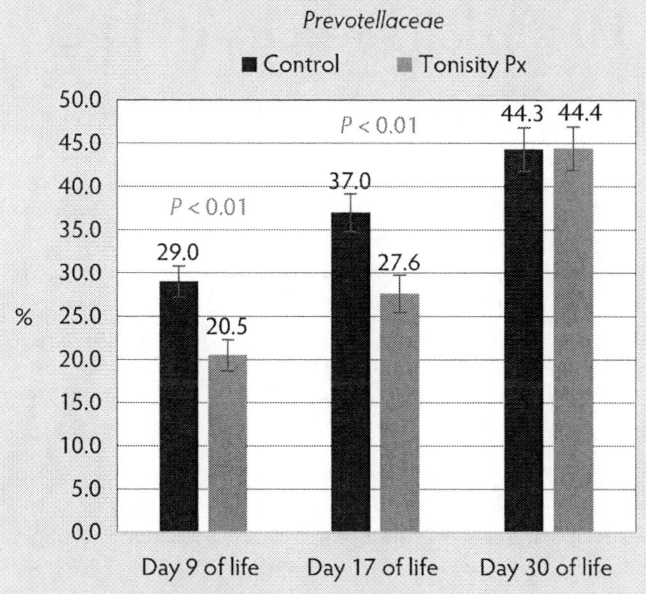

The *Actinobacillus* (*Actinobacillus pleuropneumoniae*) are mainly known for their implication in respiratory pathology in pigs, and their role in the intestine is yet to be established. However, in spite of the increase observed at day 30 of life, no respiratory pathology was observed in any of the pigs of this trial.

The ***Clostridiales UncCl291*** tended to be lower at day 17 in the TPX group compared to Control (0.4% vs 0.8%; $P = 0.10$; data not shown). The odds of detecting ***Helicobacter*** in the Control group at day 9 were 4.3 times higher than in the TPX group ($P = 0.08$), with no further differences observed. The family ***Spirochaetaceae*** which includes the known pig pathogen ***Brachyspira hyodysenteriae*** was found to be lower in the TPX group compared to Control at day 9 (0.04% vs 0.15%; $P < 0.05$). No members of the genus ***Salmonella*** were detected in any of the samples.

Escherichia is a potentially pathogenic genus, but no treatment differences were observed in the microbial sequencing analysis at the 3 time points. Overall average incidence of *Escherichia* was 0.6% for both treatment groups (SEM = 0.27%). While the pigs from the TPX group had a lower frequency of *E. coli* in the culture-based analyses, this was not reflected in the microbial sequencing analysis. This is likely due to the fact that the sequencing only allowed assignment of the data to the genus level and did not provide any information below this, at the species/subspecies level.

Conclusion

Since the pressure is high to reduce antibiotics in animal production, it is important for producers to have alternative strategies for maintaining the intestinal health of their animals. Tonisity Px has been shown previously to reduce mortality and to improve growth performance in pigs, and the promotion of a more beneficial microbial profile in the intestine provides evidence of further benefits.

The results from the current study indicate that administration of Tonisity Px between days 2 and 8 of life improved the intestinal microbial profile in piglets, by increasing the abundance of beneficial bacterial populations (*Lactobacillus, Bacteroides, Veilonella, Oscillospira* and *Ruminococcus*) and reducing potentially-pathogenic bacterial populations (*Clostridiales, Helicobacter, Spirochaetaceae*) and bacteria with variable roles (*Prevotellaceae*).

These differences show that Tonisity Px has the ability to influence some of the major intestinal genera as well as known beneficial and pathogenic groups in the pig intestine. The protection provided by the *Lactobacillus* and *Bacteroides* at days 9 and 17 of life, respectively is likely to provide a better gut environment for weaning, which is a major stressor in a pig's life. While the changes observed at days 9 and 17 of life could be linked to nutritional effects of the Tonisity Px and occur in well-known beneficial populations, the post-weaning changes, at a time of great restructuring for the gut microbiota are also worth noting. The increase and recovery of fibre-fermenting populations, along with the increase in the *Veillonella*, a potential immune modulator, would be of interest for the long-term health of the intestine.

Summary

The administration of Tonisity Px had a probiotic effect in this study. The beneficial changes observed in genera from the current study are generally linked to better overall intestinal health and immune modulation, reduced inflammation and decreased intestinal pathology. Potentially pathogenic bacteria were shown to decrease in the current study.

Such changes could provide long lasting benefits, as it is known that a window of opportunity to modify gut physiology and immunology exists in early life. Development of a healthy microbiome in early life can modulate the immune system and provide lifelong benefits.[20,21]

References

References available upon request.

Benefits of live attenuated influenza vaccination in neonatal commercial pigs, observed after challenge with contemporary IAV-S at 10 and 15 weeks of age

C. Goodell[1], DVM, MS, PhD, DACVPM; S. Playter[1]; B. Fergen[1], PhD; R. Plazola[2], PhD
[1]Boehringer Ingelheim Vetmedica, Inc., Duluth, Georgia; [2]Iowa State University, Ames, Iowa

Introduction

Influenza A virus in swine (IAV-S) is an acute respiratory disease which causes substantial economic losses in pig production. With the advent of a novel commercial live attenuated influenza A vaccine (LAIV), Ingelvac Provenza™, this study set out to determine benefits to piglet performance (average daily weight gain [ADWG]) in pigs vaccinated between 3-6 days in the face of maternal antibodies (mAb) with Ingelvac Provenza™ and challenged with a contemporary heterologous IAV-S at 10 or 15 weeks of age, compared to unvaccinated cohorts.

Materials and methods

A non-IAV-S shedding sow farm was identified and routine testing confirmed the absence of IAV-S. 44 dams in 2 farrowing rooms were balanced (parity and litter birthdate). All dams of pigs used in this study were vaccinated at 5 and 3 weeks pre-farrowing with FluSure XP (Zoetis, Kalamazoo, MI) to ensure IAV-S mAb were present at birth. Treatment was assigned by farrowing room, where all pigs in one room were vaccinated with Ingelvac Provenza™ at 3-5 days of age, and no IAV-S vaccination occurred in the second. 210 pigs of each treatment were allocated to the study, and from those, ~120 were challenged with H1N1γ (A/swine/Illinois/A01158996/2013) at 10 weeks of age, and the remaining ~300 were challenged with the same H1N1γ at 15 weeks of age. Between weaning and challenge, pigs were observed daily and monitored by oral fluids and serum weekly to ensure negative status for IAV-S and PRRSV. Prior to each study, samples were tested for homologous HI to the H1N1γ challenge virus. Pigs were allocated to the week 10 challenge using weight at D-8. The week 15 challenge pigs were reweighed at D-1 to the week 15 challenge, for the starting weight. D0 was the day of challenge and weekly weights were taken at D7, D14 and termination of the project. The week 10 challenge dose was 2mL, 10^6 $TCID_{50}$ IT followed by 2mL administered IN. The week 15 challenge dose was increased to 5mL IT and 2mL IN inoculation. Statistical analysis was done by ANOVA using SAS (Cary, NC, USA) to identify differences by days post-challenge, and LSMean values were reported.

Results and discussion

LS Mean S/N for IAV-S antibodies confirmed maternal antibodies in piglets prior to vaccination with Ingelvac Provenza at 3-6 days of age (Figure 1). Piglet weights at allocation and weaning were similar. Median weights at start and finish of each challenge are reported in Table 1. Post-challenge, ADWG was significantly different, with vaccinates gaining 0.36 lbs more per day through week one ($P < 0.001$) and 0.27 lbs more per day through week 2 ($P \leq 0.003$), than non-vaccinates, even with vaccinates starting at a lower average weight (Figure 2). In the 15 week challenge, starting mean weights were similar (144.42 lbs and 144.08 lbs). ADWG post-challenge was significantly different, with vaccinates gaining a mean 0.13 lbs more per day from day -1 through D14 (end of study) ($P \leq 0.04$) (Figure 3). There was less variability among the vaccinated pigs at 14 days post-challenge vs. non-vaccinated pigs.

In conclusion, in this study maternal antibody positive pigs vaccinated at processing age with Ingelvac Provenza™ and challenged at either 10 or 15 weeks of age had significantly greater ADWG, and less variability, than non-vaccinated cohorts.

Figure 1: IAV-S NP ELISA by treatment group over time. The figure shows data from processing age through the end of both the 10 and 15 week challenge studies. Mab detected in young pigs declined over time (positive = S/N < 0.60). Note the rapid and robust immune response difference between vaccinated and non-vaccinated animals 7d post challenge in both groups.

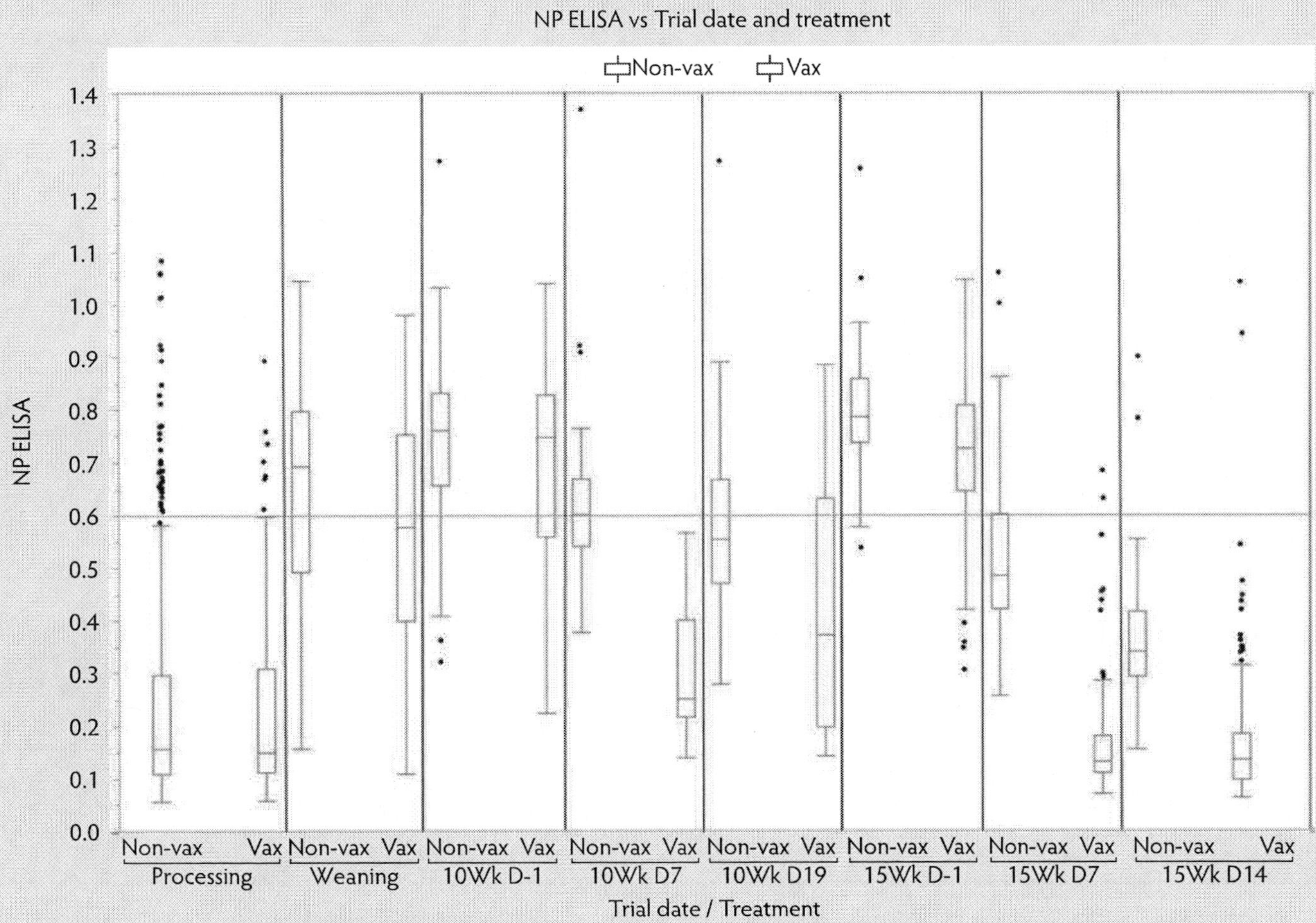

Table 1: Median start and end weight for each challenge study.

	Week 10 challenge†		Week 15 challenge	
	~63 doa Median start wt (lb)	**~90 doa Median end wt (lb)**	**~105 doa Median start wt (lb)**	**~120 doa Median end wt (lb)**
Non- vaccinated	65.20	112.65	146.00	172.00
Vaccinated	58.55	113.25	146.00	172.50

† 10wk study was used to pilot challenge model (IT and IN), and reduce the overall variability in pig weights for the 15wk study. Therefore, median weight started significantly different between the treatment groups, and distribution of weights was greater for the10wk challenge.

Figure 2: 10wk challenge: Weekly ADG of non-vx (blue) and vaccinates (red) from D-8 through D19 (end of study) post challenge. Box and whisker plot is representative of raw data.

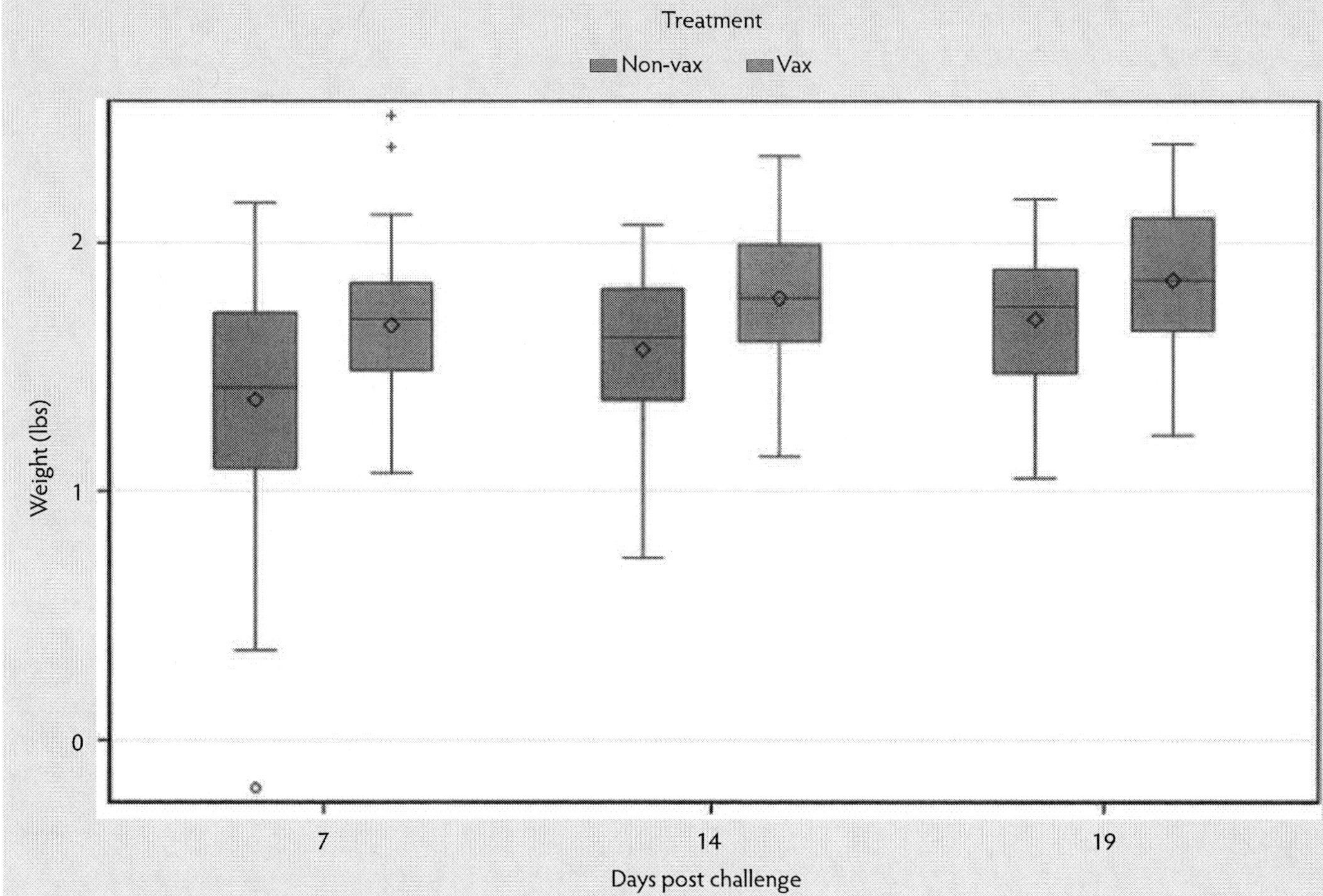

Figure 3: 15wk challenge: Weekly ADG of non-vx (blue) and vaccinates (red) from D-1 through D14 (end of study) post challenge. Box and whisker plot is representative of raw data.

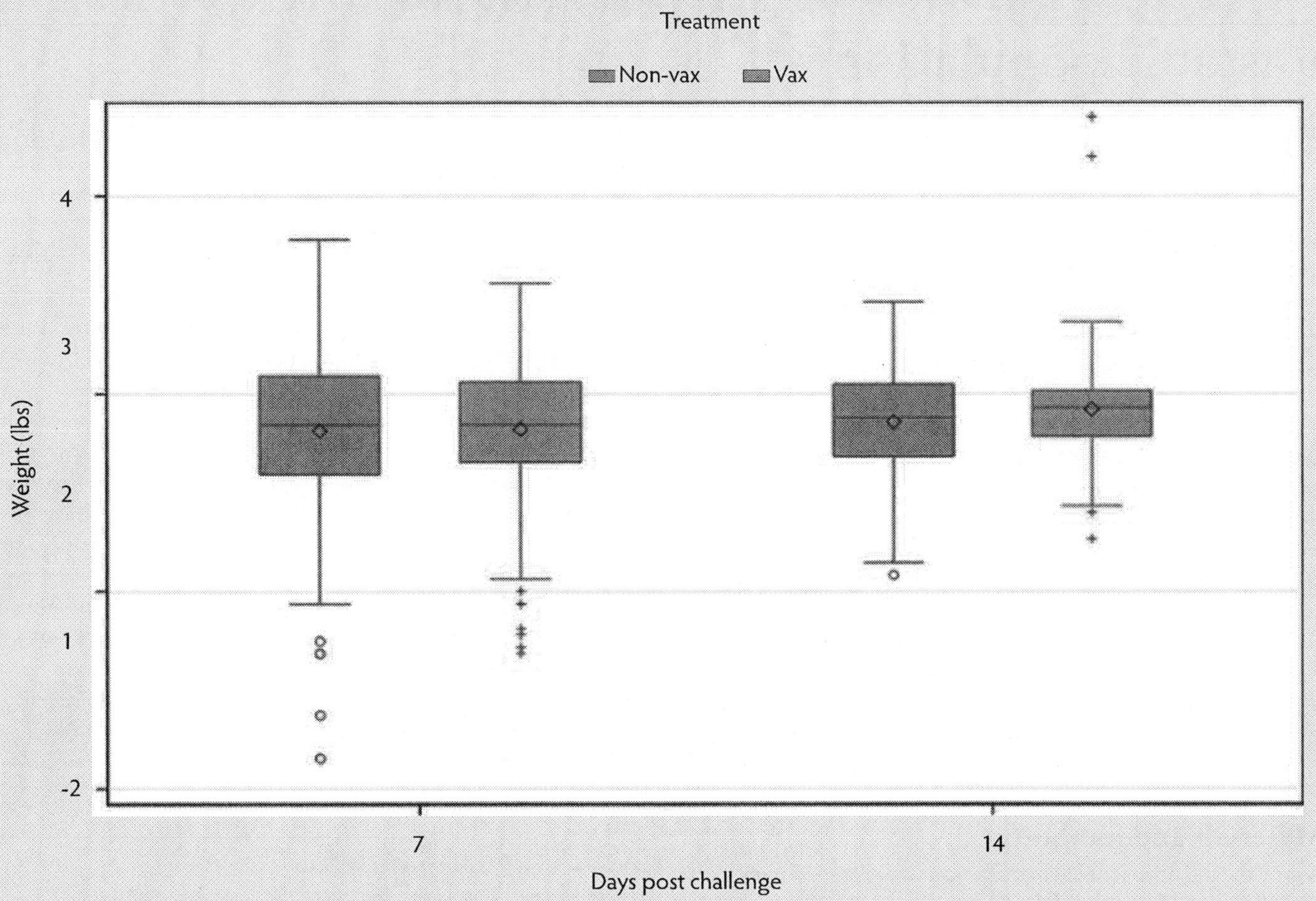

The impact of vaccination with Ingelvac Provenza™ against IAV-S on mortality in a known PRRS-unstable population

Cara Dykhuis Haden[1], DVM; Brent Sexton[1], DVM; Kellie Cicconi-Hogan[2], PhD; Christa Goodell[2], DVM, PhD
[1]Pipestone Systems, Pipestone, Minnesota; [2]Boehringer Ingelheim Vetmedica, Inc., Duluth, Georgia

Introduction

Influenza A virus in swine (IAV-S) is significant and costly to the swine industry, especially when paired with additional respiratory pathogens. Previous work estimated that the cost of uncomplicated IAV-S was more than $3.00 per pig, and when present with additional respiratory infections, cost was more than $10.00 per pig.[1] Recent data suggests losses from respiratory coinfections with IAV-S can reach $18.00 per pig.[2] By preventing IAV-S, the impact of respiratory co-infections on pig health and productivity can be reduced.[1,3] Until recently, there have been limited commercially available vaccines to provide protection against IAV-S. Ingelvac Provenza™ (Boehringer Ingelheim Vetmedica, St. Joseph, MO, USA) is an intranasally administered live attenuated influenza vaccine (LAIV) that can be given to pigs as early as one day of age. The objective of this study was to assess the value of Ingelvac Provenza™ in pigs co-infected with PRRSV and IAV-S in the nursery, compared to pigs not vaccinated for IAV-S in a known PRRS unstable population.

Materials and methods

A growing pig flow, sourced from five commercial sow farms located within a 6 miles of each other in the upper Midwestern United States, was included in this study. Wean to finish mortality for this flow was typically 3.0% - 4.5% prior to vaccination. Historically, all pigs received a commercial PRRS MLV vaccination at weaning. Cough due to IAV-S typically began in September annually. All 5 farms were confirmed IAV-S positive via PCR from oral fluids (OF) collection in September of 2017. In November 2017, three of the commercial farms broke with abortions storms and classic respiratory PRRS. Despite implementing PRRSV stabilization by means of mass vaccination, farms continued to leak wild type PRRSV to downstream growing pigs, detected by PRRSV PCR from serum at weaning, and later, processing fluids. Pigs were placed sequentially from mid-December 2017 through February 2018, into Sites A through E. PRRSV 1-7-4 was detected at all sites throughout the study timeframe. The five wean to finish sites ranged from 4,451 to 17,853 pigs per site, where each site contained either 4 or 8 × 1000 head rooms. One site double stocked at weaned pig placement. Mortality data was collected weekly. Of the five sites, Sites A, B and D were left unvaccinated for IAV-S. Site C contained pigs vaccinated with Ingelvac Provenza™ at processing, while Site E was vaccinated with Ingelvac Provenza™ at weaning. Mortality was summarized by week in the nursery for each treatment group.

Results and discussion

There were 17,934 Ingelvac Provenza™ vaccinated and 31,113 non IAV-S vaccinated pigs enrolled in this study. Combined weekly mortality by IAV-S vaccination status is shown in Figure 1. By week 7, average cumulative nursery mortality for the unvaccinated

Figure 1: Cumulative weekly nursery mortality by week of production (%)

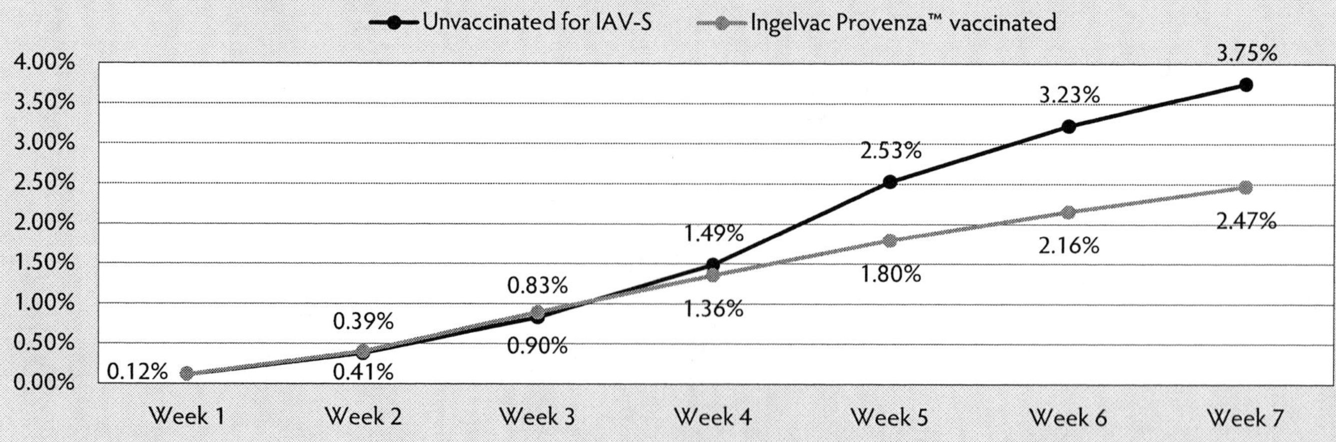

sites (Site A, B, D) was 3.75% compared to average cumulative mortality of 2.47% in Provenza™ vaccinated sites (Sites C and E). These results demonstrated that Ingelvac Provenza™ was a valuable tool to improve the health and performance of pigs co- infected with both IAV-S and PRRSV in the nursery phase.

References

1. Dyhkuis Haden C et al. Assessing production parameters and econmic impact of swine in influenza, PRRS and *Mycoplasma hyopnuemoniae* on finishing pigs in a large production system. *AASV*. 2012; 75-76.

2. Stika R et al. Measuring the respiratory pathogen burden in growing pigs to estimate the impact of disease. *Allen D. Leman Swine Conference*. 2018.

3. Vincent et al. A brief introduction to influenza A in swine. In: Spackman E, ed. *Methods in Molecular Biology: Animal Influenza Virus, Part III*. 2014; 243-258.

Approaching IAV-S control with a new unique tool

J. Morgan, DVM; E. Mondaca, DVM, PhD; E. Lowe, DVM; C. Goodell, DVM, MS, PhD, DACVPM; K.M. Cicconi-Hogan, PhD
Boehringer Ingelheim Vetmedica, Inc., Duluth, Georgia

Introduction

Influenza A virus in swine (IAV-S) is significant and costly to the swine industry, especially when paired with additional respiratory pathogens. Until recently, there have been few tools commercially available to provide protection against IAV-S or its lasting impact on health and performance in growing pigs. Ingelvac Provenza™ (Boehringer Ingelheim Vetmedica, St. Joseph, MO) is an intranasally administered live attenuated influenza vaccine (LAIV) that can be given to pigs as early as one day of age. This paper reports the results of two studies conducted to assess measures performance in growing pigs from three sow farms (10,000 total sows), within two different integrated operations in the Midwestern United States, before and after the introduction of Ingelvac Provenza.™

Materials and Methods

For both cases, all piglets of the enrolled farms received Ingelvac Provenza™ at processing (3-5 days of age). **Case 1:** Nursery and finishing closeouts of the growing pigs from a 5000 sow farm were obtained from January 2017-March 2018. From July 2017 through January 2018, pigs from this farm were vaccinated with Ingelvac Provenza.™ **Case 2:** This case study involves a two sow farm flow (5000 total sows). All piglets in both farms initiated Ingelvac Provenza™ vaccination starting July 2017, and maintained existing protocols of PRRSV, Mhp and PCV2 vaccination at weaning (approximately 20 days of age); data from January 2014 through May 2018 was obtained to determine performance mean and variance for each key performance indicator (KPI). For both cases, KPIs, such as mortality, average daily gain (ADG) and feed conversion, were entered into Statistical Process Control (SPC) software (Minitab 17 Statistical Software, State College, PA) to determine performance mean and variance for each KPI. Averages were calculated for time periods prior to and after the intervention of Ingelvac Provenza.™

Results and discussion

A process change was noted on the SPC charts for the intervention of Ingelvac Provenza™ for both cases. Several performance changes were observed throughout nursery and finishing. In case 1, a significant increase in ADG of 0.12 lbs/day was observed in the nursery, with the ADG reported as 0.81 lbs/day prior to the intervention, 0.93 lbs/day during vaccination ($P < 0.05$) (Figure 1, upper left). A significant reduction in nursery mortality was also observed, from 3.23% prior to vaccination to 2.27% after ($P < 0.05$) (Figure 1, lower left). In the finisher, there was a significant ADG increase, from 1.88 lbs/day before vaccination to 1.93 lbs/day after ($P < 0.05$) (Figure 1, upper right). An increase in the mortality of the finisher after vaccination can be observed (Figure 1, lower right), but is due in part to PRRSV activity on the farm (exclusion of those data points reduces the mortality rate by 0.15%). Similar changes in KPIs were noted in case 2 in both nursery and finishing. In the nursery phase, there was an increase in ADG of 0.08 lbs/day and a decrease in feed conversion of 0.03. There was also an increase in finisher ADG of 0.09 lbs/day in addition to a decrease in feed conversion of 0.02. SPC charting is a useful tool to visualize performance data, and allows easy notation to document interventions. These case studies observed measurable differences in pig performance KPIs after the initiation of Ingelvac Provenza™ in 3-5 day old pigs and suggest that standard KPI measures can be used to support field evaluation of Ingelvac Provenza.™

Figure 1: SPC charts of growing pigs' KPIs from Case1, by group closing date. Vertical lines indicate the date where Ingelvac Provenza™ vaccination was started in farrowed pigs. Upper charts represent average daily gain for nursery (left) and finish (right) while lower charts represent group mortality for nursery (left) and finish (right).

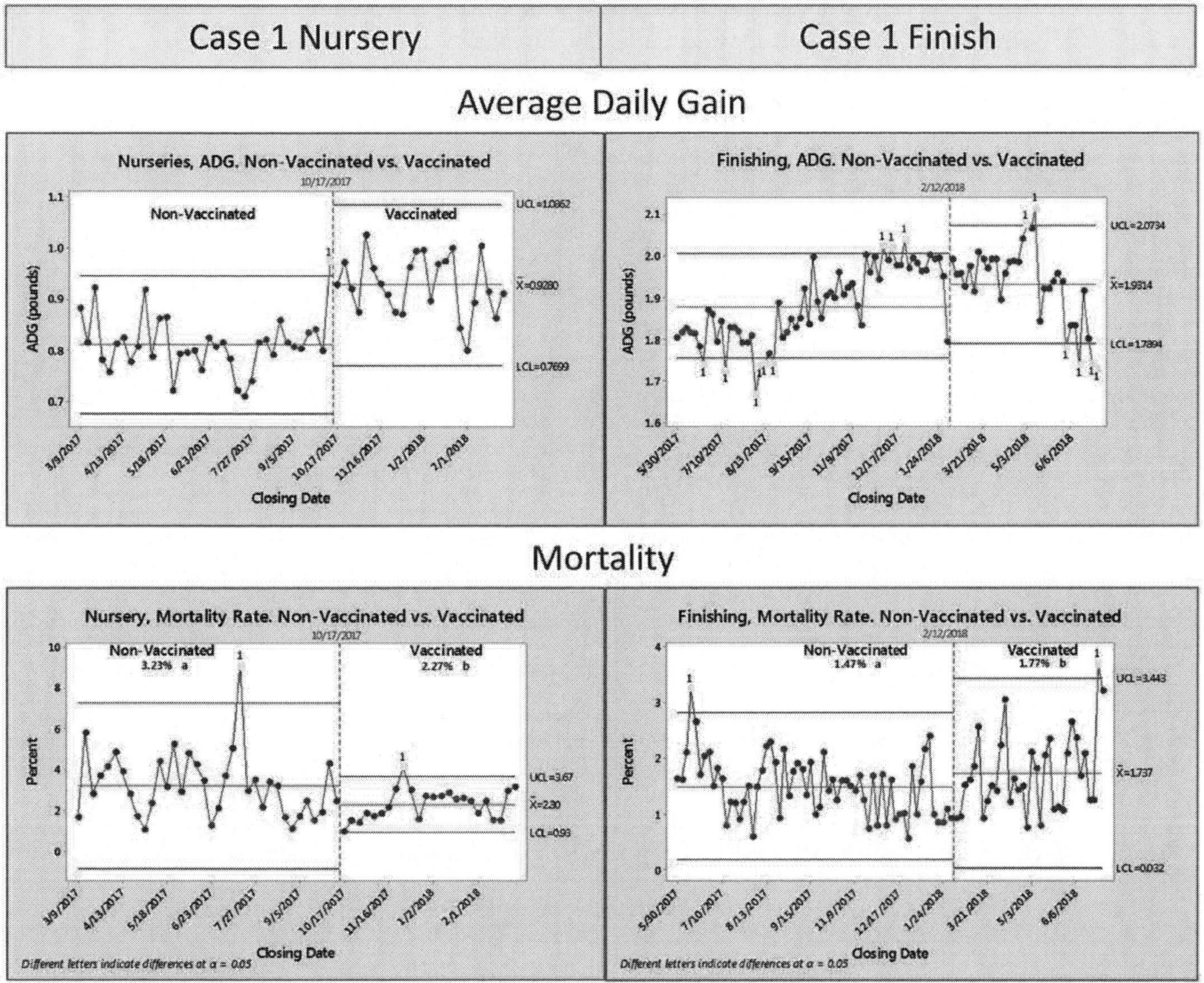

Determining optimum application length of organic acids in drinking water of nursery pigs

N.S. Ferguson[1]; M. Swantek[2]; N. de Groot[3]; D. Woods[1]
[1]Trouw Nutrition, Guelph, Ontario, Canada;[2]Trouw Nutrition USA, Highland, Illinois;
[3]Trouw Nutrition, Amersfoort, Utrecht, Netherlands

Introduction

The use of organic acids by feed or water in piglet performance has been well established, particularly in the first weeks after weaning. This is when the HCl production in the stomach of piglets is not yet sufficient for proper digestion. By decreasing the pH in the stomach, we can aid in this process. This trial also investigated multiple lengths of use of organic acid blends not only to determine the best performance but also the optimal economic returns when using the product.

This study was conducted using 768 weaned pigs (n = 12) allocated to 4 treatments. Treatments were control and then 1.5mL/L drinking water acidifier, Selko ProHydro®(PH) for 2 weeks (PH2), 3 weeks (PH3), and 4 weeks (PH4), ProHydro reduced water pH from 7.2-7.8 to 3.5-3.7. During the 42 day period, the animals were monitored for weight and feed intake was measured to calculate average daily gain, average daily feed intake, and feed conversion rate. The economic returns included the cost of the feed, the cost per kg of gain, the cost of the organic acid blend and the benefits of antibiotic reduction when using the product.

Water intake

Laboratory results confirmed the dosage level of the ProHydro to lower pH from 7.1-7.8 initially in the drinking water down to 3.5-3.7 during the trial period. ProHydro increased water intake by 48%, which is similar to the expected 38% increase observed from a meta-analysis of all ProHydro nursery studies (Figure 1).

Growth performance

During the 1st week and 3rd week there were significant benefits to using PH, but none were observed during Week 2. During week 1, PH was applied to all the PH treatments and therefore it is more appropriate to combine the effect of PH2, PH3 and PH4 when comparing with the No PH treatment. It was evident that the application of PH increased ADG (16%, $P = 0.026$), and ADFI (+7%, $P = 0.075$) while reducing FCR (-0.11, $P = 0.050$). Considering these were all low zinc diets, the addition of PH could be a strategy to replace or partially replace high ZnO in the diet immediately post weaning.

It is apparent from the data that over the first 3 weeks, PH had a significant positive effect on animal performance irrespective if provided for 2 or 3 weeks (Figure 2). This would suggest that providing PH for 2 weeks is sufficient to sustain better performances during Week 3. However, feed efficiency was numerically better when applying PH for 3 rather than 2 weeks (1.38 vs 1.41, respectively).

There were significant changes in overall FCR when combining PH treatments vs No PH, the results indicate a 0.03g/g reduction in FCR ($P = 0.046$) over the whole 6-week period, mainly due to the improvements in the first 3 weeks.

Figure 1: The average daily water intake (l/piglet/day) for the 28-day test period.

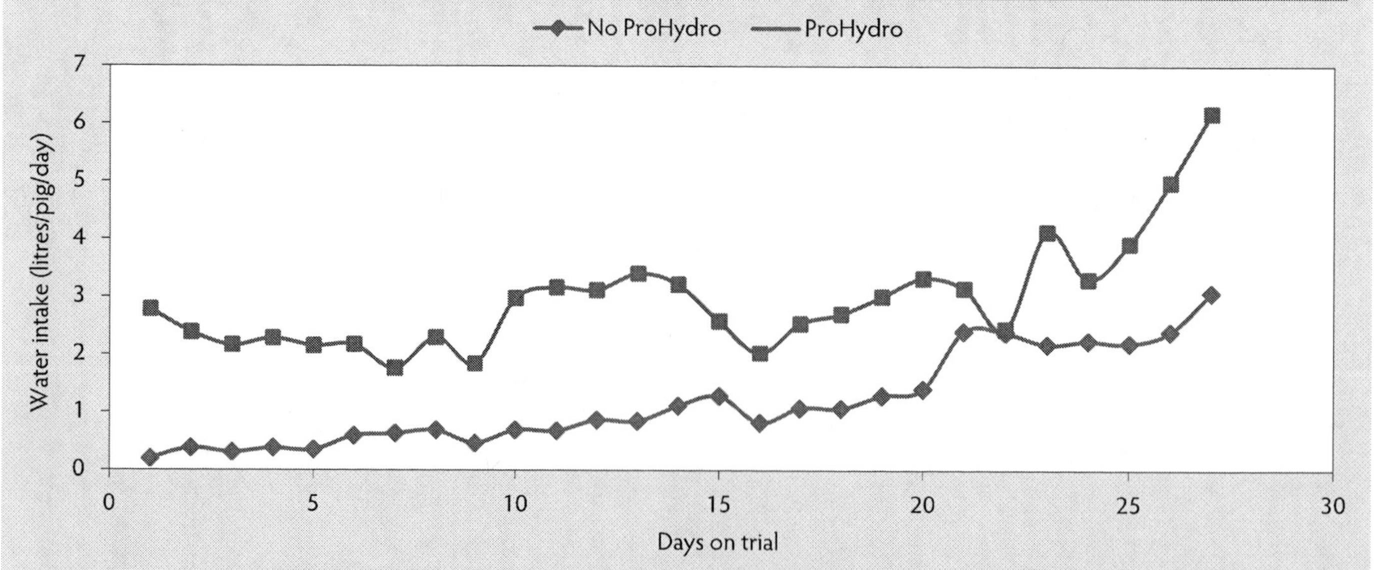

Figure 2: Effect of ProHydro on Average Daily Gain (ADG), Feed Intake (ADFI), Feed Conversion Ratio (FCR) and End weight for the cumulative period Week1-3.

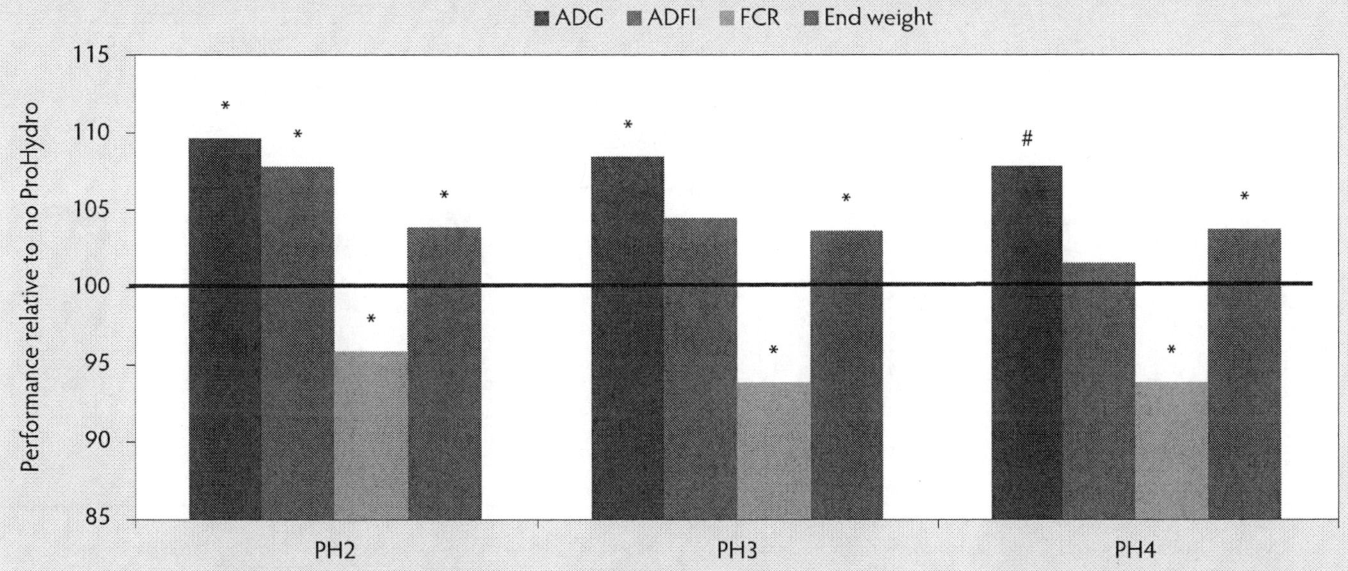

* = $P < 0.50$ versus No ProHydro treatment

From Figure 3 it is evident from Treatment PH2, that when PH was no longer provided (Week 3 onwards) FCR increased numerically. This did not occur with PH3 and PH4 treatments, suggesting PH should be available for a minimum of 3 weeks. The large increase in FCR during Week 3 is likely a result of the change of diet phases from Diet 2 to Diet 3, which occurred around Day 12, as well as some health challenges (increased diarrhea). It would appear that PH had a positive effect during these dietary and health challenges, especially PH3 and PH4.

Weight variation

There were no differences between treatments in the variation of body weight at the start of the trial, with an average within-pen weight CV of 15.7% and deviance from mean body weight of 0.81kg. However, by the end of the study, the average within-pen body weight variation was 14.6% and deviance from mean body weight was 2.7kg. However, there were significant treatment effects on body weight variation. For the most part PH had a positive effect on improving body homogeneity at the end of the

Figure 3: Changes in feed conversion (FCR) between ProHydro treatments over time.

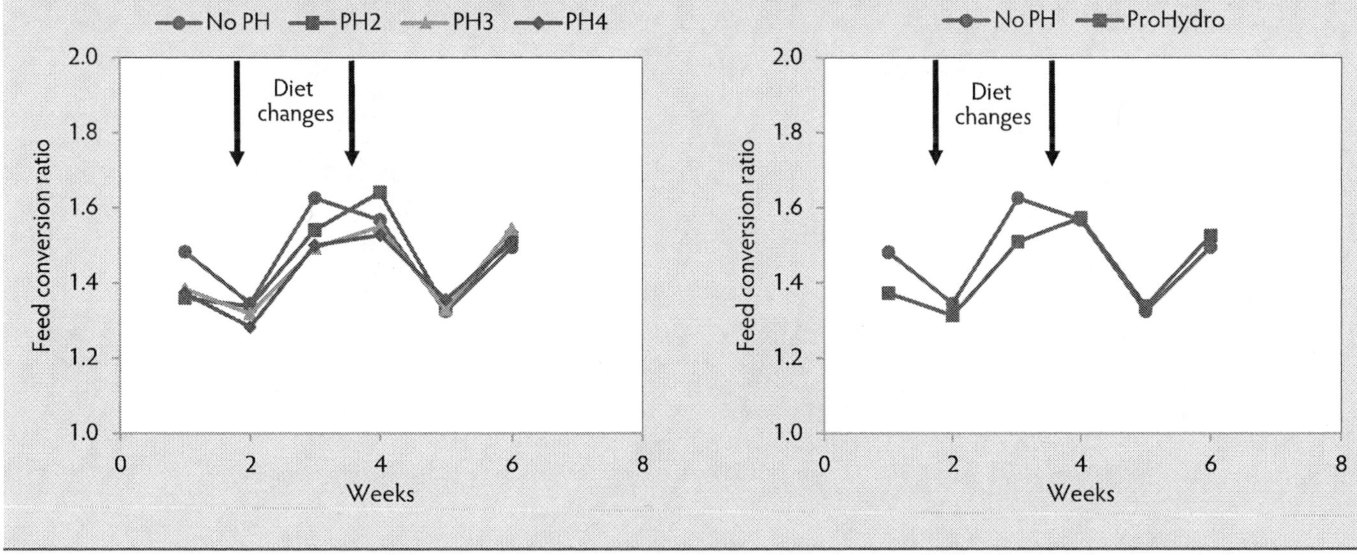

Figure 4: Effect of length of ProHydro application on variation in within-pen body weight and variation from mean body weight.

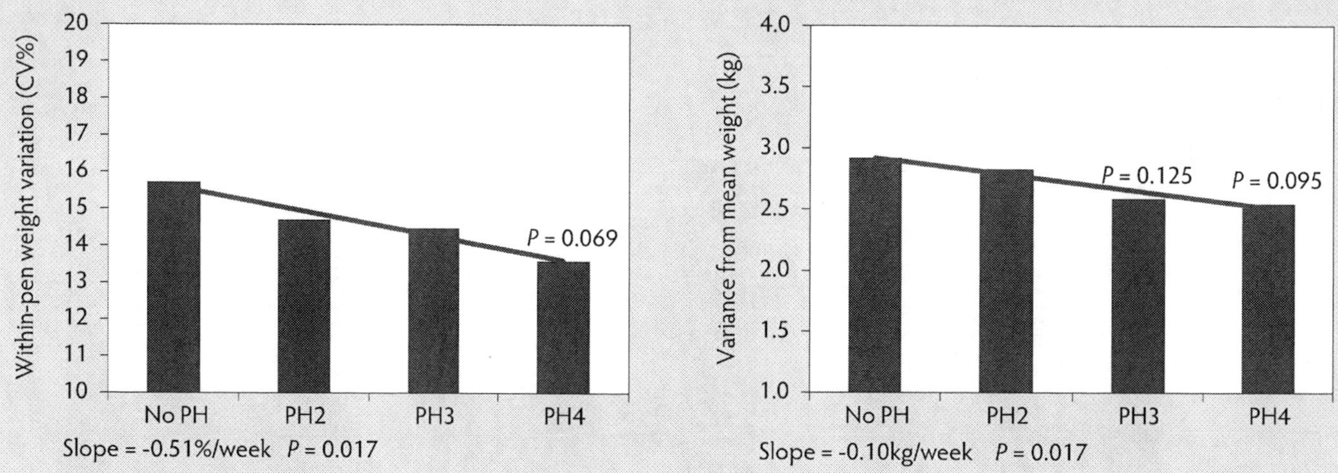

nursery, with a linear ($P < 0.05$) reduction in weight variation as the length of application increased (Figure 4).

Most of the improvement in variation was due to a reduction in the number of smaller/lighter pigs, where Small is defined as approximately 1 standard deviation less (<20kg) than mean weight (23.2kg) and Large as one standard deviation above the mean (>26kg) (Figure 5). The response appeared to be linearly related to the length of PH application.

Scouring

The cumulative (day 7, 14, 21 and 28) percentage of pens showing signs of diarrhea in this study was more than double the "normal" (14% vs 6%, respectively). No PH had the highest incidence with 19% of pens (especially Day 14), probably due to low dietary Zn and no water antibiotic medication. There was an observed increase in scours during Week 3, which could be associated with a change in diet or an endemic outbreak of enteric disease. There were numeric differences in the total number of pens with scours between PH treatments, with PH3 having the greatest effect on reducing scouring down to 8% or -44% reduction ($P = 0.15$) (Figure 6). It is important to note that in this study no water medication was applied in order to maintain treatment integrity rather than not being required. While the incidence of scouring increased during the first 3 weeks, there

Figure 5: Effect of ProHydro application on the proportion of different body weight categories.

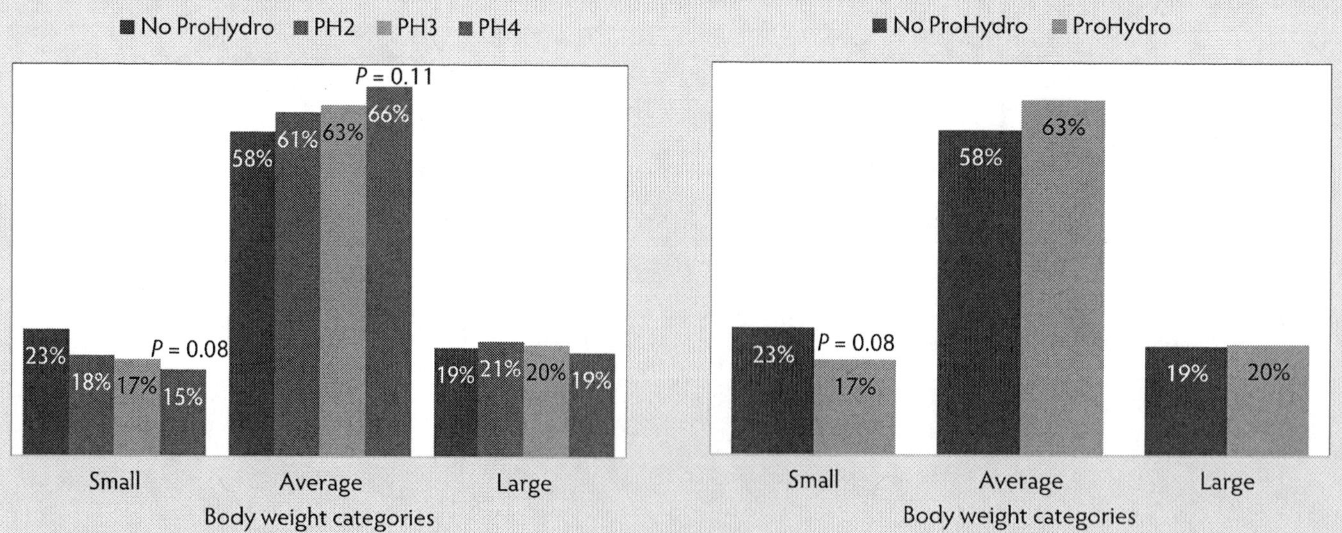

Figure 6: Effect of ProHydro application on the **cumulative** incidence of scouring, expressed as % pens showing scouring on Day 7, 14, 21 and 28 combined.

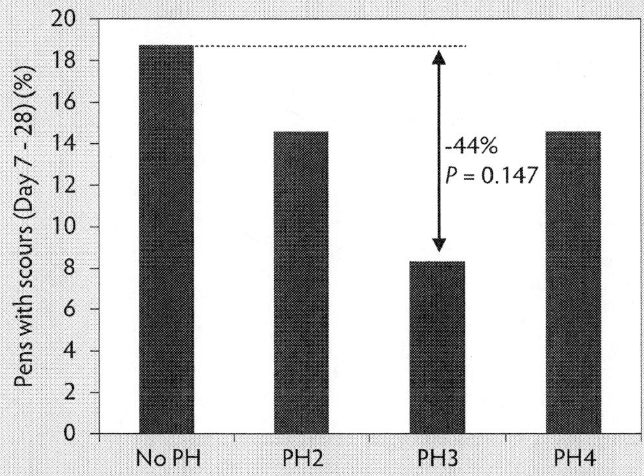

were no significant increases in mortalities or removals, especially those on PH for 3 weeks.

Feed costs

Similar to previous studies, the use of PH, particularly for 3 or 4 weeks, resulted in reduced corrected -24kg feed costs. However, the magnitude of savings was small (\$0.11-\$0.16/pig). The additional cost of PH causes all PH treatments to have higher feed + PH costs. However, with the slight improvement in end weights, only the application of PH for 4 weeks causes a significant increase in total feed + PH costs (\$0.40/pig, $P < 0.05$) and cost+PH/kg gain (\$0.024/kg, $P < 0.05$) (Figure 7). Regression responses show a trend for linear decrease in corrected-24kg feed costs with increasing length of applying PH (-\$0.042/week; $P = 0.17$), while Cost/kg gain tends to remain flat (-\$0.002/ week; $P = 0.21$). However, adding PH to these costs resulted in a linear increase in both Corrected-24kg feed costs (\$0.094/week; $P = 0.029$) and Cost+PH/kg (\$0.006/week; $P = 0.040$). From the data it was evident that applying PH for 3 weeks costs as much as 2 weeks, both of which are not significantly higher than providing no PH. It can therefore be deduced that the economic optimum length of applying ProHydro in the nursery is 2 to 3 weeks.

Potential water medication savings

No water medication for therapeutic usage was applied to maintain the integrity of the trial. A consequence being that the incidence of pens showing scouring more than doubled (14% vs 6%, respectively). For those not receiving PH, more than tripled (19% vs 6%). "Normal" is defined as the scouring incidence in pigs at the commercial facility after receiving high ZnO and two courses of water antibiotics (Trim-Sulfa@\$0.16/pig and Gentamycin @\$0.25/pig). Pigs on treatments PH2 and PH4 also had higher levels of diarrhea than "normal" (14% vs 6%), while pigs on PH3 had incidence levels close to "normal" (8% vs 6%, respectively) and significantly lower than No PH (8% vs 19%; $P = 0.15$). This suggests that the application of PH for 3 weeks could reduce the need for water-mediated antibiotic treatment for diarrhea and

Figure 7: Effect length of ProHydro application on feed costs (corrected to 24kg) and cost/kg gain with and without ProHydro costs for the whole 6-week period.

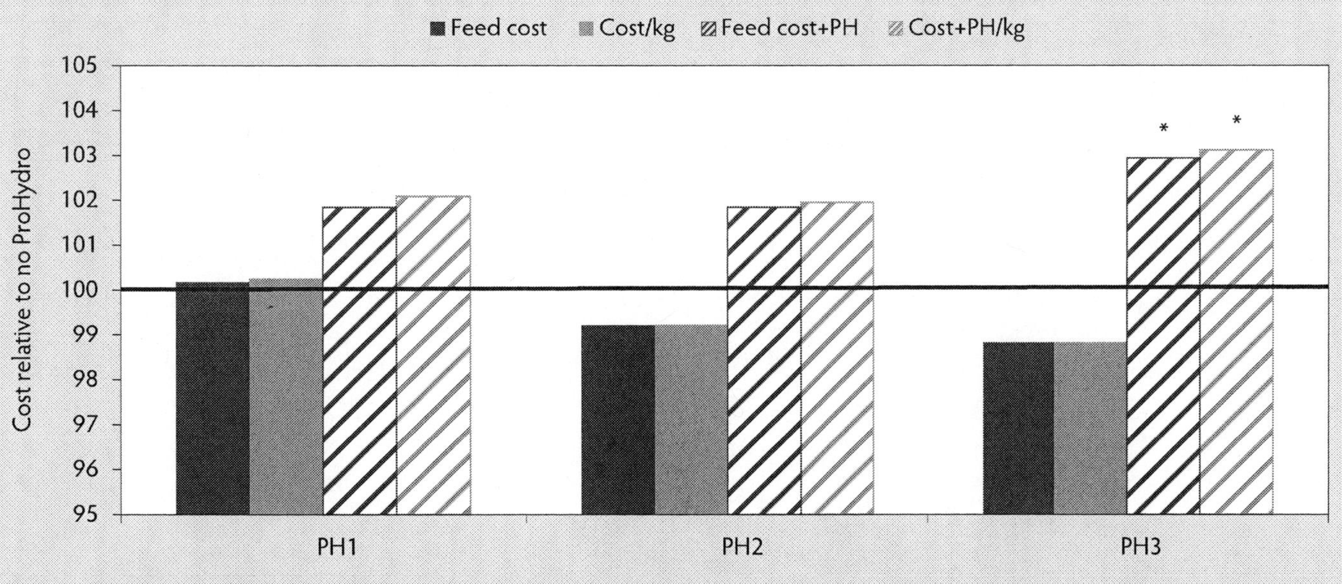

* – $P < 0.05$ versus No ProHydro

poor-doer pigs and therefore provide the producer with a net saving per pig of $0.06.

Conclusions

The application of ProHydro increased water intakes and, with diets containing low Zn, significantly improved feed intakes and growth rates in the first week post weaning. This positive effect continued through to the 3rd week post weaning but thereafter had no beneficial effect on performance, other than to reduce end of nursery weight variation. Given the higher costs associated with applying PH for 4 weeks and the lack of additional benefit compared to 3-week application, the recommended optimum length in a commercial nursery should be 3 weeks.

Evaluation of Pulmotil® AC (tilmicosin aqueous concentrate) use in lactating sows to improve herd health and productivity in herds with swine respiratory disease

E. B. Byers[1], DVM; J. L. Bargen[2], MSc; J. P. Gooding[2], MSc; J. M. Hammer[2], DVM
[1]Prestage Farms, Clinton North Carolina; [2]Elanco, Greenfield Iindiana

Introduction

Pulmotil® (tilmicosin phosphate) is approved for the use in swine at a rate of 181-363 grams per ton of complete feed through a veterinary feed directive for the control of swine respiratory disease (SRD) associated with *Actinobacillus pleuropneumoniae* and *Pasteurella multocida*. Previous studies of Pulmotil® fed to sows during lactation have been shown to improve piglet performance and subsequent nursery performance through reduced disease exposure and improved piglet health during lactation.[1-5] Pulmotil® fed at a rate of 363 grams/ton in lactation diets showed improvements of 0.5 pigs weaned/litter weaned, 0.44 pig/sow, 0.13 kg heavier weaning weight, and a 4 percentage unit improvement in pre-weaning mortality.[2] The latter study also reported an improvement in nursery mortality of 1.4 percentage units.[2] Another study demonstrated improved weaning weight (0.4 lb) with Pulmotil® fed at 181 grams/ton in lactation.[3] These studies substantiate the impact of lactation antibiotics, specifically Pulmotil®, on piglet health as demonstrated by improved productivity. Pulmotil® Aqueous Concentrate (PAC) is approved for the use in swine for control of SRD associated with *Pasteurella multocida* and *Haemophilus parasuis* and for the control of SRD associated with *Mycoplasma hyopneumoniae* in the presence of porcine reproductive and respiratory syndrome virus (PRRSV). However, the effect of administering PAC via drinking water to sows during lactation has not been reported.

Objectives

The goal of this study was to evaluate sow herd and nursery performance following PAC administration to lactating sows through drinking water as an alternative to feed-based delivery. The objective of the lactation program was to replicate the effects of a health upgrade at the commercial sow herd level as cited. In the previous studies, the positive effect of Pulmotil® administration during lactation appears to be dose-dependent. Higher doses in lactation affected pre-weaning mortality and weaning weight, but lower doses affected only weaning weight. To achieve both pre-weaning death loss improvement and weaning weight increases, the study evaluated 2 grams/sow/day, which is comparable to 363 grams/ton in feed. The objectives of the study were to evaluate the primary health upgrade impact of the lactation medication program as defined by breeding herd performance (key performance indicators cited below) and to evaluate subsequent nursery performance.

Materials and methods

Four farrow-to-wean farms (approximately 2,600 sows each) having a recent history of PRRSV infection and clinical SRD were enrolled. All sow farms sourced their replacement gilts from the same replacement gilt source and managed similarly. All farms had similar parity structure and were under common management and nutrition programs. The PRRSV outbreaks were determined to be in different stages of recovery based on date of initial outbreak and the prognosis of the herd veterinarian. Three treatment and three control groups consisting of approximately 100 sows per week per farm totaling of approximately 1,200 sows per treatment were included. A modified McREBEL program[6-9] limiting cross-fostering practices was established following PRRSV diagnosis on all farms and was continued throughout the duration of the study. Any piglet or sow movements were managed within treatment groups. An every-other-week treatment strategy was chosen to facilitate observations in the nursery on a weekly basis and to minimize farm effect in the nursery performance evaluation. Treatment groups received PAC via drinking water at a rate of 2 grams/sow/day beginning at stocking (typically 3 days prior to first expected farrowing) and continuing for the entire lactation period (average 20.7 ± 1.7 days). Farrowing rooms were treated independently, and stock solution was mixed daily to determine the appropriate intake of stock solution in medicated rooms. Control groups received fresh water only. Administration of additional individual sow and piglet treatments were via standard operating procedures. Weaned pigs from the four sow farms were commingled by site at nurseries consisting of eight hotel-style rooms that filled over a 2-week period (sow farm flow integrity was maintained each week by room). Each nursery was filled with 4 rooms of treatment and 4 rooms of control groups. Site design allowed for physical separation between treatment and control groups at each nursery, however, labor was shared between treatment and control groups.

On farm data management (Porcitec) was used to collect sow performance data by litter farrowed (i.e. pigs born alive, pigs weaned, mortality, stillborn, and mummies). Percent combined stillborn and mummies, percent gilts farrowed, and percent death loss (sum of deaths/sum of liveborn) were calculated from the data. A litter weaning weight on the entire farm and room was collected. A litter wean weight was calculated at the farm and room level using all the piglets weaned within the room (i.e. sum of total daily wean weight by each room/piglets weaned in the room). Wean weights were adjusted to standardize lactation length to 21 days and parity.[10,11] Key performance metrics evaluated for the first objective were pre-wean death loss and adjusted weaning weight. Treatments were assigned to the sow farm for the entire week and then switched the following week; therefore, the experimental units for this study were the combination of farm and week (4 farms across 6 weeks = 24 observations). Analyses were performed in SAS 9.4 (SAS Institute, Cary, NC) using a randomized, complete-block design with block defined as each two-week increment within farm.

The pre-wean performance variables were evaluated using generalized linear mixed models (PROC GLIMMIX) with farm and block within farm included as random effects and treatment (i.e. PAC vs. control) and potential covariates handled as fixed. Covariates tested were percent combined stillborn and mummies and percent of gilts, along with all possible interactions with treatment. The final models were developed using a backward selection procedure that retained covariates significant at $\alpha \leq 0.1$ level. Distributions were assumed binomial for death loss and normal for adjusted weaning weight. Effects of treatment across covariate percentile were described for significant outcomes. Percent combined stillborn and mummy and an interaction between treatment and percent combined stillborn and mummy was evaluated for covariation and retained in the final model when significance was $\alpha \leq 0.05$. Analysis of death loss assumed a binomial distribution; normal distribution was assumed for all other performance metrics. Nursery performance will be analyzed when the data becomes available.

Results

All farms were confirmed positive for PRRSV by polymerase chain reaction including ORF5 sequencing. The four farms were in different stages of outbreak progression during the study. PRRSV increases the percentage of stillbirths and mummies combined and reduces the percentage of pre-wean viability, among other clinical signs, such as abortion and sow mortality.[12-14] Percent combined stillborn and mummy data presents clinical evidence of PRRSV within the data.

Overall data summary by farm and treatment for all sow performance variables are presented in Table 1. Percent total death loss, percent combined stillborn and mummies by treatment and week, and piglet weaning weights are presented in Table 2. Percent combined stillborn and mummies ranged from 7.46% to 12.17% for controls and from 5.76% to 13.15% for treatments. Least-square mean (LSM) estimates of pre-wean death loss (Table 3) were 13.1% for treatment (PAC) versus 14.6% for control ($P < 0.01$). The treatment (PAC) group had a percent combined stillborn and mummies of 10.3% vs 9.7% for controls ($P = 0.08$) and the treatment (PAC) group had a percent gilts at 26.0% vs 23.3% for controls ($P = 0.15$) (Table 3). The treatment (PAC) group tended higher for percent combined stillborn and mummies warranting further evaluation in the final model. The final model of pre-wean death loss and adjusted weaning weights incorporated the covariate of percent combined stillborn and mummy (Table 3), therefore comparison of treatment versus control was done at similar percent combined stillborn and mummy rates.

The final pre-wean death loss model contained an interaction between treatment and percent combined stillborn and mummies ($P = 0.001$) and significant main effects for treatment ($P = 0.004$) and percent combined stillborn and mummies ($P < 0.001$). A comparison of death loss by treatment across percent combined stillborn and mummies (between the 10th and 90th percentile values) is presented in Table 4. Percent combined stillborn and mummy ranged from 7% to 12%, and comparisons showed that PAC had an advantage in death loss over controls as the percent combined stillborn and mummy was 9.5% or higher. At a 9.5% combined stillborn and mummy rate, the PAC treatment pre-wean death

Table 1: PAC in lactation: Overall data summary by farm and treatment for all sow performance variables

Farm	Treatment	No. rooms	No. sows	Avg parity	No. gilts	Total born	Live	Stillborn	Mummies	Deaths	Avg adj wean weight (lb.)	Death loss	Pct combined stillborn & mummies
Sow A	Control	15	364	3.28	83	5278	4789	277	212	835	12.99	17.4%	9.3%
	PACtx	12	290	3.42	65	4329	3814	264	251	614	13.54	16.1%	11.9%
Sow B	Control	9	230	3.37	57	3202	2910	187	105	486	13.42	16.7%	9.1%
	PACtx	9	232	3.25	68	3308	2997	239	72	469	12.68	15.6%	9.4%
Sow C	Control	10	290	3.05	75	3791	3418	292	81	370	12.82	10.8%	9.8%
	PACtx	12	352	3.15	104	4644	4208	325	111	435	13.45	10.3%	9.4%
Sow D	Control	12	331	3.55	69	4583	4072	283	228	584	13.62	14.3%	11.1%
	PACtx	10	241	3.42	56	3257	2915	218	124	314	14.42	10.8%	10.5%

Table 2: PAC in lactation: Summary data by treatment by week

Variable	Treatment	Level	N	Average	Min	Max
Death loss	Control	Week	12	14.70%	8.25%	18.43%
	PACtx	Week	12	13.21%	8.18%	17.97%
Pct combined stillborn & mummies	Control	Week	12	9.76%	7.46%	12.17%
	PACtx	Week	12	10.29%	5.76%	13.15%
Avg adj weaning weight (lb.)	Control	Week	12	13.24	11.94	14.28
	PACtx	Week	12	13.52	12.49	15.12

Table 3: PAC in lactation: Analysis with no covariates

Variable	Control	Treatment	Diff	*P*-value	Odds ratio	95% CI lower	95% CI upper
Death loss pct (deaths/born live)	14.6%	13.1%	1.5%	0.0011	1.137	1.060	1.221
Pct combined stillborn & mummies	9.7%	10.3%	0.6%	0.0820	0.935	0.865	1.009
Pct gilts	23.3%	26.0%	2.7%	0.1478	0.865	0.707	1.057
Average wean weight (lb.)	13.09	13.34	0.25	0.3810	Not applicable		
Average adjusted wean weight (lb.)	13.24	13.52	0.28	0.3145	Not applicable		

loss advantage was 1.3% (12.83% vs 14.14% for PAC vs control, respectively). At 10.0% combined stillborn and mummies, the treatment vs control advantage increases to 1.95%; at 12.0% combined stillborn and mummies, the treatment advantage was 4.89% (Table 4). There was no evidence to suggest a difference in pre-wean death loss between PAC and control when combined stillborn and mummies was ≤ 9% (Table 4).

The final adjusted average weaning weight model included no significant interaction between treatment and percent combined stillborn and mummies ($P = 0.07$). Additionally, the main effects of treatment ($P = 0.109$) and percent combined stillborn and mummies ($P = 0.071$) were not significant. Even though these main effects and interaction were not significant, Pulmotil® AC showed a linear trend that had an advantage in adjusted weaning weight over control when the percent combined stillborn and mummies was 11% or higher.

The nursery performance analysis was not available at the time of abstract submission and will be published at a future date.

Discussion

Ideally, a study design to evaluate health improvements would include the total isolation of the treated pigs from the other pigs to maintain health including filtered air spaces. However, facility limitations in commercial operations often dictate a room-by-room comparison with treated and untreated groups housed in close proximity without air space separation. This configuration complicates the evaluation of health and performance due to a likelihood of disease cross-contamination between treated and untreated groups.

The study demonstrates that PAC at 2 grams/sow/day during lactation lowered pre-weaning mortality and tends to improve weaning weight. In this study, when considering disease status (i.e. percent combined stillborn and mummies), a decrease of the pre-weaning mortality was observed when the percentage of both stillbirths and mummies increased. Application of PAC during periods of high combined stillborn and mummies (≥ 9.5%) can lower pre-wean death loss. Further studies will substantiate these findings and delineate the impact on weaning weight.

References

1. Almond G, et al. 2008. Reduction of pneumonic lesions at weaning using Pulmotil® in lactation. International Pig Veterinary Society. P08.057.

2. Almond GW, et al. 2006. Assessment of the therapeutic effect of tilmicosin in lactation feed. International Pig Veterinary Society. P34-25.

3. Kelly J, et al. 2008. Use of Pulmotil® during sow lactation to improve herd health and productivity. International Pig Veterinary Society. P08.043.

4. Lehe K, et al. 2008. Evaluation of Pulmotil® use in lactation and nursery diets in a herd with swine respiratory disease. American Association of Swine Veterinarians. Pg 229–232 .

5. Carvajal M, et al. 2010. Evaluation of the Pulmotil® premix effect on production parameters of sows and their litters in a commercial farm on the sisal zone in Yucatan, Mexico. International Pig Veterinary Society. Pg 1016.

6. McCaw MB. 2000. Effect of reducing crossfostering at birth on piglet mortality and performance during an acute outbreak of porcine reproductive and respiratory syndrome. Swine Health and Prod. 8:15–21.

7. McCaw MB. 2000. Impact of McREBEL (minimal cross fostering) management upon nursery pig sale weight and survival under production conditions in PRRS asymptomatic herds. International Pig Veterinary Society. Pg 333.

8. Dion K, et al. 2010. The importance of McREBEL procedures in successful elimination of PRRSv from large breeding herds. American Association of Swine Veterinarians. Pg 75–76.

Table 4: PAC in lactation: Least square means comparison of treatment by percent combined stillborn & mummies

Pct combined stillborn & mummies	Control death loss	PACtx death loss	Difference	Odds ratio	Probability
7.0%	10.93%	12.47%	-1.54%	0.86	0.104
7.5%	11.52%	12.55%	-1.03%	0.91	0.214
8.0%	12.13%	12.62%	-0.49%	0.96	0.492
8.5%	12.77%	12.69%	0.08%	1.01	0.892
9.0%	13.44%	12.76%	0.68%	1.06	0.195
9.5%	14.14%	12.83%	1.30%	1.12	0.008
10.0%	14.86%	12.91%	1.96%	1.18	< 0.001
10.5%	15.62%	12.98%	2.64%	1.24	< 0.001
11.0%	16.41%	13.05%	3.36%	1.31	< 0.001
11.5%	17.23%	13.13%	4.10%	1.38	< 0.001
12.0%	18.09%	13.20%	4.88%	1.45	< 0.001

9. Polson D, et al. 2010. McREBEL as a key component for the successful elimination of PRRS virus from very large swine breeding herds. International Pig Veterinary Society. Pg 267.

10. Wood CM, et al. 1990. Factors to adjust litter weight of pigs to a standard 21 days of age. J Anim Sci. 68:2628–2633.

11. National Swine Improvement Federation. Available at `https://www.nsif.com/guidel/onfarm.htm`. Accessed 24 July 2018.

12. Loving CL, et al. 2015. Innate and adaptive immunity against Porcine Reproductive and Respiratory Syndrome Virus. Veterinary Immunology and Immunopathology. 167:1–14.

13. Karniychuk UU, et al. 2013. Pathogenesis and prevention of placental and transplacental porcine reproductive and respiratory syndrome virus infection. Vet Res. 44:95.

14. Lunney JK, et al. 2016. Porcine Reproductive and Respiratory Syndrome Virus (PRRSV):Pathogenesis and Interaction with the Immune System. Annual Review of Animal Biosciences. 4:129–154.

Use of spray-dried plasma in feed for grow-finish pigs to reduce reliance on growth-promoting antibiotics

G. Lima[1], PhD; J. Polo[3],PhD; J. Campbell[4], PhD; J. Crenshaw[4],PhD; F. Faleiros[5], PhD; G. Gattas[5], PhD; L. Rangel[2], MSc

[1]EMBRAPA, Concordia, SC, Brazil; [2]APC Inc., São Paulo, Brazil; [3]APC-Europe, Granollers, Spain; [4]APC Inc., Ankeny, Iowa, [5]Gattas Consultoria, Brasil

Introduction

A common practice for some swine producers in Brazil is to use growth-promoting and (or) therapeutic doses of antibiotics in feed for growing-finishing swine. Therapeutic doses of antibiotics are typically used during strategic periods of anticipated disease stress, including at placement (65-78 days of age), during the grower stage (93 to 106 days of age) and during the finisher stage (121-134 days of age). Typically, growth promoting levels of antibiotics are used in feed between periods of feeding the therapeutic antibiotics. However, legislative changes are under discussion to prohibit the Brazilian swine producer access to growth-promoting antibiotics for use in farm mixed feed and only therapeutic use of antibiotics in feed or water will be allowed. Thus, swine producers in Brazil are seeking effective alternatives to growth-promoting antibiotics. Research has shown that spray-dried plasma in feed for nursery pigs is a powerful option for producers to reduce their reliance on antibiotics and medications.[1] Past research has demonstrated that dietary spray-dried plasma in feed for grow-finish pigs reduced mortality associated with multi-systemic disease by 50%.[2] The bioactive components of spray-dried plasma used in feed are known to maintain and support an efficient common immune system response to stressful events which can help pigs survive and recover more quickly.[3] The objectives of this study were to evaluate the effects of spray-dried plasma in feed for grow-finish pigs as an alternative to growth-promoting and (or) therapeutic antibiotics on clinical health indices, survival and performance economics under commercial conditions in Brazil.

Materials and methods

The study was done at a commercial farm in Minas Gerais State, Brazil. Growing pigs at approximately 65 days of age were distributed to a randomized complete block design to control differences in initial body weight and sex, with 5 treatment groups and 10 replications per treatment group with pen as the experimental unit. Each treatment group had 5 pens of females and 5 pens of males. Each pen had 30 pigs for a total of 1,500 pigs started for the study. Pig body weights and feed consumption was measured at d 65, 107 and 148 of age. Incidence of mortality, injectable pig medications, coughing, sneezing, and diarrhea were noted and recorded for each pen during the study.[2]

Description of feed treatment groups and phase of production are presented in Table 1 and were as follows: 1) therapeutic antibiotic use at placement, grower 2 and finisher 2 phases; 2) same as treatment 1 plus growth promoting antibiotic used in grower 1, finisher 1, and pre-slaughter phase; 3) same as treatment 1 plus 2% spray-dried plasma (SDP) included in placement phase; 4) same as treatment 1 with 1% SDP in grower 1, 0.75% SDP in

Table 1: Description of feed medication or spray-dried porcine plasma by treatment group and phase of experiment

	Feed treatment groups				
Phase (d of age)	**1**	**2**	**3**	**4**	**5**
Placement (65-78)	ATP	ATP	ATP + 2% SDP	ATP	1.20% SDP
Grower 1 (79-92)	--	AGP	--	1% SDP	--
Grower 2 (93-106)	ATP	ATP	ATP	ATP	0.90% SDP
Finisher 1 (107-120)	--	AGP	--	0.75% SDP	--
Finisher 2 (121-134)	ATP	ATP	ATP	ATP	0.50% SDP
Pre-Slaughter (134-148)	--	AGP	--	0.50% SDP	--

ATP = Antibiotic therapeutic pulse dose; AGP = Antibiotic growth promoting; SDP = percentage of spray-dried porcine plasma used in feed to provide approximately 20 g SDP per day based on estimated feed intake by phase. ATP in placement feed provided 300 mg Doxyxycline/kg feed (50% Doxyxycline at 0.060 kg/100kg feed); 100 mg Tylosin/kg feed (Tylosin 25% at 0.040 kg/100 kg feed) and 3 mg Ivermectin/kg feed (Ivermectin 0.6% at 0.045 kg/100 kg feed). AGP in grower 1, finisher 1 and pre-slaughter feed provided 12 mg Flavomycin /kg feed (Flavomycin 8% at 0.015 kg/100 kg feed). ATP in grower 2 feed provided 450 mg Amoxillin/kg feed (Amoxillin 50% at 0.090 kg/100 kg feed); 240 mg Leucomycin/kg feed (Leucomycin 30% at 0.080 kg/100 kg feed); 36 mg Abamectin/kg feed (Abamectin 6% at 0.060 kg/100 kg feed). ATP in finisher 2 feed provided 150 mg Florfenicol/kg feed (Florfenicol 10% at 0.150 kg/100 kg feed). All finisher 2 feed contained 10 mg Ractopamine /kg feed (Ractopamine 2% at 0.050 kg/100 kg feed). All pre-slaughter feed contained 15 mg Ractopamine /kg feed (Ractopamine 2% at 0.075 kg/100 kg feed).

finisher 1 and 0.50% SDP in pre-slaughter phase, and 5) had no inclusion of therapeutic or growth promoting antibiotics, but had 1.2% SDP in placement, 0.90% SDP in grower 2 and 0.50% SDP in finisher 2 phases. All feed by phases were formulated to meet nutrient requirements and diets with or without spray-dried plasma were iso-nutrient within phase.[4] The level of SDP used in feed by phase was determined based on estimated feed intake of pigs to provide approximately 20 g SDP intake per day which was similar to SDP intake of grow-finish pigs in a previous study.[2]

Pen data was analyzed by Tukey multiple comparison test with probability of $P < 0.05$ as significant. Clinical observations, injectable medications, runt hogs and mortalities were subjected to Chi-Square frequency test with probability of $P < 0.05$ as significant.

Results and discussion

Clinical indices such as coughing, sneezing, diarrhea or individual pig injections did not significantly differ among treatment groups during any phases of the study (data not shown). Pig performance results presented in Table 2 indicate pigs fed treatments 1 or 3 had higher average BW and daily gain at 107 d of age than pigs fed treatments 2 or 4 and treatment 1 pigs had higher BW and daily gain than treatment 5 pigs. Daily feed intake by 107 d of age was higher for treatments 1, 2 and 3 versus treatment 4 and for treatments 1 and 3 versus treatment 5. Feed efficiency did not differ among treatments at d 107 of age. Over the entire study from 65 to 148 d of age, pigs fed treatment 3 had the highest final BW and daily gain compared to treatments 2, 4 and 5, with no significant difference for treatment 1 compared to all other treatments. Daily feed intake by d 148 of age was higher for pigs fed treatments 1 and 3 compared to treatments 2, 4, and 5 and feed intake was lower for treatment 4 compared to treatments 2 and 5. However, pigs fed treatment 4 had improved feed to gain ratio compared to treatments 1, 2 and 5 and was numerically improved compared to treatment 3. Mortality and mortality plus runt hogs did not differ among treatment groups.

As expected, pigs fed treatment 1 had satisfactory body gain which supports the positive effects of therapeutic antibiotics during strategic periods of the grow-finish phases under commercial production conditions of Brazil. However, pigs fed treatment 2, which used therapeutic and growth promoting antibiotics during all the study period had lower body gain and feed intake than treatment 1, which suggests that excessive use of antimicrobials in grow-finish pig feed may possibly be detrimental to liver function, immune development, or result in a less favorable microbiome balance in the gastro-intestinal system.

Pigs fed treatment 3, which was the same as treatment 1 but with 2% SDP added during the placement period had the same gain and feed intake as treatment 1, but numerically had better feed efficiency than treatment 1 at 107 and 148 days of age. The early use of SDP during a high stress period of transition from nursery to placement in the finisher may support a more efficient immune response to transition stress resulting in better nutrient utilization.

Table 2: Effect of feed medication or spray-dried plasma on grow-finish pig performance.

		Feed treatment group[1]						
Variable	**Day**	**1**	**2**	**3**	**4**	**5**	**SEM**	**Trt (*P*)**
BW, kg	65	24.6	24.6	24.6	24.6	24.6	0.84	0.9976
BW, kg	107	54.4[a]	51.9[c]	54.2[ab]	50.9[c]	52.5[bc]	1.69	0.0010
BW, kg	148	100.2[ab]	98.9[b]	102.7[a]	98.4[b]	99.0[b]	2.14	0.0099
Gain, kg/d	65-107	0.710[a]	0.649[c]	0.705[ab]	0.626[c]	0.664[bc]	0.023	0.0011
Feed, kg/d	65-107	1.647[a]	1.562[ab]	1.612[a]	1.473[c]	1.505[bc]	0.049	0.0017
Feed:gain	65-107	2.322	2.412	2.292	2.362	2.269	0.039	0.0687
Gain, kg/d	65-148	0.901[ab]	0.885[b]	0.930[a]	0.879[b]	0.886[b]	0.017	0.0111
Feed, kg/d	65-148	2.176[a]	2.103[b]	2.185[a]	2.014[c]	2.083[b]	0.048	<0.0001
Feed:gain	65-148	2.417[a]	2.379[a]	2.348[ab]	2.291[b]	2.353[a]	0.033	0.0293
Mortality[2]	65-148	0.40	0.40	0.90	1.00	0.20	0.298	0.1500
Mortality + runts[2]	65-148	2.60	3.40	3.60	4.30	3.70	0.846	0.3300

[abc] Within rows means with uncommon superscripts differ ($P < 0.05$).

[1] Means represent the average of 10 pens per treatment group. No differences ($P > 0.10$) associated with sex or interaction of treatment group and sex. Treatment 1 = therapeutic antibiotic use at placement, grower 2 and finisher 2 phases. Treatment 2 = same as treatment 1 plus growth promoting antibiotic used in grower 1, finisher 1 and pre-slaughter phase. Treatment 3 = same as treatment 1 plus 2% SDP included in placement phase. Treatment 4 = same as treatment 1 with 1% SDP in grower 1, 0.75% SDP in finisher 1 and 0.50% SDP in pre-slaughter phase. Treatment 5 = no use of therapeutic or growth promoting antibiotics but with 1.2% SDP in placement, 0.90% SDP in grower 2 and 0.50% SDP in finisher 2 phases.

[2] Probability of Chi-Square with maximum SEM.

Pigs fed treatment 4 was like treatment 2, with the exception that SDP was used to replace growth-promoting antibiotics. The body weight gain was similar for treatment 2 and 4 at both d 107 and 148 of age. However, feed intake at d 107 and 148 was lower for treatment 4 versus 2, resulting in the best feed efficiency for treatment 4 among all treatments. Although final BW for treatment 4 was less than treatment 3, the use of SDP in place of growth promoting antibiotics at all phases suggests SDP enhances nutrient utilization by supporting a more efficient immune response and possibly better microbiome in the gastrointestinal tract than provided by growth-promoting antibiotics. A recent publication suggest SDP promotes a more favorable ratio of microbes in the gastrointestinal tract of nursery pigs.[5]

Pigs fed treatment 5, which was absent of antibiotics but used SDP in place of therapeutic antibiotics (treatment 1), had lower final BW than pigs fed treatments 1 or 3. However treatment 5 pigs had improved feed efficiency compared to treatment 1.

Economical results are presented in Table 3. Producer price of SDP, antibiotics and other feed ingredients were used for calculating feed cost. Pigs fed treatment 1 and 3 had higher feed cost, but also higher revenue per pig compared to treatments 2, 4 and 5. Although not statistically significant, pigs fed treatments 3 and 5 had about a R$4.7 advantage in income over feed and medication cost. In addition, results from treatment 5 suggest that SDP can be considered an alternative to replace ATB for the entire growing-finishing phase without affecting growth performance or mortality and with improved economic return over cost for the producer.

In conclusion, under the conditions of this study strategic use of SDP to reduce reliance on growth promoting and therapeutic antibiotics in feed can support adequate economical production performance of grow-finish pigs while maintaining similar health status.

References

1. Torrallardona, D. Spray dried animal plasma as an alternative to antibiotics in weanling pigs – A review. *Asian-Austr. J. Anim. Sci.* 2010;23:131.

2. Messier, S et al. Dietary spray-dried porcine plasma reduces mortality attributed to porcine circovirus associated disease syndrome. *Proc. Amer Assoc Swine Vet* 2007;147.

3. Pérez-Bosque, A et al. Spray dried plasma as an alternative to antibiotic in piglet feeds, mode of action and biosafety. *Porcine Health Management* 2016;2:16.

4. Rostagno, H et al. *Brazilian Tables for Poultry and Swine. Composition of Feedstuffs and Nutritional Requirements.* Universidade Federal de Vicosa-Departamento de Zootecnia. 2017.

5. Tran, H et al. Effects of spray-dried porcine plasma on fecal microbiota in nursery pigs. J Anim Sci 2018;96:1017.

Table 3: Effect of feed medication or spray-dried plasma on production economics.

	Feed treatment group[1]						
Variable	**1**	**2**	**3**	**4**	**5**	**SEM**	**Trt (*P*)**
Feed cost, R$/pig	161.51[ab]	157.12[bc]	165.67[a]	155.63[c]	152.62[c]	3.67	<0.0001
Injectable meds, R$/pig	0.084	0.153	0.090	0.090	0.126	0.02	0.1370
Feed cost/gain, R$/kg	2.135	2.115	2.120	2.107	2.052	0.03	0.1498
Revenue, R$/pig	350.82[ab]	346.15[b]	359.48[a]	344.52[b]	346.38[b]	7.49	0.0099
IOFMC, R$/pig[2]	189.23	188.89	193.72	188.80	193.64	4.60	0.4113

[abc] Within rows means with uncommon superscripts differ ($P < 0.05$).

[1] Means represent the average of 10 pens per treatment group. No differences ($P > 0.10$) associated with sex or interaction of treatment group and sex. Treatment 1 = therapeutic antibiotic use at placement, grower 2 and finisher 2 phases. Treatment 2 = same as treatment 1 plus growth promoting antibiotic used in grower 1, finisher 1 and pre-slaughter phase. Treatment 3 = same as treatment 1 plus 2% SDP included in placement phase. Treatment 4 = same as treatment 1 with 1% SDP in grower 1, 0.75% SDP in finisher 1 and 0.50% SDP in pre-slaughter phase. Treatment 5 = no use of therapeutic or growth promoting antibiotics but with 1.2% SDP in placement, 0.90% SDP in grower 2 and 0.50% SDP in finisher 2 phases.

[2] IOFMC = Income minus the sum of feed and injectable medications per pig.

Virus neutralizing antibodies produced after vaccination with a killed PRRS virus vaccine

B. Hause, PhD; J. Park, DVM, PhD; G. Bosch, DVM; R. Simonson, PhD
Cambridge Technologies, Worthington, Minnesota

Introduction

Porcine reproductive and respiratory syndrome virus (PRRSV) is responsible for the most significant disease plaguing pig production worldwide. Soon after the discovery of PRRSV in the late 1980's, both commercial and autogenous vaccines entered the markets in the early 1990's to control porcine reproductive and respiratory syndrome (PRRS). Despite numerous permutations of inactivated and modified-live vaccines, vaccine efficacy is perceived as insufficient. Modified-live vaccines are generally considered to provide superior protection to challenge with genetically similar viruses as compared to inactivated vaccines however largely fail to confer protection to genetically divergent viruses. Importantly, the safety of modified live vaccines has been questioned as they are prone to mutation and recombination with field strains which can result in reversion to virulence. Inactivated vaccines have a superior safety profile however improvements in efficacy are needed. Numerous publications the past few years have demonstrated that inactivated PRRSV vaccines generated with optimized processes, including antigen-friendly viral inactivation and new adjuvants, can generate neutralizing antibodies and provide significant protection to challenge. Importantly, heterologous prime-boost immunization strategies have been shown to induce heterologous immunity. The objective of this study was to evaluate the ability of an experimental inactivated autogenous vaccine to bolster and broaden the neutralizing antibody response to sows primed with a modified live virus vaccine.

Materials and methods

Ten gilts and 20 multiparous sows were randomly assigned to one of two groups. All animals received a commercial MLV vaccine at 26 weeks of age. Animals included in the autogenous vaccine group received a 1 cc intramuscular injection at 9 and 13 weeks of gestation. The autogenous vaccine contained a single strain of PRRSV with a restriction fragment length polymorphism type (RFLP) 1-8-4. All animals were bled within 24 hours of farrowing. Serum neutralizing antibodies were measured against the homologous virus as well as four heterologous strains with RFLP types (2-11-2, 1-4-2, 1-4-4, 2-6-2) using the fluorescent foci neutralization (FFN) assay. The geometric mean titer (GMT) was calculated for each group and differences were analyzed by ANOVA.

Results and discussion

Control gilts and sows receiving only commercial MLV at 26 weeks of age had FFN GMT of 4.0-5.7 to all five viruses tested. In contrast, animals vaccinated with the autogenous vaccine demonstrated a robust FFN GMT of 37.6 to the homologous virus. Neutralizing antibody GMT's of 11.6, 12.3, 23.8 and 24.4 were measured to the four heterologous viruses tested, all of which were significantly higher than control animals ($P < 0.05$).

These results demonstrated that two doses of a monovalent autogenous PRRSV vaccine administered to MLV-primed animals can stimulate robust neutralizing antibody titers which recognize both parental virus and genetically diverse field strains. Numerous studies have established neutralizing antibodies as a correlate of immunity, with titers ranging from 8-32 demonstrating immunity in different infection models. While controlled vaccination and challenge models are needed to verify protection, the results of this study using an experimental autogenous PRRSV vaccine suggest that significant efficacy can be achieved using a combination MLV-autogenous vaccine program.

Fate of injected fat-soluble vitamins in pigs, gilts, and sows

Robert L. Stuart, Ph.D.
Stuart Products, Inc. Bedford, Texas

Introduction

Fat-soluble vitamins A, D, E are vital nutrients for all swine. These vitamins are routinely added to swine feeds at or above National Research Council (NRC) recommended levels. Supplementation via drinking water and/or injection are utilized to strategically enhance fat-soluble vitamin status during critical times in the production cycle of swine. The primary emphasis for NRC dietary requirements is minimum daily requirements for animal gain and feed conversion. Besides animal gain, research has shown that under certain conditions, higher levels of fat-soluble vitamins may impact other factors besides gain.

Previous research has demonstrated that either water-supplemented or injected fat-soluble vitamins dramatically enhances serum levels. For vitamin E, serum C_{max} is attained very quickly post-injection and also diminishes within 24 hours post-injection, while serum 25-OH-D_3 status is maintained for a much longer time compared to either vitamin E or vitamin A. The purpose of this paper is to review experiments that measured the fate of injected fat-soluble vitamins in pigs, gilts, and gestating sows.

Serum levels of fat soluble vitamins in newborn pigs either injected or orally administered VITAL E-Newborn

Jang, et al., 2014[1], compared the utilization of the same product either injected vs oral administered vitamins A, D and E (VITAL E-Newborn). Injected vitamins were utilized approximately twice that of the orally administered vitamins. The data also confirmed that serum retention times were different among the three fat-soluble vitamins. Alpha-tocopherol had the shortest retention time and 25-OH-D_3 had the longest.

Serum and tissue levels of fat soluble vitamins in weaned pigs

Hidiroglou et al., 1995,[2] found that pigs injected with d-alpha-tocopherol (VITAL E-500) had dramatically higher plasma alpha-tocopherol concentrations and higher alpha-tocopherol concentrations in spleen, liver, heart, kidney, lung and muscle compared to non-injected pigs. Injected pigs also had higher antibody titers to key-hole limpet haemocyanin (KLH) compared to non-injected pigs.

In a recent experiment, Jang et.al.[3] investigated distribution of injected fat-soluble vitamins in plasma and tissue of nursery pigs (Tables 1 and 2). A total of 16 pigs were allotted to 2 treatments at d 7 post-weaning and fed a corn-soybean meal-based diet with no supplemental vitamin A. Treatments were non-injected control (no vitamins) and intramuscular injection of 3 mL VITAL E-A+D having retinyl palmitate (300,000 IU), d-alpha-tocopherol (900 IU), and vitamin D (30,000 IU) per pig. All pigs were bled at d 0, 3, 7, and 14 post-injection for E, A, and D analysis. Tissue samples were collected in three pigs per treatment at d 7 and 14 post-injection. Due to cost of analysis, tissues were not analyzed for vitamin D. Plasma 25-hydroxycholecalciferol was dramatically increased (3.44 vs. 71.00 ng/mL at d 3 post-injection) and retinol (0.09 vs. 0.18 μg/mL at d 3 post-injection) concentrations increased from d 3 to 14 post-injection ($P < 0.05$) (Table 1). Plasma retinyl palmitate was detected only in the vitamin treatment at d 3 and 7 post-injection (115.51 and 4.97 μg/mL, respectively). Concentrations of liver retinol (0.63 vs. 19.10 μg/g) and liver retinyl palmitate (43.33 vs. 199.13μg/g) at d 7 post-injection also were increased on d 14 post-injection ($P<0.05$) whereas neither were detected in other tissues. The alpha-tocopherol concentrations increased in plasma ($P < 0.05$) at d 3 (0.99 vs. 7.14 μg/mL) and d 7 post-injection and in liver (4.30 vs. 11.17 μg/g; $P < 0.10$), heart (4.60 vs. 24.80 μg/g; $P < 0.10$), and muscle (3.07 vs. 8.30 μg/g; $P < 0.05$) at d 7 post-injection. Brain was the only tissue that did not have an increase in vitamin E status (Table 2). This experiment revealed that a fat-soluble vitamin injection in pigs increased plasma and liver status of alpha-tocopherol, retinol, and retinyl palmitate, and plasma 25-hydroxycholecalciferol concentrations. Alpha tocopherol was distributed to all tissues examined but retinol and retinyl palmitate were found only in the liver. Although plasma vitamin E levels decreased, tissue levels of vitamin E and A increased in tissues except brain (Table 2). This experiment demonstrated that even though there were reductions in serum levels, there was a corresponding increase in various tissue levels, especially for vitamin E.

Fat-soluble vitamin status in serum and ovaries of gilts were dramatically improved following an injection of VITAL E-Repro

Vitamin E and vitamin A have been shown to be involved in reproduction.[4-6] Due to the high levels of polyunsaturated fatty acids (PUFA's) in ovaries, an experiment was conducted to determine if vitamin E, A, and D status in serum and ovaries would be enhanced after an injection (VITAL E-Repro). Treated gilts were injected with 5 mL VITAL E-Repro and harvested 72 hr post-injection. Each value represents the mean of eight gilts, 6.5 months old. Serum levels for vitamin E and 25-OH-D_3 increased for 72 h (Table 3). An interesting finding was that ovary levels of vitamins

Table 1: Effects of VITAL E-A+D injection on serum vitamin levels in weaned-pigs (Jang, 2018)

Vitamin	Control	VITAL E-A+D	SEM	*P*-value
25-OHD3 (ng/mL)				
d 0	4.33	4.01	0.432	0.058
d 3	3.44	71.00	6.979	<0.001
d 7	4.79	36.35	3.256	<0.001
d 14	7.24	15.50	1.220	<0.001
Retinol (µg/mL)				
d 0	0.11	0.08	0.007	0.005
d 3	0.09	0.18	0.010	<0.001
d 7	0.10	0.13	0.008	0.03
d 14	0.14	0.17	0.006	0.002
Retinyl Palmitate (µg/mL)				
d 0	ND	ND		
d 3	ND	115.51	13.693	
d 7	ND	4.97	1.089	
d 14	ND	ND		
Alpha-tocopherol (µg/mL)				
d 0	1.69	1.64	0.119	0.77
d 3	0.99	17.14	1.188	<0.001
d 7	0.80	2.35	0.137	<0.001
d 14	0.66	0.99	0.275	0.41

Table 2: Tissue vitamin E levels in VITAL E-A+D injected weaned pigs

Tissue	Control	Vital E A+D	SEM	*P*-values
Brain, µg/g				
d 7	8.20	7.57	1.333	0.14
d 14[1]	7.40	8.33	0.566	0.32
Heart, µg/g				
d 7	4.60[e]	24.80[d]	3.975	0.06
d 14	2.17[b]	4.57[a]	0.594	0.047
Lung, µg/g				
d 7	2.17	18.80	5.905	0.13
d 14	9.60	10.93	1.717	0.31
Muscle, µg/g				
d 7	3.07[c]	8.30[a]	0.655	0.01
d 14	2.80	3.90	0.734	0.23

[a-c] $P < 0.05$
[d-e] $P < 0.10$.
[1] One sample was removed from the control treatment.

E and A were enhanced. At 72 h post-injection, ovary vitamin E status was increased by 76 % ($P < .01$) and vitamin A status was increased by 27% ($P < .05$) (Table 3).

Fat-soluble vitamin levels in serum and milk of sows and serum of newborn pigs after pre-partum injection of VITAL E-Repro in sows

The recommended program for gilts and sows is to inject 5 mL VITAL E-Repro one week pre-partum and again at weaning. The pre-partum injection has been shown to reduce farrowing time and improve vitamin status of the nursing pigs. The injection at weaning is to improve embryo quality and uniformity.

An experiment in North Carolina involved injecting sows seven days pre-partum with VITAL E-Repro.[7] Sows had significantly higher serum levels of vitamins E, A and D. Colostrum and milk levels for vitamins E and A were also higher ($P < 0.05$). Besides having an impact on uterine function, this experiment showed that the injected vitamins enhanced vitamin content of colostrum and milk resulting in higher vitamin levels available to the nursing pigs. In another North Carolina field study, birthing assistance was reduced from 19% to 9.5% in sows injected with VITAL E-Repro one week pre-farrowing compared to non-injected sows.[7]

Table 3: Serum and ovary levels of vitamin E and vitamin A in gilts injected with VITAL E-Repro.

Parameter	Non-injected	VITAL E-Repro (5 mL)
Serum alpha-tocopherol (μg/mL)		
Initial	2.72	2.98
24 hrs	2.52	23.21**
48 hrs	2.43	8.63**
72 hrs	3.10	4.66
Serum retinol (μg/mL)		
Initial	.289	.304
24 hrs	.277	.362
48 hrs	.224	.312
72 hrs	.284	.336
Serum 25-OH-D$_3$ (ηg/mL)		
Initial	28.05	28.00
24 hrs	27.20	88.30**
48 hrs	27.30	108.80**
72 hrs	29.00	116.60**
Ovary alpha-tocopherol (μg/g)	4.38	7.70**
Ovary retinol (μg/g)	.495	.630*

*($P < 0.05$) **($P < 0.01$)

Importance of bioavailability of injectable fat-soluble vitamins

University experiments and field demonstrations have shown that a 5 mL-injection of VITAL E®-Repro (Stuart Products, Inc.) reduces farrowing time approximately 50 min. Instead of utilizing VITAL E-Repro, some have utilized a competitive formulation instead of VITAL E-Repro. The competitive product contained an equal amount of vitamin E, lesser quantity and different source of vitamin A, and lesser quantity of vitamin D compared to VITAL E-Repro. To determine serum fat-soluble vitamin status post-injection, six gestating sows of similar parity and breed were utilized in a small study to measure plasma concentrations of fat-soluble vitamins.[8] Three sows were injected with either VITAL E-Repro or Vitamin E+AD (Durvet, Inc.). Plasma samples were collected prior to injection (d 0) and d 1, 3 and 7 post-injection. Plasma was analyzed for 25-OHD$_3$, retinol, retinyl palmitate (source of vitamin A in VITAL E-Repro), retinyl propionate (source of vitamin A in Vitamin E+AD), and alpha-tocopherol. Table 4 shows the results obtained. VITAL E-Repro-injected sows had dramatically higher 25-OHD$_3$ on days 1,3, and 7 post-injection ($P < 0.0001$) compared to the competitive product. Although small, the plasma retinol levels were numerically higher in the VITAL E-Repro-injected sows ($P < 0.09$). As expected, retinyl palmitate levels were also dramatically higher in VITAL E-Repro injected sows. When retinyl propionate was injected, none was found for any of the plasma sampling days. The lack of retinyl propionate in serum has been confirmed in other studies in pigs and calves. For vitamin E status, even though an equal amount of vitamin E was injected (1500 I.U.), VITAL E-Repro-injected sows had 211% higher plasma status on day 1 ($P < 0.003$); and on day 3 the difference was 61% higher ($P < 0.021$). There was no difference in plasma levels on day 7. This experiment and other experiments demonstrate the importance of bioavailability of injectable fat-soluble vitamins and the importance of form of vitamin A activity in an injectable product.

Conclusions

Of the three fat-soluble vitamins, vitamin E appears to be the most deficient in all post-weaning phases of swine production, while vitamin D supplementation may also be critical in the nursing pig due to low levels of vitamin D in milk. Reduced farrowing time due to pre-partum injection of fat-soluble vitamins, reduction in mulberry heart,[9] and reduced innate immunity are areas affected by inadequate fat-soluble vitamin intake.[10,11] Esterification of fat-soluble vitamins A and E is done to improve stability in complete feeds, but there are situations in swine production when it is best not to esterify fat-soluble vitamins, especially vitamin E in nursery pigs.[12] Stability of esters will be superior, but at the expense of bioavailability.

In general, the NRC recommended levels for fat-soluble vitamins fat-soluble vitamins are too low, especially vitamin E for the gestating sow.[13] For newly-weaned pigs, the bioavailability of vitamin E is more critical than inclusion levels. The need to supplement with a biologically-available source of vitamins E, A and D as found in EMCELLE TOCOPHEROL, EMCELLE ED$_3$ Liquid, or VITAL E-injectables is critically important to be assured that fat-soluble vitamin levels are more than adequate.

References

1. Jang, Y.D., M.D. Lindemann, H.J. Monegue, and R.L. Stuart. 2014. The effects of fat-soluble vitamin administration on plasma vitamin status of nursing pigs differ when provided by oral administration or injection. Asian-Australas J Anim Sci. 5:674-82.

2. Hidiroglou, M., T. R. Batra, E. R. Farnworth, and F. Markham. 1995. Effect of vitamin E supplementation on immune status and α-tocopherol in plasma of piglets. Reprod. Nutr. Dev. 35:443–450.

3. Jang Y. D., M. J. Rotering, and R. L. Stuart. 2018. Distribution of injected fat-soluble vitamins in plasma and tissues of nursery pigs. Leman Swine Conference, St. Paul, MN.

4. Aten, R. F., T. R. Kolodecik, and H.R. Behrman. 1994 Ovarian vitamin E accumulation: evidence for a role of lipoproteins. Endocrinology.135:533-9.

5. Britt, J. 2014. Vitamin A injection improves embryonic development resulting in uniformity of piglet size at birth for Gilts and Sows. News to Use February, 2014. Stuart Products, Inc.

6. Robertson, J. A. 1997. Investigations of the action of vitamin A and beta-carotene on reproductive performance in pigs. Ph.D. Dissertation Victoria Univ. of Tech., Victoria, Australia.

Table 4: Relative bioavailability of VITAL E-Repro compared to Vitamin E+AD in gestating sows

Vitamin	VITAL E Repro	Vitamin E+AD	SEM	*P*-values	TRT
25-OHD3, μg/mL					
d 0	31.87	33.73	2.664	0.629	0.0007
d 1	53.43	34.83	2.727	0.0001	
d 3	79.77	35.93	2.639	<.0001	
d 7	76.63	34.83	2.202	<.0001	
Retinol, μg/mL					
d 0	0.15	0.13	0.011	0.220	0.0606
d 1	0.15	0.12	0.011	0.098	
d 3	0.18	0.12	0.012	0.006	
d 7	0.16	0.13	0.009	0.061	
Retinyl palmitate, μg/mL					
d 0	0.05	0.03	0.009	0.094	0.0155
d 1	9.72	0.05	2.184	0.009	
d 3	9.21	0.03	1.446	0.001	
d 7	0.06	0.04	0.006	0.043	
Retinyl propionate, μg/mL					
d 0	ND	ND	-	-	-
d 1	ND	ND	-	-	
d 3	ND	ND	-	-	
d 7	ND	ND	-	-	
Alpha-tocopherol μg/mL					
d 0	0.73	0.80	0.153	0.764	0.0135
d 1	9.23	2.97	1.211	0.003	
d 3	3.43	2.13	0.346	0.021	
d 7	1.30	1.37	0.158	0.771	

ND = not detected.

7. Becera, Roel. 2017. Improving sow productivity and nursery pig performance by injecting sows with d-alpha-tocopherol, retinyl palmitate, and cholecalciferol. Master's Thesis. North Carolina State University.

8. Jang, Y. 2018. Unpublished.

9. Pallares, F.J., M. J. Yaeger, B. H. Janke, G. Fernandez and P.G. Halbur. 2002. Vitamin E and selenium concentrations in livers of pigs diagnosed with mulberry heart disease. J Vet Diagn. Invest 14:412–14.

10. Copeland, K. R., D. B. Scales, G. M. Hill, J. E. Link, K. C. Turner, C. J. Rozeboom, and R. L. Stuart. 2017. Effects of vitamins E and D on performance and antioxidant enzymes in nursery pigs. Midwest ASAS Meeting, March 13, 2017.

11. Jensen, M., C. Fossum, C., M. Ederroth, R.V.J. Hakkarainen, R.V.J. 1988. The effect of vitamin E in the cell-mediated immune response in pigs. J. Vet. Med. B. 1988;35:549–555.

12. Specht, T.A., D.C. Mahan, N.D. Fastinger and R.L. Stuart. 2003. Effects of d-alpha-tocopherol alcohol or acetate in water soluble or emulsified form to the drinking water of weanling pig. J. Anim. Sci. 81 (Suppl. 2): p 88 (Abstract No. 356).

13. Jensen, S. K. and C. Lauridsen. 2017. Fat Soluble Vitamins - availability and recommendations for highly prolific sows and their progeny. 78th Minnesota Nutr. Conf. Sept. 20-21, 2017. pp 165-175.

Elemental impurities in injectable iron products for swine

S. Ensley, DVM, MS, PhD
Kansas State University

Introduction

Injectable iron is routinely administered to hundreds of millions of baby pigs across the world on an annual basis. In fact, it is one of the few injectable products that nearly every pig raised in modern production systems will be given in their life time. A single 200 mg dose of iron given in the first few days of life has long been common;[1-3] however, it has been shown recently that an additional 200 mg dose during lactation results in improved post-weaning growth performance.[4-7]

As with any product used in swine production, producers and consumers should expect all inputs used align with the overall goals of ensuring pig health and safety while resulting in a clean food product.[8] The first tenet of the Iowa Pork Producers Association We Care® initiative states, "Producing safe food" and the first bullet point under Producing Safe Food is, "Use management practices consistent with producing safe food."[9] Regulatory oversight by the US Food and Drug Administration Center for Veterinary Medicine (CVM) and equivalent agencies in other countries, exist to aid in this process and ensure producers have access to pharmaceutical products which align with these efforts. While all products reported in this project are approved for use in swine by the regulatory authorities in their respective countries, no inclusion limits for elemental impurities have been established for parenteral veterinary products. This is not to imply that the presence of elemental impurities are acceptable, rather is result of a lack of published data supporting the level of risk associated with injection of such impurities in swine. For these reasons, this project aimed to evaluate parenteral veterinary iron products for the presence of impurities that would be undesirable and at minimum, unintended, for intramuscular injection in food-producing animals and have no official inclusion limit established by the CVM.[10]

Materials and methods

A complete description of the sample submission and analytical testing process can be found in the original publication.[10] Briefly, 16 injectable iron products from eight countries, each approved for the treatment of iron deficiency anemia in swine, were evaluated by the Toxicology and Nutrition Laboratory at the Veterinary Diagnostic Laboratory at Iowa State University. Fifteen of the 16 samples were also evaluated for the same analytes at an independent Current Good Manufacturing Practice-compliant contract laboratory. Samples were analyzed for the elements arsenic, cadmium, chromium, cobalt, lead and mercury using inductively coupled plasma mass spectrometry.

Results

The results of this testing showed that a 200 mg injection of any of the iron products tested, with the exception of Uniferon®, contain a concentration of one or more elements, arsenic, chromium or lead that exceed the PDE limit established for humans (Table 1). Many of which exceeded human PDE limits by greater than 25% on a per kilogram basis. The presence of cadmium, cobalt, and mercury, were either not detected, or were detected at levels well below PDE limits for humans and are therefore not reported. To aid in interpretation, results in Table 1 are reported as the detected concentration of each respective element in micrograms (µg) per 200 mg dose of product. For all products with a concentration of 200 mg/mL, the reported concentration in µg/200 mg dose (i.e. µg/1.0 mL of product) is equivalent to parts per million (PPM). For products with a concentration of 100 mg/mL, the results reported by each laboratory in PPM were doubled to standardize the results to a common 200 mg dose of product per pig.

Discussion

While at the time of this writing, no formal guidance exists in the US governing acceptable concentrations of elemental impurities such as arsenic, chromium and lead in animal drugs; the presence of these elements can still pose serious risks to animal and human health when not properly controlled. The CVM has therefore recommended that sponsors of veterinary drug products are to apply their own internal control strategies for these impurities to establish appropriate acceptance criteria on a product by product basis. When the CVM was consulted to determine appropriate handling of drugs which carry the risk of containing these impurities, the response was:

> *CVM expects sponsors of veterinary drug products to apply a risk-based control strategy for elemental impurities and establish appropriate acceptance criteria and test methods for elemental impurities where necessary. In cases where elemental impurities such as As, Pb or Cr are likely to be present in significant amounts based on a risk assessment, CVM would recommend the USP <232> and ICH Q3D limits for elemental impurities, which are for humans, as a starting point for establishing a suitable limit for the animal drug; however, we would consider species or dosage specific justifications for adjusting those limits. Ensuring that elemental impurities in the final drug product are controlled within safe limits remains the responsibility of the sponsor of a veterinary drug product.*[11]

Table 1: Detected content of arsenic, chromium and lead in parenteral iron products for swine, tested at two laboratories.** (adapted from Radke et al, 2018)[10]

Brand Name	Manufacturer†	Country	Molecule	Concentration (mg/ml)	Arsenic‡ ISU µg/200mg dose§	Arsenic‡ Lab 2 µg/200mg dose§	Arsenic‡ PDE¶ µg/kg	Chromium ISU µg/200mg dose§	Chromium Lab 2 µg/200mg dose§	Chromium PDE¶ µg/kg	Lead‡ ISU µg/200mg dose§	Lead‡ Lab 2 µg/200mg dose§	Lead‡ PDE¶ µg/kg
Aspen Anem-X® 100	Sparhawk	US	Iron dextran	100	3.4	2.0		30.2	27.0		< 0.1	< 0.1	
Durvet Iron-100	Sparhawk	US	Iron dextran	100	4.0	1.9		36.2	32.9		< 0.1	< 0.1	
Ecotin 200	Iven Laboratories	Spain	Iron dextran	200	0.2	0.4		36.0	49.5		4.9	5.8	
FerroForte®	Bimeda	Canada	Iron dextran	200	1.7	1.6		12.0	35.0		0.3	< 0.1	
Ferrohipra 200	Hipra	Belgium	Gleptoferron	200	< 0.1	< 0.1		25.0	24.1		2.0	1.1	
GleptoForte™	Ceva	US	Gleptoferron	200	0.9	< 0.1		32.4	29.6		0.6	0.5	
Gleptosil®	Ceva	Germany	Gleptoferron	200	2.2	1.2		21.0	28.9		< 0.1	< 0.1	
Gleptosil®	Sogeval	UK	Gleptoferron	200	1.4	0.5		18.0	27.0		< 0.1	< 0.1	
Prolongal®	Bayer	Belgium	Gleptoferron	200	< 0.1	< 0.1	0.3	33.2	28.9	22.0	2.6	0.6	0.1
Uniferon® 200	Pharmacosmos	US	Iron dextran	200	< 0.1	< 0.1		0.4	0.7		< 0.1	< 0.1	
Ursoferran®	Serumwerk	Germany	Gleptoferron	200	< 0.1	< 0.1		25.0	36.0		0.4	0.4	
Ursoferran®	Serumwerk	Russia	Gleptoferron	200	0.1	< 0.1		19.0	21.9		1.1	1.2	
VetOne®	Sparhawk	US	Iron dextran	100	2.0	2.2		19.4	36.7		1.2	1.5	
Viloferron®	iron4u	Denmark	Gleptoferron	200	< 0.1	< 0.1		29.0	29.0		3.1	0.3	
Xue Duo Bang	Guangxi Research Institute of Chemical Industry	China	Iron dextran	100	1.4	1.8		65.4	39.8		0.6	< 0.1	
Xue Wei Bao***	Guangdong Wens Dahunong Biotechnology	China	Iron dextran	100	1.8	NA††		28.6	NA††		< 0.1	NA††	

**All values are rounded to the nearest one significant figure. Yellow highlighted cells indicate the element was present at ≤ 25% higher than the daily limits established for humans. Red highlighted cells exceed the human daily exposure limit by > 25%.

† Marketing Authorization holder/NADA owner.

‡ Values reported as < 0.1 µg/200 mg dose were below the limit of detection for the assay.

§ For all 200 mg/ml products the reported elemental concentrations in µg/200 mg dose are equivalent to parts per million. For 100 mg/mL products, detected concentrations in parts per million were doubled to represent a typical 200 mg dose.

¶ Permitted daily exposure is the published daily exposure limit for an adult human. Values were converted to µg/kg assuming 50 kg as a conservative adult human body weight and using inclusion limits reported in USP < 232 > and ICH Q3D for human pharmaceutical products.

***Sample was not available for testing at both laboratories.

†† Not applicable.

ISU = Iowa State University Veterinary Diagnostic Laboratory; Lab 2 = independent laboratory; PDE = permitted daily exposure

With these recommendations in mind, the detected quantities of impurities presented in Table 1 are compared to human PDE limits on a weight-adjusted basis. Simply, the PDE for an adult human was converted to a per kilogram basis by dividing the PDE in µg/day by a conservative adult human bodyweight of 50 kg. This conversion is justified given the magnitude of difference in bodyweight between an adult human and a newborn piglet, the target species of the products in question. In the time since the original research was submitted for publication in June 2017,[10] the CVM has issued Draft Guidance #255 *Elemental Impurities in Animal Drug Products Questions and Answers* which states:

> *CVM considers the Permitted Daily Exposures (PDEs) established in ICH Q3D acceptable for the safety of animal drug products. The calculated elemental impurity daily exposure should result in a level equal to or below the ICH Q3D PDE; however, CVM recognizes that the PDEs in ICH Q3D are for humans. If the calculated elemental impurity content of the animal drug product exceeds the PDE in ICH Q3D, the animal drug product sponsor may provide a justification or contact CVM to discuss the product and veterinary-specific considerations.*[12]

While this is a draft guidance, which is in a public comment period at the time of this writing, CVM is proposing that the PDE limits for an adult human referenced in Table 1 would be applicable to veterinary products should this guidance go into effect. When considering species and dosage, it is logical that a weight conversion be applied such as done in Table 1.

Though we do not fully understand the potential risk(s) associated with parenteral exposure to these impurities in swine, it goes without question that there are no therapeutic benefits to their presence in these products. Increasingly, it becomes clear that for heavy metals such as lead, even low levels of exposure can impact human health. For example, low levels of lead, which were once thought to be safe, have recently been implicated as a risk factor in serious human medical conditions including cardiovascular disease mortality.[13] Altogether, we recognize that further research is necessary to fully characterize the severity of the concentrations of arsenic, chromium and lead detected in the veterinary pharmaceutical products reported. The combination of limited information and potential for risk should not be taken lightly by producers or veterinarians especially considering exposure can be avoided. As such, veterinarians are implored to think critically about the risks these unintended impurities may or may not pose to baby pigs, and should consider if willing introduction of arsenic, chromium and/or lead into the food chain, albeit in low concentrations, is in accordance with the high standards of production expected in the swine industry globally.

References

1. Kernkamp H et al. Preventing iron-deficiency anemia in baby pigs. *J Anim Sci.* 1962;21:527–532.

2. Ullrey DE et al. Oral and parenteral administration of iron in the prevention and treatment of baby pig anemia. *J Anim Sci.* 1959;18:256–263.

3. Zimmerman DR et al. Injectable iron-dextran and several oral iron treatments for the prevention of iron-deficiency anemia in baby pigs. *J Anim Sci.* 1959;18:1409–1415.

4. Bhattarai S and Nielsen JP. Association between hematological status at weaning and weight gain post-weaning in piglets. *Livest Sci.* 2015;182:64–68.

5. Dunnigan S et al. The effects of hemoglobin levels on growth rates in pigs. *AASV.* 2015;305–306.

6. Perri AM et al. An investigation of iron deficiency and anemia in piglets and the effect of iron status at weaning on post-weaning performance. *J Swine Health Prod.* 2016;24:10–20.

7. Fredericks L et al. Evaluation of the impact of iron dosage on post- weaning weight gain, and mortality. *AASV.* 2018;315

8. National Pork Board. We Care. 2008. Available at: `https://www.pork.org/about/we-care/`. Accessed 30 Oct. 2018.

9. Iowa Pork Producers Association. The We Care® Responsible Pork Initiative. 2017. Available at: `http://www.iowapork. org/we-care-responsible-pork-initiative/`. Accessed 13 Nov. 2018.

10. Radke S et al. Elemental impurities in injectable products for swine. *J Swine Health Prod.* 2018;26:142-145.

11. AskCVM. Oral communication. May 9, 2017

12. US Food and Drug Administration. Animal & Veterinary Guidance by Number. 2018. Available at:`https://www.fda.gov/AnimalVeterinary/GuidanceComplianceEnforcement/GuidanceforIndustry/ucm042450.htm`. Accessed 7 November 2018.

13. Lanphear BP et al. Low-level lead exposure and mortality in US adults: A population-based cohort study. *Lancet Public Heal.* 2018;3e177-e184.

What is IDA? Experience and success factors used to eliminate iron deficiency anemia and achieve peak performance that lasts a pigs lifetime

T. Gillespie, DVM
Performance Health, PC

Introduction

Iron deficiency anemia (IDA) has been a health issue in the swine industry for decades.[1] An early suggestion was to provide additional iron even when piglets were produced where soil was avialable to them. Veterinarians in European countries that have clients using outdoor production practices have sampled piglets for their hemoglobin (Hb) levels around weaning to find the presence of iron deficiency anemia in a high percentage of piglets.[2] While not a new topic, recent research has shown that prevention of clinical even sub-clinical IDA is rarely achieved when using the long-standard single 200 mg dose of iron.[3-5] This has led to questions asking what the true impact of IDA is and what dose and timing of injections are necessary to achieve peak performance. It is therefore my goal to update readers on clinical practices that have been successful in preventing IDA and to outline the beneficial results that can occur if the proper amount of iron is provided to meet the piglets' demands.

Discussion

When discussing IDA with those in the swine industry, many common questions arise. It seemed logical then, to compile a list of questions and provide practical, science-based answers for each. By doing so, it is my hope that this information will not only be useful to you as a practitioner, but also to the clients and producers you work with on a daily basis in the field.

Why is it important to give piglets an injection of iron?

The piglet's low iron storage at birth (40-50 mg) and the low amount of iron available in the sow's milk (1 mg/day), put piglets at an increased risk of developing an iron deficiency during lactation.[1] Confounding these factors is the rapid growth rate of modern day piglets, which quickly depletes their already low iron stores. The resulting iron deficiency can rapidly lead to an anemia characterized by a lack of iron-containing hemoglobin (Hb) in the red blood cells. When Hb levels are reduced, there is a decreased oxygen carrying capacity of the blood and tissue is not properly perfused with oxygen leading to poor piglet performance seen clinically as reduced growth rates.[3-5]

How does one calculate the iron needs of a piglet?

Equations and recommended daily values are available for estimating the iron demand of a growing piglet.[6-8] While these data provide a range of suggestions, they may not accurately reflect the needs of modern, fast-growing piglets. This is evident by the number of piglets seen with IDA at weaning both in my clinical experience and in a number of research reports.[4,5,7] Perhaps the best way to evaluate if enough iron is being given is to test blood Hb concentrations at weaning. A growing piglet should have a blood Hb concentration over 110 g/L at weaning which has been shown to significantly impact post-weaning growth rates (Table 1). Measuring a piglet's Hb level is easily achieved by using a portable HemoCue® to obtain individual animal data on-farm. Measuring a subset of piglets within a day or two prior to weaning provides an accurate picture of the iron status of the herd. If after testing, you are unsatisfied with the number of clinically or sub-clinically anemic piglets at weaning, consideration should be given to modifying your iron supplementation program to avoid an "iron gap".

Can you explain this "iron gap" term?

Fast growing piglets in modern swine units where proper Day 1 Pig Care is provided, achieve tremendous weight gains from birth to weaning. An all milk ration increases the risk of developing clinical or sub-clinical IDA. This combination of factors contributes to a potential "iron gap" which is a period of time when a piglet has insufficient iron stores (Figure 1). The "iron gap" occurs when all available iron stores, including that from an iron injection early in the pigs life, have been depleted before the piglet can adequately absorb iron from their diet post-weaning. The "iron gap" is characterized by a piglets blood Hb concentration falling below optimal levels during lactation and the early post-weaning period.

Should I be giving our pigs more than 200 mg of injectable iron?

As discussed in reference to a piglets' iron demand and the "iron gap", the requirement for iron in nursing piglets even when provided with a parenteral injection of 200 mg of iron may not be satisfatory. Many ask if increasing the dosage from 200 mg to 300 mg would be sufficient in meeting iron demands and improving piglet performance. Several studies have evaluated this question and in all cases, the 300 mg injection was not detrimental to piglets' health but only resulted in non-significant numerical improvments, if any, in hematologic profiles.[10-12] None of these, or any other study, has found a 300 mg dosage of injectable iron to provide a stastically significant improvement in weight gain.

Table 1: Impact of blood hemoglobin concentration at weaning on post-weaning growth performance. Data adapted from Fredericks et al, 2018.[4]

Hb status at weaning[3,9]	Blood Hb concentration (g/L)	Live weight (lbs.)	
		8-weeks post wean	18-weeks post wean
Deficient	< 90	64.7	217.1
Sub-clinical	90-110	68.5	220.7
Optimal	> 110	72.4	226.9

Figure 1: Hypothetical model of the "iron gap" assuming 7 kg of weight gain over a 28 day lactation. Figure adapted from Van Gorp and von der Recke, 2012.[8]

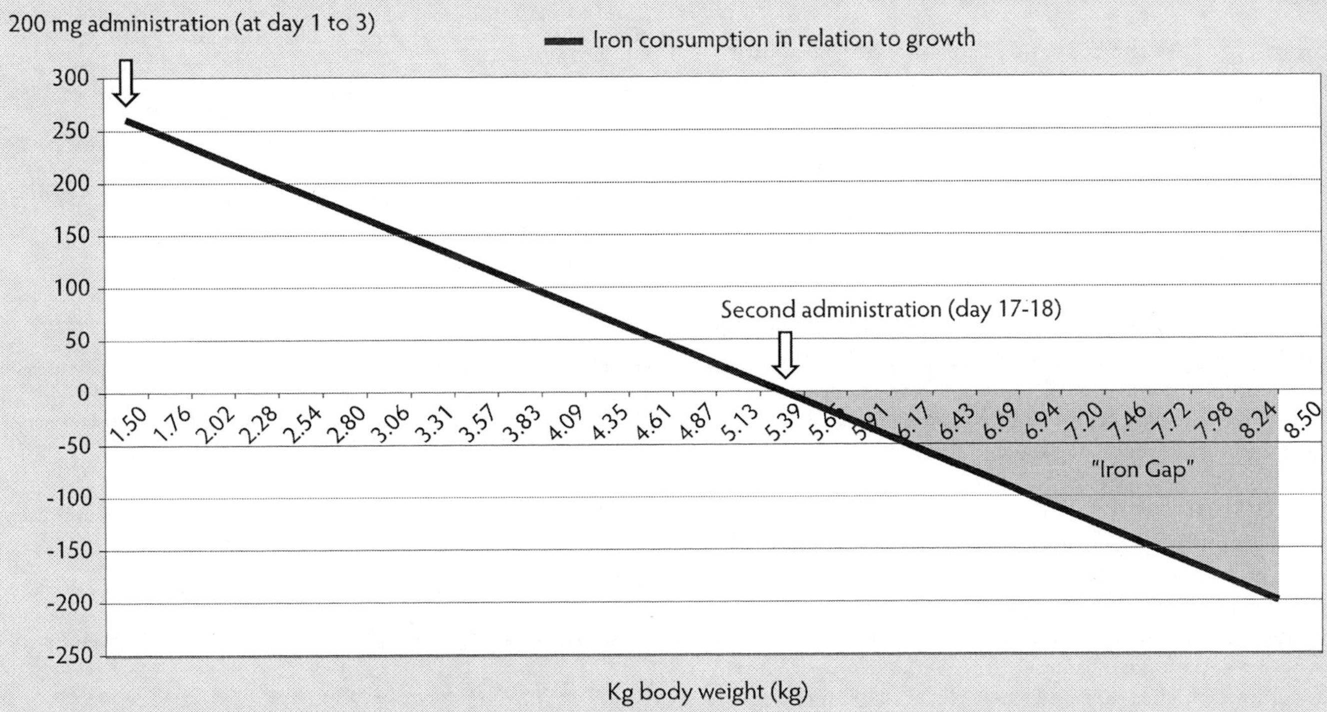

On the other hand, it has been shown that giving a second 200 mg injection of iron (400 mg total in two injections) during lactation does result in significantly improved Hb status and weight gain.[3-5] When giving 400 mg, it must be administered in two injections several days apart. I do not recommend administering a single 400 mg injection to piglets as no research has evaluated the safety of this dose in young piglets.

Are there any lasting effects of IDA and when should I give a second dose of iron?

While more research is needed to characterize any lasting detrimental affects of early-life iron deficiencies on body tissues and performance, some information is available. One recent study does indicate the importance of preventing piglets from entering the iron gap.[13] The authors showed that neuroimaging can detect undesirable structural and physiological changes in the brain of piglets due to early-life iron deficiency. This included several parameters, including grey and white matter volumes, iron contents, as well as reduced subcortical white matter integrity, despite a subsequent period of dietary iron repletion. In another peer-reviewed study by the same authors, they concluded that despite nutritional repletion of iron status after a period of iron deficiency, growth never fully recoverd relative to pigs that never entered an iron deficient state.[14] In a separate study it was documented that a statistically significant decrease in circulating B-lymphocyte numbers was found in non-supplemented animals. Iron deficiency seemingly has a negative influence on the immunocompetence in piglets.[15]

After deciding to give a second injection of iron, careful consideration must be given on the proper timing. Weaning is a tempting time to consider this activity given pigs are already being handled. However, waiting this long risks piglets entering the "iron gap" leading to reduced or no improvement in growth performance.[7,16] From an efficacy standpoint a second dose of injectable iron is best if given around 10 days of age before piglets become Hb deficient.[7] Ultimately, the goal is to provide a second injection 5-10 days after the initial injection to provide the piglet with enough iron to provide perfusion of its tissues, maximize weight gains, and prevent potential structural and immunological changes. Practically, this activity can be combined with another piglet handling procedure such as processing or vaccination if the timing and age of piglet allows. One program that has been successful is to administer the first injection on day 1 or 2 and follow with the second injection during processing at day 5-7.

How can I justify the cost and labor of giving a second 200 mg dose of iron?

A second injection of iron dextran does require more cost and labor so several things must be considered when determining if and when to give a second dose of iron. First and foremost, as veterinarians and producers, we have an obligation to do what is best for the piglet and provide it the best opportunity to thrive. If significant IDA is present on the farm, no justification should be necessary to provide the piglet the nutrients required to thrive, in this case, iron. In non-severe cases, one must consider what the return value will be in making this investment. For some, this may be in increased wean weights, for others, it may be in improved wean-finish gain. Studies have shown improvements in both which would likely pay for the costs of a second injection and more (Table 2).

Conclusion

Modern production practices promoting Day 1 Pig Care, with increasing number of piglets born per litter, and their rapid growth during the lacation and nursery phases has caused practitioners to re-evaluate an early limiting factor, iron requirements. The "iron gap" occurs as piglets continues to grow without proper body stores of iron which is captured when Hb levels are measured. Studies have shown that additional amount of iron injected in a single or as multiple injections has been beneficial to the piglet. Additional weight gain through finishing has been documented when additional iron dextran, i.e. especially when separate injections of 200 mg each, are utilized. There are scientific studies indicating potential negative structural and immunologic changes to the piglet when proper amounts of iron dextran is not used. It is the author's experience that additional iron should become a normal practice in modern swine production systems. The ease of capturing piglets' Hb levels at or near weaning has promoted this activity and provides a simple and inexpensive method to evaluate your herd.

References

1. Venn J et al. Iron Metabolism in Piglet Anemia. *J Comp Pathol.* 1947;57:314-325.
2. Waddilove, J. The Piggery Vet Ltd. Personal Communication. 2014.
3. Bhattarai S and Nielsen JP. Association between hematological status at weaning and weight gain post-weaning in piglets. *J Livest Sci.* 2015;182:64-68.
4. Fredericks L et al. Evaluation of the impact of iron dosage on post- weaning weight gain, and mortality. *AASV.* 2018;315
5. Perri AM et al. An investigation of iron deficiency and anemia in piglets and the effect of iron status at weaning on post-weaning performance. *J Swine Health Prod.* 2016; 24:10–20.
6. Kamphues J et al. Effects of a 2nd iron injection in suckling piglets on iron retention and performance before and after weaning. *IPVS.* 1992;601.
7. von der Recke C et al. Defining, evaluating and understanding iron deficiency and anemia in modern swine production. *AASV.* 2014;173-174.
8. Van Gorp S et al. Preventing iron deficiency by avoiding an iron gap in modern pig production. *AASV.* 2012;407-408.
9. Thorn CE. Hematology of the pig. In: Weiss DJ and Wardrop JK eds. *Shalm's Veterinary Hematology.* 6th ed. Ames, Iowa. Wiley-Blackwell. 2010:843.
10. Gaddy H et al. A review of recent supplemental iron industry practices and current usage of Uniferon™ (iron dextran complex injection, 200 mg/mL) in baby pigs. *AASV.* 2012;167–171.
11. Jolliff J et al. Effect of injected and dietary iron in young pigs on blood hematology and postnatal pig growth performance. *J Anim Sci.* 2011;89:4068-4080.
12. Murphy K et al. Effects of weaning age and dosage of supplemented iron on the hemoglobin concentrations and growth rate of piglets. *J Swine Health Prod.* 1997;5:135-138.
13. Mudd A et al. Early-life iron deficiency reduces brain iron content and alters brain tissue composition despite iron repletion: A neuroimaging assessment. *Nutrients.* 2018;10:135.
14. Knight LC and Dilger RN. Longitudinal effects of iron deficiency anemia and subsequent repletion on blood parameters and the rate and composition of growth in pigs. *Nutrients.* 2018;10;632
15. Svoboda M et al. Impairment of the peripheral lymphoid compartment in iron-deficient piglets. *J Vet Med.* 2004;51:231–237.
16. Peters JC and Mahan DC. Effects of neonatal iron status, iron injections at birth, and weaning in young pigs from sows fed either organic or inorganic trace minerals. *J Anim Sci.* 2008;86:2261-2269.

Uniferon® is a trademark of Pharmacosmos A/S
HemoCue® is a trademark of HemoCue AB

Table 2: Effect of Uniferon® dose and timing on pig growth performance. Data adapted from Fredericks et al, 2018.[4]

	Iron dose		
Live weight (lbs.)	**200 mg on day 1**	**200 mg on day 1, 200 mg on day 5-7**	***P*-value**
Birth	3.41	3.39	0.07
Weaning	11.9	12.3	< 0.001
8 weeks post wean	63.6	65.2	0.001
18 weeks post wean	213.2	216.3	0.02

Integrated fly control at an eastern Corn Belt swine finishing unit

T. Gillespie[1], DVM, ABVP; K. C. Thorne[2], BA
[1]Performance Health, PC; [2]Piedmont Animal Health

Introduction

Current approaches in fly control within the farm animal business are varied and based on personal decision according to successes and failures, as well as recommendations of manager's, veterinarians, and pest control experts. There has been much improvement in pest management methodology over earlier strips, traps, and chemicals, to now include adulticides, predators (i.e. beetles), insect growth inhibitors, insect growth regulators, as well as other methods. Most experts now agree that an integrated pest management approach is the key to fly control.[1] While farmers are aware of the reductions in production and potentially serious economic losses flies can create on the growth and performance of farm animals,[2] what is not commonly known is the extent of the negative impact under various production practices.

Materials and methods

A study was recently conducted in the Eastern Corn Belt, to evaluate the efficacy and convenience in applying Flynexx® (Cyromazine) an insect growth regulator from Piedmont Animal Health, within a swine finishing unit. The site consisted eight finishing rooms in four barns, housing 1070 head per room. Each room was approximately 8,000 square feet each. The goal of the study was to demonstrate significant reduction of adult fly populations in treated versus untreated finishing rooms. Diagnostically, oral fluid samples collected from ropes, plus adult flies were also collected, and were tested for bacteria and viruses. In this study, every room exhibited heavy to extremely high fly burdens, along with extremely high levels of house fly larvae in the manure pits. Manure handling was all-in/all-out by room, via pull-plug to the lagoon. No pumping of the pits was done during the life of the study. One room was not populated at the beginning of the study, and was intentionally left empty and untreated, allowing for repairs to be completed, with cleaning and disinfecting performed prior to loading with nursery aged animals. All other rooms housed animals for the duration of the study. Fly traps were deployed in hallways and in all rooms. Fly counts were made pre and post treatment. Flynexx larvicidal granules were mixed with water and used as a spray at the labeled rate (one pound per 200 square feet), and was applied evenly across the pit. No other fly control products were used.

Results and discussion

It is important to understand the fly life cycle and realize only 15-20% of a fly population exists as adult flies at any one time,[3] thus treatments that focus on elimination and/or reduction of the adults in a population is generally not an effective way to eliminate the population,[4] as was evidenced in this study. Fly species identified were *Musca domestica Linneaus* and *Fannia canicularis*, the common house fly and lesser common house fly. The highest increase in fly counts were recorded in the room unpopulated at the first counting, and repopulated with new pigs from the nursery before the second counts were made (Table 1). The data illustrates a rapid onset of fly activity within a short period of time after introducing new populations into a "clean room". There is evidence to indicate that despite proper management procedures, this room quickly sustained high fly infestations due to the activation of key trigger element; heat, humidity, carbon dioxide from exhalation, etc. This observation is a significant key finding given the extensive clean up prior to re-population. The rapid and massive increase of flies in this room shows that the application of Flynexx is a valuable tool in managing fly control populations during the "turn" and reducing the fly pressure as new pigs are introduced (Figure 1).

Current farm management's health protocol is to monitor the site monthly using oral fluid samples and serology for the presence of PRRSV, *Mycoplasma hyopneumoniae*, and Influenza type A virus. At the time this site was selected for evaluation, there were no clinical signs of major pathogen activity, although testing of the adult flies did confirm the presence of several less major bacteria. PRRSV, *Mycoplasma hyopneumoniae*, and Influenza type A virus were not detected on the adult flies that were submitted. Adult houseflies can potentially transmit swine pathogens between commercial pig farms, however several studies have shown that adult flies can serve as biological and mechanical vectors under experimental conditions.[5,6]

Conclusions

The data gained in this study indicates that testing of the bacteria-types and counts is conclusive enough to indicate that adult flies need to be considered as a method of transmission throughout a room. There is evidence to conclude that using Flynexx as an insect growth regulator (IGR) is an effective method of cing and breaking the house fly life cycle, and thus reducing population burdens within a swine operation. It is recommended as the foundation of any fly control program in a swine finishing facility where pigs are located on slats with manure pits located below the pigs. Furthermore, application when houses are emptied, and during each turn, is the most convenient and effective time to reduce fly populations.

Table 1: FLYNEXX with 1-2-3 PIT applicator pig trial

Treatment	Location	Count 1	Count 2	Count +/-	% Change
Untreated – no pigs	**Room 1**	**50**	**1068**	**1018**	**2036%**
Fully treated	**Room 2**	**548**	**251**	**-297**	**-54.2%**
Not treated	Room 3	248	556	308	124.2%
Partially treated	**Room 4**	**533**	**551**	**18**	**3.4%**
Not treated	Room 5	146	735	589	403.5%
Not treated	Room 6	148	234	86	58.1%
Not treated	Room 7	217	389	172	79.3%
Not treated	Room 8	177	648	471	266.1%
Not treated	Hallway 1	1026	1539	513	50.0%
Not treated	Hallway 2	1178	1811	633	53.7%
Not treated	Hallway 3	1045	1972	927	88.7%

Figure 1: FLYNEXX with 1-2-3 PIT applicator pig trial

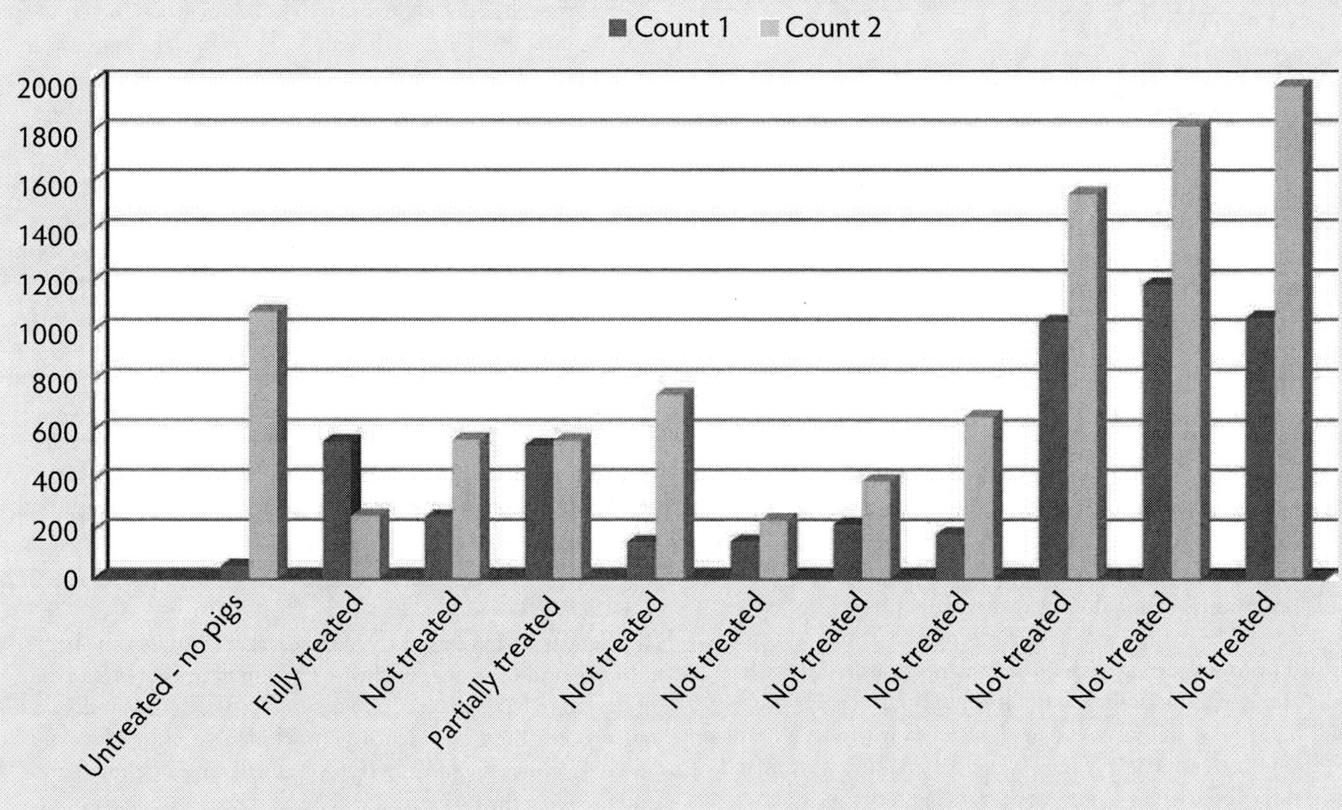

References

1. Westenbroek. 2008. Integrated Pest Management for Fly Control in Maine Dairy Barns.

2. Stork, M.G. 1979. The epidemiological and economic importance of fly infestation of meat and milk producing animal in Europe. Vet Rec., v. 105, p. 341-343.

3. Novartis, 2008. Anti-fly program, an integrated approach. Retrieved from `www.flycontrol.novartis.com/antifly/en/index.shtml` October 17, 2008.

4. Ellis, R. 2002. Mosquito and fly prevention and control on Manitoba Farms.

5. Satoshi Otake, Scott A. Dee, Roger D. Moon, Kurt D. Rossow, Carlos Trincado, Carlos Pijoan. Evaluation of mosquitoes, Aedes vexans, as biological vectors of porcine reproductive and respiratory syndrome virus. The Canadian Journal of Veterinary Research, 2003; 67:265-270.

6. Satoshi Otake, Scott A. Dee, Roger D. Moon, Kurt D. Rossow, Carlos Trincado, MacDonald Farnham, Carlos Pijoan. Survival of porcine reproductive and respiratory syndrome virus in houseflies. The Canadian Journal of Veterinary Research, 2003; 67:198-203.

Field experience with an adjuvanted autogenous hard-to-grow PRRSV vaccine

Mark Wagner[1]; Tim Miller[2]; David Nolan[2]
[1]Fairmont Veterinary Clinic, Fairmont Minnesota; [2]Huvepharma, Peach Tree City, Georgia

Introduction

Our understanding of how best to stimulate and maintain a broad cross protective immunity to PRRS virus in the previously primed (PRRS ELISA positive, PRRS PCR negative) sow continues to be a challenge. Available for stimulating immunity include several commercially available modified live vaccines, killed autogenous vaccines, and live virus inoculation. The purpose of this study was to evaluate the use of an adjuvanted autogenous killed PRRS vaccine given pre-farrow in previously primed pregnant animals challenged late gestation with a heterologous PRRS field virus that was similar to the killed autogenous vaccine PRRS isolate and the impact on pig performance during the nursing and nursery periods.

Materials and methods

Twelve sows divided into two groups of six, matched by parity were acquired from a commercial farm. These sows had historically seen and been acclimated to two farm specific field virus with RFLP's of 1-7-4 and 1-26-2. A three strain adjuvanted killed PRRS vaccine was given at 35 days of gestation. A second dose was given at 70 days of gestation. The RFLP's of vaccine virus included a 1-7-4, 1-3-4, and 1-8-4. The challenge virus had a RFLP of 1-8-4. Table 1 summarizes the difference in PRRS isolate sequence data.

At 95 and 102 days of gestation, all sows were challenged intra-muscularly with a 2cc dose of PRRS field virus with a RFLP of 1-8-4. Challenge dose had a CT value of 22 and Quantitative PRRS PCR value of $10^{\wedge 6}$. Blood samples were collected at time of challenge and 7 days post second challenge. All pigs were processed between 3-5 days of age with processing fluids collected between treatment groups. Pigs were weaned at 24 days of age at which time each pigs was identified by tag, weighed, and serum collected. PRRS PCR testing was done on each individual pig. Pigs were weaned into a conventional nursery and co-mingled. At 16 days post wean all pigs were re-bled and re-weighed. PRRS PCR testing was additionally completed. Data was also collected on mortality.

Results

PRRS ELISA results prior to challenge averaged 1.44. Post challenge averaged 2.05. Additionally sows tested PRRS PCR negative at seven days post challenge. No sow abortions or mortality occurred post challenge.

Table 2 summarizes individual sow response.

Table 3 summarizes farrowing performance.

Table 4 summarizes nursery performance.

Summary/Discussion

The challenge virus used in this study clinically had more impact on the pig as compared to the sow. Data is pending that is analyzing serum neutralization titres in the challenged and vaccinated sows against the challenge virus to determine if this had any impact on the clinical picture in sows. It is possible that since the sows had naturally seen two different field viruses that this positioned these sows to have better cross protection to the heterologous challenge virus. What is of interest in this study is the differences in the performance of the control offspring as compared to the vaccinated offspring. Often one of the main problems in the field is how to best improve the quality of the PRRS viremic weaned pig. The data shows there were fewer pigs weaned from the vaccinated sows that became infected with the

Table 1: Field experiences with an adjuvanted autogenous hard to grow PRRS vaccine

Virus description	% similar to 1-7-4 farm virus	% similar to 1-26-2 farm virus	% similar to 1-7-4 vaccine virus	% similar to 1-3-4 vaccine virus	% similar to 1-8-4 vaccine virus	% similar to 1-8-4 challenge virus
Farm 1-7-4	100%	90.2%	97.8%	86.7%	87.7%	86.5%
Farm 1-26-2	89.8%	100%	89.2%	85.3%	85.9%	83.6%
Vaccine 1-7-4	97.8%	90.2%	100%	87.1%	87.7%	86.5%
Vaccine 1-3-4	85.4%	85.3%	86.7%	100%	87.1%	86.9%
Vaccine 1-8-4	88.1%	85.9%	87.7%	87.1%	100%	99.5%
Challenge 1-8-4	86.5%	83.6%	86.5%	86.9%	99.5%	100%

field virus after being co-mingled in the nursery. One possibility for this is a difference in maternal immunity. Additionally there was a improvement in weight gain and reduced nursery mortality in the vaccinated sow offspring. Findings in this study suggest utilization of a killed autogenous vaccine pre-farrow may improve weaned pig performance.

Table 2: Individual sow response

Sow ID	PRRS ELISA value day of challenge	PRRS ELISA value 7 day post challenge
60415	1.27	1.82
61585	.79	1.71
61904	1.96	2.05
62588	.98	1.89
60356	1.67	1.85
62481	.24	1.91
61722	1.84	2.05
60664	2.42	3.10
62538	1.43	1.68
60516	1.98	2.51
61719	1.82	2.12
62231	.44	2.01

Table 3: Farrowing performance

Performance feature	Control sow offspring	Vaccinated sow offspring
Total born	13.0	12.4
Stillborn / mummies	1.5	1.1
Live born	11.5	11.3
Processing fluid samples PRRS PCR results	Positive (ct = 23)	Negative
Wean average	10.3	9.7
Pre-wean mortality	10.4%	14.1%
Average weight at weaning	15.0lbs	17.5lbs
Number of PRRS PCR positive pigs at weaning	11 of 62 (17.8%)	0 of 58 (0%)

Table 4: Nursery performance

Performance feature	Control sow offspring	Vaccinated sow offspring
Number of PRRS PCR positive pigs at weaning	11 of 62 (17.8)	0 of 58 (0%)
Average weight at weaning	15.0lbs	17.5lbs
Average weight at 16 days post wean	19.6lbs	24.0lbs
Nursery mortality (Day 1 – day 16 post wean)	6 of 62 (9.6%)	4 of 58 (6.9%)
Number of PRRS PCR positive pigs at weaning	16 of 56 (28.6%)	3 of 54 (5.5%)

Efficacy of prime-boost vaccination protocols in pigs challenged with influenza A viruses

M. Culhane[1], DVM, PhD; M. Torremorell[1], DVM, PhD; C. Li[1], DVM, MS; M. Jansen[2], DVM; M. Mellencamp[2], PhD; D. Amodie[2], MS; L. Galina-Pantoja[2], DVM, PhD
[1]College of Veterinary Medicine, University of Minnesota, St. Paul, Minnesota; [2]Zoetis, Parsippany, New Jersey

Introduction

There is an urgent need to evaluate new vaccine protocols against emerging influenza A virus (IAV) strains in pigs.

The heterologous prime-boost vaccination protocol has been evaluated in many species against a variety of pathogens.[1-3] This approach involves presenting antigen to the immune system in more than one way; i.e. using different strains of a pathogen, different adjuvants, different inoculation routes, and/or different types of vaccines (modified live vs. inactivated). Heterologous prime boost vaccination with whole inactivated virus (WIV) influenza vaccines has been demonstrated to broaden the antibody response and increase protection in pigs against H3N2 strains.[4] We conducted a vaccination-challenge experiment to evaluate the efficacy of different heterologous prime-boost vaccination protocols.

Materials and methods

This project was approved by the University of Minnesota Animal Care and Use Committee. Porcine reproductive and respiratory syndrome (PRRS) virus, *Mycoplasma hyopneumoniae* and IAV-S negative pigs (n = 90) were weighed and randomly allocated by body weight into treatment groups (Table 1) and placed into their respective isolation rooms. Within vaccine type, pigs from each treatment group were comingled and vaccinated as per allocation (Table 1). Vaccine strains varied in the percent homology to challenge strains based on HA amino acid comparison (Table 2). Pigs were weighed, bled, and received their assigned vaccine at 3 weeks of age (priming dose; Day 0 of the study) and 21 days later (booster).

Two weeks after the booster vaccine, pigs were challenged using a seeder pig model. The seeder pigs received an intratracheal and intranasal challenge with either H1N1 (A/Swine/Minnesota/PAH-618/2011 H1N1 Clade 1A3.3.3 gamma) (n = 7) or H3N2 (A/Swine/Minnesota/2015 H3N2 Cluster human-like) (n = 7). One seeder pig for each IAV-S strain was placed in each room and had unrestricted nose-to-nose contact with the pigs in the treatment groups.

The following measurements were taken at prescribed times during the study: core body temperature, presence of clinical signs, serum antibody titers, gross lung lesions, and IAV detection by PCR and/or virus isolation from specimens such as nasal swabs, oropharyngeal swabs, lungs, and bronchoalveolar lavage fluid (BALF).

Table 1: Study design and treatment allocations

Treatment group number[1]	Vaccination[2] (3 wk /6 wk)	Number of pigs in each treatment group/room
T01	COM[3]/COM	12 (10 contacts and 2 seeders)
T02	AUT[4]/AUT	12 (10 contacts and 2 seeders)
T03	AUT/COM	12 (10 contacts and 2 seeders)
T04	COM/AUT	12 (10 contacts and 2 seeders)
T05	LAIV[5]/None	12 (10 contacts and 2 seeders)
T06	LAIV/COM	12 (10 contacts and 2 seeders)
T07	None/None	12 (10 contacts and 2 seeders)
T08	None/None	6 (not challenged)

1 Treatments 1, 2 ,3 ,4, and 7 were in Rooms 3, 5, 7, 9, and 11; Treatments 5 and 6 were in Rooms 8 and 10; Treatment 8 was housed in Room 1.
2 Treatments were applied at 3 and/or 6 weeks; according to label.
3 COM = Commercial vaccine: FluSure XP® a whole, inactivated commercial multivalent vaccine which includes two H3N2 strains (Cluster IV-A and IV-B), one H1N1 strain (Clade 1A.3.2 gamma-2) and one H1N2 strain (Clade 1B.2.2.2 delta-1b).
4 AUT = Autogenous vaccine; a whole, inactivated autogenous multivalent heterologous vaccine which includes one H3N2 strain (Cluster human-like), one H1N2 strain (Clade 1B.2.2.1 delta-1a) and one H1N1 strain (Clade 1A3.3.3 gamma).
5 LAIV = Ingelvac Provenza™, a live-attenuated vaccine with one H3N2 strain (Cluster I) and one H1N1 strain (beta-gamma2-like Clade 1A.2.3-like) that have truncated non-structural influenza (NS-1) proteins.

Table 2: Hemagglutinin amino acid identity (%) between challenge and vaccine IAV-S strains.

Challenge virus	Vaccines used in this study[1]		
	COM	AUT	LAIV
H1N1	95.1%	96.5% and 97.3%	89.2%
H3N2	87.1% and 88.2%	99.1%	89.2%

[1] COM = Commercial vaccine: FluSure XP® a whole, inactivated commercial multivalent vaccine which includes two H3N2 strains (Cluster IV-A and IV-B), one H1N1 strain (Clade 1A.3.2 gamma-2) and one H1N2 strain (Clade 1B.2.2.2 delta-1b).
AUT = Autogenous vaccine; a whole, inactivated autogenous multivalent heterologous vaccine which includes one H3N2 strain (Cluster human-like), one H1N2 strain (Clade 1B.2.2.1 delta-1a) and one H1N1 strain (Clade 1A3.3.3 gamma).
LAIV = Ingelvac Provenza™, a live-attenuated vaccine with one H3N2 strain (Cluster I) and one H1N1 strain (beta-gamma2-like Clade 1A.2.3-like) that have truncated non-structural influenza (NS-1) proteins.

Statistical analyses

Statistical analyses were performed by comparing groups T01, T02, T03, T04 and T07 and separately comparing groups T05 and T06. Group T08 was summarized but not statistically analyzed. PCR results from nasal swabs and antibody responses were analyzed by a generalized linear mixed model (GLM) approach for repeated measures. . Percent lung lesions and BALF were analyzed using a GLM approach with a model that considered the fixed effect of treatment and the random effects of room and the residual error. Comparisons of LSMeans for all variables were performed by the two-sided Student's t-test at the 5% level of significance.

Results

No significant differences were observed between any treatment groups for temperature, clinical signs, or lung lesions. In the WIV groups (T01-T04), all vaccinated pigs shed less virus in nasal swabs compared to non-vaccinated controls (T08), even though there were no differences in the percentage of time nasal swabs were positive for IAV (Table 3). Virus was not isolated from nasal swabs collected from pigs vaccinated with AUT/COM. In the LAIV groups (T05-T06), percent of pigs that tested positive (nasal swabs) was similar for LAIV alone or when given in combination with commercial WIV vaccine. However, adding commercial vaccine as a booster showed significant decrease in percentage of days shedding post challenge (45.92% vs. 16.31%, respectively; $P \leq 0.05$) (Table 4). Moreover, pigs in LAIV alone shed significantly more virus compared to pigs vaccinated with LAIV followed by COM. For the WIV groups, pigs vaccinated with either COM or AUT had reduced % positive BALF samples by PCR compared to non-vaccinated controls (Table 3). No virus was isolated from BALF in groups that received a heterologous prime boost protocol (AUT/COM and COM/AUT). Serum hemagglutination inhibition antibody titers before challenge were increased in all COM and AUT vaccinated pigs compared to non-vaccinated controls. HI titers for AUT/COM and COM/AUT were higher than COM/COM and AUT/AUT for H1N1 (Table 3). The HI titers for HIN1 and H3N2 were increased in LAIV/COM pigs compared to LAIV/None pigs (Table 4).

Discussion

This study shows the advantage of vaccinating pigs for IAV-S with a heterologous prime boost protocol for both WIV and LAIV. Although the AUT vaccine has greater homology to the challenge strains than COM, a heterologous boost with COM was advantageous for reducing viral shedding and increasing H1 antibody response. For the LAIV, results showed that the prime boost vaccination protocol reduced the percentage of days shedding. There were significantly lower IAV-S levels in BALF and higher antibody titers for both H1 and H3 in the animals that received the prime boost protocol. It is important to repeat these studies and understand how to apply this vaccination approach in commercial swine production.

References

1. Steensels M, Bublot M, Van Borm S, De Vriese J, Lambrecht B, Richard-Mazet A, Chanavat-Bizzini S, Dubouef M, Le Gros FX, van den Berg T. Prime-boost vaccination with a fowlpox vector and an inactivated avian influenza vaccine is highly immunogenic in Pekin ducks challenged with Asian H5N1 HPAI. *Vaccine*. 2009; 27:646-654.

2. Woodberry T, Gardner J, Elliott SL, Leyrer S, Purdie DM, Chaplin P, Suhrbier A. Prime boost vaccination strategies: CD8 T cell numbers, protection, and Th1 bias. *J mmunol*. 2003;170:2599–2604.

3. Wu L, Kong WP, Nabel GJ. Enhanced breadth of CD4 T-cell immunity by DNA prime and adenovirus boost immunization to human immunodeficiency virus Env and Gag immunogens. *J Virol*. 2005; 79:8024–8031.

4. Van Reeth K, Gracia JCM, Trus I, Sys L, Claes G, Versnaeyen H, Cox E, Krammer F, Qiu Y. Heterologous prime-boost vaccination with H3N2 influenza viruses of swine favors cross-clade antibody responses and protection. *NPJ Vaccines*. 2017; 2:1-11.

Table 3: IAV-S in nasal swabs and brochoalveolar lavage fluid (BALF) from WIV-vaccinated groups (T01-T04) or negative control (T07).

Sample	LSMeans + Standard error				
	T01 COM/COM	T02 AUT/AUT	T03 AUT/COM	T04 COM/AUT	T07 None/None
Nasal swab					
% pigs positive[1]	65.96 ± 34.05[a]	65.96 ± 34.05[a]	82.1 ± 22.92[a]	82.1 ± 22.92[a]	92.42 ± 11.45[a]
% days shedding post challenge[1,2]	3.53 ± 1.42[b]	3.97 ± 1.42[b]	4.20 ± 1.42[b]	4.09 ± 1.42[b]	6.48 ± 1.42[a]
Virus titer, TCID50/mL[3] Day 42[4]	0.12 ± 0.079[b]	0.07 ± 0.079[b]	0.03 ± 0.079[b]	0.08 ± 0.079[b]	0.49 ± 0.079[a]
BALF					
% positive[1,2]	60.65 ± 4.23[ab]	28.90 ± 28.89[bc]	3.65 ± 5.81[c]	28.90 ± 28.89[bc]	91.44 ± 13.01[a]
Virus titer, TCID50/mL[3]	0.32 ± 0.094[a]	0.13 ± 0.094[b]	0.03 ± 0.094[b]	0.11 ± 0.094[b]	0.45 ± 0.094[a]
H1N1 geometric mean HI titer	20	17	61	40	5
H3N2 geometric mean HI titer	92	226	368	121	7

LSMeans within a row with same letter grouping are not significantly different ($P \leq 0.05$)
1 As determined by PCR (positive defined as Ct ≤ 38)
2 % days shedding post challenge was calculated by [(number of positive nasal swabs)/(number of days sampled)]*100.
3 Virus titer-positive defined as ≥ 1.75; negative results were reported as "< 1.75" and for analysis purposes assigned a value of 0.875
4 Day 42 was 6 days post exposure to infected seeder pigs

Table 4: IAV-S in nasal swabs and brochoalveolar lavage fluid (BALF) from LAIV-vaccinated groups (T05-T06).

Sample	LSMeans + Standard error	
	LAIV/none	LAIV/COM
Nasal swab		
% pigs positive[1]	94.31 ± 16.21[a]	76.13 ± 53.75[a]
% days shedding post-challenge[1,2]	6.78 ± 2.54[a]	4.04 ± 2.54[b]
Viral titer, TCID50/mL[3] Day 42[4]	0.40 ± 0.097[a] (1.49)	0.13 ± 0.097[b] (0.34)
BALF		
% positive[1*]	100[a]	50[b]
Virus titer, TCID50/mL[3]	0.61 ± 0.141[a]	0.20 ± 0.141[b]
H1N1 geometric mean HI titer	2	34
H3N2 geometric mean HI titer	2	49

LSMeans within a row with same letter grouping are not significantly different ($P \leq 0.05$)
* Fishers Exact Test
1 As determined by PCR (Positive defined as Ct ≤ 38)
2 % days shedding post challenge was calculated by [(number of positive nasal swabs)/(number of days sampled)]*100.
3 Virus titer-positive defined as ≥ 1.75; negative results were reported as "< 1.75" and for analysis purposes assigned a value of 0.875
4 Day 42 was 6 days post exposure to infected seeder pigs

Comparison of growth performance and mortality of pigs vaccinated with Fostera® GOLD PCV MH and Circumvent® PCV-M G2 in a commercial Iowa production system

N. Garbes, DVM; K. Vonnahme, PhD; D. Amodie, MS; M. Mellencamp, PhD; E. Greiner; L. Galina, DVM, PhD
Zoetis, Parsippany, New Jersey

Introduction

Porcine circovirus (PCV), a member of the genus circovirus, family circoviridiae, is a small, nonenveloped, single-stranded DNA virus with a circular genome of 1.8 to 2.3 kb.[1] PCV is divided into three different types, PCV1, PCV2, and PCV3. PCV1 was first discovered as a non-cytopathic contaminant on the continuous porcine kidney cell culture PK15 in 1974, and is non-pathogenic in pigs. PCV2, although present in pigs since 1969, was first reported to be associated with porcine multisystem wasting syndrome (PMWS) in the 1990s. The pathogenicity of PCV3 is still unknown (Franzo et al., 2018).[2]

Mycoplasma hyopneumoniae (MH) is the causative agent of Enzootic Pneumonia. *M. hyopneumoniae* primarily infects epithelial cilia and epithelial cells on the surface of trachea, bronchi and bronchioles and causes cilia and epithelial cell damage. Pigs of all ages can be infected through aerosol and direct contact, but clinical disease generally becomes apparent in growing and finishing pigs.

PCV2-associated diseases (PCVAD) have become the most economically important emerging disease worldwide affecting pigs primarily between 5 to 18 weeks of age with clinical signs of progressive weight loss, dyspnea, tachypnea, jaundice, anemia and mortality up to 30%. Similarly, MH has been continuously causing significant economic losses to swine industry through growth retardation and delayed time to market. As a major contributor in Porcine Respiratory Disease Complex (PRDC), clinical swine mycoplasma pneumonia cases are routinely identified as the mixed infections with other pathogens such as PCV2.[1]

Porcine circovirus 2 has the highest mutation rate reported for single-stranded DNA viruses.[3] The genetic gap between commercial vaccines and field viruses is widening, with prevalence of field viruses shifting from PCV2b to PCV2d. Homology studies have shown that PCV2d is more similar to PCV2b than to PCV2a.[4] Fostera® GOLD PCV MH is a new product that contains both PCV2a and PCV2b genotypes.

The objective of this study was to compare Zoetis PCV 2a2b MH vaccine (Fostera® GOLD PCV MH) with Circumvent® PCV-M G2 (Merck) throughout the nursery and finishing phases. The hypothesis was that growth performance and mortalities would be similar between pigs vaccinated with Fostera® GOLD PCV MH or Circumvent® PCV-M G2.

Materials and methods

This project was approved by the Zoetis Ethical Review Board prior to initiation. It was also conducted in compliance with the producer's animal care and handling guidelines. Rations were formulated according to the farm's usual ration for nursery and finisher pigs and exceeded NRC requirements. Only clinically healthy animals were included in this study.

Pigs (n = 2648) originated from the same sow farm and were transported to a nursery site on 2 consecutive days (~ 1370 per day). The herd health of the sow farm was characterized as PRRS naïve, MH negative, PCV stable, and Influenza (IAV-S) positive/active at the start of the study. All pigs were housed in one room with 72 pens with 38 piglets per pen. Upon entry to the nursery, pigs were weighed and ear tags were placed. Pigs were randomly allocated to one of two treatment groups in this randomized complete block trial. Although male and female pigs were housed separately, animals from both treatments were commingled in a single pen. Therefore, pig was the experimental unit. Blocking was based upon each pig's weight obtained upon arrival at the nursery. Treatments included: 1) Fostera® GOLD PCV MH- pre-license serial (Zoetis; n = 1322) and 2) Circumvent® PCV-M G2 (Merck Animal Health; n = 1324) with one 2 ml dose injected intramuscularly at 3 weeks of age (day of weaning). The vaccinations were given in the nursery as pigs were allocated to pen. All piglets were vaccinated for PRRS (Boehringer Ingelheim). After 6.5 weeks, all pigs were individually weighed and transported to the finisher. After 130 days on feed (prior to first marketing cut), all pigs were individually weighed. The general health status of the pigs was monitored daily by a qualified individual and any abnormalities were recorded. Mortalities and removals from their designated pen due to illness or physical abnormalities were recorded throughout the study.

Diagnostics were performed for PRRSV, IAV-S, PCV2, and MH. In addition to oral fluid collections upon changes in health, oral fluids were collected at 6, 11, and 16 weeks of age. Blood samples were also collected at the time of allocation in the nursery (day of entry), prior to exiting nursery, midway through finishing and at end of trial.

Statistics

Body weights and average daily gain were analyzed by a General Linear Mixed Model approach. Using the Proc Mixed Procedure (SAS 9.4, Cary, NC), these variables were analyzed with a model that considered the fixed effects of treatment, sex, and the interaction of treatment-by-sex and the random effects of pen(sex), block(allotment_day*pen*sex), and the residual error. Allocation weight was used as a covariate in the analysis of ADG. Treatment least squares means (LSMeans) were calculated for each group and compared by the two-sided Student's t-test at the 5% level of significance. Treatment and treatment by sex interaction were assessed at the 5% level. If the treatment by sex interaction was significant than treatment comparisons were assessed for each sex, separately. Mortality was defined as a binary variable (yes = 1, no = 0) and analyzed by a generalized linear mixed model approach. Using the SAS Glimmix procedure, responses were analyzed with a model that considered treatment, sex and treatment by sex as fixed effects, and pen(sex), block(allotment_day*pen*sex) and the residual error as random effects. The analysis utilized a binomial error and logit link. The 5% level of significance was used to assess statistical differences.

Results and discussion

Trial pigs were coughing upon entry to nursery and diagnostic testing performed prior to weaning revealed an IAV-S, subtype H3N2. The oral fluids samples collected in the nursery were positive for IAV-S, subtype H1N1, at 8 weeks of age (5 weeks on feed). All serum sample pools were positive for PRRSV at the end of nursery period, with all samples sequenced as vaccine virus. This trial was conducted in a low PCV2 challenge environment. Oral fluid samples collected in the nursery and finisher phases identified PCV2 but positive samples had Ct values > 30; therefore the PCV2 genotype could not be determined (Table 1). *Streptococcus suis* was confirmed to cause a small spike in death loss 4 weeks into the nursery phase. No other major health challenges were reported.

T cell epitope analysis comparison showed that a broader coverage could be expected from Fostera® GOLD as compared with a baculovirus derived PCV2a vaccine.[5] Epitope comparison has been previously used as a prediction tool of efficacy for Swine Influenza vaccine.[6] As broader coverage provided by Fostera® GOLD could not be demonstrated in the present study, these results could be a consequence of the low PCV2 challenge observed during the development of this field trial. As hypothesized, body weights and ADG were not affected ($P > 0.05$; Table 2) by treatment within the nursery or finisher phases. Moreover, mortality rates were similar in the nursery [Fostera® GOLD n = 12; Circumvent® n = 12] and finisher [Fostera® GOLD n = 27; Circumvent® n = 22] with an overall mortality rate of 2.95% in the Fostera® GOLD treated pigs and 2.57% in the Circumvent® treated pigs. At the end of the finisher, there were 5 (0.38%) and 9 (0.68%) pigs that were considered moribund in the Fostera® GOLD and Circumvent® treatments, respectively.

In an environment where there was minimal PCV2 or MH challenge, growth performance and mortality were similar for pigs receiving one dose of Fostera® GOLD or Circumvent® PCV-M G2 at weaning.

Table 1: Oral fluid and serology results for PCV2 and MH during the nursery and finisher phases for pigs receiving one dose of Fostera® Gold or Circumvent® vaccine at weaning.

			Positive test results*	
Age	**Sample**	**Location (Time on test)**	**PCV2**	**MH**
3 wk	Serology	Nursery (d 0)	PCR: 0/3, Merck; 0/3, Zoetis	ELISA: 9/15, Merck; 9/15 & 1/15 suspect, Zoetis
6 wk	Oral fluids	Nursery (3 wk)	PCR: 2/6	PCR: 0/6
9 wk	Serology	Nursery (6 wk)	PCR: 0/3, Merck; 0/3, Zoetis	ELISA: 0/15, Merck; 0/15 Zoetis
11 wk	Oral fluids	Finisher (8 wk)	PCR: 2/5	PCR: 0/5
14 wk	Serology	Finisher (11 wk)	PCR: 0/3, Merck; 0/3, Zoetis	ELISA: 1/15, 1/15 suspect, Merck; 0/15, Zoetis
16 wk	Oral fluids	Finisher (13 wk)	PCR: 3/6	NT**
19 wk	Serology	Finisher (16 wk)	PCR: 0/3, Merck; 0/3, Zoetis	ELISA: 1/14, 1/14 suspect, Merck; 2/15 Zoetis

* PCV2 = porcine circovirus 2; MH = *Mycoplasma hyopneumoniae*; Samples were pooled prior to analysis
** NT = not tested

Table 2: Body weights and average daily gain for pigs receiving one dose of Fostera® Gold or Circumvent® vaccine at weaning.

Item	Fostera® Gold PCV MH	Circumvent® PCV-M G2
Allocation wt, lb (range)	11.40 ± 0.32[a] (4.3-21.4)	11.39 ± 0.32[a] (4.8-22.2)
End of nursery wt, lb (range)	41.74 ± 0.19[a] (14.4-64.9)	41.73 ± 0.19[a] (16.8-72.6)
End of finisher wt, lb (range)	225.74 ± 1.01[a] (127.5-313)	226.89 ± 1.01[a] (140-299)
Average daily gain, lb/d		
Nursery	0.79 ± 0.01[a]	0.78 ± 0.01[a]
Finisher	2.02 ± 0.01[a]	2.04 ± 0.01[a]
Allocation to end of finisher	1.66 ± 0.01[a]	1.67 ± 0.01[a]

[a] LSMeans ± SEM within a row are not statistically different; $P > 0.05$

References

1. Baekbo P., et al. Porcine circovirus diseases: a review of PMWS. Transbound Emerg Dis. 2012. Suppl 1:60-7.

2. Franzo G et al. PCV3 a threat to the pig industry? Vet Rec. 2018 Jan 20;182(3):83.

3. Firth et al. Insights into the evolutionary history of an emerging livestock pathogen: porcine circovirus 2. J Virol. 2009 83:12813-21.

4. Xiao CT et al. Global molecular genetic analysis of porcine circovirus type 2 (PCV2) sequences confirms the presence of four main PCV2 genotypes and reveals a rapid increase of PCV2d. J Gen Virol. 2015 96:1830-41.

5. Pig Health Today. Epitope-comparison study illustrates broad coverage of Fostera® Gold PCV MH against PCV2 viruses **https://www.zoetisus.com/products/pork/fostera-gold-pcv-mh/docs/discoveries-epitope-comparison-study-fostera-gold-pcv-mh-pig-health-today-article.pdf**. Accessed 10/29/2018

6. Gutierrez A et al., T-cell epitope content comparison (EpiCC) of swine H1 influenza A virus hemagglutinin. Influenza Other Respi Viruses. 2017. 11:531-542.

Experimental safety and efficacy of a unique MLV PRRSV vaccine: PRRSGard®

F. Chamba[1], DVM, MS, PhD; J. Sui[1], PhD; B. Conarchy[1], BS; X. Zhang[1], PhD; L. Kesl[2], DVM; S. Ma[1], PhD; D. Ruth[1], DVM; A. Venegoni[1], DVM

[1]Pharmgate Animal Health, Wilmington, North Carolina; [2]Veterinary Resources Inc., Ames, Iowa

Introduction

PRRSV is the most economically important pathogen in US pigs due to significant losses in reproductive and growing pigs' performance.[1,2] PRRSV control and elimination is mainly based on herd closure and/or immunization with either live virus inoculation (LVI) or modified live virus (MLV) vaccines.[3-5] To aid in the control and elimination of PRRSV in the US, a unique MLV vaccine (PRRSGard®) has been developed. PRRSGard® is a chimeric PRRSV MLV with the backbone of a proprietary attenuated isolate within lineage 1 and the structural proteins from a high virulence isolate (MN184) within lineage 1 too.[6] MN184 is the ancestor of the current dominant 174 PRRSV circulating in US pigs.[7] PRRSGard® has a genetic marker that will allow the molecular detection and differentiation from wild-type viruses through RT-PCR and sequencing. The purpose of this publication is to summarize the experimental safety and efficacy studies that have been completed with PRRSGard® to inform US swine veterinarians and producers. This information may aid in US PRRSV control and elimination programs.

Materials and methods

Experimental efficacy studies

A total of 3 separate studies were conducted at different times to test PRRSGard® efficacy on viremia, fever, clinical signs and lung lesions. In each study a placebo was compared against a PRRSGard® vaccinated group. For all 3 studies, pigs were vaccinated at weaning with 1 mL of PRRSGard® and then challenged at either 5 (study 1) or 7 weeks post-weaning with a heterologous PRRSV strain (NADC-20/RFLP142/Lineage 8-9). The challenge virus was administered as 2 mL intranasally containing 2.5 x 10^4 virus particles/mL (study 1) and 1.5 x 10^4 virus particles/mL (studies 2 and 3). Viremic pigs after vaccination and challenge were detected weekly either with virus isolation. Viremia after challenge was measured using a $TCID_{50}$ assay at 1- and 2-weeks post challenge. Antibody response after vaccination and challenge was determined weekly with a commercial ELISA PRRSV kit (IDEXX PRRS X3 Ab Test, IDEXX Laboratories, Inc., Westbrook, Maine). Rectal temperatures were measured daily for 14 days after both vaccination and challenge (studies 1 and 3). Clinical signs were assessed daily in 3 categories: respiratory (0=normal, 1=panting/rapid and 2=dyspnea), cough (0=cough, 1=soft or mild intermittent cough and 2=harsh or severe and repetitive cough) and behavior (0=normal, 1=mild to moderately lethargic and 2=severely lethargic or recumbent). A pig was considered clinically affected if it had a total score of 1 or higher (studies 1 and 3). At the end of the study (2 weeks after challenge), pigs were necropsied to assess macroscopic lung lesions and collect lung tissue samples for microscopic lung lesion and immunohistochemistry testing and scoring (study 3). Body weight was also measured at vaccination, challenge and end of the study (study 3).

Experimental transmission and tissue tropism

A group of 16 PRRSGard® vaccinated pigs were housed together with 4 commingled and 4 adjacent sentinel pigs to assess horizontal PRRSV transmission for 4 weeks after vaccination. Blood samples were collected weekly from all pigs to detect viremia by qRT-PCR and virus isolation. Clinical signs were assessed daily similarly to the efficacy studies described above. Nasal and fecal swabs were also collected weekly, and 4 vaccinated pigs were necropsied weekly to detect PRRSGard® in different tissues and score macroscopic lung lesions. Serum samples were tested by ELISA to detect antibodies against PRRSV. All samples were tested by RT-PCR to detect PRRSGard® strain.

Experimental in-vivo genetic stability and safety in pregnant sows

To assess PRRSGard® genetic stability after multiple passages in PRRSV-negative pigs, a group of 3 pigs intramuscularly were injected with 1 mL of PRRSGard® (passage 1). Then, serum was collected from those piglets 14 days post-vaccination and used to intranasally inoculate (2 mL) another group of 3 pigs (passage 2). This process was repeated three additional times to reach passage 5. Clinical signs were recorded daily as described above in all passages. Sanger sequencing of ORF5 was performed to compare PRRSGard® and the isolate from passage 5. Viremia was measured with a TCID50 assay at 7- and 14-days post vaccination/inoculation in all 5 passages. Additionally, to assess the safety of passage 5, a group of 8 piglets were inoculated intranasally with 2 mL of serum (passage 6) from pigs infected in passage 5 with another 5 pigs as negative controls. Clinical signs and rectal temperature were recorded daily as described above. Body weight was also measured and pigs were necropsied at 21 days after inoculation to assess lung macroscopic lesions. Finally, a group of 8 pregnant sows were intranasally inoculated with 2 mL of serum from passage 5 at 3 weeks prefarrow to evaluate experimental safety in pregnant sows. Five pregnant sows served as negative controls. Clinical signs in sows scored as explained above were recorded daily from inoculation to weaning. Sow rectal temperature was recorded daily during 14 days post inoculation, weekly sow serum samples were collected from inoculation to weaning, and sow reproductive performance and body weight of the suckling piglets were recorded too.

In-vitro genetic stability and recombination potential

A total of 15 passages in MARC-145 cells were completed in-vitro to assess PRRSGard® genetic stability. ORF5 of PRRSGard® strain (passage 0) and passage 15 were Sanger sequenced to assess nucleotide and amino acid differences between them. Finally, MARC-145 cells were co-infected once with PRRSGard® strain and an experimental infectious chimera to assess PRRSGard® recombination potential after 5 days in culture. RT-PCR primers were designed to specifically detect the genomic region between ORF3 to ORF7 of each strain.

Results

PRRSGard® was experimentally efficacious against a heterologous challenge

PRRSGard® significantly reduced PRRSV viremia after challenge with a heterologous strain. Figure 1 shows viremia levels as $TCID_{50}$ of infectious virus at 7 and 14 days after challenge. Moreover, PRRSGard® significantly reduced the number of PRRSV viremic pigs at 7 and 14 days after challenge in studies 2 and 3. Figure 2 summarizes the percentages of virus isolation positive pigs after vaccination and challenge. PRRSGard® significantly reduced the percentages of macroscopic lung lesions 2 weeks after challenge in studies 1 and 3. Figure 3 shows the mean percentage of macroscopic lung lesions 2 weeks after challenge. In addition, PRRSGard® significantly improved weight gain after challenge (study 3) and reduced the duration of viremia, fever (studies 1 and 3) and microscopic lung lesion scores (study 3). Percentage of ELISA PRRSV positive pigs were significantly higher in the PRRSGard® groups at 7 days post challenge and all pigs tested positive at 14 days post challenge. Table 1 summarizes the details of the evaluated parameters in each study.

PRRSGard® had limited experimental transmission and natural tissue tropism

Four weeks post-vaccination, one out of four commingled sentinels tested PRRSV VI and RT-PCR positive and none of the adjacent sentinels tested positive. All sentinels tested PRRSV ELISA negative on the weekly samplings including the sentinel that tested positive. All pigs vaccinated with PRRSGard® tested RT-PCR positive on the weekly blood samples, nasal and fecal swabs. PRRSGard® was detected in cervical lymph nodes, muscle, thymus, heart, large intestine, lung, liver, mandibular lymph node, spleen, stomach and kidney by RT-PCR. PRRSGard® was not detected in urine or small intestine. All sentinels were negative in all their tested tissues. The broad tissue detection of PRRSGard® confirmed its natural tropism similar to the wild type PRRSV parental strains. Finally, there were not observable clinical signs and lung lesions in any of the studied pigs.

PRRSGard® was experimentally stable (ORF5) in-vivo and safe in pregnant sows

PRRSGard® ORF5 passage 5 had just one amino acid change at position 23 [S23F] (GP5 signaling peptide) when compared to the PRRSGard® ORF5 master reference strain. At passage 6, pigs

Figure 1: PRRSGard® efficacy on viremia reduction after challenge.

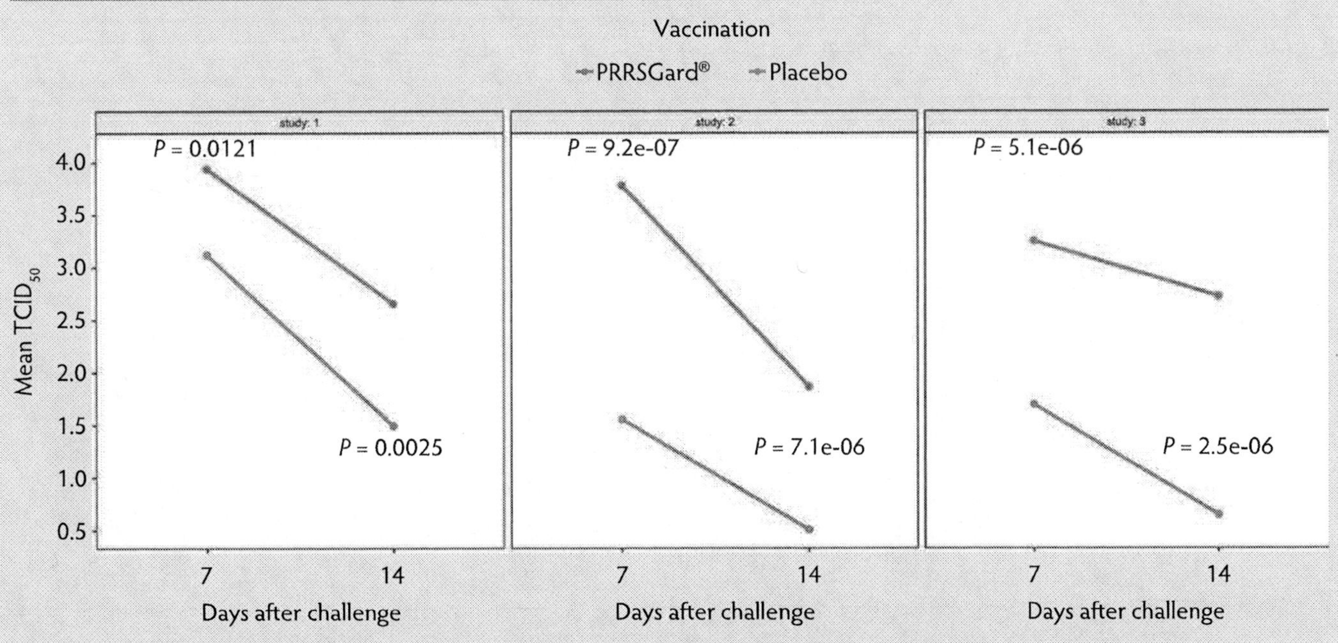

and pregnant sows did not show worse clinical signs than passage 5. There were no abortions in the inoculated pregnant sows. Piglet weight gain and sow reproductive performance was not affected by inoculation/vaccination with PRRSGard®. Table 2 summarizes the results from the PRRSGard® in-vivo passages and its safety in pregnant sows.

In-vitro, PRRSGard® was stable (ORF5) and had limited recombination

After 15 in-vitro cell passages, PRRSGard® ORF5 had one amino change at position 23 [S23C] (GP5 signaling peptide) in relation to the PRRSGard® ORF5 master reference strain. Recombination was not observed after one simultaneous infection with 2 PRRSV chimeras (PRRSGard® and one experimental PRRSV chimera).

Figure 2: PRRSGard® efficacy on reducing the number of viremic pigs after challenge.

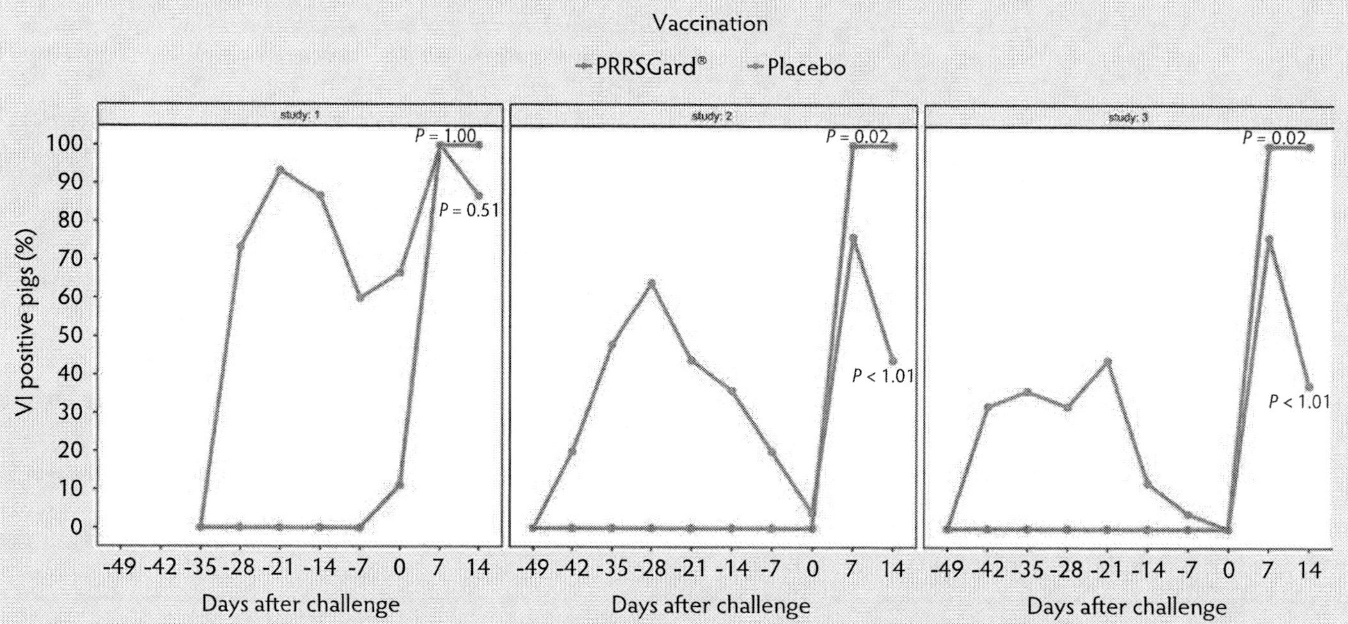

Figure 3: PRRSGard® efficacy on reducing macroscopic lung lesions 2 weeks after challenge.

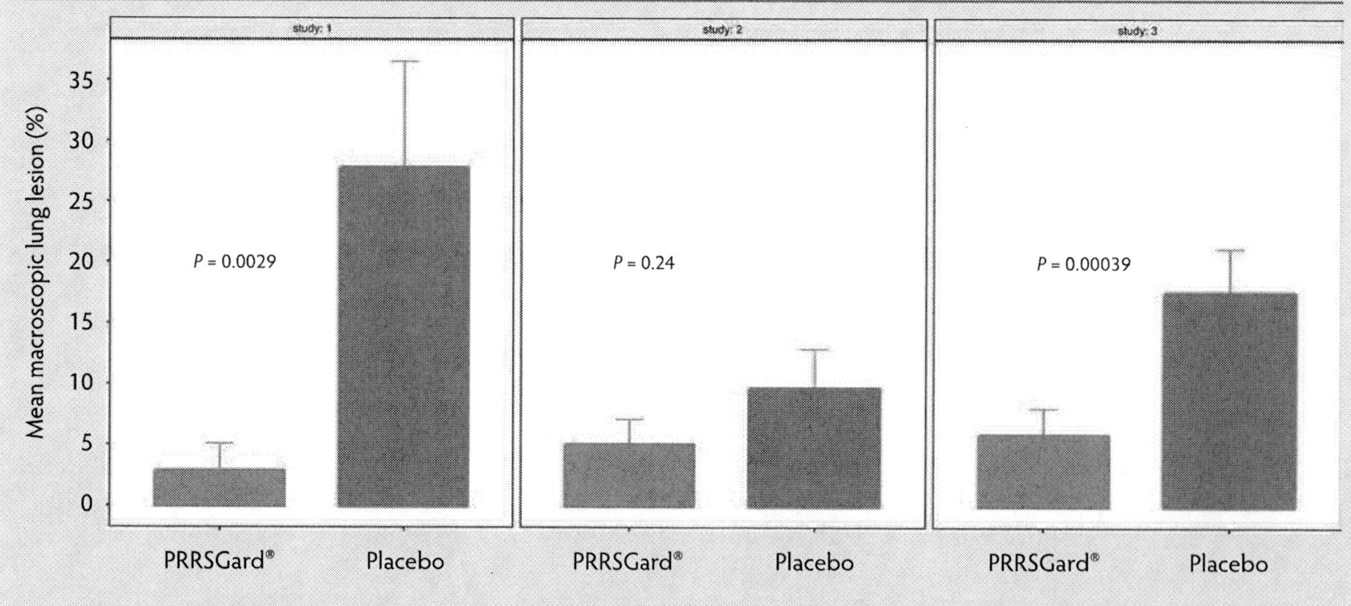

Table 1: PRRSGard® efficacy study results after challenge with a heterologous strain.

Parameter	Study 1			Study 2			Study 3		
Study group	**PRRSGard®**	**Placebo**	***P*-value**	**PRRSGard®**	**Placebo**	***P*-value**	**PRRSGard®**	**Placebo**	***P*-value**
No. pigs	15	11		25	25		25	25	
Duration of viremia, weeks, median	1.9	2.0	0.29	1.2	2.0	<.01	1.1	2.0	<.01
ELISA positives at 7 dpc, %	100	56	0.02	100	21	<.01	92	67	0.04
ELISA positives at 14 dpc, %	100	100		100	96	0.49	100	100	
Duration of fever, days, median	2.5	5.8	0.01				4.5	6.6	0.03
No. pigs with clinical signs	4/15	4/9	0.41				10/25	12/25	0.78
Duration of clinical signs, days, median	0.5	1.0	0.36				0.8	1.3	0.32
Micros. lung lesion score (0-4), median							2.3	2.8	0.08
IHC lung score (0-4), median							0.9	1.6	0.02
ADG, lb/day, median							0.72	0.43	0.03

Gray boxes mean that data was not collected or statistical comparisons were not made. IHC=Immunohistochemistry. ADG=Average daily weight gain. ELISA=Enzyme Linked Immunosorbent Assay. Micros.=Microscopic. dpc=days post challenge.

Conclusions and discussion

Experimentally, PRRSGard® was efficacious against a heterologous challenge and safe in piglets and pregnant sows. Experimentally, PRRSGard® had limited spread to naïve commingled pigs and natural tissue tropism. Experimentally, PRRSGard® ORF5 was stable and had limited in-vitro recombination. New PRRSV vaccines are extremely important to aid in the control and elimination of this disease in US pigs. PRRSGard® needs to be tested under field conditions to understand its unique features.

References

1. Holtkamp, D. J. Assessment of the economic impact of porcine reproductive and respiratory syndrome virus on United States pork producers. (2013).

2. Neumann, E. J. *et al.* Assessment of the economic impact of porcine reproductive and respiratory syndrome on swine production in the United States. *J Am Vet Med Assoc* **227**, 385-392 (2005).

3. Linhares, D. C., Johnson, C. & Morrison, R. B. Economic Analysis of Vaccination Strategies for PRRS Control. *PloS one* **10**, e0144265, doi:10.1371/journal.pone.0144265 (2015).

4. Linhares, D. C., Cano, J. P., Torremorell, M. & Morrison, R. B. Comparison of time to PRRSv-stability and production losses between two exposure programs to control PRRSv in sow herds. *Preventive veterinary medicine* **116**, 111-119, doi:10.1016/j.prevetmed.2014.05.010 (2014).

5. Rathkjen, P. H. & Dall, J. Control and eradication of porcine reproductive and respiratory syndrome virus type 2 using a modified-live type 2 vaccine in combination with a load, close, homogenise model: an area elimination study. *Acta Veterinaria Scandinavica* **59**, 4 (2017).

6. Shi, M. *et al.* The spread of type 2 Porcine Reproductive and Respiratory Syndrome Virus (PRRSV) in North America: a phylogeographic approach. *Virology* **447**, 146-154, doi:10.1016/j.virol.2013.08.028 (2013).

7. Murtaugh, M. P. in *2015 NA PRRS Symposium* (ed NA PRRS Symposium) (NA PRRS Symposium, Chicago, IL, 2015).

Table 2: Clinical findings after PRRSGard® inoculation in piglets and pregnant sows.

Parameter	Passage					Piglets			Pregnant sows		
Study group	**P1**	**P2**	**P3**	**P4**	**P5**	**P6**	**Control**	***P*-value**	**P6**	**Control**	***P*-value**
No. pigs	3	3	3	3	9	8	5		8	5	
No. viremic pigs	3/3	2/3	3/3	2/3	9/9	8/8	0/5	< .01	8/8	0/5	< .01
Weeks tested for viremia	2	2	2	2	2	3	3		6	6	
Duration of viremia, weeks, median	1	2	1	2	2	3	0	< .01	1.5	0	< .01
Viremia at 14 dpv, TCID50, median	2.8	4.9	2.5	4.1	4.8						
No. pigs with clinical signs	0/3	0/3	0/3	0/3	1/9	5/8	1/5	0.27	3/8	0/5	0.23
Duration of clinical signs, days, median	0	0	0	0	0	1.5	0	0.11	0	0	0.17
No. pigs with fever	3/3	2/3	3/3	3/3	9/9	8/8	2/5	0.04	4/8	0/5	0.11
Duration of fever, days, median	1	1	5	2	3	4.5	2	0.04	0.5	0	0.27
Weaning weight, lb, median						12.9	12.5	0.88			
Weight at 21 dpv, lb, median						29.0	26.0	0.22			
ADG at 21 dpv, lb/day, median						0.74	0.51	0.13			
Macro. lung lesion, %, median						0.1	0.0	0.09			
Total born, No. pigs, median									13	13	0.94
Born alive, No. pigs, median									11.5	11	1.00
Born dead, No. pigs, median									1	2	0.55
Weaned, No. pigs, median									10	10	0.71
Preweaning deaths, No. pigs, median									0	1	0.40
Birth weight, lb, median									3.4	3.5	0.72
Weaning weight, lb, median									12.4	14.9	0.42
Preweaning ADG, lb/day, median									0.45	0.54	0.22

Gray boxes mean that data was not collected or statistical comparisons were not made. dpv=days post vaccination. P =Passage. $TCID_{50}$=Tissue Culture Infectious Dose that infects 50% of the cells. lb=pounds. No.=number. ADG=Average Daily weight Gain. Macro.=Macroscopic. dpv=days post vaccination.

Anti-inflammatory properties of Aivlosin® (tylvalosin) and its implications for swine respiratory disease treatment

A. Hintz[1], DVM; J. Mora[2], DVM
[1]Pharmgate Animal Health, Wilmington, NC; [2]ECO Animal Health, London, UK

Introduction

Swine respiratory disease (SRD), a disease complex of numerous bacterial pathogens, continues to be one of the major health concerns challenging veterinarians in the North American and worldwide swine industries. Concurrent infections with porcine reproductive and respiratory syndrome virus (PRRSV) and swine influenza virus (SIV) exacerbate the disease caused by those bacterial organisms. Aivlosin® (62.5% w/w tylvalosin) water soluble granules, a second-generation macrolide antibiotic, is labeled for use in the drinking water for the control of SRD associated with *Bordetella bronchiseptica, Haemophilus parasuis, Pasteurella multocida,* and *Streptococcus suis* in swine buildings experiencing an outbreak of SRD.

Recently, there has been increased interest in the effects of inflammation and inflammatory mediators and their role in clinical disease outcomes due to bacterial and viral lung infections and how to minimize the negative effects. Macrolide antibiotics have demonstrated in vitro[1,2] and in vivo[2] anti-inflammatory and immunomodulatory properties along with their anti-microbial effects. This affords macrolides a two-pronged approach for the treatment and control of disease caused by pneumonic pathogens. This paper will briefly touch on the mechanisms of inflammation in pulmonary tissue and introduce the anti-inflammatory and immunomodulatory properties of the macrolide antibiotic, tylvalosin (Aivlosin®).

Mechanisms of inflammation and resolution

Infection by bacterial pathogens and their associated virulence factors drives the host animal's immune response. Alveolar macrophages detect the presence of abnormal bacteria and present antigens to other immune cells. They release inflammatory mediators to trigger the immune response.[3] Neutrophils are recruited by the inflammatory mediators (especially Interleukin-8 and Leukotriene B_4)[4] to the site of infection to neutralize the bacteria. Overabundance of neutrophils with concurrent necrosis, or uncontrolled cell death, of the neutrophils leads to the release of cytotoxic compounds. These compounds include enzymes, hypochlorous acid, nitric oxide, and other inflammatory mediators. While these compounds target and help neutralize the bacterial infection, they also lead to inflammatory damage of surrounding tissues.[5]

To quell the inflammatory process, prostaglandin D_2 and prostaglandin E_2 are released by apoptotic neutrophils, initiating the process of resolution. These prostaglandins activate the release of pro-resolution mediators such as lipoxins, resolvins, and protectins, which function by activating specific resolution mechanisms. These mechanisms include stopping the recruitment of neutrophils to the site of inflammation, activating macrophage phagocytosis of bacteria and apoptotic neutrophils, and increase removal of the engorged phagocytic macrophages, among other effects. Resolution is becoming accepted as a separate pathway versus the anti-inflammatory pathway.[6]

Macrolide antibiotics are proposed to drive neutrophils away from necrosis and towards a pathway of apoptosis, or programmed cell death. They also are thought to induce the resolution of inflammation by activating pro-resolution compounds. During apoptosis, neutrophils are destroyed, however their membranes remain in-tact until they are cleared by macrophages, preventing the release of the cytotoxic compounds. The clearance of apoptotic neutrophils by alveolar macrophages is known as efferocytosis.[7,8] Without the release of cytotoxic compounds, the neutrophils can ingest and neutralize invading bacterial pathogens and clear the infection with minimal inflammatory damage to neighboring pulmonary tissues.

Anti-inflammatory and immunomodulatory effects of tylvalosin

Recent research has shown tylvalosin suppresses nuclear factor kappa-light-chain-enhancer of activated B cells (NF-KB). NF-KB is a transcription factor that leads to the synthesis of pro-inflammatory mediators such as Interleukin-8 (IL-8) and others. These inflammatory mediators recruit an overabundance of neutrophils to the site of infection, leading to an increase in inflammatory damage.

In mouse lung tissue treated with lipopolysaccharide (LPS) (in vitro inflammation model), as well as piglets challenged with PRRSV, tylvalosin markedly decreased IL-8 and other important inflammatory mediators such as IL-6, IL-1β, and tumor necrosis factor alpha (TNF-α).[9] The pigs were clinically observed for 21 days post-PRRSV challenge and subsequently necropsied after the study. Lung lesions were markedly reduced in pigs treated with tylvalosin. This indicates a reduction in the inflammatory damage caused by neutrophil infiltration of the pulmonary tissue when treated with tylvalosin.[10] Clinical signs including dyspnea, erythema, lethargy, and inappetence were reduced in the treatment group. Rectal temperatures were lower in the pigs treated with tylvalosin and average daily gain during the treatment period was increased in the treatment group.

More recently, research indicates tylvalosin promotes apoptosis of porcine neutrophils while driving the removal of dead cells by efferocytosis in vitro. This effect was concentration dependent, with higher levels of tylvalosin (10 μg/mL) demonstrating significantly higher percentage of apoptotic cells versus the untreated control. Tylvalosin also demonstrated an increase in macrophage efferocytosis of apoptotic neutrophils without altering macrophage phagocytosis.[10] Together, these effects would be expected to reduce the inflammatory damage caused by bacterial infection and concurrent uncontrolled immune response.

The study also demonstrated a reduction in IL-8 and IL-1α released from LPS-stimulated macrophages and Leukotriene B_4 production in neutrophils activated by calcium ionophore. LPS-stimulation is an accepted model for inflammation, while the calcium ionophore was used to reflect the high levels of activated neutrophils in an inflammatory response. Along with the reduction in pro-inflammatory mediators, an increase in the pro-resolution compounds Lipoxin A_4 and Resolvin D_1, both of which are considered powerful contributors to the reduction of inflammation.[11] The effects in the studies were independent of the anti-microbial properties of tylvalosin.

Discussion

Dual-effect anti-microbials such as tylvalosin possess immense potential in the treatment of clinical respiratory disease caused by bacterial infection. Advanced research into the mechanisms of action of these anti-inflammatory properties is ongoing, though the current findings are promising for the identification of novel therapies for the treatment and control of bacterial and quite possibly viral pneumonia.

In addition to its anti-microbial properties, tylvalosin has demonstrated anti-inflammatory and immunomodulatory benefits for the control of swine respiratory disease. Further research needs to be completed to quantify and qualify the effects of tylvalosin, though the initial data are encouraging. The ongoing research will help to gain a better understanding of the underlying mechanisms to help direct and develop targeted therapeutics to control swine respiratory disease.

References

1. Culic O et al. Anti-inflammatory effects of macrolide antibiotics. *Euro J Pharmacol.* 2001;429:209-229

2. Bulska M et al. Immunomodulatory and anti-inflammatory properties of macrolides. *Curr Issues Pharm Med Sci.* 2014;27(1):61-64.

3. Mosier DA et al. Bacterial pneumonia. *Vet Clin North Am Food Anim Pract.* 1997;13:483-493.

4. Di Paolo NC et al. Interleukin 1alpha and the inflammatory process. *Nat Immunol.* 2016;17:906-913.

5. Buret AG. Immuno-modulation and anti-inflammatory benefits of antibiotics: The example of tilmicosin. *Can J Vet Research.* 2010;74:1-10.

6. Serhan C et al. Resolving inflammation: dual anti-inflammatory and pro-resolution lipid mediators. *Nat Rev Immunol.* 2008;8(5):349-361.

7. Jorgenson I et al. Programmed cell death as a defense against infection. *Nat Rev Immunology.* 2017;17:151-164.

8. Fadok V et al. Macrophages that have ingested apoptotic cells *in vitro* pro-inflammatory cytokine production through autocrine/paracrine mechanisms involving TGF-beta, PGE_2, PAF. *J Clin Invest.* 1998;101:809-818.

9. Zhao Z et al. Tylvalosin exhibits anti-inflammatory property and attenuates acute lung injury in different models possibly through suppression of NF-kB activation. *Biochem Pharmacol.* 2014;90:73-87.

10. Moges R et al. Anti-Inflammatory Benefitsof Antibiotics: Tylvalosin Induces Apoptosis of Porcine Neutrophils and Macrophages, Promotes Efferocytosis, and Inhibits Pro-Inflammatory CXCL-8, IL1α, and LTB4 Production, While Inducing the Release of Pro-Resolving Lipoxin A4 and Resolvin D1. *Front Vet Sci.* 2018;5:57-67.

Trends of mycotoxin contamination in US corn

E.G. Hendel,*[1] VMD, PhD; S. Curry[1], PhD; P.N. Gott[1], PhD; U. Hoffstetter[2]; G. R. Murugesan[1], DVM, PhD
[1]BIOMIN America, Inc., Kansas City, KS, USA, [2]BIOMIN Holding GmbH, Getzersdorf, Austria

Introduction

Mycotoxins are harmful secondary fungal metabolites frequently found in various feed ingredients. Historically discussion of impacts on animal health have emphasized classic symptoms such as reduced feed intake, vomition and reduced productivity. Producers are often unaware of subclinical impacts of mycotoxin exposure. Even at low levels, mycotoxins act as predisposing factors for enteric and respiratory diseases by causing inflammation,[1] modulating the gastrointestinal environment,[1] and immune suppression[2] which can contribute to vaccine failure.[3] Additionally, mycotoxins cause reductions in reproductive performance and may contribute to sow pelvic organ prolapse. BIOMIN annual surveys have indicated increasing trends of prevalence and co-occurrence of toxins produced by *Fusarium* spp. fungi in US corn in prior years.[4] The data herein presents the most current information regarding mycotoxin contamination of samples from the 2018 corn harvest in comparison to the previous five years.

Materials and methods

Corn samples from the United States used in multiple livestock industries (cattle, swine, and poultry) were collected and submitted for analysis utilizing the liquid chromatography with tandem mass spectrometry (LC-MS/MS) method. Samples were marked with harvest year, with the 2018 harvest starting from August of 2018. Six major mycotoxin groups were analyzed: aflatoxins (Afla), type A trichothecenes (A-Trich), type B trichothecenes (B-Trich, i.e. DON; vomitoxin), fumonisins (FUM), zearalenone (ZEN) and ochratoxin-A (OTA). The data of this sample pool of 2018 US corn were compared to previous years to examine trends of mycotoxin contamination (Figure 1, 2, and 3).

Results and discussion

A high proportion of samples of the 2018 corn harvest from the US contained at least one detected mycotoxin type (94%, 176 samples), similar to the situation in 2017 (91% positive, 733 samples). The prevalence of co-occurrence (% of samples with ≥ 2 mycotoxins) in samples thus far is higher compared to 2017 (2018: 58%, 2017: 48%). Currently in this year's sample pool, FUM is the most prevalent mycotoxin, and this year's crop appears to have higher prevalence and average contamination levels compared to last year (2018: 70%, 3375 ppb, 2017: 43%, 2166 ppb). DON, has lower prevalence and similar average contamination levels compared to last year's corn crop (2018: 64%, 974 ppb 2017: 79%, 994 ppb), while ZEN contamination levels are similar to last year both in prevalence and contamination level (2018: 34%, 235 ppb, 2017: 33%, 239 ppb). The prevalence and average contamination level of Afla is higher than the previous three years (2018: 10%, 147 ppb, 2017: 3%, 12 ppb, 2016: 5%, 19 ppb 2015: 1%, 39 ppb). The percent of positive samples above threshold levels for the three fusarium toxins (samples over 1000 pb DON, 1000 ppb FUM, and 200 ppb ZEN) are similar in DON, increased in FUM and decreased in ZEN from 2017 (2018: 25%, 59%, 27%, 2017: 25%, 37%, 32%, respectively).

Variation in the prevalence of individual mycotoxins are apparent at the state level (Table 1) where certain states have higher prevalence of DON and other have higher pressure from FUM.

The preliminary results from the 2018 corn harvest suggest a continued risk due to mycotoxins produced by *Fusarium* fungal species with increased risk of Afla and FUM compared to last year. Because of continued risk of multi-mycotoxin contamination in samples thus far, multiple mitigation strategies are needed beyond adsorption, including biotransformation and supporting the immune system and liver function.

References

1. Pinton et al. Toxicity of Deoxynivalinol and its acetylated derivatives on the intestine: differential effects on morphology, barrier function, tight junction proteins, and mitogen-activated protein kinases. *Toxilogical Sciences*. 2012; 130(1): 180-190

2. Pierron et al. Impact of mycotoxin on immune response and consequences for pig health. *Animal Nutrition*. 2016; 2: 63-68.

3. Savard, C. et al. Deoxynivalenol (DON) naturally contaminated feed impairs the immune response induced by porcine reproductive and respiratory syndrome virus (PRRSV) live attenuated vaccine. *Vaccine*. 2015; V. 33, Issue 32, pp. 3881-3886.

4. E.G. Hendel, P.N. Gott, V. Starkl, U. Hofstetter-Schahs, and G.R. Murugesan. Mycotoxin survey in 2017 US Corn, *Biomin World Mycotoxin Survey 2017*. 2017.

Table 1: Prevalence of mycotoxins in corn from specific states

	State								
TOXIN	IA	MN	IL	WI	OH	MO	GA	MS	TX
Afla	-	-	9%	10%	-	15%	21%	19%	25%
FUM	56%	25%	100%	40%	38%	90%	100%	100%	63%
B-Trich	94%	100%	55%	70%	100%	25%	36%	-	-
ZEN	56%	63%	45%	50%	69%	-	7%	-	-

Figure 1: Prevalence of mycotoxins

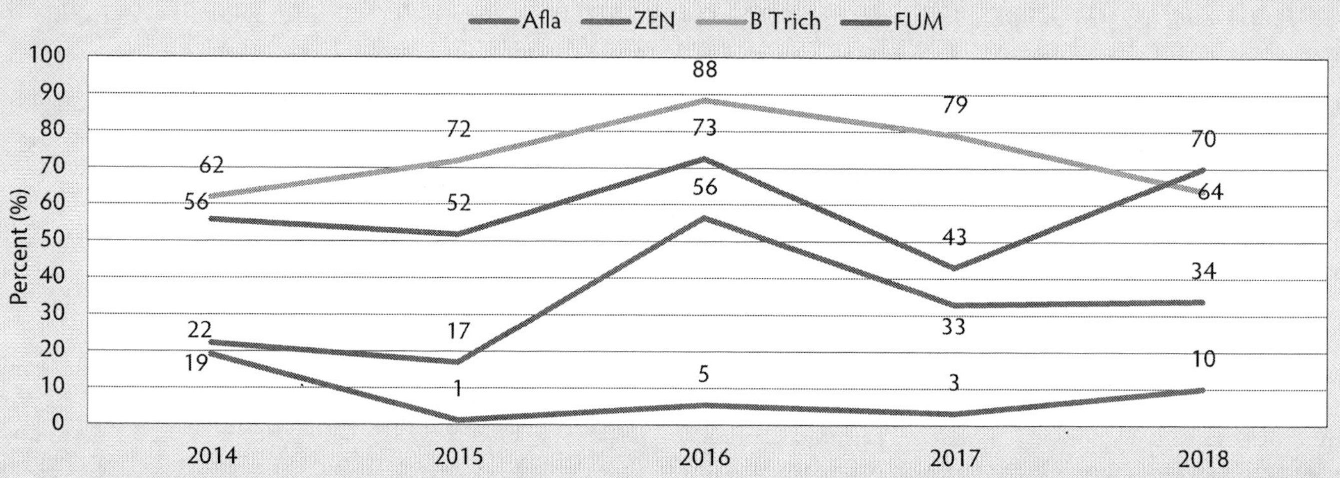

Percent positive samples for Afla, ZEN, DON and FUM are presented by year (due to small numbers of positive samples, T2 and OTA are not represented graphically).

Figure 2: Co-occurrence of mycotoxins

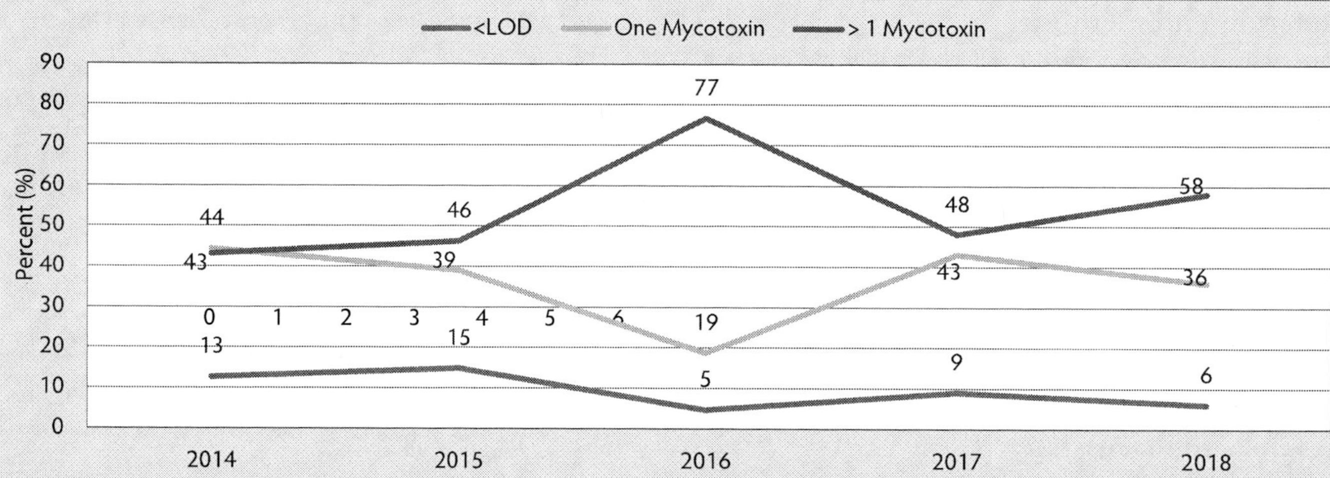

Percent of samples below the limit of detection, with one, or more than one mycotoxin for each year.

Figure 3: Sample number per year

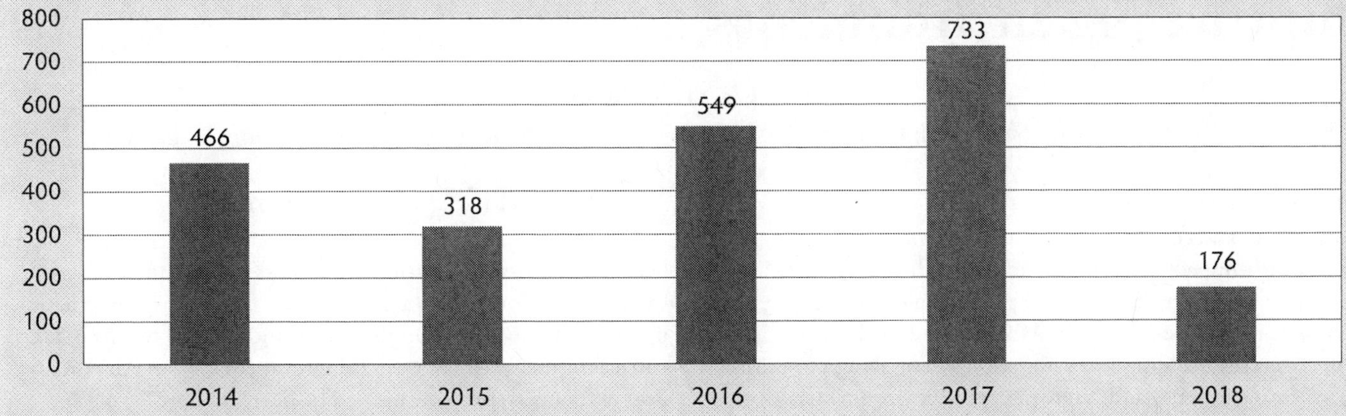

Number of US corn samples for each year from 2014 through the 2018 harvest.

Use of heat exchangers to lower production costs and improve pig environments

J. Baker, DVM, WVC
Legacy Veterinary Services, Inc. Boonville, Indiana; AirWorks Ventilation Systems, Albion, Indiana

Introduction

In moderate and cold climates, heating and propane costs can significantly impact total production costs through higher heating costs and health effects during times when producers try to minimize propane expenses. AirWorks heat exchangers are integrated into the building and ventilation system design to minimize heating expense and improve the pig environments during cold weather.

Discussion

Producers try to manage heating costs at all times but especially during times when propane prices significantly increase. To save heating costs, producers often minimize pig space air exchange to save heating costs (propane) which can create health challenges in all production buildings and increases maintenance costs.

AirWorks heat exchanger systems contribute to lowering energy costs (propane use) and improving environments and pig health in production systems. Energy costs as a percent of total production costs vary widely primarily because of fluctuating propane or Natural Gas costs. Propane costs range from 3 to 6% of total production costs in the upper Midwest and higher in the Northern Midwest and Canada.[1,2]

Methods

AirWorks heat exchanger systems are integrated into the building systems and require no daily, weekly or annual maintenance. It is a true heat exchange system and not an air blending system like those provided in other ventilation designs. The heat exchangers are designed to last the life of the building without maintenance or replacement. High quality materials with specific coefficients of thermal expansion are required to withstand the harsh environments and conditions in which the heat exchangers work. The AirWorks heat exchangers perform their function with no special management or increase in labor requirements.

Results and field observations

AirWorks heat exchangers help lower heating costs and maintain healthy air flows even during the coldest weeks and months of the year. AirWorks heat exchangers also temper large fluctuations in outside temperatures in the spring and fall and results in more consistent management of pig environments. With AirWorks heat exchangers, propane use can be decreased by 70 to 90% while maintaining and even improving air exchange in cold weather (Figure 1).

The improvement in air exchange results in fewer respiratory problems, more uniform room temperatures, drier environments, and less death loss during cold and variable weather. The fresh air coming from the heat exchangers is tempered before it enters the pig space and results in fewer temperature variations within the pig space.

Payback to AirWorks heat exchangers varies from region to region. Paybacks from propane savings and improved health varies from 1 to 4 years. The longer the winter season and the greater the health challenge on the farm the quicker the payback for the heat exchangers (Table 1).

Production budgets can be better predicted because the highly variable energy costs can be minimized to the point that it is an insignificant production cost.

References

1. Iowa State University Extension and Outreach, PM 3063E February, 2016. Farm Energy: Case Studies/Benchmarking Energy Usage for Swine Producers.

2. Larry D. Jacobson. Energy Use in Swine Production, 53rd Annual Midwest Rural Energy Council, Bloomington, MN. February 19-20, 2015.

Figure 1: AirWorks heat exchanger's tempering affect.

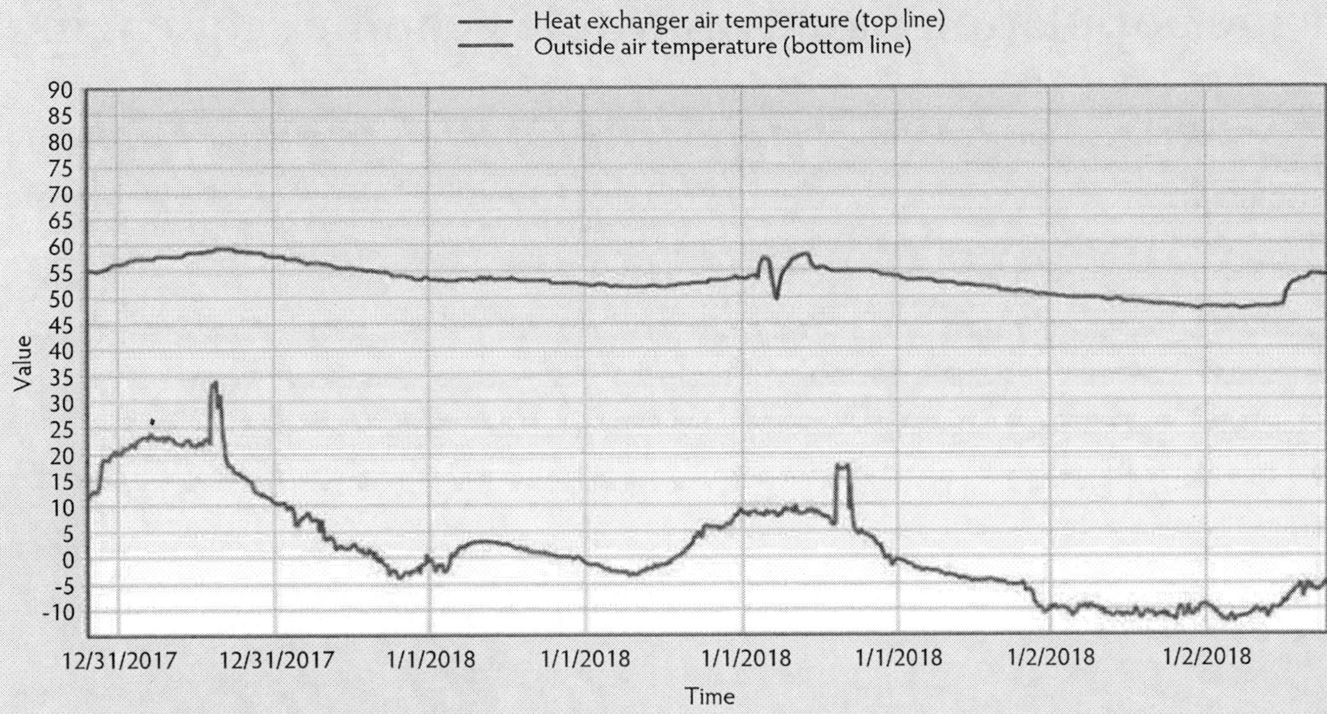

Table 1: AirWorks heat exchanger payback by region (North America and China).

Total cost of HE		**$9,550.00**			**Propane cost/ gal**		**$1.26**
	Gal of propane/pig		**Cost savings/pig**		**HE payback in years***		
	Tunnel	**AirWorks**	**Propane**	**Health**	**Propane**	**Health**	**Together**
Canada	4	0.7	$4.14	$4.00	2.55	2.66	1.16
North US	3	0.5	$3.14	$3.00	3.65	3.88	1.59
N Central US	2	0.3	$2.13	$2.50	6.44	5.04	2.22
Central US	1.5	0.2	$1.63	$2.00	10.40	7.17	3.01

Total cost of HE		**¥59,688**			**Propane cost/kg**		**¥4.08**
	Kg of propane/pig		**Cost savings/pig**		**HE payback in years***		
	Tunnel	**AirWorks**	**Propane**	**Health**	**Propane**	**Health**	**Together**
DongBei, China	7.69	1.35	¥25.89	¥25.00	2.55	2.66	1.16
North, China	5.77	0.96	¥19.61	¥18.75	3.65	3.88	1.59
Central, China	3.85	0.58	¥13.34	¥15.63	6.44	5.04	2.22
S Central, China	2.88	0.38	¥10.20	¥12.50	10.40	7.17	3.01

* Heat exchanger payback comes from both propane savings and improvement in health.
Heat exchanger payback varies by region, the colder it is the faster the payback.

Epidemiological survey of *Streptococcus suis* in the US swine population based on whole genome sequencing

Paulraj Lawrence, PhD
Biological Development and Research Department, Newport Laboratories Inc., Worthington, Minnesota

Introduction

Streptococcus suis is a gram positive bacterium that causes septicemia, bronchopneumonia, arthritis, and fatal meningitis in young piglets. Piglets are usually infected through exposure to asymptomatic sows. Thirty-five different serotypes of *S. suis* have been identified,[1] however, recent work has suggested that only 29 unique serotypes exist.[2,3] In addition, many isolates remain non-typeable using traditional methods. Of the 35 serotypes described, serotype 2 is most commonly associated with disease, but serotypes 1, 1/2, 3, 4, 5, 7, 8, 9 and 16 are also commonly isolated from diseased swine.[1] Newport Laboratories has developed a comprehensive, unambiguous method of capsular typing based on whole genome sequencing and bioinformatics. This method of typing is useful to type field and there are no non-typeable serotypes.

Materials and methods

The broth or plate cultures of *S. suis* were used for genomic DNA was isolation using a Bacterial Genomic DNA Purification Kit (Edge Biosystems, Gaithersburg, MD), following the manufacturer's recommendation, suspended in Tris EDTA buffer, quantified and purity checked on a Nanodrop. Genomic DNA libraries from all strains were prepared using a Nextera XT kit (Illumina). Whole genome sequencing was done on illumina MiSeq equipment based on 2 × 150 base pair chemistry. The raw reads were filtered for quality using a custom Python script. Low quality reads were discarded (any read containing 5 or more bases with PHRED Quality < 20 was considered low quality). Following quality filtering, reads were mapped against reference sequence, a consensus sequence generated by mapping a Newport reference serotype isolate and all available Genbank *S. suis* genomes. Reads were mapped using Bowtie2 v 2.0.6 (`http://bowtie-bio.sourceforge.net/bowtie2/index.shtml`). The capsule locus was typed based on the maximum homology to reference *S. suis* capsule region. The report is generated as percent locus coverage.

Results and discussion

Based on the whole genome based typing and surveillance results from 2015-2018, indicate serotypes 1, ½ and 2 make up almost 50% of the serotypes in the US swine belt with active *S. suis* infection (Figure 1). Among these isolates, 91% were from systemic infection from brain, or internal organs, 9% were from nasal swabs or commensal. Most of the commensal serotypes are 27, 2 and 3, which make up 45% of isolation. Among systemic infection majority was 1, ½, 2 and 7. We also observed farm specific serotype distribution, with few systems showing preponderance of a specific serotype.

Figure 1: *S. suis* serotype distribution in the US swine population.

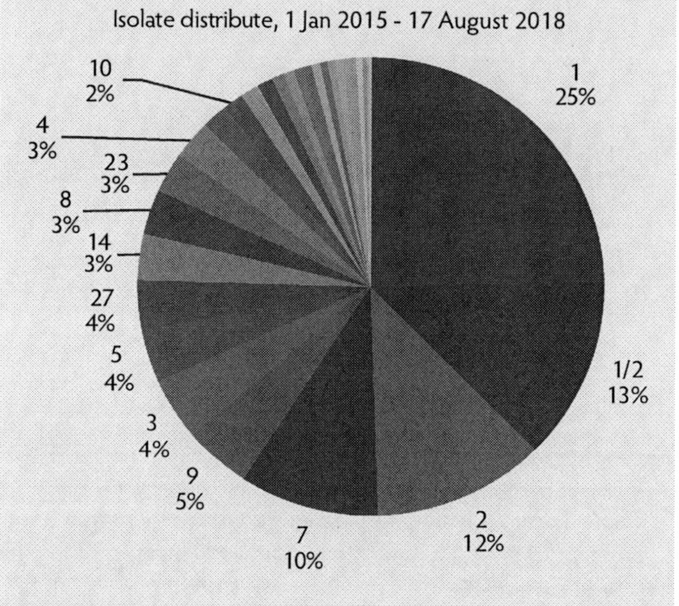

References

1. Goyette-Desjardins G et al. *Streptococcus suis*, an important pig pathogen and emerging zoonotic agent—an update on the worldwide distribution based on serotyping and sequence typing. *Emerg. Microbes Infect.* 2014;3:e45.

2. Gottschalk et al. Susceptibility of *Streptococcus suis* to penicillin. *J Vet Diag Invest.* 1991;3:170-172.

3. Food and Drug Administration. Veterinary Feed Directive Fact Sheet.2017. Available at `www.fda.gov/animalveterinary/developmentapprovalprocess/ucm449019.htm`. Accessed 20 October 2017.

Use of whole genome sequencing for identification and selection of isolates for infection chain control

Keith Wilson, DVM; Paulraj Lawrence, PhD
Biological Development and Research Department, Newport Laboratories Inc., Worthington, Minnesota

Introduction

Streptococcus suis is a gram positive bacterium that causes septicemia, bronchopneumonia, arthritis, and fatal meningitis in young piglets. Piglets are usually infected through exposure to asymptomatic sows. Thirty-five different serotypes of *S suis* have been identified,[1] however, recent work has suggested that only 29 unique serotypes exist.[2,3] In addition, many isolates remain non-typeable using traditional methods. Of the 35 serotypes described, serotype 2 is most commonly associated with disease, but serotypes 1, 1/2, 3, 4, 5, 7, 8, 9 and 16 are also commonly isolated from diseased swine.[1] Newport Laboratories has developed a comprehensive, unambiguous method of capsular typing based on whole genome sequencing and bioinformatics. We identified specific serotype associated with particular genetic stocks which were vertically transmitted. This allows us to design vaccines for flows to prevent vertical transmission of *S suis*.

Materials and methods

The genomic DNA of *S suis* isolates in our collection was isolated using a Bacterial Genomic DNA Purification Kit (Edge Biosystems, Gaithersb+urg, MD), following the manufacturer's recommendation, quantified and purity checked on a Nanodrop. Genomic DNA libraries from were prepared using a Nextera XT kit (Illumina). The whole genome sequencing was done on illumina MiSeq equipment based on 2 × 150 base pair chemistry. The raw reads were filtered for quality using a custom Python script. Low quality reads were discarded (any read containing 5 or more bases with PHRED Quality < 20 was considered low quality). Following quality filtering, reads were mapped against reference sequence, a consensus sequence generated by mapping a Newport reference serotype isolate and all available Genbank *S suis* genomes. Reads were mapped using Bowtie2 v 2.0.6 (`http://bowtie-bio.sourceforge.net/bowtie2/index.shtml`). The capsule locus was typed based on the maximum homology to reference *S suis* capsule region. The report is generated as percent locus coverage.

Results

The data presented in Table 1 as Genetic Flow 1 is from a system that stocks farms from a similar multiplier. A *S suis* Type 1 was isolated from clinically affected pigs born from this multiplier. The same *S suis* type 1 was found in new farms stocked from this multiplier.

The data presented in Table 1 as Genetic Flow 2 is an example of a *S suis* type 1/2 being isolated from the nursery of a genetic nucleus and the same *S suis* type 1/2 being isolated from nursery pigs from a new genetic multiplier stocked from gilts from this nucleus.

Discussion

Since the adoption of whole genome sequencing for *S suis* serotyping, we have been able to identify similarities between the different *S suis* isolates more accurately. These similarities are expressed as a percent locus coverage of the primary and secondary serotypes embedded within a particular *S suis* bacteria capsule. Since the *S suis* capsule is a mosaic of different capsular serotypes, it allows for a unique identification of the capsule and comparison of this mosaic capsule across isolates. As we compare these isolates, we have seen an epidemiological link to the flow of pigs. This presentation will discuss two examples of how *S suis*, serotype 1 and 1/2, have been transmitted vertically. We will also discuss how vaccination of the sow can be effective on reducing the clinical signs in the piglet.

References

1. Goyette-Desjardins G et al. *Streptococcus suis*, an important pig pathogen and emerging zoonotic agent—an update on the worldwide distribution based on serotyping and sequence typing. *Emerg. Microbes Infect.* 2014;3:e45.

2. Gottschalk et al. Susceptibility of *Streptococcus suis* to penicillin. *J Vet Diag Invest.* 1991;3:170-172.

3. Food and Drug Administration. Veterinary Feed Directive Fact Sheet.2017. Available at www.fda.gov/animalveterinary/developmentapprovalprocess/ucm449019.htm. Accessed 20 October 2017.

Table 1: *S suis* isolated from two different genetic flows

Client/site	System	Isolation date	Tissue of origin	Capsular types (% locus coverage)	Other capsular types (% locus coverage, >70%)
Genetic flow 1					
1	**A**	**17-Nov-17**	**Brain swab**	**1 (97.4%)**	**14 (98.7%), 2 (79.0%), 1/2 (70.5%)**
2	A	31-Oct-17	Brain	1 (97.3%)	14 (97.1%), 2 (78.4%), 1/2 (70.3%)
3	A	8-Nov-17	Brain	1 (96.5%)	14 (96.7%), 2 (79.0%), 1/2 (68.4%)
4	A	5-Mar-18	Joint swab	1 (98.3%)	14 (98.1%), 2 (79.8%), 1/2 (74.1%)
5	A	4-Jan-18	Brain	1 (96.6%)	14 (96.1%), 2 (82.5%), 1/2 (75.5%)
6	A	13-Mar-18	Brain	1 (98.5%)	14 (98.3%), 2 (80.9%), 1/2 (71.7%)
7	A	30-Mar-18	Brain	1 (98.5%)	14 (97.8%), 2 (78.7%), 1/2 (70.0%)
8	A	15-May-18	Meningeal swab	1 (98.6%)	14 (98.3%), 2 (80.6%), 1/2 (73.4%)
9	A	28-Jun-18	Brain	1 (92.1%)	14 (91.5%), 2 (75.2%), 1/2 (60.7%)
10	A	16-Jul-18	Brain	1 (97.9%)	14 (98.0%), 2 (80.1%), 1/2 (70.7%)
Genetic flow 2					
1	**A**	**2-Feb-16**	**Joint swab**	**1/2 (93.0%)**	**2 (90.0%), 1 (69.0%)**
1	**A**	**28-Jan-16**	**Meninges**	**1/2 (92.8%)**	**2 (91.3%), 1 (72.4%)**
2	B	20-Jan-17	Brain swab	1/2 (93.3%)	2 (90.7%), 1 (69.2%)
3	C	6-Apr-18	Brain & heart	1/2 (92.2%)	2 (90.3%), 1 (68.9%)

Inspiring a verified culture of care post-identification of abuse on an undercover video using a baseline assessment of expressed competencies compared to post training measurement of expressed competencies

Sarah Probst Miller[1,2], DVM; Seth Krantz[3], DVM
[1]AgCreate Solutions, Inc, [2]Pork Avenue Training Portal, [3]Tosh Pork

Introduction

Sometimes the need to change human behavior is determined from internal evaluation, and sometimes the need to change is uncovered and stimulated from the outside. This case study presents a company's training response to an undercover video taken by a swine farm employee simultaneously working for Mercy for Animals. The video identified unacceptable behavior outside of company protocol on a sow farm (Farm A). Two animal handling actions captured on video were animal abuse which resulted in immediate dismissal of the accountable employees. This case study details the training plan created for remaining employees potentially influenced by the expression of these unacceptable behaviors, and shares expressed competency results of employees after re-training compared with the baseline expressed competency results recorded the day of the release of the undercover video.

Materials and methods

Immediate actions and baseline competency measurements

The same day of the undercover video release all pigs on site were assessed by a 3rd party veterinarian to ensure all pigs were safe and to help assess needed actions. To measure human behavioral impact on pig welfare, the 3rd party applied the Common Swine Industry Audit (CSIA)[1] inspection guidelines to every pig and pig housing location on Farm A. The veterinarian observed animal handling events as well. Opportunities for immediate actions as well as employee training (Table 1) were identified from both the video and from the on-farm assessment via measurement of baseline pig handling competencies per employee. Baseline competencies to measure were chosen based on the video and from missing competencies observed in the on-farm assessment (Table 2). A medical record outlining needed change was created. The system's production team and veterinarian discussed with the 3rd party veterinarian and agreed on an action and training plan.

Training plan

This targeted response plan included first a Farm A specific re-training plan on animal handling followed by some lesson specific topics that worked to address missing expressed competencies (Table 1). Farm A's successful completion would be followed by a system-wide re-training plan on animal handling. This plan was initiated with a live manager training meeting to explain the situation, and its ramifications, to achieve an understanding of the importance of successful completion. Farm A's training was implemented immediately utilizing Pork Avenue Training Portal (PATP).[2] The program components per learning experience includes online training, graphical SOP's and in-barn verification which employed the *See it. Do it. Teach it.* methodology.[3-9] Training delivery occurred via learning experiences delivered on PATP's online learning management software. Simultaneously, an internal and external verification plan was detailed and executed post re-training to measure expressed competency results using the lesson's In-Barn Verification Checklists as guidance documents (Table 1, 2, 3). All training accomplishments were recorded on PATP's online learning management software and a transcript of learner accomplishment created per learner.[7]

See it.

Farm A chose to administer most videos and quizzes as a group and have a discussion after each lesson on skills they needed to express differently on farm. Groups were formed based on preferred language of English or Spanish. Learners were informed prior to the videos that they needed to learn the topics well enough that they could teach it back to a verifier. The See it portion of the experience was marked as complete on the individual employee's transcripts.

Do it.

Then the learners continued to complete skills which were manager or mentor observed until verified. If desired, pictorial SOP's were available for review in English and Spanish.

Teach it.

Internal verification included having individuals teach what they learned back to an In-barn Verifier. In-barn Verifiers presented themselves to each employee as "their student" and encouraged them to "pretend I am a new employee and teach me how to do this task." While they spent most of the verification experience listening, verifiers utilized the In-barn Verification Checklists to make sure all key competencies were taught back successfully according to the farm protocol and asked about a key competency if it was not brought up specifically. When individuals successfully passed their In-barn Verifications, Verifiers marked this part of the learning experience complete on their transcript.

Table 1: Action list and training calendar for Farm A – opportunities for immediate actions as well as employee training were identified from both the video and from the on-farm assessment by an external veterinarian

Action	Who responsible	Timeline goal	Completion
Avoiding willful acts of abuse	JS	7/19/2018	7/19/2018
Avoiding willful acts of abuse--in barn verification	JS	7/19/2018	7/19/2018
Euthanasia training (live intro)	SK	7/18/2018	7/18/2018
CO_2 euthanasia with the Euthanex AgPro lesson	SK, JS	Week of 8/6/18	8/10/2018
Pig behavior and YOU--lesson	Group lesson scheduled and discussed	8/16/2018	8/1/2018
Pig behavior and YOU-- internal verification of each employee	JS	8/16/2018	8/1/2018
Training and handling boars--lesson	Group lesson scheduled and discussed	8/16/2018	8/17/2018
Training and handling boars-- internal verification of each employee	JS	week of 8/27/18	9/5/2018
Training and handling boars-- external verification of process	SPM	9/12/2018	9/12/2018
Moving pregnant sows and gilts--lesson	Group lesson scheduled and discussed	8/16/2018	8/10/2018
Moving pregnant sows and gilts-- internal verification of each employee	JS	8/16/2018	8/10/2018
Moving pregnant sows and gilts-- external verification of process	SPM	9/12/2018	9/12/2018
Weaning sows--lesson	Group lesson scheduled and discussed	8/16/2018	8/17/2018
Weaning sows--internal verification of each employee	JS	week of 8/27/18	9/5/2018
Weaning sows--external verification of process	SPM	9/12/2018	9/12/2018
Weaning pigs--lesson	Group lesson scheduled and discussed	8/16/2018	8/17/2018
Weaning pigs--internal verification of each employee	JS	week of 8/27/18	9/5/2018
Weaning pigs--external verification of process	SPM	9/12/2018	9/12/2018
Loading market hogs--lesson	Group lesson scheduled and discussed	8/16/2018	8/17/2018
Loading market hogs--internal verification of each employee	JS	week of 8/27/18	9/5/2018
Lessons to strategically assign by task or job description			
Castration	JS, KS	By end of week of 8/27/18	8/24/2018
ID and Tx sick sows	JS, KS	By end of week of 8/27/18	9/5/2018
ID and Tx sick pigs	JS, KS	By end of week of 8/27/18	9/5/2018
Rupture repair	JS, KS	By end of week of 8/27/18	9/5/2018

Table 1 continued on next page

Table 1: cont'd

Action	Who responsible	Timeline goal	Completion
Detectable needle best practices	JS, KS	By end of week of 8/27/18	9/5/2018
Euthanasia decision	JS, KS	By end of week of 8/27/18	8/24/2018
CO_2 euthanasia with the Euthanex AgPro	JS, KS	By end of week of 8/27/18	8/24/2018
Captive bolt euthanasia--penetrating	JS, KS	By end of week of 8/27/18	8/24/2018
Captive bolt maintenance	JS, KS	By end of week of 8/27/18	8/24/2018
Fit to load lesson series	JS, KS	By end of week of 8/27/18	9/5/2018
Providing adequate space and housing	JS, KS	By end of week of 8/27/18	9/5/2018

Results

The baseline expressed competencies for the training curriculum chosen was at a 25.8% competency expression rate on farm for chosen handling competencies. Post external verification of training, staff averaged a 95% handing competency expression rate (Table 3). This is a 69.4 -percentage point difference and a 268.75% increase in handling competency expression on farm. Retraining throughout the entire system is occurring throughout Fall and Winter of 2018/19. External verification of other farms occurs in December 2018 and January 2019, and results will be reported in this presentation.

Conclusion and discussion

In 2017, this farm set clear accountability goals per key performance indicator and made an accountability plan to achieve those goals. For example, one of their 2017 goals for 2018 was to move the system from 29 P/MF/Y to over 30. They succeeded. In fact, at the time of the video release, the farm where the behavior outside of company protocol occurred averaged 32.5 P/MF/Y. Amid this production success, the animal abuse video occurred. The farm has owned this and has made a team commitment to never lose site that their number one priority is excellent pig care. Post retraining on animal handling and this animal care commitment, the farm of concern is still averaging over 32.5 P/MF/Y.

Using the "See it. Do it. Teach it." methodology[3-8] has effectively improved competency expression in other studies from a 59% level of competency expression to an 85% competency expression level.[10, 11] This case study also achieved improved levels of expressed competency. Specifically utilizing this educational methodology alongside external verification, it increased expressed competency 69.4 percentage points which was a 268.8% increase. Of note, in this study the baseline competencies (Table 2) were specifically chosen to represent those missing on the video and the assessment which makes the baseline lower than a random assessment of competencies. However, it accurately represents the missing competencies that the system wanted to ensure living on farm. Therefore, the percent change accurately measures the intentions of the intervention.

References

1. Common Swine Industry Audit, 2018. Available at https://lms.pork.org/CommonIndustryAudit.

2. Pork Avenue Training Portal. 2018. Available at www.porkavenuetraining.com.

3. Bear, Anne, Technology, Learning, and Individual Differences, Journal of Adult Education, 2012.

4. Cercone, K. (2008). Characteristics of adult learners with implications for online learning design, AACE Journal, 16(2), 137-159.

5. Dale, Edgar, Audio visual methods in teaching. Holt, Rinehart and Winston, 1965.

6. Dale, Edgar, Cone of Experience, Education Media, 1969.

7. Geaetano, Kristian, 5 ways adults learn differently than children, Learn Kit, 2016.

8. Probst Miller, Sarah, Applying the science and physiology of how adults learn to our on-farm approach as practitioners, AASV Proceedings, 2016.

9. Ryland, Bruce, Fostering Critical Reflection in Adulthood, by Jack Mezirow & Associates, San Francisco: Jossey-Bass, 1990.

10. Probst Miller, Sarah, Impact of two training implementation plans on post-training competency expression of personnel on farm, AASV Proceedings, 2018.

11. Probst Miller, Sarah, The impact of trained and untrained personnel on product implementation and subsequent pig health, AASV Proceedings, 2016.

Table 2: Baseline competencies to measure were chosen based on the video and from missing competencies observed in the on-farm assessment

Observe the following and score yes/no for key competencies	7/17/2018 — Baseline	9/12/2018 — Follow-up
Training and handling boars		
Handler properly positioned sort-board between self and boar		
Two people present if boar in pen with people		
Farm has plan for boar culling		
Farm safely loads boar to boar movement equipment		
Moving pregnant sows and gilts		
Hallway is prepped and distractions removed		
No dead pigs in hallway		
Sows are moved calmly and slowly		
Sows are moved in groups of 3 to 5		
Pressure is appropriate		
Noise is intermittent calm and not constant		
Employees use appropriate language when moving animals		
Weaning sows		
Hallway is prepped, and distractions removed		
No dead pigs in hallway		
Sows are moved calmly and slowly		
Sows are moved in groups of 3 to 5		
Pressure is appropriate		
Noise is intermittent calm and not constant		
Employees use appropriate language when moving animals		
Weaning pigs		
Hallway is prepped, and distractions removed, not slippery		
No dead pigs in hallway		
Pigs are moved calmly and slowly		
Pigs are moved in properly sized groups per farm protocol		
Pressure is appropriate		
Noise is intermittent calm and not constant		
Employees use appropriate language when moving animals		
Euthanasia		
No pigs with BCS of 1 observed in random sample of farrowing		
No pigs with BCS of 1 observed in random sample of gestation		
Farm has approved method of euthanasia in place		
There is a log for cleaning of captive bolt gun		
Captive bolt gun is cleaned and disinfected and stored properly when not in use		
Employee designated to euthanize can demonstrate how to determine if a pig is truly dead.		

Table 3: Baseline versus post-training competency assessment

	7/17/2018 Baseline competency assessment		9/12/2018 Post-training competency assessment
Number of expressed competencies	8		29.5
Total competencies measured	31		31
Percent of competencies expressed	25.8%		95.2%
Percentage point difference		69.4%	
Percent increase		268.8%	

The effect of various levels of an antioxidant support technology on the performance of health-challenged nursery pigs

S. Ebarb, MS; K. Saddoris-Clemons, MS, PhD; K. Price, MS, PhD; J. Jolliff, MS, PhD; S. Williams, MS
Cargill Premix & Nutrition, Brookville, OH

Introduction

The weaning transition is a stressful period for pigs due to nutritional, social, and environmental changes. If facing other stressors such as pathogens, the physiological systems designed to cope with stress can become overwhelmed. It has been reported that serum vitamin E (**VIT-E**) concentrations drop dramatically in the first 5-12 days post-weaning.[1] The combination of weaning and suboptimal environmental conditions has been reported to increase measures of oxidative stress at 12 days post-weaning in combination with reduced serum VIT-E concentrations.[1] Neutralizing free radicals produced during stressful periods of time may help reduce damage to both cells and tissues. Promote®AOX™ (**AOX**) is an antioxidant blend and has been shown to support antioxidant status in multiple species. The objective of this study was to determine the effects of dietary VIT-E and AOX on the performance of health challenged nursery pigs.

Materials and methods

A total of 1,848 naturally health-challenged (determined by mortality being ~2% higher than the past yearly average) pigs (13.0 ± 0.9 lb) were housed in two barns (14 pens/treatment and 22 pigs/pen) which were blocked separately by pen location and randomly allocated to treatments. Pigs were split sexed, and pens contained equal numbers of heavy, medium, and light barrows or gilts to allow for similar initial body weight for all pens. The study was a complete randomized block design and organized as a 2 × 3 factorial with three levels of AOX and two levels of VIT-E. Treatments were: 1) 16 IU/lb of VIT-E with 0% AOX; 2) 16 IU/lb VIT-E with 0.02% AOX; 3) 16 IU/lb VIT-E with 0.13% AOX; 4) 45.5 IU/lb VIT-E with 0% AOX; 5) 45.5 IU/lb with 0.02% AOX; 6) 45.5 IU/lb with 0.13% AOX. A description of dietary treatments can be found in Table 1.

Table 1: Description of dietary treatments

Treatment number	Promote®AOX™, %	Vitamin E, IU/lb
1	0	16
2	0.02	16
3	0.13	16
4	0	45.5
5	0.02	45.5
6	0.13	45.5

Pigs were assigned to treatments on d 0 post-weaning (~19 days of age). Pen weights were obtained on d 0, 7, 21, and 40 post-weaning. Feed intake was calculated by subtracting feed weigh backs from feed additions. Mortality and removals were recorded throughout the trial. Performance and health parameters were determined for average body weight (**BW**), ADG, ADFI, feed efficiency (**GF**), cost/lb of gain, closeout return over feed cost (**ROFC**; calculated assuming $1/lb and accounted for reduced animal numbers due to mortality or removal), and probability of mortality and removals. Statistics were analyzed using PROC GLIMMIX in SAS 9.3 (SAS, Cary, NC). Differences were considered significant at $P < 0.05$ and a tendency at $P < 0.10$. Pre-planned constructed contrasts were: 1) 0% AOX vs additional AOX (aggregate effect of 0.02 and 0.13% AOX); 2) 0.02 vs 0.13% AOX; 3) 16 vs 45.5 IU/lb VIT-E; 4) additional AOX × VIT-E level interaction; 5) 0.02 vs 0.13% AOX × VIT-E level interaction.

Results and discussion

Phase 1 and 2 Combined (d 0 to 21)

Results are shown in Table 2. Interactions were observed between additional AOX and VIT-E for d 21 BW, ADG, ADFI, GF, and cost/lb of gain ($P < 0.01$). When additional AOX (0.02 or 0.13%) was added to the 16 IU/lb VIT-E diet, d 21 BW increased by 1 lb, ADG increased by 7.2%, ADFI increased by 3.8%, GF increased by 3.9%, and cost/lb of gain decreased by 3.4% compared to the 0% additional AOX treatment. Oppositely, adding additional AOX to the diet with 45.5 IU/lb VIT-E resulted in similar d 21 BW, reduced ADG by 3.0% and ADFI by 3.1%, and had similar cost/lb of gain. Interactions were observed between AOX level, when increased from 0.02 to 0.13%, and VIT-E level for d 21 BW, ADG, and ADFI ($P < 0.01$). For diets with 16 IU/lb VIT-E, increasing AOX to 0.13% increased d 21 BW by 0.3 lb, increased ADG by 4.0%, and increased ADFI by 3.2% compared to 0.02% AOX. However, when AOX level increased from 0.02 to 0.13% in diets containing 45.5 IU/lb VIT-E, pigs had reduced d 21 BW by 0.8 lb, reduced ADG by 4.8% and reduced ADFI by 5.1%. There was an interaction between AOX, when increased from 0.02 to 0.13%, and VIT-E level for probability of mortality and removals ($P < 0.05$). In diets with 16 IU/lb VIT-E, probability of mortality and removals was reduced when AOX increased from 0.02 to 0.13% in the diet. When VIT-E was included at 45.5 IU/lb in the diet, increasing AOX did not further reduce the probability of mortality and removals.

Overall (d 0 to 40)

Results are shown in Table 3. Interactions were observed between additional AOX and VIT-E level for BW out of the nursery (d 40 post-weaning), ADG, ADFI, and ROFC ($P < 0.05$). Including additional AOX (0.02 or 0.13%) in diets containing 16 IU/lb VIT-E increased final BW by 1.1 lb, increased ADG by 3.5%, increased ADFI by 2.0%, and increased ROFC by $0.87 per pig. When additional AOX was supplemented in diets containing 45.5 IU/lb VIT-E, final BW was decreased by 1.1 lb, ADG decreased by 3.6%, ADFI decreased by 3.0%, and ROFC was reduced by $1.17/pig. Interactions were observed when AOX level increased from 0.02 to 0.13% and VIT-E level for ADG, ADFI, ROFC, and probability for mortality and removals ($P < 0.05$), and BW out of the nursery ($P < 0.10$). When VIT-E was included at 16 IU/lb, increasing AOX from 0.02% to 0.13% resulted in increased ADG by 4.5%, increased ADFI by 4.4%, increased ROFC by $1.59 per pig, and reduced probability of mortality and removals. Increasing AOX level from 0.02 to 0.13% in diets containing 45.5 IU/lb VIT-E resulted in reduced ADG by 2.6%, reduced ADFI by 3.1%, reduced ROFC by $0.78 per pig, and similar probability for mortality and removals. Overall, cost/lb of gain and GF were not impacted by AOX or VIT-E level ($P > 0.10$).

It is hypothesized that including elevated levels of both AOX and VIT-E may have resulted in a net pro-oxidative response that led to poorer animal performance. The antioxidant

Table 2: Effects of Promote AOX and vitamin E on growth performance and health of nursery pigs (Phase 1 and 2 combined, d 0 to 21)

	Promote AOX level, % (16 IU/lb Vitamin E)			Promote AOX level, % (45.5 IU/lb Vitamin E)			
Phase 1 and 2 (d 0 to 21)	0	0.02	0.13	0	0.02	0.13	SEM
Initial BW, lb	12.92	12.94	13.03	12.99	12.96	12.96	0.13
Final BW[1,2], lb	25.22	26.10	26.39	26.54	26.72	25.93	0.26
ADG[1,2], lb/d	0.589	0.619	0.644	0.650	0.646	0.615	0.015
ADFI[1,2], lb/d	0.701	0.716	0.739	0.745	0.741	0.703	0.018
Gain to feed[1], lb/lb	0.837	0.867	0.873	0.873	0.874	0.873	0.006
Cost/lb of gain[1], USD/lb	0.335	0.323	0.324	0.320	0.319	0.322	0.003
Probability of mortality and removals[3], %	4.94	8.61	2.90	5.20	6.69	6.26	1.71

1 Interaction: additional AOX × vitamin E level, $P < 0.01$
2 Interaction: 0.02 vs 0.13% AOX × vitamin E level, $P < 0.01$
3 Interaction: 0.02 vs 0.13% AOX × vitamin E level, $P < 0.05$

Table 3: Effects of Promote AOX and vitamin E on growth performance and health of nursery pigs (Overall, d 0 to 40)

	Promote AOX level, % (16 IU/lb Vitamin E)			Promote AOX level, % (45.5 IU/lb Vitamin E)			
Phase 1 and 2 (d 0 to 21)	0	0.02	0.13	0	0.02	0.13	SEM
Initial BW, lb	12.92	12.94	13.03	12.99	12.96	12.96	0.13
Final BW[2,5], lb	48.59	49.21	50.18	50.73	50.18	49.12	1.48
ADG[1,4], lb/d	0.869	0.880	0.919	0.930	0.908	0.884	0.033
ADFI[2,3], lb/d	1.140	1.138	1.188	1.204	1.186	1.149	0.055
Gain to feed, lb/lb	0.770	0.772	0.776	0.774	0.767	0.774	0.009
Cost/lb of gain, USD/lb	0.239	0.239	0.239	0.236	0.239	0.239	0.002
Closeout ROFC[1,3], USD/pig	25.91	25.99	27.58	27.93	27.15	26.37	1.43
Probability of mortality and removals[4], %	10.38	13.56	6.45	8.44	10.00	9.73	2.41

1 Interaction: additional AOX × vitamin E level, $P < 0.01$
2 Interaction: additional AOX × vitamin E level, $P < 0.05$
3 Interaction: 0.02 vs 0.13% AOX × vitamin E level, $P < 0.01$
4 Interaction: 0.02 vs 0.13% AOX × vitamin E level, $P < 0.05$
5 Interaction: 0.02 vs 0.13% AOX × vitamin E level, $P < 0.10$

system is robust, and the balance can be shifted by feed derived antioxidants through many different pathways.[2] In this study, providing antioxidants through either AOX or VIT-E resulted in improved performance. Mortality and removal benefits were present when AOX was included at 0.13% in a lower VIT-E diet. Plant polyphenols and VIT-E can reduce lipid peroxidation, but polyphenols have additional benefits, such as protecting both proteins and DNA from free radical damage and reduced inflammation.[3] The additional benefits of plant polyphenols on various health parameters could impact the effectiveness of AOX compared to adding only VIT-E.

Weaned pigs exposed to various stressors in the nursery may benefit from the addition of dietary antioxidant support products. Including Promote AOX in diets containing 16 IU/ b VIT-E resulted in improved growth performance parameters and reduced the probability for mortality and removals in naturally health-challenge pigs. Further benefits were observed when Promote AOX was increased from 0.02% to 0.13% in diets containing 16 IU/lb of VIT-E.

References

1. Buchet, A., C. Belloc, M. Leblanc-Maridor, and E. Merlot. 2017. Effects of age and weaning conditions on blood indicators of oxidative status in pigs. PLoS ONE 12(5):e0178487. doi:10.1371/journal.pone.0178487

2. Bouayed, J. and T. Bohn. 2010. Exogenous antioxidants – Double-edged swords in cellular redox state. Oxidative Medicine and Cellular Longevity. 3:4, 228-237. doi: 10.4161/oxim.3.4.12858

3. Hussain, T., B. Tan, Y. Yin, F. Blachier, M. Tossou, and N. Rahu. 2016. Oxidative stress and inflammation: What polyphenols can do for us?. Oxidative Medicine and Cellular Longevity. dx.doi.org/10.1155/2016/7432797

The effects of PROMOTE® Defusion® Prime on the performance of nursery pigs fed quality challenged diets: A combined analysis

Kari Saddoris-Clemons, PhD; Sara Ebarb, MS; Sabrina Williams, MS; Don Giesting, PhD
Cargill Premix & Nutrition, Brookville, Ohio

Introduction

Deoxynivalenol (DON), a mycotoxin, is produced by *Fusarium graminearum,* that is commonly found in corn, other grains, and their processed co-products, and less commonly by *Fusarium culmorum.* DON is most prevalent when wet conditions are present before and during harvest. *F. graminearum* (Gibberella zeae) and *F. culmorum* produce both DON and zearalenone, and contaminated feedstuffs may contain both mycotoxins. DON contamination of grains can lead to reductions in feed intake and performance when fed to swine. Levels as low as 1 ppm of DON negatively impact feed intake (ADFI) and daily gain (ADG),[1] with intake being reduced linearly as levels of DON increase further.[2] In addition, DON negatively impacts intestinal barrier function, impairs the immune system, and increases oxidative stress.[3,4]

Defusion® Prime is a complex product formulated to support animal growth and well-being. Several internal studies have established the efficacy of this product in swine diets containing quality challenged corn and dried distillers grains and solubles (DDGS), specifically with DON concentration used as an indicator of quality. Using a combined analysis, the objective of this trial was to determine the impact of 0.25 or 0.50% Defusion® Prime in diets with less than 1.1 ppm of DON and in diets with greater than 2.0 ppm DON on ADG, ADFI, gain/feed (GF), and cost/lb of gain across 6 studies in 15-25 lb pigs and 12 studies conducted in 25-50 lb pigs.

Materials and methods

A total of six studies (n = 6) were included in the 15-25 lb pig body weight (BW) range analysis and a total of twelve studies (n =12) were included in the 25-50 lb pig BW range analysis. In the 15-25 lb BW range, a total of 122 pens (~22 pigs/pen) were included in the analysis, for a total of 2,696 pigs. In the 25-50 lb BW range, a total of 293 pens (~22 pigs/pen) were included in the analysis, for a total of 6,305 pigs.

A complete randomized block design was utilized in each of the individual studies. Pigs were split-sex housed. Barns were blocked by initial body weight and pens were randomly allocated to treatment groups within block. The raw data from each study was combined and treatments were standardized by dietary DON level (CLEAN; <1.1 ppm DON or CONTAMINATED; >2.0 ppm DON) and inclusion rate of Defusion Prime (0%, 0.25%, or 0.50%). Table 1 describes the number of pens in each standardized category for each body weight range.

All diets were analyzed for DON content at Trilogy Laboratory (Washington, MO) by HPLC. Body weights were measured and ADG was calculated. ADFI was determined by subtracting feed weigh backs from total feed additions, and GF was calculated as a ratio of ADG:ADFI. Mortality and removals were recorded throughout the trial. Data were analyzed using PROC MIXED in SAS 9.3 (SAS, Cary, NC). Trial was considered a random effect. Differences were considered significant at $P < 0.05$, and trends were reported at $P < 0.10$. Pre-planned contrasts included:

Table 1: Description of standardized treatments and sample sizes in the 15-25 lb and 25-50 lb BW ranges.

	15-25 lb BW range		25-50 lb BW range	
Description	Pens/ treatment	Pigs/ treatment	Pens/ treatment	Pigs/ treatment
CLEAN diet[1]	36	790	46	994
CONTAMINATED diet[2]	24	535	62	1,333
CONTAMINATED + 0.25% Defusion® Prime	45	989	130	2,794
CONTAMINATED + 0.50% Defusion® Prime	17	382	55	1,184

1 CLEAN Diet = <1.1 ppm DON
2 CONTAMINATED Diet = >2.0 ppm DON

1) CLEAN vs CONTAMINATED, 2) No Defusion® Prime vs Defusion® Prime in CONTAMINATED diets, and 3) 0.25% Defusion® Prime vs. 0.50% Defusion® Prime in CONTAMINATED diets.

Results and discussion

In the 15-25 lb BW pig study group, the average analyzed DON level of the CLEAN diets was 0.29 ppm with a range from 0-0.5 ppm; and the average analyzed DON level in the CONTAMINATED diets was 2.92 ppm with a range from 2.5-3.5 ppm. In the studies with 25-50 lb BW pigs, the average analyzed DON level of the CLEAN diets was 0.39 ppm with a range from 0 ppm to 1.1 ppm; and the average analyzed level of the CONTAMINATED diets was 3.80 ppm with a range of 2.0-6.0 ppm.

In 15-25 lb BW pigs, feeding CLEAN diets resulted in significantly higher final BW ($P < 0.001$), and increased ADG ($P < 0.001$) and ADFI ($P < 0.001$) compared to pigs fed CONTAMINATED diets (Table 2). Pigs exhibited 11.3% higher ADG, 15.0% higher ADFI, and were 1.05 lb heavier after only 14 d on feed. The addition of Defusion® Prime to diets contaminated with DON significantly improved ($P < 0.001$) ADG, ADFI, and final BW compared to pigs fed CONTAMINATED diets without Defusion® Prime (Table 3). Pigs fed either 0.25 or 0.50% Defusion® Prime had 13.1% higher ADG, 12.9% higher ADFI, and were 1.37 lb heavier at (final BW) compared to pigs fed CONTAMINATED diets with no Defusion® Prime. No differences ($P > 0.10$) were detected in GF between pigs fed Defusion® Prime and pigs not fed Defusion® Prime in CONTAMINATED diets. Increasing the inclusion rate of Defusion® Prime from 0.25% to 0.50% improved ($P < 0.05$) ADG by 5.4% and ADFI by 4.6%, and tended to increase ($P = 0.08$) final BW by 0.52 lb. Overall, including 0.25% Defusion® Prime in diets with an average of 2.9 ppm DON improved final BW, ADG, and ADFI.

Pigs fed CLEAN diets from 25-50 lb of BW had higher final BW ($P < 0.001$), ADG ($P < 0.001$), ADFI ($P < 0.001$), and GF ($P = 0.002$) compared to pigs fed CONTAMINATED diets (Table 4). Pigs were 2.0 lb heavier at the end of the nursery period, and had 8.96% higher ADG, 5.4% higher ADFI, and 3.2% increased GF when fed CLEAN diets compared to pigs fed CONTAMINATED diets. The addition of Defusion® Prime to CONTAMINATED diets significantly improved ADG ($P < 0.001$), ADFI ($P < 0.001$), final BW ($P < 0.001$), and tended to improve GF ($P = 0.07$) compared to pigs fed CONTAMINATED diets without Defusion® Prime (Table 5). Pigs fed Defusion® Prime were 1.63 lb heavier at the end of the nursery period, exhibited 5.5% higher ADG,

Table 2: The effect of feeding CONTAMINATED diets to nursery pigs from 15-25 lb BW on ADG, ADFI, and feed efficiency.

	Treatment			*P*-value
Variable	CLEAN Diet[1]	CONTAMINATED Diet[2]	SEM	CLEAN vs. CONTAMINATED
Initial BW, lb	14.87	14.89	0.44	0.92
Final BW, lb	26.00	24.95	0.73	<0.0001
ADG, lb/d	0.860	0.773	0.024	<0.0001
ADFI, lb/d	1.048	0.912	0.037	<0.0001
Gain:Feed	0.823	0.851	0.022	0.014

[1] CLEAN Diet average = 0.29 ppm DON (0 – 0.5 ppm)
[2] CONTAMINATED Diet average = 2.92 ppm DON (2.5 – 3.5 ppm)

Table 3: The effect of feeding Defusion® Prime in CONTAMINATED diets (2.9 ppm average DON) to 15-25 lb BW pigs on pig performance.

	Defusion® Prime level, %				Contrast *P*-value	
Variable	0	0.25	0.50	SEM	0 vs. Any Defusion® Prime	0.25% vs. 0.50% Defusion® Prime
Initial BW, lb	14.86	14.94	14.86	0.45	0.69	0.51
Final BW, lb	24.04	25.15	25.67	0.76	<0.0001	0.08
ADG, lb/d	0.711	0.783	0.825	0.024	<0.0001	0.018
ADFI, lb/d	0.840	0.926	0.969	0.037	<0.0001	0.012
Gain:feed	0.847	0.850	0.854	0.023	0.603	0.738

Table 4: The effect of feeding CONTAMINATED diets to nursery pigs from 25-50 lb BW on ADG, ADFI, and feed efficiency.

	Treatment			*P*-value
Variable	CLEAN Diet[1]	CONTAMINATED Diet[2]	SEM	CLEAN vs. CONTAMINATED
Initial BW, lb	26.22	26.12	0.53	0.76
Final BW, lb	51.27	49.27	1.32	<0.0001
ADG, lb/d	1.252	1.149	0.049	<0.0001
ADFI, lb/d	1.744	1.654	0.075	<0.0001
Gain:feed	0.718	0.696	0.011	0.002

[1] CLEAN Diet average = 0.39 ppm DON (0 – 1.1 ppm)
[2] CONTAMINATED Diet average = 3.80 ppm DON (2 – 6 ppm)

Table 5: The effect of feeding Defusion® Prime in CONTAMINATED diets (3.8 ppm average DON) to late nursery pigs (25-50 lb BW) on pig performance.

	Defusion® Prime level, %				Contrast *P*-value	
Variable	0	0.25	0.50	SEM	0 vs. Any Defusion® Prime	0.25% vs. 0.50% Defusion® Prime
Initial BW, lb	25.93	26.16	26.26	0.53	0.11	0.62
Final BW, lb	48.18	49.26	50.36	1.32	<0.0001	0.001
ADG, lb/d	1.108	1.143	1.195	0.049	<0.0001	<0.0001
ADFI, lb/d	1.600	1.656	1.705	0.075	<0.0001	0.002
Gain:feed	0.691	0.693	0.703	0.011	0.067	0.034

a 5.0% improvement in ADFI, and a 1.0% improvement in GF. Increasing the inclusion rate of Defusion® Prime from 0.25% to 0.50% in CONTAMINATED diets significantly increased ADG ($P < 0.001$), ADFI ($P = 0.002$), final BW ($P = 0.001$), and GF ($P = 0.03$). Compared to the 0% inclusion level of Defusion® Prime, 0.25% Defusion® Prime numerically improved ADG by 3.2% and ADFI by 3.5%, which resulted in a 1.1 lb heavier final BW. Increasing the Defusion® Prime from 0.25% to 0.50% further improved ADG and ADFI by 4.6% and 2.9%, respectively, which further improved final BW by an additional 1.1 lb. Overall, including 0.50% Defusion® Prime in diets contaminated with an average of 3.80 ppm DON improved final BW by 2.2 lb due to significant improvements in ADFI and ADG compared to pigs fed CONTAMINATED diets without Defusion® Prime.

In conclusion, analysis of 6 large scale studies including over 2,600 pigs in the 15-25 lb pig weight range, and 12 large scale studies, including over 6,300 pigs, in the 25-50 lb pig weight range, indicate that under the conditions of these studies, Defusion® Prime was effective in improving pig performance when fed quality challenged diets.

References

1. Diekman, M., T. Coffey, R. Hill, E. D. Purkhiser, D. Reeves, and L. Young. Mycotoxins and swine performance. Pork Information Gateway. PIH-129. 07-06-05.

2. Döll, S. and Dänicke, S. 2011. The *Fusarium* Toxins deoxynivalenol (DON) and zearalenone (ZON) in animal feeding. Preventive Veterinary Medicine 102:132-145.

3. Pierron, A., I. Alassane-Kpembi, and I. Oswald. 2016. Impact of mycotoxin on immune response and consequences for pig health. Animal Nutrition. 2: 63-68.

4. Weaver, A. C., M. T. See, and S. W. Kim. 2014b. Protective effect of two yeast-based feed additives on pigs chronically exposed to deoxynivalenol and zearalenone. Toxins 6:3336-3353.

Meta-analysis of enteric pathogen genes identified in 122 swine flows across the United States

K. Leistikow, BS; S. Son, BS, MBA; M. King, PhD

Introduction

Several enteric bacterial and viral pathogens require control to prevent disease, reduce morbidity and mortality, and improve the efficiency of pig production.[1] Health improvement strategies to combat these pathogens may improve the profitability of the swine enterprise by increasing both throughput (e.g., pigs sold per sow per year) and margin (e.g., reduced cost per pound marketed or increased premium per genetic animal).[2] To better understand the relationship of enteric disease and environmental pressures, a large sampling effort was conducted to determine the enteric pathogen gene profiles of various swine populations from 122 flows across the United States. Utilizing a rectal sampling technique in conjunction with qPCR, this study analyzed the enteric gene profiles from over 3700 animals across twelve states in an attempt to understand how geography, seasonality, and production stage influence the enteric pathogen gene profile of these animals.

Materials and methods

A meta-analysis was conducted on enteric pathogen gene data gathered between 2016 and 2018. Rectal samples were taken and qPCR was used to characterize the enteric pathogen gene profile of animals throughout the production process including sows, newborn and late lactation piglets, nursery, and grow finish animals. Marker and virulence genes belonging to *E. coli*, *C. perfringens*, *C. difficile*, *Salmonella*, Rotavirus and *L. intracellularis* species were compared using statistical analysis for patterns associated with production stage, season, and geography.

Results and discussion

Results suggest that enteric pathogen gene profiles are influenced by a multitude of factors including stage of production, geography, and season. Despite moderate detection of *Clostridium* in sows, *Clostridium* toxin genes, *cpa*, *cpb2*, *tcdA*, and *tcdB* were detected in higher quantities in lactation-aged piglets ($P < 0.05$) and gradually decreased in quantity as the animals aged. *Salmonella* genes, though present in all populations, were more abundant in sows and animals 0-7 days post weaning, while virulence genes associated with enterotoxigenic *E. coli*, namely *STa*, *K99* and *F18*, correlated uniquely to nursery animals two weeks post wean and older. Several *E. coli*, *Salmonella*, and *C. perfringens* genes tended to be more abundant in animals sampled in the fall and spring. Rotavirus A genes were detected in larger quantities during the winter months ($P < 0.05$). This meta-analysis supports clinical observations in the field and is likely influenced by the changing production processes throughout pig development in addition to the management changes instituted throughout the year to mitigate disease pressures.

References

1.Pluske, J., Pethick, D., Hopwood, D., & Hampson, D. (2002). Nutritional influences on some major enteric bacterial diseases of pig. Nutrition Research Reviews, 15(2), 333-371. doi:10.1079/NRR200242

2.Dufresne, L. (2006, June 3). Economics of Pig Health Improvement. Retrieved October 20, 2018, from `http://porkgateway.org/resource/economics-of-pig-health-improvement/`

Genomic and metabolomic approaches to improving behavior of pigs

E. Dervishi[1], PhD; L. van der Zande[2], MSc; I. Reimert[3], PhD; P. Mathur[2], PhD; M.S. Lopes[2], PhD; E. F. Knol[2], PhD; G. Plastow[1], PhD

[1]Livestock Gentec, University of Alberta, Edmonton, Canada;

[2]Topigs Norsvin Research Center B. V., The Netherlands;

[3]Wageningen University & Research, Department of Animal Sciences, Adaptation Physiology Group, The Netherlands

Introduction

Pork consumers are demanding pork quality together with high animal welfare standards. Traits like behavior in relation to housing types, social interaction, and different stress factors, are becoming economically important in breeding programs. Breeding pigs with less aggressive behavior is attractive not only from an animal welfare standpoint, but also for the experience of animal caretakers. A new breeding method including Indirect Genetic Effects (IGE), to reduce aggression in group housing animals has been proposed.[1] Pigs selected for a positive effect on the growth of their group members, performed less non-reciprocal biting behavior and less aggression after regrouping test.[2] The purpose of this pilot study was to identify important compounds related to pig behavior such as tail biting, tail damage score and tail blood score in addition to identifying regions of the genome associated with IGE.

Materials and methods

Animals involved in the present study were part of a much larger study carried out at Wageningen University and Research Center.[2] Here we used 48 Tempo x Topigs-20 finisher animals divided over two batches balanced for gender, selected for positive (n = 27) or negative (n = 21) IGE for growth. Blood samples collected at 22 weeks of age were used to determine metabolic profiles at The Metabolomics Innovation Centre (University of Alberta) using Nuclear Magnetic Resonance (NMR), with the Absolute-IDQ platform kit (Biocrates AbsoluteIDQTM p180 Kit, Biocrates Life Science AG, Innsbruck, Austria). Finally, a catecholamine assay was performed to quantify dopamine, epinephrine, tyramine, metanephrine, normetanephrine and phenylethylamine. A total of 249 compounds were quantified. Other physiological parameters included: platelet number, blood platelet serotonin level, whole blood serotonin level and blood platelet serotonin uptake velocity.[3] In addition, observations of behavior traits such as "biter" information, tail damage score and tail blood score, were collected for all animals. A single, single nucleotide polymorphism (SNP) genome-wide association study (GWAS) was performed to identify regions of the genome associated with IGE.

Results and discussion

Preliminary results show that serum levels of serotonin and epinephrine were significantly associated with tail damage score at week 22. Acetone, ketone bodies and epinephrine were associated with tail blood score. Interestingly, concentration of vitamins B2, B5, and B7 and serotonin uptake velocity and platelet number were significantly associated with tail biting behavior. Biter animals had greater concentration of vitamins B2, B5 and B7 ($P < 0.05$), greater serotonin uptake velocity ($P < 0.05$) and lower number of platelets ($P < 0.05$). Additional results from GWAS showed that a SNP on chromosome 13 was associated with IGE. The SNP on chromosome 13 is located within a protein coding gene synapsin II (*SYN2*) and it is involved in neurotransmitter secretion. In humans and mice, variations in this gene have been associated with autism, schizophrenia, and bipolar disorder by affecting social behavior, social communication and repetitive behaviors.[4,5] In summary, preliminary results indicate that at the finisher stage, serum serotonin levels and serotonin uptake are associated with tail damage score (propensity to being bitten). Vitamins B2, B5, B7, serotonin uptake and number of platelets are associated with tail biting behavior (performing tail biting), which might be under some level of genetic control, including by genes located on chromosome 13. Currently, these metabolites are being tested on a larger number of samples.

This study is part of the "Genomic and metabolomics approach to improving new economically important traits in pigs" project, funded by the Alberta Agriculture and Forestry, AB, Canada and Mitacs Canada, in collaboration with Topigs Norsvin.

References

1. Muir, W.M. 1996. Group selection for adaptation to multiple-hen cages: selection program and direct responses. Poult Sci. 75(4): 447-458.

2. Camerlink, I., Turner, S.P., Bijma, P., and Bolhuis, E.J. 2013. Indirect genetic effects and housing conditions in relation to aggressive behaviour in pigs. PlosOne, e65136.

3. Ursinus, W.W., Van Reenen, C.G., Reimert, I., Bolhuis, J.E. 2014. Tail biting in pigs: Blood serotonin and fearfulness as pieces of the puzzle? PLoS ONE 9(9): e107040.

4. Greco, B., Managò, F., Tucci, V., Kao, H.T., Valtorta, F., Benfenati, F. 2013. Autism-related behavioral abnormalities in synapsin knockout mice, Behav. Brain Res. 251:65-74.

5. Corradi, A., Fadda, M., Piton, A., Patry, L., Marte, A., Rossi, P., Cadieux-Dion, M., Gauthier, J., Lapointe, L., Mottron, L., Valtorta, F., Rouleau, G.A., Fassio, A., Benfenati, F., Cossette, P. 2013. SYN2 is an autism predisposing gene: loss-of-function mutations alter synaptic vesicle cycling and axon outgrowth. Human. Mol. Gen. 23: 90–103.

A computer vision system for real-time group and individual pig behavior tracking

J. Labrecque, P. Eng.; F. Gouineau, MSc; J. Rivest, PhD
Conception Ro-Main inc., St-Lambert-de-Lauzon, Quebec

Introduction

Monitoring livestock behavior in real-time promises to open new possibilities to achieve early disease and lameness detection, better reproduction decisions and farm management, aggression and cannibalism monitoring, welfare assessment, and even more accurate environmental control, among others.[1] Collecting real-time behavioral metrics in commercial farms is no simple task. A good livestock behavior tracker would be both precise and affordable. In the last years, researchers and companies have tried different types of behavior data acquisition systems, but the resulting solutions are still not widely adopted. Sectors in which animals have a greater value such as dairy have partially adopted accelerometer and indoor positioning systems while other sectors like swine or poultry have not been able to integrate those in their business model due to the high cost of the systems and the difficulty to implement them at a large scale. Radio Frequency Identification (RFID) has also been tried[2,3] with different species to monitor visits to some predetermined areas, giving interesting results, but its physical implementation is complex, and the generated data is somehow poor. In fact, RFID requires antennas at different locations and tells when an animal is detected at those positions only. Therefore, we have no information about the behavior of the animal in between detections at the antennas.

Recent advances in artificial intelligence and specifically in computer vision open new ways to analyze livestock behavior. Real-time detection and multi-object tracking have reached unprecedented accuracy levels and now allow group and individual animal tracking. We propose an affordable 2D camera system to track pigs in real time. We use the individual tracks to compute behavioral metrics such as individual position, velocity, and posture, group activity level, and animal interactions, among others. This data is recorded in real-time and made available through a simple user interface. Our system can be used as a research tool to better understand animal behavior, as a diagnosis tool for veterinarians, and as an affordable farm management tool to alert pig producers in case of abnormal behavior changes.

Description of the Ro-Main smaRt Pig Tracking system

The Ro-Main smaRt Pig Tracking system (patent pending) consists of an IP camera network overlooking animal pens, a server that processes the images and a user interface that allows to visualize data and to export it (Figure 1). Each wide-angle camera (2) is installed on the ceiling such that it can see all the pigs of a given pen (1) and all cameras are connected to the dedicated local area network (LAN) through Ethernet and Power over Ethernet switches (3).

A computational server (4) receives the video streams from all connected cameras and processes the individual images of the video stream. Each pig is automatically detected in the image with the use of a deep neural network carefully trained to recognize pigs of all sizes and positions independently of the environment or lighting conditions. For every image, the output of the system is the XY coordinates of the center of every detected pig in the image (Figure 2). For individual tracking, frequent images are necessary to associate the subsequent detections of a same pig in the video and link them to create an individual path. The individual tracking algorithm uses a statistical approach based on prior knowledge about the animal position and velocity to associate each detection at a time t with its most likely position at time t+1. The output of the individual tracking algorithm is a track for each pig and associated position and velocity at all data points.

Finally, a local web application allows the user to access the data wirelessly from any platform (mac, Windows, iOS, Android, Linux, etc.) through a web browser (5). The user can visualize individual and group data from the application and the raw data can be exported to standard database formats such as CSV and TXT (6).

Use cases

Our tracking system can be used for group or individual tracking in multiple ways. A few examples of applications will be presented here to demonstrate what the system can do.

First, position data at a given frequency can be used without tracking the animals individually. This requires less computational resources as less frames from the video stream need to be processed. However, this can provide valuable data that has never been available so far at the commercial level. Given the position of all pigs at a given frequency, the user can easily know the exact number of pigs in all pens and be warned when a pig has been detected at the same position for an extended period. This can be used to identify dead or sick pigs or to monitor dead animal removal and help identify the number of pigs transferred to hospital pens on a day-to-day basis. Animal positions alone can also be used to monitor the number of visits to pre-determined areas – set through software interface - such as feeding, drinking, resting, and playing areas. A change in the number of visits to specific areas can help identify problems early.[1] The system can also provide information about animal density in pre-determined zones. Animal density could be used to identify tail biting,[4] ventilation,

Figure 1: Ro-Main smaRt Pig Tracking system architecture

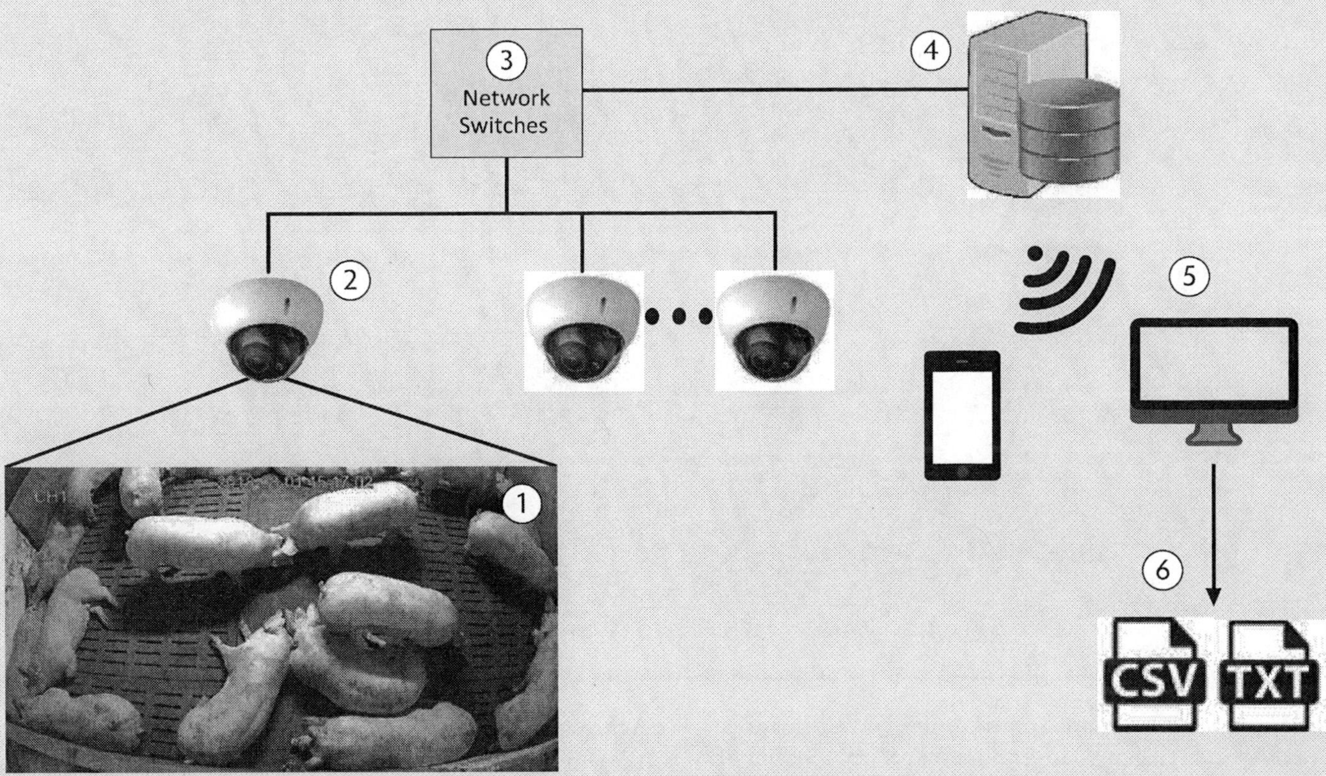

Figure 2: Output of the detection algorithm

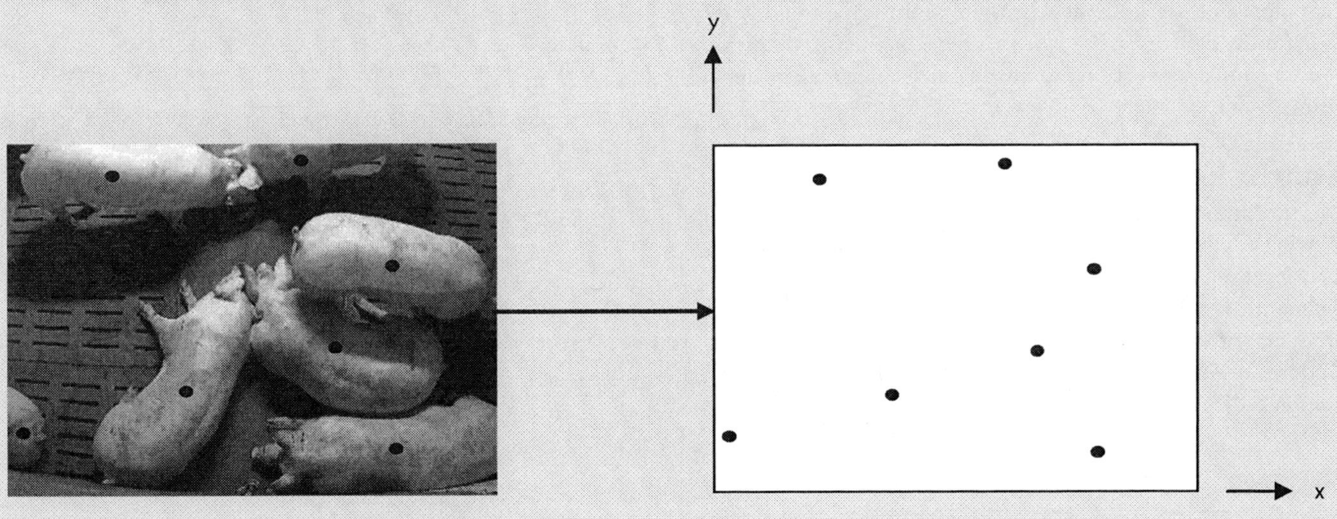

Figure 3: Example of a pig density map obtained with the Ro-Main smaRt Pig Tracking data

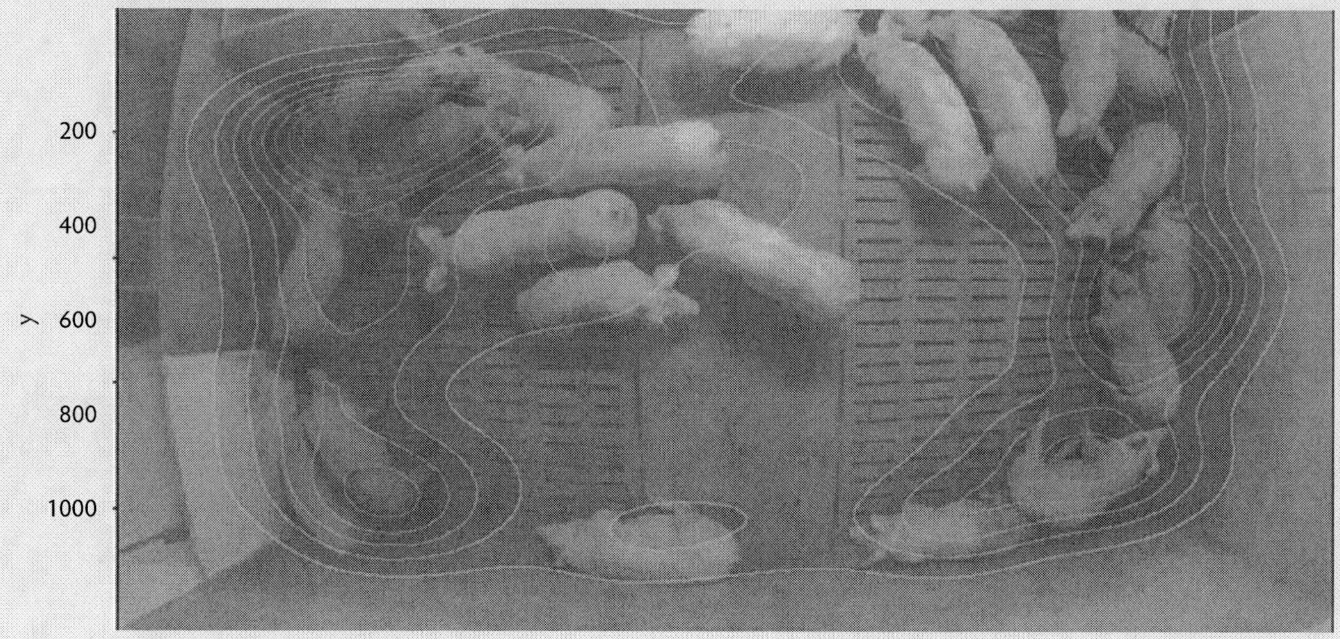

and barn design issues, among others. Animal density maps can be easily plotted (Figure 3) to visualize the general behavior of the group with respect to its environment. This could eventually help understand ventilation and barn design problems.

Tracking each animal individually is much more computation intensive but can help researchers understand better the need of individual animals. Individual behavior monitoring could lead to earlier individual disease or lameness detection, estrus detection, or even animal interaction monitoring such as aggression and tail biting.

Conclusion

Our tracking system can be used for group or individual tracking to help the research community to better understand pig behavior. Moreover, it allows producers to get valuable insights about their animals and equipment companies to better evaluate their products with respect to animal behavior. Moreover, it can be used by veterinarians as a diagnosis tool or by engineers as a calibration tool.

Acknowledgements

The authors would like to acknowledge all Ro-Main and Avantis/Olymel staff who contributed to the development of the system.

References

1. Matthews, S. G., Miller, A. L., Clapp, J., Plotz, T., Kyriazakis, I. 2016. Early detection of health and welfare compromises through automated detection of behavioral changes in pigs. The Veterinary Journal, 217, 43-51.

2. Adrion, F., Kapun, A., Eckert, F., Holland, E.-M., Staiger, M., Götz, S., and Gallmann, E. 2018. Monitoring trough visits of growing finishing pigs with UHF-RFID. Comput. Electron. Agric., 144, 144–153.

3. Brown-Brandl, T. M., Jones, D. D., & Eigenberg, R. A. 2016. Modelling Feeding Behavior of Swine to Detect Illness. In Int. Conf.Proc. International Conference on Agricultural Engineering, CIGR – AgEng 2016.

4. Edwards, S.A. 2006. Tail biting in pigs: Understanding the intractable problem. The Veterinary Journal, 17, 198-199.

Tetracore Inc., 20 years of leadership in animal health laboratory technology

J. Callahan, MT (ASCP), PhD

Introduction

In 1992, the US Navy's Biological Defense Research Program (BDRP) was founded on the grounds of the National Naval Medical Center in Bethesda, MD. The program was initiated to replace the technology fielded in the Gulf War which had major reliability and false positive issues. The initial focus was on immunodetection then later a second parallel effort focused on molecular methods. Four members of the Navy's BDRP program ultimately left to start Tetracore in 1998.

The founding members were Gary Long, Ph.D, Former Head, Biological Defense Research Program (BDRP); Beverly Mangold, Ph.D, Former Head, NMRI, molecular diagnostics; William Nelson, M.D., Commander US Navy, Gulf War Forward Laboratory Lead; and Tom O'Brien, Ph.D, Former Head, NMRI, Immunodiagnostics. While working on a Government Request for Proposals outside of their normal work hours, the four founders won a large contract which allowed Tetracore to establish itself as an LLC. What began as a company of four founders and six technical staff members has grown in 20 years to a Corporation with 100 employees specializing in multiple disciplines within the biothreat, human public health and veterinary diagnostic space. The Tetracore facility has a BSL-3 laboratory and select agent permits to support research efforts on especially dangerous pathogens.

Technical leadership

Scientists at Tetracore have made a number of technical innovations that have combined novel technologies that link point-of-care instruments and the latest high quality chemistries. Tetracore provides solutions that enable field testing in remote locations that are often far removed from normal laboratory logistical support. Innovations include custom manufacturing to provide test configurations that best match the needs of the customer, dried PCR chemistries and an advanced, portable, field-capable real-time PCR instrument.

Dried real-time/real-time reverse transcriptase PCR (rPCR/rRT-PCR)

Shipping PCR mastermix in liquid form requires a continuous cold chain from the point of manufacture to the point of delivery at the laboratory which makes shipping to remote testing sites logistically challenging. For dried PCR, the cold-chain for test delivery is broken; dried PCR mastermix from Tetracore is stable at ambient room-temperature for years and then is easily re-hydrated as needed just before use. This is the best and most practical option for testing in remote locations. The African Swine Fever Virus (ASFV) real-time PCR that was the basis of the Zsak et al., 2005[1] paper was first evaluated in a dry format at ARS at the Plum Island Animal Disease Center (PIADC) in the year 2000. Tetracore has been manufacturing dried real-time PCR tests for over 18 years. Testing and validation of a dried foot-and-mouth disease virus (FMDV) real-time PCR assay began at PIADC in 2001 and the results of this work were published in 2002.

At the height of the 2001 UK FMD outbreak, Tetracore scientists traveled to the UK to demonstrate the feasibility of testing with dried FMDV reverse transcription real-time PCR (rRT-PCR) tests using a portable device at Oxford University. Attending the demonstration was David King, the scientific advisor to Tony Blair. The Tetracore team included Dr. Floyd Horn ARS Administrator and Fred Brown, PhD, FRS., PIADC, Bill Nelson and Johnny Callahan. Other rPCR/rRT-PCR tests from Tetracore that were successfully validated in collaboration with PIADC included tests for classical swine fever virus (CSFV), swine vesicular disease virus (SVDV), vesicular exanthema of swine virus (VESV), rinderpest virus (RPV) and vesicular stomatitis virus (VSV).

The PRRS story

With a track record of success at PIADC it was suggested by a noted veterinary colleague, Professor Fernando Osorio; that the veterinary laboratory community was in need of a commercial test for porcine reproductive and respiratory syndrome virus (PRRSV), the most economically important viral disease of swine. Professor Osorio introduced us to Dr. Jane Christopher-Hennings of South Dakota State University (SDSU) and in 2002; Tetracore began an R&D partnership with SDSU to begin work on a PRRSV rRT-PCR. Tetracore had the test design expertise while SDSU had all other critical necessities for the effort including: clinical knowledge of the disease, an established in-house nested PCR and bio-assay capacity to bench-mark the new test, swine samples of known disease status, and the required test performance specifications that needed to be met by any new test. This industry-university partnership led to the first commercially available rRT-PCR reagents for the detection of PRRS viral RNA. The collaboration with SDSU remains productive to this day, a key partnership that enables the development of new products inspired by the expertise and contributions of both industry and academia.

High-throughput testing

As described above, dry tests have many advantages, but they also have limitations. Dry tests are not well suited for high-throughput testing. Tests from Tetracore can be configured and provided as either dry or liquid mastermix depending on the testing requirements. Liquid mastermix is shipped on dry ice and stored at -20°C, then thawed just before testing. For large orders, Tetracore custom manufactures to meet specific test configuration requirements.

Manufacturing and surge capacity

Tetracore can manufacture serials (batches or lots) of tests in quantities of up 50,000 reactions to meet surge capacity demands. The cGMP manufacturing facility is both USDA and FDA certified.

PCR chemistries

The latest innovative and commercially available offerings for PCR buffers and enzyme systems are monitored and evaluated in order to keep the performance of Tetracore PCR's in the top-tier of commercially available products. Tetracore serves as a beta test site for several manufacturers of PCR raw materials.

Nucleic acid extraction chemistries

Continual monitoring of the latest commercial offerings for nucleic acid extraction methods utilizing multiple sample matrices allows Tetracore to be ready to engage and support multiple customers around the globe; customers who use multiple nucleic acid extraction approaches on various test matrices. It has been demonstrated that some sample matrices can be tested directly on portable PCR instruments without performing a nucleic acid extraction procedure prior to PCR testing. Examples of successful direct sample testing in the field include influenza A virus, African swine fever virus and foot-and-mouth disease virus.

The T-COR 8™, a portable real-time PCR instrument

The T-COR 8 instrument (Figure 1) is a multiplex PCR platform developed in-house by Tetracore engineers. The instrument has 4 independent dye-channels, and will soon feature 5 dye-channels for increased flexibility of PCR multiplex applications. The under 10 lb. platform has 8 independently programmable reaction wells and many advanced electronic features including innovative software that is run from a web browser, no software installation required, auto-analysis of run data, viewing of the data in real time, and a cloud-based environment that is accessible by smart phone. This instrument is ruggedized for field use and is easily transported in a carry-on/back pack type luggage for mobilization to remote locations.

The flexibility of a small company

Large companies with multi-layered business structures often do not have the flexibility to quickly develop a new product to address an emerging market. The porcine epidemic diarrhea virus (PEDV) outbreak which began in April of 2013 is a good example of how a small company like Tetracore was able to recognize the market demand and launch a commercial multiplex real-time PCR assay to detect and distinguish three clinically similar swine coronaviruses, PEDV, transmissible gastroenteritis virus (TGEV) and porcine delta coronavirus (PDCoV). By 2014, before the PEDV outbreak was resolved, Tetracore had a commercial PEDV/TGEV/PDCoV multiplex rRT-PCR available that was utilized in research to study the transmission and control of disease.[2,3] Recently, Tetracore received SHIC funding for the development of a test for the detection of a new emerging swine coronavirus discovered in China known as Swine Acute Diarrhea Syndrome Coronavirus (SADS-CoV), which has raised concerns in the US considering how PEDV was spread to the US. The SADS-CoV PCR test could be multiplexed with other coronavirus PCR tests in custom configurations to meet testing requirements. Tetracore has the expertise and flexibility to take on new challenges and deliver high quality products for research purposes.

Other notable rPCR/rRT-PCR assays for detection of disease agents

Tetracore has developed and deployed several assays for studies on Senecavalley A virus (SVA); a multiplex influenza A/B/C/D; Burkholderia mallei/pseudomallei; and a dengue/Zika/Chikungunya virus multiplex assay that is currently being evaluated in Sarawak, Malaysia, an area where dengue is hyper-endemic.

Tetracore has a 20 year history

Tetracore has a 20 year history of collaborative research with government and university partners, and a strong track record of research and development. Great efforts are made to validate and then publish papers describing the innovative tools that are made

Figure 1: Tetracore instrument

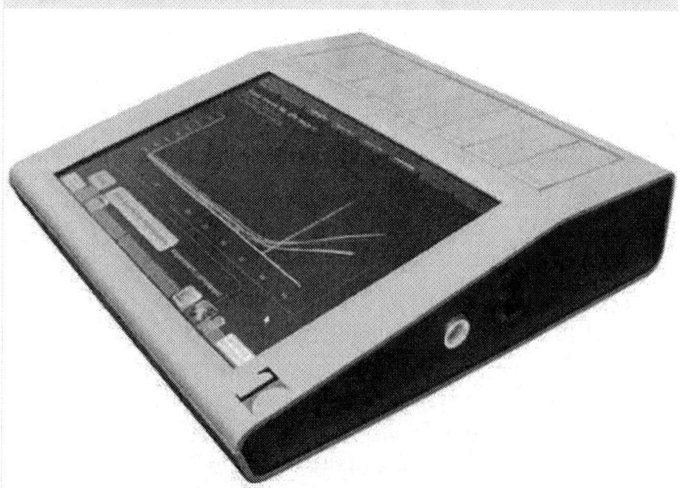

T-COR 8™ Real-time PCR Thermocycler

available on the commercial market. Evidence of scientific accomplishments are chronicled in the Tetracore publications which include Tetracore authors and co-authors, and/or publications where Tetracore products were used to support the study.

References

1. Zsak, L., et al., Preclinical diagnosis of African swine fever in contact-exposed swine by a real-time PCR assay. J Clin Microbiol, 2005. 43(1): p. 112-9.

2. Dee, S., et al., An evaluation of contaminated complete feed as a vehicle for porcine epidemic diarrhea virus infection of naive pigs following consumption via natural feeding behavior: proof of concept. BMC Vet Res, 2014. 10: p. 176.

3. Dee, S., et al., An evaluation of a liquid antimicrobial (Sal CURB®) for reducing the risk of porcine epidemic diarrhea virus infection of naive pigs during consumption of contaminated feed. BMC Vet Res, 2014. 10: p. 220.

Tetracore bibliography

Foot-and-mouth disease virus

1. Howson, E.L.A., et al., Direct detection and characterization of foot-and-mouth disease virus in East Africa using a field-ready real-time PCR platform. Transbound Emerg Dis, 2018. 65(1): p. 221-231.

2. King, D.P., et al., Detection of foot-and-mouth disease virus: comparative diagnostic sensitivity of two independent real-time reverse transcription-polymerase chain reaction assays. J Vet Diagn Invest, 2006. 18(1): p. 93-7.

3. Boyle, D.B., T. Taylor, and M. Cardoso, Implementation in Australia of molecular diagnostic techniques for the rapid detection of foot and mouth disease virus. Aust Vet J, 2004. 82(7): p. 421-5.

4. Schumann, K.R., et al., Genetic characterization and molecular epidemiology of foot-and-mouth disease viruses isolated from Afghanistan in 2003-2005. Virus Genes, 2008. 36(2): p. 401-13.

5. Paixao, T.A., et al., Diagnosis of foot-and mouth disease by real time reverse transcription polymerase chain reaction under field conditions in Brazil. BMC Vet Res, 2008. 4: p. 53.

6. Callahan, J.D., et al., Use of a portable real-time reverse transcriptase-polymerase chain reaction assay for rapid detection of foot-and-mouth disease virus. J Am Vet Med Assoc, 2002. 220(11): p. 1636-42.

African swine fever virus

7. Zsak, L., et al., Preclinical diagnosis of African swine fever in contact-exposed swine by a real-time PCR assay. J Clin Microbiol, 2005. 43(1): p. 112-9.

8. Liu, L. et al, Overcoming the challenges of pen-side molecular diagnosis of African swine fever to support outbreak investigations under field conditions. 2018. Transboundary and Emerging Diseases. 2018;00:1–7.

Classical swine fever virus

9. Risatti, G., et al., Diagnostic evaluation of a real-time reverse transcriptase PCR assay for detection of classical swine fever virus. J Clin Microbiol, 2005. 43(1): p. 468-71.

10. Risatti, G.R., et al., Rapid detection of classical swine fever virus by a portable real-time reverse transcriptase PCR assay. J Clin Microbiol, 2003. 41(1): p. 500-5.

Senecavirus A

11. Maggioli, M.F., et al., Adaptive Immune Responses following Senecavirus A Infection in Pigs. J Virol, 2018. 92(3).

12. Joshi, L.R., et al., Detection of the Emerging Picornavirus Senecavirus A in Pigs, Mice, and Houseflies. J Clin Microbiol, 2016. 54(6): p. 1536-1545.

13. Joshi, L.R., et al., Pathogenesis of Senecavirus A infection in finishing pigs. J Gen Virol, 2016. 97(12): p. 3267-3279.

14. Sun, D., et al., Emergence and whole-genome sequence of Senecavirus A in Colombia. Transbound Emerg Dis, 2017. 64(5): p. 1346-1349.

15. Resende, T.P., D.G. Marthaler, and F.A. Vannucci, A novel RNA-based in situ hybridization to detect Seneca Valley virus in neonatal piglets and sows affected with vesicular disease. PLoS One, 2017. 12(4): p. e0173190.

16. Fernandes, M.H.V., et al., Pathogenicity and cross-reactive immune responses of a historical and a contemporary Senecavirus A strains in pigs. Virology, 2018. 522: p. 147-157.

Porcine reproductive and respiratory syndrome virus

17. Topla k, I., et al., Identification of a genetically diverse sequence of porcine reproductive and respiratory syndrome virus in Slovenia and the impact on the sensitivity of four molecular tests. J Virol Methods, 2012. 179(1): p. 51-6.

18. Olsen, C., et al., Probability of detecting Porcine reproductive and respiratory syndrome virus infection using pen-based swine oral fluid specimens as a function of within-pen prevalence. J Vet Diagn Invest, 2013. 25(3): p. 328-35.

19. Biernacka, K., et al., Detection of porcine reproductive and respiratory syndrome virus (PRRSV) and influenza A virus (IAV) in oral fluid of pigs. Res Vet Sci, 2016. 109: p. 74-80.

20. Kittawornrat, A., et al., Porcine reproductive and respiratory syndrome virus (PRRSV) surveillance using pre-weaning oral fluid samples detects circulation of wild-type PRRSV. Vet Microbiol, 2014. 168(2-4): p. 331-9.

21. Wernike, K., et al., Porcine reproductive and respiratory syndrome virus: interlaboratory ring trial to evaluate real-time reverse transcription polymerase chain reaction detection methods. J Vet Diagn Invest, 2012. 24(5): p. 855-66.

22. Wasilk, A., et al., Detection of U.S., Lelystad, and European-like porcine reproductive and respiratory syndrome viruses and relative quantitation in boar semen and serum samples by real-time PCR. J Clin Microbiol, 2004. 42(10): p. 4453-61.

Influenza virus

23. Goodell, C.K., et al., Probability of detecting influenza A virus subtypes H1N1 and H3N2 in individual pig nasal swabs and pen-based oral fluid specimens over time. Vet Microbiol, 2013. 166(3-4): p. 450-60.

24. Panyasing, Y., et al., Influenza A Virus Surveillance Based on Pre-Weaning Piglet Oral Fluid Samples. Transbound Emerg Dis, 2016. 63(5): p. e328-38.

25. Goodell, C.K., et al., Ring test evaluation of the detection of influenza A virus in swine oral fluids by real-time reverse-transcription polymerase chain reaction and virus isolation. Can J Vet Res, 2016. 80(1): p. 12-20.

Coronavirus

26. Hain, K.S., et al., Immunogenicity of a recombinant parapoxvirus expressing the spike protein of Porcine epidemic diarrhea virus. J Gen Virol, 2016. 97(10): p. 2719-2731.

27. Dee, S., et al., An evaluation of contaminated complete feed as a vehicle for porcine epidemic diarrhea virus infection of naive pigs following consumption via natural feeding behavior: proof of concept. BMC Vet Res, 2014. 10: p. 176.

28. Dee, S., et al., An evaluation of a liquid antimicrobial (Sal CURB®) for reducing the risk of porcine epidemic diarrhea virus infection of naive pigs during consumption of contaminated feed. BMC Vet Res, 2014. 10: p. 220.

29. Clement, T., Singrey, A., Lawson, S., Hennings, J., Measurement of neutralizaing antibodies against porcine epidemic diarrhea virus in sow serum, colostrum, and milk samples and piglet serum samples after feedback. Journal of Swine Health and Production, 2016. 24(3): p. 147-153.

Mycobacterium avium ssp. *paratuberculosis* (Johne's disease)

30. Alinovi, C.A., et al., Real-time PCR, compared to liquid and solid culture media and ELISA, for the detection of *Mycobacterium avium* ssp. *paratuberculosis*. Vet Microbiol, 2009. 136(1-2): p. 177-9.

31. Aly, S.S., et al., Correlation between Herrold egg yolk medium culture and real-time quantitative polymerase chain reaction results for *Mycobacterium avium* subspecies *paratuberculosis* in pooled fecal and environmental samples. J Vet Diagn Invest, 2010. 22(5): p. 677-83.

The relationship between early weight measures and gilt reproductive performance

C. Bruns, PhD; B. McNeil; J. Sonderman, PhD; T. Rathje, PhD
DNA Genetics, Columbus, NE

Introduction

The positive impact of birth weight on lifetime growth and survivability has been well documented. Increased birth weight has been associated with heavier body weights throughout the pig's life[1] and improved pre-weaning and nursery survival.[2,3] However, the impact between early weight measures and gilt reproductive performance has had limited investigation. Heavier gilts at birth are more likely to farrow their first litter after they have been entered into the sow farm compared to lighter gilts.[4] The aim of this study is to determine the impact of birth and weaning weight on subsequent gilt reproductive performance. Understanding the influence of early weight measures on economically relevant traits, such as conception rate and litter size, would be beneficial to improve gilt selection methods in a multiplication system.

Materials and methods

Data for this study were obtained from 4 sow farms with similar genetics and management protocols. To determine the relationship between early weight measures and a gilt's probability of being selected to return to a nucleus herd following completion of the testing program 1,821 Yorkshire and 2,519 Landrace gilts that completed performance testing between February and August 2018 were observed. All gilts were weighed at birth (BWT), weaning (WWT, ~21 days of age), and the completion of performance testing (ENDWT, ~23 weeks of age). Four binomial traits of interest were defined as whether or not the gilt was selected (SEL), whether or not a heat-no-service was recorded on a selected gilt (HNS), whether or not a selected gilt was mated (MATED), and whether or not a gilt conceived on her first mating (CONC). All four binomial traits were analyzed using two generalized linear mixed models with a random effect of contemporary group and a covariate of either BWT or WWT. Additionally, the effects of BWT and WWT on ENDWT were estimated after adjusting for age at ENDWT and contemporary group.

A second dataset comprised of litter records from 1,495 Yorkshire and 1,767 Landrace gilts farrowing between September 2017 and September 2018 was used to determine the relationship between a gilt's early weight measure and her first parity reproductive performance. Stillborns and born alive were recorded for all litters. Total born was defined as stillborns plus born alive. Additionally, number weaned and live pigs at day 5 from the gilt's birth litter were recorded. The percent of pigs in the gilt's birth litter that survived to weaning was also analyzed. A birth weight on all pigs born alive and the total stillborn weight were obtained to determine the average birth weight of live born pigs and the total birth weight of the litter. The effect of BWT and WWT on reproductive performance measures were estimated after adjusting for contemporary group and service sire.

Results and discussion

An increase in BWT of 1 kg was associated with an increase in ENDWT of 15.05 ± 0.52 in Yorkshire ($P < 0.0001$) and 15.88 ± 0.41 ($P < 0.0001$) in Landrace. The effect of a 1 kg increase in WWT on ENDWT was smaller than the effect of BWT for both breeds (3.67 ± 0.11 for Yorkshire and 4.34 ± 0.11 for Landrace, $P < 0.0001$). For Yorkshire gilts, BWT was not associated with a higher probability of SEL, HNS, or CONC ($P > 0.1$); however selected gilts with heavier BWT were more likely to be mated compared to gilts with lower BWT (Odds Ratio = 1.76, $P = 0.04$). Additionally, selected Yorkshire gilts with heaver WWT were more likely have an HNS event (Odds Ratio = 1.22, $P < 0.0001$), be mated (Odds Ratio = 1.18, $P = 0.001$), and conceive (Odds Ratio = 1.24, $P = 0.02$) compared to gilts with lower WWT. For Landrace gilts, heavier BWT were associated with higher SEL (Odds Ratio = 2.29, $P < 0.0001$), HNS (Odds Ratio = 3.77, $P = 0.0001$), and MATED (Odds Ratio = 1.87, $P = 0.008$). The relationship between BWT and CONC was not significant for Landrace gilts. The relationships between WWT and HNS and MATED were not significant, but selected gilts with heavier WWT had a higher probability of an HNS event (Odds Ratio = 1.35, $P < 0.0001$) and tended to have a higher conception rate (Odds Ratio = 1.14, $P = 0.06$) compared to gilts with lighter WWT.

The results for the relationship between 1st parity reproductive measures and early weight measures are displayed in Tables 1 and 2. For both breeds, gilts that were heavier at birth had litters with heavier total weights and heavier average weights of live born pigs; however, there was no significant relationship between a gilt's weaning weight and the birth weight of her litter. Pre-weaning survival of a gilt's birth litter was not associated with the gilt's BWT or WWT for both breeds. Yorkshire gilts with heavier birth weights had higher total born, stillborns, born alive, and number weaned, while Landrace gilts with heavier birth weights was only significantly associated with higher total born and stillborns. The magnitude of the relationships between birth weight and litter size were higher in magnitude for the Yorkshire gilts compared to Landrace gilts. Weaning weight associations with litter size were lower in magnitude compared to birth weight.

Table 1: Effect and standard error (SE) of gilt birth weight on parity 1 reproductive performance

	Yorkshire			Landrace		
Trait	Estimate	SE	*P*-value	Estimate	SE	*P*-value
Average birth weight[a]	0.046	0.020	0.018	0.041	0.017	0.018
Total birth weight[b]	1.965	0.406	<0.0001	1.587	0.323	<0.0001
Survival[c]	0.025	0.021	0.225	-0.013	0.016	0.445
Born alive	0.755	0.363	0.038	0.137	0.285	0.631
Total born	1.198	0.379	0.002	0.979	0.288	0.001
Stillborn	0.447	0.193	0.021	0.841	0.199	<0.0001
Live pigs at day 5[d]	0.575	0.363	0.114	-0.129	0.272	0.637
Number weaned[d]	0.748	0.340	0.028	-0.027	0.268	0.919

[a] Average birth weight of pigs born alive
[b] Total birth weight of live born and stillborn pigs
[c] Percent of pigs in the gilt's birth litter that survived to weaning
[d] Measured from the gilt's birth litter

Table 2: Effect and standard error (SE) of gilt weaning weight on parity 1 reproductive performance

	Yorkshire			Landrace		
Trait	Estimate	SE	*P*-value	Estimate	SE	*P*-value
Average birth weight[a]	0.001	0.037	0.873	-0.000	0.004	0.947
Total birth weight[b]	0.200	0.077	0.009	0.227	0.071	0.001
Survival[c]	-0.004	0.004	0.358	0.001	0.004	0.885
Born alive	0.120	0.068	0.078	0.016	0.062	0.793
Total born	0.190	0.071	0.008	0.212	0.063	0.001
Stillborn	0.070	0.036	0.053	0.193	0.043	<0.0001
Live pigs at day 5[d]	0.022	0.068	0.745	-0.027	0.060	0.654
Number weaned[d]	0.019	0.064	0.763	0.028	0.059	0.634

[a] Average birth weight of pigs born alive
[b] Total birth weight of live born and stillborn pigs
[c] Percent of pigs in the gilt's birth litter that survived to weaning
[d] Measured from the gilt's birth litter

The results of this study suggest that gilt BWT and WWT are phenotypically associated with future reproductive performance in purebred gilts; however this does not necessarily extend to F1 crossbred gilts. Further research should be conducted to ensure similar relationships are found for F1 crossbred gilts. Another limitation of the current study is that sow weights at breeding and farrowing were not measured. It would be important to know whether the birth weight effect exists if sows are mated and farrowed at the same weight. Additionally, it is important to note that phenotypic relationships do not suggest that selecting gilts with higher birth weights will ultimately improve the genetic potential for birth weight within the herd.

References

1. Fix et al. Effect of piglet birth weight on body weight, growth, backfat, and longissimus muscle area of commercial market swine. *Livest Sci* 2010;127:51.

2. Fix et al. Effect of piglet birth weight on survival and quality of commercial market swine. Livest Sci 2010;132:98.

3. Panzardi et al. Newborn piglet traits associated with survival and growth performance until weaning. *Prev Vet Med* 2013;110:206.

4. Knauer. 2016. Effects of preweaning factors on sow lifetime productivity. National Pork Board project #11-146. Available at: https://www.pork.org/wp-content/uploads/2016/04/11-146-KNAUER-NCSU.pdf.

PEDV positive bioassay reveals houseflies *(Musca domestica)* can transmit infectious PEDV to pigs

Grant Allison[1],DVM; Phil Gauger[2],DVM, MS, PhD; Jianqiang Zhang[2],MD, MS, PhD; Gene Spellman[3],BS
[1]Walcott Veterinary Clinic, Walcott IA; [2]Vet Diagnostic and Production Animal Medicine, Veterinary Diagnostic Laboratory, Iowa State University, Ames Iowa; [3]Central Life Sciences, Schaumburg IL

Introduction

Despite extensive investments in biosecurity measures and protocols, farms still have unexplained cases of viral diseases caused by porcine epidemic diarrhea virus (PEDV), porcine reproductive and respiratory syndrome virus (PRRSV), and more recently Seneca Virus A (SVA). Many potential sources of virus for these diseases have been studied in the past. Insect vectors such as flies have been investigated as a possible source of PRRSV, PEDV and SVA.[1-3] Regarding PEDV, less is known about the potential role of insects in the transmission of this virus. Given the high cost of the disease, estimated at $400 per sow,[4] the role of flies as a vector for PEDV warrants further investigation. The purpose of this paper is to answer the question if flies are a vector for the transmission of PEDV to susceptible swine.

Materials and methods

Real-time reverse-transcription polymerase chain reaction (rRT-PCR) positive PEDV environmental fly samples (average Ct 24.5) and trap liquid samples (average Ct 26.5) were collected with Captivator® liquid traps (Starbar®) (Figure 1) from a 2,400 head finisher in Eastern Iowa three days after a positive diagnosis for PEDV by fecal swab (Ct 16.5 pool of five samples) as the virus source for the bioassay. Liquid from the negative control trap (a trap with entry openings screened to prevent fly entry) (Figure 2) was PCR negative. A 10% whole-fly homogenate was created as the inoculum for a bioassay using three grams of flies in 27 mL of cell culture media at the Iowa State University Veterinary Diagnostic laboratory (ISU-VDL) per laboratory procedures. Liquid trap samples were not diluted for the bioassay. The inoculum was evaluated again by PEDV rRT-PCR. The fly homogenate inoculum demonstrated a Ct of 27.1 and the trap inoculum demonstrated a Ct of 26.9. The negative fly trap inoculum was confirmed rRT-PCR negative. Inoculum was frozen at -80°C until the bioassay. The bioassay was conducted at Iowa State University under strict environmental conditions and research protocols. Four experimental groups of three-10 day old pigs, negative for PEDV were used as follows: Group 1: negative control trap liquid; Group 2: PEDV ISU reference strain known to be able to infect pigs dosed at 1×10^4 $TCID_{50}$/mL (Ct 14.0); Group 3: PEDV rRT-PCR positive flies; Group 4: PEDV rRT-PCR positive trap liquid. Each pig in their respective groups received 10mL of the inoculum through an 8 French feeding tube passed into the stomach at 0 days post inoculation (DPI) followed by 10 mL of air to gently force remaining inoculum in the tube into the stomach. Animal monitoring for clinical signs and detection of PEDV by rRT-PCR was conducted at 3, 5, and 7 DPI. Pigs were necropsied at 7 DPI.

Figure 1: Captivator fly trap

Figure 2: Negative control fly trap

Results

Pigs in Group 4 became rRT-PCR positive for PEDV at 3 DPI and remained positive at 5 and 7 DPI exclusively from the liquid flytrap attractant. The positive control group became rRT-PCR positive at 3, 5, and 7 DPI although pigs in the fly homogenate group remained rRT-PCR negative throughout the study. Negative control pigs remained rRT-PCR negative. No cross-contamination occurred during the study (Table 1).

Discussion

Under the conditions of this study, flies may be a fomite with the ability to indirectly transmit PEDV to swine. These data suggest infectious PEDV was present within or on flies that entered the liquid traps transferring PEDV to the attractant. The negative control trap liquid, in the same airspace, was negative and did not infect the pigs further suggesting flies were responsible for transporting PEDV into the liquid fly attractant. The exact route of indirect transmission is unknown; however, it is speculated flies exposed to PEDV in feces from infected swine may transport the virus to feed when pigs may be exposed to the virus or the flies may be consumed by swine (Figure 3). As flies reproduce in warm, moist conditions, the interface between PEDV positive feces and flies may provide the ideal environment for virus exposure either on the flies or what may be consumed by flies. In 2016, Tun et.al.[5] reported PEDV could survive up to 9 months in infected lagoons. Likewise Verma et. al.[6] determined that PEDV could remain infectious $\geq$ 28 days in a manure slurry stored at various temperatures. Seasonal variation in pit temperature has been reported in standard 1,000 head finishers in North Central Iowa; late winter 9°C (low) – to late summer/fall 22°C (high).[7] Likewise, temperature variation in manure slurry has been reported in fall/winter/spring as 1.5 – 11°C in earthen reservoirs versus 0.5 – 8°C in above ground tanks. A peak temperature of manure slurry was reported as 16°C for above ground tanks and 17°C for earthen storage.[8] PEDV has been shown to remain infectious at lower storage temperatures.[9] The propensity of PEDV to survive at colder temperature suggests a relationship between PEDV and the colder temperature found in swine manure pits in winter. The minimum infectious dose for PEDV in feed corresponds to a PCR Ct value of 37.0.[10] The average pooled rRT-PCR of the liquid flytrap attractant in this project was 26.9. Relative to feed as a source of infectious PEDV, liquid attractant, which contained flies, had sufficient viral load to serve as an infectious PEDV source in this bioassay. However, it remains unknown if the infectious PEDV was present on the surface of the flies or consumed by flies that was subsequently transferred to the liquid attractant. It is also unknown why the fly homogenate did not infect the pigs in this bioassay in contrast to the liquid attractant. This may suggest the liquid environment of the attractant supported infectious PEDV. These data suggest flies may indirectly transmit PEDV to feed, water, or surfaces in contact with swine as opposed to pigs consuming quantities of flies sufficient to transmit PEDV. However, flies are continually present on feed and/or water in swine facilities (Figure 3) or physically present on the backs, ears and noses of swine in large numbers (Figure 4). Therefore, transmission of PEDV via flies through contact with feed, water or physical contact with swine may serve as the vector to transmit the virus and mimics the bioassay positive liquid attractant. The intimate relationship between flies and completion of their life cycle in manure cannot be ignored regarding the potential transmission of PEDV in swine barns. Perhaps fly control should become part of the discussion on biosecurity as it relates to control of viral disease in general and to PEDV in particular. Flies do not read biosecurity manuals (Figure 5) and may spread PEDV farm-to-farm as they can fly up to four miles.[11] Previously, controlling flies has been haphazard, mostly ineffective, and labor intensive. With new options offering labor-free in-feed technology capable of breaking the fly's lifecycle, the fly-biosecurity question can now be more thoroughly investigated and discussed.

Table 1: Bioassay results. Pooled fecal swab rRT-PCR results from suckling pigs administered an oral dose of fly homogenate and liquid attractant collected from a Captivator Liquid Fly Trap and confirmed rRT-PCR positive for PEDV.

		Days post inoculation			
Group	**Inoculum**	**0**	**3**	**5**	**7**
1	Negative fly trap liquid	>36*	>36	>36	>36
2	PEDV positive control	>36	22.4	26.4	32.3
3	PEDV positive fly homogenate	>36	>36	>36	>36
4	PEDV positive liquid fly attractant	>36	12.1	15.6	23.3

* Pooled rRT-PCR PEDV cycle threshold value

Figure 3: Flies in feed

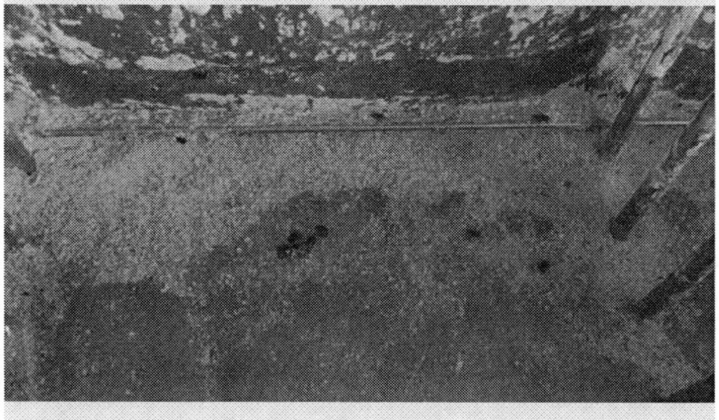

Figure 4: Flies on pigs

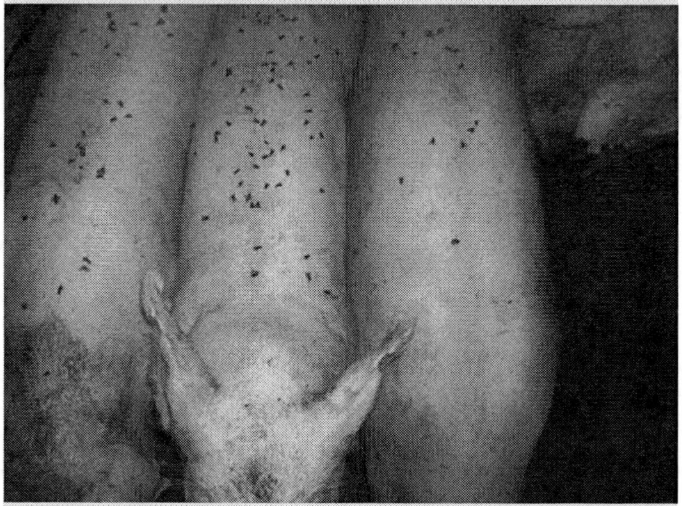

Figure 5: Fly in farm vehicle

References

1. Otake S, Dee SA et al. Transmission of porcine reproductive and respiratory syndrome virus by houseflies, (Musca domestica Linnaeus). *Vet Rec* 2003. 152: 73-76.

2. Allison G, et al. PEDV, PRRSV, and Seneca Valley virus PCR positive flies (Musca domestica). What now? *Proceedings 49th AASV 2018*. 394-396.

3. Joshi LR, Mohr KA, et al. Detection of the emerging picornavirus Senecavirus A in pigs, mice, and houseflies. *J. Clin Microbiol.* 2016. 54(6):1536-1545.

4. Ketchum R, Rix M et al. Hog producers can take steps to survive in low markets. *National Hog Farmer* 11-07-16

5. Tun HM, Cai Z, et al. Monitoring Survivability and Infectivity of Porcine Epidemic Diarrhea Virus (PEDV) in the Infected On-Farm Earthen Manure Storages. *Frontier Microbiol.* 2016:7:265.

6. Verma H, Erber J, et al. Survival of Porcine Epidemic Diarrhea virus in environmental samples. *Proceedings Allen D. Leman Swine Conference.* 2014:20.

7. Batista D, Paula A, et al. Impact of Temperature and Mixing on Methane Production Rates of Swine Manures obtained from Deep Pit Storages. *Agricultural and Biosystems Engineering Conference Proceedings and Presentations.* 2013:324.

8. Arrusa K, Holley R, et al. Influence of temperature on Salmonella survival in hog manure slurry and seasonal temperature profiles in farm manure storage reservoirs. *Journal of Livestock Science.* Vol. 102, Issue 3 July 2006, pg. 226-236.

9. Pujols J, Segales J, et al. Survivability of porcine epidemic diarrhea virus (PEDV) in bovine plasma submitted to spray drying processing and held at different time by temperature storage conditions. *Vet. Microbiol.* 174, 427-432, doi: 10.1016/j.vetmic.2014.

10. Schumacher LL, Woodworth JC, et al. Evaluation of the minimum infectious dose of porcine epidemic diarrhea virus in virus-inoculated feed. *AM J Vet Res.* 2016. 77(10):1108-1113.

11. Nazni WA, Luke H et al. Determination of the flight range and dispersal of the house fly, Musca domestica (L.) using mark release recapture technique. *Trop Biomed* 2005 Jun;22(1):53-61.

Impact of sow vaccination during lactation on piglet performance

Balderrama V.[1]; Tecpa Z.[1]; Tribó F.[2]; Hernández K.[2]; Jordà R.[2]
[1]Carroll; [2]HIPRA, Amer (Girona), Spain

Introduction

The optimum vaccination plan for sows against Parvovirus, Erysipelas and Leptospira is during lactation, thus preventing future infections during pregnancy. Furthermore, the use of reproductive vaccines is common on most pig farms around the world due to the high prevalence of these pathogens amongst swine herds.[1,2]

The humoral immune response and difference in safety between commercial reproductive vaccines has already been reported,[3,4] but the impact that these could have on sow feed intake, milk production and piglet performance up to weaning is still being evaluated.

The aim of this study was to compare how the difference in safety between two reproductive vaccines can affect the production performance of sows and piglets during lactation period.

Materials and methods

This study was conducted in a swine herd of 4,800 sows in Mexico, operating on a one-week batch system, with approximately 240 farrows per batch. A total of 80 healthy multiparous sows from one batch were selected and randomly assigned to two groups (Group 1 [G1] and Group 2 [G2]) of 40 animals each. G1 was administered Vaccine A and G2 was administered ERYSENG® PARVO/LEPTO (HIPRA) by the intramuscular route (2 mL). Both vaccinations were performed 10 days after farrowing following the manufacturer's instructions. Piglets from selected sows were individually identified with numbered ear tags and two different colors indicating the group they belonged to (891 piglets in total). Vaccine safety was measured indirectly through the absence/presence of fever (mean group rectal temperature [MRT] in °C) at vaccination (0 h), and 6, 12 and 24 hours post vaccination (hpv). Sow feed intake was measured two days before vaccination (Day -2, -1), on vaccination day (Day 0) and three days after vaccination (Day +1, +2, +3). Piglets were weighed individually at farrowing day, at 9 days of age (one day before vaccine application) and at weaning (21 days of age).

Statistical analysis software (SAS) was used to check differences between the groups. Data are expressed as mean.

Results and discussion

Regarding the farrowing parameters, no statistically significant differences were observed between the experimental groups in terms of the number of piglets born alive, mean litter weight, mean body weight per piglet, number of piglets per litter at farrowing day (Table 1).

At Day 9 of the study, which was one day prior to vaccination, no statistically significant differences were observed between groups in terms of the number of piglets, litter weight, mean body weight per piglet, average daily gain, percentage mortality (Table 2) and mean feed intake per sow (Figure 2).

After vaccination, statistically significant differences were observed in MRT measured at 6 hpv (40.3 °C G1 vs. 39.5 °C G2) and 12 hpv (39.9 °C G1 vs 39.3 °C G2) between both groups (t-student test, $P < 0.0001$) (Figure 1).

Furthermore, a statistically significant difference was also observed in mean feed intake per sow between the experimental groups (t-student test, $P < 0.0001$) (Figure 2). Thus, G2 consumed, on average, 1.4 kg more than G1 on vaccination day (6.7 kg/day G2 vs. 5.3 kg/day G1); the next day, G2 consumed

Table 1: Production parameters between groups at day 1 of study.

Variables	Group 1	Group 2	*P*-value
Number of piglets born alive, sow	11.13	11.12	0.98
Litter weight, kg	15.92	15.76	0.80
Mean body weight per piglet, kg	1.46	1.45	0.35
Number of piglets per litter, n	10.84	10.91	0.55

Table 2: Production parameters between groups at day 9 of study.

Variables	Group 1	Group 2	*P*-value
Number of piglets, sow	10.25	10.40	0.44
Litter weight, kg	28.48	28.98	0.67
Mean body weight per piglet, kg	2.76	2.79	0.50
Average daily gain between days 1 and 9, kg	0.169	0.174	0.18
Mortality, %	5.37	4.49	0.60

0.5 kg more than G1 (6.3 kg/day G2 vs. 5.8 kg/day G1); and two days after vaccination, G2 consumed 0.4 kg more than G1 (6.9 kg/day G2 vs. 6.5 kg/day G1) (Figure 2). Consequently, statistically significant differences were observed between groups in terms of mean body weight per piglet at weaning (Day 21 of study) and average daily gain between Days 9 and 21 (t-student test, $P < 0.05$) (Table 3).

Conclusions

The present study was intended to explore differences in safety and the consequent impacts of different reproductive vaccines on sow and piglet performance during lactation. The results indicate safety concerns regarding the vaccine administered to G1 due to a significant increase in mean rectal temperature, which negatively affected sow feed intake during lactation. Reduced sow feed intake during lactation caused reduced milk production,[5] which in turn, significantly affected piglet performance (mean body weight and average daily gain) up to weaning on G1.

Further studies should determine how the lack of safety of some reproductive vaccines could affect the reproductive performance of sows during subsequent insemination and gestation, since previous studies have already reported the negative impact of food restriction during lactation on ovarian development, as well as on ovulation rate and total number of piglets born.[6]

In future, reproductive vaccines should not only be evaluated by seroconversion of their respective antigens, but should also take into account safety issues that can negatively affect piglet and sow performance.

References

1. Adler, B. and De la Peña Moctezuma, A. 2010. *Vet Microbiol.* 140:292-4.
2. Mengeling, WL. et al. 1979. *Am J Vet Research.* 40(2):204-7.
3. Boix, O. *et al.* 2017. *The Asian Pig Veterinary Society Congress.* P 258.
4. Camprodon, A. et al. 2017. *The Asian Pig Veterinary Society Congress.* P 261
5. Pluske, JR. et al. 1998. *Journal of Animal Science.* 76:1165-71.
6. Quesnel, H. et al. 1998. *Journal of Animal Science.* 76:856-63.

Table 3: Production parameters between groups at day 21 of study. Different superscripts at each time point show statistically significant differences ($P < 0.05$).

Variables	Group 1	Group 2	*P*-value
Number of weaned piglets, sow	9.75	9.75	1.00
Litter weight, kg	57.63	58.97	0.57
Mean body weight per piglet at day 21, kg	**5.92^{a}**	**6.04^{b}**	**0.03**
Average daily gain between days 9 and 21, kg	**0.223^{a}**	**0.231^{b}**	**0.04**
Mortality between days 9 and 21, %	4.73	6.33	0.29

Figure 1: Mean rectal temperature (°C) on vaccination day (0 h), 6 h, 12 h and 24 h after vaccination. Different superscripts at each time point show statistically significant differences ($P < 0.05$).

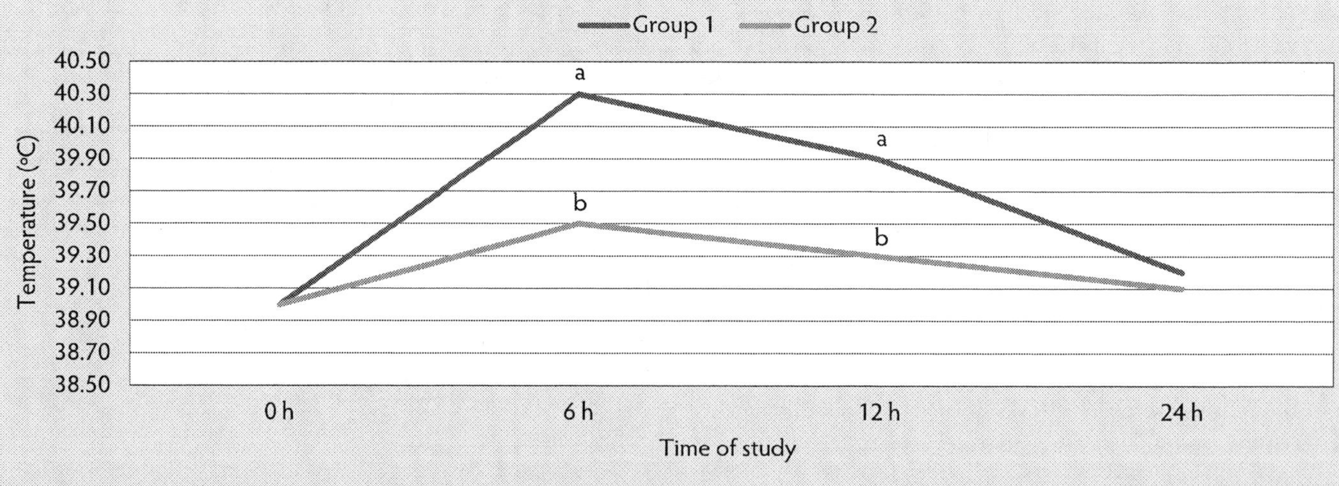

Figure 2: Mean feed intake per sow, two days before vaccination (Day -1 and Day -2), on vaccination day and three days after vaccination (Day +1, Day +2 and Day +3). Different superscripts at each time point show statistically significant differences ($P < 0.05$).

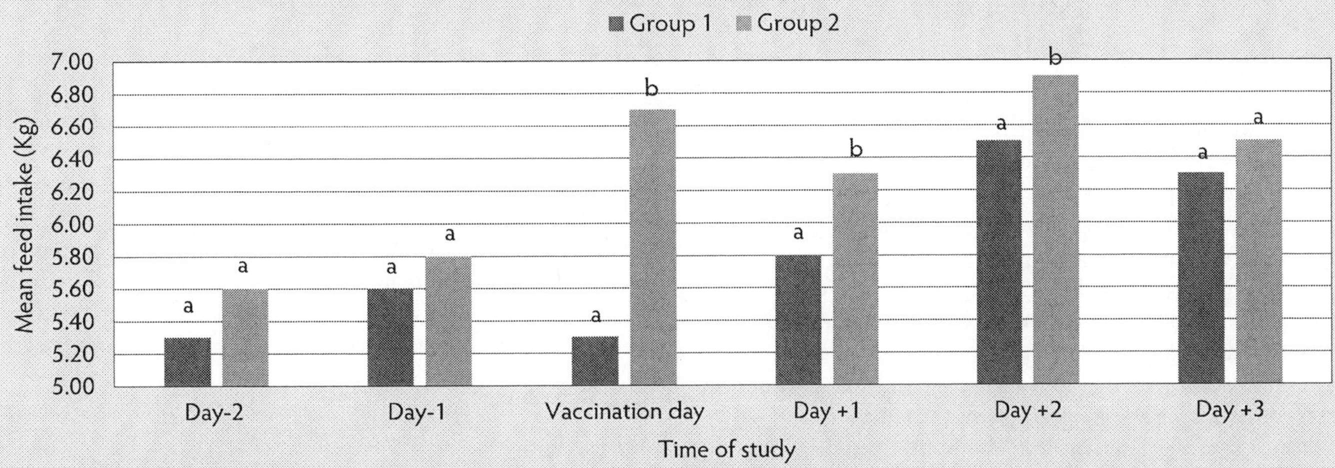

Figure 3: Measurement of rectal temperature after vaccination. Vaccine safety was measured indirectly through the absence/presence of fever (mean group rectal temperature [MRT] in °C) at vaccination (0 h), and 6, 12 and 24 hours post vaccination (hpv).

Fact vs. fiction: The truth behind disinfectant claims

Jeff Odle
Technical Marketing Manager, Lanxess Corporation, Business Line Actives and Disinfectants, North America

Introduction

Choosing who or what to trust is one of the most important decisions we make in life. Making the wrong decision can lead to financial loss and even worse irreparable harm; harm to your financial stability and your professional (and sometimes personal) brand. Given the impact of making the 'wrong decisions' it's astonishing how quickly most decisions are made. As a matter of fact, a recent study[1] published in the Journal of Neuroscience showed the human brain, responding through visual rapid responses to facial characteristics, takes a mere three hundredths of a second to decide whether to trust or not. This isn't a lot of time when you're talking about decisions that can impact your personal and/or professional success. This holds true when choosing which disinfectant to use, as part of an effective biosecurity program. Disinfectants vary greatly in performance depending on the microorganisms present, as well as the chemical and physical environment in which they are used.

You may be asking yourself, "Where does one look to find data supporting disinfectant efficacy against the pathogens of concern and can that data be trusted to be accurate?" The objective of this paper is to highlight where data, supported by scientific analysis, can be found, as well as to provide examples of misbranding or misleading statements which, if trusted, could lead to financial loss and potentially irreparable harm to your swine operation.

The regulation of disinfectants

In order to protect the consumer and the general public from unregistered, hazardous, or ineffective products, the US Environmental Protection Agency classifies disinfectants as antimicrobial pesticides and therefore the claims made for disinfectant products are highly regulated and must be supported by data in order to be registered. Every pesticide must have labeling which is accepted by EPA before the pesticide can be sold or distributed in the US. Labeling is defined under FIFRA (Federal Insecticide, Fungicide, and Rodenticide Act) as meaning labels and all other written, printed or graphic materials which accompanies the product. This includes all marketing literature whether hardcopy or electronic. Companies wishing to register a disinfectant need to submit chemical characterizations, safety, and efficacy data to EPA as well as pay registration fees. EPA reviews and approves each product label and assigns an EPA Registration Number for the product. One can find the approved EPA product label for registered disinfectants by searching the EPA Product and Label System by choosing either product name, manufacturer, or EPA Registration Number at `https://iaspub.epa.gov/apex/pesticides/f?p=PPLS:1`.

After a disinfectant is approved by the EPA, the product must also be registered in each state in which the product is intended to be sold – all of which must occur before the product is distributed into the channels of trade. Companies that sell unregistered or misbranded pesticides are subject to enforcement action by the EPA. Penalties for non-compliance with these requirements can be significant, as Medline Industries found out in September 2018 when EPA fined them $4.93M for nearly 1300 alleged violations. Amazon Services LLC was also fined $1.2M in 2018 for non-compliance. It is important to note, Amazon wasn't even the product manufacturer, yet the EPA found them in violation of FIFRA for selling and distributing products not licensed for sale in the US.

Advertising and collateral literature are considered an extension of the product label. In other words claims made through advertising (both written and verbal) must be supported by similar claims on the product label. A pesticide is considered misbranded if it's label or marketing contains any statement, design, or graphic representation which is false or misleading. Under FIFRA, it is unlawful to distribute or sell any pesticides which are misbranded.

Common examples of misbranding include:[2]

- A false or misleading comparison with another pesticide
- A false or misleading statement concerning the effectiveness of the product
- A true statement used in such a way to give a false or misleading impression to the purchaser
- Any statement directly or indirectly which implies the pesticide is recommended or endorsed by an agency of the Federal Government
- Safety claims with or without a qualifier such as when used as directed
- Statements that imply the product prevents or controls disease
- Statements which imply heightened efficacy like powerful or hospital grade

Efficacy claims

Each registrant must ensure and validate, through testing, that their product is efficacious against the pathogens in which claims are made when the product is used in accordance with label directions and commonly accepted pest control practices. This is typically accomplished by having an independent laboratory, not the manufacturer or registrant, test the product under standard EPA test protocols using Good Laboratory Practices. The standard EPA protocol for disinfectants used on hard, non-porous surfaces calls for a 10 minute contact time against a specified pathogen when exposed to the label recommended use dilution at room temperature with the sample containing 5% organic challenge and 400 parts per million hard water. If the EPA approved label lists a pathogen then one can be fairly confident there is test data to support the claim. The EPA reserves the right to review efficacy test data and routinely reviews data when a product bears a claim for organisms that pose a threat to human health.

Claims of efficacy which are not supported by testing or are inconsistent with the testing data are not acceptable. Examples include:

- Making claims for pathogens not listed on the approved label
- Claiming a 30 second contact time when testing and/or use directions specify a 10 minute contact time
- Claims of having "unique" characteristics because of composition or formula are also not acceptable
- Claims for use rates or use applications not specified on the label

Reality and field experience

Everyone is looking for a competitive advantage! However, bending the rules or ignoring them completely in order to win market share is not acceptable. Using products that do not have scientific test data that supports efficacy under defined conditions not only puts the grower at risk of an outbreak, but places the entire industry at risk of significant economic loss with lost domestic sales and suspension of export markets not to mention costs associated with containment and eradication efforts. Here are five examples to beware of:

1. Unregistered products sold as cleaners, deodorizers or absorbents, in order to circumvent EPA registration requirements, but are used in applications clearly intended for disinfection such as boot baths. If a product is used for disinfection purposes or claims of disinfection are made it must carry an EPA Registration Number on the front of the label. This is proof that a dossier, including efficacy data, has been submitted and reviewed by EPA. There are very few exceptions to this rule and reserved for products containing natural ingredients where all components are listed on the label.
2. Marketing communications which promote efficacy for pathogens not listed on the approved label, possibly supported by field trials or university studies. Often times these trials do not mimic recommended label use dilutions or contact times. The reason EPA specifies a set protocol for testing is to ensure all testing is conducted under strict adherence to Good Laboratory Practices. Non-approved protocols for field tests could be flawed in design or data interpretation.
3. Products that don't completely list all pathogens on the product label but are described as virucidal, bactericidal, or fungicidal often leave the consumer with the impression they kill all viruses, bacteria, and fungi. This couldn't be further from the truth. Enveloped viruses have a delicate lipid envelop which leaves them vulnerable to environmental factors such as high temperature or chemical attack by disinfectant chemistries. Non-enveloped viruses, on the other hand, are surrounded by a hard protein coating as well as a lipid layer which helps prevent disinfectants from entering the cell making it much more difficult to kill. For example, a product that can kill avian influenza may not be able to kill PEDV, or will require a higher concentration or longer contact time in order to provide an effective kill.
4. Products where marketing literature promotes use dilutions, contact times, or efficacy at temperatures not specified on the EPA approved label.
5. Products that promote a "one-step clean and disinfect" where cleaning is not required prior to disinfection should be avoided. Nowhere in the swine production or transportation process is the surface clean enough to presume disinfection can be effective without cleaning first.

Conclusion

Prior to selecting a disinfectant product, review the product master label filed with the EPA for the complete picture. If the marketing literature or the verbal claims differ from the master label be wary and ask to see the independent studies that supports its proposed use. Most companies will provide either the final signed efficacy report from the lab that conducted the study or a summary of the test results. Disinfectant manufacturers and distributors should not have to use misbranding or misleading statements in order to sell their products. The product should sell itself based on testing, conducted using established scientific protocols, proving efficacy under the conditions in which it will be used.

References

1. Amygdala Responsivity to High-Level Social Information from Unseen Faces by Jonathan B. Freeman, Ryan M. Stolier, Zachary A. Ingbretsen and Eric A. Hehman Journal of Neuroscience 6 August 2014.
2. *EPA Label Review Manual* – Chapeter 12 Labeling Claims, Revised November 2013, https://www.epa.gov/sites/production/files/2017-10/documents/chap-12-nov-2013.pdf

Impact of NutriQuest Swine Awake™ administration to weaned pigs on pig livability

K. Bretey[1], DVM; R. Song[2], PhD; P. Chang[2], PhD
[1]Standard Veterinary Services, LeMars, Iowa; [2]NutriQuest, Mason City, Iowa

Introduction

Porcine reproductive and respiratory syndrome virus (PRRSV) often infects sow herds and has devastatingly negative effects on the health and survivability of the downstream piglets from birth through market. While the virus can affect piglets during gestation, it is commonly spread to the nursing piglets from the dam and littermates during the nursing phase. The virus rapidly infects and suppresses the immune system, resulting in not only damage from the PRRSV itself, but also infection and disease from a myriad of opportunistic pathogens in the environment.

NutriQuest Swine Awake™ has been shown to improve livability and health in commercial wean-to-finish pigs experiencing PRRSV challenges.[1] In a controlled challenge study, pigs receiving Swine Awake™ had improved growth performance, reduced microscopic lung lesion scores, reduced neutrophil counts, increased lymphocyte counts, and increased circulating IL-1β and IL-12 on 21 days post-infection with PRRSV.[2]

This field study was conducted to evaluate the effects of NutriQuest Swine Awake™ on livability of weaned pigs coming from a PRRSV positive sow herd.

Materials and methods

Pigs utilized in the study were from a sow farm that experienced a PRRSV break (1-4-4 RFLP) in February 2018 and was undergoing a PRRS stabilization effort with live virus inoculation. Regular PCR testing confirmed that three-week-old due-to-wean pigs were PRRSV positive throughout the study. Pigs were transported to an offsite wean-to-finish facility at 3 weeks-of-age. Three 1,200-head rooms, designated rooms A, C and D were utilized in the study. Room C was double stocked with a total of 2,549 pigs between 5/21/18 and 5/29/18. Room A was double stocked with a total of 2,589 pigs between 5/29/18 and 6/4/18. Room C served as the treatment group receiving Swine Awake™ administered through the water per label directions beginning at placement and continuing for 21 consecutive days. Room A served as the control group and did not receive Swine Awake™. Room D received overflow pigs from control and Swine Awake™ rooms starting at 4 weeks post-placement, and pigs were kept separated by treatment.

Blood samples were collected from 30 pigs across pens on Swine Awake™ treatment 2 weeks post-placement. Samples were pooled for a total of 6 samples that were submitted to Iowa State University Veterinary Diagnostic Lab in Ames, IA, for detection of PRRSV and Influenza A by PCR.

Mortalities data was collected daily. A chi-square probability analysis was conducted to evaluate differences in mortality by treatment. Growth parameters including average daily gain (ADG), average daily feed intake (ADFI), and feed efficiency (FE) were not collected.

Results and discussion

All blood samples collected from pigs in the Swine Awake™ treatment were positive for PRRSV, with Ct values under 24, and all samples were negative for Influenza A.

Pig mortality during the nursery period (8 weeks post-placement) was significantly reduced in Swine Awake™ treated pigs compared to control pigs (24.8% vs. 29.5%, $P < 0.01$; Table 1).

In control pigs, mortality was highest 7 weeks after placement (n = 183), but on weeks 3, 4 and 5 after placement, over 100 pigs

Table 1: Weekly mortality of pigs placed in wean-to-finish barn in control and NutriQuest Swine Awake™ treatments.

Item	Control	Swine Awake™[1]	*P*-value[2]
Total pigs placed	2,589	2,549	
Total mortality nursery phase[3]	764	632	
% Mortality nursery phase[3]	29.5	24.8	< 0.01
Total mortality grow-finish phase[4]	129	106	
% Mortality grow-finish phase[4]	5.0	4.2	0.17

[1] NutriQuest Swine Awake™ was administered to pigs via water for 21 days post-placement per label directions.
[2] Chi-square test for probability.
[3] Nursery period consisted of day of placement to 8 weeks post-placement.
[4] Grow-finish period consisted of 8 weeks post-placement to market.

were euthanized or died weekly, accounting for 47% of the total mortality (Figure 1). In Swine Awake™ treated pigs, mortality peaked on week 4 post-placement, with 165 pigs dead/euthanized (Figure 1).

Given the differences in mortality, it is expected that growth performance would likely be improved with Swine Awake™ treatment. Pig ADG and feed efficiency have been shown to improve in a PRRSV challenge study[2] conducted with Swine Awake™ by 21% and 13%, respectively, and are currently being assessed in other commercial field settings.

Conclusions

The data from this study suggest that treating pigs with NutriQuest Swine Awake™ for 21 days post-placement was effective in improving livability of PRRSV positive pigs in a commercial wean-to-finish facility.

When analyzing the economic impact of these results, 132 pigs saved during the nursery phase valued at $50/pig, minus cost of product, provides a net value of approximately $4,000, with a return over investment of 2.6.

References

1. Holck, T., R.E. Musser, R. Song, and K.G. Friesen. 2018. The impact of Swine Awake™, a new, non-antimicrobial protein, in natural disease conditions in newly weaned pigs. 49th Annual Meeting of American Association of Swine Veterinarians Proceedings: 276-278.

2. Song, R., T. Holck, S.M. Sinn, R.E. Musser, and K.G. Friesen. 2018. A novel, non-antimicrobial protein improves growth performance and immune responses of pigs experimentally infected with porcine reproductive and respiratory syndrome virus. J. Anim. Sci. 96, suppl.2 (abstract): 127.

Figure 1: Weekly mortality of nursery pigs placed in wean-to-finish barn in control and NutriQuest Swine Awake™ treatments.

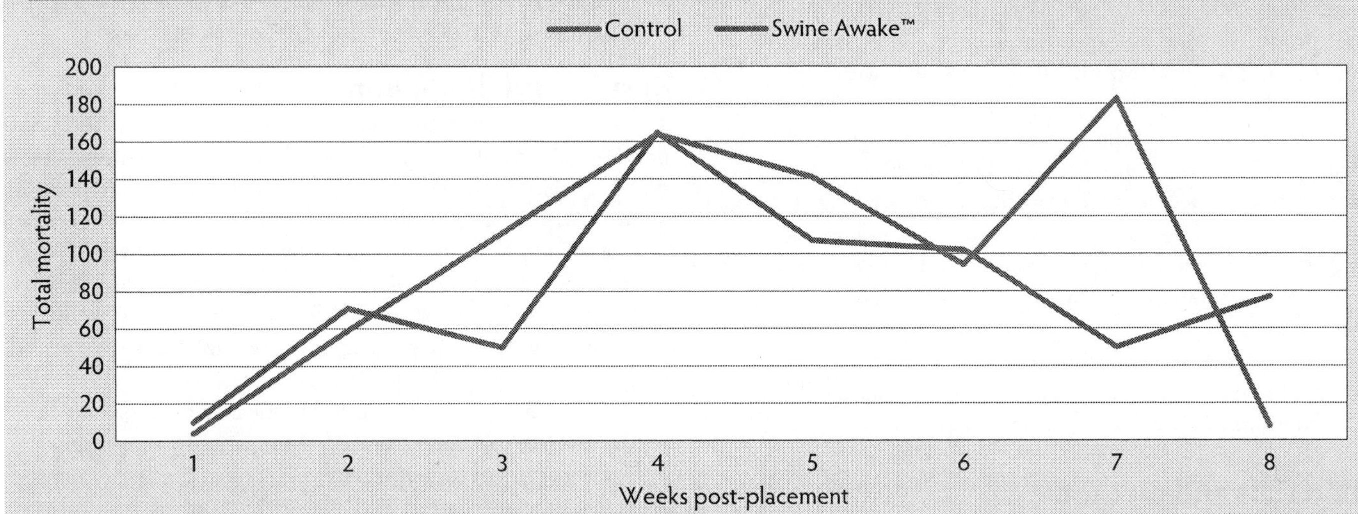

Analysis of sow mortality using big data from PigCHAMP

T. Gillespie, DVM, DABVP
Performance Health, PC

Introduction

Sow mortality in North America has increased over the past several years as established in numerous publications. The increasing incidence rate has not fully recognized underlying causes, although numerous groups and production systems have and are exploring this concern.[1-3] Specifically a sow mortality project called Pelvic Organ Prolaspe directed by Iowa Pork Industry Center, Iowa State University is an industry wide collaberation that recently reported on their findings.[4] The project's approach is a comprehnsive survey that incorporates 104 breeding herds in 15 states. The farms associated with this survey range from independent producers to production systems and totalling approximately 400,000 sows. A list of risk factors associated with the site's production and management practices, types of facilities, nutritional strategies, and animal measurements were captured on the participating farms. A total of 62 farms in the participating group were visited by trained personal to collect specific individual animal measurements. This work is ongoing with the goal of indentifying specific risk factors that are associated with pelvic organ prolapsing. This author elected to take a different approach by using a retrospective examination of combined data from thirteen data bases utilizing the same production software managed by PigCHAMP. The author's goal is to illustrate and capture key points to further help the swine industry by illustrating specific aspects, i.e. period in the reproductive life of the sow, potential reasons for death, parity comparision, etc.

Materials and methods

The database consists of the production years 2015, 2016, and 2017. A combined database is made up of 371 farms where some farms are a part of a production system and others are independent. The number of farms and size of each farm in each respective database was not considered important to this analysis. The location of the farms are all across the swine belt with the majority located in the Midwest and Eastern regions of the US. Sow housing types consist of both individual stalls and various types of open sow housing. As expected the list of causes utilized by the farms' production staff was rather numerous and diverse. Therefore, the categories used in this analysis were grouped into a shorter list as illustrated in chart titled, "Total Deaths combining years 2015, 2016, and 2017" (Figure 1). The grouping of the farms' reasons into each category that is used for this analysis were all associated with that category, i.e. lameness category dealt only with reasons listed for death that could be associated with a lameness. The total born levels did increase for nearly all of the participating farms during the three production years. The database contains over 600,000 sows that were used in this analysis.

PigCHAMP database preparation

1. Female Removal Analysis export trace list by year.
2. Performance Trend Analysis by year, by month for our database statistics and total born per sow.
3. Lookup Item Management (Dictionary): for death reason consolidation refer to `https://www.youtube.com/watch?v=W8_-m0Vyfk8&index=6&list=PLD980wxuUE1fBPbQP5TTaJmX2-IxeDDKX`.
4. Data Item Management: for subcategory euthanized, refer to `https://www.youtube.com/watch?v=WY3_TeX6Cek&list=PLD980wxuUE1fBPbQP5TTaJmX2-IxeDDKX&index=3`.

Results and discussion

A shocking number of sows died each production year with 51,550 animals in 2015, 63,275 animals in 2016, and 76,412 animals in 2017. An interesting trend in the three year time period is the growth of the average inventory of approximately 3% each year as illustrated in Table 1. The inventory increase in 2016 over 2015 was 18,851 animals and 2017 over 2016 was 20,521 animals with a two year increase over 2015 of 39,372 total animals. Equally interesting is the percentage of deaths during this same time period as captured in Table 2. The percentage increased in mortality rate of 2016 over 2015 is nearly identical to the increase mortality rate of 2017 over 2016, i.e. 1.6% and 1.7% respectfully, when using average inventory of each year is used. A look at the 2-year increase in percent mortality rate of 2017 over 2015 clearly is dreadful at 32.54%. This is due to an increasing average inventory year over year with a nearly flat mortality rate. If one is to examine the increased mortality rate using removals, a different picture develops. The percentage is nearly the same each production year with 2015 showing 16.42%, 2016 showing 19.14%, and 2017 capturing 18.78%. The 2-year percentage based on removals is equally shocking at 26.79%. Table 2 shows the number of removals and deaths per production year. Further analysis of the data including, but not limited to, the categories on reasons for death was completed.

The first step in furthering the analysis was to examine a parity breakdown as shown in Figure 2 titled "Total Deaths by Parity 2015 – 2017," comparing each parity illustrates the number of deaths for each production year. The year over year increase in mortality by parity is especially striking in the young to middle age sows. The production year of 2017 is highly outstanding across most parities. A line chart (Figure 3) illustrating monthly deaths for each production year which clearly demonstrates a

Figure 1: Total deaths by reason 2015-2017

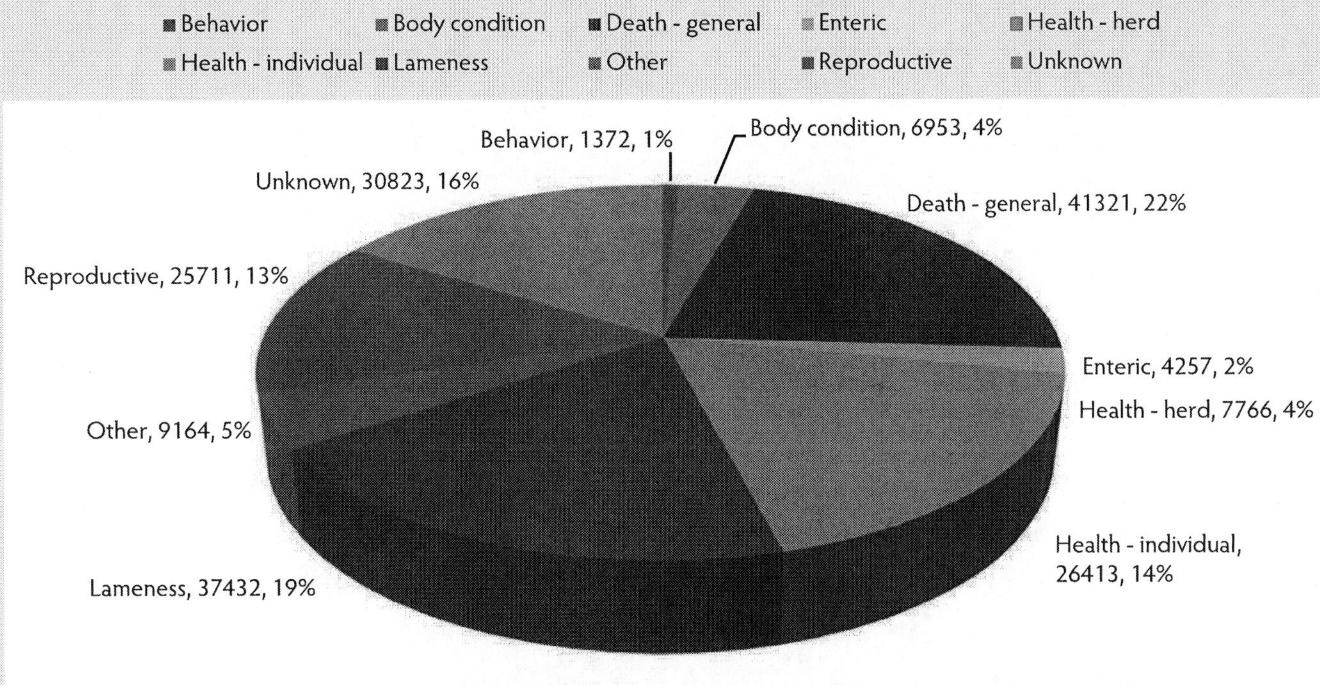

Table 1: Total database herd inventory 2015 - 2017

	2015	2016	2017
Total female herd inv.	598,210	617,061	637,582
% of growth		3.05%	3.22%

spike in March and April time period of 2017. Figure 4 illustrates a breakdown for this specific time period, i.e. March and April, 2017. The majority of deaths were listed as associated with PRRSV when an in-depth look of the reasons used by the farms for each year was performed. The category titled "health – herd" in Figure 4 is when a major pathogen is considered by the farm's staff to be the primary cause of death. The higher rate in 2017 which was listed as PRRSV for the cause of death, although co-infection with other pathogens is unknown. The category listed "death – general" is defined as a sudden death. One might wonder if the category listed as "death – general" and "health – herd" should be combined since a high pathogenic strain of PRRSV might create animals that could be listed in either category. The author chose to use the reason for death as what the farm's staff listed in the production records.

The category listed as "health – individual" consist mostly of animals that prolapse. In Figure 5, a detailed chart illustrates the type of prolapse are shown. The category listed as "death – general" consist mostly of what the farm's staff considers the animal having a sudden death. The category listed as "reproductive" is all deaths related to a reproductive reason but not associated with a prolapse.

Another technique to dig a bit deeper into when sows die in their life cycle is best illustrated in Figure 6 showing the days since last served. Sows of all parities are dying in amazing numbers around the time of farrowing and for approximately one week after farrowing. Previously, other authors have also found a high number of sows to die in the lactation phase.[5] Table 3 captures the numbers of sow deaths by parity and days since last service. This indicates that all parities are experiencing a similar increase in sow mortality during the critical period around farrowing and for up to seven days post-farrowing. In PigCHAMP, a sow changes her production status from lactation to weaning when all of her piglets are removed. For example, a sow must wean zero number of piglets if she dies after farrowing but prior to a normal weaning time. Additionally, in some cases, sows are dying very soon after weaning.

Conclusions

This exercise in analyzing large amounts of data consisting of 13 data bases, plus over 600,000 sows in a fairly short time period, i.e. three production years, has been enlightening to the author. The most outstanding features found in this analysis are the number of young to middle age sow deaths, how a major pathogen's activity influences sow mortality rate over a short period of time, the relatively low number of prolapses although still important event for our industry, the increasing total born numbers being achieved in the US industry, and the tremendous

Table 2: Total number of female deaths 2015 -2017

	2015	2016	2017
Total female cull/sales	262,476	267,351	330,420
Total female deaths	51,550	63,275	76,412
Death rate %	8.62%	10.25%	11.98%
% of death of removals	16.42%	19.14%	18.78%
% Increase of deaths over previous year		18.53%	17.19%
% increase of deaths over past 2 years			32.54%

Figure 2: Total deaths by parity 2015-2017

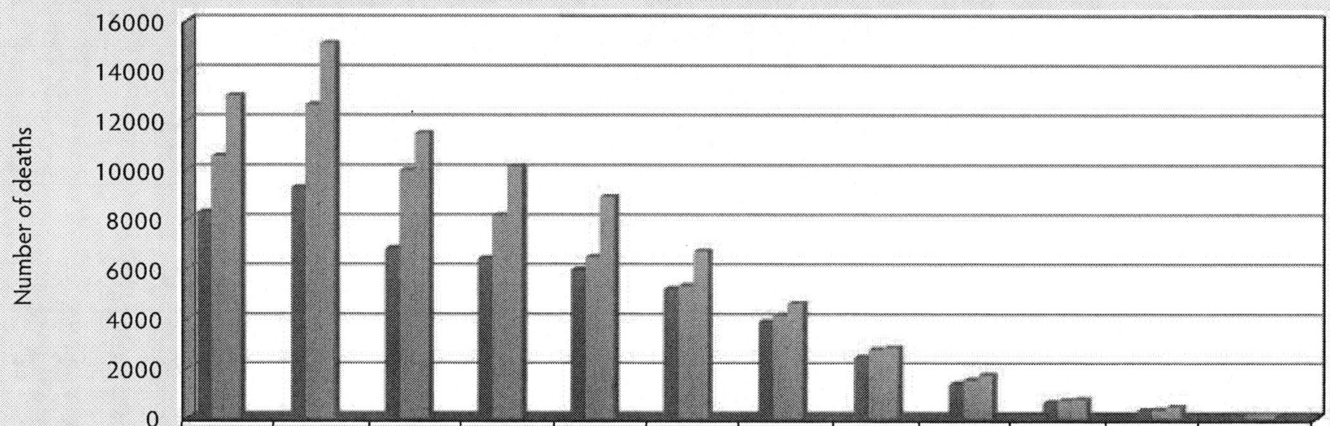

	1	2	3	4	5	6	7	8	9	10	11	12
■ 2015	8310	9296	6864	6452	6001	5224	3908	2468	1386	640	338	115
■ 2016	10579	12645	10016	8177	6507	5359	4159	2783	1552	726	356	169
■ 2017	13008	15119	11496	10171	8923	6733	4634	2851	1758	771	467	162
Total deaths	31897	37060	28373	24800	21431	17316	12701	8102	4696	2137	1161	446
Percentage %	16.72	19.42	14.87	13	11.23	9.07	6.66	4.26	2.46	1.12	0.61	0.23

number of sows that die around or shortly after farrowing. The fast pace that occurs in modern farrowing units due to large number of sows farrowing on any given day, plus the skill level of the farm's staff need to be considered as risk factors. The accuracy on why a sow dies cannot be considered absolutely accurate unless a necropsy is performed; therefore, the author acknowledges this limitation in this analysis. As practitioners, we need to continue to critically evaluate all production practices, daily events, as well as environmental and nutritional influences. One should not speculate or presume the causes of sow deaths. So many aspects in the daily routines of the farm's staff need to be examined to improve sow viability. Our industry needs to consolidate and collaborate on a shorter list of reasons for death. The author's effort to combine and consolidate the reasons utilized in this analysis should be considered as an initial attempt. A clear understanding of the risk factors that cause a sow to die is needed before a successful strategy of reducing mortality can be implemented.

References

1. Straw, B., Causes and control of sow losses. *Mod Vet Pract.* 1984; 65:349-353.
2. D'Allaire, S., Stein, T., Leman, A., Culling patterns in selected Minnesota swine breeding herds. *Can J Vet Res* 1987; 51:506-512.
3. D'Allaire, S., Drolet, R., Chagnon, M., The causes of sow mortality: A retrospective study. *Can Vet J* 1991; 32: 241-243.
4. Ross J, et.al. Pelvic Organ Prolapse: An industry-wide collaboration to identify putative contributing factors. Allen D. Leman Swine Conference proceedings, 2018;31.
5. Geiger, J.O., Irwin, C., Pretzer, S., Assessing sow mortality. Allen D. Leman Swine Conference proceedings, 1991;84-87.

Figure 3: Monthly deaths 2015-2017

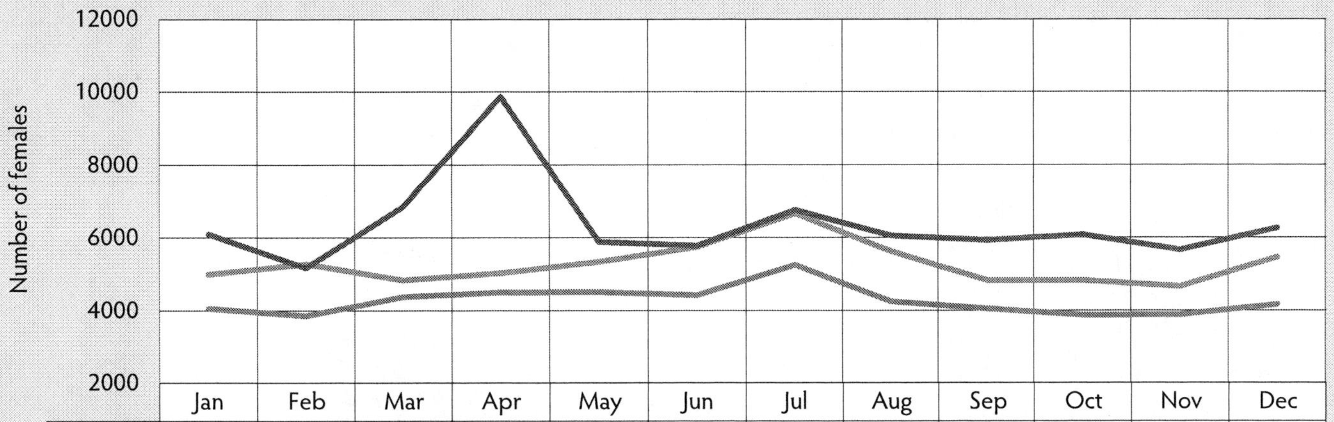

	Jan	Feb	Mar	Apr	May	Jun	Jul	Aug	Sep	Oct	Nov	Dec
2015	4052	3857	4377	4494	4502	4420	5245	4243	4047	3876	3884	4166
2016	4994	5276	4837	5028	5345	5741	6657	5617	4827	4820	4663	5469
2017	6094	5172	6844	9873	5881	5778	6757	6054	5932	6075	5667	6262

Figure 4: Deaths, March - April 2017

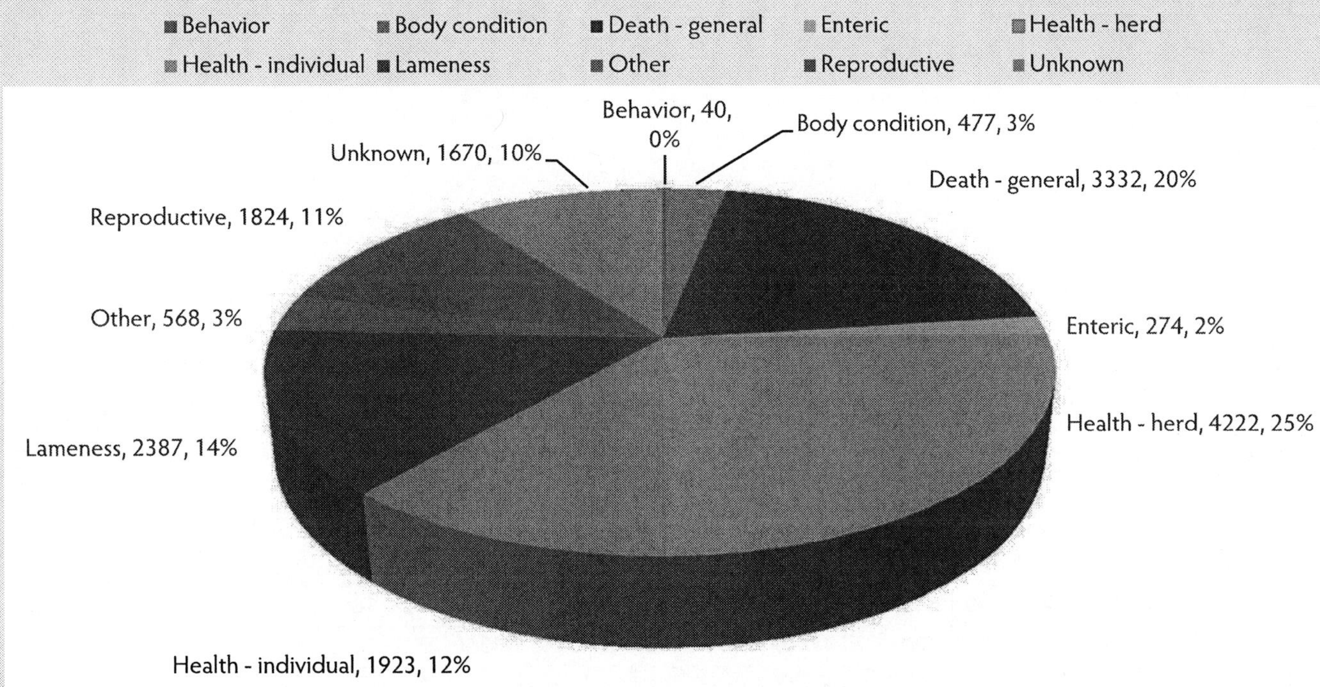

Figure 5: Prolapse by year 2015-2017

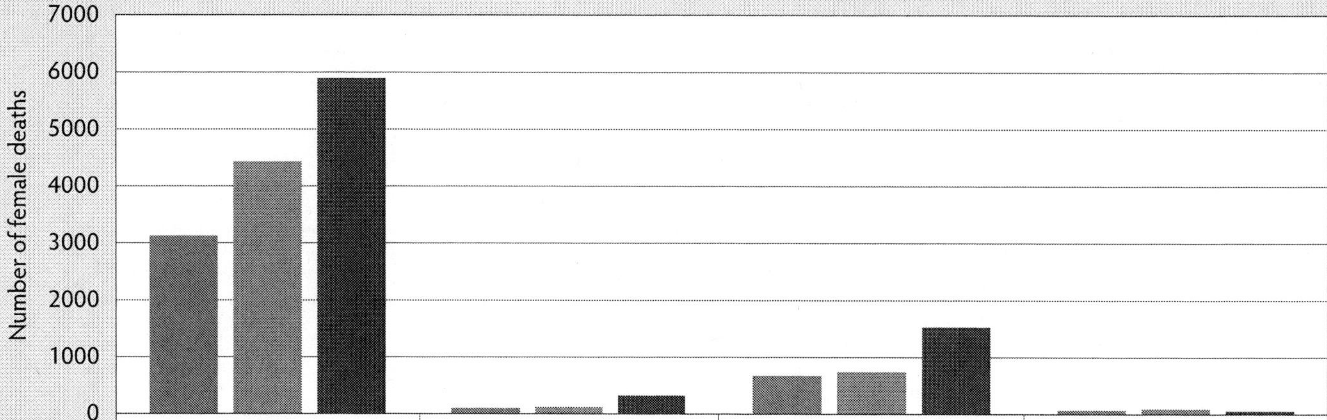

	Prolapse	Rectal prolapse	Uterine prolapse	Vaginal prolapse
2015	3123	99	667	70
2016	4431	117	736	93
2017	5883	323	1524	58

Figure 6: Day of death since service

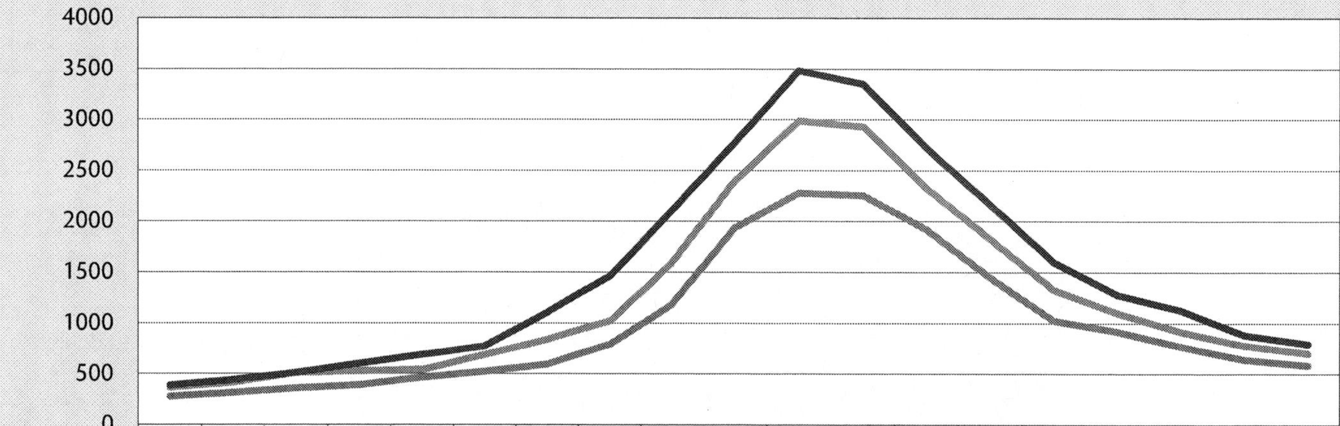

	107	108	109	110	111	112	113	114	115	116	117	118	119	120	121	122	123	124	125
2015	278	317	362	389	466	524	597	795	1183	1935	2279	2259	1918	1446	1029	921	771	645	588
2016	367	418	510	529	543	687	843	1026	1586	2405	2990	2932	2329	1822	1330	1105	916	782	707
2017	388	442	511	603	691	771	1109	1464	2105	2779	3483	3350	2718	2163	1600	1280	1128	884	798

Table 3: Day of death since service by parity 2015-2017

Days	Parity												Total
	1	2	3	4	5	6	7	8	9	10	11	12	
107	161	136	118	112	88	48	30	19	8	3	2	3	**728**
108	195	167	140	103	84	40	21	18	7	3			**778**
109	235	191	165	148	111	58	41	18	8	6	2		**983**
110	257	250	182	125	102	79	55	22	9	1	3	1	**1,086**
111	299	247	205	158	111	75	51	17	13	6	2	2	**1,186**
112	347	330	241	186	154	90	60	31	9	6	2	1	**1,457**
113	423	365	313	258	221	138	75	34	15	4	7	2	**1,855**
114	604	530	436	307	236	164	114	48	15	16	2	3	**2,475**
115	813	801	655	547	425	330	155	96	46	12	7	4	**3,891**
116	1,084	1,081	1,053	903	690	540	315	179	55	31	14	18	**5,963**
117	1,333	1,267	1,301	1,182	949	676	397	240	91	53	14	19	**7,522**
118	1,313	1,240	1,227	1,113	946	659	440	230	103	61	18	18	**7,368**
119	1,146	998	997	942	748	568	354	217	97	53	21	15	**6,156**
120	938	814	692	705	581	431	270	212	107	45	20	15	**4,830**
121	785	556	551	472	420	313	236	119	47	41	13	13	**3,566**
122	697	503	423	400	362	287	170	103	57	9	10	5	**3,026**
123	628	397	379	327	261	228	171	79	51	26	11	11	**2,569**
124	531	366	309	273	233	173	112	70	31	16	6	3	**2,123**
125	483	307	271	234	218	171	115	79	29	17	4	6	**1,934**
Total	**12,272**	**10,546**	**9,658**	**8,495**	**6,940**	**5,068**	**3,182**	**1,831**	**798**	**409**	**158**	**139**	**59,496**

Pursuing American Board of Veterinary Practitioners (ABVP) certification in Swine Health Management

Jeffrey A. Husa, DVM, Diplomate ABVP
Swine Health Management representative to the Council of Regents

Introduction

Are you interested in board certification? How might ABVP credentials benefit you and your clients? A 2017 survey of Swine Health Management (SHM) Diplomates revealed that a majority of our members pursued certification out of a desire for "personal challenge" and "increased knowledge." What's more, these same respondents reported that completing the journey to credentialing "made me a better practitioner." No matter what sparked your interest in this topic, we encourage you to take the ABVP plunge. I have found the entrance, credentialing, and maintenance processes to be rigorous and rewarding. Becoming a Diplomate required that I push my professional limits harder than ever, and has been far more valuable to me, my clients, and the animals I care for than I ever expected.

Applicant requirements*

To be eligible for certification, a veterinarian must:

A. Be a graduate of an AVMA accredited / approved college of veterinary medicine, hold a certificate from an Educational Commission for Foreign Veterinary Graduates (ECFVG), or be legally qualified to practice in the US or other country.

B. Demonstrate unquestionable moral character and ethical professional behavior.

C. Have a background in swine medicine and production in one of the following ways:

a. At least 5 years of clinical practice experience prior to first credential submission, and at least 6 years in practice prior to sitting the credentialing examination. The first year need not be in swine practice, however the applicant must have worked in a predominantly swine-oriented practice during the 5 years prior to examination.

OR:

b. Complete an ABVP-approved Residency or MS degree program, plus at least one year of swine practice.

OR:

c. Complete a swine-related PhD. The applicant must provide evidence of swine industry consulting.

The path to credentials begins with our entrance exam

So how do you become credentialed? Board certification in Swine Health Management begins by successful completion of an entrance examination. This multiple choice test is offered each Spring at the AASV Annual Meeting. We begin with this exam to assess your basic knowledge of key topics such as reproduction, disease, pathology, economics, epidemiology and more. This test also helps to set the stage for your success throughout the remainder of the credentialing process. To sit the exam, you must submit an application and fee, due January 15th of the year in which you intend to test. Additional details about the examination, schedule, topics covered, and more can be found at www.abvp.com. The Applicant Handbook is available at this address, and is an excellent guide throughout the entrance and credentialing processes. Applicants who fail may retake the exam in succeeding years through the same process (application plus fee).

The certification process

After you pass the entrance exam, you will select an ABVP advisor, subject to the approval of the SHM Regent or their delegate. You will then complete a plan of study developed in collaboration with your advisor, and you will complete the requirements for your credentials application. You will write two publications: either two case reports written in the ABVP format (Applicant Handbook 2018, pg 22),[1] or one case report and one peer-reviewed article accepted for publication. You will also begin studying for the qualifying examination during this time. Many Diplomates have found these studies to go hand-in-hand with daily clinical experiences. The credentialing process will push, stretch, and energize you!

Then in September of the year preceding your intended credentials submission, you will complete an on-line application at `www.abvp.com`.[2] This alerts us of your impending submission. Your advisor and other Swine Health Management Diplomates will be available to you throughout the journey. In particular, we recommend review of your materials prior to submission to better ensure compliance with ABVP guidelines. You will submit your credentials not later than 11:59 PM on January 15th. Then, it's time to study for the next 14 months! The certification exam is administered during our AASV Annual Meeting. It includes written and oral evaluations with short answer and problem solving components. A suggested credentialing timeline is shown in Table 1.

Maintenance of Certification*

Once you have successfully obtained credentials, they will remain valid for 10 years with submission of annual dues. Between the 8th and 10th year after you initially credential, you will recertify through the accumulation of points or by reexamination. Points

* Note: For current application, credentialing, and maintenance requirements, see the ABVP website for handbooks and other guidance.

Table 1: Suggested timeline for board certification in Swine Health Management through the American Board of Veterinary Practitioners

	Jan	Feb/Mar	Apr	May	Jun	Jul	Aug	Sep	Oct	Nov	Dec
Year 1	Submit entry exam application by 11:59PM CST, Jan 15	Pass entry exam at AASV annual meeting		work		work		work work work		work!	
Year 2	work	work work work		work	work		work	work work	work!		
Year 3	work	work work work		work		work!		Submit credentials application by 11:59PM CST, Sep 1		work!	
Year 4	Submit credentials by 11:59PM CST, Jan 15	study study		study		study!		Submit certification exam application by 11:59PM CST, Sep 1		study!	
Year 5	study!	Sit for certification exam at AASV Annual Meeting									

are earned through a variety of options including CE attendance, refereed publication, presenting lectures, writing case reports, and more. All topics attended, presented, or published must be relevant to swine practice and meet ABVP guidelines. Complete details about either pathway can be found in the Maintenance of Certification Handbook,[3] also available at the ABVP website under Forms and Documents on the Members-Only page. Credentials may be maintained for the life of the Diplomate through this rolling 10 year process. The Council of Regents urges all Diplomates to submit in their 8th year to best ensure continued success.

Conclusion

So why should you consider certification in Swine Health Management through the American Board of Veterinary Practitioners? It will stretch you! You will be recognized among your peers as an expert in this field. The journey will help you to hone new skills and enhance old ones. You will be asked to demonstrate not only your knowledge, but also your ability to solve critical problems...problems which may make all the difference for you, your producers, their pigs, and even consumers world-wide. Come join us!

References

1. Applicant Handbook: `https://abvp.com/veterinary-certification/why-get-abvp-certified`

2. Website: `www.abvp.com`

3. Maintenance of Certification (MOC) Handbook: `www.abvp.com` -> Member Login -> Forms and Documents

Are you ready for an FDA inspection? Ensuring a valid VCPR for prescriptions and treatment plans

C. Rademacher[1], DVM; T. Holck[2], DVM; P. Gorden[1], DVM; E. Brockhoff[3], DVM
[1]Iowa State University, Ames, Iowa; [2]GlobalVetLINK, Ames, Iowa; [3]Prairie Swine Centre, Saskatoon, Saskatchewan

Introduction

One of the major points of emphasis in the implementation of recent FDA guidances calls for more veterinary oversight of medically important antibiotics used in food producing animals. Only veterinarians with valid veterinarian-client-patient-relationship (VCPR) can prescribe medically important antibiotics. With this emphasis, there is a need to clarify the roles and responsibilities of veterinarians in establishing a valid VCPR, providing clients with compliant treatment plans, prescriptions, and veterinary feed directives (VFDs) while utilizing best practices for streamlining the documentation process through web-based solutions.

US regulatory requirements for a valid veterinarian-client-patient-relationship (VCPR) (Figure 1)

A VCPR is present when all of the following requirements are met:

1. The veterinarian has assumed the responsibility for making clinical judgments regarding the health of the patient and the client has agreed to follow the veterinarians' instructions.
2. The veterinarian has sufficient knowledge of the patient to initiate at least a general or preliminary diagnosis of the medical condition of the patient. This means that the veterinarian is personally acquainted with the keeping and care of the patient by virtue of a timely examination of the patient by the veterinarian, or medically appropriate and timely visits by the veterinarian to the operation where the patient is managed.
3. The veterinarian is readily available for follow-up evaluation or has arranged for the following: veterinary emergency coverage, and continuing care and treatment.
4. The veterinarian provides oversight of treatment, compliance, and outcome.
5. Patient records are maintained.

Veterinarian instructions and treatment plans

Veterinary instructions should be documented for each client. Treatment plans, or protocols, are an excellent tool to communicate the veterinary instructions for health intervention and serve as a compliance document for the 1st element of the VCPR. Treatment plans will vary by species (and within species) depending upon whether the focus of treatment is at the individual animal level (ie. dairy, beef) or group/system level (ie. poultry, swine).

Treatment plans should provide the animal owner or caretaker with clear instructions on when treatment should be initiated based upon clinical signs, including the nature of the clinical signs, percentage of animals affected, timing of disease onset, and any gross lesions if necropsies are performed. In addition, treatment plans should provide clear instructions on drug selection, route of administration, dose, duration of use and the appropriate withdrawal time. These treatment plans should be consistent with any necessary prescriptions or VFDs for the specific drugs recommended. It is a good general practice for the veterinarians and the clients to frequently review these plans and update them as necessary based on diagnostic results and efficacy of the products being used.

Figure 1: US veterinarian - client - patient relationship

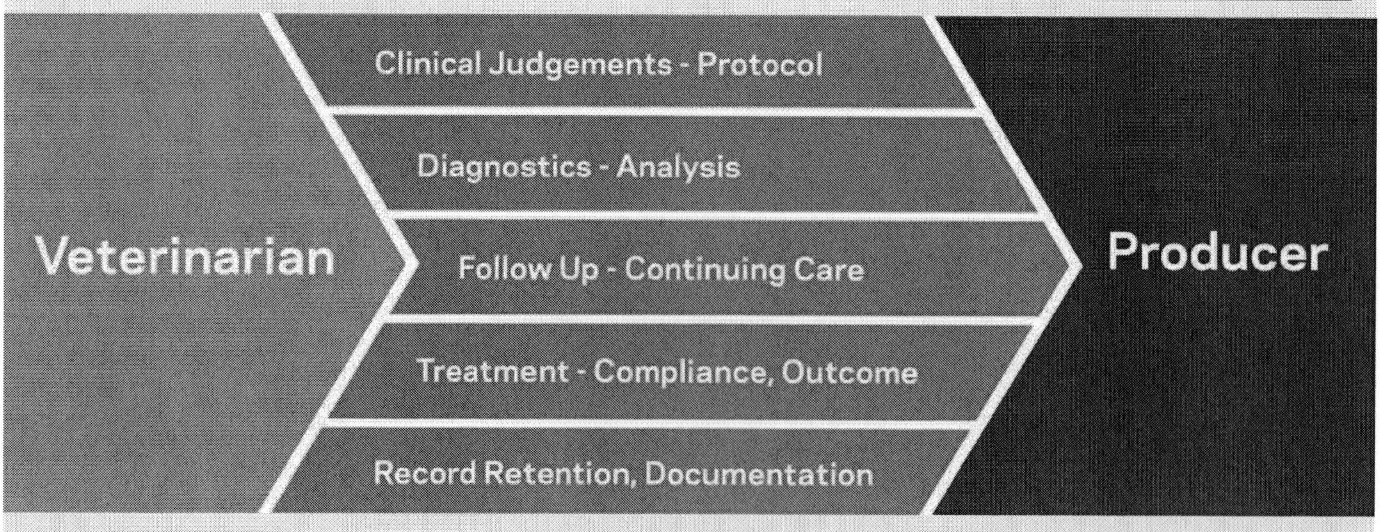

General or preliminary diagnosis

The second element of the VCPR describes the requirement for a general or preliminary diagnosis of the medical condition requiring treatment. While laboratory diagnosis is not required in the general VCPR definition by the AVMA,[1] it is generally good veterinary practice to frequently perform laboratory diagnostics for animal health intervention requiring antibiotics. As FDA continues to push for increased regulations surrounding antimicrobial stewardship in livestock, food animal veterinarians should be prepared to provide laboratory confirmation of diseases for which they are authorizing prescription or VFD drug use.

Veterinarian oversight of treatment, compliance, and outcome

Veterinary authorization of treatments are regulated by both prescriptions (for water soluble and injectable routes of administration) and veterinary feed directives (for medications provided in the feed) in the US.

US prescriptions have specific requirements and are regulated by the state, therefore these requirements may vary from state to state. But a valid prescription does require a VCPR and veterinarians have the flexibility to prescribe most antibiotics based on the Extra Label Drug Use (ELDU) regulation. The prescription label requirements are listed in Table 1.

Note: Requirements for placement of labeling on individual packaging (sachets, packs, bottles, jugs, etc.) on prescription drugs will vary by state. Please consult your State Board of Pharmacy to get the specific requirements for the state you are practicing in.

Inventory control of prescription drugs is regulated by the individual states per their State Board of Pharmacy. Prescription products will be either secured at the warehouse/distributor or delivered by qualified representative of the warehouse/distributor. Only designated personnel have access and must be present to dispense prescription medications. Antibiotic inventory on the farm generally follows one of two paradigms. The first option is that no antibiotic inventory can be maintained on the farm. It is important to finish all dispensed medication unless otherwise directed by the prescribing veterinarian. Remaining product must be returned to the warehouse after treatment. A second option may be that antibiotic inventory can be maintained on the farm. It will be important that the drugs must be properly labeled from the distributor/warehouse and administration follows the established treatment protocol. It is considered to be good practice to have some sort of documentation of medication inventory on farm. In all situations, adverse reactions and treatment failures must be immediately reported to the prescribing veterinarian.

Note: some states may require that the warehouse/distributor have a pharmacist or other specified qualified individual on staff/site. Depending on the state's statutes regarding what constitutes a "pharmacy" and "dispensing", there may be a need for a pharmacist or some sort of trained individual at the dispensing location (e.g. company warehouse).

Documentation is critical. Maintaining copies of all prescriptions is legally required. Electronic software is available to assist both veterinarians and producers to remain compliant. Based on training of the animal owner or animal caretakers by the Veterinarian of Record (VOR), they can then distribute and administer the scripted product based on protocols that were established through the VCPR.

Veterinary Feed Directive (VFD)

As of January 1, 2017, Veterinary Feed Directives (VFDs) are required to provide medically important antimicrobials via the feed in the U.S. Similar to prescriptions, VFDs have specific requirements but, unlike prescriptions, VFDs are regulated by the Food & Drug Administration of the federal government. VFDs and prescriptions can only be authorized under a valid VCPR.

Table 1: US requirements for veterinary prescription labels

	Prescription label requirements
1.	Name, address, telephone number, and license number of veterinarian
2.	Name, address, and telephone number of grower
3.	Identification of animals treated, species and numbers of animals treated
4.	Date of treatment, prescribing, or dispensing of drug
5.	Name, active ingredient, and quantity of the drug
6.	Dosage and duration
7.	Route of administration
8.	Number of refills
9.	Cautionary statements, as needed
10.	Expiration date if applicable
11.	Withdrawal period prior to slaughter
12.	Signature of the veterinarian

Each VFD can only contain one medicated feed additive, with the exception of FDA approved drug combinations. Required information for a VFD to be legal are listed in Table 2.

Canadian regulatory requirements for a valid veterinarian-client-patient relationship

Similar to the US, a VCPR is only considered to exist if medical records of the veterinarian and practice contain sufficient evidence of relevant and timely interaction between the veterinarian, animal owner and the animal(s). The VCPR is documented evidence that the veterinarian has undertaken the steps necessary to establish medical need and consequently prescribe and subsequently dispense medications. Each provincial and territorial veterinary statutory body has their own definition of VCPR in provincial legislation. No significant differences exist between these provincial definitions.

In Canada all medically important antimicrobials as described by Health Canada are prescription only. Once a drug has been authorized, Health Canada issues a Drug Identification Number (DIN). Only Category IV antimicrobials, the ionophores, are not prescription. This includes all presentations of antimicrobials including those used in feed. The Compendium of Medicated Ingredients Brochure (CMIB) is the document that lists those medicated ingredients permitted by Canadian regulation to be added to livestock feed. Veterinary prescriptions are required for all label and extra label use of all medically important antimicrobials. Beginning December 2018, medications that contain medically important antimicrobials will include a new logo on the label to help make them more easily recognized. (See Table 3)

Software solutions to enable a valid VCPR

Utilizing software can simplify many of the processes for completing veterinary prescriptions, VFDs and other animal health documentation. When it comes to regulatory compliance, veterinary software solutions are particularly helpful in establishing and documenting the validity of a VCPR.

Not all software solutions provide the necessary components for supporting evidence of a VCPR. In evaluating what system is best for your practice, critical factors to consider include:

1. Ability to consolidate past and current veterinary diagnostic results, as well as tools for analyzing, charting and communicating results with clients.
2. Scope of animal health documentation options (scripts, VFDs, CVIs, etc.), and features to check those for completeness, accuracy and regulatory compliance.
3. Additional tools and services for communicating and providing more detailed protocols, treatment plans, notes and messages.
4. Unlimited access to database retention of all communications, including documents and records, message exchanges, dates, locations, client data and animal information.

GlobalVetLINK (GVL® provides a comprehensive software platform for veterinary prescriptions, VFDs, CVIs and veterinary diagnostic results, which includes importing historical data. Veterinarians have unlimited access to an account with a searchable database of all records and information, as well as additional messaging features for communicating with clients.

Table 2: US requirements for Veterinary Feed Directives (VFD)

	VFD label requirements
1.	Name, address, telephone number of veterinarian
2.	Name, address, and telephone number of client
3.	Animal location(s)
4.	Species
5.	Number of animals to be fed
6.	Production class
7.	VFD medication or active ingredient
8.	Dosage and feed concentration
9.	Duration
10.	Indication of use
11.	Cautionary statements, as needed
12.	Special instructions
13.	Withdrawal period prior to slaughter
14.	Expiration date if applicable
15.	Combination drug approval, if applicable
16.	Signature of the Veterinarian

Table 3: Canadian requirements for feed prescriptions

	Feed prescription requirements
1.	Species
2.	Production type
3.	Age
4.	Weight
5.	Number of animals
6.	Location of animals to be treated
7.	Type of feed to be medicated (complete/supplement/macro/micro)
8.	Name of feed
9.	Total amount of feed (MT)
10.	Withdrawal
11.	DIN product/brand name/active ingredient
12.	Grams of active ingredient/kg of DIN product
13.	Grams of active ingredient/MT of complete feed
14.	Grams of DIN product/MT of complete feed
15.	Manufacturing instructions
16.	On farm mixing & feeding directions
17.	Warnings
18.	Cautions

There are also several integrations from production and feed management software providers with the GVL platform that offer veterinarians, producers and feed distributors additional capabilities for tracking and managing prescriptions and VFDs.

The GVL SmartEngine technology integrated throughout the platform is a feature that automatically checks health documents for regulatory compliance and accuracy. This technology can result in extreme time and resource savings in reducing the need for follow-up on errors, incomplete information and incorrect data.

Overall, the GVL platform offers the most comprehensive feature set that assists in documenting a valid VCPR (see Figure 2).

Conclusion

Veterinarians provide a critical role in the oversight of antibiotic usage in food producing animals. New regulatory requirements and increased consumer expectations for oversight place further demands on the veterinarian to ensure compliance to federal and state regulations.

The use of web-based certification solutions can enable veterinarians to write and maintain compliant documents and provide secure storage with ease. In addition, web-based solutions can easily facilitate communication of documents to all relevant stakeholders to ensure seamless and compliant implementation.

GVL is committed to continuing to build and maintain web-based solutions for veterinarians to establish a "one-stop-shop" for all regulatory documents and requirements.

Reference

1. AVMA. Principles of Veterinary Medical Ethics of the AVMA, 2016. Available at: https://www.avma.org/KB/Policies/Pages/Principles-of-Veterinary-Medical-Ethics-of-the-AVMA.aspx.Accessed 17 October 2018.

Figure 2: GlobalVetLINK software platform features

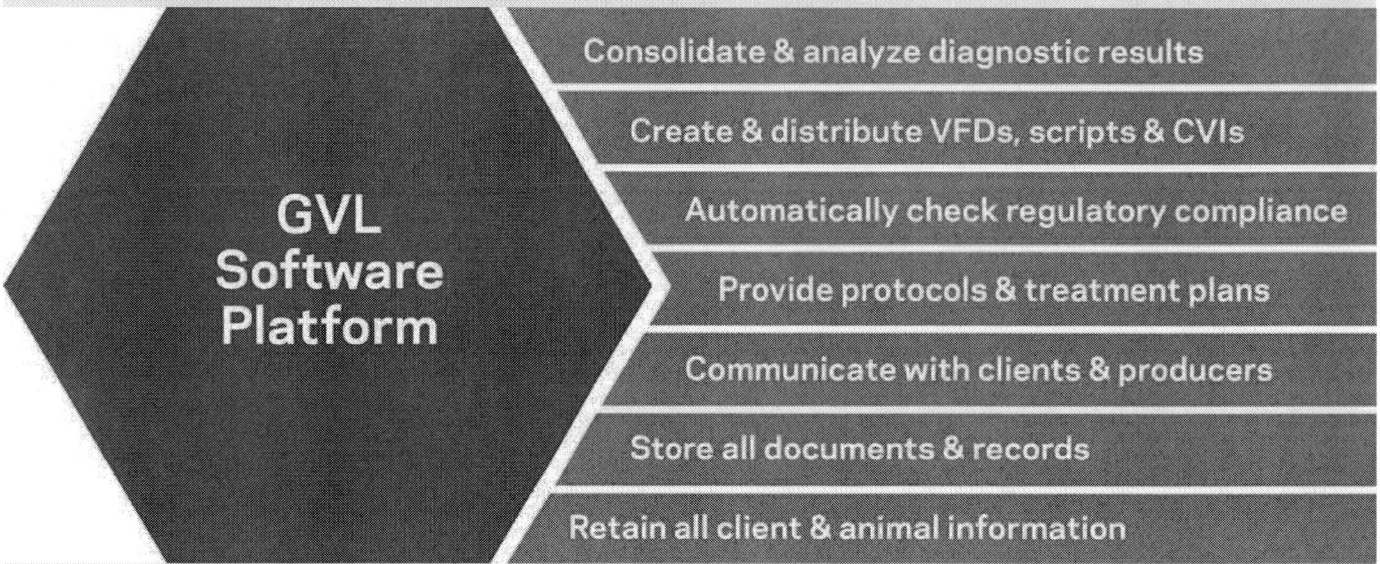

Impact of No Antibiotics Ever/Raised Without Antibiotics production on animal welfare

J. Wishnie[1], MSc, MPH, DVM, DACVPM; R. Singer[2], PhD, DVM; D. Thomson[3], PhD, DVM; A. Beaudoin[4], PhD, DVM, DACVPM; M. Gage[5], BS; L. Porter[4], MPH

[1]Wishnie Consulting, LLC and California Polytechnic University; [2]Mindwalk Consulting Group, LLC and University of Minnesota; [3]Thomson Livestock Consulting, LLC and Kansas State University; [4]Mindwalk Consulting Group, LLC; [5]Gage Group Consulting, LLC

Introduction

The use of antibiotics in food animal production has come under scrutiny in recent years in the United States because of the rise of antibiotic resistance globally and increased consumer interest in food production systems.[1-3] Antibiotics are utilized in food animal production in the US to prevent, control, and treat diseases in animals. Recent US Food and Drug Administration rule changes for feed and water-based antibiotic usage has led to new guidelines for drug labeling and a need for increased veterinary oversight.[4-6] Parallel with these changes have been attempts to monitor antibiotic usage in food animal production and to conduct research aimed at understanding the impact of food animal antibiotic usage on antibiotic-resistant infections in people.[7,8]

Restaurants, grocers and other retailers of eggs, poultry, turkey, beef, pork and dairy products have implemented niche programs centered on eliminating antibiotics throughout the life of the animals raised in their supply chain.[9-12] These "raised without antibiotics" (RWA) programs center on the premise that they will improve food safety and decrease antibiotic resistance in animals and humans, while providing a more wholesome food product.

Several studies have shown that raising chickens without antibiotics can be less efficient than conventional programs, potentially associated with negative impacts on gut health and key productivity measures.[13-15] These effects can lead to economic and environmental strain as RWA programs try to match production output of conventional programs. Demand for RWA products is growing in the US, but there are concerns that RWA practices might negatively impact animal health and welfare.

Little information is available regarding the impact of shifting from a conventional to RWA approach on production systems and animal health and welfare outcomes. As a first step to better understand these impacts, this study seeks to assess producer and veterinarian experiences and perceptions in regard to RWA programs across food animal commodities.

Methods

The survey tool was developed by study co-authors, who include veterinarians with expertise in antibiotic use and stewardship and each of the commodities studied (beef cattle, dairy cattle, swine, turkey, broilers). The tool was reviewed by industry experts in each commodity for clarity, completeness, and usability. The study included questions related to RWA program experience, disease and welfare challenges, and beliefs about RWA impacts on animal agriculture industries, antibiotic resistance, and animal welfare. Questions were minimally modified to reflect each of the five commodity types. The survey included no identifying information and was entirely anonymous. An electronic version of the survey was created in Qualtrics for online administration.

A hyperlink to the online survey was distributed by various professional organizations and commodity groups such as American Association of Avian Pathologists (AAAP), National Chicken Council (NCC), National Turkey Federation (NTF), U.S. Poultry & Egg Association (USPOULTRY), American Association of Bovine Practitioners (AABP), Academy of Veterinary Consultants (AVC), Animal Agriculture Alliance, National Pork Producers Council (NPPC), National Pork Board (NPB), American Association of Swine Veterinarians (AASV), California Department of Food and Agriculture (CDFA), and Pig Improvement Company (PIC). Announcements were also made at various professional and commodity meetings, and in key trade magazines. The survey was open during February 15 to March 23, 2018.

A total of 894 responses were received across the five commodities. Surveys were characterized as incomplete (n = 329) if the respondent left the survey without formally submitting his/her responses. Incomplete surveys were excluded from analysis. Data from international respondents were also excluded from this analysis because of low response numbers and potential for varying regulatory and production systems to influence responses. Data analysis was conducted using Stata 15.1 (College Station, TX). Respondent role (e.g., veterinarian, producer) and RWA experiences were compared with two-sample Wilcoxon rank-sum (Mann-Whitney) test.

Results

The results discussed below refer to swine respondents unless otherwise indicated.

General

Five hundred and sixty-five completed responses were received across five commodities (swine [148], broilers [69], turkey [23], beef cattle [244] and dairy cattle [81]). Ninety-six percent of swine respondents (n = 142) were located within the US.

Six international swine respondents were excluded from the analysis because of small numbers and the potential for varying regulatory and production systems to influence responses. Most swine respondents were producers (n = 70, 47.3%), practicing veterinarians (n = 56, 37.6%) and technical services professionals (n = 6, 5.4%). Producers were self-identified as manager/producer/grower/rancher/owner. Just over half of the respondents currently work with (n = 50, 33.8%) or have previously worked with (n = 30, 20.3%) animals being raised without antibiotics ("RWA respondents"). The remaining respondents (n = 68, 46.0%) had never worked with RWA animals ("Conventional respondents").

Factors contributing to decision to raise animals RWA or conventionally

RWA respondents were asked why they decided to work with swine where the use of antibiotics was not allowed. The most commonly identified reasons were market-driven, including the need to fulfill a client/customer request (n = 52, 69.3%), to increase sale price of animals/product (n = 41, 54.7%) and to gain market entry into a retail program (n = 30, 40.0%).

Conventional respondents' primary reasons for deciding not to work with swine where the use of antibiotics was not allowed were concerns about negative impacts to animal health and welfare (n = 51, 76.1%) and that they were already raising animals in a responsible use program (n = 49, 73.1%) were commonly identified reasons.

Disease challenges and broad health and welfare challenges

Respondents were asked to rank the disease challenges and broad health and welfare challenges that are problematic when raising swine in their respective production systems.

Post weaning hemolytic *E. coli* and *Actinobacillus suis, Haemophilus parasuis* and *Streptococcus suis* were identified as the most problematic swine disease challenges by both RWA respondents and Conventional respondents. Porcine reproductive and respiratory syndrome (PRRS) accounted for 10/12 responses in the "Other" category from Conventional respondents and 3/14 responses from RWA respondents.

Respiratory system disorders were by far the most problematic health and welfare disorder when raising swine in both RWA and Conventional programs. Digestive system disorders were ranked second by both groups; although with less frequency.

Effective management tools

Respondents were asked if there are effective tools (i.e. vaccine, feed/water additive, management) to manage the disease challenges they ranked in the previous questions without the use of antibiotics. The majority of respondents in both RWA and Conventional systems believe that there are effective tools to prevent post-weaning hemolytic *E. coli*, one of the top two most challenging diseases in both production types (38% management; 34% vaccine; 28% feed/water additive). Conversely, the majority of RWA and Conventional respondents did not believe there were effective tools to manage *A. suis, H. parasuis* and *S. suis*, which accounted for the other most challenging set of diseases in both production systems.

Impacts on production and management

Respondents were asked if raising swine without antibiotics would impact key production parameters (e.g. feed efficiency, morbidity, mortality, age at slaughter, weight at slaughter). The majority of both RWA and Conventional respondents believed that feed efficiency (greater than 50% across groups), morbidity and mortality (greater than 70% across groups), and age at slaughter (greater than 60% across groups) was/would be negatively impacted when switching to RWA. The majority of respondents in both RWA and Conventional groups also thought that weight at slaughter would either not change or would be decreased when switching to raising swine without the use of antibiotics. Although the majority of respondents from both groups indicated that all of these factors would be negatively impacted by switching to RWA systems, the proportion of Conventional respondents holding this view was greater than the proportion of RWA respondents across all categories.

Both RWA and Conventional veterinarians were more likely than producers to report that feed efficiency was/would be decreased when switching to raising swine without antibiotics ($P = 0.0005$ and $P = 0.07$, respectively). This was the only response for which there was a statistically significant difference between veterinarians and producers.

Management strategies

Respondent were asked if RWA, or switching to RWA, would impact their management or facilities. Both RWA and Conventional respondents identified a broad range of production practices to be modified when moving from Conventional to RWA production systems. The majority of both respondent groups identified weaning age, biosecurity, space and personnel as changing in an RWA setting (greater than 60% for both groups across all changes). Less than a quarter of respondents from both groups thought that bedding needed to be changed when switching to raising swine without the use of antibiotics.

A greater percentage of RWA veterinarians (n = 23, 69.7%) than producers (n = 10, 41.7%) thought that a change in space was needed when switching to raising swine without the use of antibiotics, although this difference was not statistically significant ($P = 0.08$).

Food safety and animal health and welfare

Respondents were asked their perception of the impacts on food safety and animal health and welfare of raising swine without the use of antibiotics. Almost half of RWA respondents (n = 35, 46.7%) believed that raising swine without antibiotics would slightly worsen or significantly worsen food safety, whereas the majority of Conventional respondents (n = 38, 56.7%) believed that switching to raising swine without antibiotics would slightly or significantly worsen food safety.

Most RWA respondents (n = 57, 76.0%) and Conventional (n = 66, 98.5%) respondents believed that switching to raising swine without antibiotics would slightly or significantly worsen animal health and welfare. However, when stratified by Role, a significantly greater number of Conventional producers (n = 25, 56.8%) compared to RWA producers (n = 6, 23.1%) believed that raising swine without antibiotics slightly or significantly worsened animal health and welfare ($P = 0.004$).

Respondents were then asked to identify how retailers/restaurants/food services think raising swine without antibiotics would impact food safety and animal health and welfare. The majority of RWA (n = 66, 82.7%) and Conventional (n = 54, 80.6%) respondents believed that retailers/restaurants/food services thought that raising swine without the use of antibiotics would improve food safety. The majority of both RWA (n = 56, 74.7%) and Conventional (n = 44, 65.7%) respondents believed that retailers/restaurants/food services thought that raising swine without the use of antibiotics would improve animal health and welfare. Similar results were observed when stratified by Role; no statistically significant differences were observed between RWA and conventional veterinarians or producers.

Impacts on cost and demand

Respondents were asked how/if they thought RWA production impacts the cost of swine production. The majority of RWA (n = 65, 86.7%) and Conventional (n = 66, 98.5%) respondents believe that raising swine without the use of antibiotics will slightly or significantly increase the cost of production. While not significantly different, when looking at Producers stratified by RWA experience, results suggest more Conventional producers (n = 43, 97.7%) than RWA producers (n = 20, 76.9%) believe that there will be an increased cost associated with raising swine without antibiotics ($P = 0.0988$). Veterinarians from both RWA and Conventional systems generally agree that there will be an increased cost of production.

Respondents were asked how/if they thought RWA production impacts demand for pork overall by consumers. The majority of RWA (n = 70, 94.6%) and Conventional (n = 61, 91.0%) respondents believe that raising swine without the use of antibiotics will either have no impact or will slightly increase overall demand for pork by consumers. However, when stratified by Role, more Conventional producers believed that raising swine without the use of antibiotics will have no benefit on consumer demand for pork overall ($P = 0.01$).

Respondent views on RWA label priority and auditing needs

Respondents were asked if there are times that they believe maintaining the RWA label on a product takes priority over herd health and welfare. More RWA veterinarians (n = 25, 75.8%) than Conventional veterinarians (n = 10, 52.6%) believe that there are times that maintaining a raised without antibiotics label has priority over herd health and welfare, although the difference was not statistically significant ($P = 0.08$). When stratified by Role, the majority of RWA producers agree (n = 19,52.8%), while Conventional producers disagree (n = 25. 56.8%), that there are times when maintaining a raised without antibiotics label has priority over herd health and welfare ($P = 0.01$). When comparing RWA respondents to Conventional respondents, Conventional respondents were more likely to believe that RWA increases the cost of swine production ($P = 0.0084$).

Respondents were asked whether more stringent health and welfare auditing is needed when raising swine without the use of antibiotics. A majority of RWA and Conventional respondents believe more auditing is needed for swine raised without antibiotics. When stratified by Role, a greater number of Conventional producers than veterinarians believe that more auditing is needed for swine raised without antibiotics, although the difference was not statistically significant ($P = 0.06$).

Respondent opinions on antibiotic resistance

To complete the survey respondents were asked to what degree they agreed or disagreed with several statements regarding antibiotic use and resistance. The following results are cumulative across species groups (swine, poultry, turkey, beef and dairy). The first statement related to the impact of antibiotic use in the swine industry on problems in human medicine; the majority of respondents agreed or strongly agreed that antibiotic use in swine does not cause problems in human medicine. The second statement related to the impact of antibiotic use in the swine industry on the ability to treat swine infections in the future; respondents were mixed in their opinion of this statement. The third statement stated that antibiotic use in the swine industry would make human infections more difficult to treat; the majority of respondents disagreed or strongly disagreed with the statement. The fourth statement related to the willingness of respondents to use alternatives to antibiotics if they were equally effective; the majority of respondents agreed or strongly agreed with this statement.

Discussion

RWA and Conventional respondents appeared to experience similar disease challenges and health and welfare challenges. While respondents felt that there were effective tools to manage post weaning hemolytic *E. coli* (one of their top diseases challenges) they did not believe that their other top disease challenges, *A. suis*, *H. parasuis* and *S. suis*, could be managed effectively without the use of antibiotics.

The potential ability to manage some of the top disease challenges without the use of antibiotics indicates opportunity for further exploration of effective tools to minimize the use of antibiotics across species and production types. A more detailed understanding of the management requirements for sustainable RWA production are necessary. However, respondents' beliefs that some diseases cannot be managed without the use of antibiotics warrants a full understanding of the impact of these diseases on animal health and welfare and food safety in systems that do use antibiotics.

Respondents generally tended to indicate that raising animals without antibiotics negatively impacts several production parameters, including morbidity, mortality, feed efficiency and age at slaughter. Conventional respondents tended to believe there would be a greater negative impact on production parameters than RWA respondents when raising animals without the use of antibiotics. Differences in RWA and Conventional respondents' beliefs about the degree of impact that raising animals without the use of antibiotics has on production parameters could be attributed to true experience of RWA respondents versus the perception of Conventional respondents of what may happen. However, further study to quantify the true impacts (type of impact and degree of impact) on production parameters when switching to raising animals without the use of antibiotics is needed.

Veterinarians were more likely than producers to report (RWA veterinarians) or believe (Conventional veterinarians) that raising animals without antibiotics has a negative impact on select production parameters. As this study did not assess production data collected by producers and veterinarians, it is not within the scope of this study to define the reasons for the observed differences between veterinarians and producers. Next steps could include assessing production data and further examining the reasons behind the discrepancies between veterinarian and producer understanding of the impacts on production parameters.

Respondents across commodity groups identified several production practices that need to be changed when moving from conventional to RWA production. There is a need to understand the specific characteristics of RWA systems that are able to maintain animal health and welfare without the use of antibiotics. Specifically, identifying aspects of baseline herd health, biosecurity, veterinary care, health programs, nutrition, and other management practices that are associated with successful RWA systems and successful transitions from conventional to RWA, should help reduce the need for antibiotics, even in those systems that continue to use them to prevent, control and treat disease.

References

1. World Health Organization. Antibiotic Resistance Factsheet. 2018. Available at http://www.who.int/news-room/fact-sheets/detail/antibiotic-resistance. Accessed 29 November 2018.

2. Cargill. Cargill Feed for Thought Survey: U.S., Brazilian consumers more likely to buy beef raised without antibiotics. 2016. Available at https://www.cargill.com/news/releases/2016/NA31966656.jsp. Accessed 29 November 2018.

3. Ellison B. et al. Which livestock production claims matter most to consumers? *Agric Hum Values.* 2017;34:819–831.

4. United States Food and Drug Administration. Guidance for Industry #209. 2012. Available at https://www.fda.gov/downloads/AnimalVeterinary/GuidanceComplianceEnforcement/GuidanceforIndustry/UCM216936.pdf. Accessed 29 November 2018.

5. United States Food and Drug Administration. Guidance for Industry #213. 2013. Available at https://www.fda.gov/downloads/AnimalVeterinary/GuidanceComplianceEnforcement/GuidanceforIndustry/UCM299624.pdf. Accessed 29 November 2018.

6. United States Food and Drug Administration. Veterinary Feed Directive. Final Rule. 2015. Available at https://www.gpo.gov/fdsys/pkg/FR-2015-06-03/pdf/2015-13393.pdf. Accessed 29 November 2018.

7. National Antimicrobial Resistance Monitoring Program (NARMS). 2015 Integrated Report. Available at https://www.fda.gov/animalveterinary/safetyhealth/antimicrobialresistance/nationalantimicrobialresistancemonitoringsystem/. Accessed 29 November 2018.

8. United States Food and Drug Administration. 2016 Summary Report on Antimicrobials Sold or Distributed for Use in Food-Producing Animals. 2016. Available at https://www.fda.gov/downloads/ForIndustry/UserFees/AnimalDrugUserFeeActADUFA/UCM588085.pdf. Accessed 29 November 2018.

9. Chick-fil-A. Chick-fil-A's "No Antibiotics Ever" Commitment. Available at https://www.chick-fil-a.com/About/Great-Food/No-Antibiotics-Ever. Accessed 29 November 2018.

10. Panera Bread. Panera Bread's Food Policy Statement. 2014. Available at https://www.panerabread.com/content/dam/panerabread/documents/nutrition/panera-bread-food-policy.pdf. Accessed on 29 November 2018.

11. The Kroger Family of Companies. Animal Welfare Policy. 2018. Available at https://www.thekrogerco.com/wp-content/uploads/2018/07/The-Kroger-Co_AnimalWelfarePolicy_2018-July.pdf. Accessed 29 November 2018.

12. Whole Foods Market. Farm Animal and Meat Standards. Available at https://www.wholefoodsmarket.com/farm-animal-meat-quality-standards. Accessed 29 November 2018.

13. Gaucher et al. Impact of a drug-free program on broiler chicken growth performances, gut health, Clostridium perfringens and Campylobacter jejuni occurrences at the farm level. *Poultry Science.* 2015; 94:1791–180.

14. Karavolias et al. Raised without antibiotics: impact on animal welfare and implications. *Transl. Anim. Sci.* 2018;2:337–348.

15. Smith, J. Experiences with drug-free broiler production. *Poultry Science.* 2011;90:2670–2678.

Field evaluation: vaccinating pigs in PRRSV infected flows with Endovac-Porci®

K. Bretey, DVM
Standard Veterinary Services, Le Mars, Iowa

Introduction

Porcine reproductive and respiratory virus (PRRSV) often infects sow herds and has devastatingly negative effects on the health and survivability of the downstream piglets from birth through market. While the virus can affect piglets during gestation, it is commonly spread to the nursing piglets from the dam and litter-mates during the nursing phase. PRRSV rapidly infects and suppresses the immune system, allowing not only damage from the PRRS virus itself, but also infection and disease from a myriad of opportunistic pathogens in the environment. "Immune-Plus" is an ingredient in Endovac-Porci® that is thought to stimulate the immune systems of pigs that are in populations undergoing significant field virus PRRS infections. This same concept has been successfully used extensively in the cattle feeders' world on groups of cattle undergoing bovine respiratory disease train wrecks (shipping fever). Moreover, this product contains a "core-antigen" bacterin that offers protection against gram-negative bacteria. This protection has potential to reduce the negative effects of gram-negative pathogens that may be secondary to PRRSV infections. The purpose of this study was to evaluate the potential health, production and economic benefits of vaccinating PRRS field-virus positive piglets per label with Endovac-Porci®.

Materials and methods

The pigs utilized in the study were from a sow farm that experienced a PRRSV break (1-4-4 RFLP) in February 2018 and was undergoing a PRRS stabilization effort utilizing Live Virus Inoculation. Regular testing confirmed that three-week-old due-to-wean pigs were PRRSV positive throughout the study. The pigs were transported to offsite nursery-grow-finish facilities. Beginning the week of 5/29/18 piglets from every other week's production received a full 1 cc dose of Endovac-Porci® at processing (2-5 days of age). These same piglets received another full 1 cc dose of Endovac-Porci just prior to weaning at 18-21 days of age. All treated piglets were ear notched for identification as being Endovac-treated. The control groups every other week received no Endovac-Porci®. All piglets in the study received all other normal treatments at the sow farm which included a 200 mg iron injection at processing and 1 cc Draxxin® 25 just prior to weaning. All piglets in the study received 3FLEX® (PCV2/M hyo/PRRS MLV) vaccine at 1 week post weaning as well as Enterisol Ileitis® vaccine at 4-6 weeks post weaning. A regimented water medication schedule based on experience with previous pigs from the flow was equally administered to all treatment and control groups. Therapeutic individual injections were administered and recorded by blinded caretakers based on their daily observations and evaluation. Four groups were included in the study (2 treated groups and 2 controls). Group sizes, performance parameters measured and results are listed in Table 1. To evaluate the performance of pigs vaccinated with Endovac-Porci® several key economic production metrics were measured. In the nursery phase data for mortality and number of antibiotic treatments by injection was collected. In site 2, nursery average daily gain was also measured. In the finishing phase mortality, ADG, F/G and percent full value pigs were measured but unavailable in time to publish.

Results and discussion

See Table 1 for nursery results. Preliminary analysis indicates a performance, health and survivability benefit to using Endovac-Porci® even during the event of a PRRS challenge.

Endovac-Porci®- Registered trademark of Endovac® Animal Health

Table 1: (Nursery phase)

	Site 1		Site 2		Total	
	Control	Endovac®	Control	Endovac®	Control	Endovac®
Head in	1,188	1,368	2,085	1,979	3,273	3,347
Head out	1,134	1,315	2,037	1,944	3,171	3,259
Duration: days	41	41	33	44	36	42
Mortality	54	53	48	35	102	88
% Mortality	4.54%	3.87%	2.30%	1.77%	3.12%	2.63%
ADG	NA	NA	.64	.78	NA	NA
Inj. treatments	240	143	1,165	1,106	1,405	1,249
Inj. per pig out	0.21	0.11	0.57	0.57	0.43	0.38

What swine producers want from their veterinarian: A market research study

R. Siakel; C. Grass
CF Grass Consulting

Introduction

Swine veterinarians are critical to the success of US pork producers. Therefore, it is important to have an in-depth understanding of what producers need from their veterinarian, so they can best help producers succeed. The aim of this project is to provide an in-depth analysis of what US swine producers need from their veterinarian(s). A similar research study[1] was presented to AASV in 1972. Our updated research study includes a comparison to the 1972 findings and expands on the scope and methods of the 1972 research.

Materials and methods

An online survey was used to collect the attitudes and opinions of a representative sample of US swine producers (n = 100, data are still being accepted) during November 2018. A monetary incentive was provided by CF Grass Consulting to all survey participants to reduce non-response bias. Key elements of the online questionnaire were developed and refined by conducting qualitative research phase among swine producers (7), AASV members (2) and industry consultants (3). The heart of the online survey consists of a test of the importance of 30 characteristics of swine veterinarians. In order to reduce order bias, this list of characteristics was randomized for each respondent. The 30 characteristics were identified during the qualitative phase of this research.

Results and discussion

The results are presented in the following discussion. Readers should bear in mind that the 1972 research was limited in scope by sample size (n = 5), number of questions asked (1), and geography (only Illinois producers).

The 1972 assessment was made by asking respondents to "Please complete the following sentence. A swine veterinarian should..." The discussion with the researcher resulted in 39 statements. In the 2018 survey, we also began our assessment by asking respondents to, "Please complete the following sentence. A swine veterinarian should..." We selected the first 39 statements received from our producer respondents so that we had a similar amount of raw qualitative information to analyze from each year. The open-end responses from 1972 and 2018 were then coded and included in one of 4 categories. These categories include the following: have specialized knowledge/expertise, prevent/treat/diagnose disease, teach/train/inform, and provide service (not health specific). This analysis suggests that it is more important that veterinarians have specialized knowledge/expertise in swine production now than in 1972 when a majority of veterinarians practiced multispecies.

More than 85 percent of swine producers rate the overall value provided by their veterinarian as 'good' or 'excellent.' This level of value is statistically the same as the value provided to beef[2] producers and dairy[3] producers by their veterinarians and to swine producers and swine veterinarians by animal health technical service veterinarians.[4] Three of the 30 swine veterinarian characteristics were found to be of highest importance to swine producers. These three most important characteristics include the following: honesty, accurately diagnoses problems, and being trustworthy. Each of these characteristics is extremely important to more than 70% of all swine producers. In addition, more than 65% of swine producers rate their veterinarian as 'excellent' in being honest and trustworthy. However, fewer than 45% rate their veterinarian as 'excellent' in accurately diagnosing problems.

References

1. Author Unknown, What Illinois Producers Want from their Veterinarians, *AASP Newsletter*, 1972.

2. Grass, Corey. What beef producers want from their veterinarian, CF Grass Consulting, 2018.

3. Grass, Corey. What dairy producers want from their veterinarian, CF Grass Consulting, 2018.

4. Grass, Corey. Animal Health Swine Technical Service Value, CF Grass Consulting, 2017.

Figure 1: A swine veterinarian should... (% of total responses)

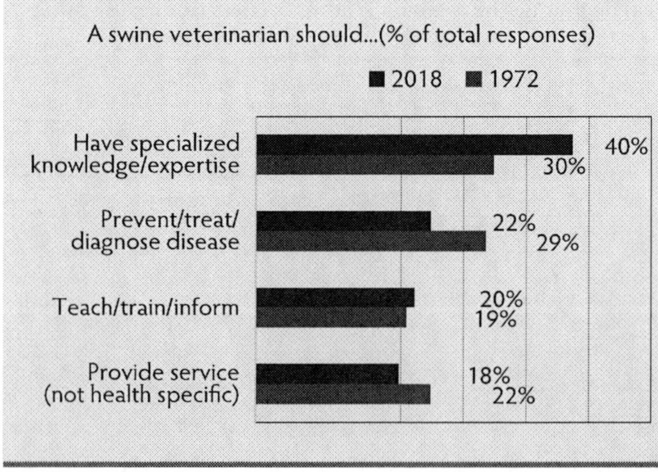

The effects of Diamond V Original XPC™ on weaning performance when fed to sows during gestation and lactation

D. Petry and K. Lehe
Diamond V, Cedar Rapids, Iowa

Introduction

The profitability of a pig enterprise is strongly influenced by sow and litter performance. A sow is usually assessed by the size of the litter, her mothering ability (pre-wean mortality and litter weaning weight), her ability to breed back quickly, and her lifetime performance. The performance of her litter is impacted by the ability to produce quality milk for the piglets. Genetics has improved litter size by over 50% in the past twenty years, while pre-weaning mortality has not changed much.[1,2] Post-weaning growth rate in piglets is correlated to weaning weight. Therefore, finding ways to help the sow produce more quality milk for her piglets is important for sustained profitability and for capturing the full genetic potential of modern sows and their progeny.

Materials and methods

In this study, farrowing and weaning performance of sows fed Diamond V Original XPC was compared to performance of sows fed a control diet. In gestation and lactation, 48 sows were either given a standard corn and soy diet (control) or control diet plus 2kg per metric ton Original XPC. No antibiotics or probiotics were added to the diets. All other nutrients were supplied at normal concentrations according to EU standards. Sows were sourced from a commercial pig herd in Segovia, Spain, and the genetics were a Danbred sow mated to a Pietrain sire. Sows were distributed equally by parity in each group. Sows were weighed and body condition scored at 107 d of gestation and 28 d of lactation. Once sows were farrowed, they were given a chance to nurse 12 ± 1 piglets within 24-48 h. A covariate of number nursed was fit in the model. Data were analyzed using general linear models in SAS v9.0.

Results and discussion

No statistical differences were detected in number born alive, stillborn, mummies, total born, or birth weight. Numerically, sows fed Original XPC had a higher body condition score, less loss of body condition, and less loss in body weight than sows fed the control diet (Table 1). A sow's ability to maintain body condition throughout lactation is correlated with her ability to breed back and to her lifetime performance. If a sow is underconditioned, the probability of breeding back quickly is lower and extra time and feed will be required to restore her condition to conceive.[3]

Sows fed Original XPC had significantly higher piglet weaning weight, higher average daily gain, number weaned and lower pre-weaning mortality (Table 2). These findings add a significant economic benefit for the producer. A 3.79% reduction in pre-wean mortality in a herd that is weaning 12 pigs per litter is an extra 0.45 pigs weaned. For the producer, that equates to $15.75 per litter weaned. Numerically, the coefficient of variation for weaning weight was lower (0.7 %) for piglets nursed by sows fed Original XPC when compared to sows fed the control diet. Sows fed Original XPC tended to have a higher conception rate and

Table 1: Sow body condition and weight pre and post lactation.

Treatment	Body condition			Body weight		
	D107G Scale 1-5	D28L[a] Scale 1-5	Loss in lactation[a]; %	D107G kg	D28L[a] kg	Loss in lactation[a]; %
Control	3.21	2.93	6.01	262.7	222.1	12.78
XPC	3.29	3.00	3.48	261.8	226.8	10.93
SEM	0.08	0.11	1.10	4.51	3.94	1.57
P	NS	NS	NS	NS	NS	NS

[a] = covariate was applied of initial measurement
N = 48 (24 sows/treatment)
SEM = pooled standard error of mean
D107G = 107 days into gestation
D28L = 28 days into lactation (weaning date)
NS = not significant, *P* > 0.05

Table 2: Nursedam performance on piglets in lactation.

Treatment	BW, birth kg	BW, 28 days of age[b] kg	ADG[bc] g/d	CV[b] %	PWM[b] %	Number weaned
Control	1.30	7.61	243.0	17.35	5.54	11.30
XPC	1.27	8.20	259.0	16.66	1.75	11.78
SEM	0.01	0.07	2.77	1.14	-	0.15
P	NS	<0.0001	<0.0001	NS	0.016	0.03

[b] = random effect of room
[c] = covariate of birth weight
N = 48 (24 litters/treatment) – nurse sows given a chance to nurse 12 ± 1
SEM = pooled standard error of mean
BW = body weight
ADG = average daily gain – grams per day
ADFI = average daily feed intake
PWM = pre-wean mortality
NS = not significant, $P > 0.05$

a shorter wean-to-1st-service interval than sows fed the control diet, however these trends were not found to be statistically significant (Table 3). This could potentially be an important observation when considering the economics of a non-productive day. A sow that has 2.4 litters per year contributes $3.20 more value to the operation annually.[4] Similarly, if conception rate is increased the economic value can be substantial.

The results of this trial are similar to results seen in the previous studies (Table 4). Sows fed Original XPC produce heavier pigs and wean more pigs than sows fed a control diet. On average across numerous trials, piglets nursed by sows fed Original XPC weaned 0.30 more pigs and have 3.40 kg more litter weaning weight than sows fed a control diet. Over a 4.5 litter lifetime, this equates to 1.35 more pigs weaned in that lifetime, giving the producer an extra value of $44.41, assuming a weaned pig value of $35. Also, pigs that are heavier at weaning tend to start better in the nursery. It has been cited that every 0.45 kg advantage at weaning translates to 1.36 – 1.81 kg advantage at market.[5] This represents an additional value of $1.61 – 2.15 per pig marketed, assuming $0.75 per pound carcass value.

Conclusions

Sows fed Original XPC during gestation and lactation showed improvements in weaning weight, number weaned and pre-weaning mortality. This study has been confirmed by numerous other studies that included Original XPC in diets fed in gestation and lactation. The return on investment using Original XPC in gestation and lactation diets is more than 5-fold. Though not statistically significant in this trial, sows fed Original XPC appeared to hold body condition score during lactation better than sows fed a control diet. Further investigation on a larger scale would provide greater insights regarding feeding Original XPC to sows to help maintain body condition during lactation, reduced wean-to-1st-service interval, and improving conception rate. Larger field studies are in process to further validate the findings in this study.

Table 3: Sow breed back performance.

Treatment	ADFI, kg	WSI, days	Conception rate, %
Control	7.72	6.36	94.8
XPC	7.81	4.21	99.5
SEM	0.16	1.61	5.30
P	NS	0.20	NS

N = 48 (24 litters/treatment)
SEM = pooled standard error of mean
ADFI = average daily feed intake
WSI = weaning to service interval
Conception Rate = confirmation by ultrasound 28/35d of pregnant
NS = not significant, $P > 0.05$

Table 4: Summary of previous Diamond V trials.

Trial	1	2	3	4	5	6	Average
Pigs weaned	0.10	0.11	1.05	0.09	0.24	0.20	0.30
Litter weaning weight	4.46	0.90	6.00	5.71	-0.38	3.60	3.40

Notes: weighted average across 6 controlled studies representing 1355 sows.

References

1.https://www.ksre.k-state.edu/news/stories/2017/05/swine-research-advances.html Accessed 2 November 2018.

2. PigChamp and Seges 2015.

3. https://www.pig333.com/articles/weaning-to-service-interval-what-affects-it-and-how-to-control-it_7507/ Accessed 2 November 2018.

4. http://porkgateway.org/resource/non-productive-days-their-significance-and-control/ Accessed 2 November 2018.

5. https://www.nationalhogfarmer.com/nutrition/maximum-profit-starts-weaning Accessed 9 November 2018.

Smart systems in pig production

Thomas Stein, DVM, MS, PhD
Maximus Systems & Maximum Ag Technologies; Saint-Bruno, Quebec, Canada; Jenison, Michigan

This paper is written in two parts. Part I describes what modern, smart control systems are, using the Maximus modular computer control system as an example. Part II describes some of the most current and emerging smart technologies for pork production.

As pork producers look to the future, they have important decisions to make now about how to get ready for these new technologies. Which barn control platform to adopt will be one of the critical success factors for future-proofing your production system. Producers will want to have a single system that can connect to or integrate with these new technologies. My thesis is that it will be the smart barn control system – "brain in the barn" – that becomes the single management and control platform for integrating all these new in-the-barn technologies.

Why? When you look in Part II at even this short list of nine emerging technologies, ask yourself questions like: How will they be managed? How many networks will pork producers install in their barns? How many different software applications will pork producers need to work with? Will pork producers want to log-in to ten or more separate networks and apps to manage each different technology? Will pork producer employees need or want to learn ten or more separate software systems? Who will service and maintain multiple networks within a pig barn? What about the cost of installing and maintaining ten separate networks in each site or barn or room? And remember, these nine technologies are the beginning, not the end.

Which management software platform to adopt is another critical success factor for the same reasons. Producers will want a single software platform that seamlessly works with a smart control system because otherwise it will be too hard to manage. Brains in the barn will communicate in real-time and in both directions with brains in the office to spearhead the next evolution in pork production management.

1: Smart control systems in pig production

1.1 Smart control systems – Overview. Modern, intelligent barn controllers fully integrate the function of all components of climate control into a single unit, eliminating the need to program and co-ordinate several smaller controls. With this ability, tighter temperature bands can be maintained, with some controls able to regulate the temperature within a band of ± 1°C. The control should show all ventilation stages and allow you to determine which fans start at which temperature, what the inlet/curtain openings should be for each stage, the ideal static pressure for each stage, etc.

The advantage of having inlet and fan function within the same unit is in having three or four types of inlets available to the same controller, which can be transitioned through and controlled by static pressure as different fans are activated. Being able to smoothly transition from one type of inlet to another as the outdoor temperature changes throughout the day can save a significant amount of heating cost. Individual heaters can be monitored and controlled in combination with inlet control. Some controls are capable of handling up to 16 zones for great uniformity along the length and width of the barn.

Obviously, heating and ventilation savings are a huge factor in payback on these computers, but some of the tools that allow users to track, analyze and manage those aspects are key. Using your controller to collect heater run-time data (to find faulty heaters or air leaks), temperature and humidity logs by zone (including outside), feed and water consumption, mortality, and many other statistics can make all the difference in diagnosing and fixing problems that may not be immediately apparent by walking the barn.

The data-tracking and alarm systems give equipment and service providers a direct line to your controller from a remote location and provide you with premium technical service, even when they can't be there immediately. Software links you to your sites and barns 24 hours a day. You can check on the situation in the barn, no matter where you are in the world. Alarms can be sent to your PC, tablet computer or smartphone. And, if necessary, you can adjust the settings in the control computers remotely to improve the processes in the barn. The result is a complete overview of your company and insight into production costs and technical results.

1.2 Essential characteristics of a modern, intelligent barn controller (Figure 1). Many brands of controls are 'tightly bound' into proprietary systems tied into the company's product line. You don't want that. You want to isolate and separate the control from any particular proprietary brand, thus breaking the "vendor lock-in" dilemma. In other words, a smart control should be able to work with any brand: any fan, any heater, any curtain, any bin load cell, any hog sorter, any stand-by generator, any inlet, etc. Here is a checklist of 28 essential characteristics that should be found in a smart barn controller system:

1. Single product that works for any species (pigs, poultry, dairy/beef, others)
2. Single product that can be configured to run any barn/barns
3. Independent: control any brand of fans, curtains, heaters, inlets, etc.
4. Online and networked (status always known)
5. Accessible directly via Internet
6. Robust and secure data transfer to private cloud database
7. Immune to power failure
8. Robust surge protection

Figure 1:

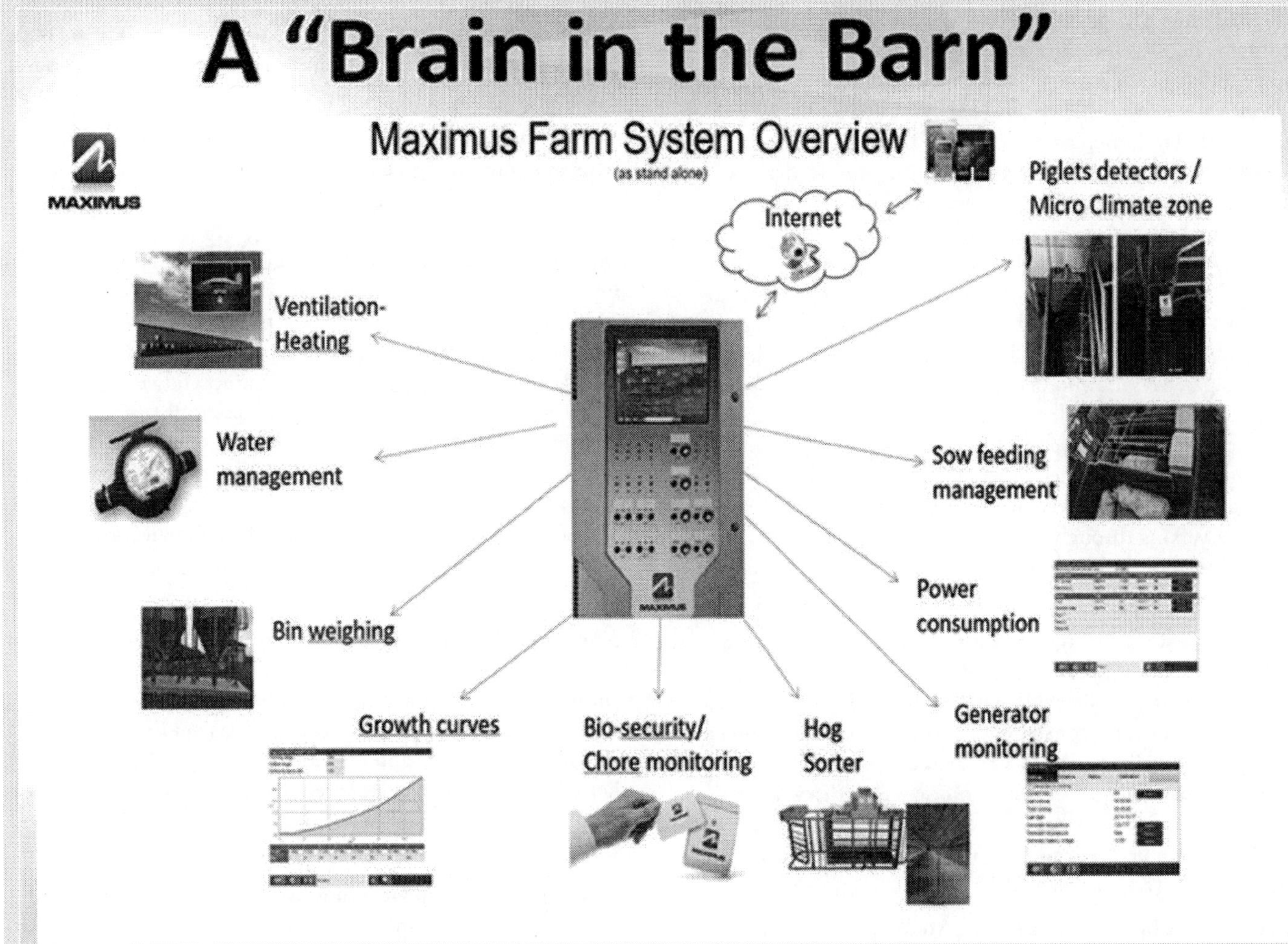

9. USB backup of all control settings for restoring the controller if power outage
10. Remote (cloud database) backup of all control settings
11. Module diagnostics (diagnose problems with a specific sensor or output)
12. Be able to create user-friendly names for all inputs and outputs
13. Location tracking (latitude and longitude)
14. Multiple languages
15. User authentication, authorization and permission system
16. Create, delete, and manage users from phone, tablet, laptop, desktop
17. Robust security to prevent unauthorized access
18. Message users by SMS text, email or phone.
19. Multiple phone and email addresses for each user/contact
20. User-specific settings for messaging availability (e.g. 24.7, 6am-8pm, etc.)
21. User-specific settings for what messages should be sent to whom
22. Be able to quickly see current conditions in the site/barn
23. Robust local data storage (at least 2 years)
24. Maintain history of interaction with controller (who set or changed what)
25. Frequent software updates (every 6 to 8 weeks)
26. User-controlled ability to manage software updates to the controller
27. Extensible (ability to add functionality in the same box)
28. Robust accessibility to settings, programs, functions (phone, tablet, laptop, desktop)

2: Smart technology – here now or coming Soon

2.1 Real-time activity and behavior monitoring to detect problems in pigs or poultry. For pigs, changes in their daily eating and drinking patterns can be used to detect sickness and provide an early warning system (Maselyne 2016). In her PhD thesis work, Dr. Maselyne put UHF RFID tags on finishing pigs along with RFID readers (antennae) on standard feeders and waterers to measure each pig's eating and drinking behavior. In addition, trained observers recorded each pig's behavior. She did five replications involving 1,000 finishers in total; every replication included four groups, each with between 35 to 59 pigs. She developed a model which was able to accurately detect changes and send out warnings about pigs based on their individual drinking and eating behavior. Among pigs detected with health problems, frequent issues included lameness, reduced growth, infections, respiratory problems or fever. Her most consequential finding was that compared with trained observers (production supervisors or barn staff), the data could detect sick pigs at least 24 hours before a human walking through the barn would be able to. There is more work to be done to make a system like this practical in finishing barns but it's clear that this is where we are headed.

In poultry, real-time image processing can detect problems in broiler houses by analyzing the birds' behavior and activity. Three or four cameras are mounted on the ceiling and give pictures of the distribution of birds in a house over time. An algorithm compares actual distribution with a predicted value for that particular time of the day. Events in a broiler house – electricity failure, blocked feeding lines, water supply problems, and climate control problems – can be detected using top-view image analysis with 95% accuracy. (Kashiha et al 2013)

2.2 Real-time behavior monitoring to detect estrus in sows PigWatch (Ro-main; Quebec, Canada) technology uses infrared sensors mounted above a gestation stall to monitor and track sow activity and behavior. Using a machine learning algorithm, it determines not only if and when a sow is in estrus but also the optimal time for insemination. A recent study (Klopfenstein et al, 2016) showed that the PigWatch system was effective at both identifying sows in estrus and providing optimal times for insemination. Unfortunately, there was no control group for comparison but sows identified by PigWatch had good reproductive performance: an average of 1.2 inseminations, a 95% conception rate, and 15.3 total pigs born.

2.3 Real-time audio signal processing to detect respiratory problems in pigs. The Pig Cough Monitor (Soundtalks NV, Belgium) uses a microphone hanging over pens of finishing pigs and a machine learning algorithm to identify and make sense of pig coughing events. It is the most robustly and formally studied audio processing sensor so far. Early research showed that it is a valid tool for automated cough detection in pig barns (Guarino et al, 2008). Over the past 10 years, farm research trials have been done throughout Europe. The trials demonstrated the tool's potential to deliver early warnings (up to two weeks earlier) compared to a situation where producers or veterinarians rely on their own routine observations without the monitor. Automated cough detection and monitoring can detect different causes of respiratory problems as well as technical problems with ventilation (Berckmans et al, 2015).

2.4 Real-time audio signal processing to prevent crushing deaths in neonatal piglets. SwineTech (Cedar Rapids, Iowa) is a company aiming to prevent piglet deaths due to crushing by sows within the first four days after farrowing. The ECHO hardware device listens for piglet sounds and squeals. When it identifies the unique sound of a piglet being crushed, it sends a vibration to a wearable patch attached to the sow. The vibration causes the sow to immediately stand up which prevents a baby pig from being crushed. Several large pork producers in the Midwest are doing field trials (2018); early results show a statistically significant reduction in crushing deaths (70% lower v controls) and an increase in Day 4 livability by 0.5 to 1.0 pigs per litter. Researchers at Kansas State (Mumm et al, 2018) showed that the ECHO system was more effective than a conventional human hand-slap at causing sows to stand up while causing a sow no more stress than the conventional method.

2.5 Voice-enabled data entry and workflow. AgVoice (TekWear, LLC; Atlanta, Georgia) is a hands-free data entry and workflow management service that helps agriculture professionals capture insights while on the move. Users include agronomists, plant scientists, veterinarians and machinery repair specialists, among others. The hands-free, voice-interactive technology that the platform offers has the potential to fundamentally shift the way food and agriculture professionals capture usable insights and data.

2.6 Real-time image processing to weigh pigs. Researchers at the French Pig & Pork Institute (at the experiment station in Romillé, France) have developed a 3D imaging prototype for weighing individual pigs in partnership with Advansee, a company specialized in image analysis and embedded electronics (Rousseliere and Marcon, 2018). For weight estimates, the objective is for a margin of error between 2% and 4%. The EyeGrow system (Fancom BV, the Netherlands) has recently been introduced (Klement et al, 2018). It uses a 3D infrared camera and machine learning to weigh pens of finishing pigs (<u>not</u> individual pig weights). It produces average daily weight gain for the pen along with variation in weights measured by standard deviation. They estimate the cost at $1,400 USD per camera with one camera needed for each pen of 20 finishing pigs.

2.7 Tracking movement of on-farm workers using Bluetooth to improve biosecurity. A new system has been tested (Piniero 2017) using small devices, called beacons, which are worn by farm staff throughout their work shift. A beacon is a small Bluetooth low energy radio transmitter and can be used as an indoor positioning system for farm staff inside the barn. Three types of staff movements are collected: (1) Correct. From higher health barns to lower (i.e., from gestation to lactation barn); (2) Wrong.

Low risk. From lower health barns to higher (i.e., from lactation to gestation); (3) Wrong. High risk. From lower health barns to higher (i.e., from finisher barns to gilts).

2.8 Geofence-based feed traceability using real-time GPS location data. FeedTrackur™ (DP Techlink, LLC; Des Moines, Iowa) is a hardware-software solution for automated and real-time traceability of animal feed (Stein, 2017). A cloud-based database contains continuously updated feed orders, load-out information, driver information, feed truck information, and real-time GPS data which enables location tracking. Cloud-based mapping software contains GPS-coordinates for all physical locations including sites, barns, and feed bins. Each is demarcated with geofences based on specific GPS coordinates. Feed truck drivers interact with a tablet computer that sits inside the cab. Whenever a feed truck crosses a geofence, logic is triggered as to whether it is the correct location for the feed delivery and, if it is, the delivery location (the feed bin itself) is GPS-stamped for traceability. If it's not the correct location, the software alerts the driver and prevents mistakes.

2.9 Smart digital monitoring of medical treatments using RFID, Bluetooth, and Wi-Fi. Automed (Canberra, Australia) and V-ETIC (Henke-Sass Wolf GmbH; Tuttlingen, Germany) have each recently introduced wifi and Bluetooth-enabled smart dosing devices that provide automatic medication delivery, data recording, and weight-based and fixed-amount dosing. Both systems record treatments, manage medication inventory, and integrate with scales, RIFD readers and with livestock management software. Both have apps which allow users to control, configure and manage the dosing devices. Because they track medication inventory and real-time usage, they enable compliance and traceability between drug inventory and use in animals.

References

References available on request.

Influenza A vaccination using the SEQUIVITY™ Technology

Channing Sebo[1], DVM; Pravina Kitikoon[1], DVM, PhD; Tara Donovan[2], DVM; Kimberly Crawford[1], DVM, MS; Mark Mogler[1], PhD; Chandra Morgan[1], BS; Heather Dempsey[1], BS; Susan Knetter[1], PhD; Brad Thacker[1], DVM, PhD; Erin Strait[1], DVM, PhD
[1]Merck Animal Health, De Soto, Kansas; [2]Hanor, Spring Green, Wisconsin

Introduction

Influenza A virus in swine (IAV-S) can be a costly pathogen for swine producers. IAV-S can cause high morbidity, increased mortality, reduced growth rates, decreased feed efficiency and increased medication costs.[1] Vaccination and prevention of IAV-S is challenging due to the vast diversity of the hemagglutinin (HA) proteins present among IAV-S circulating in the swine population. IAV-S undergoes both antigenic shift and drift leading to this extensive diversity.[1] Vaccination success is heavily dependent on sequence analysis that enables strain matching of the HA protein to the field strains that are causing disease.[2,3] To assist customers with sequence management and vaccine design, Merck Animal Health has developed a data management tool, the SEQUIVITY™ Dashboard.[4] Pairing the Dashboard with the SEQUIVITY™ alphavirus replicon technology to produce RNA particle (RP) vaccines enables veterinarians to better protect herds against IAV-S. This dashboard not only analyzes sequences using homology, dendrograms and key site analysis, but also has a geospatial mapping feature to monitor the spread of IAV-S strains throughout a production system, within a region as well as across the U.S.

Using the SEQUIVITY™ Technology in the field not only requires critical analysis for vaccine design, but also for evaluating vaccine performance post-implementation. An example of a recent IAV-S RP vaccination field experience with regard to antibody levels in sows following vaccine implementation as well as the apparent elimination of circulating of IAV-S in pigs at weaning is presented. Lastly, a comparison of detection sensitivity and sequencing IAV-S success in sow farms of nasal swabs versus udder wipes was performed.

SEQUIVITY™ Dashboard

One of the biggest challenges with IAV-S vaccination is monitoring and selection of strains for vaccine design. Merck Animal Health has developed a data analysis and management tool, the SEQUIVITY™ Dashboard, to analyze and assist customers in the design of custom, prescription RP vaccines. Sequences are analyzed in a variety of ways including homology comparisons at the amino acid level, dendrogram trees, putative amino acids (key sites), and geospatial analysis.

The most common strain detection strategy employed by veterinarians is the "reactive approach" which occurs when a new IAV-S strain enters a farm causing an increase in clinical disease. With this approach, the veterinarian designs a custom vaccine based on the virus(es) circulating within the farm. Vaccination is then typically applied only after the infection is endemic. Using the SEQUIVITY™ Dashboard, veterinarians cannot only design vaccines based on current disease challenges, they can also use the geospatial analysis to include at-risk strains into the vaccine through confidential sharing of data. This allows veterinarians to move from a reactive approach to a proactive, preventative vaccination program with the addition of area at-risk strains. Another proactive approach is ongoing monitoring of incoming gilts onto sow farms as this is one of the primary methods of IAV-S entry into sow farms. Gilt monitoring will enable the veterinarian to protect the sow farm sooner and reduce the overall clinical impact a new IAV-S introduction brings. The Dashboard can also be a valuable tool for detecting emerging strains including new IAV-S viruses or new pathogens like PCV3, tracking their diversity and understanding how they are moving through the US swine population. The SEQUIVITY™ Dashboard has over 14,000 sequences within its database for many different pathogens including IAV-S, rotavirus, PRRSV, parainfluenza, PCV2, PCV3 and PEDV. IAV-S is the most common pathogen in the database with over 9,000 sequences since 2008.

To validate the SEQUIVITY™ Dashboard IAV-S database, a comparison was performed versus the Iowa State University ISU FLUture program[5] and the Influenza Research Database.[6] This important step was taken to validate the SEQUIVITY™ Dashboard database as a true reflection of the IAV-S diversity present throughout the U.S. swine population to gain confidence in its use for analysis and vaccination selection decisions. A comparison of these three sources is summarized in Table 1. All three sources of sequence data are in agreement with each other, thus increasing the validity of the geospatial mapping and analysis that the SEQUIVITY™ Dashboard provides to our customers. Merck Animal Health continues to validate this congruence on a regular basis.

Sow herd mass vaccination field experience

Using the SEQUIVITY™ Technology, a polyvalent prescription IAV-S RP vaccine was prepared using four sequences consisting of one H1N2, delta 1a virus, and three H3N2 variants. The H1N2 virus was known to be endemic in the farm and had been included in an autogenous vaccine that was previously used on the farm. The H3N2 variants were strains detected in downstream flows but not at the sow farm.

The prescription IAV-S RP vaccine was administered twice, 4 weeks apart to all breeding age animals on the farm. Four weeks

Table 1: Prevalence of IAV-S subtypes by database (All data retrieved 01 Oct. 2018)

IAV-S subtypes	SEQUIVITY™ Dashboard	ISU FLUture	Influenza Research Database
H1Alpha	6.1%	9.4%	6.3%
H1 Beta	0.8%	0.2%	1.3%
H1 Pandemic	5.2%	4.4%	4.8%
H1 Gamma	30.6%	35.2%	36.0%
H1 Delta 1a	14.6%	13.0%	11.5%
H1 Delta 1b	3.7%	3.8%	3.3%
H1 Delta 2	8.6%	8.6%	8.7%
H3 Cluster 4	10.8%	5.5%	8.2%
H3 Human-like	19.6%	19.1%	19.9%

after the second mass vaccination, the farm began vaccinating all animals 3 weeks prior to farrowing. In order to evaluate post-vaccination serological responses, serum was obtained from 2-5 day-old pigs, at three time points as an indirect measure of sow serostatus. Serum was collected at the following three time points: 1) before RP vaccination implementation (Group A; 20 litters; 4 pigs per litter; 80 pigs); 2) after two mass sow herd vaccinations (Group B; 20 litters, 2 pigs per litter; 40 pigs); and 3) at the steady state, 4 months after the second mass vaccination (post two mass vaccinations, plus the pre-farrow vaccination; Group C; 18 litters; 6-9 pigs per litter; 139 pigs). Serum was tested by hemagglutination inhibition (HI) using an H1N2 homologous to the strain present in the vaccine and the farm. The highest dilution tested was 1:640. To detect IAV-S infection in farrowing, udder wipes were collected just prior to weaning (one per sow/litter) at the three time points listed above and tested for IAV-S by PCR. In addition, nasal swabs were obtained from Groups A and C. For Group A, 20 litters (3-4 pigs per litter) were swabbed and the swabs were tested individually. For Group C, 18 litters (8 pigs per litter) were swabbed and the samples were tested in pools of two (n = 66) or individually (n = 8). Data was analyzed by ANOVA and Fishers exact test using Statistix™ 10 (Analytic Software, Tallahassee, FL).

Following implementation of the RP vaccinations, clinical signs subsided in farrowing and decreased in the nursery as reported by the farm staff and veterinarian. Laboratory results are presented in Table 2. Antibody titers both post second mass vaccination (Group B) and pre-farrow/steady state vaccination (Group C) were significantly higher than before vaccination (Group A). Titers after the third vaccination (pre-farrow) (Group C) were numerically higher, but statistically not different to the titers induced by the two mass vaccinations (Group B). The herd had pre-existing titers to the H1N2 strain circulating on the farm and had been vaccinated with an autogenous vaccine but the titers were relatively low prior to the first RP vaccination (Group A). The boostering of these low titers demonstrates the value of using a prescription RP in order to increase the level of herd immunity. The percentage of positive udder wipes declined at each sampling time point and all were negative in Group C. Testing nasal swabs before vaccination (Group A) the percentage of positive nasal swabs from weaned pigs was 36.7% for individual pigs and 70% on a litter basis. All nasal swabs were negative in Group C. In

Table 2: Results from sow herd mass vaccination

Parameter	Group A	Group B	Group C
Number of RP vaccinations prior to sampling	0	2	3
Sampling time point	Before RP vaccination	4 weeks post 2nd mass vaccination	4 months post 2nd mass vaccination; 3 weeks after pre-farrow vaccination
HI titer to H1N2 (geometric mean)	75.9[a]	187.0[b]	276.9 [b]
Udder wipes (no. and % PCR positive litters)	7/19[a] (36.8%)	2/16[a,b] (12.5%)	0/18[b] (0.0%)
Nasal swabs (no. and % PCR positive by pig or pool)	29/79[a] (36.7%) (By pig)	Not done	0/74[b] (0.0%) (By pool or pig)
Nasal swabs (no. and % PCR positive by litter)	14/20[a] (70.0%)	Not done	0/18[b] (0.0%)

[a,b] Values with different superscripts within a row were significantly different ($P < 0.05$).

Group A, the number of litters with at least one positive swab (14/20) was higher than the number of udder wipe positive litters (7/19). However, four of the nasal swab positive litters had only one positive swab with high Cts (≥ 36.8) that would most likely yield a negative result if the swabs were pooled by litter.

Sow herd surveillance techniques

Currently, sow herd vaccination is the primary application of the SEQUIVITY™ IAV-S vaccine. One of the challenges in designing a sow farm vaccine is the ability to obtain sequences from the farm of interest. Traditionally sow herd sampling is performed through individual piglet nasal swabs with low success rates for obtaining enough virus to perform sequencing. Another issue with individual piglet nasal swab sampling is the cost of the diagnostics, as typically 30 nasal swabs from 30 different piglets are collected per farm for PCR testing. With both of these challenges, the need for better piglet population sampling methods was needed. The cost and low success rates lead us to further assess different sampling techniques to obtain a population type sample from a sow herd. Mantilla et. al found that sampling the udder skin of lactating sows was able to yield viable IAV-S and was sensitive for PCR detection.[7]

Because the SEQUIVITY™ Technology requires a sequence in order to produce vaccines for IAV-S, we sought to evaluate the success of udder wipe sampling in obtaining a sequence. For validation, a sow herd that was actively circulating virus in farrowing was chosen for sampling. Fifteen litters were selected for sampling and one udder wipe per litter was collected. Two piglets were randomly selected per litter and were individually nasal swabbed (total of 30 piglets).

The results are shown in Table 3. For the udder wipe samples, 26.6% of the litters (4/15) were positive. For nasal swabs, 23.3% of the swabs were positive (7/30) and 26.6% of the litters (4/15) were positive. Positive Cts for udder wipes ranged from 24.6-37.8 (average of 31.0) and for nasal swabs from 19.3-34.5 (average of 30.7). Two litters that had piglets with positive nasal swabs, but negative udder wipes had swab Cts ranging from 34-35.3. Both litters that had positive udder wipes, but negative nasal swabs had wipe Cts of 36.4 and 37.8. These high Cts would have a low likelihood of sequencing success by either method. Two udder wipe samples, with Cts of 24.6 and 25.2 were selected for sequencing and each attempt was successful.

Summary

The SEQUIVITY™ Dashboard and SEQUIVITY™ prescription IAV-S vaccines are both essential tools in the management and implementation of a successful IAV-S vaccination program. With the abundant diversity of IAV-S present throughout the United States swine population and lack of cross protection among strains, the need for custom prescription RP vaccines is evident. SEQUIVITY™ Technology allows veterinarians to move from a reactive, treatment-type approach to a proactive, preventative vaccination program.

The results from vaccination in a field setting demonstrated the value of sow herd IAV-S RP vaccination through elimination of evidence of IAV-S circulation in farrowing, increase of antibody titers in piglets and reduction of clinical signs throughout the nursery.

Both udder wipes and nasal swabs are adequate sampling methods for IAV-S detection and sequencing in sow herds. For sow herd population sampling, udder wipes are a valid and cost-conscious sampling technique for collection of crucial sequence information. Udder wipe sampling has since been repeated on other farms with similar sequencing success. Future trials will include an attempt to perform virus isolation from these low Ct udder wipe samples for use as HI antigen to monitor herd seroprofiles.

References

1.Vincent AL et al. Swine Influenza Viruses: A North American Perspective. *Adv V Res.* 2008: 127-154.

2.Vander Veen RL et al. Alphavirus replicon vaccines. *Ani Health Res Rev.* 2012: 1-9.

3.Tamura S et al. Mechanisms of Broad Cross-Protection Provided by Influenza Virus Infection and Their Application to Vaccines. *Jpn J Infect Dis.* 2005: 195-207.

4.Kitikoon P et al. Efficacy of swine influenza virus replicon particle vaccination including cellular and humoral responses after challenge. AASV 2018.

5.Zeller, M. A., Anderson, T. K., Walia, R. W., Vincent, A. L., & Gauger, P. C. (2018). ISU FLUture: a veterinary diagnostic laboratory web-based platform to monitor the temporal genetic patterns of Influenza A virus in swine. *BMC bioinformatics, 19*(1), 397. (data retrieved 01 Oct. 2018).

6.Zhang Y, et al. (2017). Data was obtained from the NIAID Influenza Research Database (IRD). http://www.fludb.org (data retrieved 01 Oct 2018).

7.Mantilla J et al. Novel approaches to monitor influenza in swine breeding herds. Leman 2017.

Table 3: IAV-S detection by litter from udder wipes compared to nasal swabs

		Nasal swabs (two per litter)			
		Both positive	**1 of 2 positive**	**At least 1 positive**	**Both negative**
Udder wipes (one per litter)	Positive	1	1	2	2
	Negative	2	0	2	9

Porcine circovirus type 3: Field experiences with a potentially emerging pathogen

Brad Thacker[1],DVM, PhD; Greg Armbruster[1],DVM, MBA; Emily Byers[2],DVM; Justin Cagle[1],DVM; Kimberly Crawford[1],DVM, MS; Jack Creel[1],DVM; Jamie Lehman[1],DVM, MS; Mark Mogler[1],PhD; Frank Roerink[1], MS; Channing Sebo[1],DVM; Erin Strait[1],DVM, PhD; Huiling Wei[1],DVM, PhD
[1]Merck Animal Health, De Soto, Kansas; [2]Prestage Farms, Inc., Clinton, North Carolina

Introduction

Porcine circovirus type 3 (PCV3) was first identified in the US in 2015.[1,2] Recent case studies have reported an association of PCV3 with several conditions including porcine dermatitis and nephropathy syndrome (PDNS), multi-systemic inflammation, myocarditis, congenital tremors, and reproductive failure including abortions, increased mummified fetuses and non-viable piglets, reduced litter size and delayed return to estrus.[1-5]

Merck Animal Health (MAH) recently acquired a novel vaccine platform, SEQUIVITY™ Technology RNA-Particle (RP) that utilizes an alphavirus replicon vector system to produce vaccines that contain a pathogen's gene of interest that encodes for a protective antigen. Following injection of the RP vaccine, the protective antigen is produced in host cells, which in turn elicits an immune response against the pathogen. This vaccine platform has been shown to be effective in protecting against an experimental PCV2d challenge as well as several other swine pathogens.[6,7]

In 2018, customers inquired about the possibility of producing a vaccine for PCV3. Per USDA regulations, the SEQUIVITY Technology can be used to produce vaccines under the Veterinary Prescription guidelines. With SEQUIVITY Technology and the Veterinary Prescription option, RP vaccines can be developed and manufactured under a much shorter time line than traditional fully or conditionally licensed vaccines. Importantly with PCV3, the virus can't yet be propagated in cell culture so a traditional, inactivated whole virus, autogenous vaccine approach is not feasible. In this paper, we will discuss our PCV3 related activities and experiences including the clinical pictures described by our customers, epidemiology, sampling strategies, serotesting, sequence analysis and vaccine development and use.

Clinical picture

The most common clinical problem reported by customers is reproductive failure. Lower litter size and increases in mummified or stillborn fetuses are the most common clinical signs, with abortion less common. These clinical findings coupled with detection of PCV3 in fetal tissues by PCR, and in some cases next generation sequencing, immunohistochemistry or in situ hybridization, support the diagnosis of PCV3 induced reproductive failure. Vaccine use considerations are driven by the perceived severity of the problem based on breeding herd record analysis. MAH is currently working with several operations where PCV3 has been detected in fetal tissues in order to better characterize the clinical outcomes of PCV3-induced reproductive failure. Reports of PCV3-induced disease in growing pigs are less common and have not warranted the use of vaccine in growing pigs to-date.

Epidemiology

As stated above, the first reports of PCV3 in the US were in 2015. Recently, we tested serum samples from post-marketing studies dating back to 2012. Our main objective was to identify sample sets of known PCV3 status based on PCR to use as reference sera for developing serological assays. Several of the studies contained samples that were collected periodically from the same pig from weaning to market age. We identified PCV3 PCR positive serums starting in 2014 from studies conducted in Minnesota and Virginia. In a study conducted in 2015 in North Carolina in a parity segregated system, we found a high level of PCR positives in the pigs from the farm with only parity 0-1sows, but negative or low levels in the pigs from the three, parity 2+ sow farms that received sows from the parity 0-1 farm. Additional PCR testing of individual pigs over time revealed that pigs can be viremic for at least six weeks during the nursery phase. We have also been involved in several veterinary student intern projects that dealt with various aspects of PCV3 epidemiology.

Sampling strategies

A primary goal of MAH's initial field research has been to better understand the PCV3 status of breeding herds. A convenience method for assessing PCV3 sow herd status in some systems has been to use samples collected for routine PRRS monitoring. Historically, this was done by collecting blood from pigs at weaning while recently, increasingly herds are using processing fluids. Our preliminary findings indicate the most sensitive method for detecting PCV3 in a breeding herd is testing processing fluids by PCR, either as aggregated samples such as all the litters processed in one day or as individual litters. Sampling of sows and suckling pigs for testing serum by PCR appears to be less sensitive than processing fluids. In addition, PCV3 prevalence can be impacted by parity such that parity 1 and 2 dams have in several cases been more likely to be positive compared to older parties, which may be related to sourcing of negative gilts into a positive system or to insufficient acclimation procedures. These factors should be considered when developing sampling protocols.

With regard to monitoring growing pigs, MAH's experience has been limited as this age group has not yet been targeted for vaccination. With regard to sample type collection for growing pigs, specifically serum vs. whole blood, MAH investigated whether or not there could be an impact in diagnostic sensitivity if PCV3 is cell associated. If so, serum would be void or have low levels of virus while whole blood, which would possibly contain PCV3 in white blood cells, would be a better sample. This was evaluated in a set of duplicate serum and whole blood samples. Of 31 pigs, 2 were PCV3 PCR positive and the CT values were nearly the same for serum vs. whole blood, suggesting that serum is equivalent to whole blood.

A final point is related to the PCR testing scheme. Several labs have duplex PCV3/PCV2 PCR assays available in addition to monovalent assays. MAH has found that having the PCV2 data along with the PCV3 data has been useful during diagnostic investigations, given the overlap in disease presentations that have been associated with each circovirus.

Serotesting

MAH is currently working with several research labs with regard to serological assay development. Our efforts have been related to providing samples from our R&D, market support studies and field investigations. It is of high interest to the industry to have validated antibody assays to provide a historical view of previous exposure and for measuring vaccine-induced antibody responses. MAH has used PCR to better understand exposure dynamics but PCR is limited by not having a clear understanding of how long viremia is maintained and whether or not PCR positive pigs can become PCR negative and then again positive due to re-infection or re-emergence of the initial infection.

Sequence analysis

For production of vaccine using the SEQUIVITY platform, only the sequence of the gene of interest is needed. MAH, in collaboration with the herd veterinarian, uses sequence analysis of field isolates to determine if a new sequence that exactly matches the field isolate is needed to make the prescription RP vaccine for a customer or if an existing RP is acceptable. In addition, MAH uses sequence analysis to track pathogen spread to help customers stay informed of new strains that may infect their herds, especially with diseases like swine influenza. All of the sequence and demographic data related to new field isolates is captured and managed in our SEQUIVITY Dashboard computerized database program.

Vaccine

MAH began producing PCV3 RP vaccine for customer use during the summer of 2018. As stated above, all customer requests have been based on the need to address PCV3 reproductive failure, especially in situations such as start-ups and re-populations where a high percentage of the females were gilts. The primary concern of the herd veterinarians was that young females were at the greatest risk based on their initial experiences with PCV3. It is too soon to determine the benefits of PCV3 RP vaccination based on customer experiences. One field trial in North Carolina was conducted this past summer and the results will be presented in the Student Seminar section of this meeting. Other field trials are currently underway.

With regard to producing prescription vaccines, there are 3 general requirements:

- There is a known gene of interest for the pathogen. For PCV3 we assume that ORF2 which encodes for the capsid protein is the protective antigen based on our understanding of PCV2.
- USDA Center for Veterinary Biologics (CVB) approves the gene of interest for each pathogen.
- We have a "foundation" fully-licensed RP vaccine that provides a general template for how the prescription vaccine is formulated.

Summary

The magnitude of PCV3's impact on swine production remains to be seen. At this time, reproductive failure appears to be the main problem caused by PCV3 while the impact on growing pigs is unclear. The SEQUIVITY Technology RNA-particle platform enables Merck Animal Health to quickly respond to this type of situation where a new pathogen emerges but existing, traditional vaccine technology is not capable of responding to the need for controlling the disease by vaccination.

References

1. Palinski R, Pineyro P, Shang P, et al. 2016. A novel porcine circovirus distantly related to know circoviruses is associated with porcine dermatitis and nephropathy syndrome and reproductive failure. *J Virol.* 2016; 91(1):e01879-16. https://doi.org/10.1128/JVI.01879-16

2. Phan TG, Giannitti F, Rossow S, et al. 2016. Detection of a novel circovirus in pigs with cardiac and multi-systemic inflammation. *Virol. J.* 2016; 13:184.

3. Byers E. 2018. Porcine circovirus type 3: Practitioner perspective. Proc. 49th AASV Annual Meeting, 445-447.

4. Chen GH, Mai K, Zhou L, et al. 2017. Detection and genome sequencing of porcine circovirus 3 in neonatal pigs with congenital tremors in South China. *Transbound. Emerg. Dis.* 64: 1650-1654.

5. Byers E. 2018. Allen D. Leman Swine Conference, 17 September 2018, Saint Paul, Minnesota.

6. Opriessnig T, Gerber P, Ziao C, et al. 2014. A commercial vaccine based on PCV2a and an experimental vaccine based on a variant mPCV2b are both effective in protecting pigs against challenge with a 2013 U.S. variant mPCV2b strain. *J Vaccine* 2014; 32(2):230-237.

7. Vander Veen RL, Loynachan AT, Mogler M, et al. 2012. Safety, immunogenicity, and efficacy of an alphavirus replicon-based swine influenza virus hemagglutinin vaccine. *Vaccine* 2012; 1944-1950. doi:10.1016/j.vaccine.2012.01.030.

Impact of vaccination timing on the efficacy of Porcilis™ Ileitis against experimental *Lawsonia intracellularis* challenge

Brad Thacker[1], DVM, PhD; Greg Armbruster[1], DVM, MBA; Brett O'Brien[1], VMD, MS; Robyn Fleck[1], DVM; Nathan Winkelman[2], DVM; Adam Mueller[2], DVM

[1]Merck Animal Health, Madison, New Jersey; [2]Swine Services Unlimited, Inc., Rice Minnesota

Introduction

Porcilis™ Ileitis (Merck Animal Health - MAH, Madison, New Jersey, USA) was introduced to the U.S. market in 2015 as an aid in the control of ileitis caused by *Lawsonia intracellularis,* as an aid in the reduction of colonization by *Lawsonia* and as an aid in the reduction of duration of fecal shedding.[1] The main objective of this study was to evaluate the impact of timing between vaccination and challenge along with the age at vaccination on the efficacy of Porcilis Ileitis against an experimental challenge.[2] In previous studies, the interval between vaccination and challenge ranged from six to twenty weeks.[3,4] In this study, a shorter interval of 3 weeks between vaccination and challenge was investigated.

Materials and methods

Three-week-old, cross-bred, *Lawsonia* seronegative pigs (n = 96) were used in the study. On study day -1, pigs were tagged, and their gender and litter were recorded at the sow farm. Pigs were then weaned and moved to an isolation facility for the remainder of the study. At arrival, pigs were weighed, bled and randomly allocated to treatment within litter with blocking by start weight and gender. The next day (study 0), 32 pigs were vaccinated per label directions (2 mL, IM) with Porcilis Ileitis (PI-3WOA group). On study day 20, another 32 pigs were vaccinated per label directions with Porcilis Ileitis (PI-6WOA group). Thirty-two pigs were not vaccinated at any time (control group). All pigs were weighed and bled on study days -1, 41 and 63. On study day 41, all pigs were challenged by oral gavage with 40 mL of a gut mucosal tissue homogenate containing approximately 2.04×10^9 of *Lawsonia.* Pigs were then observed every second day for general demeanor, abdominal distension and fecal consistency and color (data not shown). Fecal samples for qPCR testing were collected on 0, 7, 14 and 21 days post challenge (DPC). Fecal shedding data is reported as the sum of copies per mL for samples collected on 7, 14 and 21 DPC.

On 23 DPC, pigs were necropsied and the intestinal tracts were removed for macroscopic lesion scoring and sample collection. For macroscopic lesion scoring, jejunum, ileum, cecum and colon were scored for lesion severity (0-3 scale) and lesion length. These values were used to calculate a total lesion score that represented the sum of lesion severity multiplied by the lesion length (cm) for the four segments of the intestinal tract. For histopathology and immunochemistry (IHC) evaluations, two tissue sections were collected. Ileum was collected 1 inch proximal to the ileocecal junction from all pigs. A second tissue section was collected from an area that exhibited macroscopic lesions. If no lesions were observed, the second section was collected 10 cm proximal to the ileocecal junction. Tissues were placed in 10% buffered formalin solution immediately after collection. Both evaluations were scored on a 0-3 scale and the maximum score between the two sections was used for analysis. Because the histopath and IHC scores were the same for each pig, the results of each evaluation are reported as one variable. For tissue PCR, two areas were sampled. Ileal mucosa was scraped two inches proximal to the ileocecal junction from all pigs. A second scraping was performed from an area that exhibited lesions. If no lesions were observed, a second scraping was performed just proximal to where the second tissue section was obtained. The average of the two scrapings from each pig was used for analysis.

All laboratory testing except for serology was done at the Iowa State University Diagnostic Laboratory using standard methods and previously reported scoring criteria.[3] For serum antibody testing, the bioScreen Ileitis Antibody ELISA (Svanova, Uppsala Sweden) was performed by Merck Animal Health. Statistical analysis was performed by Kruskal-Wallis One-Way AOV, General One-way AOV, Chi Square or Fishers Exact Test depending on data type and distribution (Statistix™10, Analytic Software, Tallahassee, FL).

Results and discussion

The results are presented in Table 1. Fecal samples from all pigs were PCR negative at challenge. One pig in the PI-6WOA group died prior to challenge due to causes unrelated to the study. Compared to control pigs, both Porcilis Ileitis vaccinated groups had significantly less macroscopic and microscopic lesions and lower levels of tissue infection and fecal shedding. The percent reduction by parameter in the PI-3WOA group compared to control pigs was: total lesion score, 89.0%; histopath/IHC, 74.5%; mucosal tissue infection, 92.9%; feces at necropsy, 98.5%; and total fecal shedding, 60.3%. The percent reduction by parameter in the PI-6WOA group compared to control pigs was: total lesion score, 86.9%; histopath/IHC, 84.8%; mucosal tissue infection, 93.1%; feces at necropsy, 99.2%; and total fecal shedding, 77.8%. Weight gain from challenge to necropsy was significantly higher in the vaccinated groups compared to the control group

Table 1: Efficacy of Porcilis™ Ileitis (PI) in pigs vaccinated at 3 or 6 weeks of age (WOA) against experimental challenge with *Lawsonia intracellularis* at 9 WOA compared to non-vaccinated control pigs

Parameter	Variable		Vaccine group		
			PI - 3 WOA	PI - 6 WOA	Control
Group size	No. pigs		32	32	32
Macroscopic lesions	Total score		1.1^{a}	1.3^{a}	3.8^{b}
	Lesion length (cm)		15.8^{a}	12.0^{a}	74.2^{b}
	Total score X length		24.2^{a}	28.8^{a}	220.2^{b}
Histopath and IHC	Maximum score		0.59^{a}	0.35^{a}	2.31^{b}
	% with score	0	78.1	83.9	18.8
		1	3.1	6.5	3.1
		2	0.0	0.0	6.3
		3	18.8	9.7	71.9
Tissue infection	Copies per mL		$3,436,250^{a}$	$3,324,711^{a}$	$48,399,559^{b}$
Feces at necropsy	Copies per mL		$1,830,475^{a}$	$1,002,310^{a}$	$118,761,450^{b}$
Fecal shedding	Total copies per mL		$5,386,965^{a}$	$3,014,474^{a}$	$13,586,018^{b}$
Weight gain	Average daily gain (lb)	Start to challenge	0.828^{a}	0.842^{a}	0.808^{a}
		Challenge to necropsy	1.580^{a}	1.582^{a}	1.260^{b}
Lawsonia ELISA	Percent inhibition	At 1st Vac.	1.6^{a}	1.5^{a}	4.6^{a}
		At challenge	61.5^{a}	56.4^{a}	4.0^{b}
		At necropsy	89.9^{a}	85.9^{a}	83.4^{a}
	No. positive/ tested	At 1st Vac.	$1/32^{a}$	$2/32^{a}$	$3/32^{a}$
		At challenge	$30/32^{a}$	$28/31^{a}$	$0/32^{b}$
		At necropsy	$32/32^{a}$	$31/31^{a}$	$32/32^{a}$

^{a,b} Values with different superscripts within a row were significantly different ($P < 0.05$).

($P = 0.001$) and the percent improvement compared to controls was 25.4% for the PI-3WOA group and 25.6% for the PI-6 WOA group.

The levels of protection for each of the vaccination timings was similar to what has been reported in previous studies.[3,4] There were no differences in any parameters between the two vaccine groups. The severity of macroscopic lesions in control pigs was approximately 67% greater than our previous study reported in 2018.[4] The results reported here confirm the efficacy of Porcilis Ileitis against this relatively severe challenge model.

Porcilis Ileitis protected against this severe challenge as soon as 3 weeks after a single, 2 mL vaccination. Faster onset of immunity could be especially beneficial in antibiotic-free production systems that experience earlier onset of ileitis after weaning. Although in this study the shorter interval between vaccination and challenge was achieved by vaccinating pigs at six weeks of age, we expect a similar level of protection in pigs vaccinated at three weeks of age, the youngest age per label directions, and then challenged three weeks later.

References

1. Thacker B et al. Field experiences with Porcilis™ Ileitis: The newest tool for controlling *Lawsonia intracellularis*. AASV. 2017; 191.

2. Winkelman N et al. Validation of an ileitis mucosal homogenate seeder pig model to study the natural transmission of *Lawsonia intracellularis* in grow-finish pigs. AASV. 2017; 170.

3. Roerink F et al. Porcilis™ Ileitis: 20-week duration of immunity against *Lawsonia intracellularis* challenge. AASV. 2016; 144.

4. Thacker B et al. Efficacy of Porcilis™ Ileitis against *Lawsonia intracelluaris* challenge and post-vaccination seroresponse kinetics. AASV. 2018; 400.

Evaluation of semen production in boars given an IAV-S vaccine

J. Tubbs[1]; S. Krantz[2]
[1]Auburn University, Auburn, Alabama; [2]Tosh Farms LLC, Henry Tennessee

Introduction

Influenza A virus in swine (IAV-S) is a globally endemic disease that can have significant respiratory and performance impact in infected animals. Fever from IAV-S infection is of specific concern for boars as it can detrimentally affect semen quality and production for several weeks post pyrexia. The objective of this project was to evaluate the use of an IAV-S vaccine in boars through clinical observations and semen quality parameters. The results of this study will aid in vaccine decisions for producers with disease pressure due to IAV-S.

Materials and methods

This study was performed at a 126 head commercial boar stud located in the southeastern United States from June through August, 2018. No new boars were added to the stud during the study timeframe. Semen collection for each boar occurred once per week, with all boars being collected either Sunday evening or Thursday evening. For convenience, the Thursday night collection boars were selected for vaccinate group, the Sunday night collection boars the control group. Each group contained at least 60 boars however boar age was not balanced between collection groups. The vaccinate group was administered a single dose of an IAV-S vaccine during the collection process. Boars were observed daily. Baseline semen production data from at least 10 weeks preceding the study was compared to the subsequent 8 weeks following the vaccination event, using Computer Aided Semen Assessment (CASA) SpermVision™ Professional (MOFA Global, Verona, USA) and data management software, PRISM™ boar management software (MOFA Global, Verona, USA). Semen production parameters were assessed to determine total sperm production, morphology, and motility. Averages were calculated for time periods prior to and after vaccination.

Results

No off-feed events were noted during the study. For the reported measures of total sperm production, morphology and motility, the vaccinated and unvaccinated boars performed similarly during the duration of this study, as shown in Table 1.

Discussion and conclusions

Boar studs monitor key performance indicators because boar productivity is a way to assess boar health. Factors that are known to impact these parameters are boar age (increasing sperm production), season (reducing motility and other parameters) and diseases (highly impactful particularly with fever). In this study, boar age was not balanced between collection days, which may have contributed to the differences in starting performance parameters between groups. However thereafter, the data demonstrated the natural impact of boar maturation and season in addition to vaccination on average sperm production (modestly higher over time), and average motility (modestly lower as the summer progressed) respectively. Both vaccinated and unvaccinated boars were affected similarly.

According to a recent publication, the impact of IAV-S on semen production in boars affects semen quality and production by reducing total sperm production, percent motility and percent normal morphology.[1] The results of this study indicate that semen quality and production in boars vaccinated with an IAV-S vaccine remained comparable to unvaccinated cohorts.

Reference

1. Lugar DW, Ragland D, Stewart KR. Influenza outbreak causes reduction in semen quality of boars. *J Swine Health Prod.* 2017;25(6):303–307.

Table 1: Semen quality parameter comparison for non-vaccinates and vaccinates

	Non-vaccinates		Vaccinates	
	Pre-vaccination	Post-vaccination	Pre-vaccination	Post-vaccination
Normal morphology (%)	97.13	96.97	97.15	96.80
Total motility (%)	87.83	86.77	88.20	86.41
Sperm production (Billions/ejaculate)	73.31	79.72	76.55	76.81

Validation of the post-mortem deep bronchial swab as an appropriate sample type for influenza A virus detection in swine

A. Noel[1], BS; C. Mainquist-Whigham[2], DVM, MS; J. Creel[3], DVM; P. Gauger[4], DVM, MS, PhD
[1]College of Veterinary Medicine, Iowa State University, Ames, Iowa; [2]Pillen Family Farms, Columbus, Nebraska; [3]Merck Animal Health, Madison, New Jersey; [4]Department of Veterinary Diagnostic and Production Animal Medicine, College of Veterinary Medicine, Iowa State University, Ames, Iowa

Introduction

The United States Department of Agriculture (USDA) created the Influenza Surveillance Program in 2009 to monitor genetic evolution in influenza A virus (IAV) in swine and guide disease prevention.[1] Currently, the Program does not accept the post-mortem deep bronchial swab for virus isolation (VI) and sequencing, a sample that is easy to collect and has demonstrated detection of IAV RNA.[2] The objective of this study was to evaluate post-mortem deep bronchial swabs as an appropriate sample type to detect IAV relative to lung tissue in naturally infected swine for future validation in the USDA IAV surveillance system.

Materials and methods

Sample collection

Individual pigs, 3 to 9 weeks of age from sites with recent influenza-like illness, were selected as the experimental unit. Pigs were euthanized, necropsied, and the trachea was incised 3-4 cm cranial to the tracheal bifurcation. A rayon swab (Fisherfinest®, Fisher Scientific, Waltham, MA) was inserted in the lumen, passed into the left and right bronchi, and placed into a transport tube. Transverse sections from the left and right cranioventral lung were fixed in 10% formalin and submitted with fresh lungs and trachea.

Laboratory testing

Samples were immediately shipped to the Iowa State University Veterinary Diagnostic Laboratory (ISU VDL) and processed for analysis. Real-time reverse transcription polymerase chain reaction (rRT-PCR) for IAV using the VetMAX™ Gold SIV (ThermoFisher Scientific, Waltham, MA) kit was performed on individual lungs and swabs per manufacturer instructions and ISU VDL standard operating procedures. Histopathologic examination was conducted on fixed tissue, and subtyping, VI, and sequencing were attempted on the swab and lung tissue of one IAV positive animal from each submission.

Statistical analysis

Lung tissue was considered the reference standard for this study. Cohen's Kappa, a Chi-Square analysis, positive and negative predictive values (PPV, NPV), and diagnostic sensitivity and specificity of the swab were calculated using Microsoft Excel to assess agreement between the sample types. The rRT-PCR cycle threshold (C_T) of the paired positive samples were compared using the Wilcoxon signed-rank test to determine if there was a significant difference between mean C_T from both sample types.

Results

One hundred fourteen pigs were sampled; 18 were H3N2 positive and 89 were negative on both lung and swab. Five were positive on lung only, and 2 were positive only on swabs. The Chi-Square analysis indicated the swab and lung were not independent ($P < 0.001$). Cohen's Kappa demonstrated substantial agreement between the sample types (0.800). Significant differences between the mean C_T values were not detected ($P = 0.127$). Sequencing and VI were successful in all attempts, with swab C_T values ranging from 20.8 to 27.2. Results of IAV rRT-PCR, diagnostic sensitivity and specificity, PPV, NPV, and Cohen's Kappa are reported in Table 1.

Discussion

The statistical analyses demonstrated comparable IAV detection between lung and swabs by rRT-PCR. The low diagnostic sensitivity reported for the swab may be due to the low number of IAV positive animals or disagreement between five rRT-PCR positive lungs and negative swabs. It is possible IAV infection was subacute in these pigs as C_T values were elevated ($\mu = 35.3$) to levels where rRT-PCR detection may be inconsistent. The two rRT-PCR positive swabs might suggest early detection of infection, since bronchiolar epithelium is the site of IAV replication.[3] Consistent VI and sequencing results underscore the diagnostic potential of the swab and give further evidence of agreement between the two sample types. Based on these findings, the post-mortem deep bronchial swab could be an appropriate sample for IAV detection and potentially the USDA Influenza Surveillance Program.

Table 1: Results of influenza A virus rRT-PCR tests, diagnostic sensitivity and specificity, positive and negative predictive values, and Cohen's Kappa determined for deep bronchiole swabs collected from swine with influenza-like illness.

	Lung, positive	Lung, negative	Diagnostic sensitivity	Diagnostic specificity	PPV	NPV	Cohen's Kappa
Swab, positive	18	2	78%	98%	90%	95%	0.800
Swab, negative	5	89					

References

1. United States Department of Agriculture, Animal and Plant Health Inspection Service. National Swine Influenza Surveillance Plan, Version 3.2. 2010. Available at: https://www.aphis.usda.gov/aphis/ourfocus/animalhealth/sa_animal_disease_information/sa_swine_health/ct_siv_surveillance/. Accessed 20 September 2018.

2. Burrough E, Schwartz A, Gauger P, et al. Comparison of postmortem airway swabs and lung tissue for detection of common porcine respiratory pathogens by bacterial culture and polymerase chain reaction assays. *J Swine Health Prod.* 2018; 26(5):246-252.

3. Van Reeth K, Brown IH, Olsen CW. Influenza virus. In: Zimmerman JJ, Karriker LA, Ramirez A, Schwartz KJ, Stevenson GW, eds. *Diseases of Swine.* 10th ed. Ames, IA: Wiley-Blackwell; 2012:67-571.

The cost and causes of lameness in loose housed sows

J. Schumacher, BS; M. Pierdon, VMD, DACAW
Swine Teaching and Research Facility, University of Pennsylvania, Kennett Square, Pennsylvania

Introduction

Sow lameness is an important aspect of swine health and production, especially with the increasing transition to group-housed sow gestation. Lameness presents a major welfare concern, but also an economic concern. Premature culling of gilts or sows due to lameness can have a significant economic impact. This study had two main objectives. The first was to use analysis of a sow herd's treatment and production records over the past 4 years to determine the difference in survival time for lame sows compared to sound sows. The second objective was to determine the cause of lameness in lame sows that were euthanized.

Materials and methods

The studied herd is a 230-sow herd where gestating sows are fed by electronic sow feeders in a large 160-sow pen with gilts in a separate pen. For the first objective, treatment records were analyzed for all animals that entered the herd for the past 4 years (n = 519). A sow was defined as lame if she had been treated for lameness, euthanized for lameness, or culled for lameness. This information was combined with the removal records for all sows to calculate each sow's survival time. A Kaplan-Meier survival curve was created using STATA®v 15.0. Cox proportional hazard analysis was performed to determine the difference in incidence rate and median survival time for lame compared to sound sows. The median survival time was also compared for lame sows treated with antibiotics or flunixin meglumine with those that were not treated. From May to October 2018, 12 sows were euthanized for lameness. To determine the cause of lameness, joints of all 4 limbs were opened and examined grossly. Distal limbs were radiographed, and joints categorized as normal or abnormal.

Results

Lame sows had a significantly lower median survival time (532 days) compared to non-lame sows (874 days) ($P < .001$). The hazard ratio was calculated to be 1.6, meaning a lame animal was 60% more likely to be removed at any given time. Based on a lame sow spending 342 fewer days in the herd, she would be expected to have 2.13 fewer litters. If weaning 10.5 piglets per litter, that is 22 fewer piglets over that sow's lifetime. Treatment with antibiotics increased median survival time of lame sows (682 days) compared to lame sows not treated with antibiotics (233 days) ($P < .001$). There was no change in survival time between lame sows treated with flunixin meglumine and those not treated ($P > .05$). When the joints were examined on the euthanized sows (n = 12), the majority of the lameness was due to infectious arthritis (10/12 sows). The majority of the joint lesions were noted in the front limbs (28 lesions in 12 sows), compared to the hind limbs (10 lesions in 12 sows).

Discussion

This is a novel study that investigates the cost and causes of lameness in sows gestated in pens. In previous studies, the cost of lameness is the premature removal of sows from the herd[1], which is consistent with our findings. Understanding the cost gives producers the ability to decide how much to spend on treatment and prevention. This study indicates that infectious causes of arthritis are the most common. Likely because of this, antibiotics were beneficial in extending the survival time of lame sows in this study. While flunixin meglumine did not have an impact on survival time, the use of pain relief in lame sows is likely to improve welfare. Although infectious causes of lameness were prevalent in this study on this particular farm, further research is needed to determine if infectious causes are also most prevalent cause of lameness in loose housed sows on other farms.

Reference

1. Anil SS, Anil L, Deen J. Effect of lameness on sow longevity. *J Am Vet Med Assoc.* 2009;235(6): 734-738.

Dietary algae-derived protein as antibiotic alternative for weaned pigs infected with enterotoxigenic *Escherichia coli*

B. N. Smith; M. I. Hannas, MS, PhD, MBA; M. L. Oelschlager, MS; R. N. Dilger, MS, PhD
Veterinary Medical Scholars Program. Department of Animal Sciences. College of Agriculture, Consumer, and Environmental Sciences, University of Illinois, Urbana, Illinois

Objectives

Enterotoxigenic *Escherichia coli* (ETEC) causes high morbidity and production loss for nursery pigs. With increased regulation of antibiotic use in livestock, feed additives hold potential as antibiotic alternatives to increase herd health under pathogenic stressors such as ETEC. The objective of this experiment was to investigate the efficacy of wild-type and osteopontin-expressing algal protein sources against an ETEC infection in weaned pigs.

Materials and methods

At weaning, 108 barrows were group-housed in a BSL-2 containment facility in Champaign, IL for the duration of the 20 day experiment. Upon arrival, pigs were individually weighed and allotted to 1 of 6 treatments (2 infection levels) by body weight, resulting in 18 pigs per treatment with each treatment represented by 2 pens of 9 pigs (treatments summarized in Table 1). All diets were formulated to meet nutrient recommendations for nursery pigs and fed as a single phase. Experimental diets contained up to 1% of wild-type (WT) or osteopontin-expressing (OP) algal protein (or combination thereof), with the exception of Treatment 6 which contained antimicrobial growth promoters (AGP) as an industry positive control. Following a 10-day diet adaptation, pigs were orally inoculated on each of 3 consecutive days [days post-inoculation (DPI) 0-2] with either 3 mL of sterile PBS (sham-control) or live culture of ETEC (10^{10} CFU/3 mL dose). Growth performance was recorded on a pen basis over the course of the experiment. Individual samples of blood, feces, and intestinal sections and contents were collected throughout the 10 day post-inoculation period for measures of immunological, morphological, and microbiome responses to infection. A 2-way ANOVA (including time effect) was performed to compare Treatments 1 and 2, the non-infected and infected control groups (infection status), and Treatments 2 through 6 to compare all infected groups (algal effect within infected pigs).

Results and discussion

While ETEC infection resulted in decreased average daily gain (ADG) and feed efficiency (G:F) during the post-inoculation period when comparing uninfected and infected control groups ($P < 0.05$), there was no effect of dietary treatment on growth performance amongst ETEC-infected pigs. Across all blood cell parameters, effects were minimal for treatment and differences were typically due to ETEC-infection status alone. Serum cytokines were evaluated on 0, 7, and 10 DPI via commercially available ELISA kits, but low detection results suggest that ETEC infection and algal protein supplementation did not alter systemic inflammatory cytokine production. Flow cytometry data from 7 DPI demonstrated main effects of treatment amongst ETEC-infected groups for total proportion of peripheral mature T-cells (CD3+), mature helper T-cells (CD3+/CD4+), and mature cytotoxic T-cells (CD3+/CD8+), with greater ($P < 0.05$) percentages of total mature T-cells in pigs fed diets containing 1% algal protein. ETEC-infected pigs had higher total fecal bacterial content at 0 DPI ($P < 0.05$), though this effect was lost by 7 DPI.

Table 1: Experimental treatments

Treatment	Dietary treatments[1]	Infection status
1	1.0% wild-type algal protein	Uninfected
2	1.0% wild-type algal protein	Infected
3	0.5% wild-type algal protein	Infected
4	1.0% osteopontin+ algal protein	Infected
5	0.5% wild-type algal protein + 0.5% osteopontin+ algal protein	Infected
6	Antibiotic Growth Promoters[2]	Infected

[1] Percent inclusion of algal protein is expressed on a total mixed diet basis
[2] Diet was mixed in two phases: one containing 50 g/ton Mecadox® [Phibro] and one containing 50 g/ton Aureomycin® [Zoetis]. The phase containing Mecadox® was fed first.

Differences between ETEC-infected groups for total and *E. coli* bacterial counts were unremarkable. Treatment 4 (1% OP) had greater concentrations of total short chain fatty acids (SCFA) from the ascending colon compared with all other infected treatments at 10 DPI, a pattern than was maintained for individual SCFA concentrations of acetate, propionate, and butyrate ($P < 0.05$). Considering Treatment 4 was the only treatment containing osteopontin-expressing algal protein alone, this could indicate that this source of algal protein was more fermentable than WT algae. With minimal effects on clinical outcomes, algal protein is likely not a strong candidate as an antibiotic alternative against ETEC in weanling pigs, but algal protein type may impact microbial fermentation products. Future directions for this research include determining impact of feeding dietary algal protein, specifically osteopontin-expressing algal protein, on the microbiome and alterations of those microbial populations on gut health in pigs.

Relationship of *Mycoplasma hyopneumoniae* and *Mycoplasma hyorhinis* detection in processing fluids with growing pig prevalence

R. Stika[1]; D. Murray[2]; L. Johnson[2]; C. Vilalta[4]; M. Pieters[4]; E. Fano[3]; A. Sponheim[3,4]
[1]Iowa State University, Ames, Iowa; [2]New Fashion Pork, Jackson, Minnesota; [3]Boehringer Ingelheim Animal Health, Duluth, Georgia; [4]University of Minnesota, St. Paul, Minnesota

Introduction

It has been shown that processing fluids (PF) are a sensitive sample for detection of porcine reproductive and respiratory syndrome virus (PRRSV) at the litter level (Vilalta et al., 2018). *Mycoplasma hyopneumoniae (Mhp)* has been detected in PF, although *Mhp* is typically known to colonize the respiratory tract, and systemic dissemination of this bacterium has only been suggested under certain conditions. *Mycoplasma hyorhinis* (*Mhr*) is known to colonize the upper and lower respiratory tract and to spread systemically. PF have been partially evaluated for detection of *Mhp* and not evaluated for detection of *Mhr*. Therefore, the objective of this study was to determine the relationship between *Mhp* and *Mhr* detection by PCR in PF and prevalence of *Mhp* and *Mhr* at the end of the nursery and finishing phases.

Methods

Longitudinal sampling was conducted at three commercial sow farm flows, representing the three gilt pyramids of a Midwest swine production system. A *Mhp* negative flow (Sow 1 flow) and two *Mhp* positive subclinical infected I flows (Garza-Moreno et al., 2018; Sow 2 flow and Sow 3 flow) were followed. PF were collected for at least four weeks at each sow farm and pooled by processing day. Groups of pigs from PCR positive processing days were selected and followed based off pig movement records. Negative pigs were age-matched to the positive pigs and followed as the control group in Sow 1 flow. Two nursery sites were sampled per sow flow. One finishing site was sampled per nursery group. Deep tracheal catheter (DTC) samples were collected randomly from 55 pigs (based on detection of ≥ 7% *Mhp* prevalence) per nursery at 12 weeks of age and tested for *Mhp* by PCR. Six oral fluids (OF) per 2400 head nursery were collected at 12 weeks of age to test for *Mhr* by PCR. Sixty bronchial swabs (BS) from the accessory lung lobes were collected post-harvest from finishing pigs from each nursery group. Thirty BS were tested by PCR for *Mhp* (based on detection of ≥ 10% *Mhp* prevalence). If *Mhp* was not detected in the endemically infected flows on the first thirty BS, the additional thirty BS were tested (based on detection of ≥ 5% *Mhp* prevalence). Slaughter plant scald tank samples were collected before and after sample collection and tested for *Mhp* to access risk of cross-contamination on post-harvest samples. Genetic sequencing was performed on positive *Mhp* sow farm results and positive *Mhp* post-harvest samples to determine vertical from lateral introduction.

Results

Combined positive and suspect results for *Mhp* and *Mhr* at the three sampling time points are reported in Table 1. Scald tank samples were negative for *Mhp* before and after sample collection. Three *Mhp* sequences were obtained (Sow 2 on gilt lung tissue, Sow 3 on PF, and Sow 2 flow's Finisher B on BS). *Mhp* sequence results were placed into a phylogenetic tree rooted to US reference sequence strain 232. Sow 2 flow's Finisher B was clustered with Sow 2 flow with a 99.9% homology. Sow 2 flow's Finisher B was clustered separately with a 97.5% sequence identity to Sow 3 flow.

Conclusions

Mhp and *Mhr* were detected by PCR in PF from endemically infected farms. *Mhp* on PF, DTC and BS was not detected, as expected, in the negative flow (Sow 1 flow). *Mhp* and *Mhr* results on PF were not related to the results on nursery *Mhp* DTC and nursery *Mhr* OF. This may be due to *Mhp* prevalence being below the tested levels. Detection of *Mhp* in PF could be due to environmental contamination at processing although it could be associated with the testicle portion of the PF.[1] It was observed that positive results were detected more frequently in PF samples that were frozen over refrigeration when PF were held on farm for more than seven days. *Mhr* was observed to be less prevalent in PF when compared to *Mhp*. The low detection of *Mhr* in PF could be due to an ongoing vaccination program at all sampled sow farms. Sequencing results suggest vertical introduction of *Mhp* from the sow farm to finishing pigs in Sow 2 flow. One limitation of this study was the use of accessory lung lobes versus cranial lobes for the BS due to slaughter plant regulations.[2] Further research is needed to fully understand using PF for *Mhp* and *Mhr* testing.

References

1. Vilalta C, Sanhueza J, Murray D, Pieters M. PCR detection of *Mycoplasma hyopneumoniae* in processing fluids: A diagnostic error or a significant finding? Presented at: University of Minnesota's Allen D. Leman Swine Conference; 2018 September 15-18; Saint Paul, Minnesota, USA.

2. Fano E, Pijoan C, Dee S. Dynamics and persistence of *Mycoplasma hyopneumoniae* infection in pigs. Can Vet J (2005) 69:223–8.

Table 1: Number of *Mycoplasma hyopneumoniae* and *Mycoplasma hyorhinis* PCR positive and suspect results by sow flow and sampling time point.

		Sow farm		Nursery				Finisher	
				A		B		A	B
		Mhp (PF)	*Mhr* (PF)	*Mhp* (DTC)	*Mhr* (OF)	*Mhp* (DTC)	*Mhr* (OF)	*Mhp* (BS)	*Mhp* (BS)
Sow flow	1	0/50 (0%)	0/50 (0%) 3 S (6%)	0/55 (0%)	6/6 (100%)	0/55 (0%)	0/6 (0%)	0/30 (0%)	0/30 (0%)
	2	1/37 (2.7%) 1 S (2.7%)	0/37 (0%) 2 S (5.4%)	0/55 (0%)	6/6 (100%)	0/55 (0%)	0/6 (0%)	0/60 (0%)	9/30 (30%)
	3	15/64 (23.4%) 10 S (15.6%)	1/64 (1.6%) 4 S (6.3%)	0/55 (0%)	6/6 (100%)	0/55 (0%)	6/6 (100%)	0/60 (0%)	0/60 (0%)

Percentage of positive and suspect results are in parenthesis.
Key: PF= Processing fluids; DTC = Deep tracheal catheter; OF = Oral fluids; BS = Bronchial swabs (accessory lobe); S = Suspect results.

Impact of shoulder lesions on sow behavior and welfare

Emily Nogay; Tara Gaab, VMD; Meghann K. Pierdon, VMD, DACAW
School of Veterinary Medicine, University of Pennsylvania, Philadelphia, Pennsylvania

Introduction

Shoulder lesions in sows typically occur within the first few weeks of lactation, affecting the skin overlying the prominent tuber of the scapular spine. Post-farrowing, increased time in lateral recumbency leads to prolonged compression of the blood vessels over the sow's shoulder, resulting in lesion formation.[1] These lesions are comparable to pressure ulcers in human patients, which have been reported as painful. However, no pain relief is currently given to sows with shoulder lesions, making them a welfare concern.[1] The objectives of this study were first to observe how sow shoulder lesions develop and heal in lactation and post-weaning in gestation, and second, to determine if behavior, productivity or sow based measures, such as lameness and weight, deviate from control sows with no lesion.

Materials and methods

A total of 72 sows were evaluated for lameness, body condition score (BCS), and weight upon entering the farrowing rooms and again at weaning. A total of 12 sows developed a lesion (LES, n = 12), and each was paired with a control (CON, n = 12) based on BCS, parity, and farrowing date. Once a lesion appeared, it was photographed daily using ImageJ2 software to calculate lesion size. Both LES and CON sows' behaviors were monitored by direct observation, by the same person, for an hour at 14, 21, and 28 days post farrowing. Observations were either done directly or video footage was reviewed. Gestation location and position of each sow was recorded twice daily, as sows had the option between concrete, straw, or slatted flooring within the gestation pens. Lesion measurements occurred weekly until healing. Statistics were performed using Stata v. 15.0. All data was not normal, and a Mann Whitney u-test was used to compare farrowing room behavior, gestation location, lameness, sow weights, and productivity between LES and CON sows. Results are reported as median and inner quartile ranges. A line was fit to the lesion size by measurement day data and the slope was recorded as the rate of lesion growth and healing.

Results

On average, lesions appeared on day 22.7 post farrowing. Lesion growth trended linearly ($r^2 = 0.92$) in the first 9 days at 0.43 cm^2 per day. Healing was less linear ($r^2 = 0.77$) with a rate of 0.46 cm^2 per day. In farrowing, there were no differences in time spent lying between CON (.83 ± .19) and LES (.82 ± .21) ($P > 0.05$). There was a trend for sows with a right-side lesion to spend less time lying on their right side (0.06 ± 0.64) compared to CON sows (0.43 ± 0.64) ($P = 0.10$). Only one sow had a lesion on her left side so no comparison was made. In the gestation pen, LES sows were found in the straw bedding (.35 ± .25) more than the CON sows (0.0 ± 0.0) ($P < 0.01$). There were no differences in productivity measures, weight, or lameness between LES and CON sows.

Discussion

The trend for sows to spend less time lying on the ipsilateral side to their lesions is a possible indicator of lesion pain. This suggests that pain relief is warranted during treatment of shoulder lesions. Linear lesion growth within the first 9 days implies that earlier treatments could lead to smaller, less severe lesions. Gestation location differences between sows with lesions and their controls suggests that sows with lesions were seeking more comfortable flooring when given the option between straw, slatted, or solid flooring. Sow welfare would be improved by providing more comfortable flooring or bedding, and possibly pain relief, to sows with shoulder lesions. Further examination is required due to the small sample size of this project. Additional investigation is necessary to find the most practical means of providing comfort for sows with shoulder lesions.

Reference

1. Rioja-Lang FC, Seddon YM, Brown JA. Shoulder lesions in sows: A review of their causes, prevention, and treatment. *J Swine Health Prod.* 2018;26(2):101–107.

The influence of a porcine reproductive and respiratory syndrome virus infection on vitamin and mineral levels in serum, liver, and bone of recently weaned pigs

Jacob Baker[1], BSc; Mark Wagner[2], DVM; David Baumert[3], DVM
[1]Iowa State University College of Veterinary Medicine, Ames, Iowa; [2]Fairmont Veterinary Clinic, Fairmont, Minnesota; [3]Zoetis, Parsippany, New Jersey

Introduction

Our understanding of baseline serum vitamin/mineral levels in nursery-aged pigs is limited and primarily based on data collected over 30 years ago. Reference ranges from universities are a composite of different age and health statuses and the role of disease may confound these reference ranges. The objective of this study was to better understand the effect of a porcine reproductive and respiratory syndrome virus (PRRSV) challenge on vitamin and mineral levels and to compare those levels to reference intervals provided by the Iowa State University Veterinary Diagnostic Laboratory. By understanding what is happening on a physiological basis, diet changes or other interventions may be warranted to reduce losses caused by disease.

Materials and methods

Forty wean-age pigs from a PRRS positive-stable farm were allocated into one of two groups, a non-challenged control group and a challenged group. Four pigs selected from each of ten litters were allocated to the two groups and litters were equally represented by parity. Pigs were bled prior to weaning to confirm they were negative for PRRSV by PCR. The control group was housed in a small nursery 20 miles north of the challenge group, which was housed in a circa 1940 hog-house. Pigs in both groups were fed the same nursery diet. At seven days post-weaning, challenge group pigs were challenged via intra-muscular injection with a 1-8-4 PRRS virus to which the sows had not previously been exposed. Pigs were bled two days post-challenge to confirm PRRSV PCR-positive status. Serum samples and pig weights were collected at the time of challenge and again at three weeks post-challenge (conclusion of trial). At conclusion of trial, the control group had its PRRSV PCR-negative status confirmed via blood testing. All pigs were weighed and euthanized to collect liver tissue and second rib samples for additional testing. A standard bone panel was run on ribs, a mineral panel and vitamins A, D, and E on serum, and vitamins A and E on the liver. All sample testing was performed at the Iowa State University Diagnostic Laboratory.

Results and conclusion

The challenge pigs were found to have higher levels of vitamin A in their livers and less in serum compared to the controls. This finding could be explained by the fact the retinol-binding protein is a negative acute-phase protein, resulting in an inability of the PRRSV-infected pigs to mobilize vitamin A from their livers. Vitamin A is imperative for proper growth, and this may explain the marked difference in ADG (Control = 0.75 lb/day; Challenge = 0.40lb/day) observed between the two groups. The challenged pigs were also found to have greater calcium levels in their serum and decreased calcium levels in their ribs compared to the control pigs, suggesting the challenge pigs may have been mobilizing calcium from their bone reserves for T-cell maturation and proliferation. The challenge pigs had more vitamin E in their serum and less in their liver compared to the control pigs, suggesting vitamin E was mobilized from the liver for its immune-supporting effects. In regards to mineral levels, challenge pigs had larger increases and greater overall levels of most minerals in their blood at 21 days post-challenge. The results of this study suggest that metabolism of vitamins and minerals in PRRSV-infected pigs is altered, and, that further research to determine the impact of vitamin A, E and/or calcium interventions on the production parameters of diseased pigs is warranted.

Evaluating the correlation of *Mycoplasma hyorhinis* and *Mycoplasma hyosynoviae* detection in oral fluids and lameness scores

D. Pillman[1], BS; M. Surendran Nair[1], DVM, PhD; J. Schwartz[2], DVM; M. Pieters[1], DVM, PhD
[1]Department of Veterinary Population Medicine, University of Minnesota, St. Paul, Minnesota;
[2]Swine Vet Center, St Peter, Minnesota

Introduction

Lameness and abnormal gait are major issues affecting animal health, welfare, and economics. Different infectious and non-infectious etiologies are associated with post-weaning lameness in pigs, with bacterial arthritis bearing much attention. *M hyorhinis* and *M hyosynoviae* are often reported from different swine herds in association with lameness and arthritis.[1,2] As both these microorganisms can colonize the host as commensals or as infectious agents, mere detection will not infer their active role in pathogenesis. Oral fluids have become an increasingly popular tool to detect pathogens at a herd level due to convenience and ease of collection. However, the potential association between the detection of arthritogenic mycoplasmas in oral fluids and development of lameness in pigs has not been clearly stablished. Thus, this study was designed with the objective to detect the prevalence of *M hyorhinis* and *M hyosynoviae* in oral fluids on farms, and determine its correlation with lameness. The ultimate goal of this investigation is to improve diagnostic interpretation, in order to guide prevention and control strategies in the field.

Materials and methods

Thirty seven nursery and finisher sites in the Midwest were selected for this study. Each site housed more than 2,000 pigs and used conventional commercial management. Oral fluids were collected by pen, with 9-10 samples collected per site. Briefly, a 70-cm length of 3-strand twisted rope was untwisted and suspended at shoulder height from a metal bar in the pens. The rope was left in the pens for pigs to chew on for 20-30 min.[3] Lameness scores (0-4) were determined for pigs in each pen from where oral fluids were collected.[4] The same evaluator determined the lameness score for all pens. Species specific real-time PCR was run on each sample to determine the prevalence of *M hyorhinis* and *M hyosynoviae* at each site. The correlation of *M hyorhinis* and *M hyosynoviae* detection and Ct values in oral fluids, and lameness scores were evaluated using the Spearman's rank-order method. Results were considered significantly different when a *P* value greater than 0.05 was obtained.

Results

M hyorhinis was detected by real-time PCR in 97% of sampled sites (36/37) whereas 70% were positive for *M hyosynoviae* (26/37). The level of detection of each microorganism varied depending on the site. The proportion of sampled pens positive for *M hyorhinis* was 100% at all sites except one. However, the proportion of pens positive for *M hyosynoviae* in different sites ranged from 10 to 100%. Lameness scores (2 or higher) were observed in 3.9% of pigs across all sites sampled. No significant correlation was observed between the lameness in pigs and detection of *M hyorhinis* in oral fluid samples ($P > 0.05$). Alternatively, a significant correlation was observed between proportion of lame animals in the pen and Ct values for *M hyosynoviae* detection in oral fluids ($P < 0.05$; $r = -0.27$). Furthermore, the proportion of lame animals in a farm was positively correlated with the proportion of pens detected positive for *M hyosynoviae* in oral fluid samples.

Conclusions

Under the conditions of this investigation, *M hyorhinis* was frequently detected in oral fluids in nursery and finisher sites regardless of the clinical presentation of lameness. Therefore, detection of *M hyorhinis* in oral fluids should be carefully interpreted, avoiding diagnostic inferences in lame pig populations. On the other hand, the association of clinical presentation of lameness and *M hyosynoviae* detection in oral fluids warrants detailed investigation particularly taking the pigs' age into account.

References

1. Ross, R. F. & Duncan, J. R. *Mycoplasma hyosynoviae* arthritis of swine. *J. Am. Vet. Med. Assoc.* 157, 1515–8 (1970).

2. Barden, J. A. & Decker, J. L. *Mycoplasma hyorhinis* swine arthritis. I. Clinical and Microbiologic Features. *Arthritis Rheum.* 14, 193–201 (1971).

3. Pieters, M., Daniels, J. & Rovira, A. Comparison of sample types and diagnostic methods for in vivo detection of *Mycoplasma hyopneumoniae* during early stages of infection. *Vet. Microbiol.* 203, 103–109 (2017).

4. Nielsen, E. O., Nielsen, N. C. & Friis, N. F. *Mycoplasma hyosynoviae* arthritis in grower-finisher pigs. *J. Vet. Med. Ser. A* 48, 475–486 (2001).

Assessing the presence of infectious influenza A virus in swine in oral fluid and nasal swab samples to determine replacement gilt flow status

G. Fry[1], BS; A. Lower[2], DVM; L. Ochoa[3], PhD; J. Waddell[4], DVM, MBA
[1]Purdue University College of Veterinary Medicine, West Lafayette, Indiana; [2]Carthage Veterinary Service, Ltd., Carthage, Illinois; [3]Carthage Innovative Swine Solutions, Carthage, Illinois; [4]Boehringer Ingelheim Animal Health, Duluth, Georgia

Introduction

Replacement gilts offer a risk of influenza A virus of swine (IAV-S) introduction to recipient herds making routine monitoring an important biosecurity measure prior to every gilt entry. Gilts from IAV-S endemic source herds have been known to remain positive for up to 12 weeks post weaning as measured by oral fluids (OF) screening PCR regardless of the presence of clinical signs. The objectives of this study were to (1) assess the correlation of pen based OF samples to individual pig nasal swabs (NS) for IAV-S and (2) assess the infectivity of OF IAV-S positive samples based on virus isolation (VI). The results of this study offer guidance to industry on the duration and sampling type to reduce the risk of IAV-S direct transmission via replacement gilt introductions and improve on criterial solely based on clinical signs of IAV-S.

Materials and methods

A continuous flow finisher site with influenza-positive replacement gilt flows from two multiplication herds was used in the study. Seven pens containing approximately 60 pigs of distinct weekly age groups (ranging from 8 to 14 weeks of age [WOA]) were identified within the site (Table 1). OF samples were collected via standard methods from each pen along with NS from every pig within the pen.[1] All samples were collected, processed and handled according to previously published best practices protocols.[2] IAV-S screening PCR testing was completed on aliquots of the OF samples and NS samples were tested via pools of five at the diagnostic lab. Seventeen days after the initial sampling, age groups 12-15 WOA were resampled and tested using the same method as the initial sampling. Of these four age groups, 12 and 14 WOA were positive on OF and NS testing after initial collection, while 13 and 15 WOA were negative. Aliquots of the positive OF samples from the first collection were then submitted for virus isolation to determine IAV-S viability.

Results, discussion and conclusions

As expected, the prevalence of IAV-S PCR positive samples tended to wane with age and weeks post placement. Three of the seven age groups were IAV-S PCR OF and NS positive on the first sampling (8, 9 and 11 WOA) while the 10, 12, 13 and 14 WOA groups were negative on both OF and NS. Despite testing two previously positive age groups on the second sampling, all samples (OF and NS) were negative 17 days later due to sampling logistics. There is strong relationship between OF and NS samples as evidenced by total agreement between OF and pools of NS samples across pens/age groups, lending confidence in utilizing OF sampling in lieu of the more labor intensive and costly NS samples. However, whether each pig per pen sampled the OF rope was unable to be measured. Moreover, virus isolation was achieved from 2 of 3 positive OF samples including the OF sample from the 11WOA group with the highest average Ct (= 36.96). These results indicate that even these long-term, high Ct-value OF samples may harbor the possibility of infectious virus being transmitted by replacement gilts to receiving herds. Results of this study offer veterinarians an alternative to mere cessation of clinical signs and confidence in monitoring the IAV-S shedding status based on oral fluids.

References

1. Goodell, CK, et al. Probability of detecting influenza A virus subtypes H1N1 and H3N2 in individual pig nasal swabs and pen-based oral fluid specimens over time. *Vet. Microbiol.* 2013.

2. Foerster et al. Evaluation of the effect of shipping and handling of oral fluid samples on Influenza A PCR detection. *AASV.* 2016; 100.

Table 1: IAV-S cumulative results

Item	Barn: 1 Pen: A	1 B	1 C	2 A	3 A	3 B	3 C
First sampling period:							
Pigs per pen	66	51	62	77	49	50	69
Flow	1	1	2	2	1	2	2
Age (weeks)	14	13	12	11	10	9	8
Oral fluids:							
Number of samples	2	2	2	2	2	2	2
Positive	0	0	0	2	0	2	2
CT value[3,4]	0	0	0	24.20	0	30.05	36.96
Virus subtype	-	-	-	H1N1	-	H1N1	H1N1
Virus isolation[5]	-	-	-	Positive	-	Negative	Positive
Nasal swabs[1]							
Number of samples	14	11	13	16	10	10	14
Positive	0	0	0	16	00	4	3
CT value[3,4]	0	0	0	27.47		32.69	35.03
Second sampling period:[2]							
Pigs per pen	-	-	50	66	57	60	-
Flow	-	-	2	2	1	2	-
Age, weeks	-	-	15	14	13	12	-
Oral fluids							
Number of samples	-	-	2	2	2	2	-
Positive	-	-	0	0	0	0	-
CT value[4]	-	-	0	0	0	0	-
Nasal swabs[1]							
Number of samples	-	-	10	14	13	12	-
Positive	-	-	0	0	0	0	-
CT value[4]	-	-	0	0	0	0	-

1 Nasals swabs were analyzed in pools of 5 pigs.
2 Second samples collected 17 days after first collection.
3 Reported CT value is the average of positive samples only
4 A 0 CT value indicates all samples were negative and CT values were higher than 38
5 Three different labs were utilized and positives were yielded from one

How neonatal factors affect reproductive performance of swine replacement breeding-stock

Jordan Buchan, BSc; Alison Jeffery; Robert Friendship, MSc, DVM; Rocio Amezcua, MSc, PhD
Department of Population Medicine, Ontario Veterinary College, Guelph, Ontario

Background

Replacement gilts are important in swine production, as they keep breeding herds productive by replacing older, less efficient sows.[1] Thus, it is ideal that selected gilts produce large litters and have good breeding longevity. Gilts are currently selected based on various criteria, including quality/number of teats, vulva shape/size, and strong legs/feet.[1] However, neonatal factors affecting pigs in utero are currently not accounted for in this selection. Important aspects of future female reproductive performance depend on the environment of the fetus during pregnancy and the first few weeks post-farrowing.[2,3] This is when all the oocytes are created to last the lifetime of the female. However, factors such as litter size, neonatal competition, or the influence of a litter composed of mainly male siblings are seldom used in selection of replacement gilts. It is possible that breeding stock producers are not selecting gilts with the highest reproductive potential because they are not taking into account pregnancy and neonatal factors. This study investigates if a gilt's birth utero litter size or birth litter sex bias affects her age at first farrowing, parity 1 litter size, or longevity. Furthermore, it aims to use these relationships to develop additional criteria to help account for these factors in gilt selection to improve overall sow breeding efficiency.

Materials and methods

For each of the 1025 gilts in the data set, it was determined if she was born into a male biased (≥ 60% male), female biased (≥ 60% female), or unbiased litter. Utero litter size was also evaluated, which is the litter that a replacement gilt originated from, categorized as small (< 13), moderate (13-15), and large (> 15) sized litters. This characteristic of utero litter size was examined to determine if it affected the gilt's future reproductive performance. A statistical evaluation was then performed to see whether these neonatal factors affected her reproductive performance in the breeding herd based on her age at first farrowing, parity 1 litter size, and breeding longevity (≥ parity 3). Chi-square tests were used to compare these various factors; however, the age at first farrow variable was non-parametric, so a Kruskal-Wallis test was used.

Results and discussion

The mean age at first farrowing was 323.6 days and the mean parity 1 litter size was 12.4 pigs. Of the eight statistical tests conducted to determine significant relationships between variables, the chi-square test comparing utero litter size and breeding longevity was significant ($P = 0.04$). The difference in longevity between small and moderate ($P = 0.02$) and moderate and large litter sizes ($P = 0.05$) were both significant. Statistical results from the 7 other variable comparisons were not significant ($P > 0.05$). The most notable of these non-significant results was that gilts chosen from male-biased litters performed just as well as those from female-biased or unbiased litters. This is surprising, since male-biased litters in many other species (such as rats) produce sub-fertile females. However, the main finding from this study is that gilts born into litters of moderate size (13-15 pigs) are more likely to remain in the breeding herd until at least parity 3 compared to gilts from smaller (< 13) or larger (> 15) litters. Resultantly, utero litter size could become a criterion recommended by practitioners to improve gilt selection. However, more work should be done to see if litter size as a selection criterion will actually improve gilt breeding longevity at a herd-wide level. Additionally, the average litter size observed in this herd is much smaller compared to that of hyperprolific sows currently present in the industry. This issue of selecting gilts from very large litters may become very important to practitioners in the future if this increasing litter size trend continues. Ultimately, these results provide the starting point for future work that will provide practitioners with information to help advise producers and breeding companies to optimize breeding efficiency with easily obtained litter information that incorporates neonatal factors not currently accounted for in gilt selection.

References

1. Taylor G. & Roese G. (2006). Primefacts: basic pig husbandry, gilts and sows. *Dept of Prim Indust.* Retrieved from http://www.thepigsite.com/articles/1573/basic-pig-husbandry-gilts-and-sows/

2. Ryan B.C., Vandenbergh J.G. (2002). Intrauterine position effects. *Neurosci. Biobehav. Rev.*, 26: 665-678.

3. Flowers, W.L. (2012). Possible physiological benchmarks for sow longevity prior to puberty. *Veterinary Continuing Education.*, (39): 113-117.

Comparison of processing fluids and placental umbilical cord serum to detect porcine circovirus type 2 in sow farms

Shelby Perkins[1], BS; Jessica Davenport[2], DVM; Eva Jablonski[3], DVM, MS; Deborah Amodie[3], MS; Giovani Trevisan[4], DVM
[1]University of Missouri College of Veterinary Medicine, Columbia, Missouri; [2]JBS Pork, Greeley, Colorado; [3]Zoetis, Parsippany, New Jersey; [4]Iowa State University College of Veterinary Medicine, Ames, Iowa

Introduction

Porcine circovirus type 2 (PCV2) is a small DNA virus globally affecting swine. Recent work has proposed a stability definition as ≤ 1% of sows testing polymerase chain reaction (PCR) positive on placental umbilical cord serum (PUCS).[1] PUCS have often been used for PCV2 testing on sow farms, although a time and labor consuming technique. Processing fluids (PF) are a sample easily collected by farm staff when processing piglets. The convenience of PF makes it an attractive choice and may help in determining vertical transmission as measured by piglet viremia.[2] The objective of this study is to compare the ability of PF and PUCS to detect PCV2 in sow farms.

Materials and methods

Four farms in a large production system were enrolled. Farms B and D had confirmed PCV2 downstream, whereas farms A and C did not. Farm A had been closed for 220 days for *Mycoplasma hyopneumoniae* elimination, Farm C recently eliminated *M. hyopneumoniae* successfully, and both Farms B and C were considered antibiotic-free. On each farm, 32-35 sows and their litters were selected, with an equal distribution of parities. Litter was the experimental unit with a total of 134 litters. PUCS were collected from as many umbilical cords from each fresh placenta as possible. Each tube was labeled individually by sow. Parity, sample date, and number of cords contributing to the sample were recorded. PUCS samples that did not represent ≥ 20% of total born alive were excluded. PF were collected on the same litters at 4-6 days of age. Tails and testicles from each individual litter were collected and refrigerated overnight before extracting PF. Cross-fostered piglets were tagged and excluded from the PF sample. Precautions were taken to minimize cross-contamination including dipping tools in chlorhexidine solution and changing gloves between litters for both PUCS and PF collection. All samples were collected by researchers and tested at Iowa State University Veterinary Diagnostic Laboratory for PCV2 by PCR. The percentage of positive samples were analyzed for inter-procedural agreement considering the possibility of agreement occurring by chance (SAS 9.4, Cary, NC). Using the Proc Freq Procedure, the Kappa agreement was measured between samples with an analysis among farms and a model where PUCS and PF agree. The number of matches between procedures, along with a 95% confidence interval around those matches, was calculated.

Results and discussion

Farms A, B, and D had a greater percentage of positive PF than PUCS, whereas Farm C had more positive PUCS than PF (Table 1). Possible reasons for higher rate of detection in PF include piglets being born negative and being infected after birth or piglets born with the virus at undetectable PCV2 levels. Vertical transmission is not always in utero and may occur from sow to piglets via environment. Reasons for Farm C having more positives PUCS compared to PF include placental contamination or very low prevalence of virus.

Table 1: PCV2 PCR results agreement of PUCS and PF and percent positive by farm and sample type

		PUCS		
		POS	NEG	Total
	POS (%)	31 (23.1%)	36 (26.9%)	67 (50%)
PF	NEG (%)	9 (6.7%)	58 (43.3%)	67 (50%)
	Total	40 (29.8%)	94 (70.2%)	134 (100%)
	Farm A (n=34)	**Farm B (n=35)**	**Farm C (n=32)**	**Farm D (n=33)**
POS PUCS (%)	11 (32.4%)	20 (57.1%)	9 (28.1%)	0 (0.0%)
POS PF (%)	26 (76.5%)	32 (91.4%)	3 (9.4%)	6 (18.2%)

Agreement between sample types measured by percentage of matched results was 66.4% with a 95% confidence interval of 58.4-74.4%. Kappa agreement was calculated to be 0.143, a very poor agreement between PUCS and PF. While PF and PUCS are not directly comparable due to the age difference, the higher percentage of positives from PF may suggest it is a more sensitive sample as it represented more pigs (mean = 11.1) in the litter than PUCS had umbilical cords (mean = 5.5).

In future research, it is recommended that more herds be compared, the number and age of piglets contributing to samples within a pool be standardized, and environmental contamination be determined.

References

1. Dvorak, C., B. Payne, et al. Effect of maternal antibody transfer on antibody dynamics and control of porcine circovirus type 2 infection in offspring. *Viral Immunology*. 2018. 31(1): 40-46.

2. Lopez, W., J. Angulo, et al. Porcine reproductive and respiratory syndrome monitoring in breeding herd using processing fluids. *Journal of Swine Health and Production*. 2018. 26(3): 146-150.

Case study evaluating diagnostic tools for *Mycoplasma hyorhinis* field diagnosis

Katie Kehl[1]; Kate Dion[2], DVM; Nathan Schaefer[3], DVM; Eduardo Fano[3], MS, DVM, PhD
[1]Kansas State University – College of Veterinary Medicine, Manhattan, Kansas; [2]Hanor Company, Enid, Oklahoma; [3]Boehringer Ingelheim Animal Health, Duluth, Georgia

Introduction

Mycoplasma hyorhinis (Mhr) is a commensal bacterium commonly found in the upper respiratory tract of pigs. Under certain conditions (stress and co-infections) Mhr can become a severe systemic pathogen causing polyserositis and arthritic lesions. The mechanisms that allow Mhr to become systemic and cause disease are not well understood. Therefore, diagnostic sample selection, case interpretation, and confirmation of disease (case definition) can be challenging. The objective of this study was to assess individual and population-based diagnostic tools to measure Mhr disease dynamics within a pig population. By understanding Mhr disease dynamics, we can accurately diagnosis Mhr as a disease-causing agent within a herd.

Materials and methods

This study followed four flows, each tested weekly for four consecutive weeks. Flows 1 and 2 were selected because of a previous Mhr diagnosis with clinical lameness. Flows 3 and 4 were selected because of previous Mhr diagnosis with clinical respiratory/systemic disease. Between 6-9 weeks of age, 1-3 necropsies were performed per week on pigs with clinical lameness or respiratory/systemic disease, for a total of 10 necropsies per flow. A full set of tissues was collected from each pig and multiple serosal surface swabs were taken. Forty pigs were tagged, nasal swabs and serum were collected at weeks six and nine. Six pen oral fluids were collected weekly per flow, for four weeks.

Serum samples were tested for antibodies to Mhr by ELISA. Nasal swabs, serosal swabs, tissues, and oral fluids were tested for Mhr by PCR. All fresh/fixed tissues were evaluated via histopathology, culture, and differential PCRs at Iowa State University Veterinary Diagnostic Lab.

Results

Population diagnostic results from flows 1 and 2 (clinical lameness) were typically negative during the testing period. Conversely, results from flows 3 and 4 (clinical respiratory/systemic disease) were consistently positive on population based diagnostics. In Figure 1, one oral fluid and one nasal swab tested Mhr PCR positive in flow 1 and 2, while all oral fluids and nasal swabs were positive in flow 3 and 4. Additionally, only two pigs tested ELISA positive in flows 1 and 2 during the study, while in flows 3 and 4 initially 11% and 25% of pigs tested ELISA positive at 6 weeks and this percentage increased to 100% and 81% at 9 weeks.

Mhr was not diagnosed as a primary pathogen in any of the diagnostic cases submitted to pathology from individual clinical pigs in flows 1 and 2. In flows 3 and 4, 8/20 clinical cases were diagnosed as Mhr. Definitive diagnosis was based on Mhr PCR positive result from serosal surfaces, fibrin, and/or joint fluid in combination with histological lesions, with the most common lesions described being synovitis, epicarditis, pleuritis, and polyserositis.

Figure 1: *Mycoplasma hyorhinis* population results

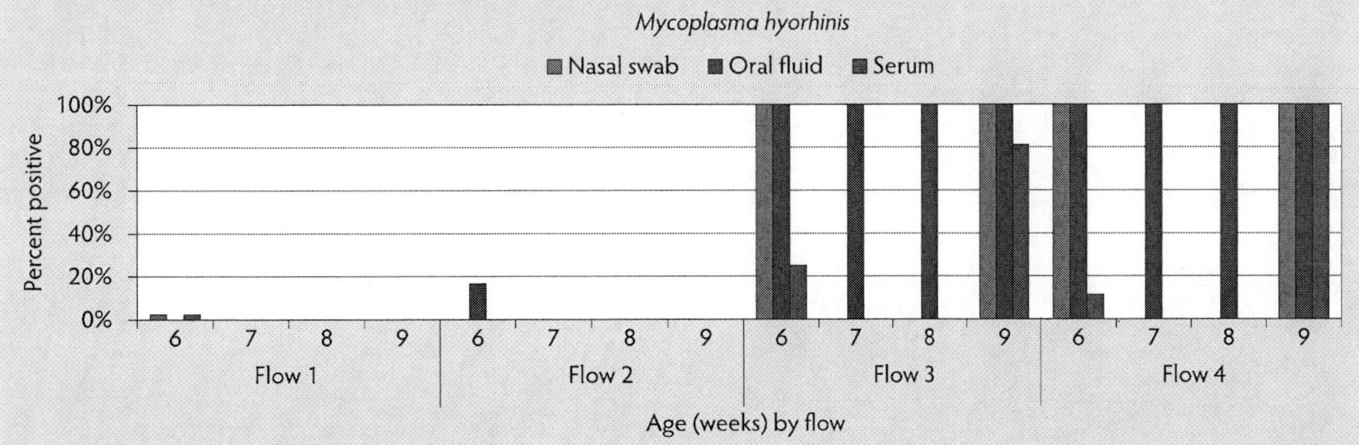

Conclusion

Population diagnostic results found Mhr positives on multiple samples types but individual pig diagnostics did not always indicate Mhr as the primary pathogen. The results of this study suggest that to definitively diagnose Mhr within groups of pigs exhibiting respiratory/systemic disease or lameness, fresh/fixed tissues and swabs from affected serosa collected from clinically acute individual pigs must be submitted to a diagnostic laboratory. Despite being a commensal bacteria, Mhr prevalence may vary by farm/flow; thus, population-based diagnostic results can be inconsistent and confusing. Additional cases are required to develop a case definition for Mhr associated disease and a risk factor analysis to measure the impact of Mhr.

Understanding bacterial resistance from clinical diagnostic submissions

T. Homann[1]; S. Wayne[2], DVM, PhD; S. Dee[2], DVM, PhD; R. Edler[2], MS; J. Nerem[2], DVM
[1]University of Minnesota; [2]Pipestone Veterinary Services

Introduction

Antibiotic resistance is one of the most important topics in human and veterinary medicine. Antibiotic use is a driver that affects the development of resistance. Therefore, judicious use of antibiotics is important to preserving their value. As there are limited data regarding the current level of resistance in common swine pathogens, the objective of this study is to develop a means to track the level of antibiotic resistance to these pathogens from clinical cases over time. Developing an average % resistance, i.e. index, is commonly used to understand CDC NARMs data and other human resistance monitoring; however, this is a new application to clinical isolates from swine populations.

Materials and methods

Data from January 2001 to July 2018 were requested from clinical swine cases submitted from one veterinary clinic to three VDLs: UMN, SDSU and ISU. Data collected included farm name, laboratory, accession ID, sample received date, the bacteria isolated, the lab-specific antibiotic sensitivity panel , minimum inhibitory concentration (MIC) values and MIC values interpretation. MIC methods (including cutoffs for MIC value interpretations) were determined to be consistent across labs.[1] Based on veterinary input, five bacteria were selected for testing: *Actinobacillis suis*, *Haemophilis parasuis* (HPS), *Escherichia coli* (*E coli*), *Salmonella* (all subtypes) and *Streptococcus suis*.

Data were organized in Microsoft Access® to standardize across VDLs, and then exported into Microsoft Excel® for analysis. A numerical value was assigned with 1 indicating "resistance" and 0 indicating "susceptible". These values were then averaged to create a resistance index, defined as the mean percentage of isolate/antibiotic combinations categorized as either "resistant" or "susceptible" for the designated time period (one month or one year).

Data were analyzed using XBar and range charts to assess total resistance index yearly variation over time.

Results

A total of 4,404 isolates were included in the analysis 325 *A suis*, 543 HPS, 1,659 *E coli*, 760 Salmonella and 1,117 *S suis*. Submission patterns were similar across years, averaging 7.6% *A suis*, 12.0% HPS, 38.9% *E coli*, 17.1% Salmonella and 25.3% *S suis*.

The overall resistance index across all five bacteria for 2018 was 44.4%, with a bacterial-specific resistance of 21.0% for *A suis*, 14.9% for HPS, 67.3% for *E coli*, 62.8% for Salmonella and 33.2% for *S suis*. The index was found to be similar across all three VDLS at 48.3% ISU, 47.8% UMN, 49.0% SDSU, as well as consistent over the years analyzed at 48.5% (Figure 1).

Discussion

Under the conditions of this study, it appeared that antimicrobial resistance to 5 common swine pathogens has remained consistent over time. This novel approach may provide swine veterinarians a means to track resistance and promote the practice of responsible use.

Reference

1. Iowa State University, Swine Susceptibility Panel, 2017, accessed via https://vetmed.iastate.edu/sites/default/files/VDL/pdf/2017-porcine-susceptibility.pdf.

Figure 1: XBar chart showing consistency of resistance index over time.

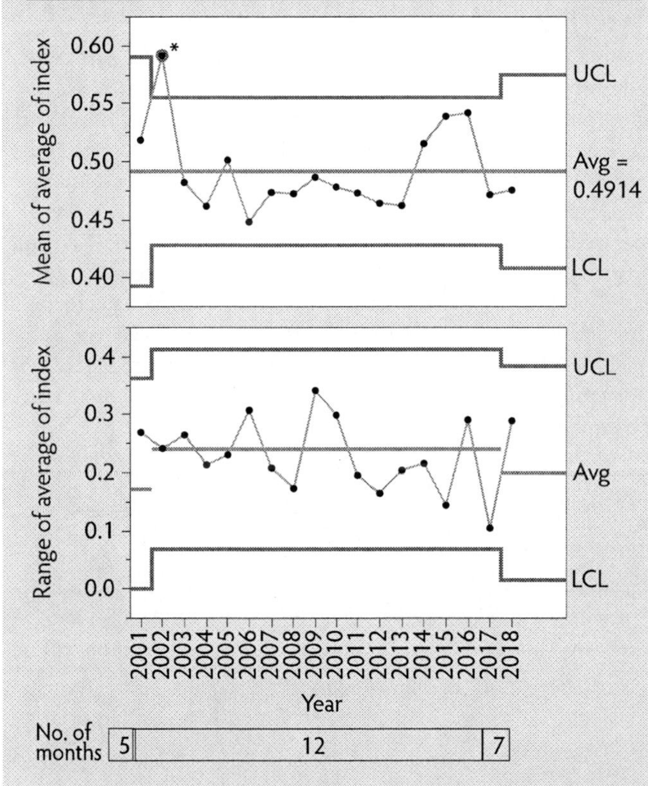

Hemoglobin concentration in sows: Correlation with the number of stillborn pigs

E. Noblett, MAS; G. Almond, DVM, PhD; J. B. Ferreira, DVM, MSc, DVSc
North Carolina State University College of Veterinary Medicine

Introduction

Iron, an essential trace element, binds to hemoglobin (Hb) and is vital for the transport of oxygen. Most iron is in red blood cells and stored as ferritin reserves which can be accessed with increased demand. Measurements of hemoglobin levels are used to determine if sows are anemic. The normal Hb range is 10-16 g/dL and less than 10 g/dL is indicative of anemia. As reported before, anemia is most prevalent at early lactation and Hb levels begin to trend upwards in late lactation.[1] In addition, sow parity also contributes to anemia; as parity increases the number of anemic sows also increases. Sows with hemoglobin levels < 9g/dL, are at risk for an increased occurrence of stillbirths.[2] A stillborn pig refers to a fetus that dies in utero prior to or during farrowing. Since these fetuses die before taking a breath, their lungs do not inflate. Performing lung flotation tests can help determine pre or post-partum death. Therefore, the objective of this study was to determine if there is a correlation between sow hemoglobin concentration and the number of stillborn piglets.

Materials and methods

Five sow farms (3,000-4,000 sows/farm) were included in this study. Sows (n = 50-150/farm) from varying parities were selected. Samples were collected in late gestation (> 110 days) and post farrowing for the Hb determinations. Blood was obtained from an ear vein prick with a 20g needle, loaded into a 10µl microcuvette, and processed in a HemoCue™ Hb 201. Sows were separated by Parity 1, 2, 3, and ≥ 4. Litter demographics/average performance were recorded for each sow. Dissections of deceased/stillborn piglets and lung flotation tests were performed. Lungs that floated in water represented post-partum deceased piglets and lungs that did not float represented intra-partum or pre-partum deaths. Pearson correlations were used to assess relationships among the various parameters.

Results

The post-farrowing Hb concentrations were less than the late gestation concentrations, resulting in a net decrease in mean Hb concentrations. Parity 1 and 2 sows had greater ($P < 0.05$) late gestation Hb concentrations than parity 3 and ≥ 4 sows. The parity ≥ 4 sows had lower ($P < 0.05$) post-farrowing Hb concentrations than sows in other parities. The total number of stillborn pigs per litter was 1.077 ± 0.07 and the number of pigs born alive per litter was 13.31 ± 0.156. As shown in Table 1, the late gestation Hb concentrations were correlated to the number of stillborn pigs, the number of pre-partum stillborn pigs, and the post-farrowing Hb concentrations. The decrease in Hb concentrations from late gestation to post-farrowing also was correlated with the number of pre-partum stillborn pigs. As parity increased, the late gestation Hb concentrations decreased, while the number of stillborn pigs increased.

Discussion

Based on the results, it is observed that Hb concentrations are associated with the occurrence of stillborn piglets, particularly, pre-partum stillbirths. The precise mechanism to explain the relationship between sow anemia and stillbirths is speculative; however, iron deficiency may contribute to impaired uterine contractions at farrowing. The low Hb concentrations in late gestation are likely due, at least in part, to the transfer of iron from the dam to the fetuses through the maternal uteroferrin - transferrin – ferritin pathway. Oxygen reserves are decreased in the fetuses of anemic sows, so there can be a delay in the piglet moving through the birth canal, thereby resulting a lack of oxygen and ultimately death. Additional studies are required to devise methods to reverse the anemia in older parity sows.

Table 1: Pearson correlations between variables.

Variable	Variable	Pearson correlation	*P*-value
Late gestation Hb concentrations	Number of stillborn pigs	-0.1538	0.0023
Late gestation Hb concentrations	Number of pre-partum stillborn pigs	-0.2307	0.0001
Late gestation Hb concentrations	Post-farrowing Hb concentrations	0.1872	0.0002
Change in Hb concentrations	Number of pre-partum stillborn pigs	0.1449	0.0041
Parity	Late gestation Hb concentrations	-0.2416	0.0001
Parity	Number of stillborn pigs	0.1930	0.0001
Number of pigs born alive	Number of stillborn pigs	-0.1738	0.0006

References

1. Castevens K, Almond G, Ferreira JB. An assessment of hemoglobin concentrations in relation to reproductive stage and parity in sows. Proceedings 49th Annual Meeting of the American Association of Swine Veterinarians, 2018. Pp 282-283.

2. Cutler R, Hurtgen JP, Leman AD. Reproductive system.In: Leman AD, Glock RD, Mengeling WL, Penny RHC, Scholl E, Straw B, eds. Diseases of Swine. 5th ed. Ames, Iowa: Iowa State University Press, 1981: 96-129.

The effect of vaccination of gilts with a pre-farrow rotavirus RNA particle vaccine following pre-breeding natural planned exposure on piglet growth and viral shedding

A. Anderson[1], BS; J. Pittman[2], DVM, MS DABVP; D. Marthaler[3], PhD
[1]College of Veterinary Medicine, Iowa State University, Ames, Iowa; [2]Smithfield Hog Production, North Region, Waverly, Virginia; [3]Veterinary Diagnostic Laboratory, Kansas State University, Manhattan, Kansas

Introduction

Rotavirus A (RVA) and rotavirus C (RVC) cause diarrhea in piglets. Due to limited cross protection between and within serogroups, natural planned exposure (NPE) is used to immunize sows prior to farrowing. In a recent study, the author found that NPE at 5, 4, and 3 weeks prior to farrowing (WPF) resulted in 1.4 pounds increased weaning weight, 4.08% lower pre-weaning mortality, and $80,000 in additional annual profit on an 1,800-head gilt farm. However, gilts exposed to NPE shed RV in farrowing while control gilts did not. For the follow-up study reported here, the goal was to limit RV shedding in farrowing and improve gilt immune status by vaccinating gilts pre-farrow with an RNA particle (RP) vaccine after initial live virus exposure prior to breeding. The objective was to compare piglet growth and viral shedding from gilts immunized using NPE or RP vaccine prior to farrowing.

Materials and methods

The study was conducted on the same farm as the previous study. NPE containing farm-specific RVA and RVC was administered to gilts twice prior to breeding. 117 gilts were randomly allocated into 4 pre-farrowing treatments; Group 1 (NPE at 5, 4, and 3 WPF; positive control), Group 2 (RP at 5 and 2 WPF), Group 3 (RP at 1 WPF), and a non-treated control group. Fecal samples were collected from gilts five times including at entry into farrowing. Piglets were weighed and tagged at 2 days of age. Three pigs per litter were randomly selected for fecal swabbing at 3, 7, 14, and 21 days of age and the samples were pooled by litter for testing. Litter diarrhea scores were recorded weekly and pigs were weighed again at weaning. Individual gilt samples and the litter pools were tested for RVA, RVB (data not shown) and RVC by PCR. Statistical analysis was done by linear regression, One-Way ANOVA and Chi Square depending on data type and distribution (Statistix™ 10, Analytical Software, Tallahassee, FL).

Results

Selected results are reported in Table 1. Prior to breeding, a low rate of shedding was found in the gilts. After pre-breeding and pre-farrowing NPE, most gilts shed RVA but few shed RVC. At farrowing, Group 1 and control gilts were shedding more RV than Groups 2 and 3. Piglets, regardless of treatment, did not shed RVA while RVC was shed at every time point, in all treatment groups. Generally, RVC shedding was greater in Group 1, while the rate of scouring litters was higher in the control group. Interestingly, average pig and litter wean weights were greatest in Group 3, but this difference was not significant.

Discussion and conclusions

The lack of differences between the treatment groups with regard to shedding, diarrhea, and growth rates in the piglets may be attributed to a low level of challenge/exposure. This could be due, in whole or part, to implementation of the previous study protocol resulting in decreased clinical disease and RV environmental load and addition of pre-breeding NPE resulting in sufficient immunity in all treatment groups. Accordingly, the pre-farrowing treatments had no additional, measurable benefit. Further studies are warranted to determine the efficacy of pre-breeding NPE and pre-farrowing RP vaccine under more severe RV challenge conditions. If efficacious, RP vaccine could potentially eliminate risks associated with NPE administration including continuous live virus on-farm, potential spread of other pathogens, difficulty of isolating on-farm strains, and labor of producing NPE material.

Table 1: Summary of RVA and RVC shedding in gilts and piglets, clinical signs, and production parameters by treatment group.

Parameter			Treatment groups Group 1: NPE 3X	 Group 2: RP 5&2 WPF	 Group 3: RP 1 WPF	 Group 4: Control
Number of sows and litters			31	27	29	30
Percentage of PCR positive gilts at farrowing by subtype		A	6.5%	0.0%	3.5%	10.7%
		C	6.5%	7.4%	3.5%	0.0%
Percentage of PCR positive litters by subtype and pig age at sampling	RVA	3 days	0.0%	0.0%	0.0%	0.0%
		1 week	0.0%	0.0%	0.0%	0.0%
		2 weeks	0.0%	0.0%	0.0%	0.0%
		Weaning	0.0%	0.0%	0.0%	0.0%
	RVC	3 days	3.2%	3.7%	3.5%	3.3%
		1 week	16.1%	11.1%	6.9%	6.7%
		2 weeks	22.6%	14.8%	6.9%	10.0%
		Weaning	35.5%	40.7%	27.6%	23.3%
Percentage of litters with diarrhea by pig age		1 week	22.6%	18.5%	20.7%	36.7%
		2 weeks	3.2%	3.7%	3.5%	16.7%
		Weaning	19.4%	3.7%	13.8%	23.3%
Pigs started per litter			12.4	13.3	12.7	12.3
Percentage of pigs died or no value			12.5%	17.5%	13.9%	14.4%
Average adjusted pig wean weight (lbs)			13.2	13.4	13.7	13.4
Average adjusted litter wean weight (lbs)			140.6	142.3	144.5	139.0

Evaluation of detergent use to improve cleaning of swine transport vehicles as measured by ATP bioluminescence

Megan E. Bloemer[1], BS; Shamus Brown[2], DVM
[1]University of Illinois College of Veterinary Medicine, Urbana, Illinois; [2]Iowa Select Farms, Iowa Falls, Iowa

Introduction

Transportation is a potential route for transmitting pathogens if not properly managed. The objective of this study was to determine if the use of detergent reduces pathogen load on trailer surfaces. The evaluation was done using ATP bioluminescence as an indicator of pathogen load. It was hypothesized that the use of a phosphoric acid detergent would significantly reduce ATP bioluminescence levels in trailer surfaces.

Materials and methods

The study was conducted at a market truck wash. Fifty trailers were enrolled over the course of six days; 25 control and 25 treated trailers. The process order was: 1) scrape out wood chips and manure, 2) high volume wash, 3) high pressure wash with hot water, and 4) foaming disinfection. Both treated and control groups were cleaned out in the same way, with the treated group having the phosphoric acid detergent applied after the wood chips were scraped out with a ten-minute contact time ensured. ATP bioluminescence levels were recorded using an ATP reader. ATP levels were measured at two different time points per trailer: 1) immediately after hot water pressure washing and 2) after disinfection, with at least 20 minutes between disinfection and sample collection. A swab sample of a 10 cm^2 floor area of was collected along with a matching wall sample at four locations (back left, front left, front right, and back right of both the top and bottom deck) on each deck for a total of eight samples per deck and sixteen samples per trailer. Samples were collected with swabs using a grid so the same areas were sampled on every trailer. In addition to recording the ATP levels, trailer use (feeder or market loads), number of loads that trailer had hauled between washes, wash bay, crew member, detergent application time and the total time it took to clean each trailer were recorded. Before each trailer was scraped a subjective dirtiness score (1-3) was recorded. ATP bioluminescence scores for each trailer were averaged to provide a composite score for each trailer. The data was transformed to log10 and then analyzed by linear regression.

Results and discussion

Results of the statistical analysis are summarized in Table 1. Variables that significantly impacted the composite ATP bioluminescence scores after the hot water phase included detergent treatment group (less with detergent), load type (market > feeder pig), number of loads prior to last wash (less was worse), and standing water present (increased). After disinfection only the trailer number and ATP level prior to disinfection were significant. ATP bioluminescence readings were lower for the treatment group after the hot water phase but were similar to the control trailers after the disinfection phase. Significant differences after the hot water phase, such as load type, were no longer significant after disinfection. Variables related to personnel, trailer age and type, and wash facility were not significant, indicating the overall wash process was done consistently. Our inability to show differences in ATP levels may be due to factors such as testing wet surfaces, rather than dry surfaces as suggested by the manufacturer. In addition, ATP levels may not be a good indicator for the level of live bacteria. Utilizing culture swabs and PCR testing may better assess pathogen load. Further studies may be warranted in order to more definitively quantify any potential benefit of detergent.

Table 1: Linear regression analysis of ATP levels

	End hot water		End disinfection	
Explanatory variable	**Significant**	**Direction**	**Significant**	**Direction**
Treatment group	Yes	Negative[1]	No	
Trailer number	No		Yes	NA[2]
Trailer model	No		No	
Trailer age	No		No	
Sample day	No		No	
Load type	Yes	Positive[3]	No	
Visual inspection score	No		No	
No. loads prior to cleaning	Yes	Negative[4]	No	
Bay number	No		No	
Wash person	No		No	
Water temperature	No		No	
Wash time	No		No	
Detergent time	Yes	Positive[5]	No	
Standing water	Yes	Positive[6]	No	
Sat with detergent	No		No	
ATP level end hot water	NA	NA	Yes	Positive[7]

1 Control > treated
2 Trailer numbers are categorical
3 Market > feeder
4 Increase number of prior loads → decreased ATP
5 Increased detergent time → increased ATP
6 Standing water increased ATP
7 Increased ATP at end of hot water → increased ATP after disinfection

Fecal microbiota transplantation shifts microbiome composition and reduces morbidity and mortality associated with PCVAD

L. A. Constance[1]; R. R. R. Rowland[1]; W. Abbas[2]; S. C. Fernando[2]; M. L. Potter[3]; M. A. Sheahan[1]; T. E. Burkey[2]; R. A. Hesse[1,4]; A. G. Cino-Ozuna[1,4]; M. C. Niederwerder[1]

[1]Department of Diagnostic Medicine/Pathobiology, College of Veterinary Medicine, Kansas State University, Manhattan, Kansas; [2]Department of Animal Science, University of Nebraska–Lincoln, Lincoln, Nebraska; [3]Abilene Animal Hospital, PA, Abilene, Kansas; [4]Kansas State Veterinary Diagnostic Laboratory, Kansas State University, Manhattan, Kansas

Introduction

Porcine circovirus associated disease (PCVAD) is a term used to describe the multifactorial disease syndromes caused by porcine circovirus type 2 (PCV-2), which can be reproduced in an experimental setting through the co-infection of pigs with PCV-2 and porcine reproductive and respiratory syndrome virus (PRRSV).[1-3] The resulting PCVAD affected pigs represent a subpopulation within the co-infected group. In co-infection studies, the presence of increased microbiome diversity is linked to a reduction in clinical signs.[4,5] In this study, fecal microbiota transplantation (FMT) was investigated as a means to prevent PCVAD in pigs co-infected with PRRSV and PCV-2d.

Materials and methods

The sources of the FMT material were high-parity sows with a documented history of high health status and robust litter characteristics. The analysis of the donated FMT material showed the absence of common pathogens along with the presence of diverse microbial phyla and families. One group of pigs (n = 10) was administered the FMT while a control group (n = 10) was administered a sterile mock-transplant. Fecal microbiomes of the transplanted and control groups were analyzed before and after FMT or mock-transplantation by a pan-microbial array (LLMDA) and 16S rDNA sequencing.

Results and discussion

Compared to the control pigs, transplanted pigs had reduced species diversity in the families *Spirochaetaceae* and *Vibrionaceae* coupled with an increase in the relative abundance of the families *Veillonellaceae*, *Lachnospiraceae* and *Ruminoccaceae*. Over the 42-day post infection period, the FMT group showed fewer PCVAD-affected pigs, as evidenced by a significant reduction in morbidity and mortality in transplanted pigs, along with increased antibody levels. Overall, this study provides evidence that FMT leads to shifts in microbiome composition and abundance of several bacterial families that are associated with a reduction in clinical signs of PCVAD following co-infection with PRRSV and PCV-2.

Acknowledgements

This work was supported by the Kansas State University College of Veterinary Medicine Success for Young Investigators Grant Program, start-up funds from Kansas State University College of Veterinary Medicine, and the State of Kansas National Bio and Agro-Defense Facility Fund.

References

1. Segales J (2012). *Virus Res* 164:10-19.
2. Trible BR., Ramirez A, Suddith A, et al. (2012). *Vaccine*. 30: 4079-4085.
3. Niederwerder MC, Bawa B, Serao NV, et al. (2015). *Clin. Vaccine Immunol.* 22: 1244–1254.
4. Niederwerder MC, Jaing CJ, Thissen JB, et al. (2016). *Vet Microbiol.* 188: 1-11.
5. Ober RA, Thissen JB, Jaing CJ, et al. (2017). *Vet Microbiol.* 208:203-211.

Implications of operating conditions in UV light decontamination chambers for items entering swine facilities

J. Kordas[1]; E. Kettelkamp[1]; B. Blair[1]; J. Lowe[1,2]
[1]College of Veterinary Medicine, University of Illinois, Urbana, Illinois; [2]Lowe Consulting Ltd., Mahomet, Illinois

Introduction

Biosecurity to limit the introduction of novel pathogens is a routine practice. With feed, pigs, supplies, personal items, and people entering and leaving daily, implementation of practices such as decontaminating personal items with UV light is common. Although commercial UV chambers are available, many farms design their own UV chambers for such use. Literature supports the bactericidal effects of UV light, however, validation of the effectiveness of UV light chambers in swine farms is lacking in the literature. The effects of wavelength and shadowing are currently undocumented and therefore the practical effectiveness of UV light decontamination under field conditions for personal item entry is poorly understood.

Objective

Measure the impacts of wavelength and shadowing on the efficacy of UV light decontamination of personal supplies entering swine farms.

Materials and methods

A model UV chamber was designed from a 30L plastic bin lined with reflective material and two light fixtures at opposite corners. Salmonella was selected as an indicator organism because of its resistance to UV sanitation. 9cm^2 square areas were marked on sterilized plastic containers (n = 10) and inoculated, inside and out, with salmonella bacteria in 20μL of fluid (Enterisol® *Salmonella T/C, BIAH*).[1] Substrates were subjected to 5, 10, or 40 minutes of UVA (315 nm), B (280 nm), or C(237.5nm) light exposure. Substrates were placed directly in the chamber with or without a covering of clear plastic, opaque plastic, paper, cotton, or polyester bags to assess UV penetration.[2] 25W UVA, B, and C bulbs were used and light intensity was measured (W/cm2) within the chamber using GENERAL UV Digital Light Meters. Post-treatment swabs were collected and then submerged in 1.0mL peptone water. Peptone samples were enriched in tryptic soy broth for 24 hours. Enriched and non-enriched samples were plated on 3M petrifilm RAC plates and read with a 3M Petrifilm™ Plate Reader. Statistical analysis (two-proportion z-tests) was conducted using R.

Results and discussion

Only UVC light reduced the number of positive sample areas following light treatment; UVA and UVB had no effect. UVC for 40 minutes decontaminated the surfaces significantly more effectively than did UVC at 10 or 20 minutes, as seen in Table 1. ($P = 0.009$). Differences in UVC penetration through the coverings resulted in varying degrees of decontamination. Opaque plastic, and clear plastic were similar to uncovered surfaces while cotton, polyester, and paper blocked the effect of UVC on the bacteria. The amount of UVC light varied depending on orientation and location across the model chamber, demonstrating the impacts of shadowing on the decontamination process. Enriched samples returned more positive results than non-enriched samples, demonstrating the inadequacy of the decontamination process. These data indicate that operational conditions of the decontamination chamber have significant implications on its effectiveness. The location and number of light sources determine the area of adequate UVC concentration. Covering and placement of items in the chamber impact the effectiveness of decontaminating all surfaces of an object. With resistant pathogens, such as salmonella, complete decontamination was not achieved. More susceptible pathogens such as PRRSv (results pending) may be more effectively killed. Even under ideal conditions, it cannot be assumed that UV light decontamination will be complete for pathogens which are more resistant to survival outside the host or novel pathogens with unknown characteristics.

References

1. Holck, A. L., Liland, K. H., Dromtorp, S. M., Carlehög, M., & Mcleod, A. (2018). Comparison of UV-C and Pulsed UV Light Treatments for Reduction of Salmonella, Listeria monocytogenes, and Enterohemorrhagic Escherichia coli on Eggs. *Journal of Food Protection, 81*(1), 6-16.

2. Dee, S., Otake, S., & Deen, J. (2011). An evaluation of ultraviolet light (UV254) as a means to inactivate porcine reproductive and respiratory syndrome virus on common farm surfaces and materials. *Veterinary Microbiology, 150*(1-2), 96-99.

Table 1: UVC disinfection summary

UV	Time (min)	Covering	Enriched	Positive (inside)	Total	Positive (outside)	Total	% Killed
C	5	none	-	2	8	1	8	81.25
C	10	none	-	4	12	3	12	70.83
C	40	none	-	0	10	0	10	100
C	10	Clear plastic	-	1	8	0	8	93.75
C	10	Opaque plastic	-	3	9	3	9	66.67
C	40	Clear plastic	-	0	9	0	9	100
C	40	Opaque plastic	-	2	8	1	8	81.25
C	5	none	+	2	8	5	8	56.25
C	10	none	+	3	12	3	12	75
C	40	none	+	0	10	2	10	90
C	10	Clear plastic	+	5	8	4	8	43.75
C	10	Opaque plastic	+	5	7	4	7	35.71
C	40	Clear plastic	+	2	9	3	9	72.22
C	40	Opaque plastic	+	2	8	2	8	75

Evaluation of PCV2/Mhp/PRRSV vaccination regimens on growing pig performance under natural exposure

Jamie Madigan[1], BS; Joshua Duff[2], DVM
[1]North Carolina State University, Raleigh, North Carolina; [2]Maxwell Foods, LLC

Introduction

Porcine circovirus 2 (PCV2), *Mycoplasma hyopneumoniae* (Mhp), and porcine reproductive and respiratory syndrome virus (PRRSV) are individually important primary pathogens, as well as important components of the porcine respiratory disease complex. Vaccines are available for all three pathogens but formulating the optimal vaccination regimen can be challenging due to the complex interactions of pathogens along with, labor and economic considerations. The purpose of this study was to evaluate four different vaccination regimens on feed efficiency (nursery only), average daily gain, mortality and post-vaccination antibody responses from weaning to market. These regimens addressed labor (single vs split dose for PCV2/Mhp vaccination) and cost (MLV PRRSV vaccination) implications for this system.

Materials and methods

The nursery phase of the study was conducted in two barns, filled one week apart. Two sow farm sources were used to fill each barn and pigs were sized and sexed at the sow farms. Pigs were sorted into pens of 29 pigs based on size with 14-15 barrows and 14-15 gilts. Sow farm sources were kept separate by pen. Each pen was assigned to either a 2mL MLV PRRSV vaccination or not at placement (3 weeks of age (WOA)) along with assignment to either 2ml PCV2/Mhp vaccine at placement or a 1ml PCV2/Mhp vaccination at placement and three weeks later, providing 4 different vaccination regimens. Pigs were tagged according to their pen's regimen. In the nursery, pens were weighed at placement, 6 WOA, and 9 WOA. Pigs were then moved by regimen to the finisher site. Each regimen group was sorted to have 21 pigs per pen with 10-11 barrows and 10-11 gilts. Pens were weighed at the start of finishing and prior to the first cut of market weight pigs.

To monitor disease status, 3 oral fluid samples were collected from each treatment group at 9, 15 and 21 WOA and tested for Mhp and PRRSV by PCR. In addition, 15 blood samples were collected from each treatment group at 3, 9, 15, and 21 WOA and tested individually for Mhp and PCV2 antibody by ELISA and PCV2 by PCR in pools of five. Sequencing was done on the 15 WOA serum and oral fluid samples for PCV2 and PRRSV, respectively. Statistical analysis was done by linear regression.

Results and discussion

Table 1 presents the growth performance and diagnostic testing results. There were no significant differences between the treatment regimens with regard to nursery or finisher growth performance and mortality/removal rates. The groups that received the split PCV2/Mhp vaccine regimen had statistically higher PCV2 and Mhp antibody levels compared to pigs that received the single vaccine regimen at 9, 15 and 21 WOA ($P < 0.05$). Sequencing of serum samples identified PCV2d. In oral fluids, a 1-7-4 PRRSV strain was identified and a MLV PRRSV vaccine like strain was identified in the non-MLV PRRSV regimen groups.

In summary, there were no significant differences between the treatment regimens in nursery or finisher performance. There were statistical differences in antibody responses between the split and single PCV2/Mhp vaccine regimens. Sequencing demonstrated that the MLV PRRSV vaccine strain had moved to the non-vaccinated pigs, which could have affected the results. Overall, it is critical for vaccination regimen evaluations to be conducted in order to guide recommendations for future vaccination protocols that would optimize labor efficiency, vaccine costs, and performance benefit.

Table 1: Growth performance and diagnostic testing results

Data type	Parameter	Unit	MLV PRRS		No MLV PRRS	
			Split PCV2/Mhp	Single PCV2/Mhp	Split PCV2/Mhp	Single PCV2/Mhp
Nursery performance	No. pigs	Count	551	551	547	548
	No. pens	Count	20	20	20	20
	Avg. start wgt.	Pound	15.3	15.3	15.4	14.9
	Avg. end wgt.	Pound	57.4	57.3	57.3	56.6
	Avg. days on feed	Count	41.2	41.3	41.3	41.4
	Avg. daily gain	Pound	0.96	0.97	0.98	0.97
	Avg. daily feed intake	Pound	1.268	1.26	1.265	1.25
	Feed/gain	Ratio	1.325	1.30	1.298	1.29
	Percent mortality	Rate	2.5	2.2	2.2	1.5
	Percent removed	Rate	2.2	1.6	1.5	1.8
Finisher performance	No. pigs	Count	514	518	514	521
	No. pens	Count	25	25	25	25
	Avg. start wgt.	Pound	59.7	59.6	59.3	58.8
	Avg. end wgt.	Pound	225.6	226.2	224.9	224.5
	Avg. days on feed	Count	89.0	89.0	89.0	89.0
	Avg. daily gain	Pound	1.86	1.87	1.86	1.86
	Percent mortality	Rate	4.7	5.0	4.3	5.2
Oral fluid PCR -NA PRRS	9 weeks of age	No. positive	6/6	6/6	6/6	6/6
	15 weeks of age		2/6	3/6	4/6	5/6
	21 weeks of age		0/6	0/6	1/6	0/6
Oral fluid PCR - Mhp	9 weeks of age	No. positive	0/6	0/6	0/6	0/6
	15 weeks of age		0/6	0/6	0/6	0/6
	21 weeks of age		1/6	0/6	1/6	0/6
Serum PCR - PCV2	Weaning	No. positive	0/6	0/6	0/6	0/6
	9 weeks of age		1/6	1/6	0/6	0/6
	15 weeks of age		0/6	0/6	0/6	2/6
	21 weeks of age		0/6	2/6	0/6	0/6
PCV2 ELISA	Weaning	No. positive	30/30	30/30	27/30	30/30
		Geom. titer	2,683	2,106	2,676	1,308
	9 weeks of age	No. positive	29/30	27/30	27/30	20/30
		Geom. titer	50,653	464	17,660	468
	15 weeks of age	No. positive	23/30	13/30	28/30	15/30
		Geom. titer	539	193	899	212
	21 weeks of age	No. positive	11/30	9/30	21/30	10/30
		Geom. titer	178	152	367	185
Mhp ELISA	Weaning	No. positive	19/30	19/30	20/30	15/30
		S/P ratio	0.626	0.691	0.779	0.584
	9 weeks of age	No. positive	26/30	1/30	24/30	4/30
		S/P ratio	0.948	0.176	0.998	0.361
	15 weeks of age	No. positive	19/30	1/30	18/30	3/30
		S/P ratio	0.576	0.104	0.521	0.157
	21 weeks of age	No. positive	25/30	7/30	24/30	7/30
		S/P ratio	0.844	0.260	0.665	0.289

Novel tools to control old pathogens: molecular characterization of *Haemophilus parasuis* field isolates by serotyping PCR (sPCR) and virulence-associated autotransporters PCR (vtaA-PCR)

T. P. Marostica[1]; N. Macedo[1], DVM, MS, PhD; O. Sahin[1], DVM, PhD; M. Gottschalk[2], DVM, PhD; A. Tucker[3], DVM, PhD; K. Harmon[1], PhD; M. J. Clavijo[1]
[1]Veterinary Diagnostic and Production Animal Medicine, Iowa State University, Ames, Iowa;
[2]Department of Microbiology, Infectious Diseases and Immunology, University of Montreal, Montreal, Quebec, Canada;
[3]University of Cambridge, Cambridge, United Kingdom

Introduction

Haemophilus parasuis (*HPS*) continues to challenge swine producers causing systemic disease in pigs. HPS isolates have been classified into 15 serotypes by gel immunodiffusion and hemagglutination.[1,2] The accuracy of these methods is limited.[3] Within serotype groups there is good protection, whereas cross-protection between different serotypes is generally poor.[1] Effective control of HPS hinges on the use of autogenous or homologous commercial vaccines, strategic antimicrobial treatment, minimizing environmental stressors and reducing commingling of multiple sources.[3] Appropriate molecular characterization of circulating field strains can aid in the development of improved HPS control efforts. The objective was to characterize HPS field strains by serotyping and virulence-associated PCR and o describe the distribution of serotypes and presence of the virulence-associated marker within HPS field isolates from different swine systems over a 2 year period.

Materials and methods

Two hundred and twenty three (n = 223) isolates from diseased pigs in 14 states were included in the study.[4] All isolates were from cases submitted to the Veterinary Diagnostic Laboratory between 2014 to 2018. Isolates originated from a variety of sample type including lung, lung surface, brain, pericardium, heart, heart surface, joint and nasal cavity. The age of the pigs varied from 3 to 22 weeks. Within the 223 isolates, a total of 123 originated from 6 different swine systems (A-F) recovered between 2016-2018. All samples were tested by serotyping (sPCR) and virulence-associated autotransporters PCR (vtaA-PCR).[5]

Results and discussion

Most isolates originated from polyserositis cases (61%), compared to respiratory cases (39%). Isolates mostly were from the lung (63%), compared to systemic sites (i.e. serosal surfaces, joint, brain). Serotypes 4 (24%; 54/223) and 7 (17%; 38/223) were the most commonly detected serotypes, followed by serotypes 1 (15%; 33/223), 2 (13%; 28/223), 13 (10%; 22/223), 5 (9%; 19/223), 14 (5%; 12/223), 12 (5%; 12/223), and 6 (1%; 2/223). Serotype 1, 13, 5 and 12 isolates originated most frequently from systemic sites, whereas serotype 4 isolates were mostly from lung samples. Similar results have been reported in the UK and Canada. Less than 1% (3/223) of isolates were non-typeable. A total of 222/223 isolates were vtaA positive. Serotypes 6 and 7, previously considered avirulent,[5] originated from systemic sites of pigs with polyserositis and were vtaA positive. There was no correlation observed between age or sample site and serotype. Most systems had at least four different serotypes detected over a 3-year period (Figure 1), suggesting the potential need for multivalent autogenous vaccines and highlighting the importance of mixing sources with the same serotype. However, in most cases, only one serotype prevailed within a flow. Serotype 4 (23.5%) and 7 (22.7%) were detected in all systems. Control of HPS requires a multifaceted approach. The first step is to accurately characterize the isolates found in diseased pigs. These tools can be useful for pathogen surveillance, identification of appropriate autogenous vaccine candidates, to improve pig flow management by commingling sources with the same serotype and to prevent introduction of pigs with potentially virulent HPS strains.

References

1. Howell KJ, Peters SE, Wang J, Hernandez-Garcia J, Weinert LA, Luan SL, Chaudhuri RR, Angen Ø, Aragon V, Williamson SM, Parkhill J. Development of a multiplex PCR for rapid molecular serotyping of *Haemophilus parasuis*. *Journal of clinical microbiology*. 2015 Sep 30:JCM-01991.

2. Lacouture S, Rodriguez E, Strutzberg-Minder K, Gottschalk M. Canada: serotyping of Haemophilus parasuis field isolates from diseased pigs in Quebec by indirect hemagglutination assay and multiplex polymerase chain reaction (PCR). *The Canadian Veterinary Journal*. 2017 Aug;58(8):802.

3. Jia A, Zhou R, Fan H, Yang K, Zhang J, Xu Y, Wang G, Liao M. Development of serotype-specific PCR assays for typing of Haemophilus parasuis circulating in southern China. *Journal of clinical microbiology*. 2017 Sep 6:JCM-00688.

4. Iowa State University, Veterinary Diagnostic data 2018.

5. Galofré-Milà N, Correa-Fiz F, Lacouture S, Gottschalk M, Strutzberg-Minder K, Bensaid A, Pina-Pedrero S, Aragon V. A robust PCR for the differentiation of potential virulent strains of Haemophilus parasuis. *BMC veterinary research*. 2017 Dec;13(1):124.

Figure 1: Distribution of HPS serotypes in six different swine systems

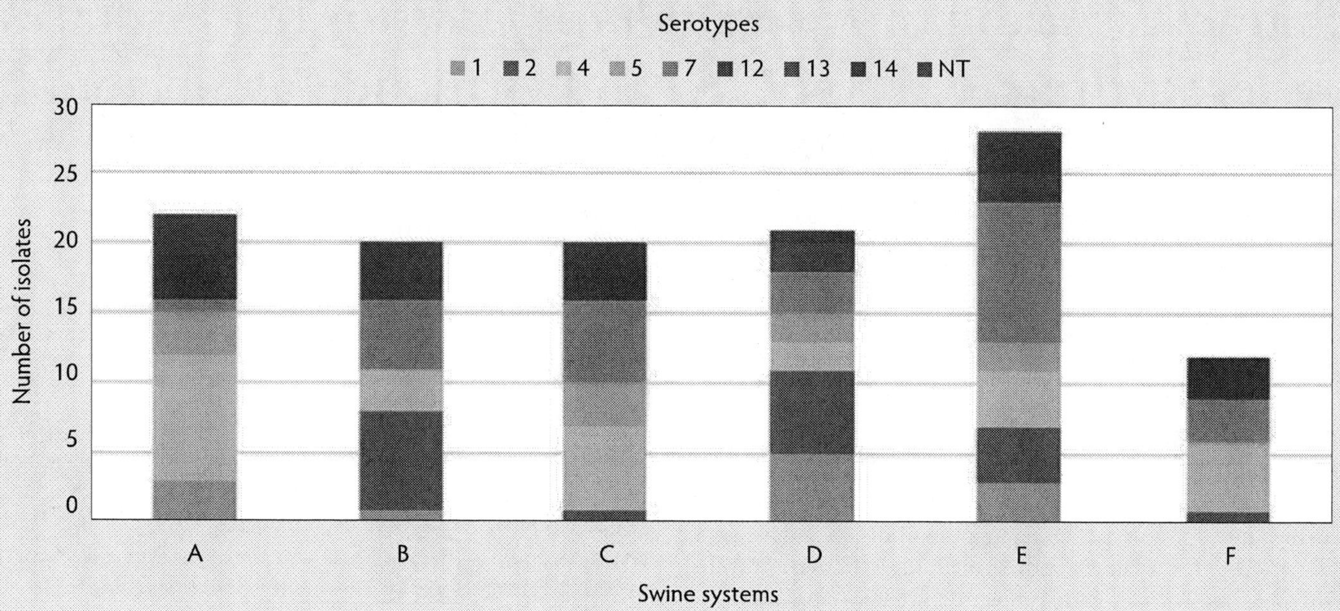

Detection of pig death patterns during the wean-to-finish period, differences in profitability, and association to management practices and presence of endemic pathogens

Samantha C. Mehling[1]; Jeremy Maurer[2]; Ed Kluber[2]; Rachel Stika[1]; Christopher Rademacher[1]; Jeff Zimmerman[1]; Luis Giménez-Lirola[1]; Chong Wang[1]; Derald Holtkamp[1]; Rodger Main[1]; David Baum[1]
[1]Iowa State University College of Veterinary Medicine, Veterinary Diagnostic and Production Animal Medicine, Ames, Iowa
[2]Smithfield Foods, Milford, Utah

Introduction

An 80,000-sow pork production system with a longstanding history of sporadic, excessive deaths near the end of the finishing period (approximately 14-22 weeks post-placement) sought to determine factors associated with these deaths. The objectives of this study were to determine if discernable patterns in number of dead pigs exist in this commercial wean-to-finish (WTF) system and to determine how such patterns, if present, affect profitability.

Materials and methods

The WTF portion of the system was used for this observational study. Sixty lots (approximately 2,000 pigs/lot) came from six 5,000-sow farms. Each WTF site consisted of six barns connected by a central hallway, with three barns on either side. The system provided production management data for each lot's entire growing period, including beginning inventory and weekly number of dead pigs (NDP). Weekly NDP was recorded for each lot from daily counts performed by barn managers. All lots' NDP were plotted by week on feed. Lots with visually observed similarities in NDP were grouped based on which week(s) on feed increased NDP appeared. The WTF period was divided into four phases: placement (weeks 1-4), nursery (weeks 5-9), early finishing (weeks 10-16) and late finishing (weeks 16 - market). These phases allowed for accurate calculation of feed consumption per phase by a linked-cell economic model, which calculated profitability for each cluster. A WTF feed model, created on a spreadsheet, calculated average daily gain (ADG) and feed conversion ratio (FCR) in each phase for each cluster. Oral fluid samples were collected bi-weekly over the finishing period using fixed spatial sampling. Sixty-six samples were collected from each lot over their first 22 weeks on feed for a total of 3,960 samples. Tests are currently being run to identify antibody against *Actinobacillus pleuropneumoniae* (APP) ApxIV, influenza A virus nucleoprotein (IAV np), porcine reproductive and respiratory virus (PRRSV), and porcine epidemic diarrhea virus (PEDV) and pathogen presence (PCR) of PRRSV, APP, IAV, PEDV, porcine circovirus type 2 (PCV2), and *Mycoplasma hyopneumoniae* (Mhp).

Results

NDP by week occurred in three detectable patterns as seen in Figure 1, referred to as clusters I-III. NDP in Clusters I and III increased after week 14 but in different patterns. Univariate analysis of NDP demonstrated significantly lower NDP in Cluster II. The economic model detected differences in economic impact that occurred during the last finishing phase: Cluster I showed the largest profit loss, cluster II the lowest loss and cluster III in-between. ADG and FCR values reported by the farms were lower in the late finishing phase relative to ADG and FCR from the WTF feed model: Clusters I-III ADG performed at 76%, 99%, and 86% of expected values, respectively; FCR 51%, 75%, and 63%, respectively. This suggests poorer biological performance than expected. Preliminary serology and PCR results show similar patterns throughout the finishing periods for each cluster but with differences noted at different sampling times post placement. The biological significance of these differences is unclear. Univariate analysis detected significant differences in weaning weight and stocking density between clusters. Cluster II lots had lower average weaning weight, more square feet per pig and the lowest NDP compared to lots from Clusters I and III.

Conclusions

Growing pigs in the US face multiple performance challenges from various environmental factors, which can reduce profit. Veterinarian-supervised health management teams are encouraged to establish a program to monitor inventory (NDP, pigs treated, and necropsy results), analyze management data in real time, and fixed spatial sampling surveillance of endemic pathogens to determine dead pig and pathogen patterns. They may then look for associations between and among NDP, treatment, stocking practices and pathogen involvement in order to allocate resources to improve production and profitability.

Figure 1: Wean-to-finish lots' number of dead pigs (NDP) by cluster

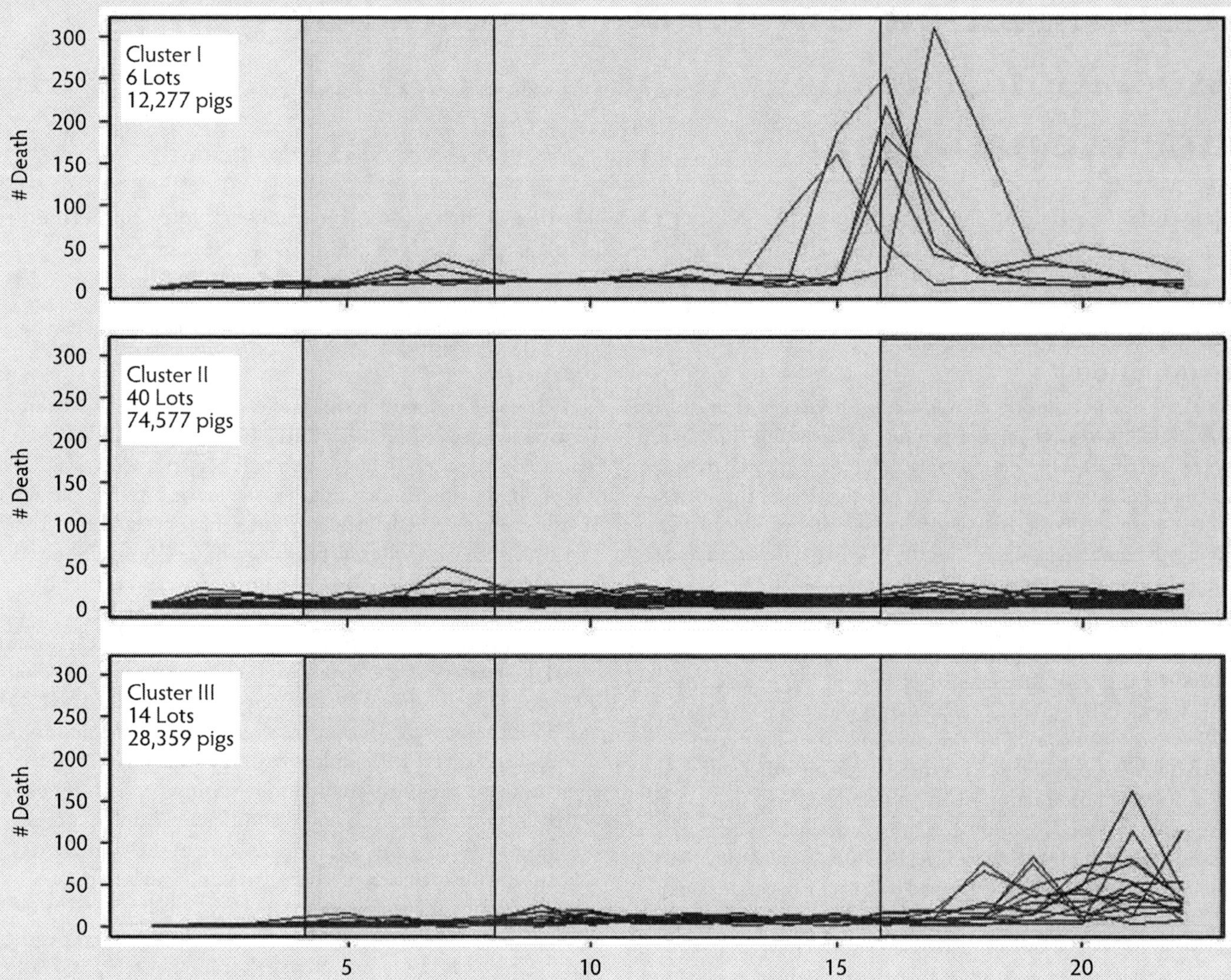

Urine concentration limits of Predef® 2X and Banamine®-S in swine

S. Miller[1], BS; D. Hendrickson[2], DVM; D. Miller[2], DVM
[1]Kansas State University College of Veterinary Medicine, Manhattan, Kansas;
[2]IVN Stoney Creek Veterinary Services and Consultation, Farmland, Indiana

Introduction

At livestock shows across the nation, urine tests are required for the selected champion animals to determine if they have any drugs in their system at the time of the show. Most shows have a zero tolerance policy, even if the drug is prescribed to the animal by a veterinarian. However, there is no data that states how long it takes for commonly used drugs to no longer be detectable in urine. While there are tissue withdrawal dates for these drugs, many livestock shows are not terminal and the urine may be the only sample from the animal being tested. There is a need to determine the period of time post-administration that these drugs are detectable in urine for exhibitors to properly follow these policies. For this study, we selected Predef® 2X, isoflupredone acetate, and Banamine®-S, flunixin, due to their anti-inflammatory properties for respiratory diseases and musculoskeletal conditions. The objective of this study is to determine the urine concentration limits of Predef® 2X and Banamine®-S in swine.

Materials and methods

Nine gilts were placed in individual crates after being weighed. Eight gilts were randomly assigned to the treatment group, while one was assigned as a control. This study was separated into two phases to analyze the concentrations of Predef® 2X and Banamine®-S in the urine. There was a 24 hour period between the last urine collection of Predef® 2X in phase A and the administration of Banamine®-S to begin phase B. During phase A, one dose of Predef® 2X was administered via intramuscular injection to the 8 treatment gilts. Urine was collected using a free catch method 24 hours, 72 hours, and 5 days post administration of Predef® 2X. During phase B, the treatment group received one intramuscular dose of Banamine®-S. Collection occurred 24 hours, 5 days, and 10 days post administration of Banamine®-S. The urine was collected in 120mL cups and sent to a diagnostic lab for qualitative testing. Any detection of either drug in the urine was considered "positive" and no detection of any amount of the drug was considered "negative" for that drug. The samples were only tested for Predef® 2X during phase A. However, since the results of phase A were not received before the start of phase B, the samples of phase B were tested for both Predef® 2X and Banamine®-S.

Results and discussion

On average, Predef® 2X was last detectable in the urine 3.38 days post-administration with a standard deviation (SD) of ± 1.86 days. All except for one pig tested negative for Predef® 2X by day 6. When comparing the results to the weights of the gilts, there was no correlation between their weight and testing negative for Predef® 2X. Banamine®-S was detected in the urine of the gilts for an average of 10 days post-administration with a SD of 0 days. The results for each collection period are shown in Figure 1a,b. Since Predef® 2X was administered prior to Banamine®-S, the urine samples were tested for both Predef® 2X and Banamine® during phase B, which is why Predef® 2X was tested for 12 days post administration. Based on these results, Predef® 2X was on average no longer detectable in urine before the withdrawal period ended. We were not able to test for Banamine®-S long enough to determine when it could no longer be detected in urine. A longer trial should be conducted to set the urine concentration limits for Banamine®-S. It should also be noted that this trial had a small sample size and future studies should include a large sample size. In clinical scenarios, an individual pig may receive multiple doses of either of these drugs over a period of a few days. Further testing should be completed to see if receiving multiple doses would prolong the concentration limit. The new insights gained about the concentration limits of Predef® 2X could allow pigs to be treated with it closer to the dates of non-terminal shows when the treatment is necessary. This study also indicates that further trials over longer periods should be conducted to determine the concentration limits of Banamine®-S and other common drugs.

Figure 1: A) Detection of Predef® 2X in urine of the gilts at 24 hours, 72 hours, 6 days, 8 days, and 12 days post-administration of Predef® 2X. B) Detection of Banamine®-S in urine of the gilts at 24 hours, 5 days, and 10 days post-administration of Banamine®-S.

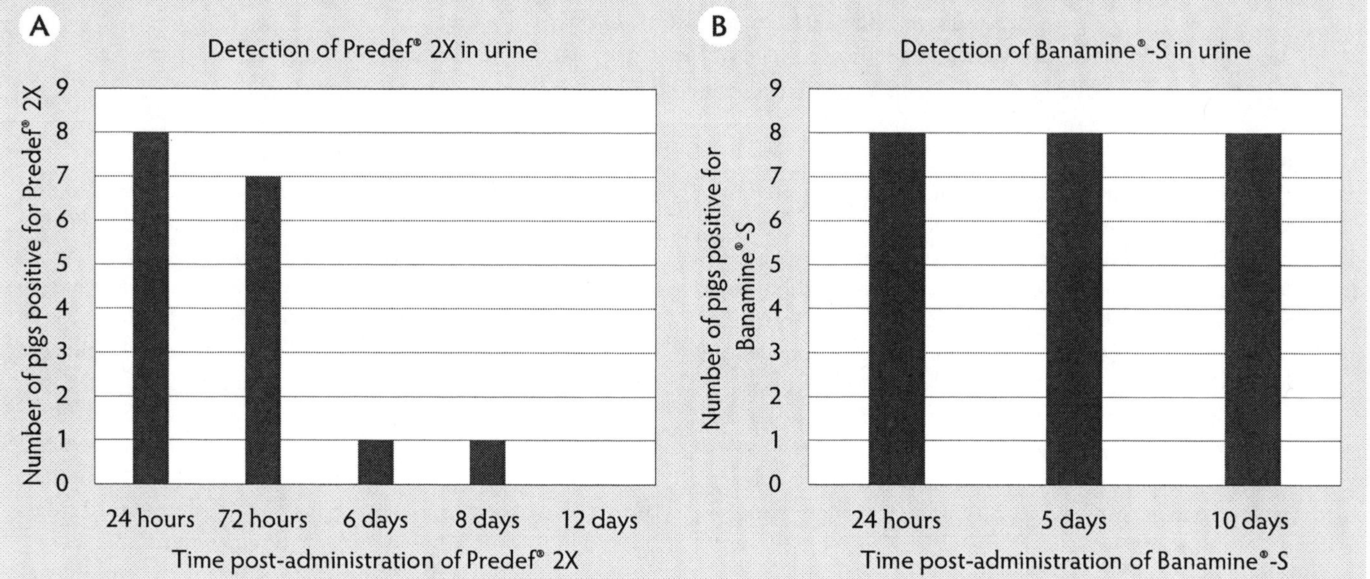

Use of highly palatable feed reward to enhance gilt training to electronic sow feeders

Haley Nelson; Meghann K. Pierdon, VMD, DACAW

Introduction

The use of electronic sow feeders (ESF) to house groups of gestating gilts and sows is becoming increasingly popular. A crucial part of effectively implementing electronic sow feeding stations is the training of gilts to properly utilize the ESF. The purpose of this study was to determine whether the addition of Tonisity Px™, a highly palatable isotonic electrolyte and protein solution often used to increase piglet weaning weights, to the training paradigm would improve the learning outcomes in gilts being trained to use ESF stations.

Materials and methods

The study took place on 2 farms. Farm 1 was a 230 sow herd where three groups of 6 month old gilts (n = 36) were randomized into Tonicity (TON, n = 18) and control (CON, n = 18) groups. On farm 2, a 750 sow herd, 6 month old gilts (n = 36) were randomized to TON (n = 18) or CON (n = 18). On both farms 500 mL of Tonicity Px™, was added to the feed of TON gilts and 500 mL of water to the feed of CON gilts. Training consisted of 3 days where TON gilts were introduced to the tonicity solution in a feed pan, not in the ESF station. The gilts were then introduced to the ESF system for 7 days of training. Tonisity was substituted for water in the ESF station for the TON gilts. For each day, time eating, amount eaten, lesion score, and lameness score were observed. Gilts were introduced to the larger gilt gestation pen on day 7 of training and monitored to determine feeding behavior, feed intake, number of visits to feeder, lameness score, lesion score, and weight change for two additional weeks. Feed rank was calculated for gilts as the order in which they entered the ESF station and ate the largest portion of their feed allotment for the day. Statistics were done using Stata V 15.0. a Mann Whitney u-test was used to compare medians between the TON and CON gilts for each outcome. The results are reported as median and interquartile range.

Results

During training, TON gilts ate faster (41.1 ± 9.0 sec) than CON gilts (45.9 ± 9.5 sec) ($P < 0.05$). On farm 1, once gilts were transferred to the large group gestation pen, there was a trend for TON gilts to eat earlier in the day and have a higher feed rank (54.4 ± 13.7) than CON gilts (61.7 ± 4.6) ($P = 0.05$). There was also a trend for TON gilts to lose less weight (2.5 ± 17.0 lbs) than CON gilts (4.0 ± 16.0 lbs) ($P = 0.08$). There was no difference in lameness score, lesion score or human approach test for TON compared to CON gilts.

Discussion

In sow farms implementing ESF systems, positive reinforcement added to the training paradigm has the potential to improve training outcomes. In this study gilts ate faster during training, tended to lose less weight and tended to eat earlier than those that did not receive the positive reinforcement. In this paradigm, a single well-trained person administered the training program to all gilts, which could account for the small differences between the groups. Having such a person dedicated to gilt training would be recommended and the addition of the Tonisity Px™ may be helpful for pig farms struggling with gilt training. Regarding Tonicity Px™, further research is recommended with a larger sample size with varying gilt training methods utilized.

Mycoplasma hyopneumoniae serological testing challenges in a negative breeding herd

C. Peterson[1]; C. Odland[2], DVM
[1]Iowa State University College of Veterinary Medicine; [2]Pipestone Veterinary Services

Problem

Routine serological testing of a herd with no history or clinical signs of *Mycoplasma hyopneumoniae* had 4 out of 15 positive results on IDEXX ELISA during routine monthly testing.

Herd and case history

The 3500 head herd is located in the Midwest, not filtered, and the closest pigs are located 2.5 miles away. Replacement gilts are born on site and no pigs are brought in from external sources. There is no herd history of *M hyopneumoniae*, no clinical signs have been observed, and the animals are not vaccinated. There is a history of *M hyosynoviae. M flocculare* status is unknown.

Monthly routine testing involves collecting serum from 10 animals in three different age groups throughout the barn (7 weeks, 24 weeks, 32 weeks) for 30 total. Of these, 15 are selected at random for *M hyopneumoniae* IDEXX ELISA testing. Any positives are then tested via Oxoid ELISA. Follow-up testing to determine presence or absence of *M hyopneumoniae* is completed via PCR.

During one summer month's routine testing, 4 out of 15 samples had positive IDEXX ELISA results. The four positive samples were then tested by Oxoid ELISA. One of the four was positive. All animals with non-negative IDEXX ELISA results, along with an additional three animals living in close proximity to the target animals, were sampled via deep tracheal swab and submitted for testing via PCR. All PCR results were negative. See Table 1 for testing results and test interpretation.

Recommendations and discussion

Cross-reactivity during testing can occur between the various swine mycoplasmas, including *M hyopneumoniae* although it is less likely to do so than with *M flocculare*.[2] This site does have a history of *M hyosynoviae* infections among the grow-finish pigs. *M flocculare* is generally considered to be of little concern and nonpathogenic. Its status in this herd is unknown.

Considering the history of the farm and complete lack of any clinical signs, I recommend continuing the current practice of using ELISA to monitor for *M hyopneumoniae*, with PCR used to follow up on any results that are positive or suspect. Serology testing for *Mycoplasma* spp. is, unfortunately, not a perfect art, but is the most economical method to use for surveillance.

Should any future samples come back positive, I highly recommend sacrificing the animal to investigate the presence of *M flocculare*. It has been shown that *M flocculare* and *M hyopneumoniae* have a very similar antigen profile and thus *M flocculare* can cause false positive results when testing for *M hyopneumoniae* via IDEXX ELISA.[2] It would be worth investigating the presence of *M flocculare* on this farm. *M hyosynoviae* shares comparatively fewer antigenic similarities with *M hyopneumoniae* and is less likely to be causing the false positives in this scenario.

This case has demonstrated the challenges of monitoring for *Mycoplasma hyopneumoniae* and highlighted the need for further knowledge and understanding in testing methods. It is important to consider clinical signs when interpreting test results and to obtain the best sample possible. While oral fluids and nasal swabs are easy to collect, they have lower sensitivity compared to other methods.[3] In the case of *M hyopneumoniae*, deep tracheal swabs are best, as they allow us to obtain a sample deeper in the airways, detect the organism for a greater period of time during infection, and are detected via PCR at a higher rate than laryngeal swabs.[1]

References

1. Clavijo, M.J., Johnson, Cl, Farkas, A., Connor, J., Cano, J.P. Longitudinal assessment of *M. hyopneumoniae* natural infection in gilts. Bacteriology & Bacterial Diseases. June 13 2018; 201.

2. Petersen, A.C., Oneal, D.C., Seibel, J.R., Poel, K., Daum, C.L., Djordjevic, S.P., Minon, C.F. 2016. Cross reactivity among the swine mycoplasmas as identified by protein microarray. Veterinary Microbiology. 192 (2016) 204-212.

3. Pieters, M., Daniels, J., Rovira, A. 2017. Comparison of sample types and diagnostic methods following detection of *Mycoplasma hyopneumoniae* during early stage infection. Veterinary Microbiology. 203 (2017) 103 – 109.

Table 1: Serology and PCR results for *Mycoplasma hyopneumoniae* testing

Animal ID	Specimen	ELISA result	S/P[a]	Oxoid result	% buffer Ctl value[b]	Specimen	Test	Results
A	Serum	Positive	0.725	Negative	64.6465	Swab (Tracheal)	M hyo	Not detected
B	Serum	Positive	0.511	Negative	60.1641	Swab (Tracheal)	M hyo	Not detected
C	Serum	Positive	0.657	Positive	49.6212	Swab (Tracheal)	M hyo	Not detected
D	Serum	Suspect	0.343	N/A		Swab (Tracheal)	M hyo	Not detected
E	Serum	Positive	0.447	Negative	60.6061	Swab (Tracheal)	M hyo	Not detected
F	Serum	N/A		N/A		Swab (Tracheal)	M hyo	Not detected
G	Serum	N/A		N/A		Swab (Tracheal)	M hyo	Not detected
H	Serum	N/A		N/A		Swab (Tracheal)	M hyo	Not detected

[a] Negative if S/P < 0.30; suspect if 0.30 ≤ S/P ≤ 0.40; positive if S/P > 0.40
[b] Positive if mean OD value < 50% of ODbuffer control

Measure and compare the pharmacokinetics of intramuscular, oral, and topical administration of flunixin meglumine in pre-wean piglets

B. E. Ratliff[1], BS; H. C. Kittrell[2], DVM; A.C. Szczotka[2], BS; J. T. Brown[2], DVM; K. P. Hayman[2], DVM; A. K. Forseth[2], DVM; K. J. Skoland[2], BS; M. E. Breuer[2], BFA; S. M. Rajewski[3], Ph.D; J. P. Mochel[4], DVM, MS, Ph.D, DECVPT; L. A. Karriker[2], DVM, MS, DACVPM
[1]University of Tennessee College of Veterinary Medicine, Knoxville, Tennessee; [2]Swine Medicine Education Center, Iowa State University College of Veterinary Medicine, Ames, Iowa; [3]Analytical Chemistry Services, Iowa State University College of Veterinary Medicine, Ames, Iowa; [4]Department of Biomedical Sciences, Iowa State University College of Veterinary Medicine, Ames, Iowa

Introduction

Routine piglet processing can be associated with pain and inflammation. The use of non-steroidal anti-inflammatory products may reduce these side effects, as well as pain associated with disease and injury. Evaluating the pharmacokinetics of oral, topical, and intramuscular flunixin meglumine (FM) in pre-wean piglets analyzes the serum concentration and the length of time those concentrations are maintained in the piglet. Topical or oral application of FM would reduce the number of needle injection sites and piglet stress during processing. Currently FM is available as a topical, pour-on liquid for cattle labeled for the control of pain and could be used as an extra-label drug with understanding of the pharmacokinetics of the product in pre-wean piglets. This study allows for future studies of the bioavailability and efficacy of FM.

Materials and methods

Twenty-four pre-wean piglets, weighing 5 to 10 pounds, with no history of treatment, were randomly allocated into three groups according to route of FM administration: oral FM (O), topical FM (T), and intramuscular FM (IM). Blood collection occurred prior to drug administration for time point 0 minutes (M), and FM was administered to each piglet according to study group. Blood collections were obtained at the following time points: 15M, 30M, 45M, 60M, 90M, 120M, 180M, 6 hours (H), 12H, 24H, 36H, 48H, 60H, 72H post drug administration. All samples were collected via jugular venipuncture. Orogastric administration of FM was accomplished with use of a 10 French red rubber catheter to ensure complete dosing. The topical group was administered FM on the dorsal surface of the piglet, between the shoulder blades. FM was injected in the lateral neck muscle using a 22 gauge by 1 ½ inch needle for the intramuscular group. All blood samples were placed in heparinized blood tubes and centrifuged at a speed of 1500g for 10 minutes. Plasma was pipetted into cryogenic vials, and stored in a -80°C freezer until analysis. Pharmacokinetic analysis included serum extraction, liquid chromatography and mass spectrometry. These processes separated the sample from serum proteins to isolate the concentration of FM at each of the listed time points. Flunixin concentration time-courses were analyzed using stochastic approximation expectation maximization algorithm implemented in Monolix v4.3.2.

Results and discussion

All three routes of administration produced measurable concentrations of FM. Intramuscular and oral administration produced concentrations expected to mitigate pain. These routes had short half-lives and clearance rates consistent with FM in older animals.[1] The topical route produced much lower concentrations that persisted beyond 72 hours post administration. This study provides information that will be useful for calculating dose regimens of FM.

Reference

1. Pharmacokinetics of flunixin meglumine in mature swine after intravenous, intramuscular and oral administration. Pairis-Garcia MD, Karriker LA, Johnson AK, Kukanich B, Wulf L, Sander S, Millman ST, Stalder KJ, Coetzee JF. *BMC Vet Res.* 2013 Aug 13; 9:165.

Prevalence of porcine circovirus type 3 based on PCR in processing fluids or piglet serum collected from Midwest sow farms

A. Riley[1], BS; A. Betlach[1,2], DVM; H. Johnson[2], DVM
[1]College of Veterinary Medicine, University of Minnesota, Saint Paul, Minnesota;
[2]Swine Vet Center, Saint Peter, Minnesota

Introduction

Porcine circovirus type 3 (PCV3) is an emerging virus that has been associated with several disease syndromes including porcine dermatitis and nephropathy syndrome (PDNS) and reproductive failure manifested as abortions, mummified and stillborn fetuses and low conception rates. PCV3 was initially detected in the United States in 2015 and is believed to be widespread around the world.[1] Within the US swine industry, the prevalence of PCV3 in sow farms remains unclear. Accordingly, the objective of our study was to determine the prevalence of PCV3 in sow farms across the Midwest by testing either pig serum obtained at weaning or processing fluids for the presence of PCV3 as detected by real-time PCR. In addition, a retrospective analysis of production records was used to assess the impact of PCV3 on reproductive outcomes on multiple sow farms.

Materials and methods

Sixty sow farms, representing 4 production systems and 8 independent producers, that were routinely collecting serum from piglets at weaning or processing fluids for porcine reproductive and respiratory syndrome virus surveillance were enrolled in the study. Processing fluids (1 to 6 samples per farm, n = 37 farms) composed of fluid from tails and testicles or serum from 30 weaned pigs per farm (5 pools of 6, n = 23 farm) were collected. All samples were held at 0-4°C and then submitted to the University of Minnesota Veterinary Diagnostic Laboratory for PCV3 and PCV2 (data not shown) detection by real-time PCR. Samples with a Ct value < 38 were considered positive and some of the positive pooled samples were further tested individually. Production records from 51 farms, from 4 production systems were obtained for analyzing the potential impact of PCV3 status on reproductive performance. Statistical analysis of reproduction data was performed by linear regression.

Results

Of the 60 sow farms sampled, 75% of farms were positive for PCV3 in processing fluids or sera. For those farms that were PCV3 positive, Ct values ranging from 19-38 and 25-38 were detected in processing fluids and wean piglet sera, respectively. Table 1 presents the PCV3 PCR results based on sample type and production system. Overall, 97.3% of processing fluids and 39.1% of serum samples were positive. The percentage of processing fluids and serum samples that had Ct values < 25, 25-38, > 38 were 36.6% and 0%, 55.5% and 13.1%, and 7.9% and 87%, respectively. Statistical analysis of reproductive data was limited to the farms that collected serum as the processing fluid farms with reproductive data were all positive. There was no correlation between the presence of PCV3 and the sow farms' reproductive outcomes.

Discussion and conclusions

Based on the conditions of this study, processing fluids seemed to be a more sensitive sample type for detecting PCV3 compared to serum. Overall, the prevalence of PCV3 in sow farms selected for this study, located in the Midwest, was 75%. This finding suggests that PCV3 is widely present across the Midwest industry. A limitation of this study was that a routine sampling program was not established across the farms, thus resulting in potential variation across farms. We were unable to evaluate the impact of PCV3 herd status on reproductive outcomes. Accordingly, additional investigations regarding the significance and role that PCV3 plays in to clinical disease are needed.

Reference

1. Kedkovid R, Woonwong Y, Arunorat J, et al. Porcine circovirus type 3 (PCV3) shedding in sow colostrum. *Vet Microbiol* 2018;220:12–17.

Table 1: Porcine circovirus type 3 real-time PCR results by system and sample type

System	Sample type	No. tested	No. positive	% positive
A	Processing fluid	3	3	100.0%
	Serum	9	4	44.4%
B	Processing fluid	12	12	100.0%
C	Processing fluid	10	10	100.0%
D	Processing fluid	5	5	100.0%
	Serum	13	4	30.8%
None	Processing fluid	7	6	85.7%
	Serum	1	1	100.0%
Overall	Processing fluid	37	36	97.3%
	Serum	23	9	39.1%

Effects of farm and harvest plant geographical locations on movement of cull sows in the United States

Z. Talbert[1]; A. Bullis[1]; B. Blair[1]; J.Lowe[1,2]
[1]College of Veterinary Medicine, University of Illinois, Urbana, Illinois; [2]Lowe Consulting Ltd., Mahomet, Illinois

Introduction

Culling rates on US sow farms currently exceed 50% resulting in over 3 million sows entering the market annually. The cull sow-marketing network is distinct from the lean hog market where pigs move directly from farm to harvest plant. Market intermediaries, local points of collection, sort and regroup sows according to size and quality prior to resale to harvest plants. This intermediary step between the farm and harvest creates a convenient point for farms to sell small batches of sows and to meet individual plant specifications for size and quality. However, this added layer of market complexity creates an opportunity for the movement of infectious disease. Descriptions exist of the biosecurity risk associated with loading and unloading of cull sows into livestock trailers.[1,2] However, there has been an inadequate investigation of the risk that the movement of cull sows across geographies pose to the swine industry.

Objective

To describe geographical movement of cull sows in the national market based on a systemic random sampling of sows at harvest.

Materials and methods

Collection of all 840 compliant tags from the USDA-APHIS-VS disease surveillance program that represents 90% of the cull sow harvest in the US from a single week per month in May, June and July 2018 served as the dataset for this study. The tags were submitted by 6 slaughter plants for each week. For each tag, the NAIS premise number, state of origin and terminal slaughter plant was recorded. The premise ID was located in a publically available database that contained street addresses. Conformation of physical locations entailed locating the street address on Google Maps and recording the geo-coordinates. Python, using the Seaborn library, served as the platform for exploratory data analysis including calculation of standard distance from farm of origin to the plant.[3] Standard distance served as a measure of the geographical distribution of sows entering the plant.

Results/discussion

Exploratory data analysis revealed that sow movement distances are similar for all harvest plants and month did not affect movement patterns. All plants purchased sows from multiple states and all plants purchased sows from large geographic regions. Sows originated from 30 states and 853 unique premises IDs. Plants in close proximity to states with higher numbers of swine breeding herds (based on USDA-NAS) tended to have a shorter transportation distance. To investigate factors that determine geographical distribution of source farms for a plant, standard distance was calculated and ranged from 97.9km to 645.2km (Figure 1). There was a weak association between the plant capacity and standard distance (correlation coefficient 0.48). The data suggests that purchasing patterns for individual plants are stable over three months in the summer and sow purchasing occurs over a large geographic region for all plants. The data set is limited in its scope. Additional data during different seasons of the year may reveal seasonal effects or market effects on sow movements. Accurate measurement of these effects will allow better predictions of the movement of disease within the US pork industry. Trace backs by the USDA for Seneca Valley Virus demonstrated that the cull sow-marketing channel is an efficient means of spreading a pathogen. This study, along with further studies, may allow for more effective decisions to limit the transmission of pathogens across large geographical regions.

References

1. Lowe J, Gauger P, Harmon K, et al. Role of transportation in spread of porcine epidemic diarrhea virus infection, United States. Emerg Infect Dis. 2014;20(5):872-874. doi:10.3201/eid2005.131628.

2. Dee SA, Deen J, Otake S, Pijoan C. An experimental model to evaluate the role of transport vehicles as a source of transmission of porcine reproductive and respiratory syndrome virus to susceptible pigs. Can J Vet Res. 2004;68(2):128-133.

3. Python Software Foundation. Python Language Reference, Version 3.4. Available at `http://www.python.org`.

Figure 1: Plant distance distribution

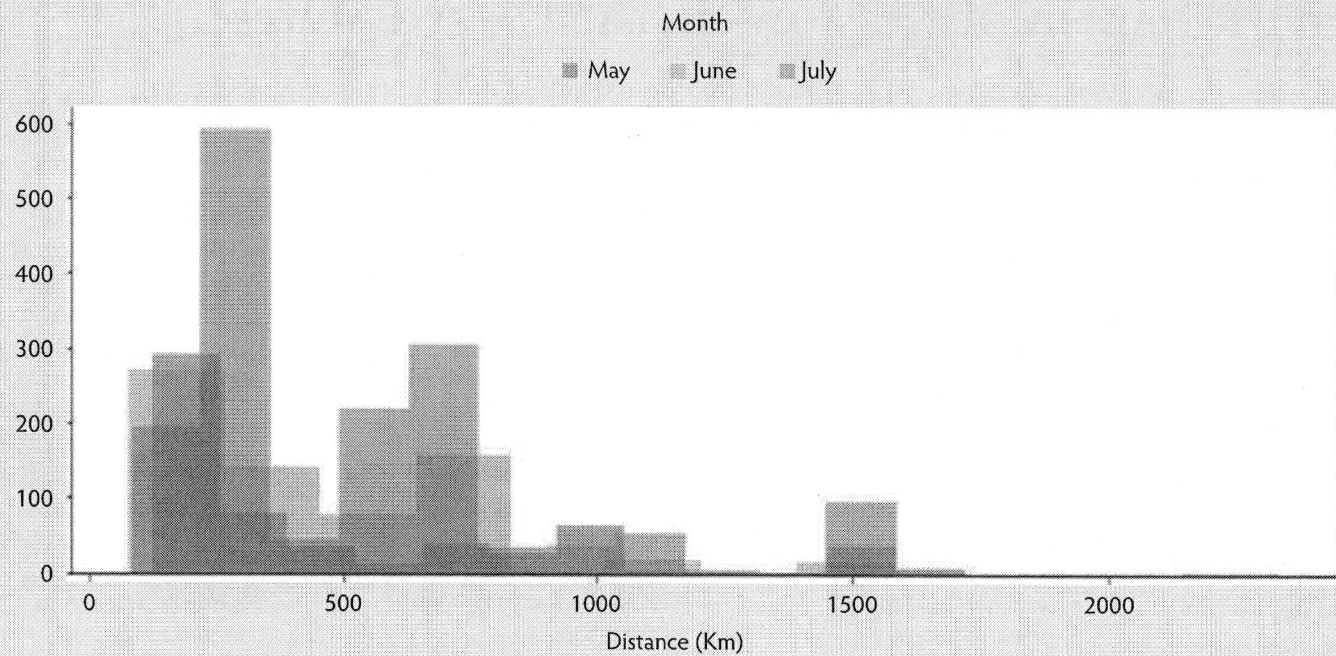

Evaluation of the immunological response of previously exposed and vaccinated pregnant sows to a killed autogenous PRRS 1-7-4 RFLP vaccine

Zoe Wolfe, BS; Jeremy Pittman, DVM; Tobias Kaeser, PhD; Andrew Kick

Introduction

Porcine reproductive and respiratory syndrome virus (PRRSV) has a significant economic impact on the swine industry, particularly in breeding herd populations.[1] PRRSV continues to be a challenge due the inherent high mutation rate, resulting in clinical and immunological variability between isolates.[2] PRRSV isolates are commonly characterized by a predicted restriction fragment length polymorphism (RFLP) from sequencing. One methodology used to control PRRSV is stabilization of the breeding herd, through maximizing immunity, using commercial modified live vaccines (MLV). However, commercial MLV vaccines are often heterologous to contemporary isolates. The objective of this study was to determine if sows previously exposed to PRRS 1-7-4 RFLP and MLV would demonstrate a homologous immunological booster response to a killed autogenous PRRS 1-7-4 RFLP vaccine.

Materials and methods

This study was conducted on a 3,200-head breed-to-wean farm, previously exposed to and actively shedding, a wild type PRRS 1-7-4 RFLP. The herd was vaccinated, semi-annually, with a commercial MLV. Sixty pregnant sows, blocked by parity and due date, were randomly assigned to a treatment or control group. The vaccine used was a killed autogenous PRRS 1-7-4 RFLP vaccine that was 98.51% similar on ORF5 Sequence between the vaccine and field isolates. The treatment and control groups were given 1 ml of an autogenous PRRS 1-7-4 RFLP vaccine or sterile water, respectively, at 5 and 2 weeks pre-farrow (wpf). Blood samples were collected from all animals at 5 wpf and at farrowing and submitted for PRRS ELISA, PCR, IFA, and FFN testing. The IFA and FFN tests utilized a lab standard heterologous isolate and the homologous vaccine isolate. A heterologous MLV isolate was also utilized for FFN testing. Statistical analysis was performed using ANOVA on ELISA S/P ratios and log transformed reciprocal titers for IFA and FFN, between treatment and control groups within time points. Significance was considered at a P-value ≤ 0.05.

Results and discussion

Prior to vaccination, no difference was observed in PRRS ELISA, IFA or FFN titers across treatment groups. Post vaccination no significant difference was observed in PRRS ELISA or FFN titers. Significant difference was observed in homologous PRRS IFA titers, demonstrating a reduced titer (1.68 vs 3.00) and lower percentage of positive sows (36.7 vs 66.7%) in vaccinated sows compared to control sows, respectively. The heterologous IFA, showed similar results with a reduced titer (0.85 vs 3.86) and lower percentage of positive animals (16.7 vs 70.0%) in vaccinated sows compared to control sows, respectively. One sow was positive by PRRSV PCR, at both samplings; confirmed to be RFLP 1-7-4 by sequence analysis. Full serological data can be found in Table 1.

The lack of difference in post vaccination PRRS ELISA titers, is expected due to immunological tolerance to this antigen, commonly seen with repeatedly vaccinated sows. The lack of difference in FFN titers may be due to a lack of response to vaccination or small sample size. The PRRS IFA results demonstrate a decrease in both percent positive and titers in the vaccinated group, which could be due to preferential activation of other immune responses (cellular immunity). The current study only assessed the humoral response of the sows. Samples were collected, processed and stored to evaluate the cellular immune response (e.g., T-cell proliferation, cytokine production) in a future study. In addition, evaluation of maternal transfer of immunity to piglets was not an aspect of this study.

References

1. Holtkamp DJ et al. Assessment of the economic impact of porcine reproductive and respiratory syndrome virus on United States pork producers. *J Swine Health Prod.* 2013;21:72.

2. Rahe MC, Murtaugh MP. Mechanisms of adaptive immunity to porcine reproductive and respiratory syndrome virus. *Viruses.* 2017;9:148.

Table 1: Serological data (ELISA, IFA and FFN assay) from sows prior to and 2 weeks post-vaccination with PRRS 1-7-4 RFLP autogenous vaccine or sterile water at 5 and 2 weeks prior to farrowing.

PRRS ELISA

	#POS/Total (%POS)		ELISA S/P	
Treatment	PRE	POST	PRE	POST
Control	21/30 (70.0)	15/30 (50.0)	0.580^a	0.546^a
Vaccinated	18/30 (60.0)	17/30 (56.7)	0.642^a	0.605^a

PRRS IFA - Heterologous - SDSU

	#POS/Total (%POS)		Titer (LOG^2)†	
Treatment	PRE	POST	PRE	POST
Control	9/30 (30.0)	21/30 (70.0)	1.56^a	3.86^a
Vaccinated	12/30 (40.0)	5/30 (16.7)	2.10^a	0.85^b

PRRS IFA - Homologous - 174

	#POS/Total (%POS)		Titer (LOG^2)	
Treatment	PRE	POST	PRE	POST
Control	19/30 (63.3)	19/30 (63.3)	3.50^a	3.00^a
Vaccinated	21/30 (70.0)	11/30 (36.7)	3.79^a	1.68^b

PRRS FFN - Heterologous - SDSU

	#POS/Total (%POS)		Titer (LOG^2)	
Treatment	PRE	POST	PRE	POST
Control	ND	29/30 (96.7)	ND	4.77^a
Vaccinated	ND	29/30 (96.7)	ND	5.33^a

PRRS FFN - Heterologous - MLV

	#POS/Total (%POS)		Titer (LOG^2)	
Treatment	PRE	POST	PRE	POST
Control	24/30 (80.0)	24/30 (80.0)	4.03^a	2.67^a
Vaccinated	26/30 (86.7)	20/30 (66.7)	4.27^a	3.43^a

PRRS FFN - Homologous - 174

	#POS/Total (%POS)		Titer (LOG^2)	
Treatment	PRE	POST	PRE	POST
Control	30/30 (100.0)	29/30 (96.7)	4.93^a	2.93^a
Vaccinated	30/30 (100.0)	25/30 (83.3)	4.87^a	3.13^a

[a,b] Numbers within column, within test with different superscripts are different ($P < 0.05$); ANOVA
† Reciprocal titers for dilution tests were transformed to Log base 2 for statistical analysis.
ND; testing not done

A new approach for benchmarking feed conversion across different swine growing system types

C. Alonso[1]; A. Fernández-Fontelo[2]
[1]Independent Data Analyst and Swine Consultant, Barcelona, Spain
[2]Departament de matemàtiques, Universitat Autònoma de Barcelona, Barcelona, Spain

Introduction

Despite the importance of the growing period to the economics of swine production, the statistical integrity of the methods used to compare the performance of growing groups of variable entry ages, entry weights, and flow types remains lacking. The largest production system in Spain is a vertically integrated system that is comparable to large scale North American production. We completed a modelling study to compare the feed conversion (FC) of their 5 growing systems (*i.e.* wean-to-finish (WTF)), 3 site production AI/AO finisher (3AF), 3 site production continuous flow finisher (3CF), 2 site production single source (2SS), and 2 site production multi-source (2MS)). Our objectives were threefold: 1) to create a practical statistical method that would allow producers to compare FC results from close-out data of animals entering in different weight ranges; 2) to assess the FC difference between growing systems; 3) and to understand the impact and interactions of production variables in the FC model.

Materials and methods

The close-out data from a production group in Spain consisting of over 200,000 sows from 2014 to 2017 was consolidated and prepared for analysis. A total of 6,188 groups from 5 different growing systems (wean-to-finish (WTF), 2 site production single source (2SS), 2 site production multiple source (2MS), 3 site production continuous flow (3CF), 3 site production AI/AO (3AF)) were analyzed. Variables of interest for each close-out, such as Mort_pct, Year (2014-2017), Season (spring, summer, fall, winter), Med_dollars, Mort_pct, PoundDowners_pct, Geographic_Area, Sow_Flow, and Genetics (A, B, C, D), were considered for the analysis. Data were modelled by means of a multivariate linear regression including both quantitative (discrete- and real-valued) and qualitative (factors with two or more levels) variables, and some of the more relevant interactions among them. Adjusted means were compared considering the Tukey's correction and assuming a 5% level of significance.[1] The models were assessed through the standardized residuals. The analysis was conducted using the R statistical software (version 3.3.2).

Results

For objective 1, a new variable called FC_Entry was calculated to reflect the impact that each incremental change in entry weight had on the final FC. For each unit of change in FC_Entry, an increment of 0.37 was observed in the final FC. Results from objective 2 are shown in Figure 1A. Overall, there were no differences between WTF and the 3 site systems (3AF and 3CF). These 3 systems performed significantly better than 2 site multi-source (2MS), and better than 2-site single source (2SS).

Results from objective 3 indicated that the FC of close-outs from the zone 4, year 2016, and fall were significantly lower (Figure1B). For every $1200 spent on pharmaceuticals there was an increase of 0.004 in FC and for every 1% Mort_pct increase there was an increase of 0.017 in FC. Genetics and Sow_Flow variables were excluded from the analysis for being highly correlated with growing system type.

Conclusions and discussion

Results from this study indicate that WTF and continuous and AIAO 3 site production systems performed equally after adjusting for the variables in the model. Further development of this model is planned in order to integrate the variable costs of each growing system (*i.e.* transport and animal space costs) to allow for better economic benchmarking. Because of the similarities between this large Spanish production system and typical large scale North American production, this information is valuable for the economics of both industries.

Acknowledgments

The authors appreciate the collaboration of Albert Vidal in the study.

Reference

1. Montgomery, D.C., Peck, E.A. and Vining, G.G. (2012). *Introduction to Linear Regression Analysis.* John Wiley & Sons, Inc, Hoboken, New Jersey.

Figure 1: A) Feed conversion results (Least squares means of feed conversion and 95% confidence interval) for all 5 grow finish systems by year after adjusting for all significant variables in the study. B) Feed conversion results (Least squares means of feed conversion and 95% confidence interval) for all 5 grow finish systems by year and geographic zone after adjusting for all significant variables in the study.

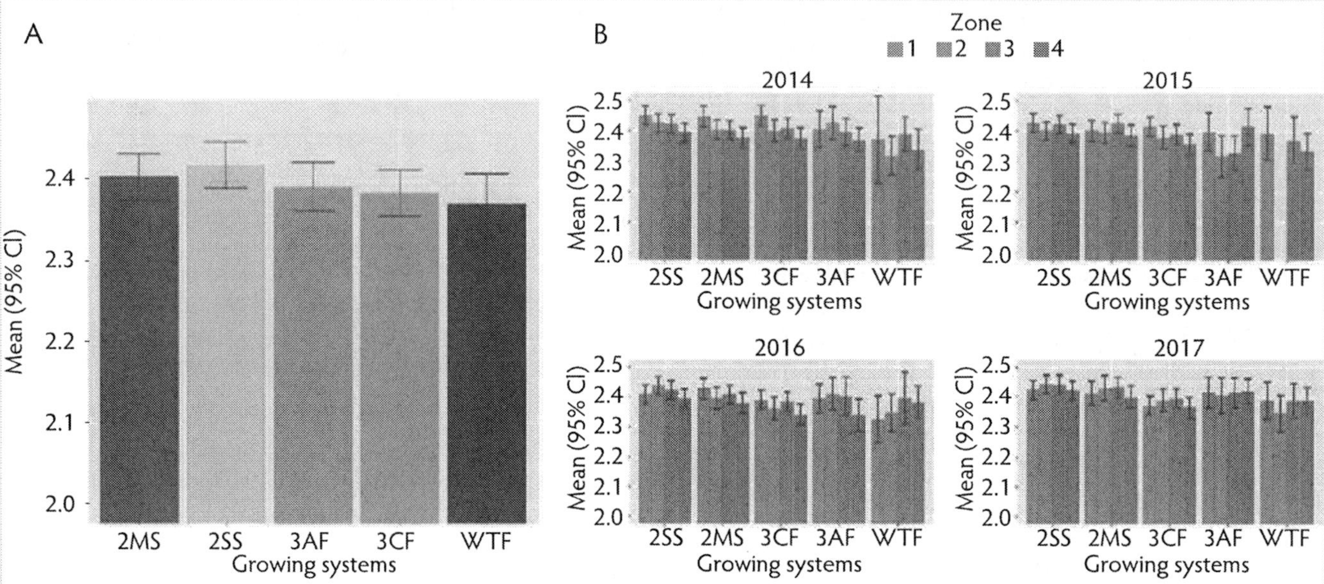

Assessment of internal biosecurity in US swine farms using innovative technology

Andréia G. Arruda, DVM, MS, PhD; Nicholas Black, MPH
The Ohio State University, Columbus, Ohio

Introduction

Porcine reproductive and respiratory syndrome (PRRS) is one of the most costly infectious diseases for the swine industry and, even though research on PRRS control and prevention has been intense during the last 20-year period, its control and eradication remain almost as challenging as 20 years ago. Even though the swine industry is aware that fomites pose an important risk for the spread and maintenance of PRRS and other infectious diseases in swine herds, detailed characterization and quantification of internal movement patterns inside farms has never been previously published in the literature. Factors that likely contribute to the lack of such information include the absence of technology to capture this type of data, and the challenges that comes with summarizing and interpreting it in a way that is meaningful to swine producers and practitioners. Such characterization would facilitate focused identification of biosecurity breaches, improving disease prevention and shortening disease duration. The goals of this study were to use an innovative tracking system to describe farm movement patterns that relate to internal biosecurity practices, and to investigate whether an increase in "risky" movements may be associated with production parameters of interest in the farrowing rooms.

Materials and methods

Three commercial farrow-to-wean farms located in the states of Indiana (n = 2) and Iowa (n = 1) were enrolled in this project. Inclusion criteria was that farms had a history of PRRS outbreaks, were willing to complete a biosecurity questionnaire, conduct PRRS sampling as requested by the investigators and coordinated with their herd veterinarian, implement and comply with an internal biosecurity system (B-eSecure®, PigCHAMP Pro Europa) for the period of one year; and share production data. Farm location, farm map (number and location of barns and rooms within the farm), worker's information, and initial PRRS PCR status by room (using oral and processing fluids) was gathered by research personnel. Internet services were optimized throughout the farm, and devices were installed in each room of the facility. These devices were able to detect the presence of Bluetooth devices ("beacons") that were individually distributed to all farm staff; and send data in real-time to a central database through Wi-Fi.

A movement was defined when one worker left a certain room and remained in a different room for at least two minutes. Movement data was aggregated at the week level, and movements were categorized into three types based on a combination of initial PRRS status, and commonly recommended biosecurity practices: "risky" (movements from either the staging nurseries or loading areas to any room except the office); "moderately risky" (movements between farrowing rooms, or from PRRS positive to negative areas), and "safe" (movements from PRRS negative to positive areas). Production data was extracted at the week level from PigCHAMP®. Statistical analysis was conducted using Stata-IC 14, and included basic descriptive statistics and univariable linear regression models using number of piglets weaned per litter and pre-weaning mortality as outcomes, and number of different types of movements during current and previous weeks as predictors. Statistical significance was declared at $P < 0.05$.

Results and discussion

Preliminary analysis included 14 weeks of data (March to July 2018) from one of the enrolled farms. This farm was a 4,400-sow filtered, shower-in facility with a total of 16 farrowing rooms, 2 nursery areas, one grower/gilt development area, 2 gestation barns, 2 load out areas, one supply room and one office; therefore, a total of 25 devices were installed for movement detection. The mean number of movements per week was 1,810.64 (SD: 344.81), with the majority of movements being "moderately risky" (mean 1,226.86, SD: 210.74), followed by "safe" (mean 408.36, SD: 89.04), and "risky" movements (mean: 175.43, SD: 77.12). The vast majority of "moderately risky" movements corresponded to movements between farrowing rooms. Preliminary analysis showed a tendency for an increase in "risky" movements in a previous week to increase pre-weaning mortality in a current week by approximately 3% ($P = 0.052$). Furthermore, an increase in "risky movements" also tended to decrease the number of piglets per litter by 0.37 ($P = 0.09$). Analysis of additional production data from this and the other two enrolled farms using an extended period is under way.

Acknowledgements

Merck Animal Health provided funding for this study and for the graduate student. PigCHAMP Pro Europa provided technology support for B-eSecure; and the means for all data collection and processing. We also would like to acknowledge and thank participating swine veterinarians, swine producers and farm staff for enrolling in this project.

PCV3, a newly discovered swine pathogen

B. Arruda[1], DVM, PhD; R. Derscheid[1], DVM, PhD, DACVP; K. Schwartz[1], DVM, MS; P. Piñeyro[1], DVM, PhD; L. Giménez-Lirola[1], PhD; S. Lierman[1]; G. Temeeyasen[1], DVM, PhD; C. Sievers[2], DVM, MS
[1]Veterinary Diagnostic and Production Animal Medicine, Iowa State University, Ames, Iowa;
[2]Swine Vet Center, St Peter, Minnesota

Introduction

Porcine circovirus 3 (PCV3) is a newly discovered species of PCV detected in mummified fetuses[1] and multisystemic inflammation.[2] PCV3-associated disease with the concurrent detection of PCV3 by PCR and *in situ* hybridization (ISH) from different production systems has not been previously reported, leaving many to question the clinical relevance of PCV3 in the US swine herd. Accordingly, this report investigates the role of PCV3 in various clinical syndromes across different production systems through the aggregation of clinical history, clinical signs, and diagnostic results. We also report preliminary findings from an experimental inoculation trial in cesarean derived, colostrum deprived (CDCD) pigs.

Diagnostic cases

PCV3 was detected in at least twenty-two cases of reproductive loss from 20 sites between January and November of 2018. Fetal tissues representing multiple litters from each of these cases were positive for PCV3 by quantitative PCR (qPCR) with Ct values commonly at 20 or below with some as low 9.4, suggesting a very high viral load (billions of genome copies/mL). Fetal tissues were PCR negative for other known causes of porcine reproductive failure including porcine parvovirus, porcine reproductive and respiratory syndrome virus, porcine circovirus 2, and *Leptospira*. In addition, next-generation sequencing (NGS) on fetal tissues identified PCV3 alone in six out of six cases assayed, and PCV3 was detected by ISH in fetal tissues in five out of five cases. Histologic evaluation of a subset of fetal tissues demonstrated myocarditis with concurrent detection of PCV3 by ISH similar to previously reported cases of PCV2-associated reproductive failure (Figure 1). When a clinical history is provided, case submissions commonly indicated a spike in mummies and stillborns. PCV3 has also been detected by qPCR at low Ct values (7.5-13.3) in the cerebrum of 1-day to 10-day-old pigs from two different production systems reporting weak-born piglets. These piglets had encephalitis and/or periarteritis and myocarditis consistent with a viral etiology with concurrent ISH detection of PCV3. PCV3 has also been strongly associated with periarteritis and arteritis in the heart, kidney, and/or mesocolon or mesentery in growing pigs in twelve cases from ten different sites. PCV3 qPCR on lung was consistently positive (Ct ~20s) in these cases and non-affected pigs are commonly PCV3 qPCR negative or Ct values are in the low 30s. Again, PCV3 is detected by ISH in cardiac myocytes, smooth muscle cells of arteries, and endothelial cells.

Experimental inoculation

A tissue homogenate inoculum was prepared from PCV3 qPCR positive material. This material was screened and found negative for other agents by NGS. CDCD pigs with (n =6) and without (n = 6) immunostimulation (adjuvant) were inoculated. Serum was collected at twelve time points. Viremia was consistently detected in both PCV3 and PCV3+adjuvant groups starting at 3 days post-inoculation (DPI), which continued until the end of study (42 DPI). Clinically, all animals were unremarkable. At necropsy at 10 or 42 DPI, PCV3 was detected by qPCR in every tissue assayed. PCV3 has not been detected in any of the negative control pigs. Histopathologic examination was pending at the time of this report.

Summary

This report suggests PCV3 is a putative cause of reproductive failure as well as multisystemic inflammation in suckling and grow-finish pigs derived from multiple US swine farms, with some features similar to those reported with PCV2-associated PCVAD. Retrospective investigations have reported that PCV2 infection resulted in sporadic systemic disease as early as 1985, long before devastating epidemics were observed in the late 1990s and early 2000s. It remains to be seen if PCV3-associated disease will follow a similar course. Experimentally, PCV3 can cause a prolonged viremia which suggests that the production of neutralizing antibodies may occur after 42 DPI.

References

1. Palinski R et al. A Novel Porcine Circovirus Distantly Related to Known Circoviruses Is Associated with Porcine Dermatitis and Nephropathy Syndrome and Reproductive Failure. *J Virol.* 2016;91:pii: e01879-16.

2. Phan TG et al. Detection of a novel circovirus PCV3 in pigs with cardiac and multi-systemic inflammation. *Virol J.* 2016;13:184.

Figure 1: A) Lymphocytic myocarditis and cardiac myocyte loss in a fetus in which PCV3 only was detected by qPCR and NGS. B) Detection of PCV3 (brown labeling) in cardiac myocytes of the same fetus by an RNAScope® in situ assay.

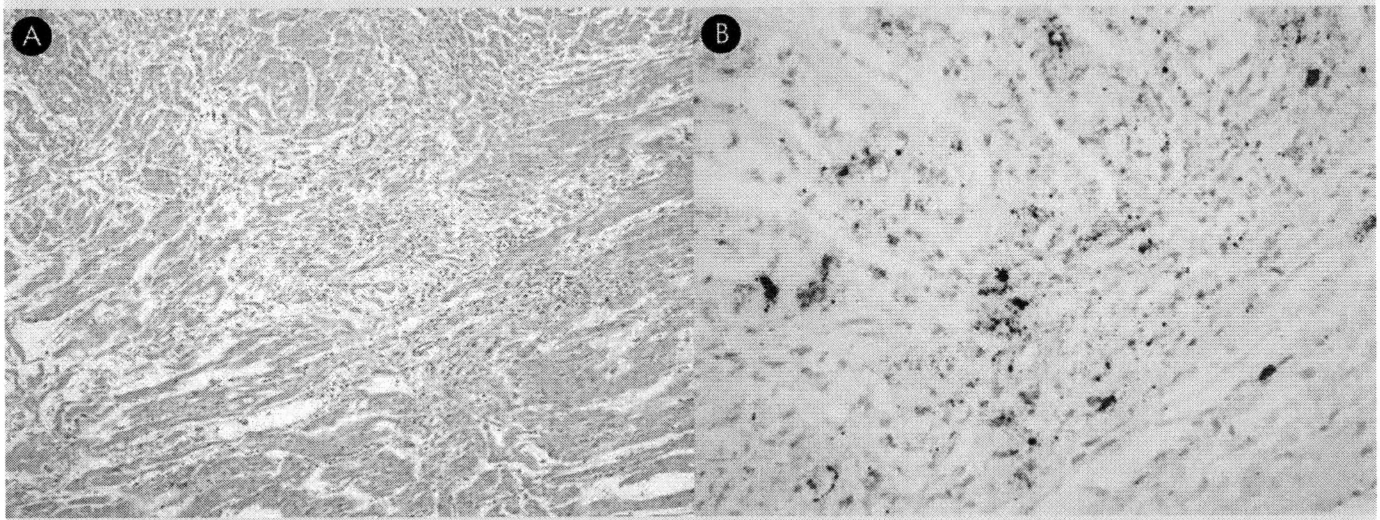

Mycoplasma hyopneumoniae cell-mediated immune responses after vaccination at weaning

F. Chamba[1], DVM, MS, PhD; M. Surendran Nair[3], PhD; V. Krishna[3], PhD; J. W. Hall[1], PhD; B. Conarchy[1], BS; N. Winkelman[2], DVM, MS; M. Baarsch[1], MBA, PhD; M. Pieters[3], DVM, PhD; M. C. J. Cheeran[3], PhD
[1]Applied Research, Pharmgate Animal Health, Wilmington, North Carolina; [2]Swine Services Unlimited Inc., Rice, Minnesota; [3]Veterinary Population Medicine Department, University of Minnesota, St. Paul, Minnesota

Introduction

Infection with *Mycoplasma hyopneumoniae* (Mhp) results in respiratory disease, reduced weight gain and an increase in morbidity and mortality, either alone or in combination with other pathogens. Mycoplasma induced cell-mediated immunity has been suggested to be associated with protection from disease, and limited data is available for cell-mediated immunity to Mhp after vaccination. We evaluated cell-mediated immune responses to Mhp vaccination at weaning using 5 different Mhp/PCV2 combinations of commercially available vaccines.

Materials and methods

One hundred eighty pigs were randomly selected, ear-tagged and vaccinated at weaning. Six groups of pigs were vaccinated intramuscularly with one of five Mhp/PCV2 vaccines or a saline control (30 pigs/group). All study pigs were housed in the same room in a nursery barn and all groups were equally represented in 8 pens. Group 1 received 1 mL of vaccine A, groups 2, 3, 4 and 5 received 2 mL of vaccines B, C, D and E, respectively. Group 6 received 2 mL of saline. Administration of vaccines was randomly assigned and personnel were blinded during vaccine administration, sample collection and testing, and data analysis. Blood samples were collected from all pigs at weaning for baseline analysis. At 14 and 28-days post vaccination, blood samples were collected from 50% of pigs in each group. Delayed-Type Hypersensitivity (DTH) test was performed in 15 pigs in each group at 14 days post vaccination, and in the remainder 15 pigs at 28 days post vaccination. To measure DTH responses, each pig received a 0.1-mL intradermal injection of Mhp 232 strain (moderate virulence), Mhp ISU-MY0051-10 strain (high virulence), Phytohemagglutinin (positive control) and Saline (negative control) at separate sites surrounding the umbilicus (Figure 1). DTH reaction was calculated as the product of two midpoint diameters, ($A = (1/2\ d1)\ (1/2\ d2)\ [\pi]$) of the wheal for each reaction site, ~32 hours after injection. Lymphocyte Stimulation index (SI) and interferon gamma (IFN-γ) production in response to Mhp 232 and Mhp ISU-MY0051-10 cell lysates were measured in peripheral blood mononuclear cells (PMBC) at 14- and 28-days post vaccination. Rectal temperature and clinical scores were recorded at ~8 hours post-vaccination. Injection site reactions were assessed at ~24 hours post-vaccination. Pigs were weighed at weaning and at the end of the study (28 days post-vaccination).

Results

DTH reactions of vaccine groups A, C, D and E, and A and C were significantly larger than the saline group at 14 (Figure 1) and 28 days post vaccination respectively for both Mhp strains (Wilcoxon P-values < 0.01). Further, the vaccine A group showed significantly larger DTH reactions than the other vaccines (Wilcoxon P-values < 0.001) at both time points. PMBC stimulation index (SI) lacked difference among groups for both Mhp strains at both time points. Interferon gamma (IFN-γ) production for the Mhp ISU-MY0051-10 strain was significantly higher in the vaccine A and C groups at both time points in relation to the saline group. Vaccine A group showed significantly higher IFN-γ production than the other vaccines at 14 days post vaccination. Weight gain was similar among all groups. No clinical signs or injection site reactions were observed in vaccinated pigs. Rectal temperatures were higher in all vaccinated groups compared to the saline group, except in the vaccine B group.

Conclusions and discussion

DTH response is a measure of $CD4^+$ T_h1 lymphocyte response to a specific antigen. $CD4^+$ T_h1 lymphocyte stimulation has been suggested to be associated with protection against disease produced by mycoplasmas in other species. $CD4^+$ T_h1 lymphocytes modulate both cell-mediated immunity to intracellular pathogens and humoral immune responses against pathogens such as mycoplasmas. Our findings suggest that all Mhp vaccines induced variable and measurable cell-mediated immune responses to cognate antigen. Vaccine A and C induced higher levels of DTH response and gamma interferon production to Mhp ISU-MY0051-10 strain (high virulence) compared to the saline group.

Figure 1: Mhp ISU-MY0051-10 strain (high virulence) DTH reactions 14 days after vacc.

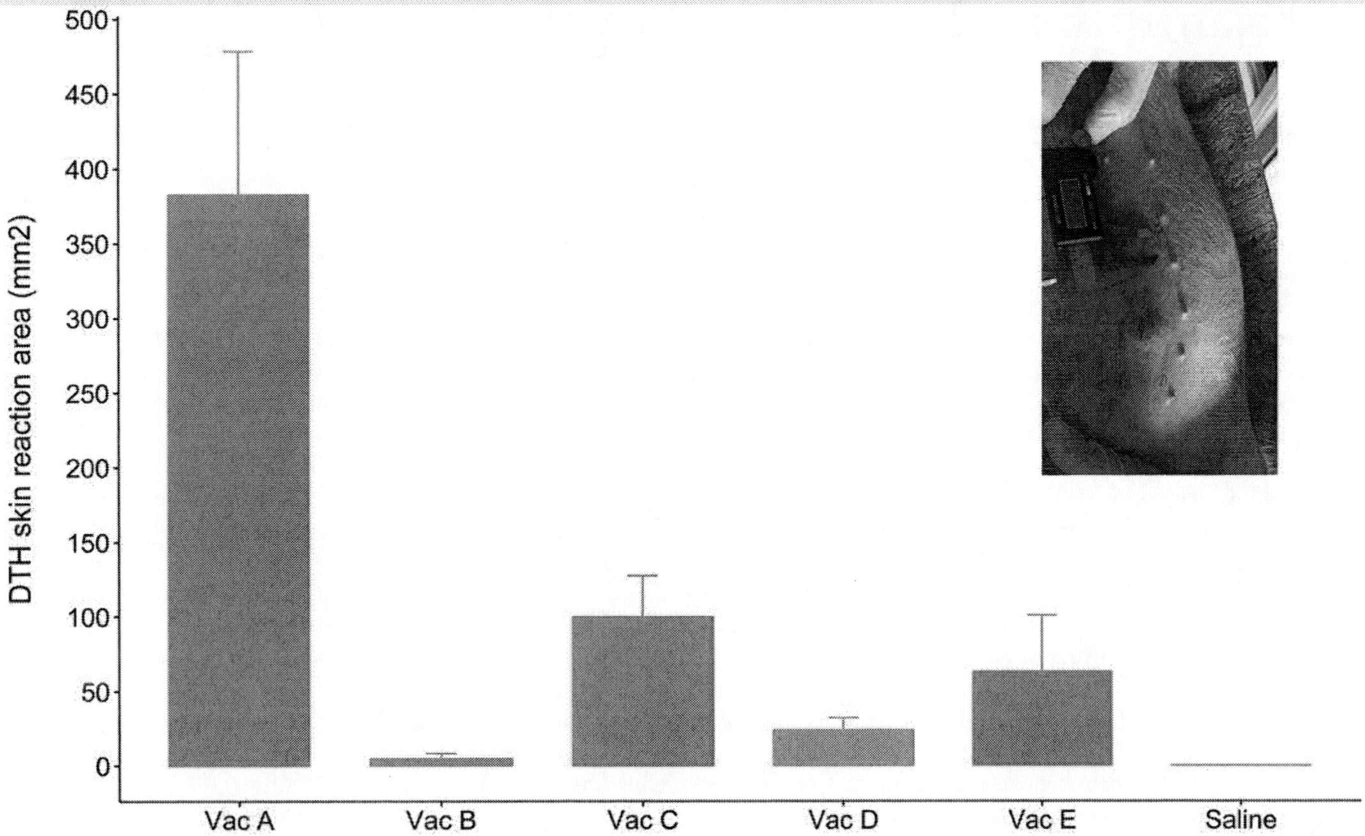

Effect of plant polyphenols as antioxidant alternatives to vitamin E in diets offered to pigs

Theo van Kempen[1,2], PhD; Neil Ferguson[2,3], PhD; Nienke de Groot[2], DVM; Nestor Gutierrez[2], PhD
[1]North Carolina State University, Raleigh, North Carolina; [2]Trouw Nutrition R&D, Boxmeer, The Netherlands; [3]Trouw Nutrition Canada, Guelph, ON

Introduction

Oxidative stress is the product of an imbalance between free radicals and antioxidants that can damage cell components, and is associated with compromised growth, disease, and the ageing process.[1] However, free radicals can also serve a regulatory role in metabolism (e.g., free radicals participate in induction of hyper-activated sperm motility and acrosome reaction).[2] Vitamin E is a key antioxidant, and is typically dosed well above NRC requirements[3] to reduce half-life of free radicals.[4] However, overdosing antioxidants may also impede regulatory processes dependent on free radicals. A more balanced antioxidant approach should be favored: rather than adding one antioxidant at a very high level, dosing a range of molecules with antioxidant properties but with different radical affinities and different distributions in the body. This allows the body to better control the antioxidant defenses without impeding the regulatory actions of free radicals. Plant polyphenols possess the same radical-quenching basic structure as vitamin E, and may complement the antioxidant defense of the animal.[5] Two research trials were conducted to determine the effect of vitamin E in combination with a polyphenol blend vs. a high dose of only vitamin E on oxidative stress and performance of pigs.

Materials and methods

In a first trial, weanling pigs were housed in normal environmental conditions and allotted to 3 dietary treatments containing either 11 ppm vitamin E (11 ppm E; NRC requirement), 80 ppm vitamin E (80 ppm E; commercial level), or 11 ppm vitamin E + 69 ppm vitamin E equivalent provided by a polyphenol blend (11E+PP). After 7 days of feeding, blood samples were collected to determine lipid peroxidation as malondialdehyde (MDA) concentration. Pigs were then subjected to heat stress (40°C) for 24 h after which MDA concentrations were again determined. In a second trial, grow-finish pigs (30.8 kg) were allotted to 3 dietary treatments containing either 11 ppm Vitamin E (11 ppm E; NRC requirement), 40 ppm Vitamin E (40 ppm E; commercial level), or 11 ppm Vitamin E + 29 ppm Vitamin E equivalent provided by a polyphenol blend (11E+PP). Dietary treatments were offered up to slaughter, and growth performance and carcass characteristics were determined.

Results and discussion

In the first trial, lipid peroxidation (MDA) after 7 days of feeding was not affected by the treatments ($P = 0.174$), but MDA increased after pigs were subjected to heat stress ($P = 0.080$) when 11 ppm E were supplemented (8.41 nM; Figure 1A). Values of MDA were intermediate with 80 ppm E (7.74 nM) and significantly lower with the 11E+PP blend (7.33 nM). In the second trial, results showed that levels of Vitamin E, or equivalent, supplied by the dietary treatments had no effect on growth performance (Figure 1B). However, loin depth ($P = 0.051$), margin over feed cost (MOFC; $P = 0.089$), and carcass value ($P = 0.099$) tended to improve with the added Vitamin E in combination with polyphenol blend. In conclusion, a blend of polyphenols and Vitamin E at NRC level is safer and more effective mean of providing antioxidant protection than an equivalent antioxidant dose of only Vitamin E as it resulted in the lowest MDA levels in heat-stressed nursery pigs and in better carcass characteristics in grow-finish pigs.

References

1. Davies, K. J. A. Oxidative stress, antioxidant defenses, and damage removal, repair, and replacement. *IUBMB Life*. 2000; 50:279-289.

2. Aitken, J. A. and Fisher, H. Reactive oxygen species generation and human spermatozoa: The balance of benefit and risk. *BioEssays*. 1994; 16:259-267.

3. NRC. Nutrient Requirements of Swine: 11th rev. *Natl. Acad. Press*. 2012.

4. van Kempen, T. A. T. G. et al. Vitamin E plasma kinetics in swine show low bioavailability and short half-life of *all-rac-α*-tocopheryl acetate. *J Anim Sci*.2016; 94.

5. Rice-Evans, C. et al. Antioxidant properties of phenolic compounds. *Trends Plant Sci*. 1997; 2:152-159.

Figure 1: A) Lipid peroxidation measured as plasma MDA levels in piglets fed Vitamin E to NRC standards (11 ppm E), commercial Vitamin E levels (80 ppm E), or to 11 ppm Vitamin E supplemented with a polyphenol blend to a dose equivalent to 80 ppm total Vitamin E (11E+PP). B) Ratio of growth performance and carcass characteristics of grow-finish pigs offered diets with Vitamin E included at NRC recommendations (11 ppm Vitamin E; set to 100%) to 40 ppm Vitamin E (40 ppm E; commercial level), or 11 ppm Vitamin E + 29 ppm Vitamin E equivalent provided by a polyphenol blend (11E+PP).

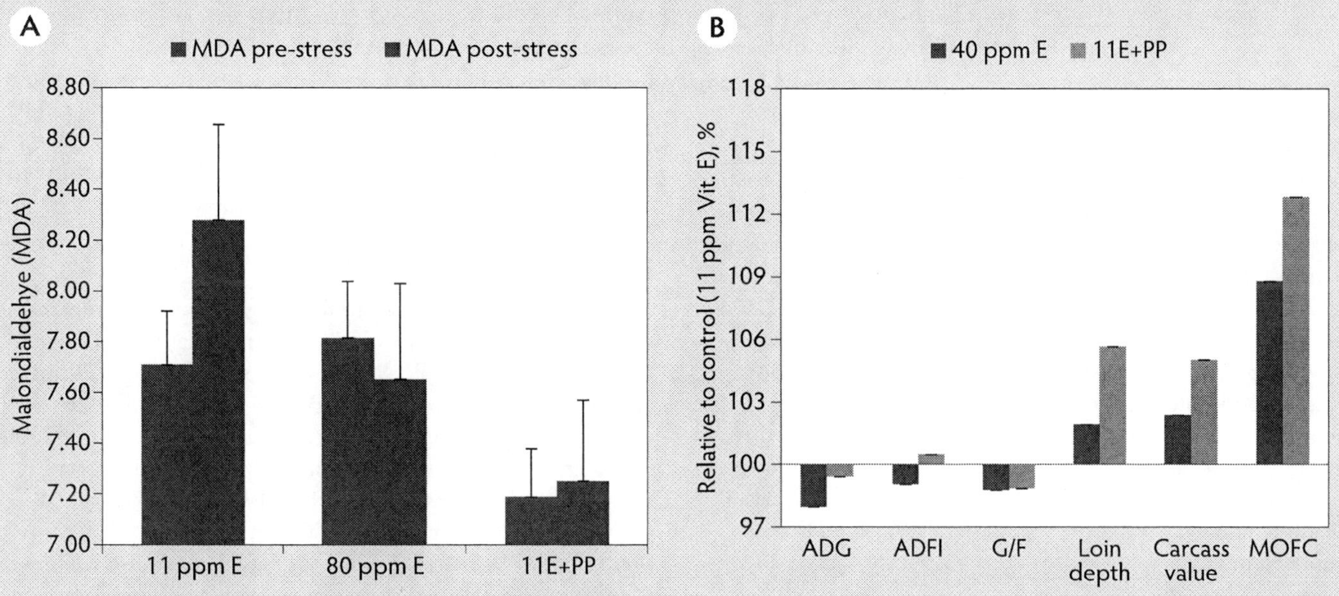

Efficacy of a PEDV intramammary vaccine delivery system

Franco Matias Ferreyra[1]; Jianqiang Zhang[1]; Balaji Narasimhan[1]; Kent Schwartz[1]; Douglas Jones[1]; Jennifer Schleining[1,3]; Paulo Arruda[2]; Bailey Arruda[1]
[1]Department of Veterinary Diagnostics and Production Animal Medicine. Iowa State University, Ames, Iowa; [2]Veterinary Resources Inc. Cambridge, Iowa; [3]Department of Large Animal Clinical Sciences, Texas A&M University, College Station, Texas

Introduction

Enteric diseases are a major cause of mortality and morbidity in pigs and result in significant economic loss. Intestinal diseases caused by viruses can be extremely severe, particularly in young animals, as was demonstrated by porcine epidemic diarrhea virus (PEDV).[1] Most of the economic losses incurred with PEDV are due to the high mortality in suckling piglets.[2] Antibodies cannot cross the placenta of a sow. As a result, neonatal pigs are born without antibodies and are unable to mount a rapid immune response to an infectious challenge. Their survival depends directly on the acquisition of passive lactogenic immunity via colostrum and milk. Protective passive immunity for PEDV from birth to weaning relies on continuous ingestion of IgA found in colostrum and milk.

There is little evidence that currently available inactivated commercial vaccines are effective in preventing neonatal mortality when used in PEDV-naïve sows subsequently challenged with PEDV at farrowing or in the first few days of life. Accordingly, there is a need to develop an effective vaccine to prevent PED in nursing piglets through the stimulation of lactogenic immunity in PEDV naïve sows.[3]

Material and methods

In this study, we developed an intramammary vaccine delivery system to induce the production of IgA in PEDV naïve sows and assessed the efficacy of this platform following inoculation of a highly virulent strain of PEDV. A polymer-based vaccine platform with adjuvant and without inactivated PEDV was administered to Group 1 (negative control: n = 2) and 2 (positive control: n = 2) sows. The polymer with inactivated PEDV and adjuvant was administered to Group 3 (vaccinated: n = 6) sows. Serum samples were collected on day 75, 85, 95, 105, and 115 days of gestation and 8 and 18 days post farrowing (DPF). Colostrum/milk and sow fecal samples were collected on 0, 4, 7, 10, 14 and 18 DPF. Piglet serum and fecal samples were collected at days 2, 4, 7, 10, 14 and 18 days of age. Individual piglet weights were measured at birth, 7, 14, and 18 days of age. Piglets in Group 2 and 3 were inoculated at 4 days of age with 1 mL of 1×10^3 TCID50/mL of PEDV. Two sows in Group 3 did not seroconvert and were removed from the study. During the experiment, no electrolyte, antibiotics or supporting therapy was administered to piglets. Clinical signs in piglets and sows were monitored daily. All piglets were euthanized 18 DPF. Statistical analysis was performed using R software (v3.5.1) using Survival package.

Results and discussion

Diarrhea and vomiting of variable magnitude were observed as soon as 1 day post-inoculation (DPI) in piglets in Group 2 and 3. At 7 DPF/3 DPI, PEDV was detected via qPCR in feces of all piglets from Group 2 and 3. By 14 DPF/10 DPI, two of the three animals that remained in Group 2 were still shedding virus, while 59% (13 of 22) of animals from Group 3 were negative by qPCR. At 18 DPF/14 DPI a 100% survival rate was observed in Group 1, 0% in Group 2 and 37% in Group 3. Preliminary ELISA results demonstrate detectable serum levels of IgA in piglets from Group 3 as soon as 2 DPF, in contrast to piglets in Group 2 where IgA was detected at 14 DPF. Log rank survival analysis for Group 2 and Group 3 was statistically significant at a level of $\alpha = 0.05$ (P = 3e-05; Figure 1).

In conclusion, piglets from dams that seroconverted in Group 3 had a higher survival rate and milder clinical signs than those in Group 2. Based on these data, this intramammary vaccine

Figure 1: Survival plot of piglets from sows inoculated with implant+PEDV, and from control sows. Horizontal black line indicates survival probability of 0.5 (50%)

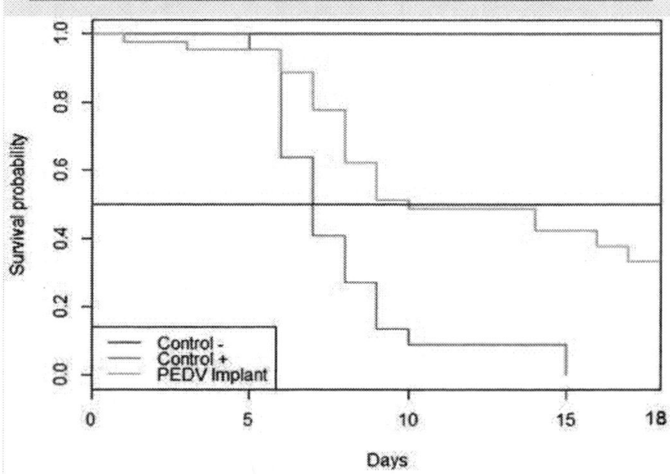

delivery system seems to be a promising alternative to increase the survivability of piglets following a PEDV challenge. In the future, this platform could be customized to include other enteric pathogens diminishing the economic losses attributed to enteric diseases in nursing piglets and decreasing or eliminating the need for feedback.

References

1. Stevenson G. et al. Emergence of Porcine epidemic diarrhea virus in the United States: clinical signs, lesions, and viral genomic sequences. *J Vet Diagn Invest.* 2013 Sep;25(5):649-54. doi: 10.1177/1040638713501675.

2. Chen Q. et al. Isolation and Characterization of Porcine Epidemic Diarrhea Viruses Associated with the 2013 Disease Outbreak among Swine in the United States. *J Clin Microbiol.* 2014 Jan; 52(1): 234–243. doi: 10.1128/JCM.02820-13.

3. Chattha K. et al. Strategies for design and application of enteric viral vaccines. *Annu Rev Anim Biosci.* 2015;3:375-95. doi: 10.1146/annurev-animal-022114-111038.

Using T cell epitopes to match the optimal vaccine for PCV2 field strains

Lucina Galina Pantoja[1*]; Dennis L. Foss[2]; Meggan Bandrick[2]; Andrés Gutierrez[3]; Anne S. De Groot[3]; Vitelio Utrera[1]
[1]Zoetis, Pork Technical Services, Parsippany, New Jersey; [2]Veterinary Medical Research and Development, Zoetis Inc., Kalamazoo, Michigan; [3]EpiVax, Inc., Providence, Rhode Island

Introduction

The constant evolution of porcine circovirus type 2 (PCV2) could impact viral genetic divergence on T cell epitope coverage of vaccine strains. EpiCC, an immunoinformatics tool that compares T cell epitope cross-conservation, can be used to determine the relatedness between vaccines and field strains and aid the vaccine selection process.[1] Improved epitope content has been reported when comparing a PCV2 bivalent vaccine and PCV2 strains obtained across time and from global geographies when compared to monovalent vaccines.[2] The objective of this study was to compare T cell epitope content of commercial vaccines and contemporary field strains circulating in the US.

Material and methods

The ORF2 of PCV2 monovalent vaccines (VacA, VacB and VacAlt), a bivalent vaccine with 2 PCV2 genotypes (VacAB, currently marketed in the US and Canada as Fostera Gold® PCV and Fostera® Gold PCV MH) and a set of field PCV2 strains were analyzed. Blind sequences from 94 strains submitted to the UMN VDL in 2017 were evaluated. Samples originated from tissues from 58 farms, from 12 different states, and with low PCV2 PCR Ct values ranging 10-20. The shared T cell epitope content between vaccines and field strains was identified using EpiCC. EpiCC scores were based on shared T cell epitope content and by the predicted SLA binding probability of each T cell epitope including both SLA class I and class II alleles. Scores were normalized by number of 9-mers in the field strains and the number of SLA alleles used for the analysis. Baseline EpiCC scores (comparison to self) were also calculated for vaccines and field strains. An average EpiCC score was calculated for the comparison of each vaccine and all the field strains. Average EpiCC scores by genotype were also calculated. The vaccine T cell epitope coverage was quantified to determine if the T cell epitope content of field strains was matched by a given vaccine.

Results

PCV2d (49%), PCV2a (37%), PCV2b (12%) and PCV2e (2%) genotypes were found. The epitope content and conservation between vaccine and field strains varied. All vaccine strains provide broad coverage of the field strains including heterologous genotypes; but none of the vaccines covered all T cell epitopes in the field strains. PCV2a vaccines had greater epitope coverage against PCV2a field isolates, but PCV2a strains showed more epitope diversity. The PCV2b vaccine strain had greater epitope coverage against PCV2b and PCV2d field strains. VacAB had, on average, the greatest epitope coverage based on EpiCC score. All vaccines showed lower epitope coverage against PCV2e (Figure 1).

Conclusion

The EpiCC analysis provides a tool to assess the possible impact of virus genetic divergence on T cell epitope coverage of vaccine strains. As multiple PCV2 genotypes are concurrently circulating in US pig farms today, the vaccine that combines PCV2a and PCV2b genotypes appears to provide broader T cell epitope coverage to strains currently present in the US.

References

1. Gutierrez et al., *BMC Bioinformatics* (2015) 16:290.

2. Bandrick et al., Proceedings of the 2018 *IPVS Congress*, Chongqing China, poster proceedings, p607.

Figure 1: Total (Class I and II) shared EpiCC scores between monovalent and bivalent PCV2 vaccines and contemporary PCV strains currently circulating in the US.

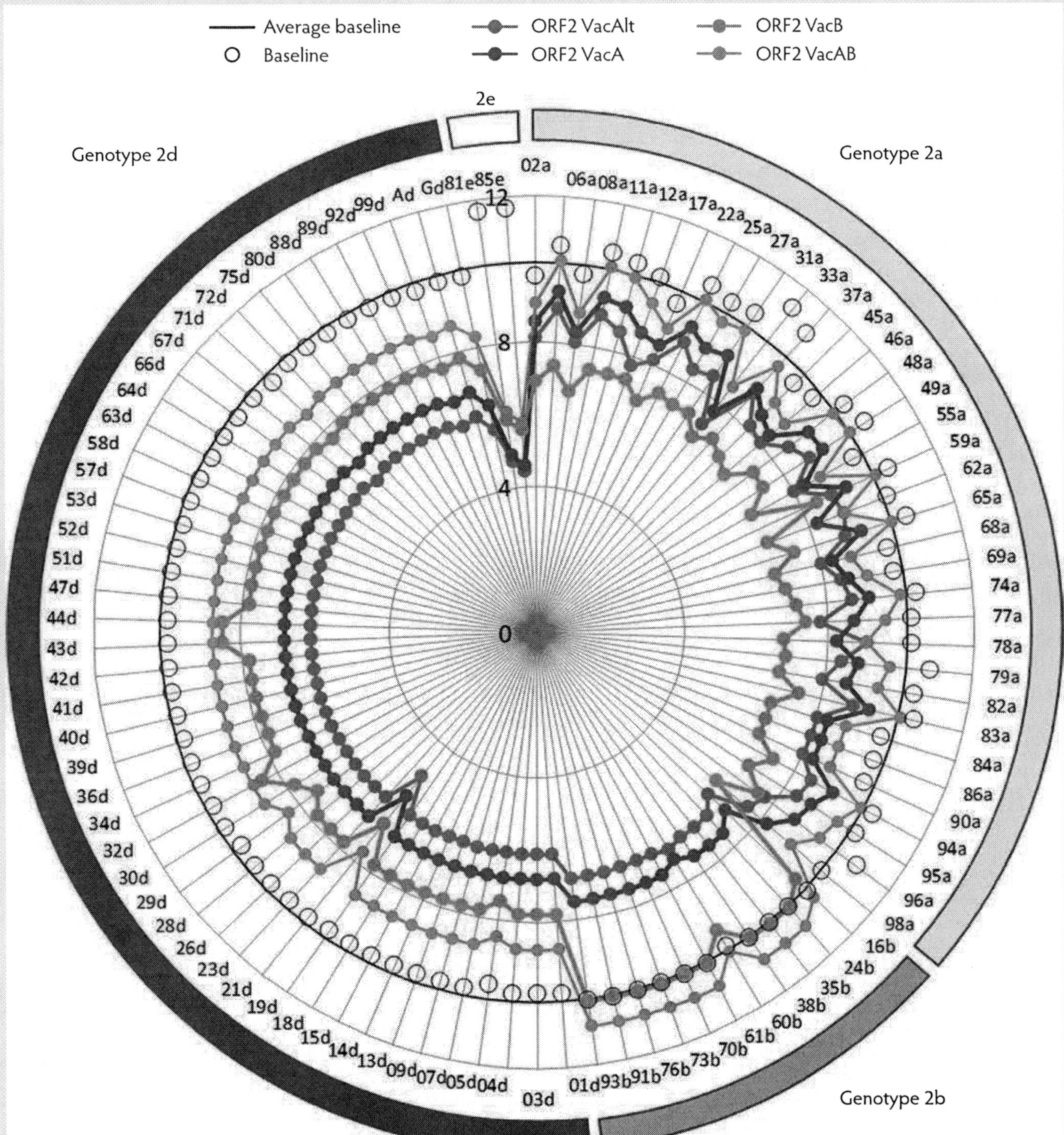

Genotypic and phenotypic characterization of *Streptococcus suis* strains isolated from healthy and diseased pigs

J. G. Santos, DVM; N. Macedo, DVM, MS, PhD; K. Harmon, PhD; L. Bradner, MS; A. Krull, DVM, PhD; O. Sahin, DVM, PhD; B. Arruda, DVM, PhD
Veterinary Diagnostic and Production Animal Medicine, Iowa State University, Ames, Iowa

Introduction

Streptococcus suis (*S suis*) is one of the most important and costly bacterial pathogens of pigs causing systemic and respiratory disease. *S suis* is also a zoonotic agent. Serotyping and genotyping are useful tools to differentiate isolates of this highly diverse species. Thus, the objective of this study is to characterize *S suis* isolates from healthy and diseased pigs using genotypic and phenotypic methods.

Materials and methods

One hundred and twenty five *S suis* isolates, obtained from 14 pigs (7 healthy and 7 with meningitis), from four different farms undergoing *S suis* outbreaks, were serotyped by PCR-based serotyping (mPCR) and coagglutination assays, and subsequently screened for 3 virulence-associated genes (*mrp, epf* and *sly*). Thirty-five *S suis* reference strains were used as positive controls.

Results and discussion

The mPCR correctly assigned all serotype reference strains, except serotypes 1 and 1/2, which cannot be distinguished from serotypes 14 and 2, respectively.[1] The sites of isolation included tonsil (44%), nasal cavity (40.8%), brain (6.4%), joint (4.8%) and bronchoalveolar fluid (4%). Serotype 1 or 14 (16%) was the most frequently identified serotypes overall, especially from brain (30%), tonsil (30%) and joint (15%) samples of diseased pigs. Thirty-one of the 63 (49.2%) untypeable (UNT) isolates by coagglutination were typed by mPCR, which agrees with a previous report demonstrating that mPCR may identify additional isolates.[1] The inclusion of up to eight isolates from the upper respiratory tract (84.8%) from the same pig may help explain the high percentage of UNT isolates in this study. Serotypes 1 (16/125, 12.8%), 8 (11/125, 8.8%), and 7 (11/125, 8.8%) were the most frequently identified serotypes by coagglutination, while serotypes 1 and 14 (20/125, 16%) were the most commonly identified ones by mPCR, followed be serotypes 7, 8, 21 (13/125, 10.4%), 16 (12/125, 9.6%) and 4 (11/125, 8.8%). Serotypes 1 or 14 by mPCR were further identified as serotype 1 (17/20), 14 (1/20) or UNT (2/20) by coagglutination. Two serotype 1 isolates by coagglutination were UNT by mPCR. Moreover, all isolates from serotypes 1 and 14 were the only serotypes positive for all three virulence genes tested, most of them (82%) from diseased pigs, including all the brain isolates and joint isolates. This finding suggests that serotype 1 and 14 strains were the main cause of streptococcal meningitis in these pigs. These serotypes were also identified in the tonsil of diseased and pen-matched healthy pigs In contrast, all isolates from serotypes 7 and 8 by coagglutination were assigned the same serotype by mPCR. mPCR further identified 2 isolates from serotypes 7 and 8 that were UNT by coagglutination. Furthermore, serotype 7 was predominately isolated from the nasal cavity, from both healthy and diseased pigs, and serotype 8 was mostly isolated from the nasal cavity of healthy pigs and tonsil of diseased pigs. Interestingly, serotype 21 was also predominately isolated from diseased pigs, but only from the nasal cavity and tonsil. Serotypes identified less frequently (< 5%) by both methods were serotypes 4, 5, 9, 12, 15, 22, 27, 28, 31 and 34. Preliminary results from this study indicate that once a virulent strain has been identified from aseptically collected samples from either the joint or brain, the herd population could be screened by tonsil swab to identify carriage status. The use of genotyping tools in addition to coagglutination could potentially aid the control of *S suis* disease in swine herds, and expands the knowledge of critical factors involved in the epidemiology and pathogenesis of infections by *S suis*.

Reference

1. Okura M et al. Development of a two-step multiplex PCR assay for typing of capsular polysaccharide synthesis gene clusters of Streptococcus suis. *J. Clin. Microbiol.* 2014; 52, 1714-1719.

Does lameness etiology affect sow-caretaker interaction and movement?

A. Forseth[1], DVM; S. Azarpajouh[2], DVM, PhD; L. Karriker[1], DVM, MS, DACVPM; K. Stalder[2], PhD; S. Millman[3,4], PhD; R. Parsons[3], MS; J. Stock[2], MS; A. Johnson[2], PhD

[1]Swine Medicine Education Center, Iowa State University, Ames, Iowa; [2]Department of Animal Science, Iowa State University, Ames, Iowa.[3]Veterianry Diagnostic Production Animal Medicine, Iowa State University, Ames, Iowa; [4]Biomedical Sciences, Iowa State University, Ames, Iowa

Introduction

Sow lameness is a major production disease affecting welfare and profitability. Lame sows require additional management and caretaker handling considerations to better assure their comfort. Little research has been completed to understand how lameness etiologies influence sow behavior and the relationship between the sow and caretaker. The objective of this study was to evaluate the effect of sow lameness etiology on the time and number of sow-caretaker interactions to traverse two obstacles in a handling course.

Materials and methods

A wooden ramp (11° slope, 4.61 m length) and wooden boards (10 boards measuring 0.08 m length × 0.76 m width × 0.03 m height) were constructed in the sow laboratory at Iowa State University College of Veterinary Medicine. Twenty-seven commercially crossbred sows (PIC 1050; parities 2 to 5) were individually removed from their home pen and walked over the ramp and wooden boards. Three color cameras were positioned over the ramp and two cameras were positioned over the wooden boards. Cameras recorded in real time (30 fps). Time (seconds; s) to traverse the two obstacles and the number of sow-caretaker interactions were reviewed from the video by one trained observer blinded to treatment. Handler interaction was used if the sow stopped making forward movement for 10 continuous seconds at any point on the handling course. Each level of interaction that occurred between the sow and the handler was completed in 20 second intervals. Handler interaction had a defined order until the sow resumed forward movement: 1) Noise maker (plastic coffee can filled with metal pieces), 2) Sorting panel used to apply pressure to the animal's hind, 3) Handler used their hands to make contact with the sow's back, 4) Feed placed in front of the sow and, 5) Rattle paddle used to make contact with the animal's back. After all interaction levels had been applied, she was left to rest for five minutes. In the event that the sow went down on two legs or laid down on the obstacle course she was left to rest for 10 minutes, without intervention. After periods of rest, animals were encouraged to continue moving through the course beginning with intervention number one. If the sow did not complete the obstacle course in 40 minutes, she was returned to her home pen. Following an antemortem evaluation, a swine veterinarian categorized sows based on the suspected lameness etiology: integumentary, musculoskeletal, or integumentary/musculoskeletal. Statistical analysis: Time and sow-caretaker interaction were shown to be highly correlated and were therefore analyzed separately using generalized linear mixed model methods (PROC GLIMMIX) of SAS. The significance level was set at $P < 0.05$.

Results and discussion

Twenty sows displayed lameness and seven sows were classified as non-lame (control etiology). Of the 20 lame sows, seven were classified as integumentary, seven were classified as musculoskeletal, and six were classified as integumentary/musculoskeletal. Lameness etiology was not a source of variation when comparing time to traverse the ramp (control 356.2 s; integumentary 105.8 s; musculoskeletal 68.9 s; integumentary/musculoskeletal 63.6 s; $P = .10$). Lameness etiology was not a source of variation when comparing time to traverse the wooden boards (control 115.3 s; integumentary sows 45.1 s; musculoskeletal 20.0 s; integumentary/musculoskeletal sows 32.8 s; $P = .10$). Lameness etiology was a source of variation ($P = .0005$) when comparing sow-caretaker interaction to traverse the ramp, with control sows needing 42.0 interactions, compared to integumentary (5.2 interactions), musculoskeletal (3.2 interactions), and integumentary/musculoskeletal (3.0 interactions). Lameness etiology was a source of variation ($P = .006$) when comparing sow-caretaker interaction to traverse the wooden boards. Control required 8.6 interactions, compared to integumentary (1.7 interactions), musculoskeletal (0.36 interactions), and integumentary/musculoskeletal (1.5 interactions). The results from this study do not suggest that lame sows with these specific etiologies need additional interaction from the caretaker to move through a ramp or wooden board obstacles, when compared to non-lame sows.

Understanding the impact of subclinical *Lawsonia intracellularis* infection on intestinal function

Fernando L. Leite[1]; Erika Vasquez[1]; Fabio Vannucci[1]; Connie Gebhart[1]; Juan E. Abrahante[2]; Jerry Torrison[1]; Adam Mueller[3]; Nathan Winkelman[3]; Zachary Rambo[4]; Richard Isaacson[1]
[1]College of Veterinary Medicine, University of Minnesota; [2]University of Minnesota Informatics Institute, [3]University of Minnesota; Swine Services Unlimited Inc; [4]Zinpro Corporation

Introduction

L intracellularis infection is estimated to occur in more than 90% of US swine farms and is known to negatively impact production performance.[1,2] *L intracellularis* infection occurs primarily in the small intestine, the major site of nutrient absorption in swine.[3,4] How intestinal function is impacted by infection is still poorly understood, yet it is known that subclinical infection has a significant negative impact on feed conversion efficiency and average daily weight gain which are correlated to lesion severity.[2] Subclinical infection is defined as infection without clinical sings, including diarrhea. A better understanding of how this pathogen can impact intestinal function is needed to be able to develop novel strategies to mitigate this infection and understand the impact that effective control methods such as vaccination can have on these measures.

Materials and methods

In this study, we evaluated total gene expression (transcriptome) of the ileum of groups of experimentally infected and non-infected swine. Animals were given an oral dose of 8.0×10^7 *L intracellularis* organisms in a mucosal homogenate and were euthanized at 14, 21 and 28 days post challenge (n = 6 per time point) to capture developing and resolving lesions. A group of non-infected animals (n = 6 per time point) was also euthanized at each time point.

Results and discussion

Infection peaked at 21 days post infection, and no animals developed clinical signs, validating this subclinical infection model. At this time point, half of the infected animals (n = 3) developed severe lesions while the other half developed mild lesions. Among the most significantly ($P < 0.05$) down regulated genes in animals with severe lesions were the genes that code for fibroblast growth factor 19 (FGF19), and the apolipoproteins APOC3 and APOA4 (Table 1). FGF19 is known to act as a hormone regulating bile acid biosynthesis and apolipoproteins play a fundamental role in the absorption of lipids.[4,5] These results indicate that subclinical infection by *L intracellularis* and development of lesions may be associated with altered lipid and bile acid metabolism which are important components to normal intestinal function, thus possibly interfering with normal nutrient absorption. These results demonstrate some of the possible mechanisms responsible for the production loss due to *L intracellularis* infection and presents targets to be investigated further.

References

1. Armbruster, G. A. et al. Review of *Lawsonia intracellularis* seroprevalence screening in the United States, June 2003 to July 2006. In 38th Annual Meeting of the American Association of Swine Veterinarians. 231–234 (2007).

2. Winkelman, N. Subclinical ileitis: Diagnostic monitoring, R^2, and economics. In 49th Annual Meeting of the American Association of Swine Veterinarians. 488-493 (2018).

3. Lawson G. and Gebhart C. Proliferative Enteropathy J. Comp. Path. 2000, Vol. 122, 77–100.

4. Laerke H.N., Hedemann M. S. The digestive system of the pig. Nutritional physiology of pigs - Online Publication. 2012.

5. Jones S.A. Physiology of FGF15/19. Endocrine FGFs and Klothos, 2012.

Table 1: Down regulated genes identified at 21 days post infection comparing animals with high (severe) and low (moderate) lesions.

Gene symbol	Fold change	Corrected *P* - value
ENSSSCT00000028967.1	-16.92	0.017351
ENSSSCT00000010414.2	-17.93	0.000002
AQP7	-20.45	0.000621
ENSSSCT00000010411.2	-20.46	0.011745
HELZ2	-22.44	0.045878
PLB1	-23.42	0.006509
APOC3	-42.21	0.000568
APOA4	-62.85	4.63E-05
FGF19	-63.33	0.037558
ENSSSCT00000022909.1	-239.38	0.049848

Assessment of the effect of different dynamics of wild-type PRRSV detection overtime on the mortality of vaccinated pigs raised at high pig dense area

C. A. A. Moura[1], DVM; R. Philips[2], DVM; G. S. Silva[1], DVM, MS, PhD; E. Mondaca[2], DVM, PhD; A. Scheidt[2], DVM, MS; Jens Kjaer[2], DVM, PhD; D. C. L. Linhares[1], DVM, MBA, PhD
[1] Veterinary Diagnostic and Production Animal Medicine Department, Iowa State University, Ames, Iowa;
[2] Boehringer Ingelheim Animal Health, Duluth, Georgia

Introduction

Porcine reproductive and respiratory syndrome virus (PRRSV) causes a considerable economic impact to the US swine industry, due to reproduction losses and worsened growth performance.[1,2] Fifty-five percent of the economic impact is on the growing phase,[1] and thus there is the critical need to better understand how the virus spreads within infected populations of growing pigs under field conditions. The objectives of this study were 1) to identify different patterns of PRRSV detection overtime on vaccinated growing pigs at high pig dense area, and 2) to compare the effect of the different detection dynamics on the growing pig mortality.

Methods

Fifty-five groups of growing pigs, from 2 PRRSV-positive and 5 PRRSV-stable sow farms, from two different US pig production systems were followed from weaning to slaughter by collecting 6 spatially distributed oral fluids samples every three weeks, totalizing 8 collection points (at 4, 7, 10, 13, 16, 19, 22, and 25 weeks of age). Samples were tested for PRRSV RNA by rRT-PCR. Moreover, at least one PCR-positive sample per time point was submitted to ORF-5 sequencing to differentiate between vaccine-like, or wild-type PRRSV. All pig batches were (A) originated from a single sow farm source, (B) vaccinated with attenuated PRRSV vaccine either at processing or at weaning, (C) placed at high pig dense areas, and (D) flowed in a wean-to-finish, or nursery followed by finishing facilities. A K-means clustering analysis, using R program, was performed to define the clusters based on the percentage of wild-type PRRSV, type 1 and 2, detection at each collection point. Groups with no wild-type virus detection over time were not included in the clustering analysis. Additionally, a poisson regression model, using R program, was used to compare differences in mortality among the clusters and the groups with no wild-type PRRSV detection.

Results

The study included 55,230 growing pigs, with 440 sampling time points totaling 2,640 oral fluids samples. Ninety-three percent of study groups were detected with wild-type PRRSV type-2 at least once in the growth phase. Three clusters among all groups were observed, representing an early (n = 18), middle (n = 21) and late (n = 12) PRRSV type-2 detection patterns (Figure 1). Four groups had no detection of wild-type virus during the study period and were classified as 'no PRRSV type-2 detection' pattern. The early challenge cluster had the highest mortality rate (13.4%). The middle challenge cluster had a mortality rate of 10.01%, and the late challenge cluster had a mortality rate of 8.86%. The groups with no PRRSV type-2 detection had the lowest mortality rate (7.3%). The difference of mortality rate among the 4 infection patterns were statistically significant ($P < 0.05$).

Fifty-six percent of study groups were detected with wild-type PRRSV type-1, at least once in the growth phase. Three different clusters among all study groups were observed, representing an early (n = 6), middle (n = 8) and late (n = 17) PRRSV type-1 detection patterns (Figure 1). Twenty-four groups had no detection of wild-type virus during the study period and were classified as 'no PRRSV type-1 detection' pattern. The early, middle, and late challenge clusters had a mortality rate of 9.5%, 10.94%, and 8.72% respectively. The groups with no PRRSV type-1 detection had a 9.58% mortality rate. The only significant difference on mortality was between 'no PRRSV type-1 detection' pattern and the late detection cluster ($P = 0.0121$).

Conclusions

Under study conditions, 93% of groups had wild-type PRRSV type-2 detection. Groups detected with type-2 PRRSV had increased growing pig mortality. The results sugested that the earlier the PRRSV type-2 detection, the worst was the growing pig mortality. Fifty-six percent of groups had wild type type-1 PRRSV detection, which was not associated with increased growing pig mortality.

References

1. Holtkamp D, Kliebenstein J, Neumann E, et al. Assessment of the economic impact of porcine reproductive and respiratory syndrome virus on United States pork producers. *Journal of Swine Health and Production.* 2013;21(2):72-84.

2. Neumann EJ, Kliebenstein JB, Johnson CD, et al. Assessment of the economic impact of porcine reproductive and respiratory syndrome on swine production in the United States. *J Am Vet Med Assoc.* 2005;227(3):385-392.

Figure 1: Clusters of wild-type PRRSV types 2 and 1, based on percentage of positive samples over time.

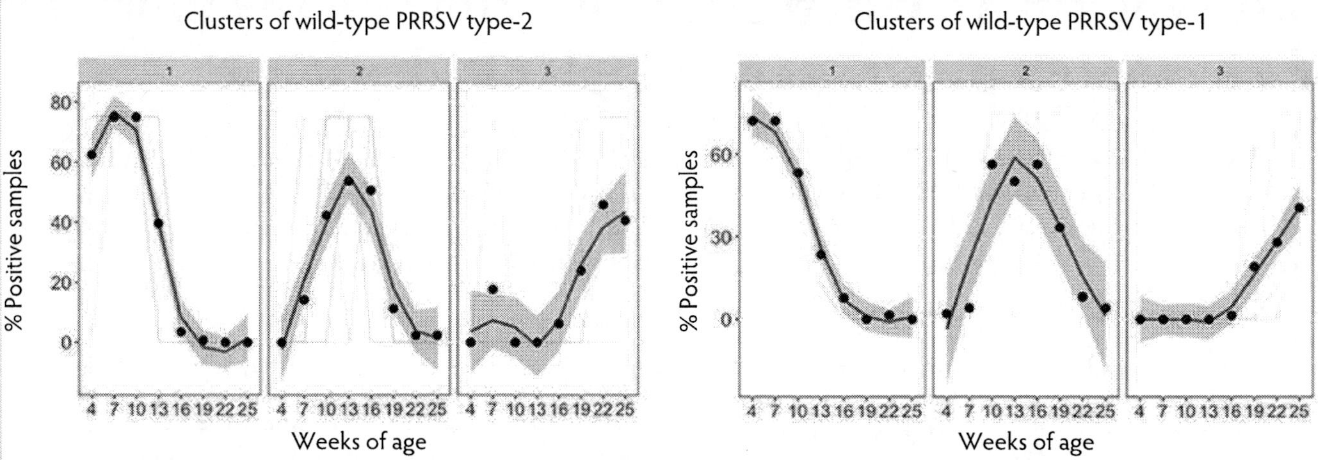

Assessment of the protective effect of influenza vaccination on morbidity and mortality in nursery pigs

E. Paladino[1]; A. Rendahl[1]; B. Chappell[2]; J. Alvarez[3]; M. Culhane[1]
[1]College of Veterinary Medicine, University of Minnesota, Saint Paul, Minnesota, USA
[2]Swine Health Professionals, Winnipeg, Manitoba, Canada
[3]Facultad de Veterinaria, Universidad Complutense de Madrid, Madrid, Spain

Introduction

Vaccination, a widely used disease control strategy,[1] can decrease transmission rates and shedding duration of influenza A virus (IAV) and also reduce the number of susceptible animals in the population.[2,3] The aim of this study is to assess the role of passive-protection on the morbidity and mortality of pigs in the nursery. We conducted a 22-month retrospective cohort study to assess mortality and morbidity over seven weeks in the nursery and assess the effect of vaccination after accounting for the possible effects of other factors, such as source, season, and wean age.

Materials and methods

Data were collected from pigs originating from three sow farm sources with endemic and on-going IAV infections but vaccinating sows routinel. Pigs from the three sow farms (E, S, and T) were commingled at weaning as batches into nursery sites. Upon arrival, piglets with low potential to thrive were either euthanized or designated as "Class B;" thus these two categories were used as a measure of morbidity. A negative binomial mixed multivariable regression model was fitted to the data, in which "batch" and "nursery site" were modeled as random effects to account for the lack of independence of the observations recorded in a same batch/nursery site, and batch and source were nested to account for within-group effects. Vaccination was assumed to result in protection of the pigs if the pigs were born to vaccinated sows within 3 to 8 weeks of sow vaccination. Pigs born outside this vaccination timeframe were assumed to be unprotected by sow vaccination. Variables were first screened using mixed bivariable models and those factors for which association with the outcome was estimated at *P*-values higher than 0.2 were selected for inclusion as candidate variables in the multivariable model. Variables that were highly correlated (as determined using Cramer's V and Variance Inflation Factor) were excluded prior to the multivariable analysis.

Results and discussion

There were 250 batches with 964,301 total pigs analyzed, with an average time in nursery of 49 days. The multivariable model included these predictors: protection, month at weaning, class B, euthanasia at arrival, weekly euthanasia, source, barn size, wean age in quartiles and days in nursery. Most pigs died during week 2 in the nursery and mortality was 4.8 times more likely to occur during the second week compared to week 1. Twenty six percent of piglets were born to vaccinated sows, thus considered protected, and a significant negative relationship (IR:0.87)was observed for this category (CI: 0.86, 0.87). Piglets that were weaned in March, May, June, September and November also had decreased mortality; however, no apparent seasonality effect can be inferred from this perspective. Lastly, there appears to be a large variability in the mortality ratesdue to sow farm source, where source "T" showed a significantly lower mortality when compared to the other two sources. The preliminary results are suggesting that sow vaccination plays a role in reducing mortality in the nursery as protected pigs had overall less mortality.

References

1. USDA. Swine 2012. Part II: Reference of Swine Health and Health Management in the United States, 2012. NAHMS Swine 2012 2017 02/23/2017 [cited 2017 March]; NAHMS Swine 2012]. Available from: `https://www.aphis.usda.gov/aphis/ourfocus/animalhealth/monitoring-and-surveillance/nahms/nahms_swine_studies`.

2. Corzo, C.A., et al., Observations regarding influenza A virus shedding in a swine breeding farm after mass vaccination. Journal of Swine Health and Production, 2012. 20(6): p. 283-289.

3. Mughini-Gras, L., et al., Control of a Reassortant Pandemic 2009 H1N1 Influenza Virus Outbreak in an Intensive Swine Breeding Farm: Effect of Vaccination and Enhanced Farm Management Practices. PLoS Curr, 2015. 7.

Influence of piglet gut microbiota on severity to *Mycoplasma hyopneumoniae* infection

M. Surendran Nair[1]; T. Eucker[2]; J. Victoria[2]; B. Martinson[2]; A. Neubauer[2]; M. Pieters[1]
[1]Department of Veterinary Population Medicine, University of Minnesota, St. Paul, Minnesota;
[2]Boehringer-Ingelheim Animal Health, Duluth, Georgia

Introduction

Mycoplasma hyopneumoniae is the causative agent of porcine enzootic pneumonia and considered as one of the most important etiologies associated with the chronic respiratory illnesses in swine production.[1] Recent research indicates that a state of compromised gut microbiota could be associated with increased incidence of respiratory infections in animals and humans.[2] In this study, we hypothesized that differences in piglet gut microbiome composition modulate susceptibility to *Mycoplasma hyopneumoniae*.

Materials and methods

Thirty conventional crossbred piglets, from five different litters, were included in this study. Piglets were fed a commercial diet *ad libitum* and were administered Ceftiofur (5 mg/kg, IM) at three weeks of age. Fecal bacterial community compositions of piglets at three weeks of age (one day after antibiotic exposure), eight weeks of age (prior to *M hyopneumoniae* challenge) and twelve weeks of age (28 days post-challenge) were targeted by amplifying the V3-V4 region of 16SrRNA gene. Illumina MiSeq generated sequences were processed using DADA2 workflow and microbiome community analyses were conducted using R software. Pearson's correlation coefficients were used to determine the association of relative abundance of amplicon sequence variants with *M hyopneumoniae* induced lung lesions. Lung lesion scores (LS) were recorded as described in European Pharmacopoeia from pig lungs at necropsy.[3]

Results and discussion

The LS ranged from 0.3% to 43% with an evident clustering of the scores observed in piglets within litters. Piglets from two litters showed the highest mean LS (20%; 95%CI: 30.32 - 10.4%) when challenged with *M hyopneumoniae*, whereas two other litters showed the lowest mean LS (2.5%; 95%CI: 4.7 - 0.31%). There were significant differences in species richness and alpha diversity in fecal microbiomes among piglets within litters at different time points ($P < 0.05$). The dissimilarity matrices indicated that at week three, the fecal microbiota of piglets was more dissimilar compared to those from weeks eight and 12. This microbial shift at week three was mainly driven by the increase in abundance of the indicator family, Ruminococcaceae in piglets with low LS ($P < 0.05$). In addition, the richness counts determined on week three was negatively correlated with LS ($r = -0.44, P = 0.013$). The taxa, *Ruminococcus*_2 having highest richness scores, correlated significantly between litters with lowest LS ($r = -0.49, P = 0.005$). These findings suggest that early life gut microflora can be a potential determinant for *M hyopneumoniae* susceptibility in pigs.

References

1. Thacker, E. L. & Minion, F. C. Mycoplasmosis. in *Diseases of Swine* (eds. Zimmermann, J. J., Ramirez, K. L. A. & S.K. J, S. G. W.) 779–797 (John Wiley & Son, Inc., 2012).

2. Marsland, B. J. & Gollwitzer, E. S. Host–microorganism interactions in lung diseases. *Nat. Rev. Immunol.* **14,** 827–835 (2014).

3. Ph. Eur. (European Pharmacopoeia). Monograph 04/2013:2448. Porcine enzootic pneumonia vaccine (inactivated). *Ph. Eur. (European Pharmacopoeia)* (2013).

Effect of an isotonic protein drink on intestinal microbe population

CallaBria Putrino[1]; Fabio Vannucci[2], DVM, PhD, MS; Ava M. Firth[3], DVM, MVS, DACVECC
[1]California Polytechnic State University, San Luis Obispo, California; [2]University of Minnesota, Veterinary Diagnostic Laboratory, St Paul, Minnesota; [3]Tonisity International, Dublin, Ireland

Introduction

The rapid development of the intestinal environment in the neonatal pig sets the stage for production performance as well as nutrient absorption in the intestine. The aim of this study was to assess the effect of an isotonic protein drink on intestinal morphology and microbial population in the first 30 days of the pig's life.

Materials and methods

15 gilts (Yorkshire × Landrace × Duroc) and their litters (161 piglets) were enrolled in the study which was performed in a dedicated high-health research facility. Litters were allocated to one of two study groups, control (Group A) or supplementation with an isotonic protein solution (Group B). Group A litters received no supplementation during the pre-weaning phase, and were given creep feed from weaning up until two days after weaning. From days 2-8 of age, group B litters were supplemented with 500 mL of a 3% isotonic protein solution once daily in an open pan. Group B litters also received the 3% solution 3 and 2 days before weaning followed by a gruel mixture of creep feed and the 3% protein solution from 1 day before weaning up until 2 days after weaning. Both groups had ad lib access to normal feed after weaning. Weaning was at ~ 21 days. A total of 56 piglets were randomly selected from each litter for euthanasia at 9 and 17 days of age and a further 20 piglets were randomly selected from each treatment group at 30 days of age. Samples taken from the ileal mucosa of each piglet were submitted for semi-quantitative aerobic bacterial culture (*E coli*) and rotavirus PCR. Samples were also taken intestinal microbiome analysis.

Frequency of positive results and odds ratios were analyzed using the Fisher's exact test in SAS 9.4 (SAS Inst. Inc., Cary, NC). Significance was declared at $P \leq 0.05$ and tendencies declared at $0.05 < P \leq 0.1$.

Results

In the preweaning phase (day 9 and 17 combined), group B pigs had significantly fewer numbers of pigs (7/30, 23%) with positive cultures for *E coli* compared to Group A pigs (14/26, 54%) ($P = 0.01$), representing a 57% reduction (Figure 1). Group B pigs had 1/30 (3.3%) cultures positive for beta-hemolytic *E coli*, compared to 4/26 (15.3%) in Group A, though this was not a statistically significant difference. Within each group, there was an even distribution of *E coli* cases between day 9 (12 cases total, 8 in Group A and 4 in Group B) and day 17 (9 cases total, 6 in Group A and 3 in Group B). Differences in the incidence of *E coli* between Group A and Group B at day 30 were not significant. Differences in the incidence of beta-hemolytic *E coli* were not significant between groups. The odds ratio of a negative *E coli* result was 3.85 times greater in Group B compared to Group A in the preweaning phase. Overall (preweaning and postweaning combined) Group B pigs tended to have fewer pigs positive for *E coli* (10/40, 25%) compared to Group A pigs (15/36, 42%) ($P = 0.06$), representing a 40% reduction.

There were no significant differences in the incidence of rotavirus at any time point. Results of microbiome analysis are reported elsewhere in these proceedings (Buzoianu, Firth, AASV 2019 Industrial Partners session).

Conclusion

The administration of an isotonic protein drink in the first week of life was associated with a 57% reduction in the preweaning phase and an overall 40% reduction in intestinal *E coli* load in this study. Enterotoxigenic *E coli* (ETEC) is an important pathogen in neonatal and post-weaning pigs as it can negatively impact production performance, and reducing it at an early stage may be of benefit to the health of pigs.

Figure 1: Percentage of cases positive for *E coli.*

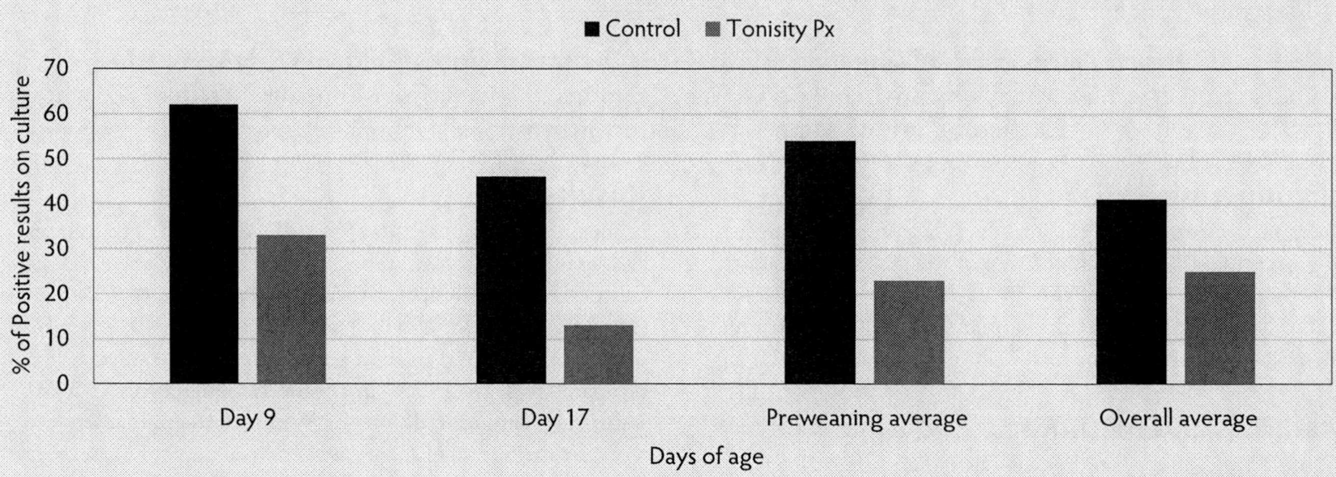

Development of a novel multiplex real-time PCR for the detection and differentiation of SADS-CoV, an emerging swine coronavirus

R. Rauh[1,2]; T. Clement[2]; J. Christopher Hennings[2]; Feng Li[2]; Steve Lawson[2]; S-L Zhai[3]; J-Y Ma[4]; J. D. Callahan[1]; E. A.Nelson[1]; Diego G. Diel[2]
[1]Tetracore; [2]Animal Disease Research and Diagnostic Laboratory, South Dakota State University; [3]Guangdong Academy of Agricultural Sciences; [4]South China Agricultural University

Introduction

A coronavirus variant known as Swine Acute Diarrhea Syndrome Coronavirus (SADS-CoV) has emerged in China where the first clinical signs in pigs were observed in late December 2016. In 2017, an outbreak in several Guangzhou farms was associated with fatal acute diarrhea in neonatal piglets. The introduction of this new swine enteric coronavirus in the US, could have significant economic consequences to the swine industry. Therefore, rapid detection methods are required for the early diagnosis of this pathogen. Here we report the development of a new real-time PCR assay for SADS-CoV that will be multiplexed with two other swine coronaviruses. The final configuration for the test in development will be PEDV/SADS-CoV/PDCoV/IC.

Materials and methods

Currently there are only a few SADS sequences available on Genbank. These sequences were downloaded and imported into the Sequencher 5.4.5 Software. The software was used to assemble the sequences into a contig. The sequences were visually inspected for conserved region. For Corona Viruses mainly the Nucleocapsid (N) and RNA-dependent RNA polymerase (RdRp) gene regions seem to be highly conserved. For the new SADS rtRT-PCR assay the N region was chosen. An In Vitro Transcript was designed which included the target sequences for the SADS assay. The process to create the IVT began with designing a syntemp oligo. The oligo includes a T7 promotor which is used by the MEGAscript® T7 Kit to transcribe the In Vitro Transcript (IVT).

Results and discussion

The IVT is then tested with primer, probes and a chosen One-Step RT-PCR kit on the ABI 7500 Real-Time PCR Instrument. The generated standard curve with the IVT has a slope of -3.321280 and R2 value of 0.984144 which demonstrates that the chosen design of the rtRT-PCR SADS assay is efficient. The next step is to evaluate the analytical specificity of the SADS-CoV PCR assay against a panel of enteric swine corona viruses at SDSU to ensure that the test specifically amplifies SADS-CoV viruses and does not cross react with closely related non-SADS swine coronaviruses. A commercial set of research reagents to detect swine coronaviruses has been on the market since 2013 from Tetracore, Inc. Rockville, MD. The PEDV-TGEV-PDCoV multiplex PCR has a three year track record of consistent performance although TGEV is very seldom seen in the US. Since the practical limit of multiplex real-time PCR is four fluorochromes; it is necessary to replace the TGEV assay with the SADS-CoV assay for a product that is a multiplex rRT-PCR consisting of PEDV, SADS-CoV, PDCoV and an internal positive control. The standalone SADS-CoV and multiplex PEDV-SADS-CoV-PDCoV-IC rRT-PCR will be tested in parallel against Chinese reference test methods with two collaborative partners, each with a collection of swine coronaviruses including SADS-CoV in China.

Conclusion

It is important to the US Pork industry that there be well validated, commercially available tests that can be mass-produced under cGMP conditions and deployed to veterinary diagnostic laboratories for an outbreak contingency should this novel swine coronavirus make its way to the US.

Porcine astrovirus type 3 is an emerging cause of atypical neurologic disease: Diagnostic cases and infection dynamics on affected flows

Gaurav Rawal; Franco Matias Ferreyra; Adam Mueller; Grant Allison; William Hedberg; Nubia Macedo; Laura Bradner; Karen Harmon; Daniel Linhares; Bailey Arruda
Iowa State University

Introduction

Astroviruses (AstVs) are viral agents in the Family *Astroviridae* that are able to infect and cause disease in a wide variety of host species.[1] Despite the broad host range and ability to cause disease, there are significant gaps in knowledge concerning the epidemiology, ecology, and pathophysiology of a majority of AstVs. Porcine astroviruses (PoAstVs) are distributed worldwide.[2] Five PoAstV lineages (PoAstV1–PoAstV5) have been identified perhaps reflecting diverse origins, interspecies transmission, and recombination events, some presumably with human strains.[3-6] In the past decade, different members of the genus *Mamastrovirus* have been associated with neurologic disease in humans,[7] bovine,[8] mink[9] and most recently, porcine astrovirus 3 (PoAstV3) in swine.[10,11]

Clinical signs

Lateral recumbency is one of the most common clinical signs reported. Others include knuckling over on forelimbs and ataxia. Animals can move their legs and seem to have more use of their hind limbs than forelimbs; however, they appear unable to bear weight and if able to move commonly crawl with their abdomen near or touching the ground (astasia) as shown in Figure 1. Note that "paddling" is NOT a characteristic sign of neurologic disease due to this type of viral agent that primarily affects spinal cord function.

Diagnostic cases

There has been an increase in the number of cases submitted to the Iowa State University Veterinary Diagnostic Laboratory. Zero

Figure 1: Astasia and knuckling of a pig with PoAstV3 polioencephalomyelitis.

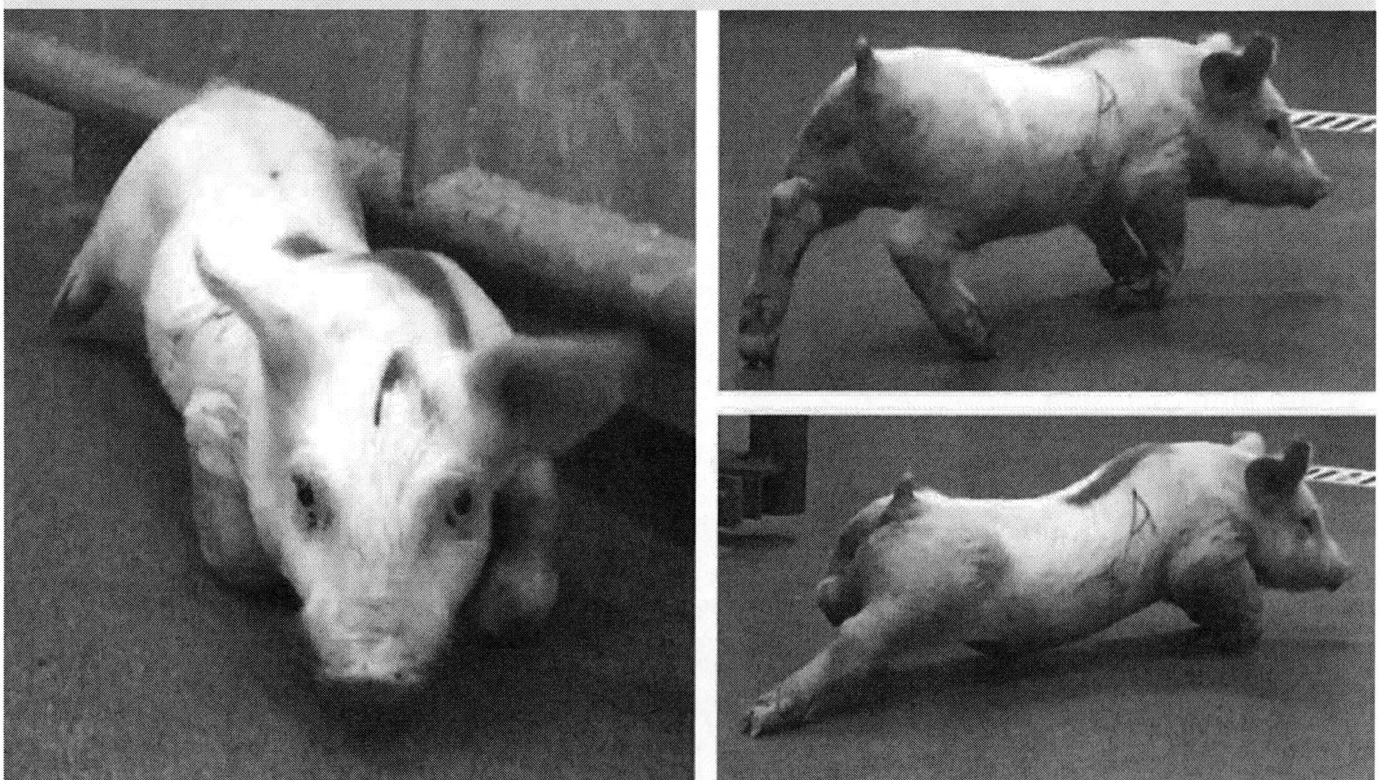

to three cases per year were identified between 2010 to 2017 with 13 cases as of November 2018 suggesting this virus is affecting more flows than in the past.[12]

Distribution of PoAstV3 on farms

To gain a better understanding of the infection dynamics and epidemiology of PoAstV3, cross-sectional studies were conducted that included gilts, sows, and suckling piglets on farms with and without a diagnosis of neurologic disease due to PoAstV3 but previous detection of PoAstV3 in feces. The overall detection frequency of PoAstV3 by PCR in a breed-to-wean farm with no previous history of PoAstV3 neurologic disease was 77% (54/70), with the highest detection frequency in suckling piglets (92%) followed by due-to-wean pigs (90%), sows (90%), neonates (73%), and incoming gilts (20%). The same farm was intensely sampled after 3 months to reassess viral prevalence. The overall detection frequency was 85% (207 of 244), with similar detection frequencies in categories reported previously. A longitudinal study was conducted following 64 individually identified newly weaned pigs to market on a farm with neurologic disease due to PoAstV3. PoAstV3 was detected by qPCR on fecal swabs intermittently over time suggesting that pigs shed PoAstV3 in their feces intermittently or can be re-infected with PoAstV3. The fecal swabs and pen-oral fluids had the highest concentration of PoAstV3 RNA compared to other sample types, and the longest duration of detection lasting for up to 126 days.

Conclusion

PoAstV3 is an emerging cause of neurologic disease affecting spinal cord functions in adult females, nursery pigs, and finishing pigs. It should be considered a possible cause of disease in pigs that are found on their side or have difficulty controlling their limbs when they walk. On farms in which it has been detected, PoAstV3 is frequently detected in the feces of pigs at various stages of production and in the environment.

References

1. Caliciviridae and Astroviridae. In: *Fenner's Veterinary Virology* (Fifth Edition), eds. 2017

2. Benedictis et al. Astrovirus infections in humans and animals. *Infect Genet Evol New York*, 2013.

3. Laurin et al. Detection and genetic characterization of a novel pig astrovirus. *Arch Virol* 2011

4. Luo et al. Multiple novel and prevalent astroviruses in pigs. *Vet Microbiol* 149: 316-323, 2011

5. Shan et al. The fecal virome of pigs on a high-density farm. *J Virol* 85: 11697-11708, 2011

6. Xiao et al. Identification and characterization of novel porcine astroviruses with high prevalence and frequent co-infection of individual pigs with multiple PAstV types. *J Gen Virol*, 2013

7. Quan et al. Astrovirus encephalitis in boy with X-linked agammaglobulinemia. *Emerg Infect Dis* 16, 2010.

8. Spinato et al. Identification of bovine astrovirus in cases of bovine non-suppurative encephalitis in Canada. *CVJ*, 2017.

9. Blomström et al. Detection of astrovirus in brain of mink suffering from shaking mink syndrome. *J Clin Microbiol* 48, 2010.

10. Arruda, B. et al. Porcine Astrovirus Type 3 in Central Nervous System of Swine with Polioencephalomyelitis. *Emerg Infect Dis* 23, 2017.

11. Boros, A. et al. Outbreaks of Neuroinvasive Astrovirus Associated with Encephalomyelitis, Weakness, and Paralysis among Weaned Pigs, Hungary. *Emerging Infect Dis* 23, 2017.

12. Rawal et al. Porcine astrovirus type 3 emerging cause of neurologic disease in swine. Available at: https://www.nationalhogfarmer.com/animal-health/porcine-astrovirus-type-3-emerging-cause-neurologic-disease-swine.

Diagnostic specificity of ASFV, CSFV and FMDV real-time PCR assays for swine pen-based oral fluid samples

Marisa Rotolo; Christina Loiacono; Kimberly Dodd; Tracy Sturgill; Diane Holder; Kate Schumann; Gregory Mayr; Jeffrey Zimmerman; Jane Christopher-Hennings; Rodger Main; Jerry Torrison; James Trybus; Maria Antognoli

Introduction

In commercial swine production systems, swine oral fluid specimens are routinely collected and tested for pathogen-specific nucleic acids (PCR) or antibodies (ELISA) to monitor domestic pathogens, e.g., porcine reproductive and respiratory syndrome virus (PRRSV), influenza A virus (IAV), porcine epidemic diarrhea virus (PEDV), and others. The ease of sample collection and high diagnostic value make oral fluids an efficient and cost-effective sample for surveillance.

The use of oral fluids in the surveillance of foreign animal diseases (FADs) in North America, e.g., African swine fever (ASFV), classical swine fever (CSFV), and foot-and-mouth disease viruses (FMDV) by real-time polymerase chain reaction (rPCR) has not been fully evaluated outside of experimental settings. In particular, a careful evaluation of diagnostic specificity is required because false positive results increase producer and program costs through delayed movement of animals and products, follow-up investigations, and additional sampling/testing. Conversely, high diagnostic specificity (few false positives) builds participant confidence in the program – an element that becomes increasingly important as the outbreak is brought under control and the majority of samples tested are truly negative. Therefore, the objective of this study was to estimate the diagnostic specificity of ASFV, FMDV and CSFV oral fluid rPCR assays.

Methods and results

The tests evaluated in the study included assays that had been validated by the National Veterinary Services Laboratories (NVSL) and employed by the National Animal Health Laboratory Network (NAHLN), as well as three non-USDA licensed commercial assays. Field oral fluid samples, including samples representative of all six USDA Veterinary Services (VS) Districts, were collected by four veterinary diagnostic laboratories (VDL) and tested for all three diseases with rPCR assays from NVSL-NAHLN and Tetracore, Inc.

Diagnostic specificity was defined as the proportion of oral fluid samples collected from pens of negative animals that tested negative by rPCR. For each VDL, the preliminary diagnostic specificity estimates and the corresponding 95% confidence intervals were calculated for each assay using statistical software (SAS 9.4, SAS Institute Inc., Cary, NC). Initial study results indicated that all rPCR assays evaluated had high diagnostic specificity, i.e., no false positive results were reported. Additional testing is in process to finalize diagnostic specificity estimates.

Discussion

In a FAD outbreak, aggregate samples, such as swine oral fluids, will be needed for efficient and rapid detection of foci of infection, to support the coordination of control efforts, to assist in the verification of disease status for premises within the control area, and help evaluate the risk associated with animal and product movement outside of a control area. The results of this study support the further evaluation of oral fluids as an effective swine surveillance sample in the event of an FAD outbreak. The next steps include the evaluation of the capability of each assay to detect FAD viruses in swine oral fluids using known positive samples.

Application of spatially balanced sampling to regional surveillance

M. Rotolo; C. Wang; R. Main; J. Zimmerman
Iowa State University, Ames, Iowa

Introduction

The long-term objective of this project is to develop efficient and cost-effective regional surveillance methods, with an emphasis on preparing the swine industry for detecting and eliminating emerging and/or foreign animal diseases (FAD). In an emergency, how can we achieve the fastest detection with the fewest samples?

The concept of "representative sampling", i.e., testing a randomly selected subset of a population, was introduced in 1895,[1] but not widely used in swine surveillance until the US PRV pilot eradication projects in the 1980's. The convention of testing 30 samples for disease detection is a lingering legacy of the PRV eradication program. The sample size of 30 is expected to provide a 95% probability of detecting ≥ 1 positive animals in a population of 1,000 animals with ≥ 10% disease prevalence.[2]

Sample sizes for representative sampling estimate the power of detection using probability distributions (binomial or hypergeometric). These distributions rely on several assumptions. Most importantly for this discussion, they require the target to be randomly distributed in the population. This is true whether the population is pigs on a site or production sites in a region. However, it is increasingly recognized that disease shows a spatial (non-random) distribution. For example, neighboring pens (or farms) are more likely to share the same disease status versus distant ones.

"Spatially balanced sampling" (not random sampling) is an approach for surveilling populations with spatial trends in the target of interest, such as disease. "Spatially balanced" means systematically distributing the samples over the population within the defined space. In essence, spatially balanced surveillance means designing the survey so that neighboring farms (likely to be of the same status) are not both sampled. A common spatially balanced approach is a method called "generalized random-tessellation stratified design" (GRTS).

Materials and methods

To evaluate different spatially balanced surveillance methods, we used PEDV test data from the ISU VDL (2014 – 2017). To protect client confidentiality, the spatiotemporal distribution of PEDV was only determined on the county level and only for the state of Iowa. The diagnostic test results were modeled to compare the probability of detection using GRTS versus a simple random sampling (SRS) approach.

Results and discussion

Our analysis of the data showed that GRTS sampling was more timely and powerful than simple random sampling (SRS) in detecting the emergence of PEDV. GRTS is just one of several statistical approaches for spatially balanced sampling. Others include the Local Pivotal Method, Spatially Correlated Poisson Sampling, Cube Method, and the Local Cube Method. These methods will also be analyzed and compared in terms of time to detection and efficiency.

References

1. Anderson LA, Black N, Hagerty TJ, et al. 2008. Pseudorabies (Aujeszky's Disease) and Its Eradication. USDA:APHIS Technical Bulletin #1923.

2. Kruskal W, Mosteller F. 1980. Representative sampling. IV: The history of the concept in statistics, 1895-1939. Int Stat Rev 48:169-195.

Effect of breeding herd PRRS status on productivity

D. Torrents[1]; J. Miranda[1]; R. Pedrazuela[1]; A. Ramirez[2]; D. Linhares[2]
[1]Laboratorios HIPRA S.A, Amer, Spain; [2] Iowa State University, Ames, Iowa

Introduction

PRRS is an endemic swine disease causing significant productive and economic loses in pig farms. Spain is one of the main global pig producers. However, data on the impact of PRRS virus (PRRSV) infection on Spanish, or other European pig farms are still scarce. The aim of this study was to establish, for the first time, a systematic monitoring system for PRRS in Spanish sow farms to classify farms according the PRRS status using terminology currently adopted in the US swine industry.[1] Furthermore, we investigated associations between PRRS status of PRRS-positive breeding herds, and selected production parameters, allowing the characterization, for the first time, of production and economic impact of PRRSV under Spanish field conditions.

Materials and methods

Between February 2017 to March 2018, 40 breeding herds (total of 93,800 sows) belonging to one large integrated group located in Spain have been enrolled voluntarily in a one-year program for the PRRSV monitoring project based on the American Association of Swine Veterinarians guidelines. PRRS status of study farms were determined by a diagnostic monitoring protocol, which consisted of monthly individual due-to-wean blood testing of 30 piglets.[2] For the evaluation of productive parameters and their association with the PRRS status, weekly Stable (ST) and Unstable (UN) PRRS-status for each positive-PRRS farm was stablished according results of the last test. At the same time weekly data from all PRRS-positive farms included in the study were collected throughout the entire monitoring period and the following weekly key performance indicators (KPI) were calculated: Abortions/1000sows (ABTHS), born alive piglets rate (BAR), Pre-weaning mortality rate (PWMR) and wean-piglets/1000sows (WPTHS). Comparative statistical analysis of KPI between different PRRS status was performed by generalized linear mixed model using SAS 9.4 analytic software. PRRS status was included in the model as fixed effect meanwhile farm and genetics were included as random effects in the model.

Results and discussion

A total of 1997 weekly data were analysed from 35 PRRS-positive farms. 782 weeks were classified as PRRS UN, meanwhile 1215 weeks were considered ST. Statistical difference (Table 1) was observed between UN and ST in weekly BAR (+1.08%), PWMR (-0.95%) and WPTHS (+26.2 piglets) parameters. However, no statistical difference was observed regarding weekly ABTHS. This lack of difference in abortions, one of the main clinical effects of PRRSV in breeding herds, could be explained by the classification system of PRRS status in this project, which was based on monitoring sampling instead of clinical diagnostic. Thus, possible peaks of abortions due to a new PRRSV circulation in ST farms may have not been associated with viremia in suckling pigs, misclassifying farms as ST. On the other hand, farms presented lower PWMR and higher BAR and WPTHS during the periods of stability when compared with instability periods, arising the productive impact of PRRSV circulation in breeding herds. Taking into account annual basis, we could estimate a reduction of 1.36 wean piglets/sow/year due to PRRS instability.

PRRS-positive breeding herds had better productive performance during stability periods than during unstable weeks, being the reduction in the number of wean piglets the main productive effect of PRRS circulation. The estimated loses of wean piglets (1.36 piglets/sow/year) due to European PRRSV circulation would be slightly lower from recent estimations in US of North American PRRSV.[3] These results provide better understanding of the productive impact of European PRRSV circulation in breeding farms, reinforcing the efforts to stabilize and maintain stable the PRRS-positive farms, especially in high-density swine production areas where elimination of PRRS virus is costly and of low risk of long-term success.

References

1. Holkamp et al. Terminology for classifying swine herds by porcine reproductive and respiratory syndrome virus status. *J Swine Health and Prod.* 2011; 19(1):44-56.

2. Torrents et al. Associations between PRRS farm status and productive parameters in Spanish sow farms. *Proc. 49th AASV.* 2017; P68:380-381.

3. Valdes-Donoso P et al. Production Losses From an Endemic Animal Disease: Porcine Reproductive and Respiratory Syndrome (PRRS) in Selected Midwest US Sow Farms. *Front Vet Sci.* 2018; 5:102.

Table 1: Weekly KPI estimated means ± SE for stable and unstable PRRS-status

	ABTHS	BAR (%)	PWMR (%)	WPTHS
Stable	0.84 ± 1.14	91.61 ± 0.78***	11.20 ± 1.05**	533.95 ± 21.57***
Unstable	0.78 ± 1.20	90.53 ± 0.78	12.15 ± 1.05	507.74 ± 21.38

** $P < 0.005$; *** $P < 0.001$

Use of aggregated veterinary diagnostic laboratory PCR results to monitor the activity of pathogens in the US swine industry

Giovani Trevisan[1]; Leticia Linhares[1]; Bret Crim[1]; Poonam Dubey[1]; Kent Schwartz[1]; Eric Burrough[1]; Rodger Main[1]; Paul Sundberg[2]; Mary Thurn[3]; Paulo Lages[3]; Kimberly VanderWaal[3]; Andres Perez[3]; Jerry Torrison[3]; Jamie Henningson[4]; Eric Herrman[4]; Gregg Hanzlicek[4]; Ram Raghavan[4]; Douglas Marthaler[4]; Jon Greseth[5]; Travis Clement[5]; Jane Christopher-Hennings[5]; Daniel Linhares[1]
[1]Iowa State University; [2]Swine Health Information Center; [3]University of Minnesota; [4]Kansas State University; [5]South Dakota State University

Introduction

Swine veterinary diagnostic laboratories (VDLs) are constantly testing enormous amount of material to detect swine pathogens. Aggregated results obtained by PCR-based tests offer useful indicator of disease megatrends affecting the US swine industry. Aggregated information from rRT-PCR results for porcine reproductive and respiratory syndrome virus (PRRSV), porcine epidemic diarrhea virus (PEDV), porcine deltacoronavirus (PDCoV), and transmissible gastroenteritis virus (TGEV), from four major VDLs (Iowa State University, University of Minnesota, South Dakota State University, and Kansas State University) are available at the Swine Disease Reporting System (SDRS). The purpose of this study was to develop an algorithm to scan the SDRS PCR results for significant changes from historic patterns of RNA detection for PRRSV, PEDV, and PDCoV using VDLs results.

Materials and methods

Standardized submission data and results of tests performed on porcine samples were retrieved by participating VDLs, with test result status determined for each accession as positive, suspect, inconclusive or negative using SAS 9.4 (SAS Institute Inc., Cary, North Carolina). Data for a 4-year period from 2015 through 2018 are fitted with 4 cycles, each corresponding to the 52 week yearly periods using PROC SQL and PROC SORT scripts on SAS 9.4. Weekly proportion of PCR-positive results over total number of accession ID cases was calculated. Three years of historical data (2015-2017) was used to fit a cyclic regression model using PROC ROBUSTREG scripts on SAS 9.4. The findings were used to predict the expected percentage of PCR-positive results for the year of 2018 using PROC SCORE. Moreover, 1 standard deviation of the expected results were calculated using PROC MEANS to represent 'upper' and 'lower' confidence limit lines around the expected values. Charts were built using PROC SGPANEL scripts on SAS 9.4. The final graphs have been reported at the Swine Health Information System (SHIC) website under Domestic Disease Monitoring reports.

Results

The number of PRRSV-positive cases were above the upper confidence limit lines (i.e. expected level based on historic data) for the spring and summer months of 2018. The relatively higher detection rate was mostly from sow farms cases, in part due to increased use of aggregates of processing fluids as a sample type. For PEDV, the observed number of PCR-positive cases was in line with the expected levels predicted by the cyclic regression model. The number of PCR-positive results for PDCoV were above expected values for all year of 2018, with higher detection rate in sow farms during the winter months (Figure 1).

Conclusions

Anonymous, aggregated data from participating VDLs used to monitor pathogen detection is valuable and informative. Charts can be built in a quasi-real time, and can report megatrends of disease detection in the USA swine industry. This tool has the ability to inform the swine industry regarding disease diagnostic trends on a national or state level. Having the ability to predict and track changes on disease detection generates early alerts for emerging challenges, and report macroepidemiological aspects associated with endemic pathogen detection. It should be noted, however, that this tool is not appropriate to determine actual prevalence or incidence because sampling is not random. Further standardization of submission data fields will be important to better answer questions regarding pathogen control. Continuous input from swine veterinarians and producers is encouraged to allow continuous improvement of this project.

Figure 1: Top: expected percentage of positive results for PRRSV RNA by rRT-PCR, with 1 standard deviation above and below the expected value. Middle: expected percentage of positive results for PEDV RNA by rRT-PCR, with 1 standard deviation above and below the expected value. Bottom: expected percentage of positive results for PDCoV RNA by rRT-PCR, with 1 standard deviation above and below the expected value.

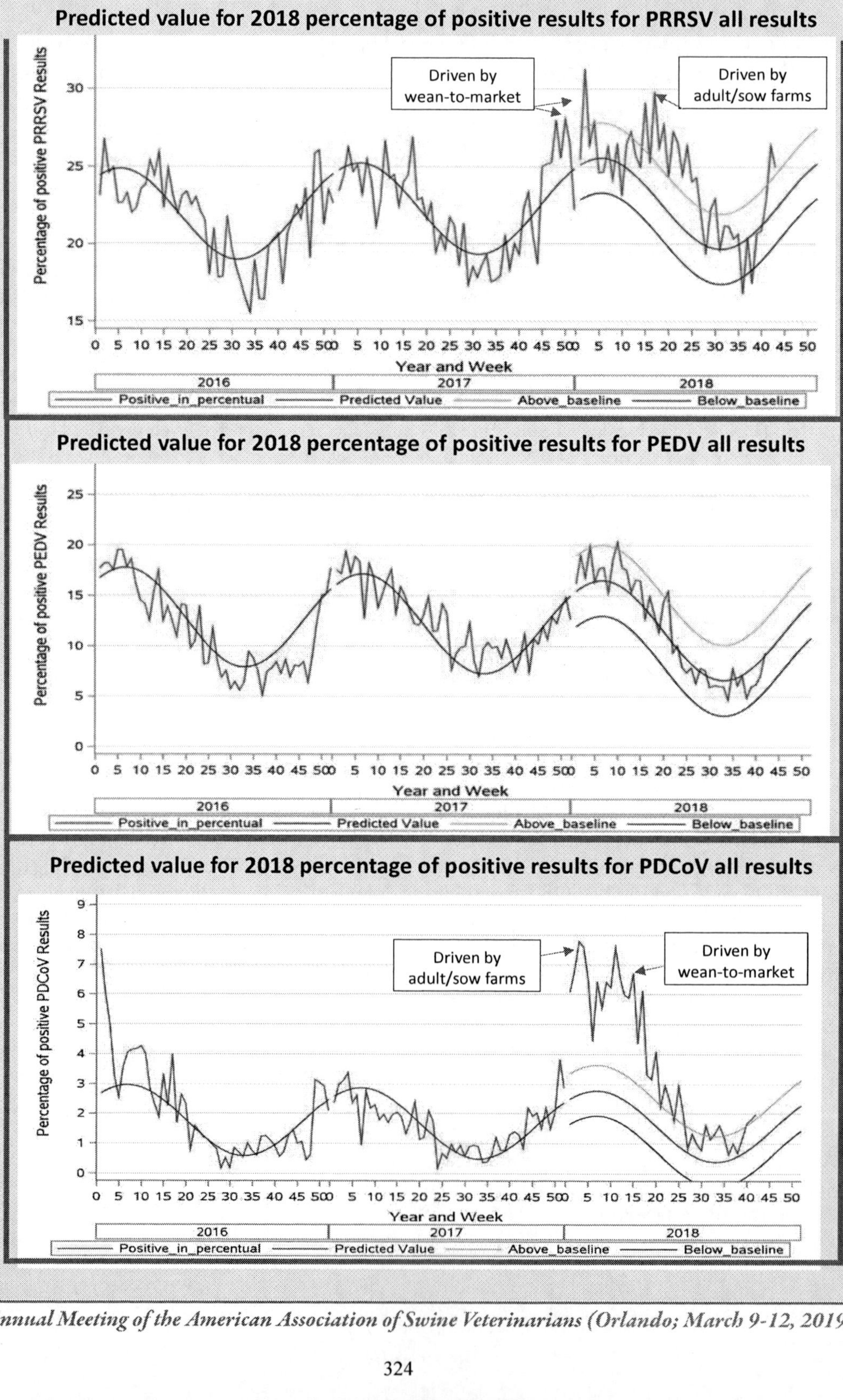

The potential of a mutant PEDV as a vaccine

A. Van Geelen[1], DVM, PhD; X. Deng[2], PhD; S. Baker[2], PhD; K. Lager[1], DVM, PhD; K. Faaberg[1], PhD
[1]Virus and Prion Research Unit, USDA-ARS-National Animal Disease Center, Ames, Iowa;
[2]Department of Microbiology and Immunology, Loyola University Chicago Stritch School of Medicine, Maywood, Illinois

Introduction

Ever since its emergence in the US swine population in 2013, porcine epidemic diarrhea virus (PEDV) continues to be a costly infection in sow farms that results in large losses in piglet production. In the absence of any vaccines, the traditional methods of "feed back" were initiated to fight the enteric virus. Although this method was efficacious, it was inherently dangerous since replacement gilts had to be exposed to wild-type PEDV to induce immunity which created considerable risk to the sow herd. An alternative to wild-type live virus immunization would be the use of a vaccine that was safe, efficacious, and had a long duration of immunity. Currently available PEDV vaccines are non-replicating virus preparations that are safe, induce a low to moderate level of protection to piglets via colostral immunity, require boosting, and the duration of immunity is variable. Immunity from a replicating virus vaccine might stimulate a better mucosal immune response compared to non-replicating virus vaccines, however, there is always the question of safety when replicating virus vaccines are used in livestock. PEDV has become endemic in the US swine herd indicating a need for improved control and prevention strategies that could include the use of a safe, replicating or modified-live virus (MLV) vaccine which may have increased efficacy and longer lasting immunity compared to non-replicating virus vaccines. This paper reports 1) the development of an infectious wild-type PEDV clone, 2) creating a targeted mutation in the clone to attenuate the virus, and 3) completion of a pilot study comparing the pathogenicity of the mutated virus to the wild-type parental and clone virus. The goal of this initial study was to evaluate the potential of a targeted mutation to generate a PEDV MLV.

Methods

At 7 days old, 8 pigs in each of 3 litters farrowed from PEDV-negative sows were all orally exposed to either the wild-type parental PEDV (WT), the wild-type virus clone (WTc), or the mutant PEDV (MUT). Each litter was left with the sow and monitored for 7 days post inoculation at which time the surviving pigs were euthanized. Daily, clinical scoring and rectal swabs were collected. Ileum was collected from each pig and placed in formalin for histology and PEDV immunohistochemistry (IHC). Daily virus load was determined by qPCR on rectal swab samples.

Results and discussion

Although all pigs in each of the 3 litters became infected, replicated virus, and developed diarrhea, the magnitude of virus load and extent of disease was variable among litters. By conclusion of the study, 5/8 and 4/8 pigs died or were humanely euthanized because of disease in the WT and WTc litters, respectively. In contrast, all 8 MUT pigs were alive. However, the MUT pigs did develop diarrhea that was moderate, but less intense than the WT and WTc litters. Similarly, the virus load was less in the MUT pigs compared to the WT and WTc litters. Microscopic lesions observed in each litter were compatible with 7-day-old PEDV lesions in young pigs, and were IHC-positive for PEDV antigen. Attenuation of wild-type PEDV through a specific mutation was demonstrated indicating the potential for developing an oral PEDV MLV. Studies are underway to select additional mutations that can further attenuate the virus and evaluate its efficacy in gilts as a PEDV MLV.

Use of a GnRH agonist for estrus synchronization in post-weaning sows

S. Williams[1], DVM, Dr Vet Sc; J. Rodriguez Pérsico[2], DVM; M. Compagnoni[1], DVM; R. Rearte[1], DVM; R. L. de la Sota[1], DVM, PhD

[1]Faculty of Veterinary Sciences, National University of La Plata; [2]Biogénesis Bagó, La Plata, Argentina

Introduction

The success of artificial insemination (AI) depends on placing an adequate number of spermatozoa at the uterine-tubal junction at a time close to ovulation. The GnRH analogues have been demonstrated as efficient inducers of ovulation in sows.[2-6] The objective of this study was to evaluate the use of a combination of eCG and eCG+buserelin (GnRH agonist) in sows at weaning on reproductive efficiency.

Materials and methods

First parity sows (n = 55) were used from a commercial herd (La Plata, Buenos Aires, Argentina). Sows were housed in individual crates and received commercial gestation diets. At weaning, sows were randomly divided into 5 groups: 1) control group (Con, n = 12) females received an intramuscular dose of a placebo saline solution at weaning; 2) eCG400 group (n = 12), received an intramuscular dose (400 IU; 2 ml) of eCG (Ecegón®, Biogénesis Bagó) the day of weaning; 3) eCG400+G group (n = 10) received an intramuscular dose (400 IU; 2 ml) of eCG (Ecegon®, Biogenesis Bagó) at weaning followed by an intramuscular dose (10.5 µg; 2.5 ml) of acetate of buserelin (GnRH analogue, Gonaxal®, Biogenesis Bagó) 48 h after eCG application; 4) eCG800 group (n = 11), received an intramuscular dose (800 IU; 4 ml) of eCG (Ecegón®, Biogénesis Bagó) the day of weaning; 5) eCG800+G group (n = 10) received an intramuscular dose (800 IU; 4 ml) of eCG (Ecegon®, Biogenesis Bagó) at weaning followed by an intramuscular dose (10.5 µg; 2.5 mL) of acetate of buserelin (Gonaxal®, Biogenesis Bagó) 48 h after eCG application. Heat detection was performed in order to calculate weaning-to estrus-interval (WEI), and sows that did not showed estrus during the first 10 days post weaning were not considered. From the onset of estrus, ultrasound scans were performed to measure number and size of follicles. AI was performed according to the program used in the farm. At farrowing, litter size (total piglets born) and average of total weight at birth were recorded. WEI was evaluated by adjusting Kaplan-Meier survival analysis models; litter size and average total weight were evaluated through linear regression models, and for the last parameter, litter size was included as a covariate. The effect of treatment on the number and size of follicles was evaluated using a regression model that takes into account the effect of the repeated measurements over time (diagnostic ultrasound days - 1, 0 and + 1; D0 = AI). All the results are expressed in mean values (LSM ± SE), and contrasts were used to estimate the difference between treatments and control groups.

Results and discussion

The WEI found showed non statically differences between groups, although in the eCG400+G group the WEI was the lowest (90 h), compared with sows treated only with eCG400 (102 h) and 24 h less than the control group (114 h). Similar differences were also reported in favor of the use of eCG+GnRH, but with minor differences and higher doses of eCG, 133.3; 135.9 and 141.5 h sows receiving eCG (600UI)+GnRH (buserelin 10µg); only GnRH or the control group, respectively.[1] In sows that received higher doses of eCG (eCG800 and eCG800+G) the WEI range was the same (108 h). Sows that received 800IU of eCG had the higher average follicular diameter ($P < 0.01$) compared to Con group. Average number of follicles was higher for groups receiving eCG800 or eCG800+G ($P < 0.05$) when compared with the control sows (Table 1). Mean weight of total piglets born and mean total piglets born showed no statically differences. However, sows treated with eCG400+G produced 1.2 more piglets than the control group (14 vs 12.8, respectively, Table 1). In spite of this, when average total weight at birth was evaluated taking litter size as covariate, eCG400, eCG400+G and eCG800+G groups had greater litter sizes than the control group (NS differences, Table 1).

References

1. Baroncello E. et al. Fixed-time post-cervical artificial insemination in weaned sows following buserelin use combined with/without eCG. *Reprod Dom Anim*, 2016, 52:76-82.

2. Brüssow K.P. et al. Studies on fixed-time ovulation induction in the pig. In: Rodriguez-Martinez, Vallet & Ziecik (Eds), *Control of pig reproduction*, 2009. vol 8:187-195. Nottingham, Univ Press.

3. De Jong E. et al. Effect of GnRH analogue (Maprelin) on the reproductive performance of glits and sows. *Theriogenology*, 2013. 80:870-877.

4. Kirkwood RN & Kauffold J. Advances in breeding management and use of ovulation induction for fixed-time AI. *Reprod Dom Anim*, 2015. 50 (Suppl 2), 85–89.

5. Knox RV et al. Design and biological effect of a vaginally administered gel containing the GnRH agonist, triptorelin, for synchronizing ovulation in swine. *Theriogenology*, 2018. 112:44-52.

6. Martinat-Botté F. et al. Induction and synchronization of ovulations of nulliparous and multiparous sows with an injection of gonadotropin-releasing hormone agonist (Receptal). *Theriogenology*, 2010, 73: 332-342.

Table 1: Least square means (LSM) ± SE of reproductive parameters in post-weaning first parity sows

	First parity sows				
Treatment	**CONTROL**	**eCG400**	**eCG400 + G**	**eCG800**	**eCG800 + G**
Follicular diameter (mm)	4.99 ± 0.25*	4.97 ± 0.26	5.30 ± 0.26	5.73 ± 0.29*	4.92 ± 0.25
Follicular number	4.70 ± 0.50**	4.93 ± 0.53	5.44 ± 0.60	6.20 ± 0.69**	6.01 ± 0.60**
Mean weight of total piglets born (kg)	18.0 ± 1.07	18.5 ± 1.02	17.0 ± 0.78	18.5 ± 1.15	18.7 ± 0.96
Mean total piglets born	12.8 ± 1.13	13.2 ± 1.29	14.0 ± 1.25	12.7 ± 1.35	13.1 ± 1.28

* In the same row, values with statistically significant differences $P < 0.1$.
** In the same row, values with statistically significant differences $P < 0.05$.

Study of the protective action of lutein in porcine semen

M. Compagnoni, DVM; C. Tittarelli, DVM, Dr. Vet Sc; M. C. Bonaura, DVM, Dr. Vet Sc; M. Marmuntti, Dr. Vet Sc; M. Gavazza, Dr. Vet Sc; S. Williams DVM, Dr. Vet Sc.
Faculty of Veterinary Sciences, National University of La Plata, La Plata, Argentina

Introduction

Boar spermatozoa are susceptible to peroxidative damage due to the high proportion of polyunsaturated fatty acids present in their membranes[1,2] and to the low antioxidant capacity of boar seminal plasma.[1] In recent years, several authors have worked on adding different antioxidants (Ax) on diluents, both for conservation at 15-17°C[3] as for cryopreservation.[4,5] Some authors have worked on various types of Ax, such as alginate, superoxide dismutase, vitamin E, glutathione, cysteine, hypotaurine and Rosemary, and on the determination of the appropriate concentrations to maximize the effect of the Ax. These substances have the ability to preserve sperm quality by neutralizing the action of the free Radicals of Oxygen Spices (ROS) on cell membranes. The high proportion of fatty acids in the boar sperm membrane turns it especially susceptible to the peroxidative damage generated by ROS. ROS are formed naturally as a byproduct of the normal metabolism of oxygen and are involved in the physiological functions of sperm. The addition of Ax in the cooling extender could minimize sperm damage generated by the ROS. The objective of this study was to determine the protective capacity of the antioxidant lutein (natural α-carotene antioxidant) against oxidative damage during the conservation of boar semen at 15°C.

Materials and methods

Ejaculates (n = 6) from 2 mature boars were collected and stored for 48 h at 15°C with and without the addition of lutein in different degrees of inclusion (0.15 or 0.25 mg/mL). In order to determine the protective capacity of lutein, semen quality (motility, percentage of live spermatozoa and percentage of cell abnormalities) was determined. In addition, to demonstrate the degree of peroxidation by chemiluminescence, the light emission was determined over a period of 120 min. quantified in counts per minute (cpm) every 10 minutes, in equipment provided with a chemiluminescence program using a Packard1900 TR liquid scintillation counter (Meriden CT, USA). The data was analyzed using the GLM procedure of SAS®.

Results and discussion

The addition of lutein to a concentration of 0.15 mg/mL had a beneficial effect on the motility and the percentage of live spermatozoa in both animals when comparing with the use of 0.25 mg/mL, although the values obtained showed no statistically significant differences (Table 1). After non-enzymatic peroxidation, the values of light emission showed statistical differences ($P < 0.005$) between the control samples in presence of ascorbate compared with spermatozoa samples incubated with lutein, for both Ax concentrations (Table 1). The addition of a natural Ax in the cooling extender at 15°C has a beneficial effect which is seen when measuring the in vitro peroxidation degree in presence of ascorbate. These results agree with those obtained by other authors who have achieved the reduction of oxidative damage with an improvement in semen quality adding antioxidants such as vitamin E (α-tocopherol), vitamin C (ascorbic acid), Rosemary (Rosmarinus officinalis) and L-cysteine in the freezing extenders.[5]

Table 1: Values (mean±SD) of motility (%), live sperm (%) and light emission after non-enzimatic peroxidation, between samples of boar semen without or with lutein in different degrees of inclusion.

Boar	Lutein (mg/mL)	Motility (%)	Live sperm (%)	Light emission (cpm) Control+ Ascorbate
1	0	75.00 ± 2.15	71.67 ± 6.88	535.33 ± 65.45[a]
	0.15	67.50 ± 6.18	82.00 ± 6.11	301.33 ± 26.41[b]
	0.25	53.33 ± 3.97	65.00 ± 3.93	315.00 ± 58.08[b]
2	0	71.67 ± 2.15	66.33 ± 6.88	685.67 ± 61.33[a]
	0.15	58.33 ± 5.05	48.00 ± 4.99	436.67 ± 29.30[b]
	0.25	50.00 ± 3.97	50.00 ± 3.93	376.67 ± 17.62[b]

[a,b] In the same column, values with statistically significant differences $P < 0.005$.

References

1. Brzezinska-Slebodzinska E. et al. Antioxidant effect of vitamin E and glutathione on lipid peroxidation in boar semen plasma. *Biol Trace Elem Res*, 1995. 47: 69–74.

2. Cerolini S. et al. Viability, susceptibility to peroxidation and fatty acid composition of boar semen during liquid storage. *Anim Reprod Sci*, 2000. 58: 99–111.

3. Zhang X. et al. The effects of different levels of superoxide dismutase in Modena on boar semen quality during liquid preservation at 17°C. *Anim Sci J*, 2016. 88(1): 55-62.

4. Jinghua H. et al. Effects of alginate on frozen-thawed boar spermatozoa quality, lipid peroxidation and antioxidant enzymes activities. *Anim Reprod Sci*, 2014, 147: 112–118.

5. Malo C. et al. Antioxidant effect of rosemary (Rosmarinus officinalis) on boar epididymal spermatozoa during cryopreservation. *Theriogenology*, 2011. 75: 1735–1741.

PRRSV oral fluid ELISA performance comparison among three commercial kits

YA. Henao-Díaz[1], DVM, MS; L Giménez-Lirola[1], PhD; J Ji[2], MS; J. Zimmerman[1], DVM, PhD
[1]Department of Veterinary Diagnostic and Production Animal Medicine, College of Veterinary Medicine, Iowa State University, Ames, Iowa; [2]Department of Statistics, College of Liberal Arts and Sciences, Iowa State University, Ames, Iowa

Introduction

Continuous and reliable PRRSV monitoring/surveillance is required to track PRRSV in commercial swine herds and the success of control/elimination programs. Oral fluid is the only diagnostic specimen that can be collected either from individuals or groups of animals and oral fluid-based ELISA is the most cost-effective method for routine antibody detection in commercial swine herds.[1-3] New commercial PRRSV oral fluid ELISA kits have appeared in the market, but the diagnostic performance of these commercial assays has not been evaluated or compared. Therefore, the purpose of this study was to compare the diagnostic performance of three commercial PRRSV OF ELISAs using samples collected from pigs of known status.

Materials and methods

Twelve (12) individually-housed pigs were vaccinated with a PRRSV MLV vaccine (Ingelvac® PRRS MLV) and a panel of oral fluids (n = 564) and serum samples (n = 132) was collected over a period of 50 days (day post vaccination -7 to 42). Serum samples were tested using the IDEXX PRRS X3 Ab ELISA (ELISA 'S'). Oral fluid samples were tested on each of 3 commercial PRRSV oral fluid ELISAs (1) IDEXX PRRS OF Ab; (2) HIPRA CIVTEST® SUIS PRRS A/S PLUS; (3) QIAGEN Pigtype® PRRSV Ab OF. Using the manufacturers' designated cut-offs, the proportion of positive results was compared among ELISAs using Cochran's Q. Numeric ELISA test results were analyzed by receiver operating characteristic curve (ROC AUC), with samples collected ≤ DPV 6 classified as antibody negative and samples collected ≥ DPV 12 as antibody positive. Thereafter, ELISA diagnostic performance was compared using a pairwise ROC Chi-square test. All analyses were considered statistically significant at $P > 0.05$.

Results and discussion

1. Data on individual pig temporal antibody responses (Figure 1) demonstrated that all animals responded immunologically to PRRSV vaccination.

2. Comparisons of diagnostic sensitivity and specificity detected no statistically significant difference between ELISAs 1 (100% - 100%) and 3 (93.8% - 100%) ($P > 0.05$), despite the fact ELISA 3 produced 23 false negative results. ELISA 2 (61.6% - 97%) was significantly different from ELISAs 1 and 3 ($P < 0.05$), producing 143 false negative results.

3. Statistical comparisons of ELISA 'S' and ELISA 1 or 3 found no difference in the proportion of positive results ($P > 0.05$), whereas ELISA 2 was significantly different ($P < 0.002$). Comparisons in the proportion of positives among oral fluid ELISAs (Figure 1) showed significant differences between ELISA 2 and ELISAs 1 or ELISA 3 ($P < 0.05$).

4. In terms of overall test performance, the three commercial PRRSV oral fluid ELISAs were ranked as: ELISA 1 ≥ ELISA 3 > ELISA 2. Thus, not all commercial PRRSV oral fluid ELISA kits demonstrated the same level of diagnostic performance.

Acknowledgments

Independent of the research reported, JZ has served as consultant to IDEXX Laboratories, Inc. on areas of diagnostic medicine. The terms of the consulting arrangement have been reviewed and approved by Iowa State University in accordance with its conflict of interest policies.

References

1. Ramirez A, et al. Efficient surveillance of pig populations using oral fluids. *Prev Vet Med* 2012;104:292–300.

2. Kittawornrat A, et al. Porcine reproductive and respiratory syndrome virus (PRRSV) surveillance using pre-weaning oral fluid samples detects circulation of wild-type PRRSV. *Vet Microbiol* 2014;168:331–339.

3. Duinhof TF, et al. Detection of PRRSV circulation in herds without clinical signs of PRRS: Comparison of five age groups to assess the preferred age group and sample size. *Vet Microbiol* 2011;150:180–184.

Figure 1: PRRSV oral fluid ELISAs performance by day post vaccination (n = 12 pigs). Left axis denotes numeric response (sample-to-positive or indice relativo por ciento). Right axis denotes percent of antibody-positive pigs based on manufacturer's recommended cut-off (dash line).

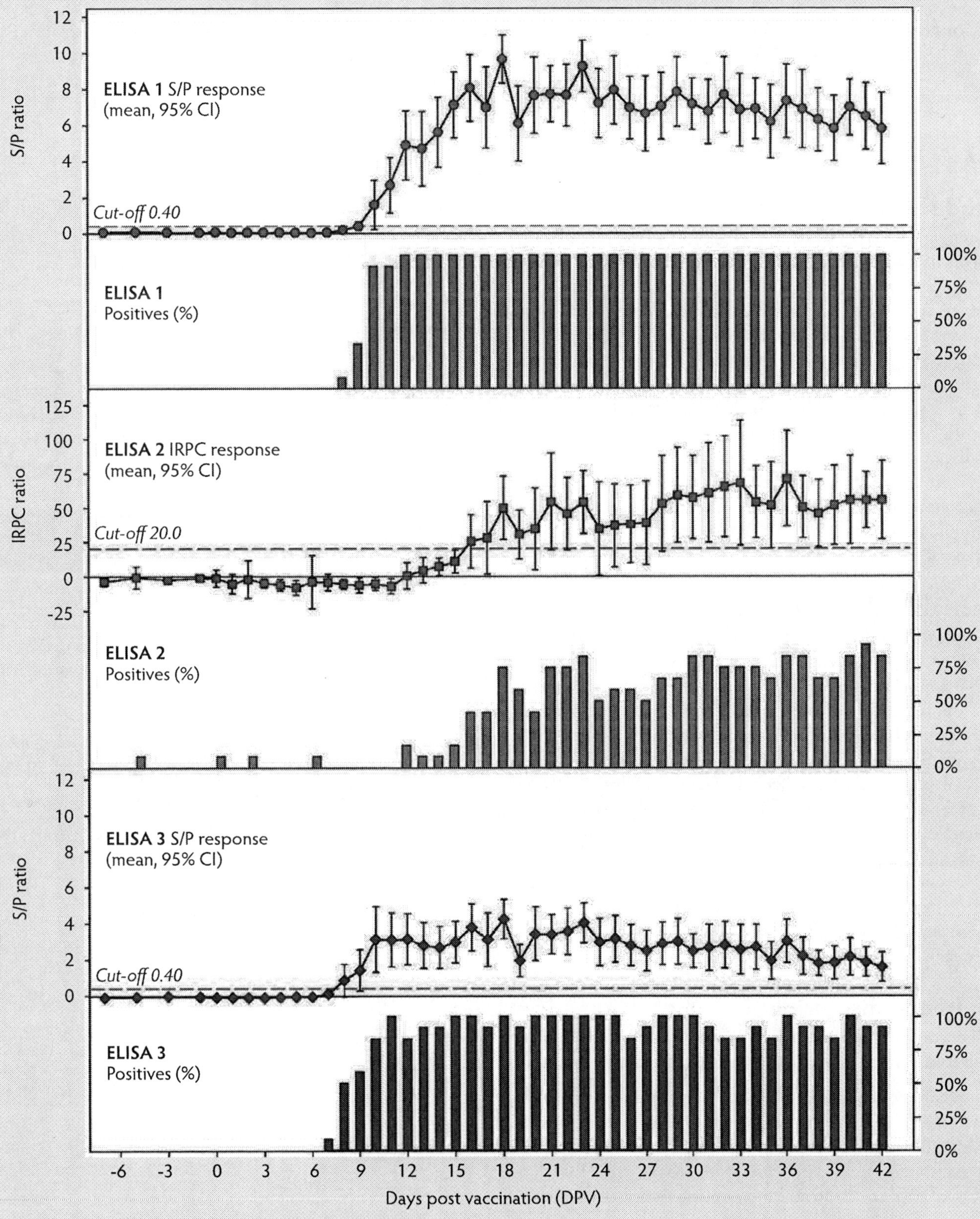

Role of collared peccary (*Pecari tajacu*) in the ecology of PRRSV?

R. Molina-Barrios[1], DVM, PhD; J. Luevano-Adame[1], DVM; Y. A. Henao-Díaz[2], DVM, MS; L. Giménez-Lirola[2], PhD; P. Piñeyro[2], DVM, PhD; R. Magtoto[2], PhD; J. Cedillo-Cobian[1], DVM; C. Diaz-Rayo[1], DVM; J. Hernández[3], DVM, PhD; J. Zimmerman[2], DVM, PhD

[1]Departamento de Ciencias Agronómicas y Veterinarias, Instituto Tecnológico de Sonora, Ciudad Obregón, México; [2]Department of Veterinary Diagnostic and Production Animal Medicine, College of Veterinary Medicine, Iowa State University, Ames, Iowa; [3]Laboratorio de Inmunología, Centro de Investigación en Alimentación y Desarrollo A.C, Hermosillo, México

Introduction

Collared peccary (*Pecari tajacu*) and pigs (*Sus scrofa*) are both members of superfamily *Suoidea*. These species coexist in the area ranging from the southwestern United States to northern Argentina and even occupy some Caribbean islands. With some exceptions, peccaries and swine share susceptibility to important parasitic, bacterial, and viral infections. This fact, in combination with the potential for contact between peccaries and domestic livestock, has stimulated discussion regarding the role of peccaries in the maintenance and/or transmission of different infectious diseases in the Americas. Because there were no data on PRRSV infections in peccaries, the objective of this experiment was to contribute to the discussion of the role of peccaries in the epidemiology of swine diseases by evaluating the susceptibility of collared peccaries to PRRSV.

Materials and methods

This experiment is fully described elsewhere.[1] In brief, adult female (1- to 2-year-old) collared peccaries (n = 4) and 3-week-old PRRSV-naïve domestic pigs (n = 8) were included in the experiment. To evaluate their susceptibility to PRRSV, one pig (Pig A, positive control) and three peccaries were exposed to a Sonoran wildtype PRRSV by intramuscular inoculation using serum from a PRRSV-viremic pig. To evaluate the transmissibility of PRRSV from peccaries to domestic swine, four pigs were placed in pens contiguous to the pens holding the inoculated peccaries on day post inoculation (DPI) 3. The remaining peccary and pigs (n = 2) served as negative controls. Serum samples collected on DPI 0, 3, 7, 10, 15 and 23 were tested by isotype-specific ELISA for the presence of PRRSV IgM, IgA and IgG and by rtRT–PCR for the presence of PRRSV nucleic acid.

Results and discussion

1. All peccaries and pigs remained clinically normal throughout the observation period, including the PRRSV positive control.

2. Cumulatively, the data showed that peccaries were susceptible to PRRSV. All animals were negative on DPI 0 for PRRSV RNA and for PRRSV IgM, IgA, and IgG (Figure 1). Serum samples collected from inoculated peccaries were PRRSV rtRT-PCR-positive from DPI 3 to 23. In addition, a marked antibody S/P response was observed on DPI 10 and DPI 15 for IgM and IgG, respectively, with a slight increase in IgA (Figure 1). It should be noted that, although the PRRSV ELISA was not optimized for peccaries, the secondary reagent (anti-pig IgG) reached peccary antibodies reflecting their common ancestry. This cross-reaction had been reported previously for different suiformes tested by IFAT.[2]

3. Pigs exposed to infected peccaries via nose-to-nose contact tested negative by PRRSV rtRT-PCR and PRRSV IgG, IgA, IgM ELISAs, with the exception of pig C, in which an increased IgM response was observed at DPI 23. The lack of definitive evidence of PRRSV transmission between peccaries and pigs should not be over interpreted. In small groups of animals, transmission of PRRSV and other arteriviruses, tends to occur because of aggressive interactions (injection or/and ingestion of body fluids) which were not observed in this study.

4. These results raise broad questions regarding the natural history of PRRSV in superfamily *Suoidea*, and more specifically, the role of non-pig species in the evolution and ecology of PRRSV.

References

1. Molina-Barrios R, et al. Collared peccary (Pecari tajacu) are susceptible to porcine reproductive and respiratory syndrome virus (PRRSV). *Transbound Emerg Dis* 2018;00:1–8. https://doi.org/10.1111/tbed.12944

2 Herrera HM, et al. The role played by sympatric collared peccary (*Tayassu tajacu*), white-lipped peccary (*Tayassu pecari*), and feral pig (*Sus scrofa*) as maintenance hosts for *T. evansi* and *T. cruzi* in a sylvatic area of Brazil. *Parasitol Res* 2008;103:619–624.

Figure 1: PRRSV-inoculated peccaries' serum testing results by day post inoculation. Left axis: Individual peccary PRRSV rtRT-PCR results (RNA copies per mL). Right axis: Grouped PRRSV IgG, IgA, and IgM ELISA, inoculated peccaries mean S/P values.

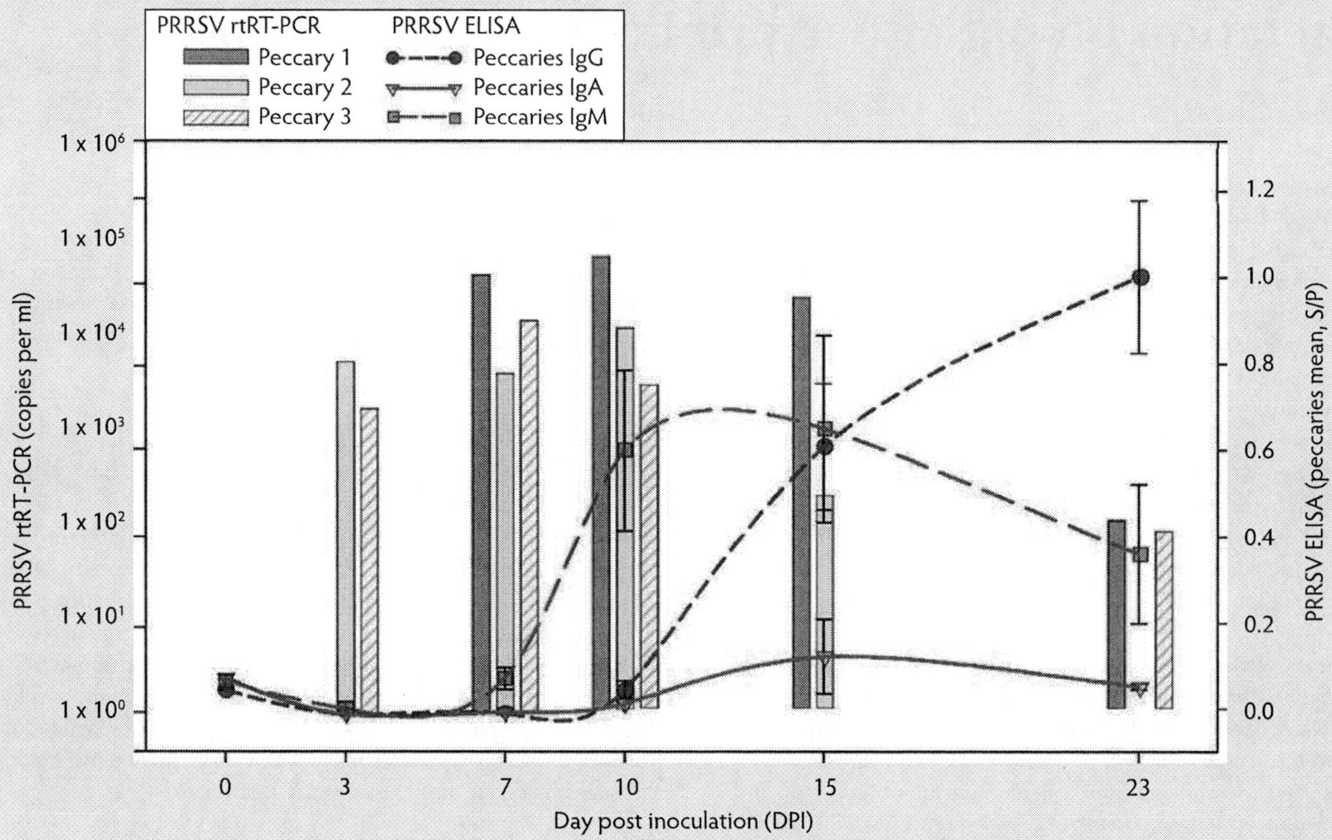

Effects of a blend of sodium diformate and glycerine monolaurate in sow feed during late gestation and lactation on piglet performance until weaning

C. Lückstädt, PhD
ADDCON GmbH, Bitterfeld-Wolfen, Germany

Introduction

Sustaining growth rate and optimizing feed efficiency in pigs are key to their economic performance through to market. With mounting pressure on the costs of pig production in general worldwide, nutrition is increasingly under scrutiny. The application of organic acids and their salts to diets for pigs has been studied extensively for more than 50 years.[1] They have proved especially effective in maintaining growth performance since the ban on antibiotic growth promoters came into effect in Europe. Several studies have also demonstrated benefits of diformates in diets for sows.[2,3] However, while the antimicrobial impact of organic acids and their salts, including potassium or sodium diformate, is mainly directed against Gram-negative bacteria. Medium chain fatty acids (C-6 to C-12), on the other hand, have been shown to have an antibacterial impact against various Gram-positive bacteria.[4] This effect is magnified if monolaurate is used, making it a promising candidate as an additive or as an alternative to antibiotics for treatment of different diseases. Despite the well-documented impacts of both additives, data on the combined impact in sows during late gestation and lactation, and their subsequent effects on suckling piglets under semi-commercial conditions had not yet been generated. This formed the impetus of this trial.

Material and methods

The study tested the efficacy of a blend of sodium diformate with the monoglyceride of lauric acid – monolaurate, in multiparous sows at a research farm in Saxony-Anhalt, Germany. 40 sows were allocated to 2 equal groups and fed a commercial lactation diet from one week before farrowing until weaning. The test diet contained 1.0% of the sodium diformate – monolaurate mixture (traded as Formi® GML, ADDCON, hereafter referred to as GML). The lactation diet was fed according to a feeding curve, from the last week before farrowing to day 13 of lactation and thereafter *ad libitum*. All piglets were weighed at birth and at weaning and performance parameters recorded. Data were analysed using the t-test and a confidence level of 95% was defined for these analyses.

Results and discussion

The number of piglets born alive, the weight of new-born piglets and the litter weight at birth were significantly ($P < 0.05$) increased (Table 1), while the number of weaned piglets differed only numerically ($P = 0.11$). At weaning, the body weight of piglets as well as the litter weight were again significantly improved ($P < 0.05$). Here, piglet weight was more than 500g greater and litter weight more than 8 kg heavier in the treated group compared to the negative control, while weaning age did not differ (25.7 days). The above findings support the use of dietary sodium diformate and monolaurate in sows as a performance enhancer for suckling piglets under European conditions.

References

1. Lückstädt C and Mellor S. The use of organic acids in animal nutrition, with special focus on dietary potassium diformate under European and Austral-Asian conditions. *Recent Advances in Animal Nutrition - Australia*. 2011;18:123-130.

2. Øverland et al. Potassium diformate in the diet of reproducing sows: Effect on performance of sows and litters. *Livestock Science*. 2009;122:241-247.

3. Lückstädt C. Effects of dietary potassium diformate on feed intake, weight loss and backfat reduction in sows: pre-farrowing till weaning. *Advances in Animal Biosciences*. 2011;2(1):145.

4. Preuss et al. Minimum inhibitory concentrations of herbal oils and monolaurin for gram-positive and gram-negative bacteria. *Molecular and Cellular Biochemistry*. 2005;272(1-2):29-34.

Table 1: Performance data of piglets from sows fed a lactation diet with or without Formi GML

	Control	1.0% GML in sows	*P*-values
Piglet born alive, per sow [n]	16.4 ± 2.8	18.0 ± 2.8	0.042
Weight of new-born piglets [kg]	1.20 ± 0.14	1.33 ± 0.22	0.018
Litter weight at birth [kg]	19.56 ± 3.26	23.67 ± 3.85	0.001
Piglets weaned, per sow [n]	11.95 ± 0.59	12.25 ± 0.89	0.11
Weight of weaned piglets [kg]	6.91 ± 0.62	7.41 ± 0.68	0.011
Litter weight at weaning [kg]	82.44 ± 7.58	90.78 ± 10.40	0.004

Investigating the transmission dynamics of *Lawsonia intracellularis*

Fernando L. Leite[1], DVM, MS, PhD; Paul Lawrence[2], DVM, PhD; Kris Kovach[1]; Greg Cline[1], DVM; Jens Kjaer[1], DVM, MBA

[1]Boehringer Ingelheim Animal Health, Duluth, Georgia; [2]Newport Labs, Worthington, Minnesota

Introduction

Lawsonia intracellularis (LI) causes proliferative enteropathy in swine. Infection has been estimated to occur in more than 90% of US swine farms and can lead to significant economic loss due to decreased production performance.[1,2] Much is still unknown regarding the transmission dynamics of LI in swine. In this study, we investigated the transmission dynamics of LI using a seeder pig model with a pure culture challenge.

Materials and methods

A herd with no previous clinical history of ileitis was used to source animals for this study. Three treatment groups were used in this study to assess LI transmission: Group 1, one seeder animal to 29 sentinel animals; Group 2, two seeder animals to 28 sentinel animals; Group 3, control non-infected animals (n = 5).

Challenge of seeder pigs was performed with a pure culture of 1.5×10^7 LI organisms per animal at three weeks of age. The seeder pigs were kept in the same pen until their introduction to sentinel pigs at 9 days post infection. Each treatment group was housed in a separate room and pen with 2.5 square feet per animal. Sentinel animals were monitored for 38 days of exposure to seeder animals. Fecal shedding of LI was assessed by qPCR twice weekly on all animals. All animals were weighed before seeder challenge at 3 weeks of age and at the end of the study when animals were 10 weeks of age.

Results and discussion

LI shedding of the seeder pig in Group 1 ranged from 0 to 1.5×10^5 organisms per gram of feces during the 38 day exposure period to sentinels, while the average shedding level of the two seeders in Group 2 ranged from 0 to 1.89×10^7 organisms per gram of feces (Figure 1). Shedding of LI was first detected in sentinels at 14 days post seeder introduction (dpsi) in both groups (Figure 1). In Group 1, 62% (18 of 29) and in Group 2 86% (24 of 28) of sentinel pigs shed LI at least once during the 38 dpsi. Infection in Group 1 peaked at the last time point of the study (38 dpsi) when 55% (16 of 29) of sentinels shed on average 3.1×10^6 LI per gram of feces (Figure 1A). A different pattern was observed in Group 2. The highest level of shedding was observed at 21dpsi when 36% (10 of 28) of sentinel animals shed on average 3.6×10^3 LI per gram of feces (Figure 1B). Average daily gain (ADG) was significantly reduced ($P < 0.05$) in both infected Group 1 (ADG = 1.00 pound) and Group 2 (ADG = 0.99 pound) compared to non-infected control Group 3 (ADG = 1.33 pound).

The results from this study demonstrate that LI infection can spread to as many as 18 (62%) pigs within a pen due to the introduction of one infected animal within a 38 day period. Had animals been monitored for a longer period, it is likely that LI would have been transmitted to all of the pigs within the pens of Group 1 and Group 2. At the peak of infection, animals in Group 1 shed on average much more LI (6 Log_{10} per gram of feces) as compared to Group 2 which did not shed above an average of 3 Log_{10} LI per gram of feces. These results indicate that the transmission dynamics of LI can vary and are not always the same. A limitation of this study was that each Group 1 and Group 2 only had one replicate pen. This study demonstrates the consequences of introducing LI infected animals to the transmission of LI within susceptible animals in a pen. These results indicate that the introduction of as little as one infected pig shedding at most 5 log_{10} LI per gram of feces during a 38 day period can result in the transmission of LI to more than half of its pen mates and lead to a significant reduction of average daily gain.

References

1. Armbruster, G. A. et al. Review of *Lawsonia intracellularis* seroprevalence screening in the United States, June 2003 to July 2006. In 38th Annual Meeting of the American Association of Swine Veterinarians. 231–234 (2007).

2. Winkelman, N. Subclinical ileitis: Diagnostic monitoring, R^2, and economics. In 49th Annual Meeting of the American Association of Swine Veterinarians. 488-493 (2018).

Figure 1: Average level of *Lawsonia intracellular* shedding in A) Group 1 and B) Group 2.

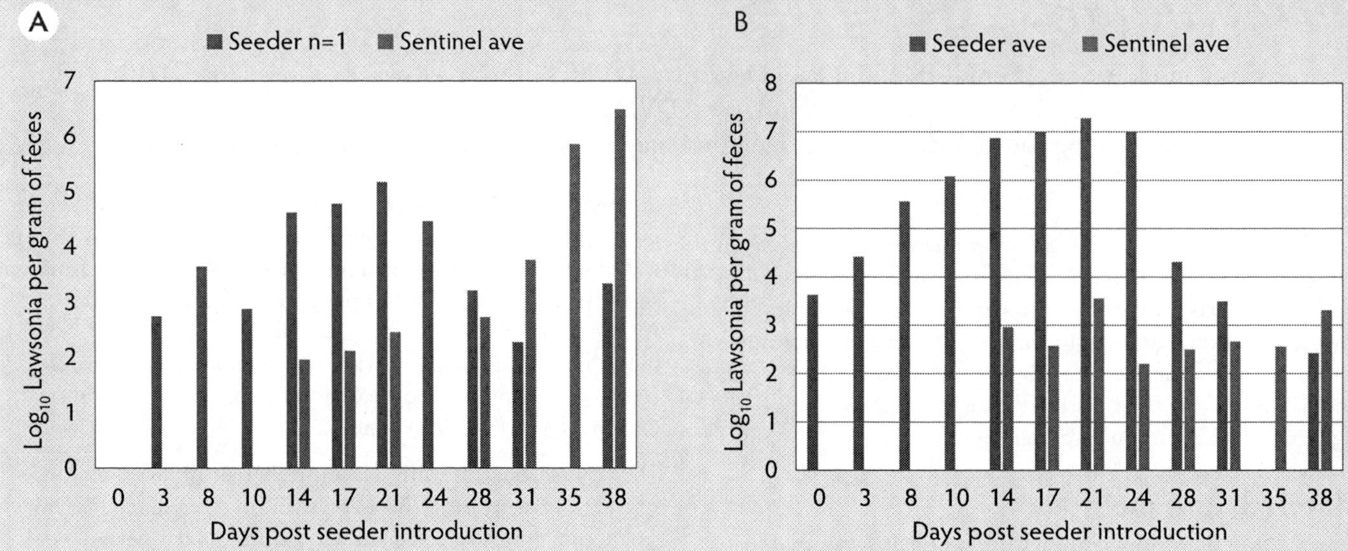

PEDV "re-breaks": Is the enemy from within? New information on manure pits PEDV positive (PCR) after prolonged lengths of time post-disease outbreaks

Grant Allison[1], DVM; Phil Gauger[2], DVM, MS, PhD; Jianqiang Zhang[2], MD, MS, PhD; DVM; Gene Spellman[3], BS
[1]Walcott Veterinary Clinic, Walcott Iowa; [2]Vet Diagnostic and Production Animal Medicine, Veterinary Diagnostic Laboratory, Iowa State University, Ames Iowa; [3]Central Life Sciences, Schaumburg, Illinois

Introduction

Despite extensive investments in biosecurity measures and protocols, farms still have unexplainable cases of viral diseases caused by porcine epidemic diarrhea virus (PEDV). Investigating and finding the source of a PEDV disease re-break is difficult and rarely accomplished. Theories and speculation is the usual end result. Deciding what would happen or change on the farm as a result of finding the source is not clear. Reducing risks of exposure to PEDV to naïve animals is the goal of biosecurity protocols and vaccination plans. Recent work has shown flies' ability to harbor and transmit PEDV[1,2] and porcine reproductive and respiratory virus (PRRSV).[3] Veterinary consultants posed the question: Where can flies encounter viable PEDV in sufficient quantity to serve as a vector? Manure pits were a natural target for investigation due to flies' ability to reproduce in manure, a substrate known to harbor PEDV. The ability of deep manure pits to serve as a possible source of PEDV for flies is the goal of this diagnostic investigation.

Materials and methods

A confirmed (fecal swab PCR positive) PEDV disease outbreak in a sow farm on 12/28/2016 was chosen for this study. The last positive PEDV sample by rectal swab on piglets was on 4/3/2017. Separate from the central farrowing rooms the site consisted of both a north and south gestation barn. Each gestation barn had separate manure pits and held approximately 1,200 sows. Five manure pit sampling locations at the center and the 12, 3, 6, and 9 o'clock positions were identified in each barn as the location to draw pit samples. Samples were collected using Sterile Amies Agar Gel Transport Swabs (JorVet J0867). A piece of twine sufficient in length to reach the surface of the pit was tied to the handle of the swab. The swab was lowered between the slats with care taken to only contact and sample the surface layer of the pit. The pit surface layer is the expected habitat for flies completing their life cycle. Samples were pooled and analyzed by real-time reverse transcription polymerase chain reaction (rRT-PCR) per routine protocols at the Iowa State University Veterinary Diagnostic Lab (ISU VDL). The pooled sample from the South gestation barn on June 5, 2018 was individually tested and reported also.

Results and discussion

Manure pit samples were PCR positive (Ct 33.5) for PEDV at least 388 days post last positive rectal swab sample in the North gestation barn; and PEDV PCR positive (Ct 33.1) at least 428 days post last positive rectal swab sample in the South gestation barn. The five individual PCR Ct samples from the South barn collected on 6/5/2018 are reported: 33.7, 33.6, 36.4, 34.6 and 32.6. Even though these manure pits were agitated and pumped in the spring of 2018, the samples were PEDV PCR positive for over a year after the last confirmed positive fecal swab was taken at the sow unit (Table 1). Gestation pits are often free of common disinfectants. Gestation manure pits tend to be coolest in winter. The combination of a cold pit containing little disinfectant may create an optimal environment for the prolonged survival of PEDV in a swine barn. More research is needed to assess the viability of PEDV in manure pits over extended periods of time. Deep manure pits harbor PEDV for prolonged periods of time and may serve as a source of PEDV for flies to transmit PEDV disease to naïve pigs.

References

1. Allison G, Gauger P, et al. PEDV, PRRSV, and Seneca Valley virus PCR positive flies *(Musca domestica)*. What now? *Proceedings 49th AASV 2018* 394-396.

2. Allison G, Gauger P, et al. PEDV positive bioassay reveal houseflies (*Musca domestica)* can transmit infectious PEDV to pigs. *Proceedings 50th AASV 2019.*

3. Otake S, Dee SA et al. Transmission of porcine reproductive and respiratory syndrome virus by houseflies, (Musca domestica Linnaeus). *Vet Rec* 2003. 152: 73-76.

Table 1: Test results of manure pit samples

Date	Days post last positive sample	Ct
North gestation pit		
April 9, 2018	371	33.7
April 26, 2018	388	33.5
June 5, 2018	428	Negative (> 36)
South gestation pit		
April 9, 2018	371	31.3
April 26, 2018	388	32.5
June 5, 2018	428	33.1*

* June 5th individuals: 33.7, 33.6, 36.4, 34.6, 32.6

Prevacent™ PRRS: A new tool for porcine reproductive and respiratory syndrome virus (PRRSV) control

J. M. Hammer, DVM; C. L. Puls, PhD; R. Evelsizer, DVM
Elanco Animal Health, Greenfield, Indiana

Introduction

Since porcine reproductive and respiratory syndrome virus (PRRSV) emerged in the late 1980's, it has grown to a costly, difficult disease to control at the farm, regional, and national levels. Understanding the onset of immunity (OOI) and duration of immunity (DOI) is useful for veterinarians to optimize immunity to PRRSV based on expected exposure, infection pressure, and demonstrated immunity. Two challenge studies are summarized with viremia, viral lung load, growth, and clinical parameters to demonstrate OOI and DOI of Prevacent PRRS. Although limited in duration after challenge, growth, viral lung load as demonstrated by broncho-alveolar lavage fluid (BALf) samples, and cumulative viremia are presented to assess health impact and immunity of the vaccine pre- and post-challenge.

Onset of immunity (OOI)

An OOI study was conducted with 40 piglets vaccinated at 14 days of age (d0) with placebo or Prevacent PRRS. Pigs were heterologously challenged 28 days later (d28) with NADC 20, lineage 8, 84.5% homologous to Prevacent PRRS, at a challenge dose of 3.98×10^4 $TCID_{50}$/mL with 2 mL delivered/nostril and 2 mL IM. The study concluded upon necropsy on day 42 (14 days after challenge).

Compared to controls, vaccination with Prevacent PRRS reduced ($P < 0.05$) lung lesions (54.3 vs. 2.7%) and viral load in BALf (8.1×10^8 vs. 2.2×10^6 Gmean RNA copies/ml) (Table 1). Vaccination increased ($P < 0.05$) ADG in the 14-day post-challenge period (0.12 vs. 0.72 lb/day; SEM = 0.05) and for the overall 42-day study period (0.65 vs. 0.89 lb/day; SEM = 0.02).

Cumulative daily clinical signs post-challenge were reduced and stable for pigs vaccinated with Prevacent PRRS compared to controls, which were experiencing escalating clinical signs.

These results indicate an OOI as early as 28 days post-vaccination. Prevacent PRRS improved the health of vaccinated pigs by reducing pneumonia and viral load in the lung while improving growth rate compared to controls in this heterologous challenge.

Duration of immunity (DOI)

To demonstrate 26-week DOI, a study of 80 PRRSV seronegative PCR-negative pigs (n = 40/group) were allotted to two treatments: 1) unvaccinated controls and 2) pigs vaccinated with Prevacent PRRS at 14 days of age. Treatments were housed separately until heterologously challenged (d181) with 10 ml NADC 20 at 2.6×10^6 $TCID_{50}$/mL (4 mL/nostril and 2 mL IM). At d181, a planned reduction in animal numbers was made to 32 pigs per treatment during randomization of two controls and two Prevacent vaccinated pigs per pen. On d196 (14 days post-challenge), pigs were euthanized and lung lesions were measured and BALf was collected for qRT-PCR.

Compared to controls, pigs vaccinated with Prevacent PRRS had reduced ($P < 0.05$) lung lesions (12.3 vs. 1.1%), reduced ($P < 0.05$) viral lung load by 98% (5.1×10^6 vs. 1.1×10^5 Gmean RNA copies/mL) in BALf samples, and reduced ($P < 0.05$) viremia in cumulative samples by approximately two logs (4.1×10^6 vs. 4.2×10^4 Gmean RNA copies/mL) after challenge (Table 1).

Table 1: Onset and duration of immunity for Prevacent PRRS

Onset of Immunty (28 days)		Lung lesions	BALf (Gmean RNA copies/mL)	ADG d28–d42 (lb/day)	ADG d28–d42 (lb/day)	Duration of immunity (26 weeks)	Lung lesions	BALf (Gmean RNA copies/mL)	Cumulative Viremia (Gmean RNA copies/mL)
	Placebo	54.3%[a]	8.1 × 108	0.12[a]	0.65[a]		12.3%[a]	5.1 × 106[a]	4.1 × 106[a]
	Prevacent PRRS	2.7%[b]	1.8 × 106[b]	0.72[b]	0.89[b]		1.1%[b]	1.1 × 105[b]	4.2 × 104[b]

[a,b] Columns with different superscripts are statistically different ($P < 0.05$); NADC 20, lineage 8, 84.5% homologous to Prevacent PRRS; OOI -challenged at d28 with NADC 20 at 3.98×10^4 $TCID_{50}$/mL with 2 mL/nostril and 2 mL IM; DOI - challenged at d181 with NADC 20 at 2.6×10^6 $TCID_{50}$/mL with 4 mL/nostril and 2 mL IM.

Conclusion

Prevacent PRRS demonstrated onset of immunity as early as 28 days post-vaccination through reduced viral lung load and pneumonia. Prevacent PRRS demonstrated 26-week duration of immunity through reduced lung lesions, viremia, and lung viral load. These results can assist veterinarians in vaccination timing and implementation to control PRRSV in their client's herds.

Simplifying veterinary prescriptions with software

Joel Stave
PrairE Systems, Spencer, Iowa

As of January 1, 2017, animal producers can no longer purchase medicated feed containing certain medically important antimicrobials over the counter. This new structure, known as the Veterinary Feed Directive (VFD), has triggered a change in the relationship between veterinarians, distributors and animal producers and has generated new recordkeeping, tracking and communication needs in the industry.

In response to the directive, PrairiE Systems, a software and precision systems developer, and GlobalVetLINK (GVL®), a leader in digital solutions for animal health, have established a collaborative web-based solution. In addition to VFDs, this integration allows all prescriptions to be tracked throughout the system, from request, to fulfillment, to end use. The collaboration between these two web-based products is meant to make the compliance process easier on vets, feed mills, diagnostic labs, and producers by establishing clear communication and documentation.

PrairiE Systems' Feed Allocation System

PrairiE Systems' Feed Allocation System (FAS) is meant to simplify and clarify the feed ordering process. This web-based application lets users place feed orders, track and enter inventory, and communicate directly with company administration. With the Feed Allocation System a user can know exactly what animals in a herd need medication and can share that information with the rest of the supply chain.

GVL veterinary prescription and VFD software

VFD and veterinary prescription software systems from GVL provide user-friendly step-by-step processes for veterinarians to create, sign and manage scripts and VFDs. This web-based platform stores information about customers, feed mills and drugs. (Figure 1)

How the software works together

1. Client is in need of a prescription.
2. Client requests script in FAS.
3. GVL platform is automatically notified of the request.
4. Vet follows through and acquires sufficient knowledge to make a decision. FAS gives the vet the ability to do consultation research. It acts as the monitoring tool answering questions like: Has this animal been sick before? What has or has not worked? Were treatments actually given?
5. Vet either writes the prescription or denies it.
6. The vet's decision is automatically sent to FAS, which notifies the producer of whether or not the prescription is filled, and to the feed mill associated with the account.

Figure 1: GVL and Feed Allocation System software integration

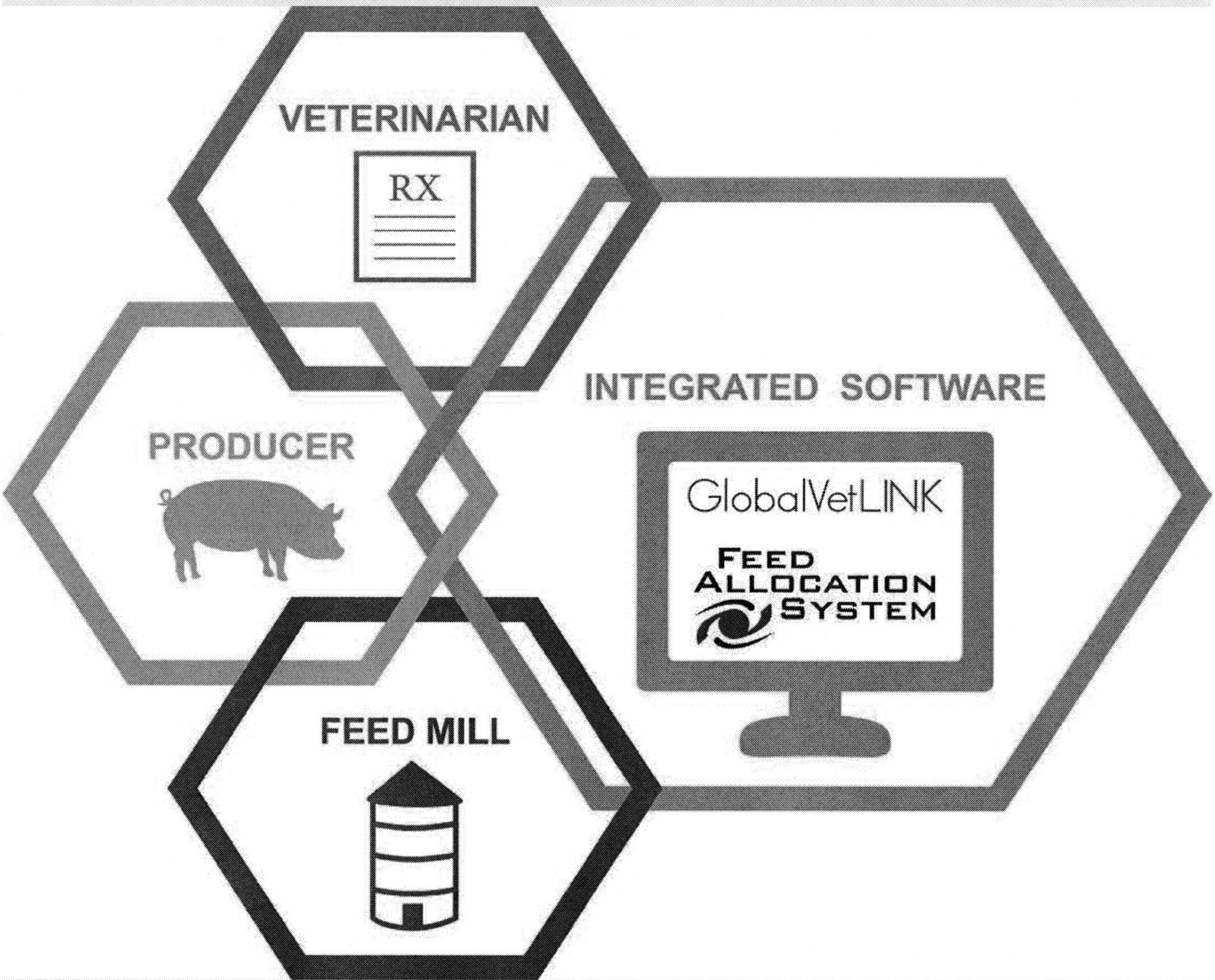

Use of KEM SAN® to impact mortality due to PEDV and bacterial coinfection in a nursery

G. Allison[1], DVM; J. Lane[2], BS; T. Marsteller[3], DVM

[1]Walcott Veterinary Clinic, Walcott, Iowa; [2]JWV Pork, Washington, Iowa; [3]Kemin Industries, Inc., Des Moines, Iowa

Introduction

The first PEDV diagnosis in the United States was made in May 2013, and the virus was found in SE Iowa in December 2013.[1] Bacterial coinfections are common after a PEDV outbreak in nursery pigs causing mortality to increase.[2] Nursery mortality was recorded in two sow farms in 2013/2014 and compared to the nursery mortality resulting from the reoccurrence of PEDV in 2018 in the same sow farms. KEM SAN® was used in the 2018 PEDV outbreaks, but not the 2013/2014 PEDV outbreaks. KEM SAN® is an EPA-registered proprietary formulation of propionic and other organic acids and is designed for the control of bacteria in drinking water for livestock and poultry. KEM SAN® has biocidal activity against a broad spectrum of bacteria.[3]

Material and methods

The outbreak of PEDV was confirmed at ISU Veterinary Diagnostic Laboratory in sow farm 1 on December 26, 2013. Nursery mortality from this 2013 outbreak was tracked on 20,410 pigs. The reoccurrence of PEDV was confirmed on farm 1 on February 23, 2018. Nursery mortality from this 2018 outbreak was tracked on 17,658 pigs. On farm 2, diagnostic confirmation of an outbreak of PEDV was made on November 5, 2014. Nursery mortality from this 2014 outbreak was tracked on 10,338 pigs. The reoccurrence of PEDV was confirmed on farm 2 on February 23, 2018. Nursery mortality from this 2018 outbreak was tracked on 6,611 pigs. The only management change between the initial (2013/2014) PEDV outbreaks and the second (2018) outbreaks was the use of KEM SAN® after weaned pig placement into the nursery sites. KEM SAN® was diluted into a stock solution and used at 0.33 oz/gallon dilution for 10-14 days. The number of days for return to baseline mortality (< 3%) was determined for each outbreak of PEDV.

Results and discussion

Without KEM SAN®, nursery mortality for farm 1 reached baseline mortality (2.5%) 111 days post-outbreak of PEDV in 2013. With the incorporation of KEM SAN® in 2018, farm 1 nursery mortality reached baseline mortality (2.0%) in 47 days post-outbreak of PEDV (Figure 1A). Without KEM SAN®, nursery mortality for farm 2 did not return to baseline (2.5%), with mortality being 3.8 % 152 days post-outbreak of PEDV in 2014. With the incorporation of KEM SAN® in 2018, farm 2 nursery mortality reached baseline mortality (1.4%) 56 days post-outbreak of PEDV (Figure 1B). The weighted average time for farms 1 and 2 to reach baseline mortality using KEM SAN® in 2018 was 1.5% mortality at 54 days post-weaning. In contrast, without using KEM SAN® in 2013/2014, nursery mortality for farms 1 and 2 did not reach baseline mortality; 3.3% mortality after 137 days. It has been shown that blended organic acids will enter the outer membrane of bacteria and lower intracellular pH. In turn, this may help control pathogenic bacteria that do not easily survive in an acidic environment.[3] It has also been suggested that lowering the pH of drinking water in newly weaned pigs will lower the *E coli* load.[4] The need for supportive treatment is critically important for the management of PEDV disease in nursery pigs. Using KEM SAN® can be an effective non-antibiotic supportive therapy to help maintain acceptable nursery mortality after outbreak(s) of PEDV.

References

1. Kwonil Jung, Linda J. Saif. Porcine epidemic diarrhea virus infection: Etiology, epidemiology, pathogenesis and immunophylaxis, The Veterinary Journal (2015), doi:10.1016/j.tvjl.2015.02.017.

2. Guggenbiller JD, et al. Impact of Neo-Terramycin, Mecadox, Denagard plus CTC and Denagard feeding programs on nursery pigs with PEDV, *Salmonella* and *E. coli*. AASV Proceedings 2015, pgs 135-138.

3. A Comparative Analysis of the Bactericidal Activity of KEM SAN® Liquid Antimicrobial Versus Other Water Acidifiers. Kemin Internal Document, 17-00104.

4. Busser De, EV et al. Effect of administration of organic acids in drinking water on faecal shedding of *E. coli*, performance parameters and health in nursery pigs. Vet J. 2011. May 188(2):184-8. doi 10:1016/j.tvjl.2010.04.006 Epub 2010 May 14.

Figure 1: A) Post weaning mortality: Days post outbreak of PED

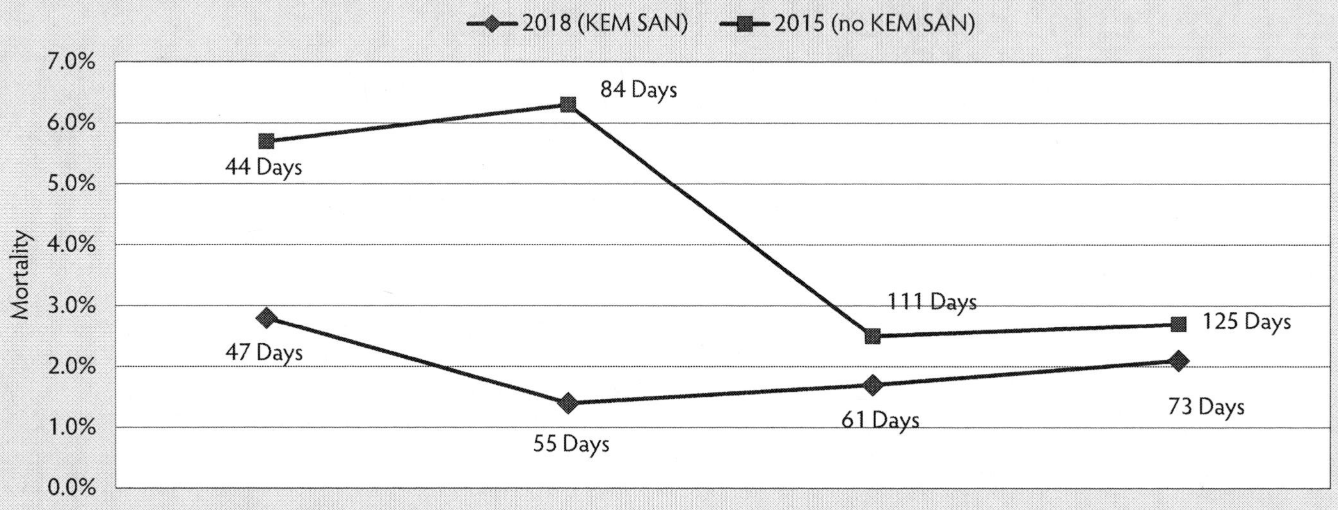

B) Post weaning mortality post PEDV diagnosis

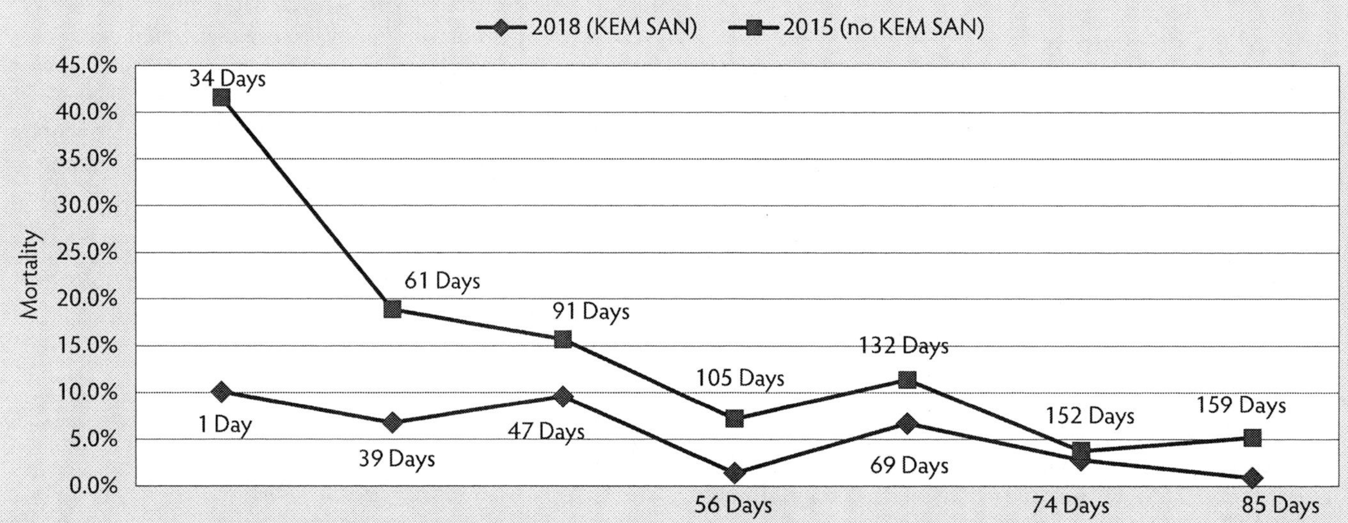

Evaluation of shedding and serological profiles of pigs following vaccination with Prime Pac® PRRS RR

Brett O'Brien, VMD, MS; Brad Thacker, DVM, PhD; Robyn Fleck, DVM
Merck Animal Health, Madison, New Jersey

Introduction

Commercially available modified live PRRS vaccines are widely used across the US because of their demonstrated protection against one of the most economically devastating swine diseases.[1,2] Prime Pac® PRRS RR is a modified-live virus vaccine labeled for "the vaccination of healthy swine 3 weeks of age or older against respiratory disease caused by PRRSV and for the vaccination of female breeding age swine against reproductive disease caused by PRRSV".[3] Serological monitoring post-vaccination of pigs administered a modified live PRRS vaccine occurs in the field regularly to measure vaccination compliance and to investigate clinical disease outbreaks. Here we report the results of two evaluations of antibody development and vaccine virus levels following vaccination of PRRSV naïve pigs with Prime Pac® PRRS RR and Ingelvac® PRRS MLV.

Materials and methods

Study 1- A long term shedding evaluation was performed in which 20 three-week old pigs sourced from a high health, PRRSV negative sow farm were vaccinated IM with 1 mL of Prime Pac PRRS RR at weaning. Pigs were housed in four pens in a traditional wean-to-finish barn at 5 pigs per pen. Sera from all pigs and oral fluids from all pens were collected immediately prior to vaccination, 3 and 7 days post-vaccination (dpv) and then weekly for 10 weeks and then biweekly until 24 weeks post-vaccination. Individual fecal samples were obtained at -2, 3, 7, 14, 21, and 28 dpv. All collected samples were tested individually by PRRS qPCR and PRRS IDEXX ELISA (S/P > 0.1 was classified as positive).

Study 2- A comparative post-vaccination serological evaluation was conducted in pigs vaccinated with either Prime Pac® PRRS RR or Ingelvac® PRRS MLV. For this study, nine average sized pigs per litter were selected from a PRRSV naïve sow farm, tagged, bled, and sexed one week prior to weaning from 15 intact litters (no cross-fostering). Pigs were weaned at 21-28 DOA into a contract research facility, weighed upon entry, blocked by litter and weight, and randomly assigned to one of three treatment groups (T01: Prime Pac® PRRS RR, N = 60; T02: Ingelvac PRRS MLV, N = 60; T03: Unvaccinated sentinels, N = 15). T01 pigs received 1 mL Prime Pac® PRRS RR IM and T02 pigs received 2 mLs Ingelvac® PRRS MLV IM three days post-weaning. Each treatment group was housed in separate rooms with the sentinel pig room located in between the two vaccinated rooms and strict biosecurity procedures employed between rooms to prevent spread of vaccine viruses. All pigs were bled one day prior to vaccination, and weekly post-vaccination for six weeks. Pigs were weighed at 28 dpv. All serum was tested individually by PRRS qPCR and PRRS IDEXX ELISA and PRRSV sequencing was performed on a subset per treatment group at every sampling. PRRS qPCR and ELISA results were analyzed using Fisher's Exact test and ANOVA.

Results and discussion

Study 1- 100% of pigs were viremic at three dpv (Average CT = 26.4) and 100% were PRRS ELISA positive by 14 dpv. Some individuals began to no longer shed virus beginning at 28 dpv and the majority of the study population was no longer viremic by 42 dpv. Fecal shedding of virus was short and detected only through 21 dpv.

Study 2- There was no cross-contamination of PRRSV strains between rooms during this study as evidenced by PRRSV sequencing and sentinel pigs remained PRRSV negative. Both vaccinated groups were 97% viremic by 14 dpv. Prime Pac® PRRS RR vaccinated pigs had significantly lower viral loads at all time points post-vaccination ($P < 0.01$) and exhibited a more transient viremia than pigs vaccinated with Ingelvac PRRS MLV (Table 1). Both vaccinated groups showed > 90% seroconversion rate by 21 dpv. However, Prime Pac PRRS vaccinated pigs averaged lower S/P values than Ingelvac MLV at each time point ($P < 0.01$). In this study, all three groups had an average starting weight of 15.1 lbs. but an average final weight of T01 = 30.1 lbs., T02 = 27.7, and T03 = 29.4.

References

1. Strait, E. et al. Efficacy and field safety of Prime Pac® PRRS RS, a new tool against porcine reproductive and respiratory syndrome virus. AASV Proceedings. 2018; 138-141.

2. Strait, E. et al. Prime Pac™ PRRS RR: Field Safety and Efficacy against Porcine Reproductive and Respiratory Syndrome Virus Reproductive Disease. Leman. 2018; 114-115.

3. Data on file, Merck Animal Health.

Table 1: PRRSV qPCR and IDEXX ELISA results in PRRSV vaccinated pigs.

Weeks post-vaccination	Prime Pac® PRRS RR				Ingelvac® PRRS MLV			
	PCR positives	Avg. qPCR	ELISA positives*	Avg. S/P	PCR positives	Avg. qPCR	ELISA positives	Avg. S/P
0	0	0	0/60	-0.008	0	0	0/60	-0.018
1	51/60	2,057,918[b]			56/60	2,529,239[a]		
2	54/60	1,903,824[b]	42/60	0.356[b]	57/60	59,500,765[a]	52/60	0.817[a]
3	56/60	287,488[b]	60/60	0.978[b]	58/60	21,235,201[a]	55/60	1.210[a]
4	35/60[b]	50,680[b]	60/60	1.019[b]	58/60a	2,642,000[a]	58/60	1.439[a]
5	42/60	51,748[b]	60/60	1.010[b]	50/60	2,016,000[a]	58/60	1.431[a]
6	25/59[b]	32,220[b]	59/60	1.116[b]	38/59[a]	255,130[a]	59/59	1.458[a]

* PRRS ELISA S/P > 0.1 is classified as positive
[a,b] Means in rows with different superscripts are significantly different ($P \leq 0.05$) for each of the four parameters.

ELISA to monitor anti-*Mycoplasma hyorhinis* IgG in serum from pigs immunized with inactivated *M. hyorhinis* vaccines

Paulraj Lawrence, PhD
Biological Development and Research Department, Newport Laboratories Inc., Worthington, Minnesota

Introduction

Mycoplasma hyorhinis is an important pathogen of swine that can often occur as a respiratory co-infection with viral pathogens but has also been shown to cause arthritis and polyserositis in infected animals.[1,2] The economic impact of disease associated with *M hyorhinis* in the United States swine industry has not been evaluated previously.[3] Nonetheless, this organism remains a significant source of infectious disease in commercially raised swine. Currently, no diagnostic assay is available to assess the serologic response to swine vaccines. In this study, recombinantly expressed *M hyorhinis* p37 protein was used to monitor the magnitude of the IgG response in vaccinated animals. The results indicate that the assay has a high degree of specificity in terms of its ability to distinguish animals vaccinated with *M hyorhinis* from those vaccinated with other important *Mycoplasma* species.

Materials and methods

The *M. hyorhinis* strains used for vaccine production were grown in Friis medium supplemented with 5% v/v yeast extract and 20% v/v porcine serum. Culture fluids were inactivated and stored at 4°C. Following inactivation, culture fluids were either formulated using medium and low dose antigen in adjuvants A or B for vaccine groups A-L.

Twenty four high health animals were obtained from a commercial supplier and housed in a clean facility until they had reached eight weeks of age. Test animals received two vaccinations of inactivated *M hyorhinis* vaccine formulated at different doses in adjuvant A, depending on the group to which they were assigned. Immunizations were given at days 0 and 21 and delivered via the intramuscular route in a 2cc dose. Serum samples were obtained from animals at day 35 for testing in the ELISA. In a second trial, one hundred nineteen high health animals were obtained from a commercial supplier and housed in a clean facility until they had reached ten weeks of age. These animals received two vaccinations of inactivated *M hyorhinis* vaccine formulated at different dose levels using commercial adjuvants A and B. Three groups (J, K, and L) were also included for comparison to groups A, B, and C to serve as control. The animals were vaccinated on days 0 and 21. Serum samples were obtained from animals at days 0, 21, and 35 for testing in the ELISA.

Purified *M hyorhinis* p37 protein was suspended in carbonate/bicarbonate buffer and 100 µl/well was coated onto the surface of Immulon 2HB 96 well plates (Corning, Corning, NY) at a concentration of 100 µg/mL. Plates were allowed to incubate overnight at 4°C and the coating solution was removed prior to blocking plates. Plates were blocked with 300 µL/well of Neptune Block (Immunochemistry Technologies LLC, Bloomington, MN) and incubated for two hours at 37°C before washing once with PBS-T. Next, serum samples and controls were diluted 1/1000 in Super Block (Thermo-Pierce, Rockford, IL), added to the plate in duplicate in 100 µL volumes, and incubated for one hour at 37°C. Plates were washed three times in PBS-T. Goat anti-swine IgG: HRP conjugate (Kirkegaard & Perry Laboratories, Gaithersburg, MD) was added to the plate at a 1/5000 dilution and allowed to incubate for one hour at 37°C. Plates were then washed four times in PBS-T and developed using a two component 3,3',5,5'-Tetramethylbenzidine (TMB) substrate (Kirkegaard & Perry Laboratories, Gaithersburg, MD). The substrate was added at100 µL/well and allowed to develop for ten minutes at room temperature. The reaction was stopped by adding 100µL of 1N sulfuric acid to each well. Plates were read at a wavelength of 450 nm and absorbance data was collected. The mean absorbance of the two assay control wells was determined and subtracted from that of every other well on the plate. Next, the mean absorbance of the duplicate samples and positive and negative serum controls was determined. Finally, the mean absorbance of every sample and the negative serum control was divided by that obtained from the positive control serum to obtain sample/positive (S/P) absorbance ratios that were used to evaluate the immune response in swine.

Results and discussion

The initial proof of concept study that was done to determine whether the ELISA could detect antibody against *M hyorhinis* successfully demonstrated that we might be able to detect an immune response in vaccinated animals that was not present in the control group. However, the immune response against this vaccine was not potent enough to achieve statistical significance (Figure 1A). Therefore, an additional study was performed using a dose titration of inactivated *M hyorhinis* vaccine containing either of two different adjuvants. The results of this study (Figure 1B) demonstrated that a much more potent response could be obtained when using a different adjuvant than that used in the initial study. Also, the response in the ELISA diminished as the dose of the antigen was decreased, indicating that the assay may have value as a means of measuring the antibody response against *M hyorhinis* semi-quantitatively.

Figure 1: A) Results of ELISA testing for *M hyorhinis* trial. B) Results of ELISA testing for *M. hyorhinis*. Yellow bars depict the mean S/P ratio for each treatment group at day 0 of the study. Red bars depict the mean S/P ratio for each treatment group at day 21 of the study. Blue bars depict the mean S/P ratio for each treatment group at day 35 of the study.

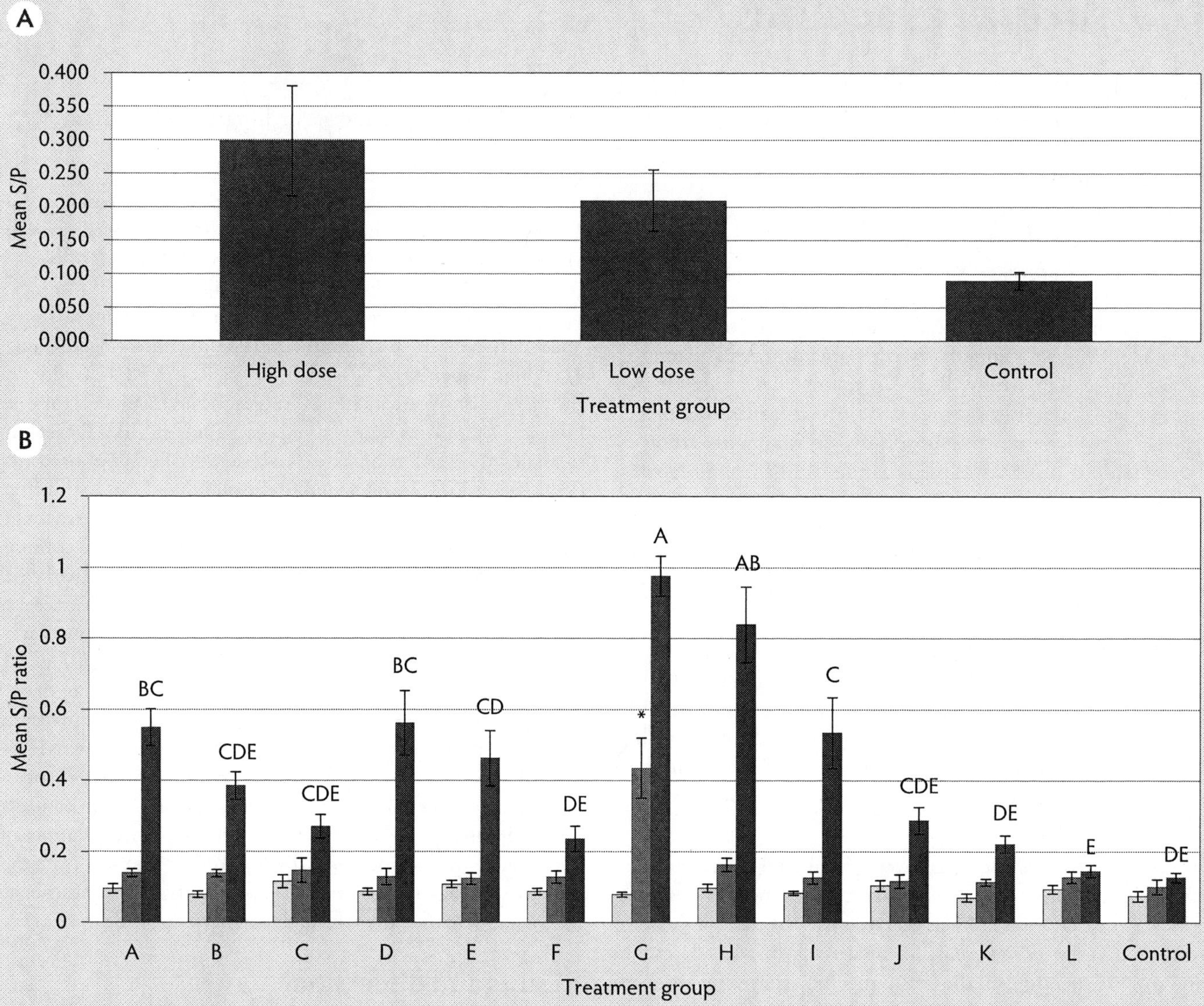

References

1. Armstrong CH et al. Serological, pathological, and cultural evaluations of swine infected experimentally with Mycoplasma flocculare. Can J Vet Res 1987;51:185-188.

2. Citti C et al. Elongated versions of Vlp surface lipoproteins protect *Mycoplasma hyorhinis* escape variants from growth-inhibiting host antibodies. Infect Immun 1997;65:1773-1785.

3. de la Fe C et al. Inactivation of Mycoplasma species involved in contagious agalactia. Berl Munch Tierarztl Wochenschr 2004;117:1-5.

The impact of NutriQuest Sow Protect™ for increasing litter size in multiparity sows

J. Tyler Holck[1], DVM, MS, MBA; Rob Musser[2], PhD; Ran Song[2], PhD; Kim Friesen[2], PhD
[1]Feed His People, LLC, Gilbert, Iowa; [2]NutriQuest, Mason City, Iowa

Introduction

NutriQuest Sow Protect (NSP) is a novel, bio-active, non-antimicrobial nutritional technology that has been shown to increase liveborn and subsequent total pigs weaned when fed in late gestation (beginning day 94 of gestation) and throughout lactation.[1] The beneficial impact of NSP is believed to be associated with positive shifts in intestinal microbiota[2] that reduce oxidative stress and subsequent tissue inflammation resulting reproductively in reduced stillbirth rates, higher liveborn and weaned pigs per litter. The objective of this study was to assess the impact of NSP fed through the entire gestation and lactation period on sow reproductive and litter performance.

Materials and methods

A total of 206 sows (PIC 1050; Parity ≥ 2) from a single, high-health sow farm in the upper Midwest were used in this experiment. Sows were housed individually in maternity pens after breeding. All sows were fed the same corn-soybean meal-based gestation diet once daily based on their body condition score via feed/water trough in gestation with water provided on an *ad libitum* basis. NSP was applied by daily top-dressing at 5.4 g/d (equivalent to 4.0 lb/ton of feed) starting on day 3 after breeding in gestation. On approximately d 112 of gestation, sows were moved into the farrowing barn and placed randomly in farrowing crates. The diet was changed to the lactation diet and fed approximately 6.0 lbs per day until farrowing and then *ad libitum* during the entire lactation period. Top-dressing of NSP was performed daily at 10.8 g/d (equivalent to 4.0 lb/ton of feed) in lactation until weaning. On the day of weaning, sows were returned to the gestation barn and monitored for wean to estrus intervals. Basal diets were based upon current standard gestation and lactation diets of the sow farm.

The number of piglets born (alive and dead), individual pig birth weight (alive and dead), individual pig weaning weight, and the number of piglets that died during lactation was recorded for each litter. Lactation length and wean to estrus interval was also recorded. Cross-fostering was allowed within 2 days after birth and only within treatments or to non-test litters. In addition, piglet management, including tail docking, injection of Fe and male piglet castration, was performed following the sow farm SOP. Piglets' access to the sow feed was not restricted.

Results and discussion

Feeding NSP resulted in significantly ($P = 0.005$) higher total born compared to control (15.0 vs. 13.7 total born per litter), an increase in 1.3 total born per litter. Of the 1.3 increase in total born per litter, 1.1 pigs were liveborn, a significant improvement in NSP litters compared to controls (13.9 vs. 12.8; $P = 0.005$). This increase was maintained through weaning with NSP litters weaning 12.7 pigs/litter, significantly ($P = 0.007$) higher than the control litters at 11.6 pigs/litter (Figure 1).

Litter weights (lbs) were also significantly higher in NSP litters at birth (42.1 vs. 38.8; $P = 0.003$) and numerically higher at weaning (151.6 vs 143.7; $P = 0.13$) compared to controls. Wean-estrus intervals (4.8 vs. 4.7; $P = 0.96$) were not impacted with the higher pigs/litter or wean weights per litter.

Parity differences were observed in this trial with Parity 2 as well as Parity 6+ sows showing significant responses to NSP (Figure 1). These differences in parity response would suggest the benefits of feeding NSP are most pronounced in higher stress parities and NSP can help mitigate typical second parity dips in performance as well as reduced performance typically seen in older sows.

Extrapolating the results of this trial, NSP provided an improvement of over 2500 pigs/1000 sows when feed throughout gestation and lactation. Using a conservative economic model this would result in an increased margin of $92,000/1000 sows based on a $40 weaned pig price.

Conclusions

Under the conditions of this field trial, feeding NutriQuest Sow Protect™ at an inclusion level of 4.0 lb/ton in the feed throughout gestation and lactation resulted in significant overall improvements in total born, liveborn, and pigs weaned per litter compared to non-treated control animals. These improvements were most pronounced in younger sows (parity 2) and older sows (parity 6+) with beneficial responses in all parity sows except for parity 3. In addition, feeding NutriQuest Sow Protect™ improved litter weaning weights in all parity sows except parity 3 while not significantly increasing wean to 1st estrus intervals. The reproductive performance improvements associated with NutriQuest Sow Protect™ in this trial would result in approximately 2500 additional pigs weaned per 1000 sows per year at a value of $92,000.

References

1. Data available upon request from NutriQuest.

2. Sinn, S.M., R. Song, D. Beckler, R. Musser, and K.G. Friesen. 2019. A bio-active, mineral-based feed additive improved growth performance and reduced severity of diarrhea in weanling pigs challenged with enterotoxigenic *Escherichia coli*. J. Anim. Sci., paper in press.

Figure 1: Parity responses to NutriQuest Sow Protect™

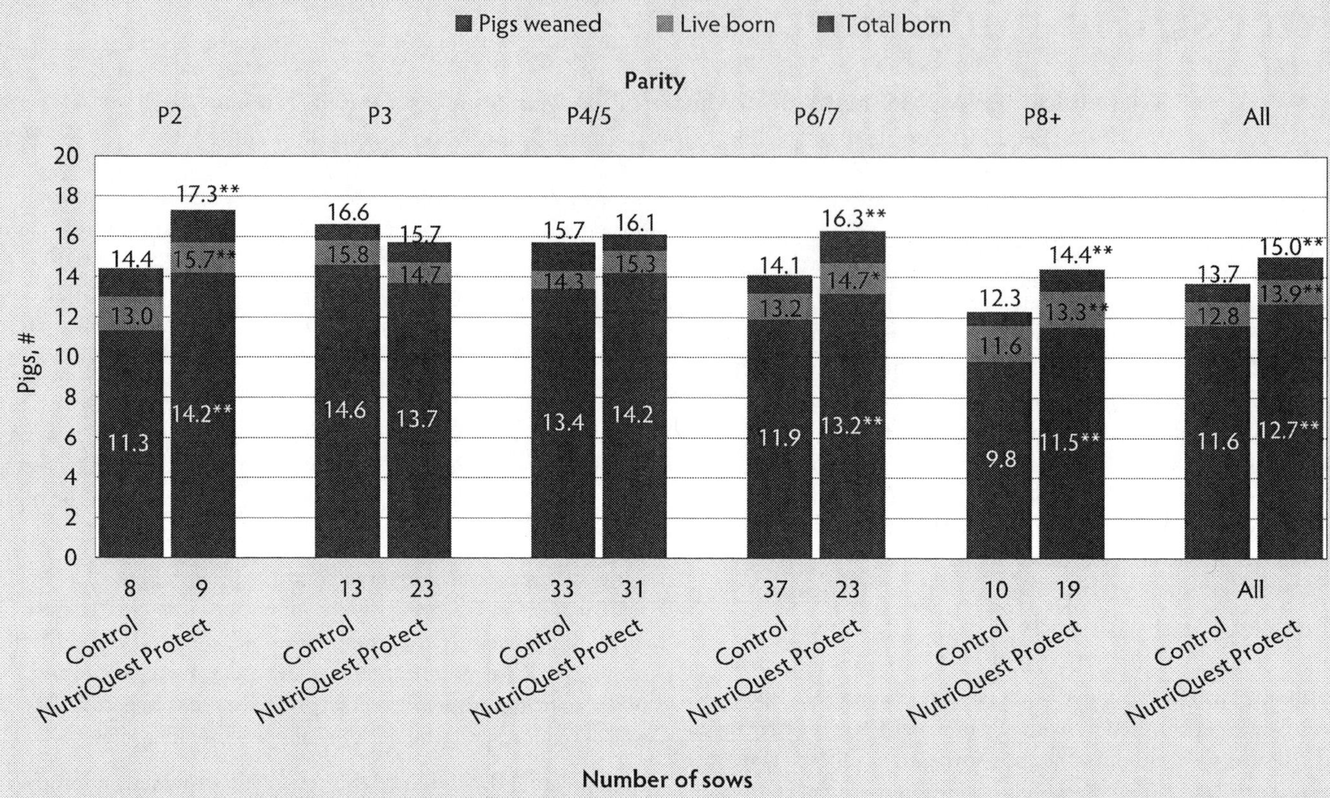

* Means tend to differ than the corresponding Control ($P < 0.10$).
** Means differ than the corresponding Control ($P < 0.05$).

The economics of iron deficiency anemia on US swine production: An annual impact of 46-335 million US dollars

C. Olsen, DVM, MS
Pharmacosmos, Inc.

Introduction

It has been well-documented that, without treatment, baby pigs are prone to developing a severe iron deficiency anemia.[1-3] Thus, the use of injectable iron is nearly ubiquitous in modern swine production. Despite this, both published research and internal data have shown that a large percentage of pigs have sub-clinical or full-scale iron deficiency anemia at the time of weaning.[4,5] This has led to studies which found that a baby pigs hemoglobin status at weaning (Hb Status) can significantly impact post-weaning growth.[4-6] Knowing this information, the objective of the current project was to evaluate the economic impact of iron deficiency anemia on swine production in the United States by using Hb status at weaning as a predictor of wean to finish average daily gain.

Methods and results

Blood hemoglobin was measured at the time of weaning in 235 healthy baby pigs and classified as follows; >110 g/L (optimal), 90-110 g/L (sub-clinical) and < 90 g/L (deficient).[6,7] To calculate wean-finish average daily gain (ADG), each baby pig was weighed individually at weaning and again at 131 days post-weaning. Using this data, Ordinary Least Squares linear regression was used in RStudio to fit a regression model for the purposes of predicting ADG for a group of pigs.[8] The model was fit to predict the outcome of ADG using Hb Status, sex and weaning weight as covariates. Hb Status and sex were included as ordinal variables while weaning weight was continuous. The final regression model was:

$$ADG = 1.158259+(0.063996^{*}Optimal.Hb)+ (0.003783^{*}SubClinical.Hb)+ (0.034634^{*}WeanWeight)-(0.074726^{*}female)$$

Using this equation, a model was built to predict the economic return associated with improvement in a herds Hb Status. The model considered the expected distribution of Hb Status when giving a single 200 mg dose of Uniferon® as a baseline and compared that against the expected improvement in Hb status associated with administration of a second 200 mg dose of Uniferon® by day 12 of the baby pigs life (Table 1A). Realistic estimates were included to account for the daily cost of production, cost and labor of a second injection of Uniferon® and the market value of finishing hogs. The model is built so that each estimated value can be replaced by farm-specific data to capture the true return for a specific production system. Calculations were made to reflect the impact improvement in Hb Status would have for both a fixed-weight or fixed-time marketing strategy and extrapolated to the US pig herd (Table 1B).[9]

Conclusions

The model developed in this study clearly demonstrates that despite long-standard recommendations to use a single 200 mg dose of injectable iron for prevention of iron deficiency anemia, a tremendous economic potential remains untapped by a failure to maximize Hb status at weaning. Depending on farm-specific input paramaters and the marketing strategy used, the total economic impact of sub-clinical and full-scale iron deficiency anemia in the US swine herd ranges from 46.3-335.7 million US dollars despite current iron injection practices.

References

1. Kernkamp H et al. Preventing iron-deficiency anemia in baby pigs. *J Anim Sci.* 1962;21:527–532.

2. Ullrey DE et al. Oral and parenteral administration of iron in the prevention and treatment of baby pig anemia. *J Anim Sci.* 1959;18:256–263.

3. Zimmerman DR et al. Injectable iron-dextran and several oral iron treatments for the prevention of iron-deficiency anemia in baby pigs. *J Anim Sci.* 1959;18:1409–1415.

4. Perri AM et al. An investigation of iron deficiency and anemia in piglets and the effect of iron status at weaning on post-weaning performance. *J Swine Health Prod.* 2016;24:10–20.

5. Fredericks L et al. Evaluation of the impact of iron dosage on post- weaning weight gain, and mortality. *AASV.* 2018;315

6. Bhattarai S and Nielsen JP. Association between hematological status at weaning and weight gain post-weaning in piglets. *Livest Sci.* 2015;182:64–68.

7. Thorn CE. Hematology of the pig. In: Weiss DJ and Wardrop JK eds. *Shalm's Veterinary Hematology.* 6th ed. Ames, Iowa. Wiley-Blackwell. 2010:843.

8. RStudio Team. RStudio: Integrated Development for R. RStudio, Inc., Boston, MA. 2015. Available at: http://www.rstudio.com/. Accessed 5 November, 2018.

9. United States Department of Agriculture, Economics, Statistics and Market Information System. 2017. Available at: http://usda.mannlib.cornell.edu/MannUsda/viewDocumentInfo.do?documentID=1096. Accessed 1 October 2018.

Table 1: Economic impact of sub-clinical and full-scale iron deficiency anemia on United States swine production

A. Regression model inputs

	Baseline*		Improved†
Hb status at weaning (%)‡			
Optimal	21		70
Sub-clinical	54		27
Deficient	25		3
Percent female pigs		50.0	
Wean-finish cost per day on feed ($US)		0.28	
Cost of second 200 mg dose of Uniferon (product and labor in $US)		0.25	
Average weaning weight (lbs)		13.0	

B. Predicted economic return resulting from improved hemoglobin status at weaning

Parameter	Marketing strategy	
	Fixed weight	Fixed days§
Targeted wean-finish gain (lbs)	260	n/a
Targeted wean-finish days on feed	n/a	165
Predicted Average daily gain (lbs)§	1.587	1.617
Average days to market		
Baseline	163.9	n/a
Improved	160.8	n/a
Weight at marketing (lbs)		
Baseline	n/a	261.8
Improved	n/a	266.8
Carcass yield (%)	n/a	75.0
Lean hog price/cwt ($US)¶	n/a	80.37
Days on feed saved per head	3.1	n/a
Weight difference at marketing per head	n/a	5.0
Net return on investment per head ($US)	0.38	2.77
Total market hogs slaughtered in the US**	121,317,200	
Annual impact on total US swine production ($US)	**46,276,152.98**	**335,710,988.44**

* Expected hemogobin distribution at weaning in baby pigs receiving a single 200 mg dose of Uniferon®.
† Expected hemogobin distribution at weaning in baby pigs receiving two 200 mg doses of Uniferon® by day 12 of life.
‡ Optimal (>110 g/L), sub-clinical (90-110 g/L), deficient (< 90 g/L).
§ Estimated using the regression formula: ADG = 1.158259+(0.063996*Optimal.Hb)+(0.003783*SubClinical.Hb)+(0.034634*WeanWeight)-(0.074726*%female)
¶ Iowa State University Extension and Outreach Historic Hog and Lamb Prices. Average lean hog price 2010-2016.
** 2017 United States Department of Agriculture, Economics, Statistics and Market Information System.

Elimination of *Mycoplasma hyopneumoniae* from a multisite production farm in Perú

José A. Suárez-Mestas[1]; Yoni Marcos-Ríos[1]; Jorge Martínez Schmiel[1]; Jenny Tonassi[2]; Andres Díaz[3]; María José Clavijo[4,5]
[1]PIC Perú; [2]PIC Andina; [3]PIC Mexico; [4]PIC USA; [5]College of Veterinary Medicine, Iowa State University

Introduction

Mycoplasma hyopneumoniae (MHP) is the etiological cause of one of the most important respiratory diseases in pigs. In North America the cost of MHP infection has been estimated in approximately 6 to 10 USD per pig marketed. Hence implementing effective strategies for the elimination of MHP in the contemporary swine industry is crucial to minimize the impact of MHP on swine health and production, especially in developing countries where it is assumed that most pig farms are infected with MHP and few or none MHP elimination protocols have been implemented. Therefore, the objective of this study was to implement a MHP elimination program in a 1500-head sow farm with three production sites in which MHP infection was endemic.

Materials and methods

Sow farm and growing sites are located in Lima, Perú. Site 1, 2, and 3 house farrow-to-wean, nursery and finisher pigs respectively. There are 3 nursery sites and 2 finishing sites located at approximately 8 kilometers of distance from Site 1. All pigs were vaccinated at 21 and 42 days of age against MHP. Additionally, gilts received a third dose of MHP vaccine at 165 days of age. Before this MHP elimination program gilts were weaned and developed on site until breeding age, while commercial pigs grew on sites 2 (until 10 weeks of age) and sites 3 (until market). The elimination program was based on controlled exposure, herd closure, partial depopulation, vaccination and medication.

Loading gilts

43 weeks of replacement gilts were selected (n = 1060) at once (week 0 of the protocol) and included 285 gilts born on site 1, and 775 obtained from an external multiplier (MHP negative source).

Controlled exposure

Two weeks before exposure all MHP vaccination and medication was withdrawn from the production sites. Exposure was achieved by two ways: a) Use of 12-week old seeder (MHP PCR positive) pigs housed with females in site 1 (1pig/50 female) for 5 weeks, and b) Inoculation of all females with 2mL of bronchoalveolar lavage fluid (PCR CT value 19) twice by either use of a long handled spoon used to deposit the content into the larynx (females > 70 days) or intranasal with a syringe (< 70 days).

Determining day zero for herd closure

Strict mapping of cough in the GDU and sow herd associated with MHP infection was performed by pen/crate and barn to estimate real exposure. Then the farm was closed for 38 weeks.

Medication

All animals in the breeding herd received medicated feed with tylvalosin (2 mg/Kg) during 5 consecutive weeks between week 35th and 39th). Piglets born during week 36 were treated with tulathromycin (2.5 mg / kg intramuscularly) at 7 and 14 days of age.

At week 37th suckling piglets were weaned at 14 days of age to create a "bubble" in site 1 and sites 2 and 3 were depopulated, cleaned and disinfected.

Diagnostics

Laryngeal swabs were collected from 450 sows at weeks 33rd, 36th and sampling times to measure population shedding.

On site weaning and entry of negative gilts: new gilts were weaned on site 38 weeks post closure.

Results and conclusions

Elimination of MHP from this site was confirmed by complete absence of clinical signs in sites 2 and 3, evaluation of lungs after slaughter and by active surveillance using ELISA (n = 300 samples distributed in 11 batches of 848 pigs with an average of 120 days of age). To date, 100% of samples collected (PCR and ELISA) have tested negative. Additionally, all gilts born after week 39 and weaned on site remained ELISA negative after 60 weeks post closure. This protocol demonstrates that it is possible to eliminate MHP in endemically infected farms using controlled exposure, to guarantee a homogeneous and specific immunity against MHP, herd closure, and pig flow management to minimize pig to pig transmission.

Genetic analysis of tail bite scores of piglets at weaning

L. E. van der Zande[1,2], MSc; P. K. Mathur[2], PhD; J. R. Dunkelberger[3], PhD; E. F. Knol[2]; PhD
[1]Adaptation Physiology Group, Wageningen University & Research, Wageningen Livestock Research, The Netherlands;
[2]Topigs Norsvin Research Center B.V., The Netherlands; [3]Topigs Norsvin USA; Burnsville, Minnesota

Introduction

Tail damage is caused by excessive tail biting from pen mates. This undesirable behavior has both animal welfare and economic implications. A common management practice used to reduce incidence of tail damage is to dock tails at a young age. However, in several countries (especially within the European Union), the practice of docking tails is also considered an animal welfare issue and a movement to ban tail docking is underway. Therefore, alternative strategies to prevent tail biting behavior must be considered. Results from previous studies show that tail biting behavior is due to many factors, such as housing, feed, and genetics (although lowly heritable, according to Breuer et al.).[1] However, a classical animal genetic model, which was used by Breuer and others to derive this heritability estimate, does not utilize the genetic information of pen mates for genetic parameter estimation.[2,3] A social genetic effects model, which estimates heritability as the proportion of total phenotypic variation explained by both direct and indirect genetic effects, due to the individual as well as its pen mates, respectively, is a useful alternative for social traits. Therefore, the aim of this study was to use a social genetic effects model to estimate the direct genetic effect (corresponding to receivers of tail damage) separately from the indirect genetic effect (corresponding to the performers) for tail damage in piglets at weaning.

Materials and methods

A total of 12,372 large white piglets, originating from a single source, were scored for the presence/absence of tail damage at weaning age. Within three days of farrowing, 24% of the piglets were cross-fostered. Cross-fostered litters (hereafter referred to as groups) varied in size from eight to fifteen piglets. Tail damage was recorded using a binary scoring system, where piglets that exhibited evidence of tail biting received a score of "1" and piglets that lacked evidence of tail biting received a score of "0". Piglets that died one week prior to weaning were included in analyses, given that these piglets could have been responsible for the tail damage of pen mates. Data was analyzed using a linear mixed model with indirect genetic effects with ASReml. Gender, month of birth, group size, litter size, and the interaction of cross-fostering by birth weight were fitted as fixed effects, whereas group ID, pig ID and group mate ID were fitted as random effects. Five generations of pedigree records were used for genetic analyses.

Results and discussion

Results from this study show that the proportion of the total phenotypic variation explained by genetics (T^2) was 0.21 ± 0.6. Of the total genetic variation, 85% and 18% of this variation was due to direct vs. indirect genetic effects, respectively, with a covariance of -3%. Therefore, extra genetic variation was identified upon applying a social genetic model for tail biting behavior, given that 18% of the total genetic variation (due to indirect genetic effects) would otherwise have been missed using a classical animal genetic model. These results indicate that both biting behavior and propensity to being bitten are heritable traits, where the latter (i.e. victim effect) is the more heritable trait of the two. Results also show that the correlation between direct and indirect variation was -0.04 ± 0.35, indicating that a victim and a biter were not necessarily the same individual. Given the experimental design used for this study, the presence of a maternal effect is likely, but not possible to separate from the group effect due to confounding. In conclusion, results from this study show that social genetic models can be used to identify genetic variation in tail biting behavior that would otherwise be missed using a classical animal genetic model. However, such models are complex, meaning that slight changes to the model can impact results, and large amounts of data are required for proper genetic parameter estimation.

References

1. Breuer, K., Sutcliffe, M. E. M., Mercer, J. T., Rance, K. A., O'Connell, N. E., Sneddon, I. A., & Edwards, S. A. (2005). Heritability of clinical tail-biting and its relation to performance traits. Livestock production science, 93(1), 87-94.

2. Bergsma, R., Kanis, E., Knol, E. F., & Bijma, P. (2008). The contribution of social effects to heritable variation in finishing traits of domestic pigs (Sus scrofa). Genetics.

3. Bijma, P., Muir, W. M., Ellen, E. D., Wolf, J. B., & Van Arendonk, J. A. (2007). Multilevel selection 2: estimating the genetic parameters determining inheritance and response to selection. Genetics, 175(1), 289-299.

Pathogen virulence factors present in sow represent potential health threats to piglet

K. Leistikow, BS; S. Son, BS, MBA; M. King, PhD

Introduction

Enteric pathogens continue to play a major role in shaping the landscape of pig production. Health improvement strategies to combat these pathogens may improve the profitability of the swine enterprise by increasing both throughput (e.g., pigs sold per sow per year) and margin (e.g., reduced cost per pound marketed or increased premium per genetic animal).[1] *E coli, C perfringens,* and *C difficile*, are widely accepted as causative agents of enteric disease within production facilities. Using a rectal sampling technique, this study investigated the transmissibility of virulence genes harbored by *E coli, C perfringens* and *C difficile* from the sow to her piglets. An additional sampling was performed on piglets 8 days post weaning to determine if the microbial virulence gene profile had been transported into the nursery.

Materials and methods

A rectal sampling technique was used in conjunction with qPCR to determine the virulence gene profile of animals at a sow unit and corresponding nursery. Four healthy litters showing no observable enteric disease and four sick litters presenting classic signs of enteric disease were sampled to determine differences in virulence gene profiles between sows and their progeny and differences between symptomatic and asymptomatic crates. Litter age ranged from 9-19 days old at the time of sampling, and each litter consisted of three piglets and their maternal sow. These twenty-four piglets were followed into the nursery phase and allowed to acclimate for eight days before a second rectal sampling was conducted. Virulence gene quantities and diversity were compared to the profiles observed in the sow unit.

Results and discussion

The results indicated that similar virulence gene profiles exist between piglets and their corresponding sows. In addition, these virulence genes were detected up to eight days post weaning in nursery animals originating from the studied sow unit. Despite a gradually decreasing presence of pathogenic *Clostridium* species, it was apparent that particular *E coli* virulence genes, namely *F18, Stx2,* and *EAST1*, persisted into the nursery and continued to negatively affect pig health. Piglets harbored larger virulence gene quantities than their maternal counterparts. *E.coli* toxin gene, *EAST1*, *C perfringens* toxin genes, *cpa* and *cpb2* were all significantly greater in piglet populations irrespective of health status ($P < 0.05$). *F18 E coli* detected in the sow unit continued to be detected in the nursery pigs. Despite the absence of an *F18* receptor in pre-weaned piglets, this adhesin might contribute to issues downstream in the nursery once the receptor develops.[2] This study provides evidence of likely transmission of enteric microorganisms harboring specific virulence genes from the sow to the piglet and subsequent carry over into the nursery.

References

1. Dufresne, L. (2006, June 3). Economics of Pig Health Improvement. Retrieved October 20, 2018, from `http://porkgateway.org/resource/economics-of-pig-health-improvement/`.

2. Frydendahl, K., Jensen, T. K., Andersen, J. S., Fredholm, M., & Evans, G. (2003). Association between the porcine *Escherichia coli* F18 receptor genotype and phenotype and susceptibility to colonisation and postweaning diarrhoea caused by *E coli* O138:F18. Veterinary Microbiology, 93(1), 39-51. doi:10.1016/s0378-1135(02)00348-6.

Evaluation of interference of Fostera® Gold PCV MetaStim combined with Fostera® PRRS

Jose Angulo[1], DVM; Lucina Galina Pantoja[1], DVM, PhD; Marcia L. Keith[2], BS; Lucas P. Taylor[2], MS
[1]Zoetis, Parsippany New Jersey; [2]Zoetis, Kalamazoo Michigan

Introduction

Porcine respiratory disease complex (PRDC) is a result of a combination of infectious respiratory pathogens in conjunction with environmental and management conditions to cause a pneumonia of mixed etiology causing significant disease and economic losses.[1] Porcine circovirus type 2 (PCV2) and porcine reproductive and respiratory virus (PRRSV) are two important components of PRDC in growing pigs and their control through the right vaccination strategies is critical to minimize economic losses. The objective of this study was to evaluate antigen interference of Fostera® Gold PCV MetaStim (circovirus vaccine type 1 – type 2 chimera, killed virus (PCV1-2a/PCV1-2b)) when used to rehydrate Fostera® PRRS.

Materials and methods

Sixty healthy three-weeks of age PRRSV-negative (by IDEXX ELISA) were housed in a BSL-2 containment facility at Midwest Veterinary Services Inc. In order to detect whether common vaccines could be mixed without a detrimental effect on efficacy, a non-inferiority study was implemented using two treatments groups, T02 and T03 (N = 24 pigs per group), with lung lesions as the primary variable and a margin of non-inferiority delta of 5%. T02 received a Fostera® PRRS vaccine reconstituted with sterile diluent and T03 received a Fostera® PRRS reconstituted with Fostera® Gold MetaStim (PCV 2a/2b MetaStim). An additional group, T01 (N = 12 per groups) was used as a non-vaccinated-challenged control group. The vaccine was administered intramuscularly (2ml) once to pigs at 21 days of age. Pigs were then challenged with virulent PRRSV isolate NADC20 at approximately 7 weeks of age (four weeks after vaccination). Upon arrival pigs were housed in three separate rooms by treatment with T01 pigs in one room and T02/T03 in 2 additional rooms (one treatment per room). Prior to the challenge phase all pigs (T01, T02 and T03) were rehoused into 1 room. Serum was harvested from all pigs at various time points pre-and post-challenge. Body weights were recorded prior to challenge and the day of necropsy. On study-day 38 (10 days post challenge), pigs were euthanized and lung lesions were scored. The primary variable in determining non-inferiority was percent lung lesions in treatment group T03 in relation to T02. Secondary variables included clinical signs attributable to respiratory disease as well as level and duration of viremia (RT-qPCR) and body weights.

Results and discussion

Treatment group T03, Fostera PRRS rehydrated with Fostera Gold MetaStim (PCV 2a/2b MetaStim) was non-inferior to treatment group T02, Fostera PRRS rehydrated as labeled using a delta of 5% (Non-inferiority decision passed as the lower confidence limit of the comparison of the test to the standard was above -5%) (Table 1). Additionally, the least squares mean percent lung with lesion for both treatment groups (T02 and T03), was significantly less ($P \leq 0.05$) than the level of lesions observed in control group (T01) (Table 1). When comparing secondary variables, clinical signs attributable to respiratory disease were not biologically different between treatment groups, in addition, level of viremia (RT-qPCR), and body weight including ADG showed significant differences ($P \leq 0.05$) between vaccinated groups (T02 and T03) compared to non-vaccinated controls (T01) post-challenge supporting efficacy against a virulent PRRSV challenge in both vaccinated groups (T02, T03) when compared to controls.

Conclusions

This study provides evidence of efficacy against a virulent PRRSV challenge in both vaccinated groups (T02: Fostera PRRS alone and T03: Fostera PRRS rehydrated with PCV 2a/2b MetaStim) when compared to controls.

Reference

1. S. L. Brockmeier et al. Polymicrobial Diseases:Porcine Respiratory Disease Complex.Washington(DC): ASM Press; 2002,Ch 13.

Table 1: Non-inferiority test and percent lung lesions following a virulent challenge 28 Days post vaccination with PRRS-MLV or PRRS-MLV rehydrated with PCV2a/2b MetaStim

Non-inferiority test			
Comparison		**95% lower bound (back transform delta)**	**Non-inferiority decision**
T02 (Fostera® PRRS) vs T03 (Fostera® PRRS rehydrated with Fostera® Fold PCV MetaStim.)		-4.45%	Passed
Lung with lesions %			
Treatment group	**N**	**Back transform Ls mean % lung with lesions**	**Standard error % with lesions**
T01 (Control)	12	32.8	5.59
T02 (Fostera® PRRS)	24	13.4	3
T03 (Fostera® PRRS rehydrated with Fostera® Fold PCV MetaStim)	24	17.1	4.82

T02. Lyophilized PRRSV was reconstituted with sterile diluent.
T01 vs T02 *P*-value = 0.0044. T01 vs T03 *P*-value = 0.0477

Your Pork Checkoff dollars at work

W. Even, JD, BS
Drake University, Des Moines, Iowa; South Dakota State University, Brookings, South Dakota

Introduction

Established by the 1985 Pork Act, the National Pork Board (NPB) has responsibility for Checkoff-funded research, promotion and consumer information projects and for communicating with pork producers and the public. Pork producers invest $0.40 for each $100 value of hogs sold through a legislative national Pork Checkoff.[1] Importers of pork products contribute a like amount, based on a formula. The NPB, consisting of an elected, 15-member group of pork producers, funds national and state programs in advertising, consumer information, retail and foodservice marketing, export market promotion, production improvement, science and technology, swine health, pork safety and sustainability and environmental management.

US pork industry update

This presentation introduces session attendees to the Pork Checkoff and shares information about its three defined goals: build consumer trust, drive sustainable production and grow consumer demand for pork and pork products domestically and abroad. The update will focus on how the NPB invests producer dollars in programs to: enhance swine science research and implementation of findings; grow US and international demand for pork through comprehensive forecasting and trend analysis; build educational programs that improve producer performance and consumer awareness. The session will define and review current marketplace dynamics including: the ongoing increase in pork production following a slight retraction due to porcine epidemic diarrhea virus in 2013; current and future carcass market value; increasing consumer demand for pork returning to pre-recession levels.[2]

Defining and delivering value

In 2017, the Pork Checkoff conducted an economic analysis to assess return on investment to producers.[3] The study quantified the financial impact of research, promotion and education programs. The results, based on 2011 to 2016 programming, demonstrated that for every $1 invested, the Checkoff generated $25 in producer value. Production research specifically returned $83 for each $1 invested. NPB leadership will continue to share program updates with producers and allied audiences, including the American Association of Swine Veterinarians, to articulate this return on investment.

Current trends driving the pork industry

This presentation will define three current trends that influence consumer awareness and activation to purchase meat – and specifically pork – products. The megatrends to be discussed include: growing global population and income; changing consumer demographics and a need to better market to millennial and multicultural audiences that are consuming more protein; an industry and marketplace focus on health/wellness and sustainability. The presenter will discuss the ethical principles driving the US pork industry and a commitment to share the need to reignite producer commitment and consumer awareness of these principles. The six principles include: produce safe food; promote animal well-being; protect public health; safeguard natural resources; provide a safe work environment; contribute to a better quality of life in producer communities.

Checkoff responds by turning insight into action

This presentation will review current market-focused research projects that move beyond the Checkoff charge to advertise and market pork and pork products. Those projects – first implemented in 2018 following a 12-month review and analysis of previously defined marketing strategy and consumer trends – take into account the changing US demographic make-up and media landscape, demand patterns linked to consumer motivation and barriers to purchase, and innovative new digital marketing platforms to reach the packer, retail and foodservice customer and, ultimately, the consumer. Specific actions implemented due to this comprehensive store-level insight include: outreach on end-point cooking temperature (145°F) consumer education; use of digital and social media platforms (Google and YouTube, specifically); creation of customizable images and digital assets for retail and foodservice customers; and outreach to grill and meat thermometer manufacturers.

References

1. US Department of Agriculture, *Pork Promotion, Research, and Consumer Information Act of 1985.*

2. Kerns & Associates, *U.S. pork production per breeding animal; Market hog inventory trend: 1963-2017.*

3. Harry Kaiser, *An economic analysis of the National Pork Board checkoff program*, Cornell University, 2017.

Landscape and outlook for US pork in the global markets

B. Stuart, MBA
Global AgriTrends

Introduction

US pork is not the largest global producer or exporter. But with US$6.2 billion in exports last year,[1] the industry is globally-significant. Competitive production and pricing make US pork affordable all over the world. Global consumers eat more than one-fifth of US pork production.[2] But the global market for pork is somewhat of a tangled web of import protocols, regulations, culture, and economics, unfortunately, often in that order. What is the future for US pork in the global markets? Let's take a look.

It's complicated

Success in global markets requires a few key factors. First the product must be safe, humanely-raised and affordable. Next, logistics must exist to move a highly-perishable, biologically-sensitive product across oceans and continents, as well as a "cold chain" at the end to get this product to consumers. Market access must be negotiated to allow entry into foreign markets without undue burdens of regulation, duties, and non-standardized sanitary requirements. And finally, the product must be demanded by the ultimate consumer, a step involving marketing, education, and other promotional efforts. Global pork trade is a complicated business requiring a vast team of professionals from varying fields with skills ranging from logistics to pathogen control to microbiological research to consumer research.

And a pig is not just a pig. It is not a single SKU, but a bundle of many SKUs. We may think of pork chops, loins and hams for meals, while consumers in China treasure pig feet, intestine, ears, tails and snouts. Global consumers also enjoy a wide variety of processed pork products.

Global competitors abound. On a global basis, pork is much more standardized in quality, taste, and flavor profile, than beef which varies widely by genetics and production system. I have seen top-notch hog farms producing high-quality pork on 5 continents. Note that genetics, feed, technology, and equipment are all globally-mobile. An example is Chile. In the early 1990's, Chile's pork was a small, backyard swine industry with limited feed grains. Investors planned and dreamed. They imported technology, genetics, equipment, and feed, and built, a world-class pork production base that exports pork all over the world today. Key US competitors also include Canada, the EU, Brazil, and Mexico.

So how does the US fit into this global pork puzzle?

Week-in, week-out, US pork is often the most affordable pork on the planet, based on comparable hog prices. (Figure 1). While China holds more than half of the world's hogs,[3] and is the world's "low-cost" manufacturer of many consumer products, they are nearly always higher cost hog producers than the US. Think of it this way: US agriculture is generally the "China" of the agriculture world. We are typically the "low-cost" producer; of corn, wheat, pork, poultry, beef, rice, and soybeans; but particularly in pork.

This significant cost advantage suggests that US pork should dominate markets everywhere. But these markets are more complex than that. Various cuts trade at various prices in various cultures. The meat case in a Walmart in Nanning, China showed pork loins at 16.8 rmb/500g (US$2.25/lb) next to pig feet at 23 rmb/500g (US$3.07/lb) and trimmed rib bones at 31.8 rmb/500g (US$4.25/lb).[4] There are major cultural variances in demand for various cuts. Another example is the STRONG US demand for pork bellies (bacon...) and ribs. Cultures and customs shape consumer demand.

Figure 1: Global hog prices

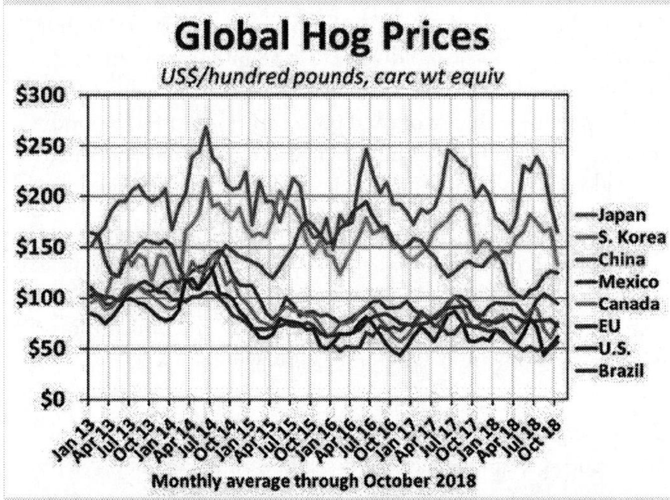

Source: USDA, Japan ALIC, Korean Meat Trade Association, Agriculture and AgriFood Canada, Agriculture and Horticulture Development Board U.K., Porcimex, China Ministry of Agriculture, Efeedlink LLC

Note the difference between US and Chinese hog prices in Figure 1. This would suggest a giant, sucking sound of US pork to China. But look at Figure 2; that scenario rarely happens. US pork accounted for 0.3% of Chinese consumption[5] (and 1.5% of US production) last year.

An actual hindrance to being the low-cost supplier is just that: the fact that US pork can undercut domestic prices in foreign lands. It works fine for consumers, but what about foreign producers. Farmers are strategically-important in any country and thus politically protected. Affordable US pork is a threat to countries seeking to maintain self-sufficiency goals. And this "Trade Policy" arena is where it gets complicated.

What about China?

This chart (Figure 2) is interesting. It shows China's monthly pork imports, both from the US, and from all other sources. Note the volatility around Chinese price cycles. Each of the "spikes" in imports revolved around high Chinese prices, either from disease-driven shortages, or financial losses driving sow liquidation. When China needs pork, China buys pork. Unfortunately the inverse is somewhat true also. And, note the big run-up in non-US pork during the last price cycle. Most of this was EU pork. The US was largely excluded.

US pork has been restricted based on China's non-scientific requirements over residues of ractopamine. The non-scientific part of that restriction is a zero-tolerance for residues that goes beyond global-accepted standards. But don't get hung up on details, China has long "managed" US pork imports through a number of means including plant de-listings, inspection delays, and other trade distorting measures. The bottom line is that the US and China have a complex, codependent, bipolar, political, and commercial relationship. And both sides jockey for power.

China is holding to pork self-sufficiency rates above 95%[6], but the market is massive and diverse; the $8 billion pork import market (including China and Hong Kong)[7] is basically a niche market serving only 5% of total pork consumption. US opportunities in China necessitate patience and continued pressure toward global standards, as well as continued market development.

China's discovery of African swine fever (ASF) will lead to pork shortages and pull on global pork supplies. However, China has shown strong resolve in boycotting US soybeans due to US trade actions (including duties) against China. Will China import US pork, or seek only non-US pork? Time will tell.

So where does US pork freely flow?

Free trade agreements have been secured allowing duty free pork into Australia, Canada, Chile, Colombia, Hong Kong Mexico, and South Korea. Figure 3 shows pork exports to these duty free nations versus all others. Note the growth in exports to the duty-free markets. There are also duty free quotas for CAFTA and other FTA nations; most of those quotas end in 2020/2021.

Figure 2: China pork imports

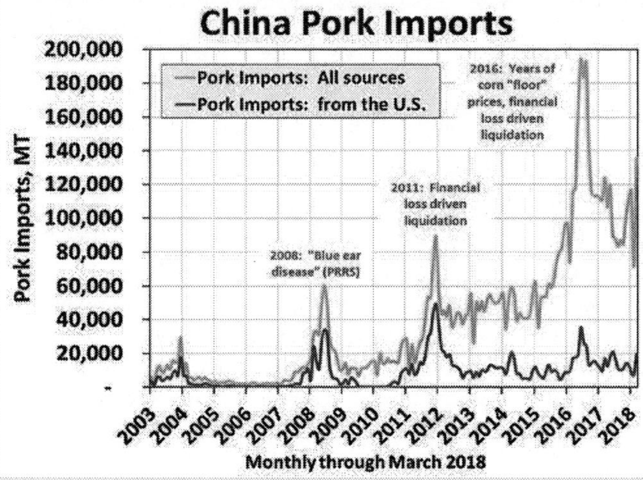

Source: Trade data monitor

Figure 3: US pork exports to duty free markets and all others

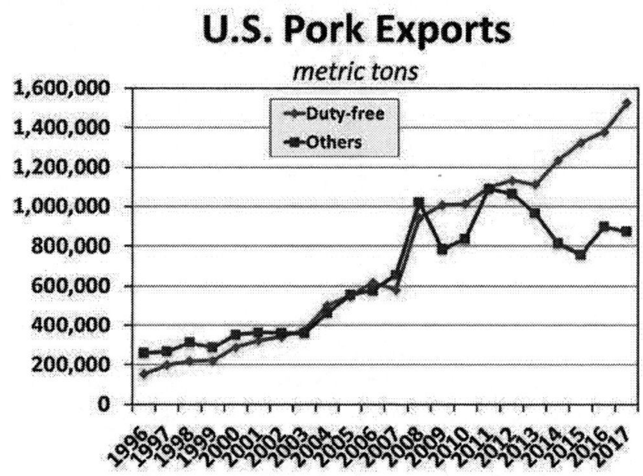

Source: US Department of Agriculture, Foreign Agriculture Service, Global AgriTrends

The FTA game

The last major US FTA was Korea, completed in 2008, altered and implemented in 2012.[8] The agreement now gives US the only duty-free pork access into the country, an enviable spot. The Korea - EU FTA will provide duty free pork access in 2021 for bellies (their most-traded item to Korea); Canada gets duty free pork to Korea in 2019.

Outside of the TPP agreement (now CPTPP) the US has not implemented a FTA since 2012 (Korea, Colombia, and Panama). Key competitors are cutting deals around us. Chile, Canada, and

Mexico now have preferential access for pork into the Japan, the #1 US pork export market by value, under the newly-ratified CPTPP agreement. Also, the EU will soon gain the same access to Japan under their bilateral agreement, leaving the US as the "odd man out" for our most lucrative pork export destination (barring a US – Japan FTA, promised by President Trump).

President Trump pledged to seek new bilateral agreements, even as he withdrew the US from the TPP (now CPTPP) agreement. But those have yet to occur outside of the renegotiation of the existing Korean agreement and the Canada/Mexico agreement. The US Trade Representative Robert Lighthizer notified Congress of his intention to begin trade talks with Britain, the EU, and Japan back in October 2018, but progress remains slow. Time is of the essence, FTAs take years to complete and often include years of tariff phasedowns. What we sow now may take a decade to reap.

Where is the future?

I am often asked, "Where is our next export growth opportunity?" The question is simple, the answer more complex. I see the next significant export growth opportunities to a basket of markets, not one single market. With our duty-free access to Colombia, Chile, Central American nations (CAFTA and the Dominican Republic), we will be strong force in those markets. Colombia has a population (48 million) similar to Korea; the CAFTA-DR group is nearly that same size (40 million).

And there are other regions of the world seeking more pork. The EU and Brazil sent a combined $374 million in pork to Africa last year (307,000 metric tons).[10] The US sent just $2.6 million. But a new US agreement with South Africa is beneficial, as would be a potential agreement with Nigeria (population 186 million), hinted to by Nigerian President Buhari in meetings with President Trump last year. Granted, Africa is a unique market, but remember that US pork prices (think hams and offal) are VERY competitive globally.

These and other markets hold the key to US export growth. And the US is well poised to capitalize on new opportunities. The US Meat Export Federation has a global presence with staff tracking consumer trends, competitor prices, and market access all over the world.

Longer term optimism

If we look from the 40,000 foot level at global pork demand, Figure 4 shows the correlation between global GDP (economic output or total global income) and meat and poultry consumption from 1970 to 2016; each dot is a year in that sequence. This shows global meat and poultry consumption rising nearly lock-step with rising GDP. The forecast line is based on World Bank and IMF forecasts of global GDP, and estimates future meat and poultry consumption levels. If these forecasts are correct, an additional US$38 billion in global GDP (forecast from 2017 to 2030) will require an additional 134 million metric tons of global meat and poultry production. The growth in meat and poultry production over the past 13 years was only 74 million metric tons. There is a hungry world rising out there as incomes rise globally.

Figure 5 breaks out the key meat species in the same analysis. Note the flatter slopes of the pork and poultry lines suggest that they are very elastic. Said another way: Rising global GDP is like rocket fuel for pork and poultry demand. Forecasts by the World Bank and International Monetary Fund suggest that global GDP growth will average $2.7 trillion annually out to 2030. Continuing that correlation calculates to 4.7 million metric tons more global pork needed annually. For reference that is about 40%

Figure 4: Global GDP, forecasts, and estimated meat and poultry consumption

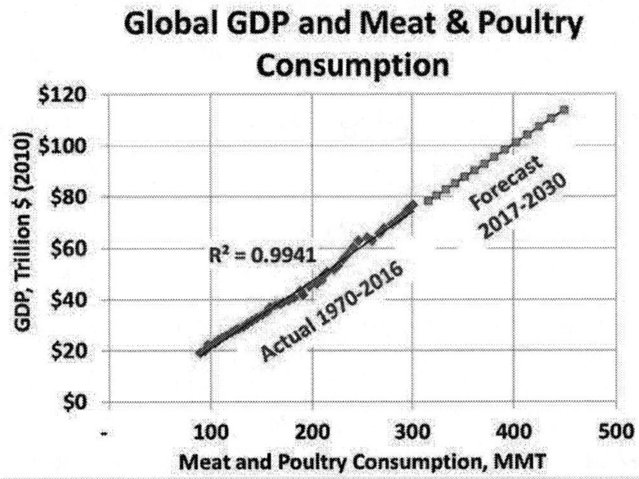

Source: World Bank, IMF, FAO, Inflation-adjusted to 2010 dollars

Figure 5: Global GDP and global meat/poultry consumption

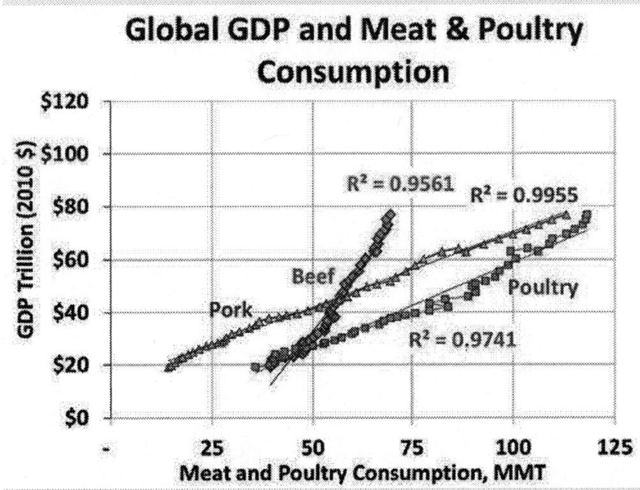

Source: World Bank, IMF, FAO, Inflation-adjusted to 2010 dollars

of US production last year. Global supplies may not grow fast enough to supply these levels of demand; the model may shift as a result. But global demand appears poised to chase global supplies over the next decade (note that is a global picture, not necessarily the US). And remember that in the short-term, production can expand faster than demand growth; we can overproduce in the short term.

Are we producing too much pork?

The final question I'll address is "Are we producing too much pork?" That's an easy question to answer because a free market sends very clear and direct signals regarding our output. The text-book definition of consumer demand is "a consumer willingness to pay at a specified level of production". And based on per-capita US supplies nearing the highest levels in 19 years (based on my 2018 forecasts), I'd say the market has been mostly fine with current levels. But the continued expansion in 2019 takes per-capita supplies to the highest levels since 1981 (37 years). Adding in beef and poultry supplies takes my US per-capita supply number to a record 205 pounds in 2019, up +21 pounds in five years. That level feels "heavy" to me.

Recent large investments into US hog production and packing infrastructure by some bold souls will be needed in coming years. The question is how fast do we build and fill these plants with hogs. I would suggest that with rising beef and poultry supplies, caution is warranted.

Bottom line

US pork is the typically the most competitive on earth. Global populations are rising by 78 million per year. Global incomes are rising and will pull nearly 3 billion into the global middle class (2009 to 2030).[11] Demand is rising but not in a straight line. US producers can over-produce. There will be a lot of up and downs on the broad global demand growth trendline. But in the end the future belongs to those who can adapt, and I would add, to those who can gain the best market access.

References

1. U.S Department of Agriculture, Foreign Agriculture Service. GATS Database. Available at: www.fas.usda.gov/gats/default.aspx.

2. US Department of Agriculture, Foreign Agriculture Service. PSD Database. Available at: https://apps.fas.usda.gov/psdonline/app/index.html#/app/advQuery.

3. Ibid.

4. Prices noted by author at a Walmart store in Nanning China in May 2017.

5. Chinese customs data as reported by Trade Data Monitor, subscription required. Available at: www.tradedatamonitor.com.

6. US Department of Agriculture, Foreign Agriculture Service. PSD Database. Available at: https://apps.fas.usda.gov/psdonline/app/index.html#/app/advQuery.

7. Chinese customs data as reported by Trade Data Monitor, subscription required. Available at: www.tradedatamonitor.com.

8. Office of the U.S. Trade Representative. Available at: https://ustr.gov/trade-agreements/free-trade-agreements/korus-fta.

9. Mexico customs data as reported by Trade Data Monitor, subscription required. Available at: www.tradedatamonitor.com.

10. Europe and Brazil customs data as reported by Trade Data Monitor, subscription required. Available at: www.tradedatamonitor.com.

11. OECD, An Emerging Middle Class. OECD Yearbook 2012. Available at: http://oecdobserver.org/news/fullstory.php/aid/3681/An_emerging_middle_class.html.

Elimination failures and LVI experiences

J. Miller, DVM; M. Hensch, DVM; C. Stoffel, DVM; B. Peters, DVM
The Maschhoffs, LLC, Carlyle, Illinois

Introduction

PRRS has long been the most economical impacting disease in the swine industry over the past decade. As management practices adapt to minimize the impact of the disease, indications show virus adaptation to longer persistence in a population of animals than previously documented. A common management practice of eliminating the virus from a breed to wean facility has been to load, close, and homogenize with either a live virus inoculation (LVI) or modified live virus vaccine (MLV). Prior to 2012 within The Maschhoffs system, LVI was a common practice until the time to negative production (TTNP) data was summarized by Linhares, et al.[1] Since this point in time, the preferred protocol for elimination has been MLV, which also has an economical benefit estimated at $20 per sow. Recently, the current epidemic disease management plans within The Maschhoffs have shown an extended TTNP and failed to remove field strain virus from the farm. This situation has led to the use of LVI in a few herds to evaluate its effectiveness in an elimination plan.

Farms 1 and 2

Farm 1 and 2 are each 1200 head mated inventory farms that are approximately 1100 feet apart and have maintained the same health status within The Maschhoffs system. The farms do not possess a gilt development unit or an isolation facility. Starting in early 2017, Farm 1 was confirmed positive via PCR with PRRS 1-1-4. Figure 1 gives a detailed timeline of events from this point forward. Farm 2, although never confirmed positive, proceeded with a load/close/homogenize protocol using vaccine only. This process appeared to be successful in Farm 1, which achieved PRRS 2B status at 36 weeks post closure. Testing at this time was via 60; four-day-old piglet blood swabs for eight consecutive weeks, as well as 60 wean piglet blood swabs for four consecutive weeks. In Farm 2, at 35 weeks post closure, a virus that was typed 6.6% genetically different than Farm 1 was identified in processing fluids. At this time, Farm 2 continued to test and managed to achieve stage 2B via the blood swab procedure described above. However, the more recent virus subtype detected (PRRS 1-1-2) continued to show up on processing fluids in Farm 2. With a plan to implement LVI at Farm 2 for a new closure along with the risk it posed for causing an outbreak at Farm 1, the decision was made to use LVI at both farms at the same time. LVI was then followed by Ingelvac MLV four weeks later. Farm 1 experienced severe clinical signs with pre-weaning mortality (PWM) up to 24%, thus leading us to believe this was a new virus introduced into Farm 1. At Week 48 of 2018, both farms had adopted processing fluids for their testing protocol and achieved stage 2B at 28 and 31 weeks post-LVI.

Farm 3

Farm 3 is a 3500 head mated inventory breed-to-wean facility in west central Illinois consisting of crated and pen gestation facilities, an isolation unit, and two farrowing units. In week 47 of 2015, the naïve farm experienced a PRRS outbreak with subtype 1-7-4, which matched an outbreak in a farm three miles to the north. The farm implemented a load, close, and homogenize protocol with Inglevac MLV, and achieved Stage 2V at 34 weeks post-closure in 2016. Testing at the time was weekly blood swab testing on 30, four day old piglets for eight consecutive weeks, and 30 weaned pigs for four consecutive weeks. The farm continued to test bi-weekly on the same blood swab protocol until week 43 of 2016. At that time, the farm tested positive on wean pig samples with an RFLP cut pattern of 1-6-4, which was 1.7% different than the previous 1-7-4. The farm was loaded with vaccinated gilts several times and closed on week 5 of 2017 when all animals received Ingelvac MLV. Figure 2 shows exact testing timeline, but at 26 weeks post-closure, it was determined that the closure was failing, and a different approach needed to be used. Management, along with health services, decided LVI would be the next option. Serum was harvested and mixed with Gentamicin and PBS to create an autogenous inoculum. Management decided the intranasal administration route would save time and be less of a work load for the farm staff. Pump sprayers were then used to administer intranasal inoculum to all P1+ and half of gilts on site. The remaining half of on-site gilts were administered an intramuscular (IM) injection of virus with little to no clinical signs observed over next two weeks. Two weeks later, blood samples of gilts were collected and revealed only one pool out of five to be positive to PRRSV. It was then decided the farm would not be homogenized and the intranasal route of administration was not effective. LVI was attempted again with all animals receiving an IM injection rather than intranasal. The farm experienced severe clinical signs of a PRRS infection including 200-250 abortions. The farm continued to have small numbers of positives using the blood swab testing protocol described above and did not achieve 2V status until week 16 of 2018, 33 weeks after the second LVI. In week 17 of 2018, the farm did a receive routine mass Ingelvac MLV vaccination. Farm 3 has maintained the Stage 2V status; however, the downstream pig flow did test PRRS PCR positive for the 1-6-4 strain in week 39 of 2018. This would indicate that Farm 3 may be leaking the PRRS virus and the LVI may not have be successful at removing the resident field strain.

Conclusions

Multiple different PRRS control methods were used above in the farms, including vaccine only, LVI/vaccine combo, and LVI alone. All resulted in slightly different results, both of success and failure. These case studies demonstrate the variety of tools available, and the wide range of unpredictable factors presented in a production setting. Because of this, it is no surprise that this would produce a variety of successes as a result. LVI was used

Figure 1: Farm 1 and Farm 2 PRRS closures, 2017 - 2018

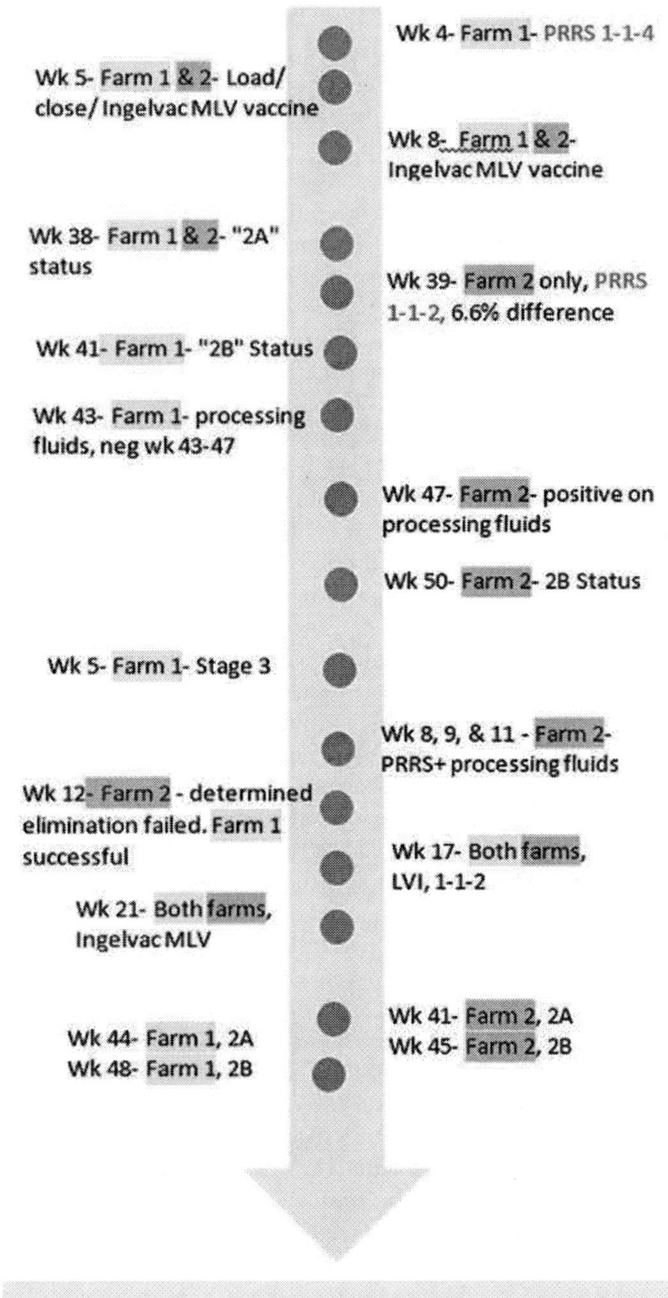

Figure 2: Farm 3, 2015 - 2018

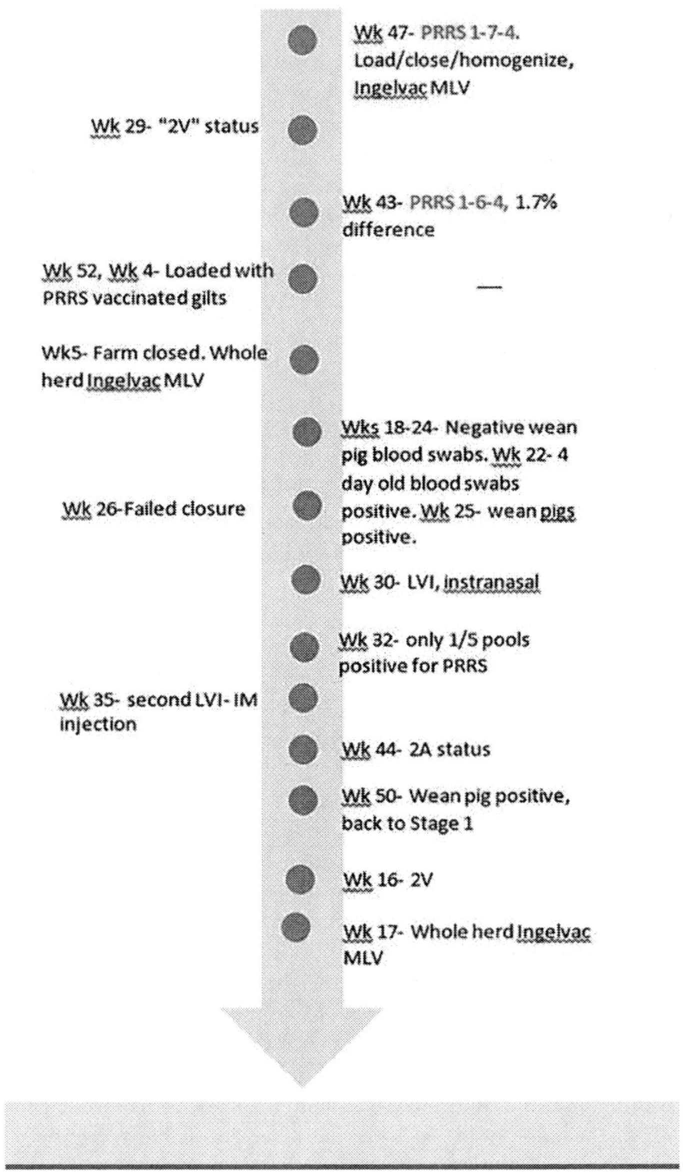

in both cases, with variable responses, production losses, and success. It can be a tool used to supplement or adjust a closure plan with, while always keeping in mind the potential for major production loss when doing so. As PRRS mutates and changes its response to traditional load, close, homogenize protocols, we must also mutate and change our response.

Reference

1.Comparison of whole herd exposure programs to produce PRRSV negative offspring from PRRSV positive breeding herds Daniel Linhares[1]; Jean Paul Cano[2]; Montserrat Torremorell[1]; Robert Morrison[1] [1]University of Minnesota, Saint Paul, Minnesota; [2]Boehringer Ingelheim Vetmedica Inc, St Joseph, Missouri

Economics of PRRS control

Dale D. Polson, DVM, MS, PhD
Boehringer Ingelheim Vetmedica Inc., Duluth, Georgia

Introduction

Classic economic analysis comes in some combination of the following:

- Actual calculated results (retrospective) or estimated anticipated results (projected, predicted)
- Short-term or long-term time horizon
- Deterministic or stochastic

The most commonly used economic analysis methods (and their corresponding key output metrics) typically used to generate actual calculated results are:

- Enterprise budgets and partial-budgets – Profit, margin-over-[feed costs or variable costs], net difference, benefit:cost ration, return on investment, return on equity, return on assets
- Cash flows – Beginning and ending cash balances, periodic net cash balance (positive or negative)
- Discounted cash flows – Net present value, internal rate of return

The levels at which economic analysis methods are typically applied are:

- Group (batch, cohort) level
- Farm/site level
- Enterprise level
- Industry level

For all methods and at all levels, analysis can be done either deterministically or stochastically.

- Deterministic models – The same set of input values always results in the same outcome values
- Stochastic models – The same set of input values results in a distribution of outcome values

"What-if" analysis is commonly used to attempt to determine

- Deterministic methods – Two commonly used methods are the Decision Tree and Payoff Matrix, and both of these methods can incorporate probabilities for each input variable level use in the Tree or Matrix to attempt to factor in the likelihood of the various input levels evaluated.
- Stochastic methods – Various types of distributions are used, but a simple and relatively commonly used one is the Triangle Distribution. Another commonly used distribution is the PERT distribution.

The basic methods described above for evaluating the cost of disease are no different for PRRS than they are for any other disease, and no different than they are for any other economic evaluation related to pig and pork production.

Economic analysis applied to PRRS control

PRRS remains as the single most costly disease faced across the North American swine industry. In 2005, the cost of PRRS (US) was estimated to be $560,000,000 per year.[1] In 2011, the cost of PRRS (US) was estimated to be $664,000,000 per year.[2] Based on these two industry-wide estimates, PRRS has cost the US industry:

- Approximately $18 billion since 1987
- 14 months of pig production (give or take)
- Roughly $6 per pig marketed every year for 31 years

For an industry worth $18 ($16-$20) billion USD in market hog sales revenue per year, that is like the entire United States not producing any pork at all for more than a year. Clearly, routine economic analysis is worthwhile at all levels – group, farm site, production business, industry – when it comes to PRRS. This is true for research projects to estimate the value potential of what is being evaluated, but is even more important when it comes to determining value in the operational setting of an ongoing business.

For business operations, the 5-Step process, adapted from the DMAIC process of Six Sigma, is a valuable tool when doing economic analysis for various purposes at the business, farm and group levels; as each of the 5 steps can be viewed in economic terms:

1. Economic goals (profit, cash flow)
2. Economic status (profit/loss, cash flow)
3. Economic constraints (low market prices, high input costs, poor cash flow, availability and/or high cost of debt capital)
4. Expected cost and benefit of solution options (net benefit, benefit:cost ratio, ROI, NPV, IROR)
5. Actual cost and benefit of implemented Solution (net benefit, benefit:cost ratio, ROI, NPV, IROR)

During the presentation, following setting the context as described above, a brief review of prior PRRS economic analyses from the literature and technical/professional meetings will be made. Then several examples of PRRS economic analysis will be described. These examples will include application of short-term and long-term economic analysis methods, as well as apply both deterministic and stochastic analytic methods.

References

1. Neumann, E.J., Kliebenstein, J.B., Johnson, C.D., Mabry, J.W., Bush, E.J., Seitzinger, A.H., Green, A.L., Zimmerman, J.J., 2005. Assessment of the economic impact of porcine reproductive and respiratory syndrome on swine production in the United States. J. Am. Vet. Med. Assoc. 227, 385–392.

2. Holtkamp, D.J., Kliebenstein, J.B., Neumann, E.J., Zimmerman, J.J., Rotto, H.F., Yoder, T.K., Wang, C., Yeske, P.E., Mowrer, C.L., Haley, C.A., 2013. Assessment of the economic impact of porcine reproductive and respiratory syndrome virus on United States pork producers. J. Swine Health and Production. 21, 72–84.

Evolution of PRRSV monitoring: 3 decades of advance and improvement

Will Lopez, DVM; Marcelo Almeida, DVM, MS; Gustavo Silva, DVM, MS, PhD; Giovani Trevisan, DVM, MBA; Daniel C. L. Linhares, DVM, MBA, PhD
Veterinary Diagnostic and Production Animal Medicine, Iowa State University, Ames, Iowa

Introduction

Porcine reproductive respiratory syndrome (PRRS) is a swine-specific disease that was first described in 1987 in North America, and in 1990 in Europe.[1] It affects pigs of all ages in most swine-producing countries since the late 80's and it has been causing significant economic loses since then. Currently, PRRS virus (PRRSV) remains the single pathogen that most impacts economics of US pig production, costing about 664 million dollars per year.[2] Control and elimination of PRRSV infection in swine herds require continuous and reliable virus monitoring and surveillance systems (MOSS). More specifically, there is the need to continuously improve methods to assess within herd virus transmission, and to correctly define the PRRSV shedding status of suckling pigs and weaned pigs. That was not the case in the early years of the emergence of the mystery swine disease. When PRRSV started to appear, there was not much access to reverse transcription polymerase chain reaction (RT-PCR) or enzyme linked immunosorbent assay (ELISA) based testing for PRRSV. The monitoring and surveillance was in most part based on clinical signs, and lesions associated with the virus. We present an overview of the evolution of PRRSV monitoring practices and methods since the emergence of the disease.

A brief look at the past

At the time PRRS emerged, clinical signs of the disease were extremely severe and different than those of any other known pathogen, and thus red flags started to raise, indicating that an infectious process was taking place and it was something new. The diagnostic and subsequent monitoring was really very dependent on having great practitioners and producers sending cases to diagnostic labs, having very well trained pathologists and diagnosticians with a very open mind that would ask questions, identify the agent and document their findings. In some cases, the diagnosticians would go and do field visits with practitioners and get more samples, bring it back to the laboratory to further investigate, and that was really key in terms of doing PRRSV monitoring and surveillance in the past. The rapid development of antigen and antibody testing after the discovery of PRRSV and its isolation, allowed researchers and practitioners to apply serologic information to understand the epidemiology of the virus and to design more effective control and elimination measures, but somehow, active monitoring and surveillance was not a very common practice for PRRSV. When asked about how much of PRRS surveillance was really done by farms or production systems in the 90's and going forward, the common answer from veterinarians and producers is that it just consisted of a protocol of "I would test when I see clinical signs", which fortunately, has been evolving and advancing through the years due to the increasing consolidation of bigger pig production systems. The decreasing tolerance to the effects of infectious diseases like PRRSV, that produce such negative health, production and economic impact to the swine industry, have been driving the efforts to monitor and surveil swine populations. In today's world, the decisions made in terms of management, control and elimination of PRRSV and other infectious diseases, affect more and more pigs and the consequences of not detecting them on a timely manner comes at a high price. Among diagnostic and confirmatory methods and tests available for PRRSV detection in swine herds in the early years of its appearance, immunohistochemistry (IHC) techniques were used at first, ELISA, the immunoperoxidase monolayer assay (IPMA) and indirect fluorescent antibody test (IFA) would provide reliable confirmation of positive ELISA tests (ELISA became the most commonly used test to screen herds) and the more specific technique, in situ hybridization. But still, sampling pig populations required an individual subject approach (sampling individual pigs), which made things uphill in terms of costs, availability to all levels of the swine industry and ease of implementation.

Current commonly used diagnostic tools

There is a diversity of diagnostic assays to detect the PRRSV, its RNA and/or anti-PRRSV antibodies. The RT-PCR is a commonly used method to detect viral RNA in samples. The ELISA is the most common test used for screening herds, having also the IFA and the IPMA as confirmatory back-up tests for unexpected positives on ELISA.

There are also other molecular techniques being more increasingly used in the field of molecular epidemiology. One example is the sequencing of the open reading frame 5 (ORF5) of PRRSV which can be used to determine patterns of transmission between herds, to track down the spread of virus across regions, and/or to understand the genetic variation that may occur over time within or between herds. Whole genome sequencing (WGS) has been gaining critical importance in biological and medical research and it is another method being more frequently used in veterinary research, becoming a trend as it may be used to identify whole families of virus, categorize predominant field virus strains and gain more understanding about the variations within and between the PRRSV field isolates, it represents a major improvement for molecular epidemiology.

PRRSV monitoring

Often, 'monitoring' and 'surveillance' are two terms used interchangeably.[3] In his paper from 2013, M. D. Salman writes about this matter and briefly explains the difference between the two terms as follows: "Animal disease surveillance is watching an animal population closely to determine if a specific disease or a group of diseases makes an incursion so that a prior plan of action can be implemented. Monitoring of animal diseases focuses on identifying a disease or a group of diseases to ascertain changes in prevalence, and determine the rate and direction of disease spread. Therefore, monitoring by definition lacks action to prevent or control a health problem." In the case of PRRSV, the term 'surveillance' seems more adequate due to the fact that veterinarians and producers often implement PRRSV control/elimination programs, once the virus is detected. Regardless of the use the these both terms, monitoring and surveillance for PRRSV have a common objective, to watch the pig population to detect virus infection. For the rest of this paper we will use both terms 'monitoring' and 'surveillance'. A great example of a PRRSV monitoring and surveillance system is the Morrison's swine health monitoring project MSHMP.[4] In regards of PRRSV, the goal of the Swine Health Monitoring Project is to monitor the disease statuses of a population of US sow farms, enrolled voluntarily in the program. The MSHMP is currently funded by the USDA, the Pork Checkoff and the Swine Health Information Center (SHIC). It was created in 2011 with retrospective data since 2009 and since then, has been a major improvement in organizing PRRSV monitoring information, offering a valuable and updated snapshot of PRRSV incidence and prevalence over time to the swine industry and academia. Similarly, Trevisan et al., has recently reported the development of the Swine Disease Reporting System, which gathers diagnostic data from the veterinary diagnostic laboratories from Iowa, Minnesota, Kansas, and South Dakota and reports patterns of PRRSV (and other pathogens) detection over time, geographical region, specimen, age group, and farm type. Then, we will briefly discuss two common approaches for monitoring and surveillance for PRRSV that are currently in use and continue to evolve and advance to provide diagnostic information and increase our knowledge about PRRSV epidemiology: Individual pig based blood sampling and population based sampling.

Individual pig-based blood sampling

This sampling approach is one of the most commonly used ways to monitor PRRSV infection status in swine herds. This sampling technique consists in drawing blood from randomly, or conveniently selected individual pigs using approved restraining methods to hold the animals. The sample size for this approach depends on the expected prevalence in the pig population and the desired confidence level, for instance, the impact that a failure to detect PRRSV would have in a genetic nucleus will differ from the impact that it will have in a commercial sow farm, in terms of economic losses and, thus confidence levels may vary and therefore sample size as well. A great example of the application of an individual pig based blood sampling protocol can be found embedded in the herd classification system developed in 2010 by the American Association of Swine Veterinarians (AASV) for describing the PRRSV status of herds. The criteria for defining this herd classification scheme for the PRRSV status of breeding herds is in part supported by information derived from a MOSS, this includes diagnostic results and information from production records, both of which should be routinely collected. Absence of clinical signs consistent with PRRS can also be used to support the herd-status classification, but clinical signs alone are not sufficient.[5] The disadvantages of this sampling approach include the need of qualified personnel to take samples, the potential risk of losing pigs during the procedure (due to its invasive nature), it is time consuming and it has limited potential for pooling high number of samples.

Population-based sampling

PRRS monitoring in breeding herds using processing fluids (PF)

Processing fluid (PF) was described as an aggregate (population) sample, formed by the serosanguinous fluid recovered from piglet castration and tail docking.[6] The development of this population based sampling method responds to the necessity of improving PRRSV detection in scenarios where the prevalence is close to zero, that is, improved PRRSV monitoring and surveillance systems. Industry current standard protocol for monitoring PRRSV in breeding herds consists of collecting individual pig serum samples from 30 weaning age animals on a monthly basis and test samples (pooled by five) for PRRSV RNA detection. But this monitoring scheme assumes that PRRSV cannot sustain clinical infection and/or remain in the population at a prevalence lower than 10% during a period of 90 days. On the other hand, near-zero PRRSV prevalence has been documented in endemically-infected breeding herds.[7-10] These findings are proof that the main assumption upon which the current monitoring plan is based is not sufficiently robust to provide reliable results. The current monitoring scheme could be improved upon by testing higher numbers of individual piglets, at a higher testing frequency. However, bleeding piglets is time consuming, requires two trained persons, and causes additional piglet stress. These practical and economic constraints render this option unsatisfactory for most commercial production systems.[6] Processing fluids sampling method represents a robust tool for PRRSV monitoring and surveillance in suckling pigs due to ease of implementation and relatively lower cost (compared to individual pig sampling). At the litter level, PF-based sampling has a moderate sensitivity when compared with bleeding all piglets from the population but has 100% sensitivity and 100% specificity at the whole room, or whole day level. Results on PF suggest that pooling litters increases the sensitivity of the method, PF is more cost-efficient to monitor PRRSV by qPCR as compared to individual pig sampling, especially at low prevalence level.[11]

PRRS monitoring in breeding herds using family oral fluids (FOF)

Family oral fluids (FOF) was recently described as an alternative way to collect aggregate samples for PRRSV monitoring from weaning age piglets.[12] What is Family Oral Fluids?, in 2008, Prickett et al. described the collection of oral fluids from growing pigs. "Oral fluid" was described as a mixture of saliva and mucosal transudate from which anti-PRRSV antibodies and PRRSV RNA could be detected.[13] Oral fluids could be collected from several pigs within a pen by hanging a cotton rope and allowing sufficient time for the pigs to chew on that rope, serving as an aggregate sample of that group of animals. Now, FOF is a population based sampling method to collect oral fluids from due-to-wean litters where both the sow and the piglets have access to the cotton rope hung in front of the crate in the opposite side of the drinker for the pigs. Ideally, we would allow sow and piglets a 30 minutes' period of rope exposure. Almeida et al. reported in 2018 that FOF-based techniques have consistently delivered a higher success rate for obtaining oral fluids from due-to-wean piglets when compared to litter oral fluids (LOF) sampling approaches, and should, therefore be used as the preferred technique for that. The first report of PRRSV RNA detection by RT-PCR in FOF was made by Almeida and others in 2017.[14] Seventy-two matching sets of FOF and blood samples from all piglets were collected from the same piglet population. FOF detected 84.4% (27/32) of all the litters as positive, where there was one or more viremic piglets within the litters (PCR positive on serum). From all the litters with 3 or more PRRSV-positive individual piglets, 100% were PRRSV-positive by RT-PCR in FOF (22/22), but from the litters with 1 or 2 viremic piglets, only 50% (5/10) were detected as positive for PRRSV by RT-PCR on FOF. Forty of the 72 litters had PRRSV-positive pigs by serum, as well as by FOF, which is really important finding regarding the specificity of the method, as no false negative results were produced when using FOF to detect PRRSV on due-to-wean pigs.[14] Almeida has also recently reported a success rate of 94% to obtain FOF in 756 attempts in 6 different sow farms. Most of those FOF were obtained from farm staff.

Production parameters monitoring

A classic situation in the face of PRRSV outbreaks in breeding herds, has been the observation of negative changes in some key production parameters, but once these parameters are altered, it is typically too late and certainly PRRSV is already running free, which may have catastrophic consequences in the downstream production flow of a given system. Changes in production parameters include: increase in the number of abortions, increased neonatal losses (stillborn and mummified piglets), and increased pre-weaning mortality. Based on the above exposed, Silva et al. reported the use of production data to monitor changes on productivity to early detect PRRS outbreaks in breeding herds.[15] Measuring changes in farm productivity before and after PRRS infection enables quantifying the production and economic impact of outbreaks. This study assessed the application of exponentially weighted moving average (EWMA A statistical process control (SPC) method was used to screen and check significant deviations on productivity and if a "deviation" was detected A statistical process control (SPC) chart is used to screen and check significant deviations on productivity. If a "deviation" in farm-specific baseline parameters is detected, it is taken as an alert of a possible outbreak which could then activate a more intense monitoring approach with the collection of samples for confirmation by diagnostic testing. To further improve this method, based on the aforementioned work of Silva et al.,[15] Cesar Moura [Iowa State University, Veterinary diagnostic and production animal medicine(VDPAM)] and collaborators developed an online application to collect production data, assess deviations in productivity and alert the users about significant changes in production parameters (via automated email). This represents a virtually no-cost and efficient tool, being particularly useful to screen farms and trigger action plans for early disease detection and therefore, a more effective and efficient management and control of PRRSV outbreaks.

Swine disease reporting system

Among new tools for disease monitoring being developed and explored, there is a new alternative for PRRSV monitoring at the aggregated level. The swine disease reporting system (SDRS), concentrates real-time, aggregated diagnostic tests results from four participating major US veterinary diagnostic laboratories (VDL) (Iowa State University, University of Minnesota, South Dakota State University, and Kansas State University). The SDRS is a user-friendly informatics tool to summarize and report routine pathogen detection from tissues submitted to participating VDLs over time.[16] SDRS collects standardized submission data and results of tests performed on porcine samples. Data is organized and analyzed at an accession ID case level using statistical software. SDRS is a user-friendly tool that provides information regarding swine pathogen and disease detection aggregated from four participating VDLs with the ability to be updated on a near real-time basis.[16] This tool has the ability to keep the swine industry informed regarding disease and pathogens detection. Also, allows users to rapid visualize disease detection results, assess trends, and perform comparisons between the aggregated datasets, from participating VDLs, with users own datasets. Although this new development represents a great tool for swine disease monitoring at the VDL level, it is not appropriate to determine actual prevalence or incidence.

Conclusion

Porcine reproductive and respiratory syndrome monitoring and surveillance systems (PRRS-MOSS) have become a major component in veterinary practice and research, and in recent years, it is prerequisite for effective PRRSV control and elimination programs. Today, management decisions taken within swine production systems affect more pigs, more farms and bigger regions than ever in the past, thus, opportune detection and identification of pathogens and dynamics of disease is critical for the profitability and sustainability of the business. There are different approaches to monitor PRRSV including laboratory diagnostic tests and/or production data and they have different advantages as well as limitations. One of the biggest limitations of ancient monitoring and surveillance protocols have its root in the method applied for its execution: individual pig-based clinical signs and/or blood sampling. This approach does not allow for PRRSV detection under low prevalence thresholds, depending on sample size. The standard monitoring protocol of 30 blood serum samples drawn from due-to-wean pigs, used to declare breeding herds stable for PRRSV[5], for instance, can only detect PRRSV at 10% or higher prevalence level. However, breeding herds attempting PRRSV control/elimination programs, have detected the same PRRSV found in the herd prior the implementation of the control/elimination program, shortly after achieving stability.[7] This constitutes evidence that the virus continues to circulate under specific prevalence thresholds for undetermined periods of time. The need for more sensitive methods is certainly clear. The swine industry has been showing some progress towards the adoption of population based sampling methods for PRRSV-MOSS, the increasing usage of oral fluids for PRRSV diagnostics in the last years and the rapid very well reception of the processing fluids are proof of that.[16] Population-based sampling methods for PRRSV monitoring in swine populations (processing fluids, oral fluids, family oral fluids, and production parameter monitoring) are the next step in the evolution of PRRSV-MOSS, these methods allow us to sample a higher number of pigs and some of them possess a high potential for sample pooling, adding more practicality to the method and lowering diagnostic costs, without loss of herd sensitivity. Evolving PRRSV-MOSS will continue to help to early detect outbreaks as well as new virus introductions to endemic herds, it could help to minimize production and economical losses due to PRRSV by making us more accurate in managing interventions according to the level of virus circulation (detection at near-zero prevalence) or by confirmation of its absence from the herd.

References

1. Keffaber KK. Reproductive failure of unknown etiology. *Am Assoc Swine Pract Newsl.* 1989;1(2):1-10.

2. Holtkamp DJ, Kliebenstein JB, Neumann EJ, et al. Assessment of the economic impact of porcine reproductive and respiratory syndrome virus on United States pork producers. *J Swine Heal Prod.* 2013;21(2):72-84. %3CGo.

3. Salman MD. Surveillance tools and strategies for animal diseases in a shifting climate context. *Anim Health Res Rev.* 2013;14(2):147-150. doi:10.1017/S1466252313000121.

4. *Dr . Bob Morrison's Swine Health Monitoring Project.*; 2018.

5. Holtkamp D, Polson D, Torremorell M, et al. Terminology for classifying swine herds by porcine reproductive and respiratory syndrome virus status. *J Swine Heal Prod.* 19(1):44-56. color:

6. Lopez WA, Angulo J, Zimmerman JJ, Linhares DCL. Porcine reproductive and respiratory syndrome monitoring in breeding herds using processing fluids. *J Swine Heal Prod.* 2018;26(3).

7. Linhares D. Evaluation of Immune Management Strategies to Control and Eliminate Porcine Reproductive and Respiratory Syndrome Virus (PRRSv). 2013.

8. Kittawornrat A, Panyasing Y, Goodell C, et al. Porcine reproductive and respiratory syndrome virus (PRRSV) surveillance using pre-weaning oral fluid samples detects circulation of wild-type PRRSV. *Vet Microbiol.* 2014;168(2-4):331-339. doi:10.1016/j.vetmic.2013.11.035.

9. Ramirez A, Wang C, Prickett JR, et al. Efficient surveillance of pig populations using oral fluids. *Prev Vet Med.* 2012;104(3-4):292-300. doi:10.1016/j.prevetmed.2011.11.008.

10. Prickett JR, Zimmerman JJ. The development of oral fluid-based diagnostics and applications in veterinary medicine. *Anim Heal Res Rev.* 2010;11(2):207-216.

11. López L. WA, Linhares D. Processing fluids sensitivity and specificity for RT-PCR assays to detect PRRS virus under field conditions. In: *Allen D. Leman Swine Conference.* Minneapolis; 2018.

12. Almeida M, Zimmerman J, Holtkamp D, et al. Comparative evaluation of family oral fluids and piglet sera to detect PRRSV RNA by PCR. In: *AASV 49th Annual Meeting.* San Diego, California; 2018.

13. Prickett J, Simer R, Christopher-Hennings J, Yoon KJ, Evans RB, Zimmerman JJ. Detection of Porcine reproductive and respiratory syndrome virus infection in porcine oral fluid samples: a longitudinal study under experimental conditions. *J Vet diagnostic Investig.* 2008;20(2):156-163.

14. Almeida M, Zimmerman J, Holtkamp D, Rademacher C, Linhares D. Maximizing herd sensitivity to detect PRRSV in due-to-wean pigs using family oral fluids sampling. In: *James D. McKean Swine Disease Conference.* Ames, Iowa; 2017.

15. Silva GS, Schwartz M, Morrison RB, Linhares DCL. Monitoring breeding herd production data to detect PRRSV outbreaks. *Prev Vet Med.* 2017;148 (October):89-93. doi:10.1016/j.prevetmed.2017.10.012.

16. Trevisan G, Linhares L, Crim B, et al. Swine Disease Reporting System: Real-Time, Aggregated Results from Major U.S. Veterinary Diagnostic Laboratories. In: ; 2018:43-44.

Field applications for population-based sampling in breeding herds

Giovani Trevisan[1]; Rebecca Robbins[2]; Jose Angulo[3]; Will Lopez[1]; Marcelo Almeida[1]; Nubia Macedo[1]; Maria Eizmendi[1]; Daniel Linhares[1]
[1]Iowa State University; [2]Seaboard Foods; [3]Zoetis, Parsippany, New Jersey

Introduction

Porcine reproductive and respiratory syndrome (PRRS) is a major clinical and economically important disease affecting North America swine herds. In United States it was first described at the end of the 1980's as the mystery swine disease.[1] Annual economic losses due to PRRS virus (PRRSV) in United States has been estimated in $664 million.[2] Due to the importance of this virus, the AASV published a guideline for classifying breeding herds in an effort to help veterinarians and producers to standardize the terminology on PRRS monitoring programs. These classification criteria relied on the presence of clinical signs, anti-PRRSV antibody detection, and detection of viral RNA in serum samples from 30 due-to-wen piglets, and incoming gilts.[3] The sample size of 30 serum samples was based on providing 95% confidence to detect at least one positive sample when prevalence is at least 10% prevalence.[4] In an effort to enhance PRRS surveillance, population-level sampling methods to detect PRRSV have been developed. In 2008 the detection of PRRSV on oral fluid was described[5] and guidelines for collection and validation of this sample to monitor herds to PRRSV was proposed.[6] In 2017, a novel specimen described as processing fluid (PF) was reported as a useful tool to monitor PRRSV in 3-5 days old piglets.[7,8] In 2018 an new approach for PRRSV detection in family oral fluids (FOF) from due-to-wean piglets was first described and added to the portfolio to surveil breeding herds.[9] These new specimens allowed sampling more piglets, more frequently, and with great confidence to detect PRRSV at low prevalence. Because of these advantages, PF-based monitoring made it practical and affordable to determine PRRSV status of whole batches of pigs (e.g. piglets from a whole farrowing room) for PRRSV status at the time of castration and tail docking. Data from the Swine Disease Reporting System (SDRS) demonstrated a significant increase in the percentage of cases using oral fluid from 2010 to 2018, and a rapidly increase in the submissions of PF samples in 2018. In the opposite way, the overall percentage of submission using of serum samples has decreased in the same period.[10] The purpose of this work is to summarize the knowledge obtained from field-based projects using PF and OF to monitor PRRSV on a population of piglets, and assess the relationship between diagnostic results and growth performance parameters downstream.

Results and discussion

Obtaining PF from 3-5 days old piglets from whole farrowing rooms is a practical, time-efficient, and affordable way to monitor PRRSV in breeding herds,[11] as it allows covering a great representation of piglets within rooms. One potential application for this specimen is to determine the PRRSV status at processing time and use this to estimate time to produce PRRSV-negative piglets at processing time, which may be associated with whole-herd time-to-stability (TTS). A field evaluation was conducted in 29 sow herds operating under 4-week batch farrowing system that became infected by PRRSV and adopted load-close-expose with PRRS Modified Live Virus (MLV) vaccine as a strategy to achieve stability and eliminate the virus.[12] Day zero was defined as the day of whole-herd exposure to PRRS MLV vaccine. Stability was achieved when PRRSV-negative piglets at processing time and weaning age were consistently produced. The herds were monitored using PF until 2 consecutive piglet batches had negative results for PRRSV RNA by rRT-PCR. Then, 30 due-to-wean serum samples were tested for PRRSV RNA by rRT-PCR. Naïve gilts were introduced after a negative result on due-to-wean serum samples. Sera from incoming naïve gilts were tested by ELISA for anti-PRRSV antibodies. Twenty eight of 29 (96.55%) of the herds achieved TTS at the end of the study. The TTS ranged from 18 to 55 weeks with a median of 27 weeks and interquartile range (25th to 75th percentile) of 25 to 30 weeks. Blood collection in due-to-wen piglets was performed at 28 farms, and 26 (92.85%) obtained negative results on the first collection for PRRSV rRT-PCR. The overall TTS was similar to previous work performed in breeding herds with farrowing barns operating in a continuous flow using serum samples of due-to-wean piglets.[13] The major difference found in the present study was the much narrower interquartile range for TTS. This work showed that PF results can potentially be used to determine TTS after PRRSV exposure in breeding herds operated in the batch-farrowing system. Also, based on results from this work one could hypothesize that the probability of achieving PRRSV-negative piglets is higher in herds implementing batch system compared to those operated in continuous breed-farrow flow.

Family oral fluids (FOF) sampling is a recently described strategy to monitor the due-to-wean pig population. In this technique, ropes are hanged in front of the sow crate in farrowing rooms, where both the piglets and the sow can chew on it, and thus the FOF is an aggregated sample from the dam and the piglets.[9] Sows consistently chew on the rope, and piglets follow the dam's behavior interacting with the rope. Almeida et al., reported a 94% success rate to obtain FOF on 750 attempts in 6 different farms.[9] To detect PRRSV at 2% prevalence in a population of 400 piglets (one farrowing room) it is necessary to sample 125 individuals.[4] With FOF sampling, sampling 10 litters would provide a similar probability of detection. PF and FOF are alternatives to monitor PRRSV at low (< 5 %) prevalence. As demonstrated recently by Almeida et al.[9] FOF can also be used to estimate PRRSV prevalence on an individual level from the due-to-wean piglet population.

Producers have been collecting PF from multiple rooms, and testing them by pools of 5 for PRRSV by rRT-PCR. This is a reasonable strategy to sample most piglets produced within a week. This information can be used to 'tag' (classify) piglet batches according to disease status. This procedure can potentially be applied to different pathogens including PRRSV and PCV2. Having the ability to know the batch status before the group reaches the nursery can provide a competitive advantage for health management strategies within production systems. We are conducting an on-farm ongoing project, which is looking for the relationship between PRRSV RNA detection at processing time, and the subsequent nursery closeout mortality results.[14] A total of 126 cohorts have been enrolled at this moment. Each cohort was classified based on PRRSV rRT-PCR Ct values as low, medium, high, or negative when Ct value was < 27, between 27 and 34, between 34 and 37, or > 37 respectively. Pig batches having low Ct had significantly higher mortality than all other groups (Tukey adjusted P-value < 0.05). Groups from the low Ct had 2.61 percentage points higher mortality than those of medium Ct; low Ct group had 3.17 percentage points higher mortality than those with high Ct; and low Ct group had 3.18 percentage points higher mortality than those of negative PCR results. There was no statistical difference among other comparison groups: medium Ct vs. high Ct = 0.56 percentage points higher mortality, medium Ct vs. negative = 0.57 percentage points higher mortality, and high Ct vs. negative = 0.01 percentage points higher mortality. A practical implication of these results would be to avoid comingling batches having significantly different Ct values (low, medium, high, negative) from PF-based monitoring.

Final comments

The ability to monitor breeding herds, and piglet batches for low prevalence PRRSV infection have significantly improved with the adoption of PF and FOF sampling. Adoption of PF by the swine industry had rapidly increased as it became incredibly practical to monitor herds under very low prevalence. Helping swine practitioners and producers to better understand PRRSV circulation dynamics in breeding herds. FOF-based sampling is another recent approach, and continues to show an increase in use to efficiently monitor piglets close to weaning time. One of the applications for PF- and FOF-based sampling is the ability to efficiently monitor disease status of batches of piglets (i.e. whole room, or whole week). This allows classifying breeding herds and piglets batches for disease status, as well as estimate predictions of growth performance downstream.

References

1. Hill H. Overview and history of mystery swine disease (swine infertility/respiratory syndrome). In Proceedings of the Mystery Swine Disease Committee Meeting, Livestock Conservation Institute. Denver, Colorado1990. p. 29-31.

2. Holtkamp D, Kliebenstein J, Neumann E, Zimmerman J, Rotto H, Yoder T, et al. Assessment of the economic impact of porcine reproductive and respiratory syndrome virus on United States pork producers. *Journal of Swine Health and Production*. 2013;21(2):72-84.

3. Holtkamp D, Polson D, Torremorell M. Terminology for classifying swine herds by porcine reproductive and respiratory syndrome virus status. *Journal of Swine Health and Production*. 2011;19(1):44-56.

4. Chase C, Polson D, editors. Sampling the herd: when is enough? 32nd Annual American Association of Swine Practitioners Meeting; 2000; Nashville, Tennessee.

5. Prickett J, Simer R, Christopher-Hennings J, Yoon KJ, Evans RB, Zimmerman JJ. Detection of Porcine reproductive and respiratory syndrome virus infection in porcine oral fluid samples: a longitudinal study under experimental conditions. J Vet Diagn Invest. 2008;20(2):156-63. doi: 10.1177/104063870802000203. PubMed PMID: 18319427.

6. Prickett JR, Kim W, Simer R, Yoon K-J, Zimmerman J. Oral-fluid samples for surveillance of commercial growing pigs for porcine reproductive and respiratory syndrome virus and porcine circovirus type 2 infections. *Journal of Swine Health and Production*. 2008;16(2):86-91.

7. Lopez W, Angulo J, Zimmerman J, Linhares D. Porcine reproductive and respiratory syndrome monitoring in breeding herds using processing fluids. *Journal of Swine Health and Production*. 2018;26(3):146-50.

8. Lopez W, Trevisan G, Zimmerman J, Harmon K, Gauger P, Gimenez-Lirola L, et al. Monitor Heard for PRRS using procesing-fluids samples. *National Hog Farmer* Nov. 06; 2017.

9. Almeida M, Allison G, Silva GS, Holtkamp D, Zimmerman J, Linhares D. Field Studies on Family Oral Fluids for PRRSv Surveillance. 2018 ISU James D McKean Swine Disease Conference, 2018; Ames, IA.

10. Trevisan G. Swine Disease Reporting System. Phase 1 Porcine Reproductive and Respiratory Syndrome Virus - PRRSv. Available at: `www.powerbi.com` username: `sdrs@iastate.edu` password: Bacon 100 APP PRRSv ed2018.

11. Vilalta C, Sanhueza J, Alvarez J, Murray D, Torremorell M, Corzo C, et al. Use of processing fluids and serum samples to characterize porcine reproductive and respiratory syndrome virus dynamics in 3 day-old pigs. Vet Microbiol. 2018;225:149-56. Epub 2018/09/17. doi: 10.1016/j.vetmic.2018.09.006. PubMed PMID: 30293648.

12. Trevisan G, Jablonski E, Angulo J, Lopez W, Linhares D. Use of processing fluid samples for longitudinal monitoring of PRRS virus in herds undergoing virus elimination. Submitted 2018.

13. Linhares DC, Cano JP, Torremorell M, Morrison RB. Comparison of time to PRRSv-stability and production losses between two exposure programs to control PRRSv in sow herds. Prev Vet Med. 2014;116(1-2):111-9. Epub 2014/06/02. doi: 10.1016/j.prevetmed.2014.05.010. PubMed PMID: 24931129.

14. Trevisan G, Robbins R, Angulo J, Dufresne L, Linhares D. Relation of nursery closeout mortality and PRRSV status at rRT-PCR in processing fluid - Research in progress. 2018.

Diagnostic methods for effective *Mycoplasma hyopneumoniae* control and elimination

Maria Pieters, DVM, PhD
Department of Veterinary Population Medicine, College of Veterinary Medicine,
University of Minnesota, St. Paul, Minnesota

Introduction

Diagnostics of *Mycoplasma hyopneumoniae* (*M hyopneumoniae*) have improved over the years; however, they are far from perfect. In the meanwhile, diagnosticians, swine practitioners, and producers face challenges when attempting to make decisions based on diagnostic schemes that result difficult to interpret.

Recent efforts have generated information regarding sample type, tests, and timing that appears to be more appropriate for specific diagnostic scenarios. Nevertheless, it is important to keep in mind that tools and assays should be carefully selected based on the specific diagnostic objective.

In this summary, three specific situations in which a selection of samples and tests for *M hyopneumoniae* detection is not standard are presented, along with the pros and cons of the selected diagnostic protocol.

Various scenarios

Naïve replacement gilt introduction into negative herds

On a regular basis, replacement gilts are introduced to breed-to-wean farms. These introductions are needed in order to improve the genetic quality of the herd, but they doubtlessly represent a potential source for introduction of *M hyopneumoniae* (or other pathogens) into high health herds. Thus, close diagnostic monitoring of gilts to be introduced is necessary for all farms. The most commonly applied protocol for sampling and testing for *M hyopneumoniae* implies blood sample collection for serum separation, which later is tested by ELISA for antibody detection. Nevertheless, it is important to remember that development of antibodies against *M hyopneumoniae* in infected pigs can be a very slow process, sometimes taking weeks to occur. Moreover, vaccination against *M hyopneumoniae* can induce antibodies, which would be equally detected with the commercially available ELISA tests. Therefore, understanding the limitations of the sampling protocol, and in light of recent information showing that early detection of *M hyopneumoniae* can be obtained using different diagnostic approaches and sampling sites, other strategies are applied on gilts prior to entering the sow farm. PCR testing in samples obtained from the lower respiratory tract are becoming more and more common, based on the high sensitivity obtained in those type of samples, which are obtained from live pigs. Thus, either laryngeal swabs or deep tracheal catheters assayed by real-time PCR are becoming more generalized when sampling gilts prior to entering *M hyopneumoniae* negative sow farms. This diagnostic approach provides greater accuracy, compared with antibody monitoring alone. However, the sensitivity still depends on many other aspects such as sample size, sample type, sample dilution (in pooling situations), etc. One drawback can be the cost of testing, which can be significantly greater when using PCR, however pooling samples can aid decreasing costs, while still keeping sensitivity high. Side-to-side sample and testing comparisons studies, incorporating economics into the scenario, would help practitioners in the decision making when dealing with this situations.

Confirming gilt exposure prior to herd closure

Herd closure and medication is a widely used protocol for eradication of *M hyopneumoniae* in sow farms. It is understood that a significant aspect for successful Mycoplasma elimination resides in achieving solid exposure of the population to *M hyopneumoniae* prior to herd closure, assuring that no susceptible individuals linger in the herd, while a strong immunity is built. Thus, establishing Day 0 has become extremely important in timing the start of the closure and increasing success chances. However, diagnostic protocols for detection of infection in exposed pigs are more complicated that it may appear. In populations that are initially naïve and in which vaccination has not been employed, detection of antibodies can be a good option. Nevertheless, most herds are vaccinated against *M hyopneumoniae*, leaving the ELISA assay use out of the question. Real-time PCR testing in laryngeal swabs or deep tracheal catheters combined with PCR have shown to be sensitive for *M hyopneumoniae* detection, but sampling time can be critical based on exposure. The pathogen exposure method significantly influences the timing and likelihood of *M hyopneumoniae* detection post-exposure. For example, seeder-to-naïve exposure will take longer time to achieve infection in susceptible pigs, compared with direct exposure via intra-tracheal inoculation or fogging.

Detection of genetic material

In the two previous scenarios, the use of PCR testing in combination with samples obtained from the lower respiratory tract has been suggested in order to achieve high sensitivity. *Mycoplasma hyopneumoniae* genetic material can be detected with a high degree of accuracy. Real-time PCR can be potentially used for quantification, for example when fitting a standard curve based on CCU/mL. However, absolute quantification can be hard to obtain, mainly due to the lack of homogeneous distribution of the pathogen in the respiratory tract and the intrinsically varied nature of the lung tissue. Another factor to consider is that genetic material detection does not imply that the pathogen is viable. Thus, a PCR positive result has to be carefully interpreted when inferring potential for shedding and transmission, especially after the use of agents that can contribute to pathogen detachment from the respiratory epithelium.

Summary

Basic and applied research are still needed in order to fine-tune the use of existing diagnostic tools to achieve high *M hyopneumoniae* detection accuracy. All assays and samples types may not be applicable and all situations and specific scenarios will require the use of certain protocols for better results. In recent years, control and eradication of *M hyopneumoniae* have become a priority in swine production and a strong diagnostic support has allowed producers move in the eradication direction. However, a comprehensive understanding of the ecology and epidemiology of this respiratory bacterium is necessary for diagnostic purposes.

Mycoplasma hyopneumoniae boar stud elimination experience

R. W. Thompson[1], DVM, MS; Lucina Galina[2], DVM, PhD, Jean Paul Cano[1], DVM, PhD; Maria Jose Clavijo[1,3], DVM, PhD
[1]PIC Health Assurance, Hendersonville, Tennessee; [2]Zoetis, Florham Park, New Jersey; [3]Iowa State University

Introduction

Expanded epidemiological knowledge of *M. hyopneumoniae* (Mhp) coupled with improved diagnostic protocols and a quick return on investment, have led to an increase in successful Mhp eliminations industry wide. While successful Mhp eliminations from sow herds have been extensively documented, few boar stud eliminations have been reported.[1] Frequent introductions of boars (~ 8 weeks) into studs is critical for continuous genetic improvement. However, this aspect, in addition to limited stud space, makes average Mhp herd closures (~35 weeks) challenging to implement without incurring significant genetic index losses. The other challenge with endemically infected populations is the lack of whole herd exposure and the frequent introduction of Mhp negative boars, impeding establishment of day zero. The objective of this document is to share a recent successful Mhp elimination from an endemically infected boar stud.

Elimination phases

Site

The 700-head stud located in KY and was stocked Mhp positive in the late 90s. The site consists of two interconnected barns, housing 350 boars in each, with a central office and detached laboratory. Every 8 weeks a new group of Mhp negative boars entered the stud. Prior to entry, boars were vaccinated with minimal clinical signs observed post entry.

Baseline diagnostics

With the objective of measuring the exposure level prior to closure, a baseline testing was carried out. Tracheal swabs and serum samples were collected from three different entry groups (25/each). PCR results showed 15 weeks post stud entry was 64% PCR positive while two older groups were 4% and 8.3% positive at 23 and 30 weeks post entry, respectively, suggesting that younger boars were getting exposed and the older boars were likely decreasing shedding. All boars were ELISA positive.

Achieving day zero

To accomplish whole herd Mhp exposure, 20 negative by-product pigs were introduced into the positive stud population to ensure exposure and serve as seeders. Four weeks later, the infected and clinical seeders were introduced into the quarantine to start acclimation of newly arriving negative boars, last group to be introduced before closure. Allowing the seeders to walk the alleyways in quarantine promoted transmission and infection of negative boars. No vaccine or prophylactic antibiotics were administered to either the sentinels or the last group of incoming boars to maximize transmission and measure exposure accurately. The boars in quarantine showed evident clinical signs and all tested 100% ELISA positive by 11 weeks post-exposure with seeders. Using this information Day Zero was estimated to have occurred 6 weeks post exposure.

Boar segregation

With the objective of segregating boars by timing of infection, all low index and non-working boars were culled from the stud and the remaining older boars were moved into Barn 2. New positive arrivals plus the most recent groups stocked Barn 1. Skipping one quarantine group lengthened our closure period with the first group of Mhp negative boars arriving to quarantine 16 weeks from Day 0. Washing and disinfection of the entire site, barns, alleyways and office along with removal of any porous material prior to negative entry was completed. To assure older boars were not shedding five weeks before negative boars entered 60 tracheal bronchial swabs were collected in Barn 2, oldest population on site with all negative PCR results. A bench system with a change of boots and coveralls separated employees to reduce the possibility of a fomite transmission back from Barn 1. Culls and older non-working boars were removed from both barns prior to entry of each negative group. As space required for new negative boars previously positive boars went back to Barn 1, eventually Barn 2 filled with negatives.

Vaccination/medication

One whole stud vaccination was given with Respisure One to booster immunity prior to negative boars entering Barn 2. The entire stud was injected with Draxxin prior to arrival of first negative boars to site. Pulmotil at 365 Gm/T was fed to the site starting one week prior to entry of the negative boars to Barn 2 for two weeks.

Validation

Before our first group of negatives entered Barn 1 (younger barn) an additional 60 tracheal bronchial swabs were collected from the youngest boars (last exposed population) in that barn, all of which were PCR negative. The first Mhp negative boars entered the stud site 175 days after Day 0. It took until Week 38 of 2018 to introduce the first negative boars to the last positive barn, 336 days from Day 0. Weekly serology in the stud have remained negative in all negative boars entered. Week 45 of 2018 the stud was officially declared negative.

Reference

1. Holst S, Yeske P, et al. 2015. Elimination of *Mycoplasma hyopneumoniae* from breed-to-wean farms: A review of current protocols with emphasis on herd closure and medication. *Journal of Swine Health and Production*. 2015 Nov 1;23(6):321-30.

What if a *Mycoplasma hyopneumoniae* naïve herd breaks?

Dyneah M. Classen, DVM
Carthage Veterinary Service, Ltd.

Introduction

Mycoplasma hyopneumoniae outbreak in a naïve breeding herd and downstream flows requires immediate action. For the purpose of this discussion naïve herd is defined as a herd that is PCR and ELISA negative for *Mycoplasma hyopneumoniae*. No killed vaccines are used prior to the outbreak. This paper will focus on control measures immediately following a *Mycoplasma hyopneumoniae* outbreak. The decision on *Mycoplasma hyopneumoniae* elimination can be made later.

Discussion

The veterinarian is often called upon to examine a herd with a clinical cough and/or increased mortality. The veterinarian short differential list for these clinical signs is porcine reproductive and respiratory syndrome virus (PRRS), Influenza A virus (IAV-S), or *Mycoplasma hyopneumoniae*. Other pathogens in the porcine respiratory disease complex (PRDC) such as *Streptococcus suis* and *Haemophilus parasuis* may be involved but typically won't cause the acute clinical cough and/or increased mortality, especially in a mature breeding herd.

The first action is to collect the necessary samples to correctly diagnosis the causative agent. The gold standard for any of the differentials above would be tissues from affected animals. If euthanizing an animal and collecting samples is not an option collect oral fluids for PCR testing and serum for ELISA testing. I would also encourage veterinarians to contact researchers, like Marie Pieters, DVM, PhD and Maria Clavijo, DVM. Sending them samples will help increase the knowledge base on *M hyopneumoniae*. The ELISA testing can be used later to determine the timing of the outbreak in epidemiological follow up work.

The gross lung lesions are not pathognomonic for *M hyopneumoniae*. Lung lesions are predominately well demarcated and cranioventral.[1]

If the herd in question is a multiplication sow herd, all replacement gilt shipments should be stopped and gilts moved to a biosecure nursery/finisher. This will protect the downstream sow herds from receiving clinical sick gilts and give the source herd ample time to determine the causative agent. If the herd is a commercial sow herd, notify those buying pigs immediately so the downstream flow can make changes. These changes may include the addition of an antimicrobial injection on arrival or in the water. The veterinarian may need to notify them of a change in health status and call back later to confirm the causative agent. Even if the downstream sow farms are mycoplasma positive or mycoplasma stable there may be differences in the *M hyopneumoniae* strain types resulting in clinical disease post weaning.[2] The authors' experience, not all mycoplasma stable herds behave the same when sent gilts originating from a multiplication herd experiencing a mycoplasma outbreak. Moving replacement gilts to a biosecure nursery/finisher can allow for clinical sick gilts to recover before the stress of acclimation at the receiving sow farm.

The clinical history, clinical signs of a non-productive cough, increased mortality, and gross necropsy lesions will start to point you in the right direction for a diagnosis. However, with the fast turnaround times of the veterinary diagnostic labs, mass action is often delayed until preliminary results are returned.

While waiting for diagnostics to return the on-farm teams will be trying to save sick sows. This would include individual injectable treatment with an antimicrobial labeled for *M hyopneumoniae* and an anti-inflammatory. If the suckling pigs are clinically ill, remember they have not been vaccinated and do not have any maternal antibodies to *M hyopneumoniae*. The timing may vary between sow farms based on when clinical signs and coughing pigs are occurring. The minimum would be to add an injectable antimicrobial at weaning. These piglets are about to undergo the stressful event of weaning. This phase of production will benefit from the use of an injectable antimicrobial labeled for *M hyopneumoniae*. This is to keep pigs alive and prevent the organism from developing life threating pneumonia.

Once the diagnostics are completed and the diagnosis of *M hyopneumoniae* is returned, now it's time to jump into mass action.

The sow herd and wean pigs should be vaccinated using a conventional killed for *M hyopneumoniae*. There is evidence that conventional, killed *M hyopneumoniae* bacterins did not alter the transmission rates of *M hyopneumoniae* in populations.[3,4] An *M. hyopneumoniae* bacterin may minimize or prevent clinical signs of *M hyopneumoniae* in pigs,[5] but it doesn't prevent infection or reduce the number of infected pigs in a herd setting.[6] Older pigs at the nursery/finisher sites that will be exposed to wean pigs should also be vaccinated for *M hyopneumoniae*. A favorite saying of the author is "vaccines don't prevent infection, just minimize the risk of dying."

Regardless of the sow herd's decision maker's (veterinarian, owner, pig buyers) thought process on elimination, a mass antimicrobial labeled for *M hyopneumoniae* should be administered to the sow herd. This could be done in either the feed or water depending on the farm set up. Medication programs should also be evaluated at the nursery/finishers and make sure *M hyopneumoniae* control is in place.

This was something learned from the author's experience. Failing to start antibiotics early in the outbreak resulted in increased sow mortality due to respiratory deaths (Figure 1). Figure 1 represents the sow mortality 20 weeks before and after the week that clinical signs were observed. Sows necropsied during the weeks following the outbreak had severe pneumoniae. The concern is that adding an antimicrobial treatment to the sows would stop the spread of *M hyopneumoniae* to herd mates and prevent 100% of the sows becoming exposed. This sow herd had made the decision that Mycoplasma elimination was necessary for their production. Antimicrobials should be started and clinically coughing animals can be moved around the barns to aid exposure.

The replacement gilt customers and herd veterinarians from a multiplication sow farm with a *M hyopneumoniae* outbreak now have some decisions to make. The commercial sow farm may be *M hyopneumoniae* positive and acclimating replacement gilts. The research community is learning more about *M hyopneumoniae* strain differences.[6] The author would advise a *M. hyopneumoniae* positive/stable commercial sow farm to have the multiplication farm hold the gilts until they are non-clinical for *M. hyopneumoniae*.

The sow herd decision-makers can now turn attention to *Mycoplasma* stabilization or *Mycoplasma* elimination projects.

Conclusion

Mycoplasma hyopneumoniae outbreaks require immediate action. Once diagnostics has confirmed *M hyopneumoniae,* killed commercial vaccination with a *M hyopneumoniae* bacterin and an antimicrobial program for all phases of production is warranted immediately. The decision to continue *M hyopneumoniae* vaccination depends on the long term goal and monitoring for that herd.

References

1. Schwartz KJ, et al. Swine Disease Manual Third Edition.
2. Pieters M, et al. Tools for molecular characterization of *Mycoplasma hyopneumoniae.* Allen D. Leman Swine Conference; 2015.
3. Pieters M, et al. An experimental model to evaluate *Mycoplasma hyopneumoniae* transmission from asymptomatic carriers to unvaccinated and vaccinated sentinel pigs. Can J Vet Res 2010; 74:157-160.
4. Meyns T, et al. Comparison of transmission of *Mycoplasma hyopneumoniae* in vaccinated and non-vaccinated populations. Vaccine 2006; 24:7081-7086.
5. Okada M, et al. Evaluation of *Mycoplasma hyopneumoniae* inactivated vaccine in pigs under field conditions. J Vet Med Sci 1999;61:1131-135.
6. Lowe J, *Mycoplasma hyopneumoniae*; 'Fix the gilts, fix Mycoplasma. Pig Health Today. Posted on September 25, 2017.

Figure 1: Sow mortality 20 weeks before and after a clinical *M. hyopneumoniae* outbreak.

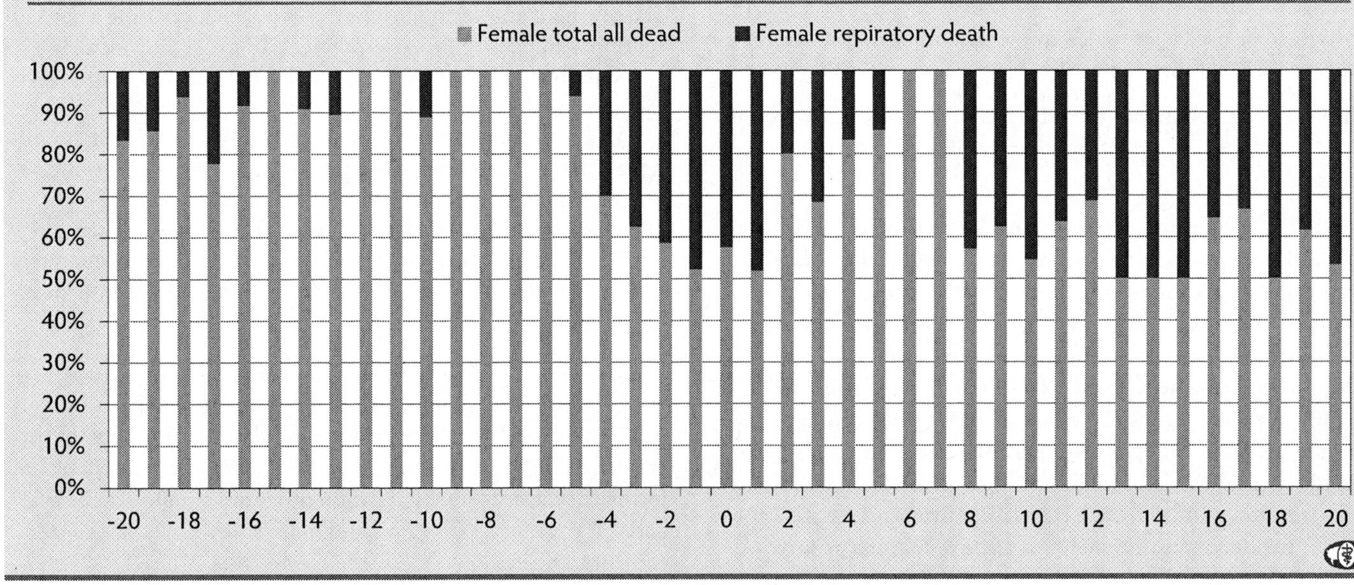

When emotions run high

Mary Battrell, DVM, MS
Smithfield Hog Production

When contacted and asked to speak on the topic of how I prepared farm staff emotionally for porcine epidemic diarrhea virus (PEDV) breaks and the exposure and clean up that follows, the first thought I had was, "You have got to be kidding me. Who did I upset?" The more I thought about it, the more I realized that someone was far more insightful than me. I have always said "the pigs are easy, it's the people that make things difficult." Although most of us chose this profession due to our love for animals, we are very much in the people business. We can develop and lay out a perfectly devised plan, but if we do not have buy in from the farm staff it is destined for failure.

The first PEDV case in North Carolina was June 21, 2013, roughly 50 days after it was identified on US soil. This lag time allowed us the opportunity to educate our production team on the clinical signs and expected economic consequences of a PEDV break. We had the opportunity to review existing biosecurity procedures and make necessary improvements. Farms tightened up on their biosecurity, and a few sow farms added drive-through disinfectant stations. Our efforts were in vain as 89% of the sow farms in my area of responsibility have experienced and recovered from a PEDV break.

PEDV brought with it a wide range of emotional responses, sorrow being the most common. Sorrow is defined as a feeling of deep distress caused by loss, disappointment, or other misfortunes suffered by oneself or others. Animal care givers, particularly those working in the farrowing house were full of sorrow during a PEDV break. It was not uncommon to be greeted by owners or workers at the farm counter with tears in their eyes. Many of those teary-eyed employees had worked in this industry for years. They care deeply for the pigs and hated to see what this disease did to them. Owners were sorry that their employees would not be seeing a bonus check for some time, and there was real concern that those employees might seek other employment opportunities. I knew people were hurting, but the local general practitioner really put it into perspective for me when he made the correlation with the timing of PEDV and the increased number of people who had recently been prescribed antidepressants. There are PRRS strains that are more devastating than PEDV. There are natural disasters, and foreign animal diseases that can put us to the test. During these challenging times in our industry, it is important to be prepared to suggest a mental health professional for those who are deeply distraught and in need of help.

The approach I have always taken is to listen to people. Give them the time to explain what they are experiencing and how they feel. Often you will hear the same story over and over. You might pick up some innovative ideas, but listening is mostly for them. They need the opportunity to get things off their chest. After giving them time to talk, it is then important to turn the conversation toward focusing on the solution and moving the farm forward. Tell them what we are going to accomplish today. Explain the exposure process in detail and explain why things are being done a certain way. Assign them specific task to assist in the exposure process and task for the days to follow. Help paint a picture of what the farm health should look like in the weeks and months to follow. Provide a timeline for improvement so they understand that things will not always be as bad as they are today. Give them specific tasks and timelines in writing. There is a lot of information to cover and their emotions make it that much more difficult to listen and focus. Written instructions allow them to review the information later when they have had time to calm down and ask questions as needed. Volunteer for the nastiest job. I ask employees to bring the pigs to me and my assistant (which is usually the production specialist). We take it from there. If anyone appears emotionally distraught, I ask the owner if they can go home for the day, or I might even suggest that they allow that person to take some vacation.

Encourage the farm manager or owner to look for potential sources of disease introduction before you leave that day. Explain the incubation period and ask them to list anything or anyone that has come near that farm within that incubation period and beyond. They may come up with the most probable cause. Hopefully, they will be willing to share that information, so you can help protect other farms. Either way, it is important to identify a possible cause, so they can focus efforts to avoid that risk in the future. It is not an appropriate time to point fingers or lay blame. It is time to focus on healing.

Frustration is the feeling of being upset or annoyed, especially because of inability to change or achieve something. This emotion kicks into high gear when farms do not meet the expected timelines for PEDV cleanup. This is an emotion that I am all too familiar with! Farm owners or workers get extremely frustrated when PEDV drags on. In some circumstances they put the blame on their veterinarian, their production specialist, the company, anyone but themselves. This is where those written instructions/protocols came in very handy. Walk through the farm together and review each bullet point. They often follow those instructions that are easily implemented and skip the more difficult ones. If deemed necessary, we added more detail to the instructions. We would change the flow of people through the farm or assign each employee a room or barn they were responsible for. The greater their involvement in finding the solution, the more successful the outcome.

Unfortunately, we have all experienced anger at some point. People were angry that it entered the United States in the first place, that our biosecurity recommendations did not protect our farms, that there was no available vaccine or treatment that was proven to be 100% effective, and that there was a shortage of pigs. As swine practitioners we cannot prevent every disease, solve every problem, or save every pig. We can listen, learn, teach, lead and coach others toward a resolution. It is trying at times. Dealing with sick pigs and emotional people does not make for a winning combination. You must truly love both to find the strength to see it through.

Persistence, control, and elimination of PEDV

L. Batista, DVM, PhD
Batista & Asociados, Mexico/Canada

Introduction

Porcine epidemic diarrhea (PED) virus is highly contagious with low infection doses. It causes an acute and highly contagious enteric disease characterized by severe enteritis, vomiting, watery diarrhea and high mortality in suckling pigs. Presently, there is no cure, except for vaccines that can enhance lactogenic immunity, therefore strict biosecurity protocols are the best way to prevent, and control the PED virus from entering or from spreading in, and to other farms. A very costly PED epidemic began in the United States of America on May of 2013, and subsequently spread to Canada and Latin-American.

Lessons learned[1-3]

1. Infection sources
   a. Infected swine.
   b. Infected transport,
   c. Area spread/neighbors.[4]
   d. Pig movement.
   e. Infection status of source.
   f. Personnel.
   g. Infected feed or material.[5]
   h. Infected facilities.
   i. Farm flow.
   j. Animal disposal and pest control.
   k. Infected manure.

2. PEDv can spread, and survive during hot weather, (Latin-American 2014-2018,[6] and Ontario experience 2017[2]).

3. How to avoid infection (I) or reinfection (RI), both (B)?

a. Working with herd veterinarian (B).
b. Establishing and following strict biosecurity protocols (B).
c. Ensure your replacement gilts are negative (I).
d. Be aware of the PED status of your area (B).
e. After infection, record, and analyze real time movement information to understand transmission in your area (I & RI).
f. Know the Infected farm status (RI).
   - Source
   - Incoming animals
g. Diagnostic monitoring (RI).[7]
h. Herd immunity (RI).
   - Post infection status
   - Feedback
   - Vaccination
   - Both

Persistence

Dictionary definition[8]

1. To continue steadfastly or firmly in some state, purpose, course of action; 2. To last or endure tenaciously.

Scientific definition[9]

1. Persistent infections last for long periods, and occur when the primary infection is not cleared by the adaptive immune response. There is no single mechanism responsible for establishing a persistent infection; a key feature is reduction in host defenses and the ability of the virus to kill cells.

In PED infection, virus persistence in the environment, different surfaces, and manure play a very important role for the control and elimination of the pathogen; therefore for its control and elimination.[10]

After infection in a farm/system, very importantly decide your objective: control or elimination; areas to consider, and then proceed to establish a protocol:

1. Economic impact, invest and return on investment (ROI).
2. Biosecurity protocol (above).
3. Replacement animal status.
4. Quarantine area/period, management.
5. Transport.
6. Feed.

Conclusions

Main goal: ELIMINATION

Disease prevention

- Biosecurity culture.
- Community communication.
- Getting things done.

Disease control

- Above.
- Objectively measuring status, and implemented actions success.

References

1. https://www.aasv.org/aasv%20website/Resources/Diseases/PorcineEpidemicDiarrhea.php.

2. https://www.manitobapork.com/images/producers/pdfs/Lessons-Learned--2017-Manitoba-PED-Outbreak---March-2018.pdf.

3. https://www.vetmed.umn.edu/centers-programs/swine-program/outreach-leman-mshmp/mshmp.

4. Vanderwall, K. 2018. Leman conference.

5. Dee SA, Bauermann FV, Niederwerder MC, Singrey A, Clement T, de Lima M, et al. (2018) Survival of viral pathogens in animal feed ingredients under transboundary shipping models. PLoS ONE 13(3): e0194509. https://doi.org/10.1371/journal.pone.0194509.

6. Personal communication.

7. https://vetmed.iastate.edu/vdpam/research/disease-topics/swine/PED.

8. https://www.dictionary.com/browse/persist.

9. http://www.virology.ws/2009/03/17/persistent-viral-infections/.

10. Sun, R. et al. 2014. Journal of Animal and Veterinary Advances 13(6):410-415 · January 2014.

General immunology: Concepts and key points for swine veterinarians

Brian Aldridge, BVSc, PhD, Dipl ACVIM, MRCVS
Integrated Food Animal Systems, College of Veterinary Medicine,
University of Illinois at Urbana-Champaign, Urbana, Illinois

The immune system as an organ of homeostasis

All life systems are comprised of components that work together to produce normalcy and therefore to ensure health. The health or success of a particular system is determined by its capacity (and by extension, the capacity of its components) to resist or adapt to external or internal perturbations away from normalcy. The capacity to resist perturbation is termed robustness. The ability to adapt to perturbation is termed homeostasis. Homeostasis is the maintenance of constancy within a system and is one of the key biological concepts in health and disease. When we scrutinize or observe health we are really performing an integrated study of normal structure and function. When we are considering a disease state, we are exploring perturbations or abnormalities in structure and function. In some ways a disease state can be viewed as a disturbance or failure to resist or adapt to biophysical or biochemical challenges. These characteristics or qualities can be applied at both a system and component level. For instance, a population is comprised of a system of inter-related individuals; an individual of interrelated organs; an organ of interrelated cells; a cell of interrelated molecular components. The overall success (or health) of any one of these systems depends on the ability of each unit to resist or adapt to any changes in its external environment or internal state. While it is possible to identify a role for almost every organ system in the process of homeostasis, the ultimate agent of mammalian homeostasis is the immune system, particularly since it operates in a continual state of response to external stimuli, particularly at the mucosal surfaces of the gastrointestinal, respiratory, and urogenital tracts. In describing the function of the immune system, many would emphasize its role in antigen-specific defense against infectious pathogens, and yet its primary role is to protect and, in the event of damage, to restore tissue structure, balance and function. It could be said that the endpoint of an immune response is not simply pathogen removal, but culminates in of healing, and that homeostasis fails when this process is incomplete.

Inflammation is our friend

In clinical discussions, a great deal of consideration is given to the adaptive immune responses and the production and activity of pathogen-specific elements of immune defense. It is important to note that the most immune defense, and 99% of pathogens, are dealt with by the innate immune defenses through the sequence of events known as inflammation. This is also the principal means by with healing occurs. Inflammation can be defined as the protective response of an organism to stimulation by invading pathogens or endogenous signals such as damaged cells, which results in the elimination of the initial cause of injury, the clearance of any dead or necrotic cells and eventually tissue repair. The cardinal signs of inflammation were first recognized in the fifth century, and have been classically described as rubor (redness), calor (warmth), dolor (pain), tumor (swelling) and function laesa (loss of function). While the importance of inflammation has been recognized for thousands of years, our understanding of the specific elements, the steps or stages, and the mechanisms that control information are rapidly advancing.

The key soluble components of the inflammatory response are mediators of intra- and inter cellular communication (cytokines and chemokines) and of neutralization and killing (antimicrobial peptides and complement). The key cellular components of inflammation are macrophages, dendritic cells, epithelial cells, neutrophils, natural killer cells and fibroblasts. The action of these cells are related to the presence of several cell-associated molecules that function in the roles of microbial or damage recognition (e.g. toll-like receptors, TLRs), immunochemical signaling, and cell activation and recruitment. Inflammation starts when host cells recognize evolutionarily-conserved structures on pathogens called pathogen-associated molecular patterns (PAMPs) or signals from internal damage or stress called danger-associated molecular patterns (DAMPs). The recognition of PAMPs or DAMPs by host cells comes through the presence of germline-encoded pattern recognition receptors (PRRS), such as TLRs, which are present on many host cell types such as monocytes, macrophages, neutrophils, and dendritic cells. PRRS are also present on cells not classically associated with inflammation, such as epithelial cells, lymphocytes and fibroblasts. Once these receptors are engaged on the host cells, the inflammatory process is initiated through the release of pro-inflammatory cytokine or chemokines. The first function of cytokines can be local and involves the offending agents or the cells that release them (autocrine effect). The second function of cytokines is on cells such as neutrophils and macrophages in the immediate region (paracrine effect). The third function of cytokines occurs when the inflammatory response is of a sufficient magnitude to overwhelm local defenses (endocrine effect). The endocrine effect includes the induction of acute phase proteins in the liver, the stimulation of production of new inflammatory cells in the bone marrow, or systemic and behavioral host responses like fever, fatigue and anorexia. Cytokines also activate endothelial cells on blood vessels.

Physiologically, this helps immune cells reach the site of inflammation; but pathologically, this can also lead to capillary leakage, vasodilation, and low blood pressure or hypotension.

Stages of inflammation

To understand the inflammatory response, it is useful to deconstruct or break down the process, and to elucidate the role of each component at each specific stage of the response, particularly in terms of their hierarchy, organization, interaction, and regulation. From a systems viewpoint, we want to specifically describe and understand the role of each component in relation to one or more key functions:

- Sensing (to detect, monitor and measure)
- Signaling (to transmit information to other components)
- Processing (to quantify, segregate and compare information)
- Capacity to control (in a central role of process coordination, or as part of control network of response feedback loops)
- As an effector (the capacity to respond to the stimulus and achieve a beneficial outcome)

Macrophages and dendritic cells can be viewed as the sensors, conductors or regulators of the inflammatory response. The key signaling components are soluble chemicals such as cytokines and chemokines, which are able to communicate locally or systemically for cell activation and recruitment, and the induction of broader inflammatory mechanisms such as fever and acute phase protein induction. The primary effectors of early stage inflammation, through neutralization and killing, are soluble mediators such as antimicrobial peptides and complement, and neutrophils which neutralize, kill, or eliminate pathogens and remove damaged cells and debris from tissues through phagocytosis. In the later stages of inflammation, the primary mediators are fibroblasts, which activate and enact tissue healing and restoration. With these components participating in the functions of sensing, signaling, control/regulation and effecting, it is useful to provide a bird's eye view of the primary sequence of events during a normal inflammatory response.

1. Recognition, in which the microbes, or tissue damage is detected by receptors on and in immunologically active cells
2. Neutralization, in which local cells produce soluble mediators that function to neutralize and localize potentially injurious agents
3. Amplification, which occurs when the nature or magnitude of microbial challenge or tissue damage exceed the capacity of local defense responses. This involves an increase in local blood flow, the recruitment of additional immune cells and soluble mediators, and the provocation of systemic immune responses (e.g. fever)
4. Killing, in which local and recruited immune cells destroy the inciting agents of pathology
5. Clearance, in which immune cells remove microbiological and cellular debris from the diseased tissue
6. Healing, in which the local pathological processes are arrested and reversed, and cellular and tissue structure and function is regenerated, rejuvenated or restored.

It is important to appreciate that while inflammation is a normal, homeostatic physiological process, when the magnitude or sequence of immune responses becomes disturbed or dysregulated in some way, the immune system can become a major source of pathology to the host tissues.

Control of inflammation

Inflammation is a complex biological response to pathogens or damage that initiates changes in vasculature, cell trafficking and movement of exudative elements. The process is tightly regulated and coordinated because it is bioenergetically expensive and if allowed to become dysregulated can cause collateral damage to the host tissues

In some ways we can think of inflammation as a fire, that when encased in a furnace or oven and maintained within certain parameters of magnitude, extent and time, can have a desired and beneficial effect. To ensure regulation within these parameters the furnace or oven possess a thermoregulatory system with a capacity for sensing, signaling processing, control, and effecting change in the fire. The regulatory system could be a simple feedback loop, or in the case of the immune system, a more complex multidimensional system, involving multilayered, negative, autoregulatory, feedback pathways, and a control system that could ignite and amplify, or reduce and dampen the flames of the immune response. Of course, if the system breaks down and the inflammatory fire became unregulated or dysregulated, the previous benefits are usurped by local or widespread damage or destruction such as occurs in lung consolidation in pneumonia, or the development of self-destructive pathologies, such as those that occur in autoimmune disease. So while the inflammatory process is hugely beneficial and an essential part of homeostasis, it needs to be controlled or regulated quantitatively, qualitatively, and temporally, within well-defined parameters of magnitude, scope, and time.

Now, the mechanisms of immune regulation are multidimensional and involve various structural and functional elements. The most important structural means of regulation is the compartmentalization of the immune system which provides a tissue specific approach to controlling the inflammatory fire. We have some tissues which are immunologically privileged. This means that immune cells cannot easily travel into these sites. Examples of this are the central nervous system, the eye, and parts of the reproductive tract such as the male testes. These are often referred to in terms of the primary underlying structures and processes which combine to limit the ability of extraneous immune factors such as cells and immunoglobulins to move from the circulation in to those tissues. Another element of this control process is the existence of unique micro environments between and within different tissue types. Just as the nature and magnitude of fire is determined by the combination of substrates and chemicals used as fuel, each inflammatory response is qualitatively and quantitatively governed by the character, combination and availability of local metabolic and regulatory factors.

The other mechanisms of immune regulation occur by functional means throughout the inflammatory fire cascade. These mechanisms exist at all levels of the immune system, and affect a wide range of cellular, neurohumoral, chemical and molecular elements involved in the detection, signaling, and effector processes of the inflammatory response. The mechanisms are tissue specific and are often targets for attack by many infectious agents which induce their disease by allowing the inflammatory fire to become dysregulated and pathogenic.

Every animal we see has an immunological phenotype status

If we visited a livestock production unit on which there is an outbreak of infectious disease in the animals, what would we observe or what could we determine about the immune system by collecting data on these animals? Could we categorize these animals in terms of their current or potential contribution to the disease scenario, and in relation to immunological homeostasis? What factors are impacting their immunological status? It is important to understand that in all occurrences of infectious disease there are three very different groups of individuals represented in terms of pathogen spread, those that are infected, those that are susceptible to infection and those that are resistant to infection. It may be very difficult to differentiate these groups visibly, but those three groups will all be present, albeit in different proportions at different times. It would certainly be useful to differentiate between these three groups, because each require a different management approach. In addition, the ultimate determinant of infectious disease trajectory is the nature and magnitude of contact and interaction between infected and susceptible animals.

So, it is important to recognize that, while it may not be obvious immediately, these three groups of animals are comprised of individuals with different pathophysiological phenotypes. There will be animals with an infected phenotype (some of whom are exhibiting symptoms of disease), animals with a susceptible phenotype, and animals with a resistant phenotype. Now, while a phenotype is not always visible, biologically we should be able to define parameters by which we could identify, measure and monitor each animal's phenotype status. After all, we already do this with diseased animals. In most cases we have clearly defined parameters by which we identify infected animals - at least a proportion of them will exhibit clinical signs on which we can perform diagnostic tests. But, how can we differentiate between animals with a resistant or susceptible phenotype status? That is an important question and to answer it, we need to understand what constitutes a resistant and susceptible phenotype, and what factors affect the presence of either status. We can define the resistant phenotype as the capacity of an individual to maintain health in the face of potentially disruptive influences. This resistance to disease can be equated with biological resilience, and results from a combination of constitutional robustness (that is - the capacity to resist perturbation) and a dynamic homeostasis (that is - the ability to adapt to perturbation), most importantly with regard to immune function. This concept is probably better understood in the world of agronomy and horticulture. The existence of drought or disease resistant plants, in the absence of an adaptive immune system, is well recognized. We also we have breeds of cattle, particularly those in the *Bos indicus* or Zebu lineage, that are resistant to heat or parasite infection. In sheep there are also genetic lines that are resistant to endoparasite load, and in pigs PRRS virus resistant populations are under development. In all of these situations we see individuals or populations that are less sensitive, or non-sensitive, to immuno-microbiological challenges, and that are able to effectively recover from pathophysiological disturbances. In combination, we see a greater consistency and stability in health and production in biologically resilient, and disease resistant populations. The definitive measure of phenotype status would be the presence of superior health and performance at a given infection challenge, and the maintenance of health and performance with increasing infection challenge. Ultimately, our understanding of immunology should help our clients to develop and deliver management inputs that cultivate and maintain a stable resistant phenotype, and prevent animals from declining in to a susceptible status.

Gastrointestinal immunology: Basic principles and factors that impact immune development and function?

A. J. Moeser, DVM, MS, PhD
Gastrointestinal Stress Biology Laboratory, Department of Large Animal Clinical Sciences, College of Veterinary Medicine, Neuroscience Program, Michigan State University, East Lansing, Michigan

Introduction

The gastrointestinal (GI) immune system constitutes the largest immune compartment in the body and plays diverse roles in gut and systemic health. The GI immune system, which includes gut epithelial and immune cells, provides a first line of defense against to the massive amount of potentially deadly microorganisms and toxins residing within the intestinal lumen. The GI immune system must be able to mount a rapid and robust immune response in response to an infectious challenge or breach in epithelial barrier function while at the same time limit its activation in response to common feed antigens and commensal microorganisms. The GI immune system also becomes highly activated during the stress response which may serve as a critical defense mechanism but if not controlled may impair performance and increase disease susceptibility. Furthermore, GI inflammatory disorders are often co morbid with diseases in distant organs such as the brain, bone, and respiratory system demonstrating that the state of the GI immune system health can impact systemic immune and organ function. The objective of this paper is to provide a broad overview of the anatomy and mechanisms of GI immune function. A specific area of emphasis here will be on postnatal development of the GI immune system in the pig and how early life stressors such as weaning can impact GI immune development and pig health across the production lifespan.

Anatomy of the GI immune system: key cellular players and organization

The GI immune system is comprised of a diverse array cell types that act in concert to carry out the diverse immune functions mentioned above. Discussed below are the major GI immune cell types and brief overview of their roles.

Intestinal epithelial cells (IECs): While not recognized as classical immune cells, IECs play a central role in GI immune function. The GI epithelium forms a permeability barrier that limits the entry and exposure of luminal contents, which contains massive amounts of potentially harmful microbial and feed antigens, to the immune system and systemic circulation. The "leakiness" of the GI epithelium is predominantly regulated by an array of tight junction proteins which are localized to the apical paracellular space. In additional support of GI barrier function, different IEC types secrete various protective substances. For example, goblet cells secrete large amounts of protective mucins and paneth cells in the intestinal crypts secrete antimicrobial peptides (e.g. defensins, cathelicidins, lysosomes, etc). Breakdown in epithelial barrier function (i.e. "leaky gut") is a central pathogenic mechanisms in many GI inflammatory diseases and if severe can lead to sepsis and death, thus highlighting the critical importance of this function. In addition to providing a physical and chemical barrier, it has become well-recognized that IECs perform direct immune functions including antigen recognition uptake and presentation. In inductive immune sites such as the Peyer's patches, specialized epithelial cells call Microfold cells (M cells) are responsible for the uptake of luminal antigens and pathogens and delivering them to antigen presenting cells such as dendritic cells. This process is an important initial step in initiating the mucosal immune response and development of adaptive or tolerogenic functions (discussed later in this paper). The GI epithelium also produces an array of immune mediators known as "alarmins" (e.g. interleukin 33 (IL-33), IL-23, defensins, etc.) which are released early in response to acute infections or epithelial injury and serve as an early warning system that activates, recruits and directs the immune system.[1-3] More recent evidence demonstrates that specific epithelial nutrient transporters, such as sodium glucose linked transporter 1 (SGLT1) play a non-classical role in the immune response to GI infections and stress.[4]

Cells of the innate immune response: In the GI system innate immune cells consist mainly of dendritic cells, macrophages, eosinophils, mast cells, neutrophils and innate lymphoid cells. This diverse array of cells carry out specific and sometimes overlapping functions such as antigen presentation (e.g. dendritic cells), immune cell recruitment (e.g. mast cells), and inflammation (e.g. neutrophils and macrophages). The innate immune cells play important roles in both early and later stages of immune responses including pathogen challenge, induction of adaptive immunity and tissue healing and repair.

Cells of the adaptive/acquired immune response: Lymphocytes are found both scattered throughout the GI lamina propria and submucosa mucosa and in organized lymphoid structures (Peyer's patches, lymphoid follicles and mesenteric lymph nodes). Lamina propria and intraepthelial lymphocytes lymphocytes include mainly CD4+ and also contain CD8+ cytotoxic and regulatory T lymphocytes. The majority of the T cells residing in the intestinal mucosa have already been in contact with antigens at inductive sites, whereas as the lymphocytes in organized lymphoid structures (e.g. Peyer's patches) are mainly naïve. Plasma cells are differentiated B-cells that produce polymeric IgA (pIgA) playing a critical role in mucosal defense.

The gut-associated lymphoid tissue (GALT). The GALT consists of organized lymphoid tissue including Peyer's patches localized in the distal ileum, organized lymphoid follicles and mesenteric lymph nodes located along the length of the GI tract. The GALT constitutes the inductive sites of mucosal immunity where luminal antigens are sampled and delivered to the underlying lymphoid tissue resulting in activation of naïve and memory T and B lymphocytes.

Nervous system and microbiome as modulators of immune function: Immune cells receive cues from the local environment and systemically which have a major impact on immune cell function and immunity. For example, the GI microbiome has been an intensely studied area and evidence has continued to mount for its influence of GI immune function. The influence of the microbiome is an intense area of study and outside the scope of this paper but several review papers are now available on this topic.[5] Another critical modulator of GI immune function is the central and enteric nervous systems. The GI tract is a richly enervated organ receiving neural input from the central nervous system (CNS) and the enteric nervous system (ENS). The ENS contains as many neurons as the spinal cord and performs many of its functions (e.g. digestion, motility, etc.) independent of the CNS. Enteric neurons release a diverse array of neuropeptides and mediators (e.g. acetylcholine, substance P, norepinephrine, corticotropin releasing factor, etc.) that bind to receptors on GI immune cells which can profoundly modulate their function. An excellent example of a classical neuro-immune pathway in the GI system is the cholinergic anti-inflammatory mechanism where vagal nerve-mediated release of acetylcholine binds to nicotinic receptors on immune cells to downregulate pro-inflammatory cytokine production and dampen intestinal inflammation.[6] In contrast, the cholinergic nervous system can be proinflammatory which is also mediated by release of acetylcholine, but acting via muscarinic receptors. The muscarinic proinflammatory response has been shown to be important for mucosal immunity and clearance of pathogens such as *S.* typhimurium.[7] The cholinergic anti- and proinflammatory responses are examples of how the nervous system divergently modulates the immune system to maintain immune homeostasis. We have recently demonstrated that pigs undergoing a challenge with *S.* typhimurium exhibit robust changes in the expression of cholinergic proteins in the ileum and colon which correlate with clinical and inflammatory disease.[8]

How does the GI immune system work to simultaneously defend against infections and develop protective immunity while limiting activation to commensal microbiota dietary antigens?

The GI immune system is exposed to a diverse array of commensal organisms, pathogenic organism and dietary antigens and thus must develop appropriate and different mechanisms to respond to each stimuli to maintain mucosal homeostasis. In response to a pathogenic challenge (or mucosal vaccine) or breach in the intestinal epithelial barrier, the GI immune system response must be rapid and robust to prevent the local and systemic spread of luminal microorganisms, antigens and toxins. This innate immune response is critical for setting up the adaptive immune response that is critical for targeted protection from subsequent pathogen exposures or protective immunity. In general, pathogen exposure and invasion often begins with the activation of pattern recognition receptors (e.g. Toll like receptors) expressed on epithelial and innate immune cells. In turn, activated cells release of a variety of chemotactic molecules (e.g. IL-8, TNF, chemokines, etc) that recruit innate immune cells such as neutrophils to the site of infection. While this response causes inflammation and surrounding GI damage contributing to the clinical disease observed in animals, a robust initial immune response is critical for preventing the spread of infection and sepsis and clearance. For example, *E. coli* infection elicits mast cell activation and release of TNF and IL6 which is critical for neutrophil recruitment and clearance of *E. coli*. Animals deficient in mast cells (mast cell deficient mice) are more likely to succumb to an infectious challenge. At inductive sites, APCs present antigen to naïve T and B cells via their MHC class I or II molecules. Activation of T cells causes differentiation into specific T cell subtypes including Th1, Th2, Th 17 and T regulatory cells and cytotoxic T lymphocytes, dependent largely by the local cytokine environment. B lymphocyte differentiation in the presence of TGFβ favors immunoglobulin class switching from IgM to IgA which take place in the Peyer's patches and allow maturation of B cells into antibody producing cells. Activated T lymphocytes, plasma cells and memory T and B cells migrate out of the lymph nodes via effect lymphatics and thoracic duct where they home to mucosal effector sites to mediate a protective immune response.

As mentioned above the GI immune system must be active to induce protective immune against pathogens, however, it must also be able to prevent inappropriate or excessive immune responses to food antigens and commensal microbiota that could result in massive inflammation. A major mechanism by which the immune system prevents over-activation to luminal antigens is via the establishment of mucosal "tolerance". The mechanisms of tolerance have been and continue to be intensively investigated and thus several detailed reviews are available for further reading.[9,10] Broadly, the mechanisms of the tolerogenic response include (1) anergy in which antigen-specific T cells become unresponsive to antigens or (2) induction of a T regulatory cells with an immunosuppressive actions. Epigenetic mechanisms, via the regulation of histone binding to tolerance-associated genes and TLR activation, has also been demonstrated play important roles in the establishment of mucosal tolerance.[11]

Long-term GI immune function is shaped in early postnatal critical periods

While the mature GI immune system can adapt and change throughout life, it is largely shaped in first 3-4 months of postnatal life. Therefore it is important to understand the basic developmental aspects of GI immune system as any significant stressors or environmental challenges that occur during postnatal GI immune development have the potential to drive long-term changes in immune function and health. Ontologically, the early life immune system has been described as suppressed, yet active. At birth, the host must adapt to microbial colonization of the lungs, intestine, and skin as well as consumption of milk antigen, without inducing massive inflammation. Research over the last 25 years has described multiple layers of both exogenous and endogenous immunosuppressive mediators that modulate neonatal immunology in order to promote an active, yet tolerogenic immune system. Maternal milk provides immune-supportive factors such as sIgA, maternal leukocytes, and milk glycans, all of which modulate and neutralize intestinal microbes. Breastmilk also provides massive amounts of anti-inflammatory cytokines and peptides, which negatively regulate neonatal TLR and inflammatory cytokine expression. As a further mechanism to limit over-stimulation of the neonatal immune system is that neonates are born with fewer lymphocytes and low co-stimulatory molecules expression. The neonatal immune system is biased toward a Th2 (T helper 2: humoral) immune response as their ability to produce Th1 (T helper1) cytokines such as IFNγ and interleukin-12 (IL-12) is suppressed while producing comparatively higher Th2 cytokines such as IL-4, IL-10 and IL-13.

Several weeks after birth, neonates overcome the neonatal immunosuppressive when lymphocytes start developing functionally as well as structurally (Figure 1). Unweaned piglets achieve a normal number of lymphocytes at about 6 weeks of age.[12] Around the same time, the secondary lymphoid organ such as jejunal and ileal Peyer's patches rapidly mature characterized by an infiltration of lymphocytes. Interaction of lymphocytes to wide variety of antigen including self-antigen, microbiota containing commensal or pathogenic bacteria, help in differentiating between self and non-self-antigen and further facilitate to achieve homeostasis. The proper interactions of lymphocytes with antigen also lead to their proliferation, immunoglobulins and balanced Th1/Th2 cytokine production.

In summary, the early postnatal immune system is a 'window of opportunity' for tolerogenic induction and for lifelong GI immune function. During this time the immune system is reactive and possesses *inflammatory capability*, however, suppressive factors predominate to prevent overstimulation. This reinforces that there may be consequences of immune-overstimulation during this quiescent period that may have long-term consequences and thus it is critical that this period remains undisturbed to support proper immune development.

Early weaning and stress change long-term GI immune development and health of the pig

As mentioned above, development of the GI (and systemic) immune system in the pig occurs predominantly in the first 3-4 months of postnatal life and the conditions under which this immune system develops shapes its long-term function. Therefore any over-activation of the GI immune system, which could be caused by infection, dietary antigens and stress can theoretically have lifelong impacts. Stressors (physical and or psychological) cause robust activation of the immune system and early life stress is a significant risk factor for the later life onset of inflammatory diseases in people. In pig production, the most significant early life management-related stressor is early weaning which occurs between 2-4 weeks of age. This is in contrast to natural weaning process, in which piglets wean from sows gradually and is completed at around between 3-4 months of age (just about the time when the GI immune system is developed). Thus weaning in production is stressful involving multiple stressors (e.g. maternal and littermate separation, changes in social hierarchy, transport, anorexia, etc.) and occurs during the peak of GI immune development. It has been established that early weaning can compromise wean to finish performance and infectious disease susceptibility[13,14] but the question remains how can we intervene to improve GI immune development without implementing major changes in weaning age? A more fundamental mechanistic understanding of how early weaning impacts GI immune development will be critical in identifying new strategies and therapeutic targets. Our laboratory has been focusing on early life stressors that impact the long-term development of GI epithelial and immune barrier function. In order to study how early weaning influences long-term GI development, we compare the immediate and long-term GI responses to between early weaned and late weaned pigs. Our studies have demonstrated that weaning age has a profound impact on immediate and long-term GI barrier function and disease susceptibility.[15-19] We have also shown that subtle differences in weaning age can dramatically impair the subsequent mucosal immune response and clinical disease to later life enteric challenges.[20] Currently we are investigating how early weaning influences immune development in pigs and the role that the ENS and innate immune system (e.g. mast cells and eosinophils) interact to modulate immune development and function in the early weaned pig.

Summary

The GI immune system represents the largest immune organ in the body. The GI immune system functions to maintain a balance between activation, required for survival and protective immunity, and suppression/tolerance to limit activation to commensal microorganisms and dietary antigens. Any disruption in this

Figure 1: The developmental changes in neonatal pig immune system. Neonatal piglets gets majority of their immunoglobulin from mother through transplacental route. Which further increase within 24-48 hours after birth following colostrum ingestion. Immunoglobulin concentration then decline until newborn start producing their own antibodies through activated immune cells. The actual activation and proliferation of lymphocytes occur after maturation of secondary lymphoid organs around two months of age Activated lymphocytes interact with wide variety of antigen (self-antigen, microbiota containing commensal or pathogenic bacteria) to differentiate between self and non-self and achieve homeostasis.

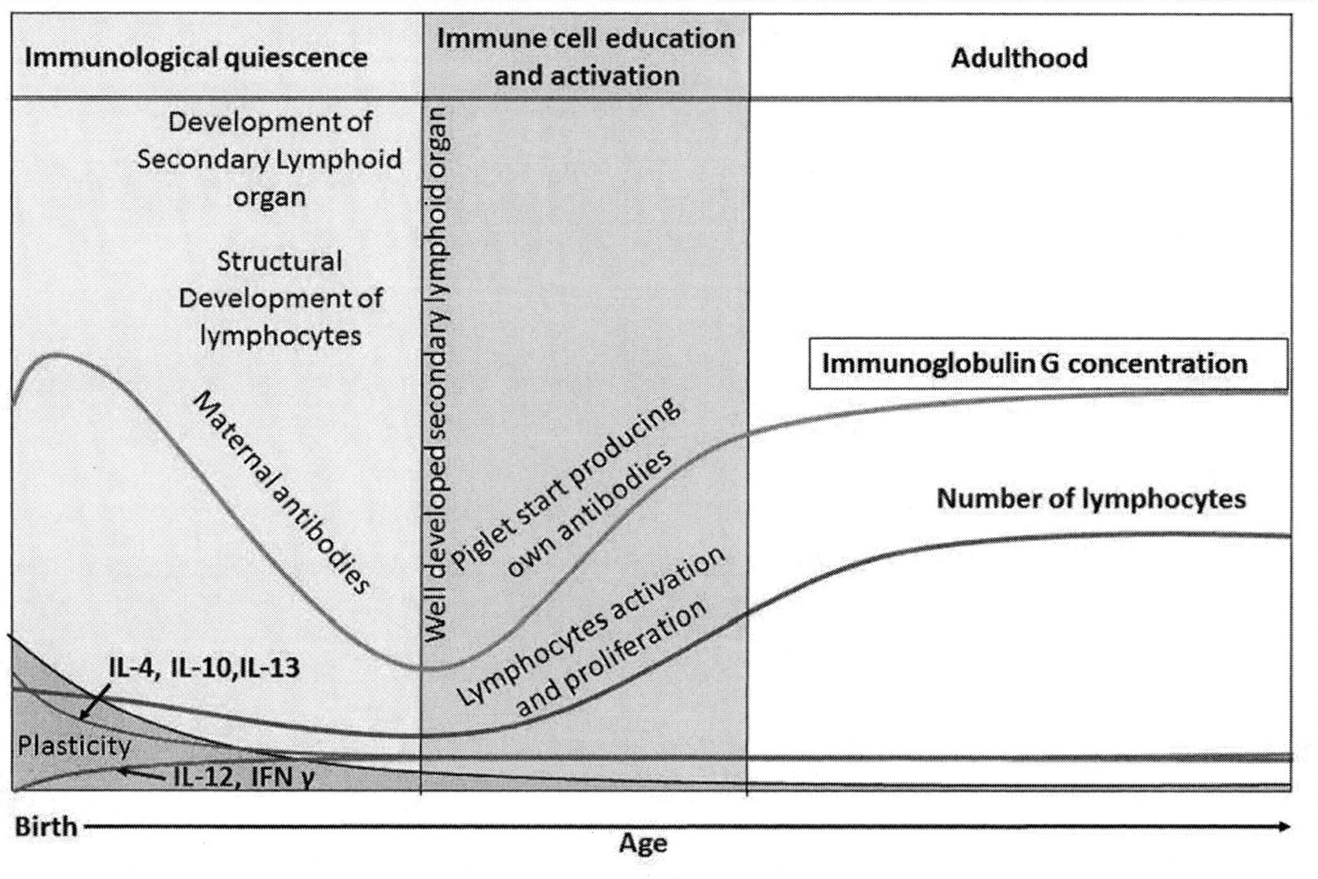

balance can lead to inflammation or immunosuppression and disease. Importantly, the specific functions of the GI immune system are largely shaped in the first 3 months of postnatal life. Unfortunately, this critical period in GI immune development coincides with the most stressful event in swine production--weaning. How inherent management stressors in pig production adversely impact long-term GI barrier and immune health remains poorly understood. This represents a critical gap in knowledge with major implications to lifetime pig health and performance as strategies and biological targets for optimizing gut immune development are currently limited.

References

1. Hodzic, Z., Schill, E. M., Bolock, A. M., Good, M. (2017) IL-33 and the intestine: The good, the bad, and the inflammatory. Cytokine 100, 1-10.

2. Schiering, C., Krausgruber, T., Chomka, A., Frohlich, A., Adelmann, K., Wohlfert, E. A., Pott, J., Griseri, T., Bollrath, J., Hegazy, A. N., Harrison, O. J., Owens, B. M., Lohning, M., Belkaid, Y., Fallon, P. G., Powrie, F. (2014) The alarmin IL-33 promotes regulatory T-cell function in the intestine. Nature 513, 564-8.

3. Bevins, C. L. (2013) Innate immune functions of alpha-defensins in the small intestine. Digestive diseases (Basel, Switzerland) 31, 299-304.

4. Wright, E. M., Ghezzi, C., Loo, D. D. F. (2017) Novel and Unexpected Functions of SGLTs. Physiology (Bethesda, Md.) 32, 435-443.

5. Thaiss, C. A., Zmora, N., Levy, M., Elinav, E. (2016) The microbiome and innate immunity. Nature 535, 65-74.

6. Matteoli, G. and Boeckxstaens, G. E. (2013) The vagal innervation of the gut and immune homeostasis. Gut 62, 1214-22.

7. Fernandez-Cabezudo, M. J., Lorke, D. E., Azimullah, S., Mechkarska, M., Hasan, M. Y., Petroianu, G. A., al-Ramadi, B. K. (2010) Cholinergic stimulation of the immune system protects against lethal infection by Salmonella enterica serovar Typhimurium. Immunology 130, 388-398.

8. Pohl, C. S., Lennon, E. M., Li, Y., DeWilde, M. P., Moeser, A. J. (2018) S. Typhimurium challenge in juvenile pigs modulates the expression and localization of enteric cholinergic proteins and correlates with mucosal injury and inflammation. Auton Neurosci 213, 51-59.

9. Joosse, M. E., Nederlof, I., Walker, L. S. K., Samsom, J. N. (2019) Tipping the balance: inhibitory checkpoints in intestinal homeostasis. Mucosal Immunol 12, 21-35.

10. Ahluwalia, B., Magnusson, M. K., Ohman, L. (2017) Mucosal immune system of the gastrointestinal tract: maintaining balance between the good and the bad. Scandinavian journal of gastroenterology 52, 1185-1193.

11. Seeley, J. J. and Ghosh, S. (2017) Molecular mechanisms of innate memory and tolerance to LPS. Journal of leukocyte biology 101, 107-119.

12. I, P. M. a. M.-D. (2011) Age-dependent changes in relative and absolute size of lymphocyte subsets in the blood of pigs from birth to slaughter Bull Vet Inst Pulawy 55, 305-310.

13. Main, R. G., Dritz, S. S., Tokach, M. D., Goodband, R. D., Nelssen, J. L. (2004) Increasing weaning age improves pig performance in a multisite production system. J Anim Sci 82, 1499-507.

14. Rosero DS, D. T., Boyd RD (2015) Impact of Weaning Age on Pig Performance: Evaluation of Close-out Data from the DN Sow Flow (Jan 2010 to Dec 2014). The Hanor Company Technical Report Report number: 2015-15.

15. Pohl, C. S., Medland, J. E., Mackey, E., Edwards, L. L., Bagley, K. D., DeWilde, M. P., Williams, K. J., Moeser, A. J. (2017) Early weaning stress induces chronic functional diarrhea, intestinal barrier defects, and increased mast cell activity in a porcine model of early life adversity. Neurogastroenterol Motil.

16. Pohl CS, F. D., Gulbransen BD, Moeser AJ (2017) Early Life Adversity in Rodent and Porcine Models Induces Enteric Cholinergic System Upregulation and Mast Cell Hyperplasia. FASEB 31, 1047.2.

17. Moeser, A. J., Pohl, C. S., Rajput, M. (2017) Weaning stress and gastrointestinal barrier development: Implications for lifelong gut health in pigs. Animal Nutrition.

18. Medland, J. E., Pohl, C. S., Edwards, L. L., Frandsen, S., Bagley, K., Li, Y., Moeser, A. J. (2016) Early life adversity in piglets induces long-term upregulation of the enteric cholinergic nervous system and heightened, sex-specific secretomotor neuron responses. Neurogastroenterol Motil.

19. Pohl, C. S., Medland, J. E., Moeser, A. J. (2015) Early-life stress origins of gastrointestinal disease: animal models, intestinal pathophysiology, and translational implications. Am J Physiol Gastrointest Liver Physiol 309, G927-41.

20. McLamb, B. L., Gibson, A. J., Overman, E. L., Stahl, C., Moeser, A. J. (2013) Early weaning stress in pigs impairs innate mucosal immune responses to enterotoxigenic *E. coli* challenge and exacerbates intestinal injury and clinical disease. PLoS One 8, e59838.

Impact of vaccination on transmission of *Lawsonia intracellularis*

Erika Vasquez[1]; Robert Valeris[1]; Dana Beckler[2]; Fabio Vannucci[1]
[1]College of Veterinary Medicine, University of Minnesota; [2]Gut Bugs, Inc, Minnesota

Introduction

Lawsonia intracellularis (LI) is an intracellular bacterium and causative agent of proliferative enteropathy (PE; ileitis) in pigs as well as a variety of other species. Although PE is one of most prevalent enteric diseases of grow-finish pigs, little is known of the epidemiology and transmission of this disease. Antimicrobial has been widely used to control and reduce the economic losses due to PE. The recent changes on regulation for the use of antimicrobials as growth promoters will directly impact well-established strategies used to control PE and other enteric diseases. This scenario has driven implementation of vaccination strategies especially for controlling bacterial endemic diseases that were previously well manageable with the use of antimicrobials. Although studies has been focused on the individual efficacy of vaccines towards the reduction of clinical signs, the impact of vaccination on transmission rates and probability of infection in naïve and vaccinated animals have proven to have great value in driving important decisions about control swine diseases, such as influenza and *Mycoplasma hyopneumoniae*.[2-4] Our broad objective is to evaluate differences in transmission patterns of PE and quantify the spread of *L intracellularis* in naïve and vaccinated pigs by estimating the transmission rate and expected probability of infection.

Experimental design

Using a seeder-pig sentinel model, 90 pigs were divided into 3 groups: Treatment 1 (T1) was orally vaccinated (Enterisol® Ileitis). Treatment 2 (T2) received intramuscular vaccine (Porcilis™ Ileitis); Treatment 3 (T3), was non-vaccinated (Figure 1). Day 21 post-vaccination, 9 seeder pigs were challenged with *L intracellularis*. Day 7 post-inoculation, seeder pigs were commingled with T1, T2 and T3. Shedding was monitored by qPCR; serological response by IPMA technique. Transmission patterns and expected probabilities of shedding were assessed with the susceptible–infectious (SI) model.

Figure 1: Experimental design.

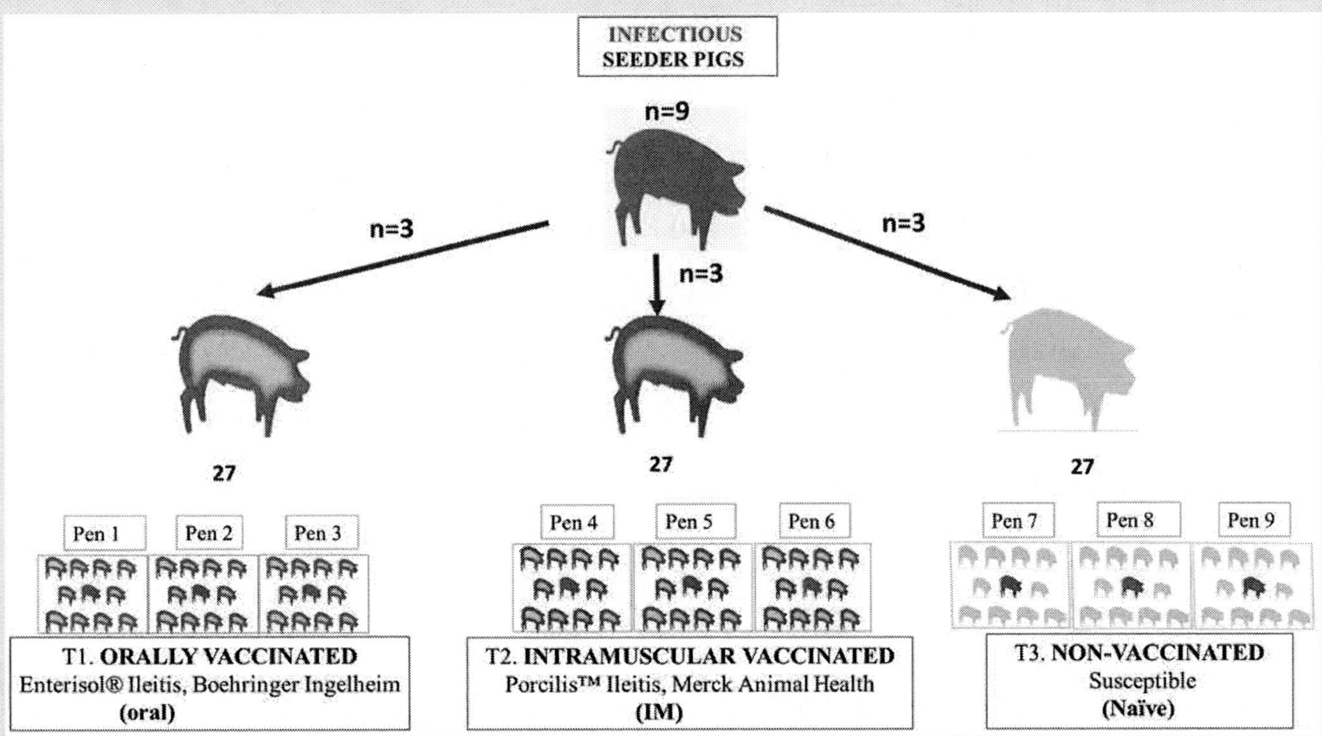

Results

- Animals were considered infectious when fecal qPCR was equal to or more than 10^3 *L intracellularis*/g of feces.[1]
- The transmission rate per week for non-vaccinated animals was 3.6, and was not significantly reduced in either vaccinated group.
- Since there is no certainty that the shedding of *L intracellularis* would have decreased below 10^3 in those animals with high shedding by the end of the study, a parametric survival regression with censoring (time from shedding *L intracelullaris* $\geq 10^3$ to shedding below that level) were used. The median length in days of *L intracellularis* shedding was 43.3% and 25.5% shorter, in the orally and intramuscular vaccinated animals respectively, compared to the unvaccinated animals.
- The adjusted predicted probability of shedding $\geq 10^3$ *L intracellularis* was higher in pigs receiving the oral vaccine compared to the intramuscular vaccine during the 4 weeks of the transmission assessment (Figure 2, A).
- After 4 weeks, there were no significant differences in the probabilities of shedding between both vaccinated animals, while the probability in the non-vaccinated group remained significantly higher ($P < 0.05$). (Figure 2, B).
- After 4 weeks of contact, both vaccinated groups showed reduced number of positive pigs by IHC compared with the non-vaccinated group (Table 1).

Conclusions

- Vaccination against *L intracellularis* has been applied over the years with successful reduction of the clinical signs and fecal shedding, which has been resulted in significant improvement of growth performance in vaccinated animals. Under the conditions of the present study, both vaccination protocols reduced the length of *L intracellularis* shedding after natural exposure through contact with infectious animals.
- Despite the shortened period of fecal shedding observed in vaccinated animals, vaccination was able to significantly decrease the transmission rate of *L intracellularis* when unvaccinated shedder pigs were introduced into a vaccinated population.
- From a practical perspective, these results highlight the importance of implementing vaccination programs at system level rather than site-specific interventions. This strategy prevents the risk of commingling batches of pigs from vaccinated and non-vaccinated sources, therefore reducing the impact of PPE in downstream flows.

References

1. Collins, A.M., et al. Immunity to *Lawsonia intracellularis*. Allen D. Leman Conference, 2001;115.

2. Velthuis AG, et al. Design and analysis of an *Actinobacillus pleuropneumoniae* transmission experiment. Prev Vet Med. 2003;53-68.

3. Romagosa A, Allerson M, Gramer M, Joo HS, Deen J, Detmer S, Torremorell M. Vaccination of influenza A virus decreases transmission rates in pigs. Vet Res. 2011;120.

4. Roos LR, et al. A model to investigate the optimal seeder-to-naïve ratio for successful natural *Mycoplasma hyopneumoniae* gilt exposure prior to entering the breeding herd. Vet Microbiol. 2016;51:8.

Figure 2: A) Weibull regression model: Adjusted predicted probabilities of shedding ≥ 10^3 *L intracellularis* by group and time. B) Generalized estimating equation (GEE) model: Proportion of pigs shedding ≥10^3 *L intracellularis* by length of time

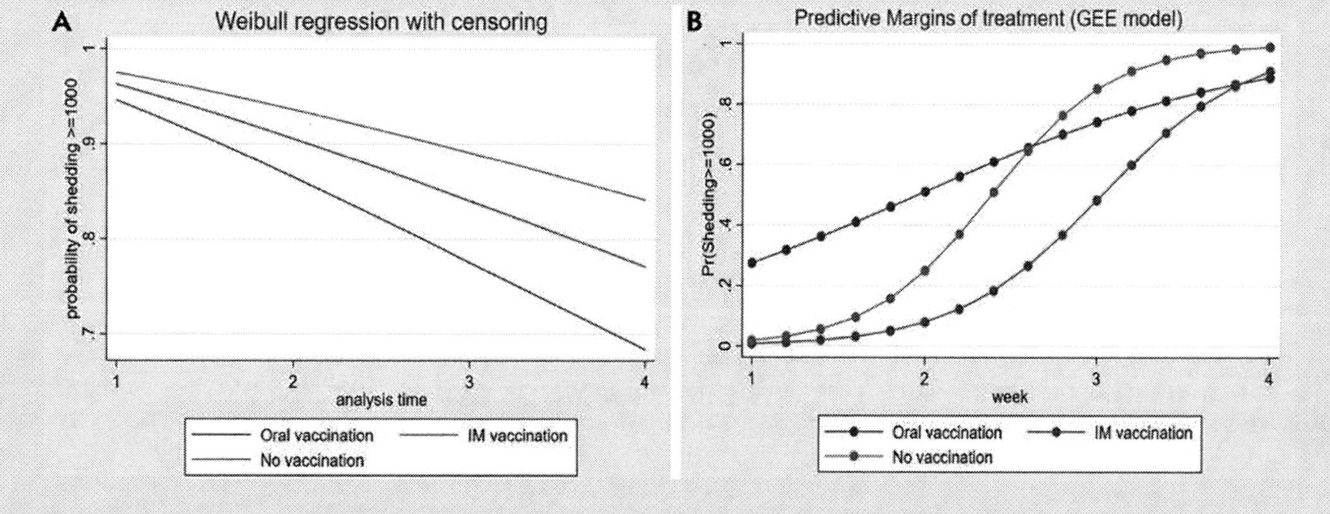

Table 1: Immunohistochemistry for *L intracellularis* among the experimental groups after 4 weeks of contact with seeder pigs.

Groups	Immunohistochemistry	
	Negative	Positive
Non-vaccinated	2	25
Orally-vaccinated	8	18
Intramuscularly-vaccinated	12	15

Farm application of ileitis vaccines

Nathan L. Winkelman, DVM
Swine Services Unlimited, Inc. Rice, Minnesota

For ileitis control, the goal of the swine veterinarian is to prevent clinical ileitis and mitigate subclinical ileitis to prevent economic loss from poor performance. Both vaccines and antibiotics have been widely used to control this wide-spread endemic pathogen. A gradual decrease in feed-grade antibiotics has occurred since VFD's were mandated in 2017, further increasing the use of commercial ileitis vaccines. Practioners need to know what to look for and how to monitor ileitis control programs to evaluate their effectiveness, and what to do when expectations are not met. In general, ileitis vaccines are moderately effective and very cost effective. However, vaccine "failure" is not uncommon if our expectations are to not have any clinical signs of ileitis in finishing pigs and to keep subclinical ileitis below an economic threshold.

The results of a recent ileitis vaccine challenge trial[1] comparing Enterisol®, Porcilis® and DLI® Immunity vaccine efficacy in the face of a *Lawsonia intracellularis* challenge will be reviewed.

Ileitis (vaccine) control: Re-evaluate your expectations

Perceptions of ileitis vaccine performance are largely linked to expectations of the producers and/or veterinarian. Often the only expectation of vaccine success is the absence of bloody diarrhea and mortality from PHE (Porcine Hemorrhagic Enteropathy). However, these expectations are much too low. Any degree of diarrhea (mild to severe) or any abnormal color from blood (orange, dark brown to melena) in formed or diarrheic feces, or loss of body condition (gauntness) is a clinical sign of ileitis. **For vaccine success, there should be <u>no</u> evidence of <u>any</u> clinical ileitis, especially in finishing pigs from about 120# to market or genetic replacement gilts.** Clinical signs in just a few pigs indicates inadequate protection. Often the majority of the rest of the pigs in that population are subclinically infected, which also causes significant performance loss.

Secondly, subclinical ileitis above a certain threshold should not be tolerated either. Subclinical ileitis is defined as a positive ileitis diagnosis via fecal PCR, intestinal IHC (immunohistochemistry) or PCR, or gross PIA (Porcine Intestinal Adenomatosis) lesion pathology in the <u>absence</u> of clinical signs. Many research trials have shown a significant economic loss from $3 to $5 per pig from subclinical ileitis.[2] It is the most ubiquitous and costly form of the disease.

Monitor your ileitis (vaccine) program - ante-mortem diagnostics

Quantitative fecal PCR ct values - Fecal samples with a PCR ≤ 31 of 40 ct (U of MN VDL) are highly correlated with gross intestinal PIA lesions, positive IHC lesions, and poor ADG (Average Daily Gain).[3] Ct values from > 33 to < 35 may or may not be significant and those ct values > 35 to 40 are considered "suspect". Therefore, swine vets now have a very useful ante-mortem diagnostic tool to monitor their ileitis control programs. (Note that *Lawsonia* PCR tests are not standardized across diagnostic labs and these numerical ct values will differ from lab to lab). Figure 1 shows an (R^2) correlation coefficient of 78%, depicting a strong negative correlation between *Lawsonia* fecal ct values and IHC pathology scores.

IPMA serology - Serology is also a useful tool to identify past exposure to *Lawsonia intracellularis*. IPMA (immunoperoxidase monolayer assay) serology measures antibody titers from 30 to 960 (or higher if requested) and is more sensitive than the ELISA antibody assay.[4] Pigs exposed to ileitis infection generally have higher titers than from vaccine alone. Field infection titers usually range from > 480 to 15,000, depending on the *Lawsonia* exposure dose and PIA lesion severity. Serological surveys can also be an indirect method of determining ileitis exposure, or vaccine compliance with Porcilis® or DLI®. Vaccine and DLI® titers are usually ≤ 480.[5] Enterisol® vaccine illicits a relatively weak IPMA immune response and no ELISA response is evident post-vaccination[4]

To detect subclinical ileitis in a pork production system at a 90% confidence level and a 10% prevalence would require sampling 30 fecal and 30 serums about every 4 weeks beginning at 6-10 weeks old. However, periodic sampling in finishing pigs via fecal PCRs, oral fluid PCRs, and/or serology will often be enough to monitor the program and detect the presence of clinical or subclinical ileitis. If ileitis is present, then consider changing the vaccination and/or feed antibiotic program.

Sampling replacement gilts to identify subclinical ileitis before it gets into the sow farm is also an important consideration. Commercial vaccines do not prevent subclinical infection and finding ileitis early may prevent a PHE outbreak later.

In a perfect scenario, grow-finish pigs would not have any clinical signs of ileitis or any subclinically positive fecal PCR's when monitoring the pig-flow for ileitis.

Ileitis vaccine "failure" – What next?

Step 1: Control clinical disease with antibiotics

Whenever clinical ileitis is diagnosed, it should aggressively be treated with antibiotics. The incubation period for *Lawsonia* is typically 2-4 weeks so clinically infected pigs are just the "tip of the iceberg," with most of the other pigs subclinically infected. If it's a PHE break, inject individual pigs with tylosin or lincomycin <u>and</u> water medicate the entire group. The risk of not treating right away at first clinical signs is that as the outbreak becomes more severe, the disease then responds less quickly to antibiotic therapy causing more mortality and morbidity.

Step 2: Determine potential reasons for vaccine "Failure"

There are many common errors which could affect a vaccine's effectiveness.[5] The herd veterinarian and sometimes the animal health technical vet can audit proper vaccination procedures. It is not the purpose of this paper to review all the potential reasons for possible vaccine "failure". A few common reasons for ileitis vaccines to "fail":

- **Not following label-directions** – e.g. giving partial vaccine doses or administering it to suckling pigs. Research has not supported these practices.
- **Antibiotic interference** – e.g. contaminated stock solution pails or antibiotics in the feed which could mitigate MLV vaccine effectiveness. Our research shows if Mecadox (carbodox) is given within 7 days of *Lawsonia* challenge, there will be no immune response.[6]
- **Poor vaccination timing** – e.g. be sure pigs are not subclinical carriers at the time of vaccination and give vaccine at least 3-4 weeks prior to anticipated exposure.

Sometimes these vaccination procedures are properly conducted but vaccines are still not controlling clinical or subclinical ileitis effectively because the animal's immunity is overwhelmed with high *Lawsonia* exposure doses, or when various stresses (environmental, gilt transportation, overcrowding, etc.) suppress the immune system. When this occurs in problem pig-flows, sites, or barns within a site or system, there are two additional options for improved ileitis control to consider:

1. Use antibiotic pulses in conjunction with vaccine
2. Use DLI® Immunity for enhanced protection

Step 3: Option 1 - Antibiotics and vaccines

Ileitis challenge trials have shown that the combination of vaccines and antibiotics are more effective than just one or the other. In randomized, controlled 2 × 2 trials, the main effect of ileitis control is from the antibiotic pulse and not from the oral ileitis vaccine.[7] Also, strategically timed antibiotic pulses (lincomycin, tylvalosin, tylosin) given about three weeks prior to anticipated ileitis breaks can be cost effective with or without vaccines.

Step 3: Option 2 - Unique intracellular immunity/DLI[R] (Diluted *Lawsonia intracellularis*) immunity

Pigs previously exposed to a *Lawsonia intracellularis* show a very high level of resistance to reinfection. Researchers have shown that pigs exposed to a relatively low oral dose (10^5) of *Lawsonia* are completely resistant to colonization, fecal shedding, clinical signs or pathology when re-exposed to a high dose (10^9) later.[8] This is the premise of the method of administering a low subclinical dose of *Lawsonia* followed by an antibiotic pulse two weeks later to control the subclinical infection, called DLI® Immunity. It has been shown to provide complete protection (sterilizing immunity) preventing both clinical and subclinical ileitis.[9] This could also be a viable option for ileitis control in the right situation.

Ileitis vaccine efficacy trial[1]

Study Objective The objective was to evaluate the comparative efficacy of two commercial ileitis vaccines and one autogenous live "vaccine" for ileitis control vs challenged controls when pigs were orally gavaged with a high dose (2.2×10^8) *Lawsonia intracellularis* (LI) mucosal homogenate at 10 weeks of age. Pigs were evaluated for clinical observations, fecal PCR shedding, IPMA (immunoperoxidose monalayer assay) antibody, performance parameters (ADG, ADFI, FCR), PPE intestinal gross lesion length and score, and IHC (immunohistochemistry) scores at necropsy.

The results indicate that all three methods of control are very effective at significantly improving all parameters measured compared to the challenged control group. However, clinical signs of ileitis and significant levels of subclinically infected carriers were still present in all three treatments indicating vaccine immunity was overwhelmed by a relatively high infectious dose.

Activity timeline: The three vaccine groups had 30 pigs each (6 Pens, 5 Pigs/Pen). A *Lawsonia* Challenge Control group (No vaccine) also had 30 pigs and a Strict Negative Control (No vaccine, No challenge) group with only 10 pigs was housed in a separate room.

Porcilis® treatment pigs were vaccinated with 2cc IM one day post-weaning on Day 0. Enterisol® and DLI® treatments were orally vaccinated at 5 weeks old on Day 14. Mecadox at 50 g/ton was fed from Day 28 to Day 42 to the DLI treatment groups. All pigs except the Negative Controls, were orally gavaged with a high dose of LI on Day 43. All pigs were euthanized for PPE intestinal scoring and IHC analysis on Day 69 (26 DPI).

Results

Table 1, Table 2, and Figure 1 and Figure 2 show some important results from the study.

Discussion

All three methods of ileitis control (DLI®, Porcilis, and Enterisol) significantly improved these parameters over the ileitis challenged controls in Table 1 and Table 2:

- Performance Parameters – ADG, FCR, and ADFI
- Significantly reduced fecal PCR shedding
- Significantly reduced IHC PPE lesion score
- Significantly reduced gross PPE severity scores
- Significantly reduced clinical diarrhea, fecal color, and abdominal appearance scores

Figure 2 shows the percentage of clinical, subclinical and ileitis negative pigs by group on Day 69. Ileitis clinical signs (diarrhea, fecal color, gauntness) were still present in the vaccinated groups in 12.5 to 19.25% of the pigs. Also, there still were 61.5 to 75% of the pigs subclinically infected 26 days post-challenge. Of those subclinically infected pigs in the vaccine treatments, 70% had fecal ct values < 31 and/or positive IHC scores indicating a significant infection is still present.

Table 1: Comparison of performance measures Days 41 to 69, 0 to 69

Treatment	Days 41 to 69			Days 0 to 69		
	ADG	ADFI	F:G	ADG	ADFI	F:G
Challenged						
Control	1.49[a]	3.44[a]	2.37[a]	1.21[a]	2.30[a]	1.91[a]
DLI	2.04[b]	4.20[b]	2.06[b]	1.49[b]	2.69[b]	1.80[b]
Enterisol	2.01[b]	4.02[b]	2.00[b]	1.45[b]	2.58[b]	1.78[b]
Porcilis	2.00[b]	4.17[b]	2.09[b]	1.43[b]	2.60[b]	1.81[b]
SEM	0.09	0.14	0.08	0.04	0.06	0.02
P	< 0.01	< 0.05	< 0.05	< 0.01	< 0.01	< 0.01
Neg C.	1.96	3.56	1.83	1.51	2.56	1.70

Different superscripts within a column are significant at a *P* value of < 0.05; Winkelman AASV 2018

Table 2: Day 69 necropsy parameters summary by treatment.

	N	Mean BW	Fecal CT value	IPMA titer	IHC	PPE severity score	Clinical obs # pig positive (%pos)
Challenged control	26	97.5[a]	25.30[a]	1:3724	2.21[a]	188[a]	15 (58%)[a]
DLI	24	121.3[b]	29.84[b]	1:1270	1.42[b]	35[b]	3 (8%)[b]
Enterisol	26	117.0[b]	30.63[b]	1:797	1.15[b]	45[b]	5 (19%)[b]
Porcilis	24	117.0[b]	33.11[b]	1:2880	1.04[b]	15[b]	4 (17%)[b]
Neg control	10	119.3[b]	0	0	0	0	0

Winkelman AASV 2018

In this trial, the *Lawsonia* challenge overwhelmed the vaccinated pigs so a combination vaccine and an antibiotic pulse may have further controlled clinical and subclinical ileitis.

Summary –key points

1. **Vaccine expectations** – To be considered effective, ileitis vaccine programs should completely prevent clinical signs and keep subclinical ileitis below an economic threshold. Ideally there should be no carrier pigs (positive PCR fecals) in mid to late finishing pigs or replacement gilts.
2. **Monitor your Ileitis program** – Ante-mortem diagnostics with routine fecal PCRs and quantative ct values will diagnose suspect ileitis diarrhea and subclinical carriers. Cross-sectional fecal and IPMA serological surveys taken every 4-6 weeks can help monitor control programs.
3. **Determine potential reasons for vaccine "Failure".**
4. **Timely feed antibiotic pulses or DLI immunity** are two ways to better control ileitis in difficult pig-flows.
5. **Understand your Ileitis control program** – Know what to expect for *Lawsonia* IPMA titers or fecal PCR's with ileitis vaccines, DLI® immunity, and/or feed medications. This will allow you to better interpret field infections.

References

1. Winkelman N., Mueller A., Domangue R.; Comparison of DLI[R] Controlled Immunity and Two Commercial Ileitis Vaccines (Porcilis[R] and Enterisol[R]) in Grower Pigs Challenged with *Lawsonia intracellularis*; AASV 2018 pp 195-199.
2. Paradis M., Gebhart C.J., Tool, et al; Subclinical Ileitis: Diagnostic and Performance Parameters in a Multi-Dose Mucosal Homogenate Challenge Model ; J of Swine Health and Prod.; 2012, 20 (3): pp 137-141.
3. Winkelman N.; Subclinical Ileitis: Diagnostic Monitoring, R^2, and Economics; AASV 2018 pp 488-493.
4. Vasquez E.; Comparison of IPMA and Commercial Blocking ELISA for *Lawsonia intracellularis*; AASV 2018.
5. Day C.; Is Vaccine Failing Me or Did I Fail Vaccine? National Hog Farmer Feb; 2018 pp 6-12.
6. Gebhart C., Winkelman N., Deen J., Gramm B.; Development of Immunity to *Lawsonia intracellularis* in Pigs Fed Carbodox (Mecadox[R]); AASV 2003 pp 141-145.
7. Kesl, L., Saltzman, R., Winkelman N., et al; Tylan[R] Premix and Enterisol[R] Ileitis Vaccine Evaluation in a *Lawsonia intracellularis* Challenge Model; Proceedings of AASV Animal Meeting pp 139-142, 2004.
8. Collins A.M., Love R.J.; Re-Challenge of Pigs Following Recovery From Proliferative Enteropathy; Vet Micro 120 (2007) pp 381-386.
9. Winkelman N.L., Mueller A., Domangue R.; Effectiveness of DLI[R] (Diluted *Lawsonia intracellularis*) Vaccine Followed by Denagard[R] LC, Aivlosin, or Mecadox in Finishing Pigs Challenged with a High Dose LI Mucosal Homogenate; AASV 2014 pp 273-278.

Figure 1: Strong negative correlation between *Lawsonia* fecal shedding and IHC score

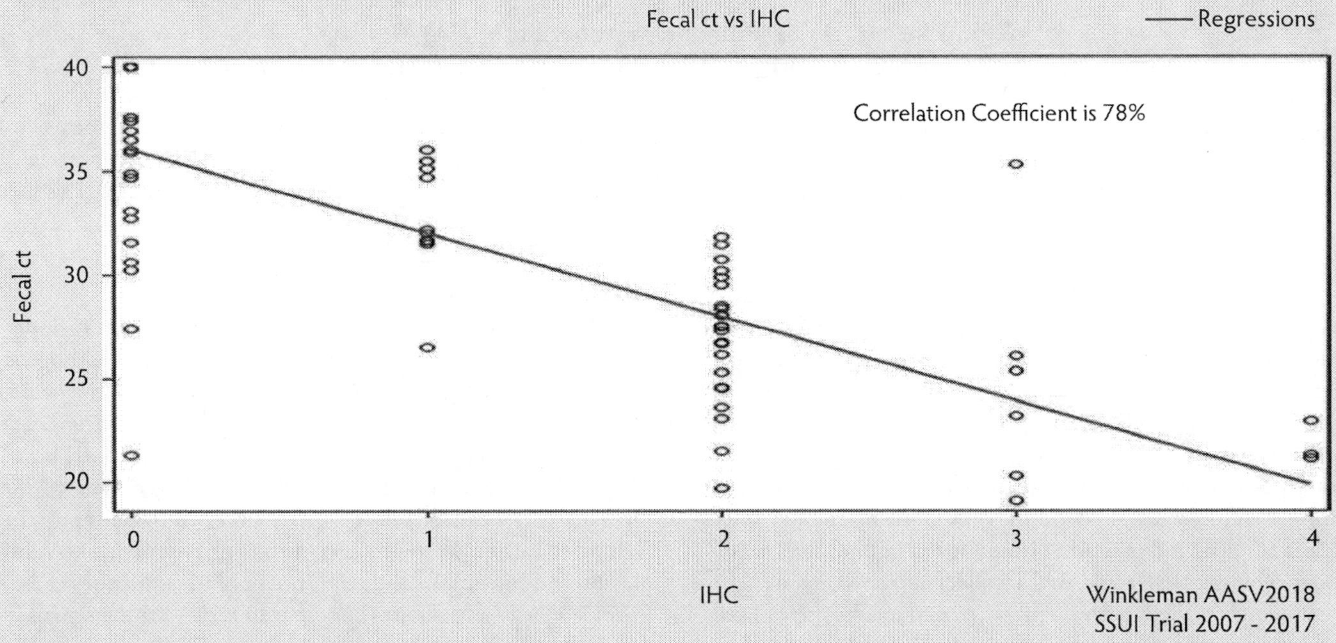

Figure 2: Day 69 (26 DPC)-% positive (suspect), subclinical, and negative PCR by group

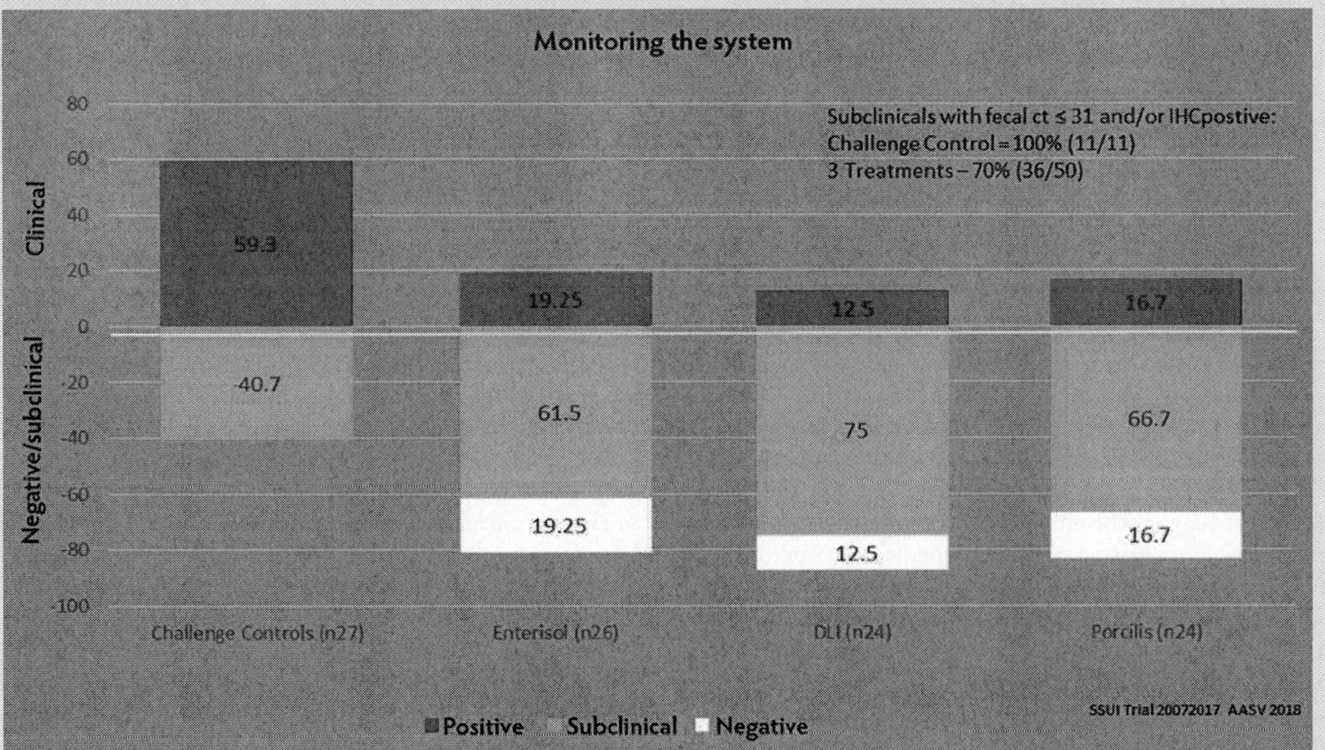

Respiratory immunology – a flu focus

A. L. Vincent, DVM, PhD; M. Wymore Brand, DVM, PhD
Virus and Prion Research Unit, National Animal Disease Center, USDA-ARS, Ames, Iowa

Introduction

The respiratory tract, like the gastrointestinal, urinary, and reproductive tracts, has the unique challenge of mucosal surfaces exposed to the external environment and its microorganisms, which may include pathogens. Unique features to protect the respiratory tract include lymphoid organs (lymph nodes and tonsils), submucosal nasopharyngeal- and bronchus- associated lymphoid tissue (NALT and BALT respectively), and the mucociliary apparatus of the ciliated lower respiratory epithelium (Figure 1).[1] Immune cells of the innate and adaptive immune system can be found throughout the respiratory tract, but are also recruited from the periphery upon pro-inflammatory and anti-viral signaling. Macrophages and dendritic cells in the respiratory tract collect antigens and present to lymphocytes in draining lymph nodes, whereas NALT and BALT collect antigens directly across the mucosal epithelium through specialized microfold (M) cells and are presented to B- and T-cells in submucosal follicles. Lymphoid tissue of the upper respiratory tract bias B-cell responses toward IgA at the expense of IgG, whereas lymphoid tissues of the lower respiratory tract tend to be biased toward IgG at the expense of IgA. Polymeric IgA tethered with a J-chain is more efficiently transported across the pseudostratified epithelium of the upper respiratory tract by the polymeric Ig-receptor, while IgG can transudate across the thin epithelial membranes of alveolar capillaries and alveoli.[2]

Respiratory viruses are particularly effective in avoiding the physiologic and immune barriers of the respiratory tract due to small particle sizes and rapid life cycles, as well as viral mechanisms to evade innate and adaptive immune mediators. Influenza A viruses (IAV) can gain access to respiratory mucosa through aerosolized particles as well as via direct contact with infectious virus in nasal secretions. IAV is in the family *Orthomyxoviridae* and has a negative sense single stranded RNA genome in 8 coding segments. In a naïve host primary response, IAV establishes infection, replicates, sheds, and clears within approximately 7-10 days, with a lag of several days before mature antigen-specific B- and T-cells are detected, and a lag of 1-2 weeks before antigen-specific protective antibodies are detected. The secondary immune response to IAV occurs much quicker and can provide sterilizing protection from homologous IAV. However, IAV evolves rapidly in the genes encoding proteins that the immune system targets, creating a continual battle between the virus and the immune system designed to guard the respiratory tract. The hemagglutinin (HA) protein of IAV mediates virus binding to the host receptors in the respiratory tract, therefore the HA protein is a major target of protective immune responses. However, the IAV surface glycoproteins continually accumulate mutations during viral replication that may lead to immune evasion call "antigenic drift," or exchange of viral RNA gene segments in co-infected cells may lead to more a dramatic change in surface glycoproteins through a process called reassortment, that can then result in immune evasion called "antigenic shift."

Innate immune response to IAV

The physical barrier for mucosa of the respiratory tract includes mucus and mucin glycoproteins to trap particles and pathogens that can then be cleared by upward movement of the mucociliary apparatus. The neuraminidase (NA) of the influenza virus enzymatically degrades mucins to facilitate HA binding to the host receptors on epithelial cells.[3] A family of proteins associated with the mucus layer called collectins, particularly surfactant protein D, bind to carbohydrates on IAV surface glycoproteins HA and NA, promoting uptake and recruitment of inflammatory cells. Infected epithelial cells are the source of early warning signals through pathogen sensing receptors and production of type I interferon and other cytokines. Type I interferon production creates an anti-viral state and recruits innate and adaptive immune cells through pro-inflammatory cytokines and chemokines. Type I interferon signals to surrounding infected and uninfected cells to reduce viral replication and promote apoptosis. Because of the importance of this innate signal, IAV evolved a mechanism to interfere with host type I interferon production through the NS-1 protein, dampening the innate immune response long enough for the short IAV life cycle to complete in the infected host and transmit to the next susceptible host.

Resident pulmonary alveolar macrophages (PAMS), dendritic cells, and neutrophils are phagocytic cells that also play a role in innate immune response to IAV. PAMs are important in the response to IAV by phagocytosis, complement activation, producing inflammatory signals, and recruitment of other immune cells. PAMs are also important in limiting lung damage by feedback to reduce inflammation and promote airway repair. PAM have also been shown to be infected by IAV, altering the response compared to PAM that phagocytose IAV particles.[4] Dendritic cells reside in the submucosa and sample the mucosal surface and airways with dendrites that extend through tight junctions between epithelial cells. Dendritic cells are also phagocytic cells and are especially important as they serve as a bridge between innate and adaptive immune responses, migrating to draining lymph nodes to present antigens from IAV detected in the lungs to T-cells. Neutrophils are recruited to lung during IAV infection to target viral infected cells and enhance the cytotoxic T-cell response. Natural killer cells also target IAV infected cells through innate receptors and signals and lyse them by releasing cytotoxic granules.

Figure 1: Respiratory tract and associated immune system in swine. The respiratory tract is divided into upper, (rostral to the trachea) and lower, (the trachea to the alveolar spaces of the lungs). Lymphoid tissues include tonsils, mucosal associated lymphoid tissues (nasopharyngeal, NALT and bronchus, BALT), and lymph nodes. The antibody isotype IgA tends to be found in higher levels in the upper respiratory tract, whereas IgG tends to be in higher levels in the lung.

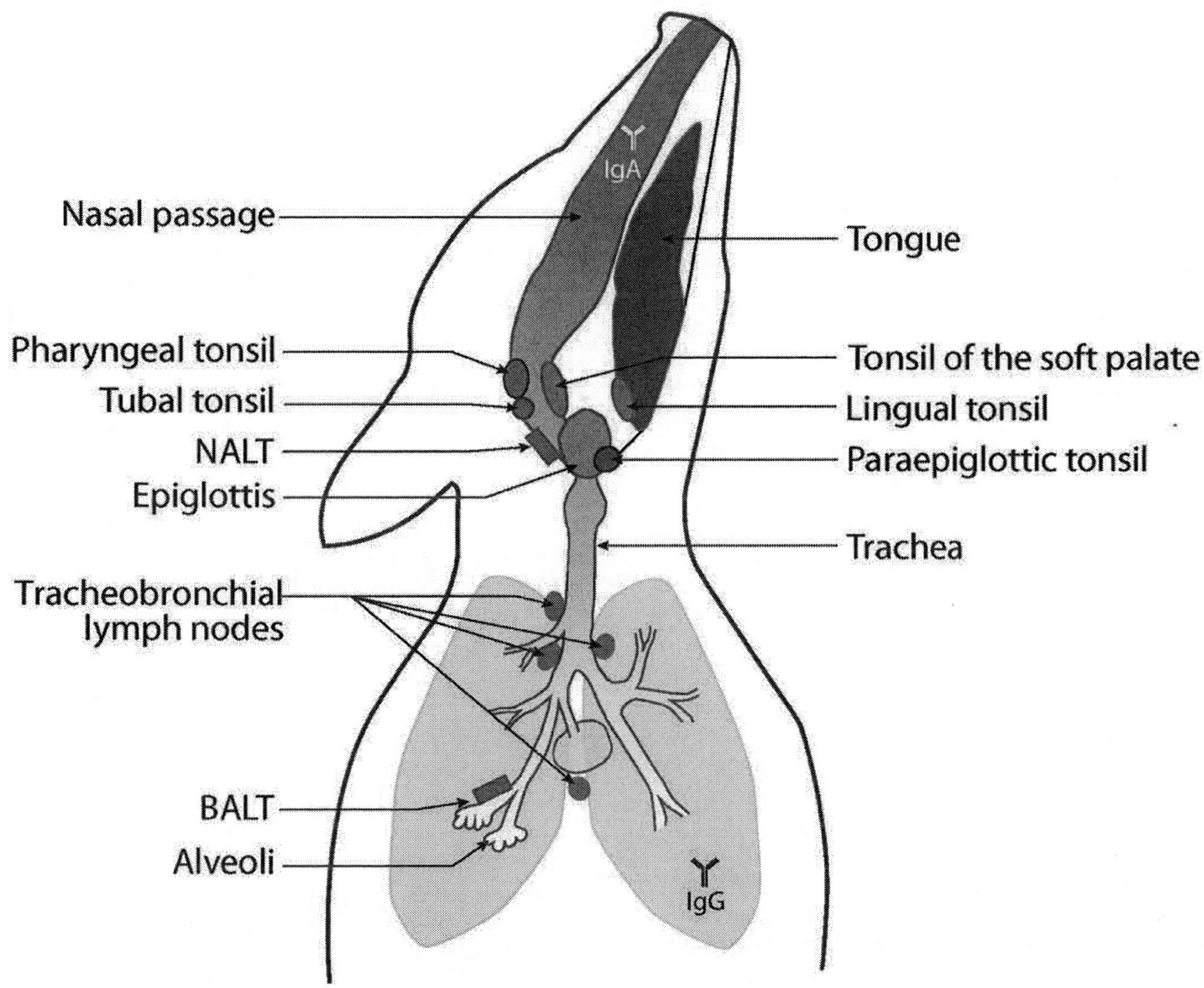

Adaptive immune response to IAV

In contrast to the generalized anti-microbial sensors and cells of the innate immune system, the adaptive immune system involves B- and T-lymphocytes that become highly specialized to particular epitopes on pathogen-specific antigens. Antibodies produced by B-cells are a focus of influenza immunity because of their highly neutralizing capacity when well-matched to epitopes near the receptor binding site on HA proteins, but also due to their diagnostic use and correlation with vaccine protection. Antibody responses to IAV in swine are primarily against the HA, NA, M, and NP proteins.[5] But, only antibodies to the globular head region of the HA can block attachment of the virus to host cell receptors and neutralize viral infectivity. These antibodies can be measured in virus neutralization or hemagglutination inhibition (HI) assays. Antibodies against the NA primarily act after infection has been established to reduce viral titers and shedding by limiting virus release from infected cells. Protective anti-NA antibodies can be measured in neuraminidase inhibition (NI) assays.

Antibodies can be divided into different isotypes, and depending on the isotype and specificity, may serve different functions. Influenza-specific IgM and IgG are the dominant antibody isotypes in serum, whereas IgA is the dominant isotype in the upper respiratory tract following natural exposure. Antibodies in the lung largely transudate from serum, and there is a predominance of virus-specific IgG in bronchoalveolar lavage fluid (BALF) samples of pigs [6] However, substantial IgA levels have also been found in BALF of IAV-infected pigs[7,8] and local antibody production in the lung parenchyma may also occur. Secretory IgA as dimers or multimers are particularly effective at neutralizing homologous IAV and have been shown to have greater cross-protection against heterologous strains.[3] Antibodies to other proteins cannot prevent infection, but they can mediate killing of infected cells by antibody-dependent mechanisms, especially IgM and IgG.[9] Antibodies bound to virions can facilitate viruses to be taken up through the Fc portion of the antibody when it binds to Fc receptors on phagocytic cells. This type of interaction leads to activation of the phagocytic cells and expression of proinflammatory cytokines. Antibodies can also bind to viral proteins being expressed on the surface of infected epithelial cells and promote killing of these infected cells through mechanisms such as complement and antibody dependent cell cytotoxicity (ADCC). Antibodies to conserved epitopes on the HA, such as the stalk domain, or to conserved internal proteins are more likely to mediate ADCC rather than neutralize viral infection. Thus, they may fail to prevent infection, but serve to reduce the magnitude and duration of the infection and illness. However, these antibodies and the cell killing they elicit may also play a role in immunopathology.[10]

Antibody expressed on the surface of B-cells, known as the B-cell receptor, can also bind to IAV, activating the B-cell and initiating the cascade of signals and interactions, including that of T-helper cells, for B-cell proliferation, differentiation, and affinity maturation. High affinity B-cells can become memory B-cells or plasma cells that migrate to the bone marrow to secrete highly specific antibodies. B-cells stimulated in mucosal lymphoid tissue that have switched to IgA antibody production leave the NALT or BALT to circulate in the blood and enter sub-mucosal surfaces where they differentiate into plasma cells to secrete IgA that can then be transported in dimeric form to the mucosal surface.[8,9] Influenza specific memory B-cells have been identified in lungs, suggesting there may also be lung resident memory B-cells.

T-cells can be divided into subsets based on surface markers and function, simplistically as CD4+ helper T-cells and CD8+ killer or cytotoxic T-cells (CTL). CD8+ CTL are essential in primary influenza infections to clear virus and are also important in secondary infections when prior antibody is no longer neutralizing due to viral shift or drift, but conserved antigens are presented on the surface of infected epithelial cells for memory CTL to encounter. Cytokines and chemokines signal T-cells to migrate to infected lungs where they kill virus infected cells by releasing cytotoxic granules and produce potent cytokines such as Type II interferon to further activate macrophages and antiviral protein expression. However, unchecked or disproportionate CTL activity can lead to immunopathology and pulmonary damage, and regulatory T-cells function to limit the extent of inflammation by controlling these pro-inflammatory immune responses. Helper T-cells are needed in influenza infections to stimulate B-cells, activate CD8+ T-cells and macrophages, and recruit other immune cells into infected lungs. CD4+ Th1-cells are potent generators of cytokines needed to enhance CD8+ responses to influenza, however disproportionate Th2 cell responses have been associated with a failure to control infection and may lead to greater inflammation. Gamma delta T-cells have both innate and adaptive immune features, and are found at higher levels in the blood of pigs compared to many other species, though their levels decrease with age. These unconventional T lymphocytes are also found in the mucosa, though their role in swine respiratory immune responses has not been well elucidated.

Cytokine-producing virus-specific CTL and helper T-cells are induced following IAV infection, working together to promote viral clearance.[11,12] A small population of these effector T-cells generated following a primary immune response to IAV survive to become memory T cells, and memory T-cells reactive to heterologous viruses have been identified in pig lungs following IAV infection.[12] Memory T-cells are primed for specific effector functions and can rapidly migrate to the lung to respond upon subsequent infection to promote faster viral clearance as compared to a primary response. When neutralizing antibody fails to prevent infection, memory helper T-cells can activate dendritic cells and B-cells to make antibodies targeting different epitopes more rapidly than what occurred in the primary response. A small proportion of memory T-cells have been shown to become resident T-cells in the lungs and important in response to secondary influenza infection.

Secondary and vaccine immunity

After a primary IAV infection in swine, protection is typically complete against reinfection with the same or a similar virus strain. Based on findings in humans, immunity may last for several years or even a lifetime, but duration of immunity has not been well-studied in the pig. However, because of the concurrent circulation of different subtypes and phylogenetic lineages and subclades of influenza viruses, pigs are typically exposed to multiple antigenically drifted and/or shifted IAV strains during their lifetime. There is no serologic cross-reaction between H1 and H3 subtype IAV strains in the HI test, nor in some cases between viruses of different clades within H1 and H3 lineages. Some experimental infection studies have shown cross-protection between influenza viruses in the absence of cross-reactive serum HI antibodies. The immune mediators of this "broad" protection need further study, but may include a combination of mucosal IgA antibodies, CMI responses, and cross-reactive VN and/or NI antibodies. In live attenuated influenza vaccine (LAIV) studies, cross-protection between H1 and H3 IAV subtypes was limited.[13,14] Reduced pathology following heterosubtypic infection was recently demonstrated with a single cycle influenza vaccine that correlated with an increased T-cell anti-viral response.[15]

Whole inactivated vaccines (WIV) are administered parenterally through intramuscular routes whereas live attenuated influenza vaccines (LAIV) are typically delivered mucosally through intranasal routes. Vectored or RNA vaccines may be delivered through either route. The route of administration as well as the manner in which vaccine antigens are presented to the immune system influences the resulting host immune response (Figure 2).[16] The immune response after intramuscular administration of whole inactivated-IAV vaccine (WIV) is fundamentally different from live virus infection. Vaccination immunity from WIV relies largely on inducing high titers of serum HI and VN antibodies to the HA of the vaccine strain(s). In contrast, mucosal IgA antibodies or CD8+ T-cells are not efficiently induced by WIV. Vectored vaccines may induce a more balanced humoral and CMI response, but do not typically stimulate the mucosal immune system if delivered by peripheral injection. There is evidence for complete or partial cross-protection between clades within subtypes of North American H1 and H3 IAV that was increased with vectored and LAIV vaccines.[17,18]

Maternally-derived serum IgG antibodies to IAV can protect young pigs against antigenically related viruses, but they interfere with the development of an active antibody response to vaccination or infection. Maternal IAV antibody levels in newborn pigs reflect the dam's immune status. Passive IAV antibody declines in the pig over a period of 4-14 weeks.[19] Research showed that pigs with maternal antibodies were not completely protected from nasal virus shedding upon challenge, but some pigs showed protection against clinical disease.[20,21] Passive antibody can also alter response to vaccine and induce immunopathology.[22,23]

Vaccine-associated enhanced respiratory disease (VAERD) has been described in pigs vaccinated with experimental whole-virus inactivated vaccine and challenged with heterologous strains of the same IAV subtype.[24-26] VAERD is characterized by greater gross and microscopic pathology in vaccinated pigs versus unvaccinated controls following challenge. It appears to be restricted to pairs of viruses of the same HA subtype which have drifted sufficiently as to no longer stimulate cross-reacting HI or VN antibodies.[10]

Figure 2: The route of administration and vaccine delivery platform influences the resulting host immune response.[16] The immune response after intramuscular administration of whole inactivated-IAV vaccine (WIV) is fundamentally different from live virus infection. Vaccination immunity from WIV relies largely on inducing high titers of serum antibodies to the HA of the vaccine strain(s). In contrast, mucosal IgA antibodies or CD8+ T-cells are not efficiently induced by WIV. Vectored vaccines may induce a more balanced humoral and CMI response depending on the platform and expressed gene(s), but do not typically stimulate the respiratory mucosal immune system if delivered by peripheral injection.

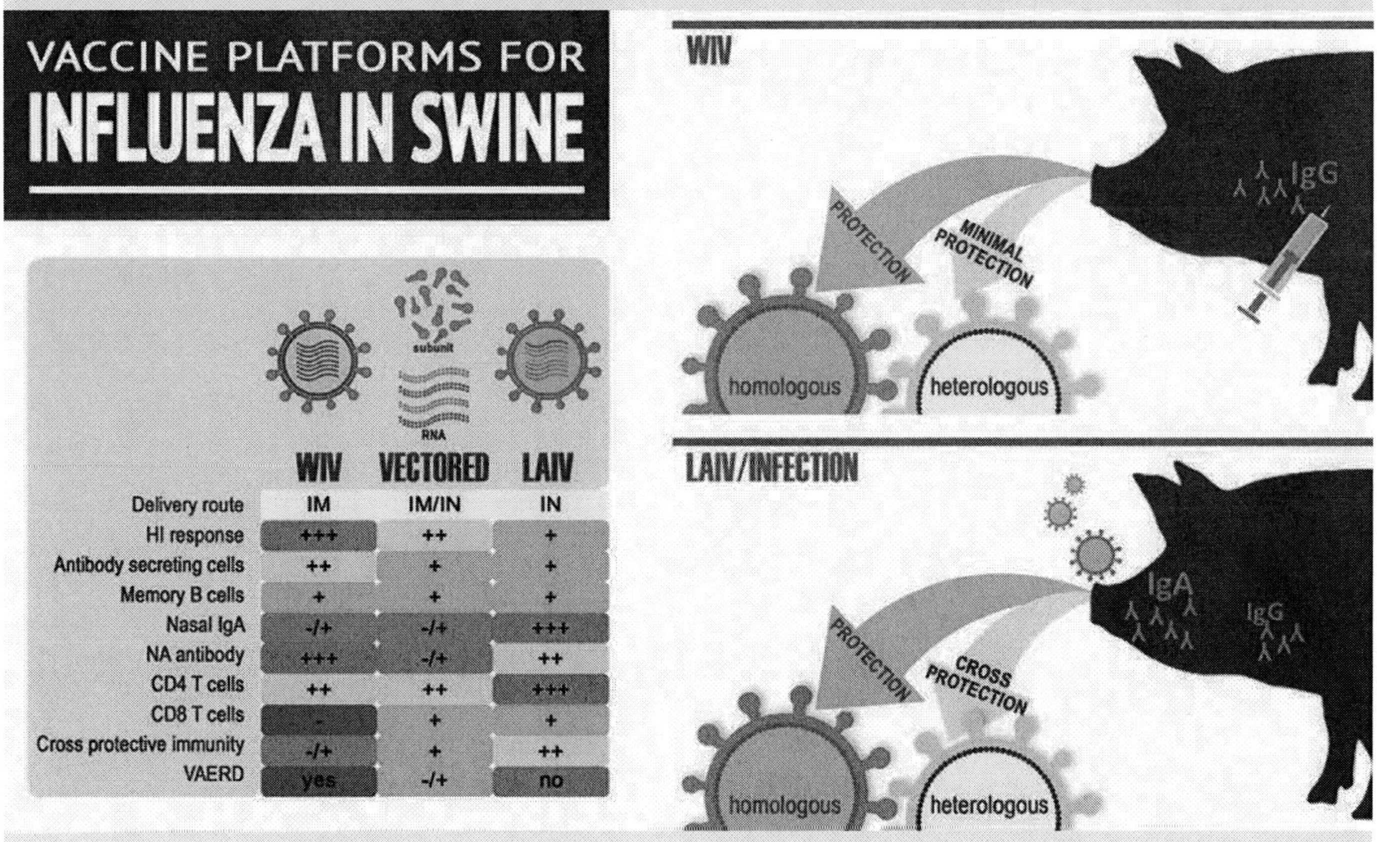

Conclusions

Because of its plasticity and complicated epidemiology in swine, IAV presents a challenge to the respiratory immune system and, thus, a challenge to control through vaccination. A greater understanding has occurred in recent years with regard to the intricate immune responses to exposure, both primary and repeated exposures in a lifetime, as well as responses to vaccines administered to immune and partially immune individuals. However, many of these types of studies are lacking in the swine host. Improved knowledge of protective responses at the respiratory mucosa and how different vaccine platforms and viral antigens alter these responses are needed in swine to improve our ability to control this important pathogen that affects so many species.

References

1. Liebler-Tenorio EM, Pabst R: MALT structure and function in farm animals. Vet Res 37:257-280, 2006.

2. Allie SR, Randall TD: Pulmonary immunity to viruses. Clin Sci (Lond) 131:1737-1762, 2017.

3. Chen X, Liu S, Goraya MU, et al: Host Immune Response to Influenza A Virus Infection. Front Immunol 9:320, 2018.

4. Marvin SA, Russier M, Huerta CT, et al: Influenza Virus Overcomes Cellular Blocks To Productively Replicate, Impacting Macrophage Function. J Virol 91, 2017.

5. Kim WI, Wu WH, Janke B, et al: Characterization of the humoral immune response of experimentally infected and vaccinated pigs to swine influenza viral proteins. Arch Virol 151:23-36, 2006.

6. Heinen PP, de Boer-Luijtze EA, Bianchi AT: Respiratory and systemic humoral and cellular immune responses of pigs to a heterosubtypic influenza A virus infection. J Gen Virol 82:2697-2707, 2001.

7. Kitikoon P, Nilubol D, Erickson BJ, et al: The immune response and maternal antibody interference to a heterologous H1N1 swine influenza virus infection following vaccination. Vet Immunol Immunopathol 112:117-128, 2006.

8. Larsen DL, Karasin A, Zuckermann F, et al: Systemic and mucosal immune responses to H1N1 influenza virus infection in pigs. Vet Microbiol 74:117-131, 2000.

9. Zimmerman JJ: Diseases of swine, in (ed 11th edition.), Vol. Hoboken, NJ, Wiley-Blackwell,, 2019, p 1 online resource.

10. Khurana S, Loving CL, Manischewitz J, et al: Vaccine-induced anti-HA2 antibodies promote virus fusion and enhance influenza virus respiratory disease. Sci Transl Med 5:200ra114, 2013.

11. Talker SC, Koinig HC, Stadler M, et al: Magnitude and kinetics of multifunctional CD4+ and CD8beta+ T cells in pigs infected with swine influenza A virus. Vet Res 46:52, 2015.

12. Talker SC, Stadler M, Koinig HC, et al: Influenza A Virus Infection in Pigs Attracts Multifunctional and Cross-Reactive T Cells to the Lung. J Virol 90:9364-9382, 2016.

13. Vincent AL, Ma W, Lager KM, et al: Efficacy of intranasal administration of a truncated NS1 modified live influenza virus vaccine in swine. Vaccine 25:7999-8009, 2007.

14. Kappes MA, Sandbulte MR, Platt R, et al: Vaccination with NS1-truncated H3N2 swine influenza virus primes T cells and confers cross-protection against an H1N1 heterosubtypic challenge in pigs. Vaccine 30:280-288, 2012.

15. Holzer B, Morgan SB, Matsuoka Y, et al: Comparison of Heterosubtypic Protection in Ferrets and Pigs Induced by a Single-Cycle Influenza Vaccine. J Immunol 200:4068-4077, 2018.

16. Vincent AL, Perez DR, Rajao D, et al: Influenza A virus vaccines for swine. Vet Microbiol 206:35-44, 2017.

17. Vincent AL, Ciacci-Zanella JR, Lorusso A, et al: Efficacy of inactivated swine influenza virus vaccines against the 2009 A/H1N1 influenza virus in pigs. Vaccine 28:2782-2787, 2010.

18. Abente EJ, Rajao DS, Santos J, et al: Comparison of Adjuvanted-Whole Inactivated Virus and Live-Attenuated Virus Vaccines against Challenge with Contemporary, Antigenically Distinct H3N2 Influenza A Viruses. J Virol 92, 2018.

19. Loeffen WL, Heinen PP, Bianchi AT, et al: Effect of maternally derived antibodies on the clinical signs and immune response in pigs after primary and secondary infection with an influenza H1N1 virus. Vet Immunol Immunopathol 92:23-35, 2003.

20. Corzo CA, Allerson M, Gramer M, et al: Detection of Airborne Influenza A Virus in Experimentally Infected Pigs With Maternally Derived Antibodies. Transbound Emerg Dis 61:28-36, 2014.

21. Allerson M, Deen J, Detmer SE, et al: The impact of maternally derived immunity on influenza A virus transmission in neonatal pig populations. Vaccine 31:500-505, 2013.

22. Rajao DS, Sandbulte MR, Gauger PC, et al: Heterologous challenge in the presence of maternally-derived antibodies results in vaccine-associated enhanced respiratory disease in weaned piglets. Virology 491:79-88, 2016.

23. Vincent AL, Ma W, Lager KM, et al: Live attenuated influenza vaccine provides superior protection from heterologous infection in pigs with maternal antibodies without inducing vaccine-associated enhanced respiratory disease. J Virol 86:10597-10605, 2012.

24. Gauger PC, Vincent AL, Loving CL, et al: Kinetics of lung lesion development and pro-inflammatory cytokine response in pigs with vaccine-associated enhanced respiratory disease induced by challenge with pandemic (2009) A/H1N1 influenza virus. Vet Pathol 49:900-912, 2012.

25. Gauger PC, Vincent AL, Loving CL, et al: Enhanced pneumonia and disease in pigs vaccinated with an inactivated human-like (delta-cluster) H1N2 vaccine and challenged with pandemic 2009 H1N1 influenza virus. Vaccine 29:2712-2719, 2011.

26. Vincent AL, Lager KM, Janke BH, et al: Failure of protection and enhanced pneumonia with a US H1N2 swine influenza virus in pigs vaccinated with an inactivated classical swine H1N1 vaccine. Vet Microbiol 126:310-323, 2008.

Universal influenza vaccines and vaccine platforms for swine

Daniela Rajao, DVM, MS, PhD; Daniel Perez, PhD
Department of Population Health, University of Georgia, Athens, Georgia

Introduction

Influenza A viruses (IAV) are among the most devastating pathogens for swine production. IAV infections in pigs lead to rapid onset of fever, lethargy, loss of appetite, and coughing.[1] Although the disease is typically mild, the high morbidity results in significant loss in production due to decreased weight gain, costs with secondary infection treatments, and sporadic abortion associated with hyperthermia in sows.[2] IAV are rapidly evolving viruses that undergo accumulation of mutations in the HA and NA genes due to the lack of proof-reading activity of the viral polymerase. These mutations lead to antigenic drift, which over time results in escape from previous immune responses. IAV can also undergo antigenic shift through reassortment, whereby an IAV with a new HA subtype enters and readily transmits in a population with no prior immunity.

Several genetically and antigenically diverse IAV strains are endemic in swine worldwide. Four major lineages of H1 and H3 IAV circulate in U.S. pigs: classical, delta (δ), 1990s H3-lineage, and 2010s H3-lineage.[3,4] The classical H1N1 lineage emerged around the 1918 human pandemic and predominated until the late 1990's, when a novel triple-reassortant H3N2 virus emerged and established the 1990s H3-lineage.[5,6] Shortly after, new introductions of human-origin H1N1 resulted in the establishment of the δ-lineage H1N1 and H1N2 viruses.[7] And recently, a novel virus derived from human seasonal H3 IAV led to establishment of the 2010s H3-lineage.[8] Since the introduction of the triple-reassortant H3N2 virus in the late 1990's, the triple-reassortant internal gene (TRIG) constellation became predominant among all viruses circulating in pigs in the U.S. There is considerable genetic diversity among the hemagglutinin (HA) gene between and within these lineages, with 16 phylogenetic clades co-circulating in North American pigs: the classical lineage is divided into α, β, γ, γ2, and 2009 pandemic (pdm09) phylogenetic clades; the δ-lineage is divided into H1-δ1a, H1-δ1b and H1-δ2 clades; the 1990s H3-lineage is divided into IV, IV-A, IV-B, IV-C, IV-D, IV-E, IV-F clades; and the 2010s H3-lineage is comprised of the human-like H3 clade.[3,4,8,9] These HA phylogenetic clades are paired with either N2 genes derived from 1998 or 2002 human H3N2 seasonal-lineages, or N1 genes from the classical lineage or from human H1N1 seasonal-lineage.[4] Different HA and NA clades predominate in different regions of the US, and the HA/NA combinations also vary.[3]

Current vaccines available for use in swine

Vaccination remains the most efficient and cost-effective means to prevent and control influenza in swine populations. Vaccines mostly rely on the effective stimulation of the immune response against the surface HA glycoprotein, the primary immunogen of influenza viruses. The HA is a highly variable protein and a major site of antigenic drift that can lead to vaccine mismatch. It was demonstrated that changes in only 6 amino acid positions in the HA account for major antigenic changes of swine H3 influenza viruses, and single amino acid changes can have a significant impact in the emergence of a new antigenic strain.[10,11] This changing nature of influenza viruses that may result in escape from earlier immune responses is one of the biggest challenges to an effective vaccination program. Frequent failure of IAV vaccines in swine production systems can lead to extreme frustration in the swine industry. Efforts towards overcoming the challenges for an effective influenza vaccination program aim at developing vaccines that induce broadly cross-protective and long-lasting immune responses. Novel vaccine technologies, the so-called "universal" vaccines, focus on stimulating better cross-protection against many or all influenza strains. However, more homogeneous vaccine platforms, that can be used in different species and scenarios, could bring benefit to the industry as well.

Typical vaccines against influenza currently on the market for swine are whole inactivated virus (WIV) vaccines combined with potent oil-in-water adjuvants delivered by the intramuscular route.[12] Inactivation is accomplished using formalin, β-propiolactone or binary ethylenimine. Originally, IAV vaccines available for use in swine were bivalent products, but some newer products contain 4-5 strains in response to the circulation of antigenically distinct viruses within the H1 and H3 subtypes.[12] Vaccination is typically done in sows to transfer maternally derived antibodies (MDA) through colostrum to their litters, in two to three doses prior to farrowing.[13] Sow vaccination will mostly protect the litter from clinical signs of disease, and MDA have been shown to persist until about 14 weeks of age in piglets born to vaccinated sows.[14] Occasionally and in herds with high virus circulation, vaccination is done in growing pigs to protect against infection once MDA have weaned. Inactivated vaccines are usually protective against influenza viruses that are antigenically the same or closely related to the viruses in the vaccine, but small antigenic changes lead to potential loss of cross-reaction in the hemagglutination inhibition (HI) assay, which is associated with loss in protection.[15]

More recently, two new vaccine platforms were licensed for swine: a technology based on a non-replicating alphavirus RNA particle[16,17] and a live-attenuated influenza virus (LAIV) vaccine. The alphavirus-vectored vaccine is based in single-cycle, propagation-defective particles generated by replacing the structural genes from the Venezuela equine encephalitis (VEE) virus with the transgene of choice rendering them unable to spread beyond the initial infected cells.[16-18] This platform was shown to reduce lung pathology and virus replication after homologous H1 and H3 challenge.[16-18] The LAIV vaccine is based on a modification in the nonstructural 1 (NS1) protein that compromises viral inhibition of the interferon I responses.[19] This vaccine was shown to reduce the number of pigs shedding after challenge when used as a bivalent.[20] Although there are several vaccines to use against IAV in swine and the new platforms show promising results, most products tend to be efficacious against homologous or related viruses, but protection drops considerably when animals are exposed to antigenically distinct viruses. Thus, the current vaccines available for use in swine, or even in humans, still do not provide cross-protection as broad as is intended by "universal" vaccines.

Given the viral diversity and variability in regional distribution, vaccine components should be selected based on surveillance taking characteristics of viral ecology under consideration. Hence, identifying circulating viruses, their spatial/temporal trends, and their antigenic relationships to other circulating and vaccine strains are crucial steps to develop an effective influenza vaccine. For human vaccines, the WHO monitors, through the Global Influenza Surveillance and Response System (GISRS), the evolution of seasonal viruses by using data obtained year-round from national influenza centers located around the globe. Meetings are held twice a year to select vaccine strains for the Northern and Southern Hemispheres. The strain recommendation is based on antigenic data from ferrets antisera, paired serologic analysis of human samples, and genetic, epidemiological and clinical surveillance data. In general, each component in the vaccine is updated every 2 to 3 years, with at least one component updated each year. A similar approach to the human vaccine selection is largely lacking in animal production systems except for the selection of vaccine strains against equine influenza.[21] However, a national swine IAV surveillance system was implemented by the USDA in 2009[4] and has generated consistent flow of genetic data that could provide input for vaccine strain selection. Still lacking are collaborative panels to continuously evaluate this surveillance data in conjunction with antigenic characteristics of predominant circulating strains. Therefore, swine influenza vaccine manufacturers make independent decisions for which strains to include in their products, and those formulations are not frequently updated. It is important to emphasize that, unlike IAVs in humans that follow common evolutionary trends, IAVs in pigs tend to adopt unique evolutionary pathways depending on the country, region, and even at the farm level. The high diversity and heterogeneous distribution of different viral lineages circulating in pigs represents a major obstacle in the development of effective vaccines against swine influenza, which need to be tailored for each country, specific areas, or production systems.[22,23] Following this trend, autogenous vaccines have become frequent products used by swine producers and veterinarians in the US.[12]

Current challenges for effective vaccination programs

The greatest challenge for vaccine manufacturing for both animal and human influenza is the high variability and rapid evolution of the virus that results on a constant chase for the vaccine virus to match the circulating strains. Since standardized programs for vaccine selection and update for swine are nonexistent, vaccine strains are often mismatched and provide suboptimal protection that can promote, rather than stop, the spread of field strains. Additionally, pigs vaccinated with adjuvanted WIV and then exposed to a heterologous virus of the same subtype can exhibit enhanced respiratory pathology due to strong stimulation of non-neutralizing antibody response, termed vaccine-associated enhanced respiratory disease (VAERD).[24] Furthermore, a suboptimal vaccine can result in selection of escape mutants that will pose an increased threat to naïve populations.

An individual's previous immunity can have a major impact on the vaccine responses. It is well accepted that the first immunological encounter with influenza, either by natural infection or vaccination, will shape up future responses in the life of an individual, a process often referred to as *original antigenic sin* and more recently as *antigenic imprinting*. The resulting secondary exposures tend to boost the antibody response to the priming exposure in total or partial detriment of the immunological response against the new strain, a process known as *back-boosting*.[25-27] Additionally, maternally derived antibodies in piglets can interfere with the active immune responses to the vaccine depending on the age at vaccination.[28]

Another obstacle for an effective vaccine is the timeline needed for vaccine manufacture and the regulations for approval. The process to license updated or new products for animal vaccines can be very time-consuming. However, recent changes implemented by the USDA Center for Veterinary Biologics now allow manufacturers to change or replace strains under an existing license without having to go through an entirely new licensing process,[29] which can be an advance on the path to implement vaccine updating programs.

Universal vaccine platforms for use in different species

In the effort to achieve better vaccine products, a myriad of platforms and technologies have been developed. An ideal "universal" vaccine approach would include not only a broadly protective vaccine methodology, but also a more standardized vaccine platform that can be applied indistinctively in multiple animal species or in different epidemiological situations, from endemic circulation to epidemics or pandemics. (Figure 1) The most widely

Figure 1: "Universal" vaccine platforms for use against influenza viruses in multiple species or epidemiological situations. Whole inactivated virus (WIV), live attenuated influenza virus (LAIV), vector, nucleic acid-based, and recombinant vaccines are used or have been successfully tested in multiple species such as pigs, humans, horses, dogs, and poultry.

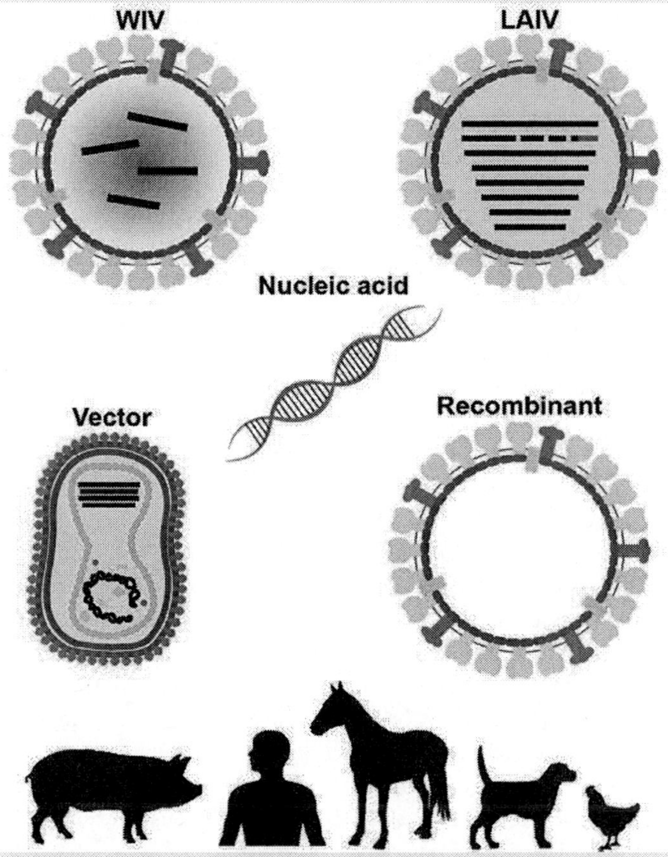

used example of an "universal" platform are inactivated influenza vaccines (IVV), which includes the WIV platform largely used in animal systems. Despite the advantage of being widely used in multiple species, the limitations pointed above highlight the need for the development of ways to improve current technologies or elaboration of alternative approaches. The addition of adjuvants to inactivated products is a method to increase the immunogenicity and efficacy of inactivated vaccines. They are widely used in poultry and swine vaccines, and have been shown to be highly effective in children and the elderly.[30,31]

LAIVs have an advantage over inactivated products that they stimulate both humoral and cell-mediated immunity, and they can induce mucosal IgA responses in the upper respiratory tract, providing cross-reactive and longer-lasting immune responses.[32,33] One perceived risk associated with the use of LAIVs is the possibility of vaccine strain reversion or recombination with circulating field strains; however, LAIV systems have been consistently shown to be safe and stable. LAIVs are already in use for swine, horses, and humans, and similar approaches have also been experimentally tested successfully in poultry and dogs. Three technologies have been tested for use in multiple species, including pigs: i) a combination of an HA tag at the C-terminus of the polymerase basic 1 (PB1) gene along with temperature-sensitive mutations in PB1 and PB2 (*att*LAIV) that limits replication to the temperature of the upper respiratory tract; ii) several amino acid deletions on the nonstructural (NS1) gene (ΔNS1-LAIV) that impedes the interferon-antagonist activity of the viral NS1 protein; iii) a mutation of the cleavage site of the HA (eLAIV) that makes the virus dependent of the presence of elastase for replication, which is absent in the respiratory tract. The *att*LAIV showed remarkable stability over multiple passages and in different IAV backbones of avian-, human-, and swine-origin. This strategy has been shown to be safe and effective resulting in protection against low- and high-pathogenic influenza strains in chickens and against swine-origin IAVs in swine[34,35] and cross protection after challenge with antigenically distinct viruses in swine.[36,37] Several ΔNS1-LAIV were shown to result in virus attenuation and heterologous cross-protection in mice, chickens, and pigs.[38-41] Additionally, this strategy has been proven to be immunogenic in young pigs, prime T cell response and result in partial heterosubtypic cross-protection.[42] The eLAIV technology has been tested and proven effective against homologous and heterologous challenge in swine.[43]

Vectored vaccines, using adenovirus, poxvirus, alphavirus, etc, can mimic natural infection when delivered through the intranasal route and are a safe approach for different species that eliminates the risk of virus recombination. Not all virus vectors can be used in common for all species, but the concept of vectored-vaccine production tends to be similar between different vector systems. Modified vaccinia virus ankara (MVA, poxvirus) is one of the most commonly used vectors for human vaccines. MVA-based vaccines have also been tested in chickens: a MVA-H9 vaccine reduced clinical signs and virus shedding against homologous challenge, but did not prevent infection.[44] A replication-defective vaccine using adenovirus serotype 5 (Ad5) has been tested in pigs and resulted in moderate protection against heterologous challenge.[45,46] As mentioned previously, a single-cycle, propagation-defective alphavirus-like replicon particle vaccine was recently approved to be used in swine.[47]

Nucleic acid-based vaccines, such as DNA or messenger RNA (mRNA)-based vaccines, allow for the expression of target antigens *in vivo* and, therefore, induce both cellular and broad antibody immunity without exposing the host to a live virus. Hence, this technology provides a rapid, stable, and safe alternative to traditional influenza vaccine production, particularly the mRNA technology, and has been effectively tested in multiple host species. DNA vaccines against influenza are often based on expression of the HA protein, but the humoral immunogenicity is considered suboptimal in humans and large animals when compared to the traditional vaccine approaches. To increase their immunogenicity, further optimization has been proposed, including the use of adjuvants, different prime-boost combinations, or different delivery methods.[48,49] Alternative delivery systems, such as intradermal delivery, have been shown to improve levels of cross-reactive antibody and cell-mediated responses in pigs and humans.[50,51] Vaccines produced using recombinant expression systems have been tested in humans, swine and poultry, and thus could be used as a universal platform. A trivalent recombinant HA vaccine has been licensed for human use in individuals 18-49 years of age (Flublok by Protein Sciences Corporation). Recombinant vaccines protected pigs[52] and chickens[53] against homosubtypic heterologous challenge. Virus-like particles (VLPs) are recombinant virus particles formed solely by viral structural proteins that do not have any genomic component and retain the morphology of the virus. VLP vaccines have been shown to be protective and induce long lasting immunity in humans,[54,55] and result in cross-reactive antibody response when used as multivalent formulations in humans and chickens.[56,57] In pigs, VLP antigens consisting of H1N1pdm09 HA, NA, and M1 proteins elicited robust levels of humoral and mucosal immune responses and protected pigs against homologous infection.[58]

Universal vaccine technologies

New approaches are being explored to overcome the drawbacks of the highly changing nature of influenza viruses and are aimed at inducing cross-protective, broadly neutralizing immunity. These so called "universal" vaccines are based on the concept of stimulating an immune response against conserved viral epitopes (Figure 2), which would reduce the frequency of vaccine update and prolong the immune response to reduce the need for re-vaccination. One of these approaches is based on stimulation of immune response against the highly conserved epitopes of the HA. The HA is formed by a globular head domain (HA1) and the stalk or stem domain (HA2). The HA stalk domain is much more conserved across HA subtypes than the head domain. Based on sequence similarities of the stalk region, IAV HA subtypes are phylogenetically classified into two groups: group 1 (H1, H2, H5, H6, H8, H9, H11, H12, H13, and H16) and group 2 (H3, H4, H7, H10, H14, and H15). Antibodies against the HA globular head neutralize infection by either preventing binding to cellular receptors or membrane fusion. Stalk-specific antibodies do not usually neutralize the virus (measured by using standard assays), but protect through inhibition of entry, inhibition of viral release, or by mediating antibody dependent cell-mediated cytotoxicity (ADCC) or complement cytotoxicity. One approach for a "universal" influenza vaccine is based on a "headless" HA as antigen and has been shown to stimulate good levels of anti-stalk broadly neutralizing antibodies (bnAbs) that protected mice against lethal homologous or heterosubtypic challenge,[59] and reduced fever after sub-lethal homologous challenge in cynomolgus monkeys.[60] In a different approach to stimulate anti-stalk antibodies, sequential immunization with antigenically different strains that belong to the same group (1 or 2) or with chimeric constructs that express the same stalk but irrelevant divergent heads protected mice against lethal heterologous challenge.[61-64] The Computationally Optimized Broadly Reactive Antigen (COBRA) technology is another approach being tested as an "universal" vaccine. Using consensus sequences that represent multiple circulating strains, it resulted in broad cross-reactive response to multiple H1 strains, protecting mice against H1N1 challenge.[65]

Other broadly protective vaccine candidates that are currently under investigation target the NA or the extracellular domain of the M2 protein (M2e) to induce cross-reactive antibodies. The influenza NA protein is antigenetically more conserved than the HA protein, and several animal studies have shown that NA-specific antibodies may induce cross-protective immunity against homologous and heterologous strains that are closely related, usually within a subtype.[66-69] Pigs previously exposed to a swine H1N1 virus showed cross-reactive neuraminidase inhibiting antibodies, were clinically protected against heterologous H5N1 challenge and showed reduced viral shedding.[70] However, when pigs were vaccinated with inactivated adjuvanted vaccine and challenged with a virus containing heterologous homosubtypic HA and NA, they showed enhanced disease characteristic of VAERD, but the enhancement was abrogated when a matched NA vaccine was used.[71] The M2e is a highly conserved protein among different influenza strains, and has been extensively explored as a "universal" vaccine antigen candidate. Anti-M2e antibodies do not neutralize the virus, instead these antibodies are thought to prevent viral budding, mediate killing of infected cells by NK cells or macrophages through ADCC, Fc-opsonization, or complement activation.[72] M2e-based vaccines have been shown to be safe and immunogenic in humans[73,74] and provide protection against heterologous, hetorosubtypic challenge in mice.[75,76] Surprisingly, pigs vaccinated with an M2e vaccine and challenged with a heterologous virus showed severe clinical signs compared to control animals.[77]

Vaccines targeting internal proteins like NP and M1 to induce cross-protective CD8+ cytotoxic T lymphocytes (CTLs) responses are potential candidates for a more "universal" vaccine. The NP and M proteins contain highly conserved epitopes that are shared by various subtypes of influenza virus and are recognized by diverse human leukocyte antigen (HLA) backgrounds.[78] A MVA-based vaccine expressing the NP and M1 proteins (MVA-NP+M1) significantly boosted T-cell response in naturally primed vaccinated humans and vaccination was shown to reduce clinical disease and virus shedding in intranasally challenged individuals.[79] MVA vectored vaccines expressing the NP protein in addition to HA stalk or HA stalk/M2e resulted in cross-protection due to the induction of broadly-reactive virus-specific CD4+ and CD8+ T-cells.[80] A single-cycle "universal" influenza vaccine candidate based on the suppression of HA signal sequence (S-FLU) was shown to induce high levels of CD4+ and CD8+ T cell immune responses when administered

Figure 2: Influenza "universal" vaccine technologies that are under development with the aim to stimulate broader cross-protective responses. The headless HA, chimeric HA or sequential exposure with different HA heads, and Computationally Optimized Broadly Reactive Antigen (COBRA) technologies are based on the stimulation of antibodies against the HA. Other vaccines are based on stimulation of antibodies against the NA or M2e proteins. Anti-stalk, -NA, or -M2e antibodies do not neutralize the virus but inhibit viral release, mediate antibody dependent cell-mediated cytotoxicity (ADCC) or complement activation. Vaccines based on conserved internal proteins, such as M1 and NP, stimulate T-cell response.

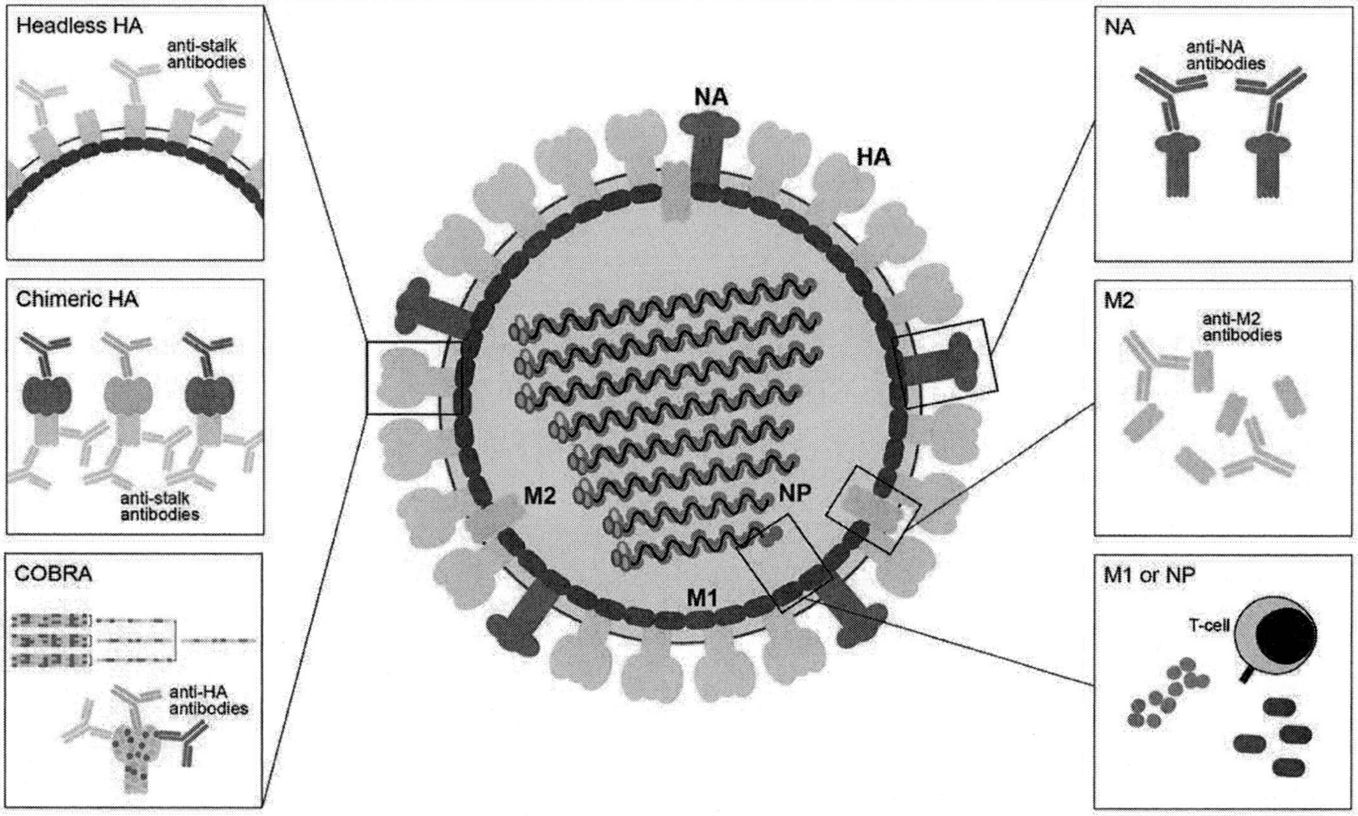

to pigs by aerosol, which was correlated with reduced viral titers following challenge with a closely related H1N1pdm09 virus.[81] Because cell-mediated immune responses do not prevent infection but rather contribute to accelerated clearance and restrict disease progression, CTL-inducing vaccines are a great aid to help reduce disease severity and mortality after heterologous infection. Hence, to reach their full potential as a "universal" vaccine they would be most efficient if used in combination with antibody-stimulating technologies.

Conclusions

Although current vaccines in use for humans and animals are continuously improving, there are several limitations to the stimulation of an ideal protection. In recent years, efforts have focused on vaccine technologies that provide "universal" broadly cross-protective that could potentially protect against several or all influenza strains. To reach a full "universal" potential these vaccines need to stimulate both antibody and T-cell immune responses. Nevertheless, successful control of influenza needs to consider its complex ecology and the virus exchange between different species and can only be achieved through collaborative efforts between human and animal health. There are no standardized vaccine platforms that can be used indistinctively in various animal species and that can easily be accessed in different epidemiological scenarios (e.g. during a season, outbreak or pandemic). In many cases, the way to reach this full breadth of the immune system is to combine platforms that are already at use, such as live-attenuated vaccines or live vectored-vaccines, with new technologies to stimulate responses against conservative epitopes of influenza viruses. A uniform approach would complement the novel "universal" technologies and could improve responsiveness to control influenza in human and animal populations, reducing infectious pressure and reaching the ultimate goal of efficacious control.

References

1. Van Reeth K *et al.* Influenza Virus. in *Diseases of Swine* (eds J. J. Zimmerman *et al.*) 557-571 (Wiley-Blackwell, 2012).

2. Janke BH. Clinicopathological features of Swine influenza. in *Swine Influenza* Vol. 370 (eds J. A. Richt and R. J. Webby) 69-83 (Springer, 2013).

3. Walia RR *et al.* Regional patterns of genetic diversity in swine influenza A viruses in the United States from 2010 to 2016. *Influenza Other Respir Viruses*. 2018.

4. Anderson TK *et al.* Population dynamics of cocirculating swine influenza A viruses in the United States from 2009 to 2012. *Influenza Other Respir Viruses*. 2013;7: 42-51.

5. Olsen CW. The emergence of novel swine influenza viruses in North America. *Virus Res.* 2002;85: 199-210.

6. Zhou NN *et al.* Genetic reassortment of avian, swine, and human influenza A viruses in American pigs. *J. Virol.* 1999;73: 8851-8856.

7. Vincent AL *et al.* Characterization of a newly emerged genetic cluster of H1N1 and H1N2 swine influenza virus in the United States. *Virus Genes*. 2009;39: 176-185.

8. Rajao DS *et al.* Novel Reassortant Human-Like H3N2 and H3N1 Influenza A Viruses Detected in Pigs Are Virulent and Antigenically Distinct from Swine Viruses Endemic to the United States. *J. Virol.* 2015;89: 11213-11222.

9. Rajao DS *et al.* Antigenic and genetic evolution of contemporary swine H1 influenza viruses in the United States. *Virology*. 2018;518: 45-54.

10. Abente EJ *et al.* The Molecular Determinants of Antibody Recognition and Antigenic Drift in the H3 Hemagglutinin of Swine Influenza A Virus. *J. Virol.* 2016;90: 8266-8280.

11. Lewis NS *et al.* Substitutions near the hemagglutinin receptor-binding site determine the antigenic evolution of influenza A H3N2 viruses in U.S. swine. *J. Virol.* 2014;88: 4752-4763.

12. Sandbulte MR *et al.* Optimal Use of Vaccines for Control of Influenza A Virus in Swine. *Vaccines*. 2015;3: 22-73.

13. Rajao DS *et al.* Pathogenesis and Vaccination of Influenza A Virus in Swine. in *Influenza Pathogenesis and Control* Vol. 1 (eds Richard W. Compans and Michael B. A. Oldstone) 307-326 (Springer, 2014).

14. Markowska-Daniel I *et al.* The influence of age and maternal antibodies on the postvaccinal response against swine influenza viruses in pigs. *Vet. Immunol. Immunopathol.* 2011;142: 81-86.

15. Vincent AL *et al.* Failure of protection and enhanced pneumonia with a US H1N2 swine influenza virus in pigs vaccinated with an inactivated classical swine H1N1 vaccine. *Vet. Microbiol.* 2008;126: 310-323.

16. Vander Veen RL *et al.* Haemagglutinin and nucleoprotein replicon particle vaccination of swine protects against the pandemic H1N1 2009 virus. *Vet. Rec.* 2013;173: 344.

17. Vander Veen RL *et al.* Safety, immunogenicity, and efficacy of an alphavirus replicon-based swine influenza virus hemagglutinin vaccine. *Vaccine*. 2012;30: 1944-1950.

18. Vander Veen R *et al.* Rapid development of an efficacious swine vaccine for novel H1N1. *PLoS currents*. 2009;1: Rrn1123.

19. Richt JA and Garcia-Sastre A. Attenuated influenza virus vaccines with modified NS1 proteins. *Curr. Top. Microbiol. Immunol.* 2009;333: 177-195.

20. Genzow M *et al.* Live attenuated influenza virus vaccine reduces virus shedding of newborn piglets in the presence of maternal antibody. *Influenza Other Respir Viruses*. 2018;12: 353-359.

21. Paillot R. A Systematic Review of Recent Advances in Equine Influenza Vaccination. *Vaccines*. 2014;2: 797-831.

22. Anderson TK *et al.* A Phylogeny-Based Global Nomenclature System and Automated Annotation Tool for H1 Hemagglutinin Genes from Swine Influenza A Viruses. *mSphere*. 2016;1: e00275-00216.

23. Lewis NS *et al.* The global antigenic diversity of swine influenza A viruses. *eLife*. 2016;5: e12217.

24. Gauger PC *et al.* Enhanced pneumonia and disease in pigs vaccinated with an inactivated human-like (delta-cluster) H1N2 vaccine and challenged with pandemic 2009 H1N1 influenza virus. *Vaccine*. 2011;29: 2712-2719.

25. Fazekas de St G and Webster RG. Disquisitions of Original Antigenic Sin. I. Evidence in man. *J. Exp. Med.* 1966;124: 331-345.

26. Fonville JM *et al.* Antibody landscapes after influenza virus infection or vaccination. *Science*. 2014;346: 996-1000.

27. Ma J *et al.* Age-specific mortality risk from pandemic influenza. *J. Theor. Biol.* 2011;288: 29-34.

28. Loeffen WL *et al.* Effect of maternally derived antibodies on the clinical signs and immune response in pigs after primary and secondary infection with an influenza H1N1 virus. *Vet. Immunol. Immunopathol.* 2003;92: 23-35.

29. Center for Veterinary Biologics. in *Notice No. 07-17* (ed USDA Animal and Plant Health Inspection Service) (Ames, IA, 2007).

30. Domnich A *et al.* Effectiveness of MF59-adjuvanted seasonal influenza vaccine in the elderly: A systematic review and meta-analysis. *Vaccine*. 2017;35: 513-520.

31. Vesikari T *et al.* MF59-adjuvanted influenza vaccine (FLUAD) in children: safety and immunogenicity following a second year seasonal vaccination. *Vaccine*. 2009;27: 6291-6295.

32. Hoft DF *et al.* Comparisons of the Humoral and Cellular Immune Responses Induced by Live Attenuated Influenza Vaccine and Inactivated Influenza Vaccine in Adults. *Clin. Vaccine Immunol.* 2017;24: e00414-00416.

33. Loving CL *et al.* Heightened adaptive immune responses following vaccination with a temperature-sensitive, live-attenuated influenza virus compared to adjuvanted, whole-inactivated virus in pigs. *Vaccine*. 2012;30: 5830-5838.

34. Pena L *et al.* Modifications in the polymerase genes of a swine-like triple-reassortant influenza virus to generate live attenuated vaccines against 2009 pandemic H1N1 viruses. *J. Virol.* 2011;85: 456-469.

35. Song H *et al.* A new generation of modified live-attenuated avian influenza viruses using a two-strategy combination as potential vaccine candidates. *J. Virol.* 2007;81: 9238-9248.

36. Gauger PC *et al.* *Live Attenuated Influenza A Virus Vaccine Protects Against A(H1N1)pdm09 Heterologous Challenge Without Vaccine Associated Enhanced Respiratory Disease* (2014).

37. Loving CL *et al.* Efficacy of inactivated and live-attenuated influenza virus vaccines in pigs against infection and transmission of emerging H3N2 similar to the 2011-2012 H3N2v. *J. Virol.* 2013;87: 9895–9903.

38. Pica N *et al.* NS1-truncated live attenuated virus vaccine provides robust protection to aged mice from viral challenge. *J. Virol.* 2012;86: 10293-10301.

39. Solorzano A *et al.* Mutations in the NS1 protein of swine influenza virus impair anti-interferon activity and confer attenuation in pigs. *J. Virol.* 2005;79: 7535-7543.

40. Hai R *et al.* Influenza B virus NS1-truncated mutants: live-attenuated vaccine approach. *J. Virol.* 2008;82: 10580-10590.

41. Shi S *et al.* Cross-clade protective immune responses of NS1-truncated live attenuated H5N1 avian influenza vaccines. *Vaccine*. 2016;34: 350-357.

42. Kappes MA *et al.* Vaccination with NS1-truncated H3N2 swine influenza virus primes T cells and confers cross-protection against an H1N1 heterosubtypic challenge in pigs. *Vaccine*. 2012;30: 280-288.

43. Masic A *et al.* Immunogenicity and protective efficacy of an elastase-dependent live attenuated swine influenza virus vaccine administered intranasally in pigs. *Vaccine*. 2010;28: 7098-7108.

44. Ducatez MF *et al.* Low pathogenic avian influenza (H9N2) in chicken: Evaluation of an ancestral H9-MVA vaccine. *Vet. Microbiol.* 2016;189: 59-67.

45. Wesley RD *et al.* Protection of weaned pigs by vaccination with human adenovirus 5 recombinant viruses expressing the hemagglutinin and the nucleoprotein of H3N2 swine influenza virus. *Vaccine*. 2004;22: 3427-3434.

46. Braucher DR *et al.* Intranasal vaccination with replication-defective adenovirus type 5 encoding influenza virus hemagglutinin elicits protective immunity to homologous challenge and partial protection to heterologous challenge in pigs. *Clin. Vaccine Immunol.* 2012;19: 1722-1729.

47. United States Department of Agriculture. (ed USDA APHIS Center for Veterinary Biologics) (Ames, IA, 2014).

48. Fotouhi F *et al.* Adjuvant use of the NKT cell agonist alpha-galactosylceramide leads to enhancement of M2-based DNA vaccine immunogenicity and protective immunity against influenza A virus. *Arch. Virol.* 2017;162: 1251-1260.

49. Crank MC *et al.* Phase 1 study of pandemic H1 DNA vaccine in healthy adults. *PLoS One*. 2015;10: e0123969.

50. Borggren M *et al.* A polyvalent influenza DNA vaccine applied by needle-free intradermal delivery induces cross-reactive humoral and cellular immune responses in pigs. *Vaccine*. 2016;34: 3634-3640.

51. Ledgerwood JE *et al.* Influenza virus h5 DNA vaccination is immunogenic by intramuscular and intradermal routes in humans. *Clin. Vaccine Immunol.* 2012;19: 1792-1797.

52. Hernandez LA *et al.* Particle and subunit-based hemagglutinin vaccines provide protective efficacy against H1N1 influenza in pigs. *Vet. Microbiol.* 2016;191: 35-43.

53. Crawford J *et al.* Baculovirus-derived hemagglutinin vaccines protect against lethal influenza infections by avian H5 and H7 subtypes. *Vaccine*. 1999;17: 2265-2274.

54. Low JG *et al.* Safety and immunogenicity of a virus-like particle pandemic influenza A (H1N1) 2009 vaccine: results from a double-blinded, randomized Phase I clinical trial in healthy Asian volunteers. *Vaccine.* 2014;32: 5041-5048.

55. Valero-Pacheco N *et al.* Antibody Persistence in Adults Two Years after Vaccination with an H1N1 2009 Pandemic Influenza Virus-Like Particle Vaccine. *PLoS One.* 2016;11: e0150146.

56. Pillet S *et al.* A plant-derived quadrivalent virus like particle influenza vaccine induces cross-reactive antibody and T cell response in healthy adults. *Clin. Immunol.* 2016;168: 72-87.

57. Li X *et al.* Influenza virus-like particles harboring H9N2 HA and NA proteins induce a protective immune response in chicken. *Influenza Other Respir Viruses.* 2017.

58. Pyo HM *et al.* Pandemic H1N1 influenza virus-like particles are immunogenic and provide protective immunity to pigs. *Vaccine.* 2012;30: 1297-1304.

59. Steel J *et al.* Influenza virus vaccine based on the conserved hemagglutinin stalk domain. *mBio.* 2010;1: e00018-00010.

60. Impagliazzo A *et al.* A stable trimeric influenza hemagglutinin stem as a broadly protective immunogen. *Science.* 2015;349: 1301-1306.

61. Ermler ME *et al.* Chimeric Hemagglutinin Constructs Induce Broad Protection against Influenza B Virus Challenge in the Mouse Model. *J. Virol.* 2017;91: e00286-00217.

62. Krammer F *et al.* Chimeric hemagglutinin influenza virus vaccine constructs elicit broadly protective stalk-specific antibodies. *J. Virol.* 2013;87: 6542-6550.

63. Nachbagauer R *et al.* Induction of broadly reactive anti-hemagglutinin stalk antibodies by an H5N1 vaccine in humans. *J. Virol.* 2014;88: 13260-13268.

64. Kirchenbaum GA *et al.* Sequential Infection in Ferrets with Antigenically Distinct Seasonal H1N1 Influenza Viruses Boosts Hemagglutinin Stalk-Specific Antibodies. *J. Virol.* 2015;90: 1116-1128.

65. Carter DM *et al.* Design and Characterization of a Computationally Optimized Broadly Reactive Hemagglutinin Vaccine for H1N1 Influenza Viruses. *J. Virol.* 2016;90: 4720-4734.

66. Krammer F *et al.* NAction! How Can Neuraminidase-Based Immunity Contribute to Better Influenza Virus Vaccines? *mBio.* 2018;9.

67. Wohlbold TJ *et al.* Vaccination with adjuvanted recombinant neuraminidase induces broad heterologous, but not heterosubtypic, cross-protection against influenza virus infection in mice. *mBio.* 2015;6: e02556.

68. Sylte MJ *et al.* Influenza neuraminidase antibodies provide partial protection for chickens against high pathogenic avian influenza infection. *Vaccine.* 2007;25: 3763-3772.

69. Wan H *et al.* Molecular basis for broad neuraminidase immunity: conserved epitopes in seasonal and pandemic H1N1 as well as H5N1 influenza viruses. *J. Virol.* 2013;87: 9290-9300.

70. Van Reeth K *et al.* Prior infection with an H1N1 swine influenza virus partially protects pigs against a low pathogenic H5N1 avian influenza virus. *Vaccine.* 2009;27: 6330-6339.

71. Rajao DS *et al.* Vaccine-associated enhanced respiratory disease is influenced by haemagglutinin and neuraminidase in whole inactivated influenza virus vaccines. *J. Gen. Virol.* 2016;97: 1489-1499.

72. El Bakkouri K *et al.* Universal vaccine based on ectodomain of matrix protein 2 of influenza A: Fc receptors and alveolar macrophages mediate protection. *J. Immunol.* 2011;186: 1022-1031.

73. Kolpe A *et al.* M2-based influenza vaccines: recent advances and clinical potential. *Expert review of vaccines.* 2017;16: 123-136.

74. Turley CB *et al.* Safety and immunogenicity of a recombinant M2e-flagellin influenza vaccine (STF2.4xM2e) in healthy adults. *Vaccine.* 2011;29: 5145-5152.

75. Lee YN *et al.* Co-immunization with tandem repeat heterologous M2 extracellular proteins overcomes strain-specific protection of split vaccine against influenza A virus. *Antiviral Res.* 2015;122: 82-90.

76. Tao W *et al.* Consensus M2e peptide conjugated to gold nanoparticles confers protection against H1N1, H3N2 and H5N1 influenza A viruses. *Antiviral Res.* 2017;141: 62-72.

77. Heinen PP *et al.* Vaccination of pigs with a DNA construct expressing an influenza virus M2-nucleoprotein fusion protein exacerbates disease after challenge with influenza A virus. *J. Gen. Virol.* 2002;83: 1851-1859.

78. Lee LY *et al.* Memory T cells established by seasonal human influenza A infection cross-react with avian influenza A (H5N1) in healthy individuals. *J. Clin. Invest.* 2008;118: 3478-3490.

79. Lillie PJ *et al.* Preliminary assessment of the efficacy of a T-cell-based influenza vaccine, MVA-NP+M1, in humans. *Clin. Infect. Dis.* 2012;55: 19-25.

80. Hessel A *et al.* MVA vectors expressing conserved influenza proteins protect mice against lethal challenge with H5N1, H9N2 and H7N1 viruses. *PLoS One.* 2014;9: e88340.

81. Morgan SB *et al.* Aerosol Delivery of a Candidate Universal Influenza Vaccine Reduces Viral Load in Pigs Challenged with Pandemic H1N1 Virus. *J. Immunol.* 2016;196: 5014-5023.

Influenza: Herd immunity and transmission

M. Torremorell, DVM, PhD
College of Veterinary Medicine, University of Minnesota, St. Paul, Minnesota

Introduction

Influenza A virus (IAV) is known to be endemic in many swine herds. The factors that affect endemicity are several but in this paper we will discuss how immunity affects IAV transmission and thus the ability of the virus to persist or become eliminated from swine herds.

Effect of immunity on transmission

Influenza is considered one the most transmissible diseases affecting people and pigs. This is evidenced by the rapid transmission of IAV during pandemics and also when new IAV strains are introduced into naïve populations. Influenza transmission is impacted by immunity, and the type of immunity (passive or active), quality of immunity (homologous or heterologous to the challenge strain) and quantity of immunity (i.e. hemagglutination inhibition (HI) titer levels) are known to influence transmission parameters.

Reproductive ratio (R) values, which are a measure of transmission and how quickly an agent will spread in a population, are impacted by immunity. R is defined as the expected number of secondary IAV infections in a population due to a typical infectious individual during the duration of the individual's infectious period. Therefore when R is > 1, each infectious individual will infect on average at least one other susceptible individual and transmission will occur, whereas when R is < 1 transmission will eventually cease. In certain situations, the goal of an intervention such as vaccination is to reduce R < 1.

R values have been estimated for IAV in vaccinated populations[1] and in populations with passive immunity[2] under experimental conditions. A summary of the R values can be seen in Table 1. IAV infection in pigs without immunity spreads rapidly, Ro = 10.66 (CI 6.57-16.46).[1] In such cases, clinical signs have an acute onset with the population being infected rapidly. In contrast, IAV spread in populations with immunity can vary. In pigs vaccinated with an autogenous vaccine that was a perfect match to the challenge strain, R could not be estimated because transmission was not observed.[1] However, this is not likely to be the most common scenario under field conditions since immunity will vary between individuals, wane overtime or the virus will mutate. It is likely that active immunity induced by autogenous vaccines most likely will resemble R values within ranges of those obtained from heterologously vaccinated populations. These values were estimated at 1 with a confidence interval of 0.39-2.09 in vaccinated population with partial immunity. In this case, IAV could spread but at lower levels than in naïve populations. Although transmission was reduced, IAV could still transmit which is central to the issue of endemicity and the ability of IAV to persist in infected herds. In other words, immunity can act as a double edge sword because it helps improve clinical disease and decrease transmission, but at the same time it appears to facilitate the persistence of IAV in populations.

Table 1: Reproductive ratio (R) values obtained from pigs with active (A) or passive (B) immunity generated with homologous or heterologous vaccines as described in Romagosa[1] and Allerson.[2]

A.

R0 No vaccine	R Heterologous vaccine
10.66	**1**
(6.57-16.46)[a]	**(0.39-2.09)[b]**

B.

R0 No vaccine	R Heterologous vaccine	R Homologous vaccine
11.01	**7.81**	**0.84**
(6.89-16.80)[y]	**(4.57-12.56)[y]**	**(0.05-3.68)[x]**

a,b and y,x differences in superscripts indicate differences with *P*-value < 0.05.

The impact of passive immunity on transmission differs slightly from the impact of active immunity. In the case of pigs that had received passive immunity generated by vaccinating dams with an autogenous vaccine that perfectly matched the challenge virus, R was estimated at 0.84 (CI 0.05-3.68).[2] The confidence interval obtained for this R value indicates that although in many cases passive immunity may result in limited transmission, in other cases it facilitates the spread of IAV within populations at a low rate. In contrast, the R value for pigs with heterologous passive immunity was estimated at 7.81 (CI 4.57-12.56) which indicated that IAV could transmit efficiently despite the presence of immunity. Often pigs with imperfect passive immunity act as "silent spreaders" of IAV because they spread IAV despite the lack of clinical signs. This is a major issue in the North American pig industry because pigs at weaning are transported to distant locations and many geographical areas. It is estimated that at least one third of weaned groups test IAV positive at weaning.[3]

When all this information is taken altogether and is used in mathematical models to estimate the number of pigs that can become infected in breeding herds,[4] piglets prior to wean were identified as a key subpopulation responsible for maintaining IAV infections endemic in breeding herds. This observation has been corroborated by others[5] and it is accepted also that the on-going recruitment of susceptible hosts (i.e. pigs born daily) is one of the main reasons why IAV persist in populations. Piglets are naïve at birth and as discussed above, although they may have passive immunity this immunity is not enough to protect the piglets from infection. Some autogenous pre-farrowing vaccines may delay IAV infections but other vaccines have little effect on preventing or delaying them.

Effect of immunity on transmission in the field

The impact of sow vaccination on IAV infection of piglets at weaning has been evaluated epidemiologically. Both, prefarrow and mass sow vaccination protocols have been shown to decrease the number of infected groups of pigs at weaning, and the prevalence within those groups. Interestingly, both types of vaccines, commercial and autogenous had similar results in a study by[3] suggesting that sow vaccination can be used to lower the level of infection at weaning. In a different study, use of sow vaccination was one of the few parameters also associated with lower levels of IAV infection at weaning.[3]

Despite the widespread use of sow vaccination, there is limited information on the impact of maternally-derived antibodies on IAV infection dynamics in growing pigs. Increase in influenza seroprevalence at weaning has been associated with low influenza prevalence at weaning and delayed time to IAV infection throughout the nursery. Piglets with IAV strain specific HI (hemagglutination inhibition) titers of 40 or higher at weaning were shown to be less likely to test IAV positive at weaning, took longer to become infected, tested IAV RT-PCR positive for fewer weeks, and had higher IAV RT-PCR cycle threshold values compared to piglets with HI titers less than 40.[6] These findings suggested that sow vaccination or infection status that results in high levels of IAV strain-specific maternally-derived antibodies may help reduce IAV circulation in both suckling and nursery pigs.

Furthermore, strain specific humoral immunity at weaning has been associated with different IAV transmission patterns in weaned pigs.[6] Figure 1 shows 4 distinct IAV transmission patterns observed in a nursery site based on their IAV prevalence and maternally derived antibody (MDA) levels at weaning. The first pattern had high IAV prevalence (97%-100%) and low seroprevalence (<5%) at weaning with a subsequent rapid decline of infection and low seroconversion. The second pattern was very similar to the first pattern and had high IAV prevalence and low seroprevalence (<5%) at weaning, a rapid decline of infection, and a recurrent infection four weeks later with the subsequent seroconversion after the second infection. The third pattern had IAV prevalence levels at weaning between 7% and 57% and high seroprevalence (~80%) with 2 peaks of infection afterwards, 1 week after weaning and 4-5 weeks later. Seroprevalence decayed to low levels (~5%) at the end of the nursery. In the fourth pattern, there were no IAV positive pigs at weaning and seroprevalence was high (~80%) although few pigs tested positive at different weeks after weaning. Seroprevalence also decayed over time. Thus this information highlights the complexity of IAV transmission patterns in populations with varying immune profiles.

Summary

In summary, the interplay of IAV transmission and IAV immunity is complex. Despite the beneficial effects of immunity at improving clinical outcomes and decreasing transmission, imperfect or partial IAV immunity appears to facilitate persistence of IAV in populations. Furthermore, there appears to be different transmission patterns of IAV in pigs after weaning

Figure 1: Influenza A virus infection patterns and strain specific seroprevalence in 10 cohorts of weaned nursery pigs (adapted from Chamba et al[6]).

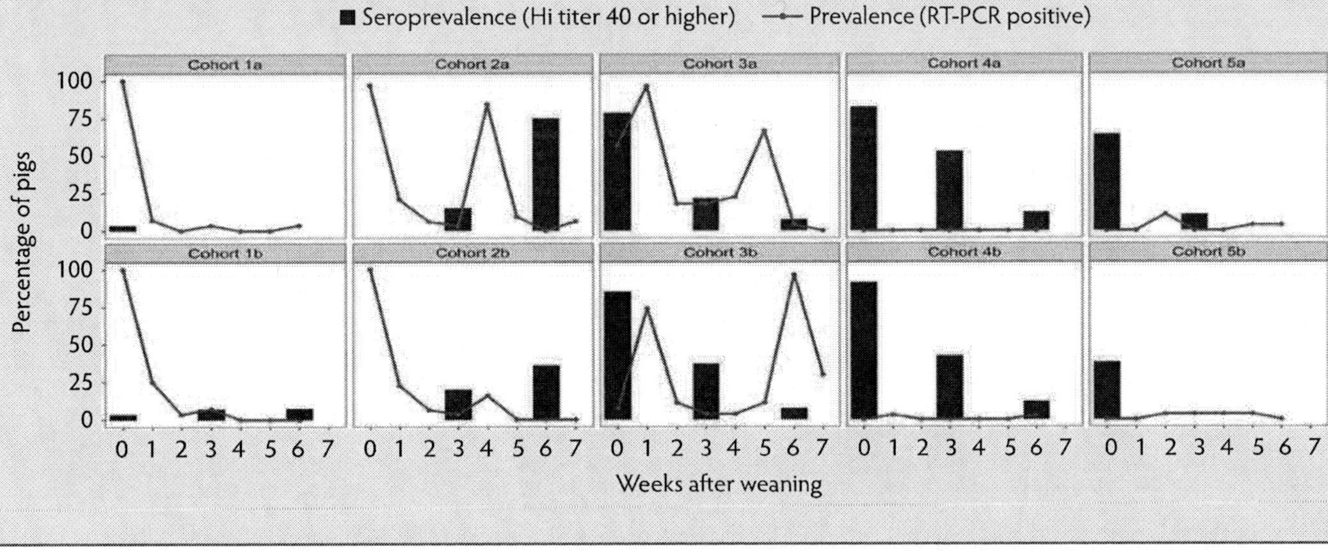

and in part these transmission patterns are driven by the quantity and quality of the immunity present against the circulating strain(s). Vaccination strategies can be helpful to decrease transmission and increase the number of groups weaned IAV negative, but vaccination can also facilitate the establishment of endemic infections. More research is needed to better understand how immunity in combination with farm management practices can be more effective at decreasing transmission to the point that IAV elimination could be considered.

References

1. Romagosa A, Allerson M, Gramer M, Joo HS, Deen J, Detmer S, Torremorell M (2011). Vaccination of influenza A virus decreases transmission rates in pigs. Vet Res. 2011 Dec 20;42(1):120

2. Allerson M, Deen J, Detmer S, Gramer M, Joo HS, Romagosa A, Torremorell M. (2013) The impact of maternally derived immunity on influenza A virus transmission in neonatal pig populations. Vaccine. doi:pii: S0264-410X(12)01616-7. 10.1016/j.vaccine.2012.11.023.

3. Chamba Pardo FO, Alba-Casals A, Nerem J, Morrison RB, Puig P, Torremorell M (2017). Influenza herd-level prevalence and seasonality in breed-to-wean pig farms in the Midwestern United States. Front Vet Sci. 4:167, https://doi.org/10.3389/fvets.2017.00167

4. Reynolds JJ, Torremorell M, Craft ME (2014). Mathematical modeling of influenza A virus dynamics within swine farms and the effects of vaccination. PLoS ONE 9(8). e106177. doi: 10.1371/journal.pone.0106177. eCollection 2014.

5. Diaz A, Perez A, Sreevatsan S, Davies P, Culhane M, Torremorell M (2015). Association between influenza A virus infection and pigs subpopulations in endemically infected breeding herds. PLoSOne, 10(6):e0129213. doi: 10.1371/journal.pone.0129213

6. Chamba Pardo FO, Wayne S, Culhane MR, Perez A, Allerson M, Torremorell M (2019). Effect of strain-specific maternally-derived antibodies on influenza A virus infection dynamics in nursery pigs. PLoS One 2019 Jan 14;14(1):e0210700. doi: 10.1371/journal.pone.0210700. eCollection 2019

Influenza vaccines: Live vs killed

Chong Li[1], DVM, MS; Marie Culhane[1], DVM, PhD; Montse Torremorell[1], DVM, PhD
[1]Veterinary Population Medicine Department, College of Veterinary Medicine,
University of Minnesota, Saint Paul, Minnesota

Introduction

Influenza A viruses of swine (IAV-S) cause respiratory disease in pigs worldwide. Pigs infected with IAV-S frequently have secondary infections with other viruses or bacteria which results in economic losses to pork producers.[1] Currently, vaccination is the primary way to control IAV-S infection in pigs. Even though much progress has been made to improve efficacy of recombinant subunit, vector and DNA vaccines against IAV in experimental studies, inactivated (killed) and live attenuated influenza vaccines (LAIV) are still the only two types of licensed swine vaccines available to control IAV-S. Here we summarize the performance of inactivated or LAIV vaccines in experimental vaccination-challenge studies in pigs. Also, we pay special attention to the potential advantages and application of these two types of IAV-S vaccines in domestic swine populations.

Inactivated influenza vaccines for IAV-S

First to be developed, the whole cell inactivated IAV-S vaccines (WIV) are widely used in sows and gilts to induce serum antibodies against the viral hemagglutinin and enhance transfer of passive immunity to newborn piglets. The vaccine strains are grown in cell lines or embryonated chicken eggs, after going through the process of purification and concentration, and finally inactivated.[2] Unlike human inactivated vaccines that contain purified viral antigens without adjuvant, most WIV for swine contain an adjuvant.[3] Through the deep intramuscular injection into the neck of pigs, antigens are presented to antigen presenting cells in the draining lymph nodes and stimulate B cells and $CD4^+$ T cells. Based on antigen delivery route, WIV usually only induce serum antibodies.[4] Additionally, WIV cannot enter the endogenous pathway during the antigen presentation process and can hardly activate the virus-specific $CD8^+$T cell and subsequent differentiation into cytotoxic T lymphocytes (CTLs), both key components of cell-mediated immunity. Compared to live attenuated influenza vaccine (LAIV), WIV are safer, less expensive, and easier to store. However, they are less immunogenic and need more booster doses to get the desired immunity.[5] Currently, most commercial WIV are multivalent, based on H1N1 and/or H3N2 and/or H1N2 IAV-S. Due to the high genetic diversity of IAV-S circulating in US pig herds,[6] it is not uncommon to develop autogenous vaccines that incorporate local strains to better match the farm or region circulating strains to provide a better protection.

Live attenuated influenza vaccines for IAV-S

Live attenuated vaccines for IAV-S (LAIV) are relatively new and most of them available experimentally although now there is one vaccine commercially available in the US market. LAIV are usually administrated intranasally to newborn piglets to bypass or minimize the limiting effect of maternal antibodies. Vaccination of newborn piglets is done to reduce the impact of IAV-S on growing pig performance. LAIV when given intranasally tend to stimulate similar immune pathways to a natural virus infection, inducing both serum and mucosal antibodies and generate more efficient cross-reactive T cell mediated protection. LAIV can elicit more rapid and efficient innate and adaptive immunity than WIV. However, producers may have safety or efficacy considerations since LAIV contain viable virus and proper storage is critical.

A LAIV for influenza was first developed for humans in the 1960s.[7] After the serial passage of influenza virus in embryonated chicken eggs under suboptimal temperature conditions, the attenuated strains became cold adapted with temperature-sensitive attenuation characteristics, replicating at approximately 25°C and not growing in the lower respiratory tract which has higher body temperature of around 37°C. As a result, the proliferation site of the LAIV is restricted to the upper respiratory tract to induce the local immune response and will not cause infection by itself. With the development of molecular engineering and reverse genetic approaches, more LAIV have emerged. One strategy is to change the trypsin-sensitive motif in the hemagglutinin cleavage site to an elastase-sensitive motif to generate elastase-dependent virus mutants.[8] The mutants have similar growing properties compared to wild virus but are highly attenuated in pigs and capable of inducing a systemic antibody response. This type of LAIV is still under development as of this writing and further study is undergoing to test the safety of this type of vaccines. Recently, a multivalent LAIV with NS1 truncated proteins became commercially available in the US market. NS1 is a nonstructural influenza protein which is not essential for virus replication; however, it is a virulence factor which can block the activation of protein kinase R (PKR) and hijack the innate immune IFN-α/β response.[9] This LAIV strategy is based on an influenza virus with modified NS1 protein as a means of virus attenuation and displays great potential to elicit both humoral and cell meditated immune response in different animal models.[10]

Our experience with live attenuated and inactivated vaccines for IAV-S

We have performed an experimental vaccination-challenge study in pigs to evaluate the efficacy of IAV-S vaccines using different heterologous prime-boost vaccination protocols. Heterologous prime-boost refers to the antigen presentation to the immune system in more than one way; i.e. using different strains of a pathogen, different adjuvants, different inoculation routes, and/or different types of vaccines (modified live vs. inactivated). In our study, pigs were vaccinated with different licensed inactivated and/or live attenuated multivalent vaccine combinations and simultaneously challenged with H1 and H3 IAV under experimental conditions. There were no serum antibody responses detected against either H1 or H3 challenge strains by hemagglutination inhibition assay (HI) in any pigs that received a single dose of LAIV. The protection efficacy for pigs that received a single dose LAIV showed no difference from the non-vaccinated challenged pigs, which suggests under our experimental conditions, that there was poor immunogenicity with a single dose of LAIV administration. Delivering a WIV boost to pigs following a single priming administration of LAIV apparently stimulated a rapid and high antibody response that was broadly cross-reactive against both H1 and H3 challenge strains. T cell immunity was also enhanced in pigs that received an LAIV prime and a booster with WIV. There were fewer IAV infected pigs and much lower levels of virus in the nasal cavity and lungs of pigs that were primed with LAIV and boostered with WIV. Regardless of vaccine type, we found the heterologous prime boost combination (e.g., one vaccine is given as a prime and a second (different) vaccine is given as the boost) provided better protection efficacy than a homologous prime boost vaccine combination. We concluded this prime boost strategy may potentiate responses to suboptimal immunogens and may elicit broadly cross-reactive responses that could eliminate the need for additional vaccinations. However, more studies are needed to validate the protection efficacy of live and/or inactivated vaccine combinations to control IAV-S under field conditions.

References

1. Morris DE, Cleary DW, Clarke SC. Secondary Bacterial Infections Associated with Influenza Pandemics. *Frontiers in Microbiology*. 2017;8:1041. doi:10.3389/fmicb.2017.01041.

2. Wong, S.S.; Webby, R.J. Traditional and new influenza vaccines. *Clin. Microbiol. Rev.* 2013, *26*, 476–492.

3. Gerdts V. Adjuvants for veterinary vaccines--types and modes of action. *Berliner und Munchener tierarztliche Wochenschrift*. 2015;128(11-12):456-63.

4. Heinen, P. P., et al. "Systemic and mucosal isotype-specific antibody responses in pigs to experimental influenza virus infection." *Viral immunology* 13.2 (2000): 237-247.

5. Wu, Nicholas C., et al. "A structural explanation for the low effectiveness of the seasonal influenza H3N2 vaccine." *PLoS pathogens* 13.10 (2017): e1006682.

6. Lewis, Nicola S., et al. "The global antigenic diversity of swine influenza A viruses." *Elife* 5 (2016): e12217.

7. Maassab,H.F. Plaque formation of influenza virus at 25°C. *Nature* 1968, 219, 645–646.

8. Masic, Aleksandar, Lorne A. Babiuk, and Yan Zhou. "Reverse genetics-generated elastase-dependent swine influenza viruses are attenuated in pigs." *Journal of General Virology* 2009;9:375-385.

9. Bergmann, Michael, et al. "Influenza virus NS1 protein counteracts PKR-mediated inhibition of replication." *Journal of virology* 74.13 (2000): 6203-6206.

10. Nogales, Aitor, and Luis Martínez-Sobrido. "Reverse genetics approaches for the development of influenza vaccines." *International journal of molecular sciences* 18, no. 1 (2016): 20.

Immunity management for the pig reared in a *Raised Without Antibiotics* program

M. Hoogland, DVM, MS
Smithfield Hog Production; Midwest Region, Algona, Iowa

Introduction

Today's commercial pig surpasses the performance of its predecessors 10-15 years ago. The continued improvements captured from genetic, nutritional, and management advancements have helped greatly. It is also exciting to see these advancements in modern hog production produce pork at a faster and more efficient pace, using less resources to achieve heavier carcass weight.

There is also additional pressure to reduce an already limited amount of antibiotic use in swine production. This pressure only continues to grow suggesting the swine industry likely will have fewer disease management tools via antimicrobials. It is important to continue to study other tools that help advance disease management that can impact positively the performance in hog production today.

The pig's immunity is one of the critical pieces we focus on in our Raised Without Antibiotic (RWA) flows. This focus has helped shed light on areas of disease management and pig health that have driven our RWA performance into the top 10% in growth and livability in our system. Our own past history 10 years ago and reports from peers within the industry would have led us to believe our RWA program would be successful if we achieved 70% or more pigs qualified for our RWA markets. We have been running well above that target pushing 90% of all pigs placed qualifying for the RWA market the last 2 ½ years with mortality trends that are below 6% wean-to-finish and growth rates exceeding 1.75 ADG consistently. This success is not met without some challenges and certainly it is not feasible in all areas of our business. We clearly recognize there is added cost into this system today. A strong integrated hog company can work through these cost and use them to improve the bottom line. We also clearly affirm that antibiotics used correctly and timely are a key part of disease control and treatment for animal wellbeing. There are some key factors learned (perhaps relearned) in our RWA program that may add to the success of a business. This paper can hopefully share some thoughts and how it relates to disease and immunity management.

The disease triad

It can be helpful to remember the disease (or epidemiology) triad when assessing all the factors that go into the health and immunity of the pig. The triad looks at the host factors, pathogen factors, and environmental factors that influence the pig's outcome. These same factors influence the immune system and how robust of a response it has. The immune response encompasses the passive immunity from the sow, the innate immunity and physical barriers of the pig, and the maturing and development of the adaptive immune response. When these areas are underdeveloped, challenged too soon, overwhelmed, or poorly managed the pig is much more likely to face illness. A successful RWA system must have these areas accounted for and well managed.

It is not always possible to control all diseases in all farms. The biggest obstacle remaining is the overwhelming of the immune system when significant disease challenges occur too early, such as porcine respiratory and reproductive syndrome virus (PRRSV). We do all we can to eliminate pathogens in the RWA flow to not overwhelm the immune system and, in the case of PRRSV, further weaken the immune system to allow bacterial infections get a hold. Key takeaway in our RWA system is to ensure all RWA flows are PRRS negative and pigs placed in the RWA program come from MLV-vaccinated farms only.

We have also experienced early challenges with vertical transmission of influenza A virus (IAVs) from the sow farm. IAVs can also compromise the respiratory tract of the pig and bacterial challenges mount a greater force when it is present. IAVs is difficult to manage with the amount of strain variation and the difficulty overcoming maternal antibodies in the piglet with an appropriate vaccine to stimulate active immune response. Our current approach is therefore to address each strain specifically at the sow farm to mount as specific immune response as possible. This has worked well and it is evident when there is a strain mismatch on the sow farms and virus leaks into the nursery phase and downstream pigs struggle a little more at placement. We have only had one group of pigs pulled from the program due to IAVs early challenge. That group also was challenged with very warm weather in shipping which further compounded the stress on those pigs weakening the pigs response to disease challenges.

Other immunosuppressive diseases and chronic diseases that can be avoided are controlled when possible. *Mycoplasma hyopneumoniae* (M hyo) act both to suppress the immune system and insidiously slow the pigs performance and growth. All pigs in the RWA flow come from older parity sows to avoid shedding of this organism and help drive better performance.

Passive immunity

The neonatal pig is born with a functional immune system. However, it is not fully developed and less efficient than the adult.[1] Key diagnostic surveillance becomes very important in management and effectiveness of passive immunity. Little to no *Lawsonia intracellularis* (LI) was detected at the onset of the RWA program using routine oral fluids in early 2016. The presence of LI and its disease timing within the population changed over the course of a year and half. By mid-to-late 2017 greater efforts were needed sooner for LI and the impact to minimize shedding from the sow farm as well as maximize consistency and peak in weaned pigs was helpful. The passive immunity from the sow is critical to manage the early challenges in life while minimizing disease load at weaning. More effort is applied on the sow farm to drive passive immunity for enteric bugs including salmonella, LI, and *Escherichia coli* (*E. coli*) than other commercial farms in our system. There is limited concern in the current RWA flows for PRRSV or Mhyo in our system. Passive immunity to IAVs is the best tool we have today for this disease and currently provides manageable immunity to the pigs to minimize the amount present at weaning being carried to the wean-to-finish barns.

Innate immunity

The innate immunity of the pig is an important function to understand in controlling infections and disease of the pig. Chase et al. describes key areas to remember are the physical, chemical, and microbial barriers of the innate immune system in Diseases of Swine 10th ed.[2]

Management

Smithfield Hog production continues to see the benefit of an older weaned pig in their system. It is not uncommon for pigs to be 24-28 days of age at weaning. This older age is especially important for pigs without antibiotics in the feed or water. Moeser et al.[3-5] has studied gut permeability repeatedly and has shown the older wean age contributes to greater gut integrity, less tight junction dysfunction and reduced mast cell infiltration in the lamina propria.The physical barrier in the gut epithelium performs better on an older pig ensuring better control of enteric pathogens and in the young pigs gut.

Minimizing stress is very important on the immune system and pig performance. There is a growing body of evidence as to how the stress impacts the immune system and its ability to respond most effectively.[6] Temperature assumptions in the newly weaned pig can lead to cold pigs and areas of the barn that are overlooked. It is important to recognize both the ambient and microenvironment temperature affecting the pig. Often it can take a week or more for concrete in a barn to warm up effectively, especially in the winter, even if the ambient temperature is set high enough. Newly placed pigs can be chilled too much during the first few days placed, leading to poorer start and health challenges to follow. Cold stressed pigs also are less effective at fighting infections, both viral and bacterial.[7,8]

Feed and water

Nutrition is critical in managing the immune system of the pig. There is evidence of key minerals and vitamins in the regulation of the immune system.[9] there are also other key factors to address in feed and nutrition of the pig. It is critical to gut health and integrity that pigs eat quickly after weaning and stay on feed. The proper diet must be freely accessible and palatable to pigs. A lot of effort is put forth to ensure that pigs have feed and are enticed to eat I our RWA program. The early period after weaning helps establish pig health and understanding that pigs adapt more quickly to soft fresh foods, grueling every pen the first week is done. This task at first sounds tedious. However, it is fairly manageable and a rewarding process that can easily be completed 2-4 times per day with some simple equipment and systems in place. Often a 5 gallon bucket, 1 liter scoop and plastic pans will get the job done.

Identifying pigs that are not eating well is easy and they can quickly be sorted and treated if necessary.

Keeping a health gut going is critical throughout the grow-finish period, but especially early. Many changes in the gut development are occurring and small disruptions have a big impact. Any out-of-feed event can lead to rapid bacterial changes in the gut, dysbiosis, and *E. coli* enteritis.

Stomach pH and bile acid production is an important area still developing at weaning. Nutritional adjustments and water adjustments can aid in the transition of pigs at weaning. We have found it helpful to focus on highly digestible protein sources that don't incite as much inflammation in the gut of the newly weaned pig. It is understood that SBM is reactive and can cause some gut inflammation and hypersensitivities.[10] Lower amounts of conventional soybean meal (SBM) are used while further processed SBMs is used in its place to avoid excess gut hypersensitivity. This ensures adequate protein is available and easy to digest. Water is monitored on each site and adjusted to ensure it is at or below pH of 5 to not further buffer the gut and help gut health. Less scours is reported with this simple step.

A host of nutritional components that aid in gut health, feed palatability, nutrient accretion, and gut flora development are helpful. Medium chain fatty acids (MCFAs), probiotics, flavor enhancers, and key enzymes help drive pigs to eat and aid in gut health and stool consistency.

Active immunity

RWA pig flows target many of the same vaccines necessary in are typical commercial flows. Porcine circovirus type 2 (PCV2) vaccine is a staple in all flows and RWA is no exception. M hyo is used but more as a precaution than necessity. PRRSV modified live vaccine (MLV) is only used 3 weeks post weaning to avoid unnecessary stress and only in geographical areas (SW Iowa) that may see some PRRSV pressure. The RWA pigs placed in South Dakota barns see little to no PRRSV pressure today and PRRSV MLV is not used.

Enteric vaccines are given a lot of attention. Salmonella and ileitis vaccines are routinely given in the RWA flows. They are given around 4-to-6 weeks of age and repeated to help boost immunity quickly. LI pressure became very evident in the RWA program. It also moved to a mid-to-late nursery age problem. Therefore, a LI vaccine was been moved to shortly after weaning. *E. coli* management involves the stated water, temperature, and feed issues above. There is no amount of vaccine that will overcome the disease challenge that can occur if one of those areas is out of whack. In addition, though, we do use competitive exclusion K88/F4 and F18 lab grown *E. coli* strains that are devoid of toxigenic genes to help control *E. coli* early. These are given at placement and repeated several times through the water during the first 10 days placed.

Summary

It is not practical and should not be our goal to attempt to raise all pigs without antibiotics ever. However, better understanding some key drivers of immunity and health can shed light on areas that may enable improved performance with less antibiotic use in a cost effective manner. Our goal is to continue to strive to improve how we raise pigs in a welfare friendly way that also brings more efficient and cost effective management strategies. External pressures will continue to push swine production to find new ways to be successful under a watchful eye of scrutiny. Those up for the challenge will remain viable long-term in the global swine industry.

References

1. Hammerburg C, Schurig GG, Ochs DL. 1989. Am J Vet Res 50(6):868-874.
2. Chase CCL, Lunney JK. Immune System: Zimmerman JJ, et al., eds. Diseases of Swine. 10th ed. Ames, IA. 2012; 227-250.
3. Moeser AJ. 2008. Predisposing factors for enteric disorders. Proceedings of the 39th annual meeting of the American Association of Swine Veterinarians (AASV). Pp 487-490.
4. Moeser AJ, Ryan KA, Kinghot PK, Blinkslager AT. 2007. Gastrointestinal dysfunction induced by early weaning is attenuated by delayed weaning and mast cell blockade in pigs. Am J Physiol Gastrointest Liver Physiol. 239:G413-G421.
5. Moeser AJ, Almond GW, Baker BB, Blinkslager AT. 2006. Does weaning age influence immediate and long-term gastrointestinal health in the pig? Proceedings of the 37th annual meeting of the American Association of Swine Veterinarians (AASV). Pp 29-32.
6. Hyun Y, Ellis M, Riskowski G, Johnson RW. J. Anim Sci. 76:721-727.
7. Shimizu M, Shimazu Y, Kodama Y. 1978. Effects of ambient temperatures on induction of transmissible gastroenteritis in feeder pigs. Infect Immun. 21:747-752.
8. Wathes CM, Miller BG, Nourne FJ. 1989. Cold stress and post-weaning diarrhea in piglets inoculated orally or by aerosol. Anim Prod. 49:483-496.
9. Borghetti P, Saleri R, Mocchegiani E, et al 2009. Vet Immunol Immunopathol 130:141-162.
10. Engle MJ. 1999. The role of soybean meal hypersensitivity in postweaning lag and diarrhea in piglets. J Swine Health and Prod. 2(4):7-10.

Sow mortality: The Danish perspective

J. M. J. Agerley, DVM
SvineVet, Moltrupvej 49a, DK-6100 Haderslev

Introduction

Sow mortality is an important welfare and production parameter, which deserves more attention also in Denmark. Reducing sow mortality in normal, stable farms, often demands long term focus and results are often first seen after a year of focus. The sow mortality in Denmark is referenced as percentage of dead sows per year sow. Sow mortality consist of culled and dead sows with roughly 50% in each category.[1] From the year 2000 an elevation in sow mortality has been recorded in Denmark. The sow mortality increased from 11% dead year sows in the year 2000 and peaked to roughly 15% dead year sows in 2005-2008[2](Figure 1). Causes where increased culling of sows due to shoulder lesions and suboptimal feeding leading to gastric ulcers, and supposed *Cl. Novyi* infections. This led to a national "sow life" campaign, which was started by the Danish Pig Research center called SEGES in 2008. The goal of the campaign was to reduce sow mortality in Denmark to 9% dead year sows in 2018. The whole Danish pig industry participated in the campaign. In 2017 Denmark reached a level of 11% dead year sows, data from 2018 is still not published.[3]

Figure 1: Danish sow mortality, 2000-2016

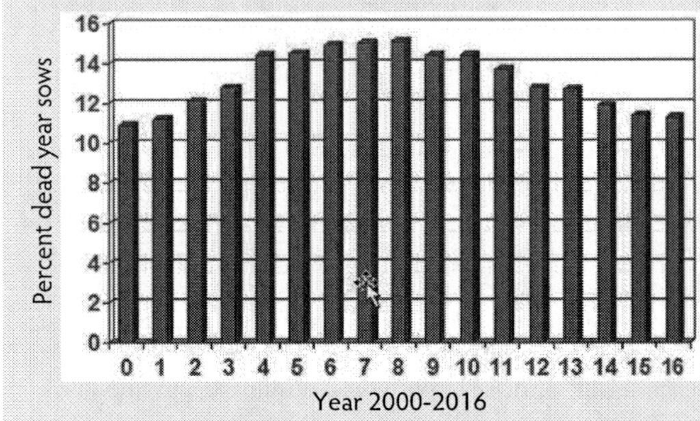

Modified from following source:
SEGES svineproduktion. Udviklingen i sodødelighed - tal fra DAKA 2016, notat nr. 1714 available at https://svineproduktion.dk/publikationer/kilder/notater/2017/1714. Accessed 28 December 2018

In 2012-2013 farms where adapting to the new laws on group housed sows in gestation. In these years we did not see any decline in sow mortality, because of more sows being culled due to leg problems. In 2016 sow mortality became part of the Danish legislation. Farms with more than 16% dead sows had an increased chance of getting control visits from the veterinarians of the Danish veterinary and food administration.

The economic impact of sow mortality was in the "sow life" campaign calculated to be 740 dollars of gross profit loos for every dead sow. This equals 7,4 dollars in gross profit loos per sow, for every percent dead sow in the herd.

Causes of culling and interventions

Causes of culling are well described in Denmark (Figure 2). A study performed in 2016 in 17 Danish farms with high sow mortality showed that lameness is the most predominant cause of culling sows. Especially in younger sows were up 70% of parity 1 sows are culled because of lameness.[1] This confirms what was found in an earlier study from 2003 in 37 Danish farms where 59% of all sows were culled because of lameness.[4] Thus, action plans on reducing lameness in sows is paramount in the attempt to reduce mortality from culling. All starts with producing the perfect gilt and inseminating her at the right age. In Denmark we aim for maximum 700 grams ADWG to ensure strong bones and joints. We aim for 32 weeks of age at first service and 130 kg's of bodyweight with a backfat level at p2 of 15-17mm. A relationship between age at first service and longevity can be shown analyzing farm data. Gilts who are older and bigger at service also become bigger sows who put more strains on their legs and joints.

Selection of gilts is another important area to focus on. Selection should focus on strong and straight legs, uniform claws and a clean gait.

Management routines, well-trained and happy staff are very important. One important experience from the "sow life" campaign was, that in farms where the staff was well trained and felt happy about their jobs, sow mortality also was lower. A Spanish study showed a relationship between low sow mortality and high productivity.[6] Also increased age at first mating was found to give a lower survival probability.

In Denmark the introduction of hospital pens combined with early response from a well-trained staff has helped to reduce culling due to lameness. In SvineVet we have had good experiences introducing claw trimming. It takes roughly 2 reproductive cycles to measure the impact of claw trimming. Labor cost is high; thus, many farms have stopped doing it again. Stables only with slatted flor tend to have bigger problems with lameness compared to stables with a combination of solid and slatted flor. Individual feeding leads to less fights and thus less sows with leg problems.

A fiber rich diet leads to more "down time". About 10% more beet pulps in the digestion diet will make sows stand up 1 hour less per day.

Figure 2: Causes for culling cows. Data recorded from 17 danish farms

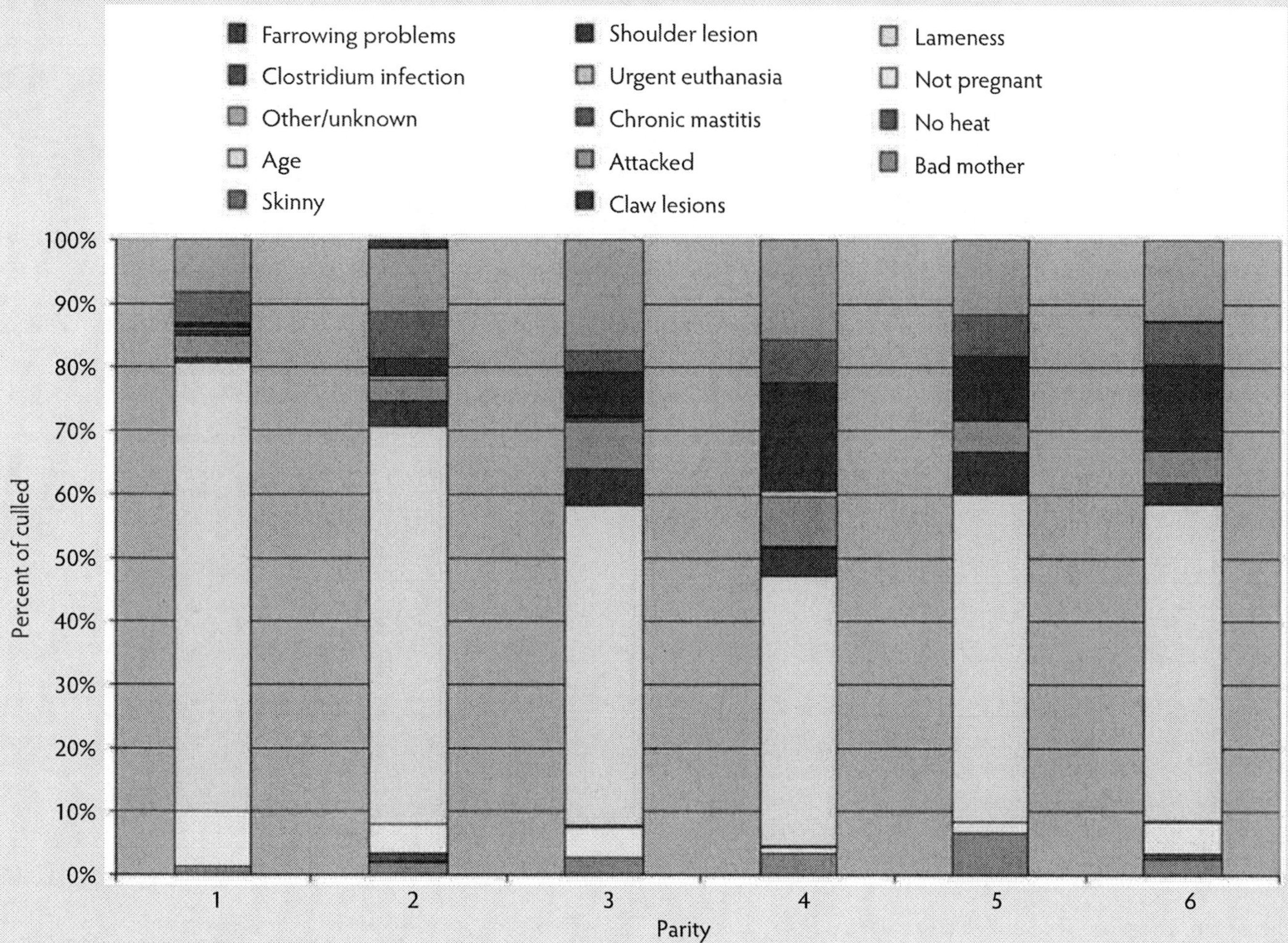

Modified from following source:
SEGES svineproduktion. Analyse af sodødelighed i 17 danske besætninger notat nr. 1604 available at https://svineproduktion.dk/publikationer/kilder/notater/2016/1604. Accessed 28 December 2018

In Denmark there are strict rules concerning transport of animals. The animals can't have any lameness at all before undergoing transport. This also leads to more sows being culled compared to other countries.

Causes of mortality and interventions

Causes of sow mortality is much less described compared to causes of culling. The reason of that is, necropsies usually are being performed in the field with obvious limitations. Studies on sow mortality were necropsies are done, are expensive and difficult to perform. If we look at the data being recorded in the Danish farms roughly 55% of the dead sows are recorded with a diagnosis of sudden death. And roughly 20 % of the dead are recorded with the diagnosis unknown (Figure 3). One larger study was performed in Denmark in 2003, where 100 dead sows from 10 farms were necropsied. Here 23% of the sows died from infections related to reproduction/farrowing, 24 % died from organ torsions, 18 % died from gastrointestinal problems, 11 % from heart problems and 9% from other reasons and on 15% no cause could be found.[4] In 2010 another study was performed based on necropsies of dead sows during lactation.[6] The study showed that 22% of the sows died from rectal and uterine prolapses. The other causes of death reported in this study was stomach ulcers and organ torsions.

Most sows die unassisted in the weeks before and after farrowing. From these sows 30% die just after farrowing.[6]

Reducing unassisted sow mortality can be difficult since many necropsies are needed to get a clear picture. Nevertheless, emphasis on gastrointestinal health can be concluded as important. Pelleted feed with very low particle size is damaging for gastric function. In Denmark we work with recommendation of min 50% of the particles being above 1000 microns. This is impossible to ensure in

Figure 3: Mortality causes. Data recorded from 17 danish farms

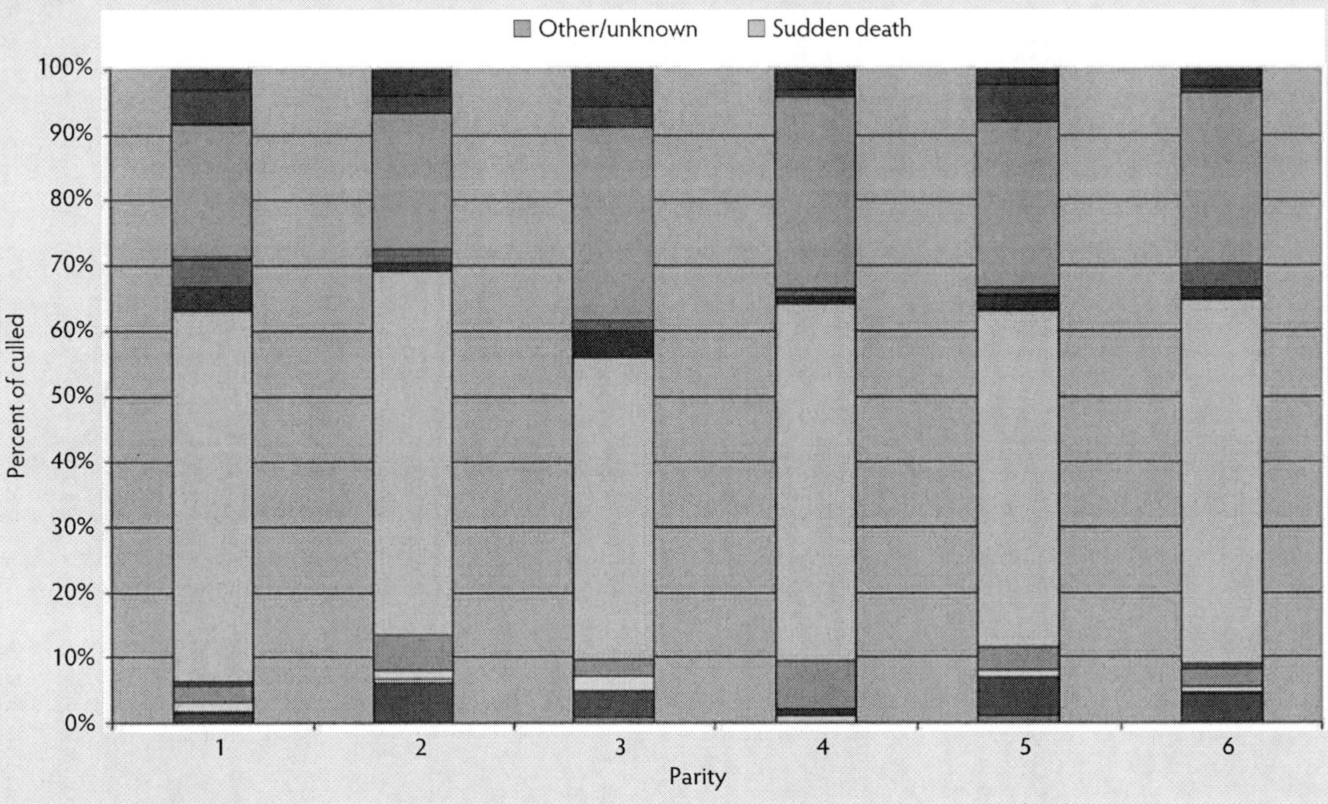

Modified from following source:
SEGES svineproduktion. Analyse af sodødelighed i 17 danske besætninger notat nr. 1604 available at https://svineproduktion.dk/publikationer/kilder/notater/2016/1604. Accessed 28 December 2018

pelleted fed. Therefore, feed mills have made never products like mixing in 10-15% smashed barley or sow feed which is processed but not pelleted. An old study in Denmark showed that sows have 0,5 parity more in farms which produce their own non pelleted feed compared to commercial feed.

More knowledge of mycotoxins and sow mortality is also needed, since a high degree of prolapses are seen.

References

1. SEGES svineproduktion. Analyse af sodødelighed i 17 danske besætninger, notat nr. 1604 available at https://svineproduktion.dk/publikationer/kilder/notater/2016/1604. Accessed 28 December 2018.

2. SEGES svineproduktion. Udviklingen i sodødelighed - tal fra DAKA 2016, notat nr. 1714 available at https://svineproduktion.dk/publikationer/kilder/notater/2017/1714. Accessed 28 December 2018.

3. SEGES svineproduktion. Udviklingen i sodødelighed - tal fra DAKA 2017, notat nr. 1815 available at https://svineproduktion.dk/Publikationer/Kilder/Notater/2018/1815. Accessed 28 December 2018.

4. SEGES svineproduktion. Afgangsårsagerhos søer – samt obduktionsfund hos aflivede og selvdøde søer. Medd nr. 656 available https://svineproduktion.dk/publikationer/kilder/lu_medd/2004/656. Accessed 28 December 2018.

5. Tani, S., Piñero, C., Koketzy, Y.(2017). Mortality and Survival Probability of Female Pigs in Southern European Commercial Breeding Herds. Journal of Agricultural Science; Vol 9. No 4; 2017.

6. Thorup, F., Pedersen, H., Olesen, A.K. 2010. Sodødelighed i farestalden. In: Meddelelsen r. 886. Videncenter for Svineproduktion, Den rullende afprøvning, pp. 1-29.

Low sow mortality: What's the secret?

Ron Ketchem, BS
University of Missouri-Columbia, Missouri; Swine Management Services, LLC, Fremont, Nebraska

Introduction

Swine Management Services, LLC (SMS) was purchased in 2002 by Ron Ketchem and Mark Rix. The company is now involved with sow data entry, consulting, financial record keeping and records analysis with a web site: www.swinems.com. SMS is most known for developing SMS Farm Benchmarking data base which has grown to 900+ farms with 1.6+ million sows. Ranging in size of farms from 125 to 10,000+ sows with most being independent pork producers in the USA, Canada and now some farms from Australia. Data is collected from 26+ different record programs from around the world with 16+ different Genetic companies represented. Our mission statement is to provide "information solutions" to the swine industry through state of the art benchmarking and analysis programs.

Review of data

In order to understand more about the issue of increasing female death loss in the swine industry in North America, I have attached several tables to show some of the history of the swine industry. Table 1 shows the last 13 years of trend lines in the SMS data base which currently is at 1.6+ million sows which is 900+ farms located mostly in the US and Canada. Most are independent producers with farm size of 125-10,000+ sows. As you see in Table 1 Pigs Weaned per Mated Female per Year for All Farms has gone from 21.3 to 26.44 pig which is a gain of 5.16 pigs and the Top 10% going from 24.7 to 31.7 pigs which is gain of 7.01 pigs in the last 13 years. Tables 2 and 3 contains data from SMS for the last 52 weeks ending October 2018 from 904 farms with 1,608,092 sows, you can see the table is broke out by Top 10%, Next 20%, All Farms and Bottom 30%. We now see the pigs weaned per mated female per year for All Farms last 52 weeks at 26.18 pigs with 91 farms in the Top 10% at 31.1 pigs dropping down to 22.4 pigs for the 271 farms making up to Bottom 30% of the data base.

A very interesting number is Total Born per Mated Female per Year which has Top 10% at 38.50 pigs with All Farms average of 34.26 pigs with Bottom 30% at 30.84 pigs. This shows that in the near future some farm could be at 35+ pigs weaned per mated female per year. We see in Table 2 and 3 that production numbers for that Top 10% which as the average of 91 farms having Farrowing Rate at 90.1% versus All Farm average of 85.1%, Replacement Rate at 61.3% versus All Farms at 59.9%, Total Pigs Born per Female Top 10% at 15.67 pigs versus All Farms average of 14.69 pigs versus Bottom 30% at 14.05 pigs, and Pigs Weaned per Female Farrowing

Table 1: SMS performance for last 13 years trend line: pigs weaned/mated female/years

	2005	2006	2007	2008	2009	2010	2011	2012	2013	2014	2015	2016	2017	Diff
10%	24.7	25.7	26.9	26.8	27.3	27.8	28.6	29.3	30.1	30.5	30.8	31.3	31.7	+7.01
All	21.3	21.9	22.5	22.9	23.3	23.8	24.3	24.8	25.3	25.2	25.4	25.3	26.4	+5.16

Table 2: SMS performance data 52 weeks average through October 2018 summary

	Top 10%	Next 20%	All Farms	Bottom 30%
Number of farms	91	181	904	271
Pigs weaned/mf/yr	31.1	28.81	26.18	22.4
Litters/mf/yr	2.46	2.42	2.33	2.19
Wean to 1st service interval	5.59	6.19	6.79	7.80
Total born/mf/yr	38.5	36.48	34.26	30.84
Percent repeat services, %	2.50	4.50	6.90	10.1
Farrowing rate, %	90.1	88.1	85.1	80.7
Female death loss, %	8.20	8.80	10.0	11.8
Replacement rate, %	61.3	55.9	59.9	66.1

has All Farms average at11.27 pig with Top 10% at 12.59 pigs versus Bottom 30% at 10.37 pigs.

As the total pigs born has increased the industry has seen an increase in Stillborn Pigs to 5.5% for Top 10%, All Farms average of 6.9% and Bottom 30% up to 7.7%, Pre-weaning death loss for All farms higher at 14.8% with Top 10% down to 12.2% and Bottom 30% at 16.8%. The industry is now moved Average Age at Weaning up to All Farms average of 20.79 days, with Top 10% at 20.40 days and Bottom 30% at 21.26 days.

If we concentrate on Female Death Loss % from Table 2 the Top 10% farms at 8.20%, All Farms at 10.0% and Bottom 30% higher at 11.8%. Table 4 has 12 years SMS data for Female Death Loss % with All Farms starting at 9.2% in 2006 down to 7.2% in 2010 going up the next 6 years to 10.2% in 2016 with a small drop in 2017 to 9.4%. The Top 10% started low in 2006 at 5.9% with some up and down to a low of 5.5% in 2014 and them a increase to 7.0% in 2016 and down a little in 2017 at 6.8%. That is an increase of 1.8% since 2013. To look at this issue of increasing female death loss we did some review of SMS data breaking out in more detail to look at this growing problem.

In Table 5 farms were sorted by size with data broke out by < 1,000, 1000 to 1999, 2000 to 2999, 3000 to 3999 and 4000+ sow per farm. That represents 348, 233, 105, 44 and 68 farms for a total of 888 farms with 1,539,149 sows in the US, and Canada. The farms along with being broke out by size were then separated by PW/MF/YR, Female Culled % and Female Death Loss % and by Top 10%(+10%), All and Bottom 30%(-30%). The table shows All Farms Female Death Loss at 9.3% with Top 10% (+10%) at 7.0% and Bottom 30% (-30%) at 11.2%. If you look at the farms by +10% which are farms that are at and average of 31.07 Pigs Weaned per Mated female had lower female death loss % across all farm sizes with range of 6.1 to 9.6%. The -30% farms had average of 22.24 Pigs Weaned per Mated Female with Female Death Loss at 11.25% with range of 10.0% to 12.8%. This shows that farms that are producing more pigs regardless for farm size had lower female death loss. Is this related to how farms are managed? We were also interested in female culling and death loss percent by parity. Table 6 takes 888 sow farm for the last 52 weeks and breaks data out by parity from P0, P1,P2, total for P0-2, P3, P4, P5, P6, and P7+ for female cull % and female death loss %. The data shows that P0 at 14.0, P1 at 18.8 and P3 at 14.5% which puts total female deaths for P0-2 at 47.3% with P3-P7+ at 52.7% with 10.6% being the older P7+ females. So this makes us wonder why there are so many younger female deaths. Do we need to look at gilts development?

Table 7 takes a deeper look at when female deaths are occurring, we broke the data out by Percent Deaths by Days Since Served by parity for 6,849 female deaths. Data was broke out by day 0-44, 45-89, 90-109, 110-119, 120-139, 140-154 and 155-200 days deaths and by parity P0-P7+. The P0 females have their own trend line with 43.4% of the female deaths being before the gilts get bred and being 13.6% of the total female deaths. P1 to P7+ all have similar trend lines with 13.5% of the deaths being days 0-44 after breeding and starting day 110 to 139 with 50.6% of all the female deaths by parity taking place, that is the time pregnant sows

Table 3: SMS performance data 52 weeks average through October 2018 summary

	Top 10%	Next 20%	All farms	Bottom 30%
Total pigs born/female farrow	15.67	15.05	14.69	14.05
Pigs born live/female farrow	14.39	13.68	13.24	12.49
Pigs weaned/female farrow	12.59	11.84	11.27	10.37
Piglet survival, %	82.4	80.2	78.3	75.6
Stillborn, %	5.50	6.40	6.90	7.70
Pre-weaning mortality, %	12.2	13.4	14.8	16.8
Average gestation length	116.0	115.8	115.8	115.9
Average age at weaning	20.40	20.63	20.79	21.26
Average parity	2.45	2.49	2.57	2.60
Mated female non-productive days	27.40	33.01	42.47	55.37

Table 4: SMS data set 12 years trend lines for female death loss %

	2005	2006	2007	2008	2009	2010	2011	2012	2013	2014	2015	2016	2017	Diff
10%	NA	5.90	6.60	7.40	7.20	6.90	6.30	5.80	6.10	5.50	5.90	7.00	6.80	-0.90
All	NA	9.20	8.60	7.90	7.60	7.20	7.60	7.70	7.60	7.90	8.60	10.2	9.40	+0.20

Table 5: SMS data set culling and female death loss by farm size last 52 weeks ending October 2018 for 888 farms

			PW/MF/YR			Female culled %			Female death loss		
Farm size	**# of farms**	**# of sows**	**+10%**	**All**	**-30%**	**+10%**	**All**	**-30%**	**+10%**	**All**	**-30%**
< 1000	348	194,059	31.16	26.06	21.97	54.7	45.2	43.3	6.0	8.0	10.0
1000s	233	319,893	30.81	26.06	22.15	52.8	43.6	38.3	6.1	9.7	12.1
2000s	195	493,588	30.86	26.05	22.45	43.1	48.7	45.3	8.2	10.1	11.6
3000s	44	147,681	30.56	25.21	22.99	48.5	50.8	50.3	6.5	10.0	11.6
4000+	68	393,927	31.51	26.63	22.45	56.0	48.7	49.2	9.6	11.1	12.8
Grand total	888	1,539,149	31.07	26.06	22.24	52.3	46.1	43.5	7.0	9.3	11.2

Table 6: SMS data set culling and death loss by parity 52 weeks ending October 2018 for 888 farms

	Parity									
	P0	**P1**	**P2**	**P0-2**	**P3**	**P4**	**P5**	**P6**	**P7+**	**Total**
Female culling %	15.7	15.3	9.3	40.4	8.5	8.5	9.3	10.5	22.8	100.0
Female death doss %	14.0	18.8	14.5	47.3	13.1	11.5	9.9	7.6	10.6	100.0

Table 7: SMS data set % deaths by days since served for 37 farms with 88,000 sows

		Percent deaths by days since served								
Parity	**Never bred**	**0-44**	**45-89**	**90-109**	**110-119**	**120-139**	**140-154**	**155-200**	**Grand total**	**% of total**
0	43.4%	16.3	11.6	8.6	16.5	3.1	0.3	0.1	929	13.6
1		10.0	10.8	5.3	21.7	32.8	14.7	4.7	1173	17.1
2		13.3	10.8	8.0	28.3	27.1	9.7	2.9	890	13.0
3		12.9	11.8	7.9	29.8	26.0	8.7	2.9	908	13.3
4		16.1	13.2	5.2	25.8	29.0	8.6	2.0	861	12.6
5		15.2	13.8	5.6	25.0	29.8	8.5	2.1	752	11.0
6		16.0	12.6	5.6	23.2	32.4	8.5	1.5	586	8.6
7+		12.0	10.8	3.6	22.9	34.8	11.6	4.3	750	11.0
Grand total	**5.9**	**13.7**	**11.8**	**6.3**	**24.1**	**26.5**	**9.0**	**2.7**	**6849**	

are moved from gestation to farrowing, going through farrowing, lactating, wean, moved to breeding, bred back and moved to gestation in a 30 days period of time. We do not have any idea how much stress this puts on these females.

To try to get more information on why females are dying and being culled we took data from 17 farms that we felt had very good reasons for female deaths and culling over a 3 year period of time and put data in to Table 8. The deaths and destroyed (euthanized) were combined so all deaths were together and culls in a separate column. Since there were 88 reasons for deaths and destroyed and 65 reasons for culls the data was reviewed and reasons reduced to 13 reasons that are listed: Other/Unknown, Prolapse, Downer, Difficult Farrow, Lame/Injured, Heart Attack, Retained Pig, Body Condition, PRRS, Age, Did Not Conceive, No Heat and Misc (all others). The most eye opening reason was 25.5% of the Female Deaths and 1.87% of culls were recorded at Other/Unknown. This makes it very difficult to figure out how to lower female deaths if that many females have no reason for their deaths recorded. Table 8 shows the next 4 reasons for female deaths being Lame/Injury, Downer, Difficult Farrowing and Misc. (all others). In our dataset prolapses contributed to 8.89% of the total deaths.

In summary looking at why female death loss has been increasing the data is showing since about 2013 this number has been increasing across the US and Canada in our data base and other ones that have been reviewed. During this time production numbers continue to improve with Pigs Weaned per Mated Female per Year improving by 5.0+ pigs per sow over the last 13 years with 91of the 904 sows in the SMS benchmarking data base the last 52 weeks at 31.10 Pigs Weaned per Mated Female per year the last 52 weeks, Farrowing Rate at 90.0+%, Total Pigs Born per Mated Female per year up to 15.67 pigs, Pigs Weaned per Female farrowed at 12.20 pigs and Females Death Loss at 8.2%. We see that the higher producing sow farms how lower female death loss regardless of sow farm size. We continue to be very concerned that 47.2% of female deaths are now the younger females P0-2 versus the older P7+ females at only 10.6% of the total. The data shows us that 50.6% of the female deaths are in a 30 days period which is the most stressful time for females from day 110 to 139 after breeding which includes the farrow, lactation, breeding and movement to the gestation area. So what are females dying from continues to be hard to figure out since 25+% of the death reasons are Other/unknown with the 4 biggest death loose reasons being Lame/injured, Downer, Difficult Farrowing and Misc.(all others).

Discussion

I have spent a lot of time in sow farms over the last several years and have seen the increasing problem of female death loss and culling of females early. A lot of these farms whether new or older continue to have labor issues such as not being able to hire enough people to get the basics done daily, lots of absenteeism, higher turnover, and lower quality workers. Due to the lack of labor some farms have had to drop doing some basic management practices such as washing all the building at the farm regularly which includes the gestation and breeding areas, GDU and grower facilities for developing gilts, and not just the farrowing rooms and hallways. This has lead to an increase in slick dirty floors, more injuries and lameness, increases in abscesses, joint infections, etc, causing more females that then have to be euthanized even after being treated. These farms are not getting minor repairs and maintenance done timely which leads to more injuries to animals and people. With more farms going to some form of loose sow housing and some issues with floor surfaces there has been an increase in sows fighting causing more toe growth and broken dew claws. There have also been more feed quality issues involving mycotoxins which can cause many issues at the farm from estrus issues, increase in non-cycling gilts, sows refusing to eat, increase in aborts etc. This may come in the feed or be in the feed tubes in farrowing or bulk bins that are not getting attended. Genetics may be some of the issue with the increased demand for gilts caused by the expansion in the swine industry the last few years. Also a lot of farms have not dedicated money and labor to improving their gilt development and selection programs on farm. Then there are farms dealing with health issues over the last few years causing long periods of time that farm is closed to entry to replacement gilts causing sows farms to rebred problem animals and increased sows deaths as the farm has older sows. There are probably more causes for the increase in female death loss that can be added to this list.

Conclusion

For the pork industry to address this issue of increasing sows deaths loss there are several areas that need to be mentioned.

Table 8: Death loss and cull reasons for SMS data-3 years 17 farms

Reason-individual farms	Deaths-55 reasons 8275 hd.		Destroyed-33 reasons 1160 hd		Combined total-9935 hd.		Culls-65 reasons 48,675 hd.		Total-58,610 hd.
	#	%	#	%	#	%	#	%	%
Other/unknown	2411	29.1	119	7.17	2530	25.5	911	1.87	5.87
Prolapses	717	8.66	175	10.5	892	8.89	248	0.50	1.95
Downer	1258	15.2	466	30.1	1724	17.4	32	0.07	3.00
Difficult farrow	1253	15.1	47	2.83	1300	13.1	271	0.56	2.68
Lame/injury	625	7.55	655	39.5	1280	19.3	2896	5.95	7.12
Heart attack	341	4.12	-	0.00	341	3.43	4	0.01	0.59
Retained pig	196	2.37	10	0.60	206	2.07	92	0.19	0.51
Body condition	117	1.41	72	4.34	189	1.90	910	1.87	1.88
PRRS	167	2.02	35	2.10	202	2.03	1	0.01	0.35
Age	27	0.33	3	0.18	30	0.30	12875	26.5	22.0
Did not conceive	10	0.12	4	0.24	14	0.14	7399	15.2	12.6
No heat	84	1.00	-	0.00	84	0.85	5041	10.4	8.74
Misc (all others)	1080	13.1	74	4.46	1154	11.6	17995	36.9	32.7

Number one on my list is labor: to help with labor there need to be more sources for labor available to pork producers, which means more contact with local organizations, state and community colleges, and being able to bring in more labor from outside the US borders. We have to create jobs that are a career in the swine industry by providing more continuous training and personal development at the farm level by online training courses, development and implementation of SOP's on farm, use of outside training groups or in conjunction with local colleges. Back to the basics in the farm of regular cleaning and sanitizing the entire farm, make sure that every animal on the farm is checked daily to make sure they have feed, clean water and good environment. That also means looking at each female for injuries or health issues daily. Each farm needs to create a health team that is responsible for daily treating of sows with issues, making sure that the vaccination program is administered correctly, looking at sows daily in crates or pens and spotting the animal that need treating and if needed removing them from the pen since they are not able to compete. Farms need to work with someone such as consultant or maybe their genetic company in developing better methods of selecting and developing replacement gilts. If doing internal multiplication, the proper feeding program for the gilts as developing replacement gilts, proper housing of these future sows, selection for proper feet and leg structure and proper handling until they are of breeding age and size. I see that with the move to more loose sows housing more research is needed on the floor these animals are going to walk around on, better selection of replacement gilts, the effect of nutrition on feet and leg structure and doing regular maintenance of sow's feet even trimming of toes and dewclaws. More nutrition research being done on how to feed the high producing females which have potential to wean 30+ pigs per sow per year and to gilt development, gestation, transition feeding before farrowing and diets in farrowing to increase quality of colostrum and volume of milk produced to pigs.

The issue of increasing female death loss is multi-factorial with many areas to be addressed, if we plan to stop the increase and start to lower the current numbers some management areas need to change. Going forward if we plan to improve production numbers more work needs to be done to get female death loss under control and lower the cost to produce a pig.

Production drivers of sow mortality

P. Mosqueira, DVM; J. C. Pinilla DVM, MS
PIC Global Technical Services Reproduction- Hendersonville, Tennesse

Introduction

Sow mortality is one of the most important indicators of performance and efficiency in modern swine production as it is a component of more complex and complete performance indicators, like sow lifetime performance. A higher sow mortality impact goes beyond the mortality rate by itself, as it can negatively impact other performance indicators such as breeding target, farrowing rate, retention rate and overall farm's throughput.
A low sow mortality could also be understood as a manifestation of good animal husbandry and a solid commitment with animal stewardship.

Over the last window of time, several sources have documented proximately three percentage points increase in sow mortality.[1] On the economic side, it is estimated a cost of $5/inventoried sow/year for every additional point of sow mortality, considering the opportunity loss of $240/sow for not selling it to the salvage market, plus $230 as added marginal cost for a needed gilt to replace the dead sow.

Drivers

Gilt selection

Optimizing gilt selection practices is a critical component in maximizing sow survivability, retention and lifetime production in the breeding herd. It is known that gilts with improper quality are prone to be early removed from herd, many times by euthanasia. A successful gilt selection process focused on sow longevity should prioritize general appearance/conformation of the gilt, proper weight/size, sound feet and leg structure, and hooves integrity.

There are still opportunities across the North American industry to improve this process and some reasons could be attributed to lack of convincement of its importance, lack of gilt availability to perform selection, economic models based solely on cost and not on cost-effectiveness as well as short-term pressure to meet breeding target.

Gilt weight at first breeding

Body weight at first breeding is arguably the single most important indicator of gilt eligibility as it reflects growth and body maturity, and it is highly correlated to health and reproductive status. Gilt weight at first breeding is one of the most neglected indicators on the sow side of the business,[2] with the body weights being essentially unknown or visually estimated, even considering that there are alternatives for estimation such as flank taping or heart girth tape. The optimum live weight at first service is 300 to 350 lb.[3] Gilts bred lighter than 300 lbs tend to have difficulties during lactation as its body may not be mature enough to cope with the lactational burden, leading in some cases to extreme loss of body condition and risk of early euthanasia. On the other side, heavier bred gilts can pose challenges on leg structure soundness leading to early removals, commonly because of lameness and/or its complications, not to mention the limited lactational feed intake shown by too heavy gilts going to farrowing.

Staffing

It has been some years that the US market is in expansion mode, which basically means new sow farms have been built or old ones have been retro-fitted. The top 40 US producers have expanded their sow inventory by 11% net by 2018 when compared with 2016 (approximately by 5% in 2018 compared to 2017, and approximately by 6% in 2017 compared to 2016).[4,5]

The expansion of the industry is very positive to the business, but it comes with short and long-term challenges as lack of specialized labor. Both understaffed farms as well as underqualified labor in a sow operation will negatively impact overall performance. The lack of experience in animal husbandry by the new people starting to work in the sow farms is often related to an increase in overall sow mortality, in some cases taking months to go through the learning curve.

Individual sow care

Sow care comprises not only the husbandry of the herd as a whole, but even more important the sow care in an individual basis. A quick and proactive identification of sows in need of intervention can save many females that otherwise would be euthanized. This management should be done in a daily basis, including weekends.

"Number of treatments initiated per week" has been used as an indicator to manage individual sow care. It is related to any type of intervention was initiated on the sows in a period of one week, as use of antibiotics, anti-inflammatory, topic medication, mats for lameness sows and others. The recommendation is to be in between 2-3% out of the average sow inventory with intervention levels under 1% or above 4%.[3]

Body condition

Body condition of the herd can determine the sow lifetime performance of the operation. It is known that fat sows tend to have more structure issues as lameness, and thin sows tend to be more fragile both on the health and management side.

Also, the optimization of lactation feed intake is highly correlated to the body condition of the individuals and proper gestation feed usage. A proper lactation feed intake will support a successful advance of the reproductive cycle of the sow and its retention.

Sows should have the condition assessed at breeding, 30, 60 and 90 days of gestation and it is accepted up to 5% of fat and 5% of thin sows in the herd.

Farm supervisors should also be encouraged to manage overall feed usage as a strategy to estimate body condition of the herd. It is recommended to have 1,550-1,650 lbs. of gestation feed usage/sow/year and minimum 750 lbs. and 1,000 lbs. for 21 and 28 days of lactation feed usage/sow/year, respectively.

Group housing

Over the last years many North American companies have been adopting group housing as a response to the demand of the market.

There is a learning curve on the implementation process inherent to this setting and the producers should be aware in advance of what are the key points to be successful. In some cases, this learning curve can lead to a reduction in performance, with sow mortality being one of the earliest indicators.

On the individual sow care perspective, the group housing farms face more difficulties to find sows in need of intervention compared to individual housing setting. In the latter, employees are able to identify a non-eater with the feed in front of her as an evidence, which usually does not happen in the former. On the body condition side, a fat herd tend to be less forgiving in a group housing setting due to the more frequent interactions for hierarchy consolidation by the sows, when the heavy structure may pose more damages to the feet and legs. Also, the proper stocking density, number of sows per pen, segregation of pens by parity/size/condition and proper maintenance of the group housing equipment/facilities comprehend key points to mitigate losses.

Parity structure

The balance of the parity structure of the herd is a combination among replacement rate, culling rate and sow mortality. Any disruption to this balance can lead to losses in performance, with the sow mortality being routinely affected. Common disruptions are related to lower replacement rates when the systems do not have enough gilts to replace the current inventory, and/or higher culling rates when the farms have been involuntarily culling more sows than the target, and/or even a higher sow mortality as a starting point. The combination of these disruptions with the pressure to meet breeding target over the weeks to support throughput may lead the farm to a negative spiral when it starts to breed marginal sows that should not be bred. These marginal sows will have more chances to die or to have an issue over the course, increasing even more the sow mortality of the farms.

Pelvic organ prolapses

Over the last several years, an increasing incidence of prolapses have been intermittently reported across many different geographic locations. These reported cases appear across a range of facility types, genetics and feed mills, making it more complex to define common factors. Moreover, some of the cases sometimes appear to be farm-specific, even within a given production system.

Although the identification of the precise etiology has been a work in process by the industry, it is known that some risk factors on the production side may predispose to its occurrence. Some examples are older parity females, low grain particle size, over conditioned sows, constipation, farrowing assistance with overuse of oxytocin as well as mycotoxins in the feed.

Considerations

Considering the production drivers of sow mortality as listed above, it can be said that different factors can influence this indicator. On top of the production drivers, the systems can also face health challenges which by itself or combined with the production challenges might lead to an increase in sow mortality.

After identifying the key drivers of the sow mortality and its impact in a given system, the next step goes through strategizing an action plan to address these findings and executing it with focus and discipline.

References

1. PigCHAMP Benchmarking Summaries. PigCHAMP. 2018. Available at `https://www.pigchamp.com/benchmarking/benchmarking-summaries`. Accessed 29 October 2018.

2. PICpro100: Lessons Learned. PIC. 2018.

3. Sow and Gilt Management Guidelines. PIC. 2017.

4. Pork Powerhouses®. Successful Farming. 2017.

5. Pork Powerhouses®. Successful Farming. 2018.

Training, motivation, and a culture of safety at Maple Leaf Agri-Farms

T. Landry, B.Sc.(Ag)

Introduction

Our goal is to always provide consumers safe, great tasting food produced in a safe work environment.

– Maple Leaf Safety Promise

This paper will explain the development stages of the safety program at Maple Leaf Agri-Farms and show how the shift of employee engagement towards a culture of safety was encouraged. It will discuss the training techniques and motivation behind the health and safety program at Maple Leaf, which helped to create a culture of safety and accountability within our teams, along with the future goals for safety within our facilities.

Maple Leaf Agri-Farms has worked hard over the years to shift our occupational health and safety program from a **reactive system**; one where action was taken in the case of an injury, to a **proactive system**; one that works as a collaborative unit to prevent injuries from happening. A system that holds its workforce, the ones performing the tasks on the front lines, accountable. This workforce must come together to keep themselves and each other safe; while management must take all necessary action to provide them with the tools to do so. Maple Leaf Agri-Farms has achieved a substantial improvement in lowering the number of injuries in our facilities since this journey began, but more importantly we have changed the mind set of the workforce from "accidents happen" to "all incidents are preventable".[1] Zero injuries within the entirety of our system is an achievable goal with the proper commitment and engagement from the team; this is the system Maple Leaf Agri-Farms continues to strive for.

History

The Maple Leaf Agri-Farms health and safety program started as a top down initiative. Safety rules came from and were enforced by management. We value our people therefore management made safety a priority in all areas. So much so that working safely was and still is, a condition of employment. This stage was critical to initiate awareness with the team, however, there was minimal credibility and very low accountability. There was room for improvement. Binders of papers with safe work procedures and rules sat in the office with little engagement from the team. Incident investigations were done because something happened, because legislation dictated the requirement. It was a reactive system, not a proactive system.

At this time, there was also a push from the Canadian government, making workplace health and safety a legislation. Legislation states that all employees have the below rights:

1. Right to know (about hazards in the workplace)

2. Right to participate (in health and safety committees and activities)

3. Right to refuse unsafe work

4. Right to work without being subject to discriminatory action for exercising any of the rights listed above or as described in the *Manitoba Workplace Safety & Health Act* or *Regulations*.

A shift began as the employees started to partake in the health and safety initiatives at their sites. They began to realize the importance of their input on safety matters. Getting home safe at the end of the day became a priority/desire for workers and less of a "rule". They began to look out for themselves and contribute to making things better in their workplace. This behavior was strongly encouraged by management, contributions were listened to and action was taken by management to follow through on suggestions, thus the ownership of the health and safety program was beginning to take form. Action and encouragement from management created a relationship of trust between the varying levels of Maple Leaf Agri-Farms. Management trusted the employees to speak up if there was a concern and the employees trusted that when a suggestion was offered to improve their workplace, management would evaluate the suggestion, develop a strategy, and take action whenever feasible.

To continue receiving these suggestions and to motivate the process to propel even further, management encouraged these suggestions with rewards. These could be in the form of a simple thank you from higher ups, recognition of a job well done, or physical rewards. For example, if the site reaches 365 days straight of being injury free the senior management team will take the staff for a nice dinner, plus the staff receives an item or gift. This practice is not only a physical reward for their hard work, but also a beneficial team building experience.

As the workforce began to feel confident in management and aware of health and safety initiatives through constant motivation, a third shift began. This is the point where Maple Leaf Agri-Farms currently sits in this process. Health and safety is a team initiative, everyone is aware and works to keep each other safe. Looking out for one another, thinking collectively and working together has been encouraged and continues to be successful through team initiatives and collective goals.

How the shifts occurred

Stage one had a struggle with a lack of accountability, "nobody taught them" the binder, there was a lack of buy in from employee's. Rules were just rules with no reason or explanations.[2] A system like this is not ideal for a variety of reasons. It is difficult for anyone to understand and follow a rule if they are not explained why this rule is in place; they will have their own ways and reasons for doing things differently, an "easier" or "faster" way. As a company, Maple Leaf recognized a strong need for employee engagement and standardization. Engagement allows people to be involved in the creation of rules, provide valuable input and hear feedback. In turn they will be involved in creating the rules they are required to follow. Thus, health and safety committees were developed further which allowed for stage 2 to begin. These committees are open to all hourly staff to discuss any concerns or ideas they have to improve their workplace. The health and safety committees also allow for engagement between all the employees, at all sites to get together and discuss any safety issues or concerns within their sites. This engagement between employees and management allowed our near miss reporting (preventative measures) to increase by almost 600%.[2] With the employees taking more ownership of the program, we began shift into the third stage. The proactive rather than the reactive stage, where we are today. Employees want to improve their workplace; most importantly employee's want to go home safely to their family's and wish the same for their teammates. We are currently at a place where employee run health and safety committees and write the majority of safe work procedures to be reviewed by management, they are the solution. It is no longer a top down delegation to be safe, our health and safety initiatives are coming directly from the employees within all facilities at Maple Leaf Agri Farm's.

Training techniques

As a baseline we use two online training systems to encourage health and safety engagement. Every employee has their own personal account that is monitored by management to ensure training is completed in a timely manner. Both systems use online videos followed by interactive quizzes.

The first system is called Alchemy Sistems. This is a mandatory program that every employee in the system must complete regardless of where you work (feed mill, logistics, barn etc.). It is specific to safety while performing MLAF (Maple Leaf Agri-Farms) tasks. It is scheduled to be completed annually every employee to refresh and stimulate our thoughts of staying safe, complacency kills!

The second system is Pork Avenue training in conjunction with our safe animal handling program. This system is specific to the MLAF production team staff who are hands on with the animals. This training is focused specifically on learning how to perform tasks on site with the animals in a safe manner for both the staff and the animals. This portion of the training involves online videos, online quizzes and a hands-on portion where the employees must perform (or "teach") their daily tasks to mentors on farm properly and safely in order to be signed off.

Another system used that could be considered training is cross audits within the sites. Managers from different sites will internally "audit" other farms within their production pyramids to help other managers find areas for improvement. This method allows for a fresh set of eyes to look over the barn to even further improve safety and encourages dialogue amongst peers around what was witnessed, possible solutions, potential gaps and lateral learnings. This method is also fantastic for increasing motivation and engagement because it inherently instills a sense of value to the person when they are chosen to go visit another site and offer their opinions.

Involvement through Maple Leaf interactive health and safety committees can be seen as further training as it is peer mentoring; having open discussions at minimum once a month. These committees encourage hourly employees to develop and record the Safe Work Procedures for their areas, do task observations to see how daily routines can be improved to make them safer, and discuss equipment and PPE trials (Personal Protective Equipment).[3]

Motivation

Proper engagement and training is an excellent tool for motivation. If employees feel they are a valued team member whose opinions are respected, they are much more willing to partake in keeping themselves and others safe. Going home safe at the end of the day is the greatest reward/motivation anyone can receive, but below are a few ways MLAF perpetuates a constant focus on health safety within the barns:

1. Daily huddles: Never become complacent with awareness of health and safety issues, it is made to be fresh on the minds of every employee at the start of each morning in MLAF facilities. Before going into the production area the team sits together to plan the day and discuss the tasks at hand. They take this time to discuss any near misses from their own farm and other MLAF sites. A discussion takes place on safety tips for the day to keep each other safe. It is critical to have daily reminders for the team to avoid complacency.

2. Daily reminders: Reminders or catch phrases are on all barn clothing (scrubs, t-shirts, and coveralls) to "Work Safe". Posters on the wall and corkboards in the office/lunch area dedicated strictly to health and safety initiatives and motivational messages.

3. Mandatory PPE: Anyone entering a production area of an MLAF barn must wear full PPE as required by MLAF standards (staff, contractors and visitors included). The full PPE required is: protective safety glasses, ear plugs, dust mask, cut resistant gloves, steel toed footwear and a personal H2S monitor on their chest AT ALL TIMES. If an individual is moving sows, entering the loose sow housing areas OR working with finisher pigs they also must wear metatarsal boots. The company provides all of the above.

4. Incentives: at Maple Leaf there is a Recordable Injury (RI) Free incentive program. This consists of every 50 days that a barn goes RI free, they receive a "gift" and recognition by the management team. Something like doughnuts, pizza party, etc. If the site goes

365 days RI free they will be taken out to a dinner by management plus receive a gift (backpack, jacket etc.). There is also an annual monetary bonus that that recognizes and focuses on teams who consistently deliver on health and safety requirements. These efforts help to prevent incidents from happening. If a team is not performing in this area they are at risk to lose their bonus. This is a strong method of motivation to increase proactive measures, and report all areas for improvement. mmm

Health and safety requires continual focus and high level of commitment from everyone at all levels of the company. Everyone needs to work together to keep not only themselves but each other safe, this is how to truly motivate a workplace culture.

Summary

Employee engagement, team building and fast action from management have been three of the key ingredients to Maple Leaf Agri-Farm's success in creating a culture of health and safety. But above all, health and safety is more than just T-shirts and posters on the wall or rewards; it is a daily discussion between all staff.[1] A commitment to doing things the safe way every time and being personally accountable along with holding others accountable to do the same. It is a constant daily focus and commitment to go home safe at the end of the day, because that is the greatest reward of all.[2] Ultimately this empowerment and awareness is what has substantially decreased the number of injuries in the workplace at Maple leaf Agri-Farms. Collectively we have changed the mind set from "accidents happen" to "all incidents are preventable".[1] It is a culture driven by the employees, it is a proactive system. Having zero injuries in a workplace is an achievable goal, it is Maple Leaf Agri Farms goal. Maple leaf Agri Farms will strive to continue being a leader in the progression of health and safety initiatives not only within the company, but globally.

References

1. Robertson, N. (2018, October 22). Email.
2. Bugera, F. (2018, September 7). Personal Interview.
3. Bugera, F. (2018, March). Christensen Farms – March 2018 PowerPoint Presentation.

Comparative lameness: What are we learning in other species?

JK Shearer DVM, MS, DACAW
Department of Diagnostic and Production Animal Medicine; Iowa State University, Ames, Iowa

Introduction

Lameness disorders of domestic ungulates are among the some of the most important of health disorders. Technically, the term "ungulate" refers to any animal with hooves. Mammals including cattle, sheep, goats and swine represent even-toed ungulates (i.e. Artiodactyla); whereas, odd-toed ungulates (i.e. Perissodactyla) includes horses, rhinoceroses, and tapirs. The causes, treatment and control of lameness in species such as horses, cattle and small ruminants are well described and documented in the veterinary literature. In the majority of cases (up to 90%), lameness is associated with conditions of the foot. While this is likely true for swine as well, there are few reports describing specific foot disorders in swine. One reason is that examination and care of foot lameness in swine is not a practical or common procedure for individual animals in a production setting. Gait abnormalities are observed infrequently in part due to individual housing, which limits freedom of movement that would otherwise facilitate lameness detection. In swine, disturbances in locomotion significantly reduce growth in feeder/grower pigs, reproductive performance in breeding gilts and sows and welfare of affected animals.[1]

Lameness is not a diagnosis; rather, a symptom of pain associated with lesions of the foot, proximal limb or in some cases the spine. Broad categories include non-infectious conditions of the hoof or claw associated with physical trauma or metabolic disease; infectious diseases of the foot and foot skin; and developmental disorders or injury of the proximal limb. The incidence and distribution of lameness-causing lesions to each of these categories is influenced by specific anatomic and weight-bearing dynamics unique to each species. Combined with housing, feeding and management factors these become the primary determinants of lameness.

The following is a brief review of lameness in common livestock species with emphasis on lesions occurring in the foot from a comparative perspective.

Relevant anatomy and weight-bearing dynamics

The equine foot and weight-bearing

The foot of the horse is composed of three bones: the third phalanx or coffin bone, the second phalanx or short pastern bone, the first phalanx and the distal sesamoid or navicular bone. Structures distal to the fetlock joint are normally referred to as the digit or foot. The distal interphalangeal joint is enclosed within the hoof horn capsule serving as the articular connection between the 2nd and 3rd phalanxes. Moving upward, roughly a finger's width above the coronet is the proximal interphalangeal joint between the second and first phalanxes. The navicular bone lies caudal to the distal interphalangeal joint serving as a fulcrum for the deep digital flexor tendon. External features of foot above the weight bearing surface includes the coronet or coronary band, periople, wall, heel, toe and quarter. Structures visible on the weight bearing surface includes the frog, heels, sole, white line, bars, quarters, wall and toe.

Most lameness in horses occurs in the front feet. The reason most often cited is increased load bearing in front limbs compared with rear limbs because of the added weight of the head, neck, shoulders and thorax. Front limbs are reported to support as much as 60-65% of the horse's weight when at rest.[2,3] Add a rider and the additional concussive forces incurred during running or jumping and the stresses on feet increase significantly. When a horse is galloping, all of the body's weight is borne on the leading foreleg. So, unlike cattle, small ruminants and swine, lameness in the horse is complicated by its athletic prowess and to some extent human demands and use.

The ruminant (bovine and small ruminants) foot and weight-bearing

In cattle and small ruminants, the foot is comprised of all structures below the fetlock joint, which includes the 1st, 2nd and 3rd phalanxes and the navicular bone times 2 making a total of 8 bones within the foot since the ruminants are even-toed ungulates. External features of the claw horn capsule are similar to that of the horse and include the coronet or coronary band, periople or cuticle, heel and wall, which is separated on the abaxial side of the hoof by the abaxial groove. Structures of significance on the weight-bearing surface include the heel, sole, abaxial and axial wall, and white line.[4]

In cattle, 90% of lameness occurs in the foot, and of that affecting the foot, nearly 90% affects the rear feet. Therefore, while similar to the horse in that most of the cow's weight is in the front of the body; contrary to the horse where the majority of problems affect front feet, most lameness in cattle occurs in rear feet. Furthermore, when lameness is hoof-related, 90% of lesions occur on the lateral claw of the hind foot. Cattle, and to a lesser extent small ruminants, bear a greater amount of weight on the lateral claws of rear feet. When subjected to hard flooring surfaces such as concrete, weight bearing results in physical and mechanical stresses that are major contributors to claw-related lameness. For example, one of the primary lameness disorders occurring in large dairies today is "thin soles" a condition associated with excessive wear of feet from long and frequent trips to the milking parlor.[4] On the contrary, small ruminants normally spend very little time on hard surfaces such as concrete. In this case, reduced rates of horn wear result in overgrowth of hoof horn, which predisposes to long claws and foot rot.

The porcine foot and weight-bearing

The porcine foot is similar to that of cattle and small ruminants; two weight-bearing digits and two dewclaws that are non-weight-bearing. Researchers have determined that more weight is borne on the front feet on average compared to the back feet (~ 58% vs. 42%). In addition, similar to cattle and small ruminants, the lateral claw of the rear feet bears the greater amount of weight, is generally larger and more often affected with lesions resulting in lameness.[5,6]

Prevalence and impact of lameness

Equine lameness

A UK report cites lameness as the most common problem affecting horses and ponies. Lameness was reported to affect 18.6% of horses with 3.8 % attributed to foot lameness and 14.8% to degenerative joint disease. The National Animal Health Monitoring System (NAHMS) Equine '98 study surveyed horse operations in 28 states throughout the US. Horse operations reported having at least one horse with lameness during the previous year and 13% reported having a horse with laminitis. On the days of the Veterinary Medical Officers visits, 3.4 (winter) to 5.4 (spring) percent of horses were affected by lameness and less than 1 percent were affected by laminitis. Leg or joint problems were perceived to be the cause of lameness in the spring and winter whereas, hoof problems were the most common cause of lameness in the summer. Laminitis, navicular disease, and sole bruises or abcesses were the most common causes of hoof or foot lameness with laminitis accounting for 7.5 to 15.7 percent of all lameness problems. Cost to the US equine industry was estimated to be from $678 million to $1 billion in 1998 ranking lameness as the most costly of diseases in equids.[7,8]

Lameness in cattle and small ruminants

The worldwide reported prevalence of lameness in dairy cattle varies from 8% to near 55%.[9,10] Where dairy cattle are confined on hard surfaces such as concrete, prevalence is usually higher. Cattle are land animals whose feet are better designed for softer earthen surfaces. Cost per case of lameness in dairy cattle is estimated to be around $500/case, which for a 100-cow herd with a 30% incidence rate of lameness amounts to $14,300.[4] This exceeds losses due to mastitis and decreased reproductive performance making lameness the single most costly clinical disease of dairy cattle. Prevalence of lameness in beef cattle is lower, but not well described in the literature. The prevalence of feedlot cattle lameness is variable, but generally in the range of 4-6%.[11]

The prevalence of lameness in sheep is reported to be between 8-10%. Observations on dairy goat farms in the UK estimated the prevalence of lameness to be between 9.1 and 19.2 %. In sheep, lameness results in poor weight gains, impaired fertility, and reduced growth in lambs. Dairy and meat goats are impacted similarly with reduced milk production, fertility and longevity.[12]

Lameness in swine

The prevalence of lameness in feeder pigs is difficult to assess accurately, but appears to be in the range of 1.8-30% throughout the finishing phase of production.[13] Farms in the US with the most severe problems reported prevalence rates near 30%. Risk factors for developing lameness in growing pigs includes heavier weights, fully slatted floors and housing of pigs in large groups. Lameness in sows is a major concern because of its impact on reproductive performance and longevity. It also contributes to higher veterinary costs for treatment or euthanasia and a higher risk for losses from condemnation of carcasses at slaughter.[14] A study comparing loose-housed sows on partially slatted concrete floors versus sows stalled on concrete revealed that 96% of cull sows from loose housing and 80% of cull sows from confined housing were observed to have at least one foot lesion.[15] Overgrown toes from a lack of exercise reduces feed consumption and impairs locomotion. Claw horn overgrowth combined with lameness reduces the number of litters per sow per year and the mean number of weaned pigs per sow per year and increases the cost per piglet weaned.

Lameness is reported to be the cause for culling of 8-15% of sows; however, this is likely to be an underestimate since many sows culled for reproductive failure are in fact lame. In consideration of the costs for decreased longevity, decreased productivity (piglets/sow), decreased salvage value, etc. economic impact of sow lameness in the US and Canada runs into the hundreds of millions of dollars annually.[16]

Common lameness disorders in ungulates

Equine lameness disorders

The top 5 causes of equine lameness are subsolar abscesses, navicular syndrome, heel pain, degenerative joint disease (DJD) and tendon and ligament injury. *Subsolar abscesses* cause extreme pain and lameness. There are multiple causes with white line disease being one of the more common causes. Subsolar abscesses will often expand beneath the sole or migrate upward beneath the wall to exit as a purulent tract at the coronet. Curettage of the hoof horn surrounding the lesion permits drainage of the purulent material and limits secondary damage to deeper structures within the foot. *Navicular syndrome* is the term used to describe heel pain associated with navicular disease that some believe is caused by an abnormal hoof-pastern angle and both front limbs are usually affected. Treatment may include shoeing that raises the heel to improve hoof-pastern angle, anti-inflammatories and rest. *Heel pain* may be associated with navicular disease; however, there are multiple other causes of heel pain including disorders affecting the ligamentous structures within the foot and lower limb. *Degenerative joint disease* occurs as horses gradually become older. It may also occur in younger athletic horses where there is damage to the cartilage in joints. Inflammation leads to continued erosion of the cartilage, the hallmark of DJD. Finally, *tendon and ligament injury* may occur in a multitude of ways. The superficial and deep digital fexor tendons and the suspensory ligaments are the most prominent and prone to injury. Treatment may be complicated and prognosis guarded with injury to these ligaments and tendons.[17]

Common lameness disorders in cattle and small ruminants

The most common causes of lameness in beef and dairy cattle affect the foot and include both infectious and non-infectious conditions. The most common infectious conditions are digital dermatitis and foot rot. *Digital dermatitis* is considered to the most common cause of lameness in dairy cattle and in recent years has become a major problem in feedlot cattle as well. Digital dermatitis lesions typically occur on the rear feet adjacent to the plantar interdigital cleft but may also occur on the interdigital skin and heel bulbs. Lesions are granular in appearance and when chronic often develop filamentous outgrowths of epithelium that look like as hairs growing out of the surface of the lesion. Treatment consists of the use of copper sulfate or formalin footbaths. *Foot rot* is mixed bacterial infection that generally starts secondary to a mild lesion in the interdigital skin. A combination of bacteria are involved in the disease process that leads to a generalized swelling of the foot and severe lameness. Systemic antibiotic therapy is the treatment of choice and usually results in a cure when instituted early. Disorders affecting the hoof or claw include ulcers of the sole, toe or heel; white line disease; traumatic lesions of the sole caused by excessive wear, treading on sharp objects or the penetration of foreign bodies encountered on walking surfaces. While these disorders may lead to the formation of subsolar abscesses in the soles of claws, the most important treatment is therapeutic hoof trimming to remove all loose and necrotic horn coupled with a foot block to relieve weight bearing on the injured or diseased claw. Lameness disorders of the foot constitute roughly 90% of the lameness disorders that occur in cattle. The remaining 10% are related to lesions of the proximal limb.[4]

In small ruminants, 90% of lameness cases are caused by footrot (Dichelobacter nodosus). Overgrowth of claw horn is a primary predisposing cause of foot rot and overgrown claws are a consistent finding in up to 90-100% of farms. Extreme overgrowth also leads to deformities in the foot that also contribute to lameness. Other lesions found on dairy goat farms include horn separation, white line lesions, abscesses of the sole, foreign bodies, and granulomatous lesions with similarities to digital dermatitis in cattle.[12]

Common lameness disorders in swine

The predominant causes of lameness in swine are non-infectious conditions including physical injury, claw lesions and osteochondrosis. Physical injury may occur from any number of traumatic events. Wounds or lesions such as skin abrasion, calluses, capped hocks, or bursitis are often the result of housing and/or flooring conditions. Fracture of a claw or dewclaw is possible on slatted flooring and prolonged lying time on a hard un-bedded surface such as concrete is likely to result in skin lesions. Claw lesions are highly prevalent with most reports suggesting nearly all sows affected. The most commonly observed lesions are wall cracks, heel erosion, heel-sole separation, white line separation/disease, and dewclaw and claw horn overgrowth. Nutrition is often referenced as an underlying cause of claw lesions in sows; however, current thinking of bovine researchers is that housing and environment likely play a bigger role in the development of lameness. One could argue that this is likely the case in swine as well. This is clearly an area in need of research to better determine the underlying causes of lameness in sows. Finally, osteochondrosis is believed to be a major cause for premature culling of sows. Osteochondrosis causes irreversible damage to the articular cartilage in the elbow joint. It is characterized by a focal or multifocal delay in endochondral ossification in the growth plate and epiphyseal cartilage of the joint. The etiology appears to be multifactorial with genetics, rapid growth rates, vascular factors, and trauma all being implicated as possible causes.[15]

In conclusion

Lameness is one of the most prevalent and costly diseases in livestock. There are a few anatomical differences between the even-toed (i.e. cattle, small ruminants and swine) and the odd-toed ungulates (i.e. horses), but the lameness disorders among them share a number of similarities. For example, white line disease and subsolar abscesses are common to all ungulates, but the prevalence is influenced by animal use, housing conditions and flooring surfaces (whether hard or soft) that contribute to abrasion, bruising, excessive wear or lack of wear and/or overgrowth. Digital dermatitis is an infectious disease of the foot skin that is unique to cattle. Foot rot in cattle is associated with abrasion of the interdigital skin, whereas foot rot in sheep and goats is associated with overgrown claws. Hoof wall cracks are usually described as vertical (sand cracks) or horizontal cracks are common disorders in horses and cattle, but they are also important causes of lameness in small ruminants and swine. Specific treatment of foot problems in individual animals is possible in horses, cattle and small ruminants, but it may not always be practical in swine. Slaughter or euthanasia may be the only real practical options for severe lameness in individual sows or gilts.Determination of the underlying causes of lameness should be the priority when dealing with here lameness conditions.

References

1.Torrison, JL. Overview of Lameness in Pigs. Merck Veterinary Manual. https://www.merckvetmanual.com/musculoskeletal-system/lameness-in-pigs/overview-of-lameness-in-pigs accessed September 16, 2018.

2. Jackman BR. The Equine Foot. The American Association of Equine Practitioners. https://aaep.org/horsehealth/equine-foot accessed September 16, 2018.

3. Sellnow L. Leg Conformation, The Horse. https://thehorse.com/14455/leg-conformation/ Jun 11, 2014, accessed September 16, 2018.

3. 4. Shearer JK, van Amstel SR. Manual of Foot Care in Cattle. Second Edition, Hoard's Dairyman, 2013.

5. Sun G, Fitzgerald RF, Stalder KJ, Karriker LA, Johnson AK, Hoff, SJ. Development of an embedded microcomputer-based force plate system for measuring sow weight distribution and detection of lameness. Applied engineering in agriculture. 2011;27(3):475-482

6. Karriker, L. Identifying, Treating and Preventing Lameness in Sows. https://porkcdn.s3.amazonaws.com/sites/all/files/documents/2013SowHousingWebinars/Lameness.pdf, accessed November 11, 2018.

7. Lameness and Laminitis in US Horses, USDA:APHIS:VS, National Animal Health Monitoring System, Equine '98, April 2000, https://www.aphis.usda.gov/animal_health/nahms/equine/downloads/equine98/Equine98_dr_Lameness.pdf accessed November 12, 2018.

8. Equine health survey highlights the prevalence of lameness. Veterinary Record. August 10, 2013 https://veterinaryrecord.bmj.com/content/vetrec/173/6/132.2.full.pdf accessed November 12, 2018.

9. Cook NB. Prevalence of lameness among dairy cattle in Wisconsin as a function of housing type and stall surface. J Am Vet Med Assoc 2003;223:1324–8.

10. M.A.G. von Keyserlingk, A. Barrientos, K. Ito, E. Galo, D.M. Weary. Benchmarking cow comfort on North American freestall dairies: Lameness, leg injuries, lying time, facility design, and management for high-producing Holstein dairy cows. J. Dairy Sci., 2012;95:7399-7408.

11. Davis-Unger J, Pajor EA, Schwartzkopf-Genswein K, Marti S, Dorin, C, Spackman E, Orsel K. Economic impacts of lameness in feedlot cattle. Translational Animal Science, 2017;1(4):467-479.

12. Groenevelt,M, Anzuino K, Smith S, et al. A case report of lameness in two dairy goat herds; a suspected combination of nutritional factors concurrent with treponeme infection. BMC Res Notes. 2015; 8: 791.

13. Ellingson JS, Karriker LA, Borgmannc MH, Buckley AC. Finishing Lameness – What Do We Know? Search and Summary of Current Evidence. Swine Medicine Education Center, July 16th, 2012.

14. Rowles C 2001. Sow lameness. Journal of Swine Health and Production 9, 130-131.

15. Zimmerman, J. Diseases of Swine 10th edition. Wiley-Blackwell. West Sussex, UK. 2012.

16. KilBride, AL; Gillman, CE; Green, LE. A cross-sectional study of the prevalence of lameness in finishing pigs, gilts and pregnant sows and associations with limb lesions and floor types on commercial farms in England. Animal Welfare, 2009;18(3)215-224.

17. Norton J. The Top 5 Causes of Equine Lameness, Horse Illustrated, September 1, 2014. https://www.horseillustrated.com/horse-health-the-top-5-causes-of-equine-lameness accessed November 13, 2018.

Structural and pathological changes in growing pig lameness compared to sow lameness

Stephanie Rossow, DVM, PhD
Minnesota Veterinary Diagnostic Laboratory, University of Minnesota

If lameness was easy to understand and correct, it wouldn't be a recurring problem. We don't know how to measure lameness. If we can't measure lameness, why do we think we can fix it? These are obvious over simplifications of a problem but I find them a more honest representation of our ignorance when compared to calling lameness multifactorial. Pigs have been telling us for a long time that we're ignorant. Pigs measure lameness as pain. Conceptually, I think pigs are ahead of us.

A basic tenant of biology is that structure equals function. Abnormal structure equals abnormal function. It's the reason for submitting formalin fixed tissues with case submissions. Correlate structural changes with other laboratory results to establish cause and effect. When we can't see something, we correlate laboratory findings with clinical abnormalities until we understand the "who, what, when, where and why" of the disease process. If we don't know what to measure and there is a clinical problem, we call it genetic (joke). In the time since I was volunteered to give this presentation, I've been looking at the structure of rib growth plates to try and understand (decrease my level of ignorance) the "who, what, when, where and why" of growth plate structure. I use the following criteria to assess rib growth plate structure: 1. Presence of periosteal bone at the growth plate. 2. The alignment of chondrocyte columns. 3. The structure of the growth plate interface with the metaphysis. 4. The degree or lack thereof of cartilage persistence into the metaphysis. 5. The degree or lack thereof of bone in the periosteum of the metaphysis. These are subjective assessments based on repetition. I'll cover the histopathology jargon in the presentation. The conclusion I've come to is that there few rib growth plates that meet these criteria whether from nursery, grow-finish or fetal pigs. The second part of this project was to try and understand the components that comprise the growth plate structure. In doing so, I've come up with a list of questions:

Do we know what normal is? What should a normal Vitamin D level be? Is there a consensus? When should we measure it? Should we measure it? Can we assess bone structure and quality? How or does sow Vitamin D status affect the piglet? Do fetuses need Vitamin D? What affects neonatal bone mass? Are there periods or opportunities to optimize peak bone mass? How are fetal calcium demands met? Does a Vitamin D deficiency always lead to rickets? Is the fetal skeleton normal when there is maternal vitamin D deficiency? Can you give too much Vitamin D? Can you just correct the diet? Is Vitamin D deficiency just about bones? Is it all just about Vitamin D? Is congenital rickets common? Is osteomalacia clinically different from rickets? Does sow supplementation of Vitamin D affect what's available for the piglet? Is bone density related to bone ash concentration? Do we feed for strong bones? Does feeding for growth equate to strong bones? Do we feed for lowest cost or for health? Do we feed for cost or the pounds of pork produced? What is a study of association? Does a study of association assume anything?

My conclusion or opinion is that sows are on the borderline or below for optimal bone health, they produce piglets that are borderline or below for optimal bone health, these piglets become growing pigs with borderline or below optimal bone health and have no reserve to deal with health issues or bad diets, we then keep gilts with borderline or below optimal bone health and return them to the breeding herd to continue the cycle of borderline or below optimal bone health and then conclude that this population of pigs with borderline or below optimal bone health is appropriate to measure and use as the definition of normal. If we can't define normal, how do we know what abnormal is?

"Every change in the form and function of bone or of their function alone is followed by certain definite changes in their internal architecture, and equally definite alteration in their external conformation, in accordance with mathematical laws"

Wolff's Law, 1892

Antibiotic use measurement – where are we at?

Dr. Peter Davies, BVSc, PhD
Department of Veterinary Population Medicine, University of Minnesota, St Paul, Minnesota

Introduction

Both nationally and internationally, governmental and other groups are advocating for improved systems to measure antibiotic use in animals. These include:

- World Health Organization and World Organization for Animal Health (OIE): policies to harmonize surveillance of antibiotic use in humans and animals, and of antimicrobial resistance
- European Union: 2011 directive to harmonize antibiotic use surveillance in food animals across EU countries
- US National Action Plan for Combatting Antibiotic Resistant Bacteria: better surveillance of antibiotic use in food animals
- US Government agencies: The 5-year plan released by FDA in September 2018 identifies evaluation of antimicrobial drug use practices as a core principle, and has a stated goal to enhance monitoring of antimicrobial resistance and antimicrobial drug use in animals (https://www.fda.gov/downloads/AnimalVeterinary/SafetyHealth/AntimicrobialResistance/UCM620420.pdf).
- Consumer groups/politicians: calls for stricter regulatory oversight of antibiotic use in food animals (e.g., Pew Charitable Trust https://www.pewtrusts.org/en/research-and-analysis/issue-briefs/2018/08/trends-in-us-antibiotic-use-2018)
- Selected EU countries (e.g., Denmark, Netherlands, Belgium, Germany, United Kingdom): systems to benchmark antibiotic use at the individual user (farm, veterinarian) level, potentially with punitive measures for high users
- Downstream food industry customers: requests for assurances related to antibiotic use in the supply chain. McDonalds announced in December 2018 that it will establish reduction targets for antibiotic use in its global supply chain for beef.
- Academic and technical experts: Identified need for reliable antibiotic use data to enable meaningful risk assessment of the impact of antibiotic use in food animals on human health and to evaluate policy options.[1,2]

In response, initiatives to report data on antimicrobial use in animals, including swine, continue to be implemented in many developed countries. A 2018 systematic review of studies of measurement of antibiotic use in pigs included 25 papers (English language only), of which 22 were published since 2010, and the others since 2000.[3] The two major data sources were antimicrobial national sales data and farm surveys. The authors described ten different metrics use to report data, and categorized them into four groups

- Use measured by milligrams of active substance per animal weight
- Use measured as daily doses per weight at treatment
- Use measured by daily doses per treatment period
- Use measured by daily dose per period-at-risk of treatment.

This variability in measurement methods limits the ability to compare data from different studies and regions. However, the natural inclination towards harmonizing methods can detract from the utility of the data for its original purpose in an industry. For example, countries use different values for weights of classes of pigs (sows, nursery, finishing) as denominators to assess use. Obviously, this makes it difficult to compare data across those countries. However, to 'harmonize' with other countries, Country X may have to employ weights that do not reflect the reality in its industry. International comparisons also overlook differences in disease risks in different geographies, and crude measures of use provide no insight into 'appropriateness' of use. In general, measurement systems should be designed to meet the specific purpose within a specific industry population. Obtaining reliable data on patterns on use and temporal trends is more useful to support stewardship efforts in a particular population than are 'apples and oranges' comparisons across disparate populations (species and countries).

International comparisons and initiatives

Data on gross sales of antibiotics by country are reported in many countries including the USA, and comparative analyses of use among countries have been published.[4] In 2017, the OIE published its second annual report on antimicrobials intended for use in animals, including an estimate that global use of antibiotics in animals was 98.97 mg/kg, with 'kg' being the estimated live weight of the animal inventory (http://www.oie.int/fileadmin/Home/eng/Our_scientific_expertise/docs/pdf/AMR/Annual_Report_AMR_2.pdf). The report acknowledges the considerable limitations in obtaining the data and cautions interpretation. Unfortunately, this approach of aggregating all antibiotics and all species into one number will deliver more confusion than enlightenment. A major limitation of national sales data is that it generally lacks granularity, and few nations can accurately parse national antibiotic use data to a species level.

In Europe, coordinated efforts to harmonize surveillance of ABU in animals have been underway via the European Surveillance of Veterinary Antibiotic Consumption (ESVAC) since 2009.[5] Differences exist among countries in the demographics of species that are included in sales data. For example, horses are considered food animals in the EU and included in sales data, and fish may be included in some countries but not others. Furthermore, the relative size of food animal industries across countries will confound interpretation of sales data at a national level. Due to differences in lifespan production conditions, and prevalent pathogens, antibiotic use can vary markedly between industry sectors within a species, such as cattle (dairy, veal, cattle at pasture, fed cattle). Consequently, simple comparisons between countries, even within a species, may be misleading if they fail to account for differences in industry demographics.

Given the lack of granularity of national sales data, numerous programs have been put in place to obtain more detailed, and species specific, data, at the farm level. The objective of the AACTING project in Europe is to develop guidelines for setting up systems to collect antibiotic use data at farm level that are applicable to guide antimicrobial stewardship. The project website (**http://www.aacting.org/**) has assembled information worldwide about systems that measuring antibiotic use at the farm level, including details of how data are collected and analyzed. The website enables systems to be selected on both a country and species basis. Some systems are run by governments (e.g., CIPARS in Canada, Sanitel-Med in Belgium) while others are industry run or part of quality assurance programs (e.g., QS in Germany, AB register in Belgium). Readers seeking more details on international programs to measure antibiotic use in food animals are referred to the AACTING website. AACTING is also convening the second International Conference on Quantification, Benchmarking and Stewardship of Veterinary Antimicrobial Usage in Bern, Switzerland, from July 2-3, 2019.

USA

Since 2009, under the Animal Drug User Fee amendments, the FDA Center for Veterinary Medicine (CVM) has published annual summaries of the amounts of antimicrobial drugs sold or distributed for food-producing species. The sales data are described by drug class, medical importance, route of administration, and label indications. The 2016 report was the first time that species level estimates were included by species, including an estimate that 37% medically important antimicrobials was intended for use in swine (**https://www.fda.gov/downloads/ForIndustry/UserFees/AnimalDrugUserFeeActADUFA/UCM588085.pdf**).

In 2017, the CVM proposed a biomass denominator to provide greater context for reported sales data in relation to the size of the livestock population. In 2016, CVM funded two cooperative agreements (one for beef and dairy, one for poultry and swine for data collection for on-farm antimicrobial use. Reports of these projects have been recently submitted to the FDA, but are not yet publically released. A goal of the agreements was to establish a framework for confidential sharing of proprietary data on antibiotic use to estimate patterns of use at a more level than possible with the national sales data. In the swine project, nine collaborating systems voluntarily provided their data for 2016 and 2017 to advance understanding of how antibiotics are being used in growing pigs in the US swine industry. The systems were distributed across most major swine producing states and collectively produced over 20 million pigs in each of 2016 and 2017. Although a number of biases limit the ability to extrapolate the findings across these systems to the overall swine industry, some of the key observations were:

- Aggregate data were similar in 2016 and 2017, although changes were evident between years within individual systems
- There was considerable variability among systems in the relative and absolute amounts of antibiotics used.
- Antibiotic use (by weight) was dominated by tetracycline use (approximately two-thirds of total weight used), with chlortetracycline alone constituting about half of total weight used.
- Antibiotics in classes categorized by FDA as critically important were around 3% of total use, and predominantly involved macrolides
- Fluoroquinolones and 3rd generation cephalosporins collectively accounted for less than 1% of total use by weight

For a range of motives, we can expect to see more private and commercial initiatives to collect and analyze antibiotic use data across food animal industries. Some of this activity will be competitive and some will be cooperative. Either individually or collectively, the common goals are to optimize antibiotic use in food animals to maintain animal health and well being, and to communicate that the industry understands and aspires to principles of sound antibiotic stewardship.

References

1. McEwen SA, Singer RS. Stakeholder position paper: The need for antimicrobial use data for risk assessment. Prev. Vet. Med. 2006;73:169-176.

2. Van Boeckel TP, et al. Reducing antimicrobial use in food animals. Science. 2017;357:1350-1352.

3. Lekagul A, et al. The use of antimicrobials in global pig production: A systematic review of methods for quantification. Prev Vet Med. 2018;160:85-98.

4. Van Boeckel TP et al. Global trends in antimicrobial use in food animals. PNAS, 2015;112:5649-5654.

5. Grave K et al. Variations in the sales and sales patterns of veterinary antimicrobial agents in 25 European countries. J Antimicrob Chemother. 2014;69):2284-2291.

The politics of antibiotic use in food-producing animals

Liz Wagstrom, DVM, MS, Diplomate ACVPM
National Pork Producers Council

Introduction

Changes to the ability to use antibiotics in food animal production have been largely driven by regulatory actions or customer specifications. Legislation restricting use at the federal level has not progressed through Congress, however there have been state and municipal actions that may have impacts.

Federal legislative activity

This proceedings paper will review the pertinent bills in the 115th Congress. If new bills are introduced in the 116th Congress prior to the American Association of Swine Veterinarians annual meeting they will be discussed there.

1. HR 1587 Preservation of antibiotics for medical treatment act of 2017 (PAMTA)

A version of PAMTA has been introduced for the last several Congresses. The current bill was sponsored by Congresswoman Louise Slaughter prior to her death and has 55 all Democrat co-sponsors. A companion bill was introduced in the Senate by Sherrod Brown.

This bill amends the Federal Food, Drug, and Cosmetic Act to require an applicant for approval of a new animal drug that is a medically important antimicrobial to demonstrate that there is a reasonable certainty of no harm to human health from antimicrobial resistance attributable to the nontherapeutic use of the drug.

Medically important antimicrobials are drugs intended for use in food-producing animals that contain: (1) specified antibiotics, or (2) certain drugs on the World Health Organization's list of critically important antimicrobials.

Two years after enactment of this bill, the Food and Drug Administration (FDA) must withdraw approval of a drug's nontherapeutic use in food-producing animals unless the FDA makes a determination that, based on the application holder's demonstration or an FDA risk analysis, there is a reasonable certainty of no harm to human health from antimicrobial resistance attributable to nontherapeutic use.

The FDA must rescind an exemption for investigational use of, or approval of a new drug application for, a medically important antimicrobial for its nontherapeutic use in food-producing animals two years after the exemption is granted or the application for approval is submitted unless there is a reasonable certainty of no harm to human health from antimicrobial resistance attributable to nontherapeutic use.

A medically important antimicrobial cannot be administered (including through animal feed) to a food-producing animal for disease control unless there is a significant risk that a disease or infection present on the premises will be transmitted to the animal.

Neither the House nor Senate held hearings related to PAMTA.

2. S629 Preventing Antibiotic Resistance Act of 2017

The Preventing Antibiotic Resistance Act was introduced by Senator Diane Feinstein and has seven co-sponsors.

This bill amends the Federal Food, Drug, and Cosmetic Act to require the Food and Drug Administration (FDA) to refuse a new animal drug application if the drug is a medically important antimicrobial used to treat humans and the applicant fails to demonstrate that the drug meets specified criteria for use in animals, including that: (1) the drug is effective, (2) the drug is targeted to animals at risk of developing a specific bacterial disease, (3) the drug has a defined duration of therapy, and (4) there is not a reasonable probability of risk to public health from microbial resistance to the drug.

The FDA must review the approvals of certain medically important antimicrobials approved for use in food-producing animals. Sponsors of these drugs must submit evidence to the FDA that demonstrates that their drug meets the criteria described above for approved indications. The FDA must withdraw approval for any indication for which there is insufficient evidence that the drug meets the criteria.

This bill declares that a veterinarian-client-patient relationship should ensure that medically important antimicrobials are used in food-producing animals in a manner consistent with best practices.

There were no hearings or votes taken on this bill.

3. Animal Drug User Fee Act (ADUFA)

The Animal Drug User Fee Act and the companion Generic Drug User Fee Act must be reauthorized every five years. The user fees paid by pharmaceutical sponsors are used to provide timely reviews by the Food and Drug Administration. The 2018 reauthorization included expansion of the conditional approval of drugs to major species, but exempted antibiotics from conditional approval. Report language was also included that would require the Food and Drug Administration to report back to Congress on progress made toward moving all remaining medically important antibiotics to prescription status. The Bill passed both houses of Congress and was signed into law.

State and local legislative activities

1. California

California passed SB27 in 2015 which required all medically important antibiotics to be under a veterinary prescription or VFD, have a veterinary visit within the last 12 months, be given to the fewest animals for the shortest duration possible and to gather data on antibiotic sales, use, resistance and livestock practices. However, advocacy groups such as the National Resources Defense Council (NRDC) have been unsatisfied with the progress made on data collection. NRDC worked with the City of San Francisco to require that retailers there report antibiotic use suppliers to the city's grocery stores. This data would include use not only in the animals whose meat would be sold in San Francisco, but on all animals that the producer has. The rules for the San Francisco ordinance have been recently published and the reporting has not yet been initiated.

2. Maryland

Maryland passed legislation in 2017 that authorized on or after January 1, 2018 the administration of specified antimicrobial drugs to cattle, swine, or poultry only if, in the professional judgement of a licensed veterinarian, the administration is necessary for specified purposes; prohibiting specified antimicrobial drugs from being administered to cattle, swine, or poultry in a specified pattern for for specified purposes; requiring the Department of Agriculture each year to collect publicly available data on the use in the State of specified antimicrobial drugs. In short, medically important drugs may not be used to promote growth, they may be used for prophylaxis when there is an elevated risk, but not be used in a regular pattern.

Other states considered, but did not pass, legislation to restrict antibiotic use in food animals.

Federal executive branch activities

Much of the activity around antibiotic use and resistance resides in the Department of Health and Human Services. The United States Department of Agriculture also has involvement in animal antibiotic activities at the federal level. Of note, the Food and Drug Administration released a five-year plan for supporting antimicrobial stewardship in veterinary settings. Areas cited include, but were not limited to, moving remaining medically important antimicrobials to veterinary oversight, considering duration of use for antimicrobials that currently do not have a label listing duration, and developing long term strategy for collection of data on use and resistance.

The Department of Health and Human Services oversees the Presidential Advisory Council on Combatting Antibiotic Resistant Bacteria. The council has been reauthorized and will concentrate on revisions to the National Action Plan on Antibiotic Resistance which focuses on both human and animal issues.

The Department of Agriculture has a role in gathering antibiotic use data via the National Animal Health Monitoring System which conducted surveys in 2017 to gather stewardship and use data to serve as a baseline prior to the implementation of Guidance 213. Subsequent surveys are planned, as well as potential longitudinal studies.

Summary

Regulatory agencies continue to work on antibiotic stewardship activities, perhaps contributing to the lack of legislative activity on the federal activity. State legislative actions have the potential to disrupt national stewardship efforts.

African swine fever: A global threat

Klaus Depner[1*]; Erika Chenais[2*]; Vittorio Guberti[3]; Klaas Dietze[1]; Arvo Viltrop[4]; Karl Ståhl[2]
[1]Friedrich Loeffler Institute, Germany; [2]National Veterinary Institute, Sweden; [3]National Institute for Environmental Protection and Research, Italy; [4]Estonian University of Life Sciences, Estonia

Internationally, African swine fever (ASF) is considered as one of the most dangerous animal diseases of pigs. The disease is affecting trade and has a serious socio-economic impact on people's livelihood. No drugs or vaccines are available to fight ASF. The most severe epidemic ever experienced outside of the African continent started in Georgia in 2007, spread throughout the Caucasus and the Russian Federation, subsequently reaching the European Union and China. In many countries, the disease has become endemic in domestic pigs and wild boar. In wild boar populations, ASF shows a pattern of habitat bound persistence lacking a tendency of dynamic spatial spread.[1]

The ASFV strain in the current epidemic is highly virulent and belongs to the genotype II. From the start of the current epidemic in 2007, until the detection of the first case inside the EU in 2014, infections were mainly seen among pig farms with generally low biosecurity, and with incidental spill over to the wild boar population. At that point in time it was predicted that the disease would spontaneously fade out from the local wild boar population as soon as the disease was under control in the domestic pig population, due to the high case fatality rate and the absence of long term carriers.[2] However, in the ecological context which prevailed in Poland and the Baltic states this epidemiological hypothesis proved to be wrong. The infection survived locally in the wild boar population independently from outbreaks in domestic pigs, with a steady and low prevalence below 5% and a local transmission speed of 2-5 km/month.[3] In addition to the local transmission within the wild boar population, long distance jumps responsible for disease incursion into areas far from known infected regions occurred. Human activities have been identified as main drivers of disease transmission in the domestic pig epidemiological cycle of ASF in other contexts.[4]

ASF is often described as a highly contagious disease with high mortality. However, field data as well as experimental studies indicate a rather low contagiosity. The case fatality rate (i.e. the proportion of infected individuals that die from a disease within a certain time period) related to highly virulent ASFV in affected populations of domestic pigs and wild boar is indeed high, often reaching 90-100%. However, the initial mortality within an epidemiological unit is rather low regardless of the high case fatality rate. In domestic swine populations the low contagiosity can be considered an advantageous feature reducing the urgency in the implementation of control measures. For wild boar however, in combination with the environmental stability of the virus and high animal densities, the low contagiosity represents a disadvantage for effective control. The qualities of the three epidemiological traits contagiosity, tenacity, and case fatality rate make ASFV efficient in both persistence and transmission.[5] The combination of high tenacity and a high case fatality rate makes the virus largely available and ensures long term persistence in the environment, meanwhile the relatively low contagiosity prevents the complete depletion of the host population. The interaction of these three parameters maximize both local persistence and geographical spread of the virus making its eradication a challenge.

Humans are recognized as the main cause of both long distance transmission and virus introduction into domestic pig farms. Thus, it has become crucial to include social science when planning prevention-, control-, or eradication-measures. By considering only the biological particularities of the disease, contagiosity, tenacity and case fatality rate, but ignoring the human aspects, the epidemic will not be controlled.

References

1. Chenais E, et al. Identification of Wild Boar–Habitat Epidemiologic Cycle in African Swine Fever Epizootic. *Emerg Infect Dis 2018*, 24(4):810.

2. Petrov A et al. No evidence for long-term carrier status of pigs after African swine fever virus infection. *Transbound Emerg Dis* 2018. 1-11.

3. Depner K, et al. Epidemiological analyses of African swine fever in the Baltic States and Poland. EFSA Journal 2017, 15(11).

4. Chenais E, et al. *Transbound Emerg Dis* 2017, 64(1):101-115.

5. Chenais E, et al. Epidemiological considerations about the African swine fever epizootic in Europe 2014-2018. *Porcine Health Management* 2018 (in pres).

A China perspective: Diseases, diagnostics, and biosecurity

Keith Erlandson, MS, MBA, DVM, DABVP (SHM)
CP Group, Beijing, China

Introduction

Having worked as a swine veterinarian in China for over three years now, I have come to several realizations about the similarities and differences between the disease pressure, biosecurity and diagnostic capabilities in the two countries. I will try to convey to you some of the challenges faced in the Chinese swine industry, as well as the opportunities.

Disease

Working in China I worry about what I call the "Big 5," or the five diseases that have the capability to severely impact the farms or production systems. The "Big 5" I worry about are foot and mouth disease (FMD), classical swine fever (CSF), pseudorabies (PR), porcine reproductive and respiratory syndrome (PRRS) and porcine epidemic diarrhea (PED). Currently, all five of these diseases can be found in China.

FMD was eradicated from the United States in 1929.[1] Classical Swine Fever was eradicated from the United States in 1976.[2] Pseudorabies was eradicated from the commercial swine herd in the United States in 2004.[3] This happens to be the same year that I graduated from veterinary school and went into swine practice, so I was hopeful I would never see any of those three disease in person. PED had been reported in European and Asian countries since the 1970's, but never in North America.[4] As we all know, that changed in 2013, when the PED virus entered the United States. PRRS was first detected in the United State in the late 1980's[5] and of course we are still fighting that disease.

When I arrived in China, I began quizzing my new colleagues as to the greatest disease challenges they were facing on farm. I was expecting them to answer that FMD, CSF, PR or some combination of those three disease were the problems, however, the consistent answer to my question was PRRS and PED. This shows some of my naiveté prior to moving to China. The PR vaccines available work well, and PR is well controlled on most of the farms I work with. The same can be said for the CSF vaccines that are available. However, when PRRS infects the farm it can throw things out balance and, well, PRRS just makes everything worse. Where we can keep PRRS under control, other diseases are easier to manage.

As all of you will be aware, I have had to add another disease to my list to make it the "Big 6," as China reported its first case of African swine fever in August.[6]

Diagnostics

One problem faced by Chinese producers is that of logistics to transport diagnostic samples to the laboratory. The area where the sample is collected may not be near a laboratory that has the capability to run the tests needed. There are shipping companies that can accommodate some sample shipment, but in some cases there is not the availability to get samples shipped to the diagnostic laboratory quickly enough to yield useful results.

Some of these problems will be have to be overcome with new laboratory construction and improved logistics. Like the cell phone allowed many developing countries to skip over the "landline" phase of telecommunication, the rapid advancements in pen-side testing, portable PCR and other technologies may allow China to skip over some of the brick and mortar construction of diagnostic laboratories.

Biosecurity

As the swine industry in China has continued to modernize, and especially since the introduction of African swine fever to the country, the level of biosecurity has increased. Boar Studs and some sow farms have HEPA filtration. Many new production facilities often have dedicated isolation areas (for both animals and people) and are shower in/shower out. Older facilities are being upgraded to increase the biosecurity. Truck washes and Thermoassisted Drying bays are being built.

Conclusion

It remains to be seen the impact that ASF will have on the Chinese industry. The introduction of the disease has increased the level of awareness about the importance of biosecurity. Diagnostic capabilities and disease control for the "Big 5" will continue to get better, so after working in swine production in the United States and now with one system in China, I can say that there is more in common between the two than I knew three years ago.

References

1. Iowa State University Swine Disease Manual. Available at: `https://vetmed.iastate.edu/vdpam/FSVD/swine/index-diseases/foot-mouth-disease`. Accessed 11 November 2018.

2. Iowa State University Swine Disease Manual. Available at: `https://vetmed.iastate.edu/vdpam/FSVD/swine/index-diseases/classical-swine-fever`. Accessed 11 November 2018.

3. Iowa State University Swine Disease Manual. Available at: `https://vetmed.iastate.edu/vdpam/FSVD/swine/index-diseases/PRV`. Accessed 11 November 2018.

4. Iowa State University Swine Disease Manual. Available at: https://vetmed.iastate.edu/vdpam/FSVD/swine/index-diseases/porcine-epidemic-diarrhea. Accessed 11 November 2018.

5. Iowa State University Swine Disease Manual https://vetmed.iastate.edu/vdpam/FSVD/swine/index-diseases/porcine-reproductive. Accessed 11 November 2018

6. OIE African Swine Fever, China (People's Rep. of). Available at: http://www.oie.int/wahis_2/public/wahid.php/Reviewreport/Review?page_refer=MapEventSummary&reportid=27442. Accessed 11 November 2018.

The foreign animal disease risk of feed

Scott Dee[1]; Megan Niederwerder[2]; Jeff Zimmerman[3]
[1]Pipestone, Diego Diel, SDSU; [2]KSU; [3]ISU

In 2013, the US pork industry became infected with porcine epidemic diarrhea virus (PEDV). Shortly thereafter, a second strain of PEDV (Indel), porcine delta corona virus and a new strain of Seneca virus A entered the county. This unique situation caused our veterinary team at Pipestone to contemplate new routes of entry, with feed coming to the forefront of the discussion as a possible vehicle for the transboundary movement of pathogens. Through a series of experiments, we demonstrated proof of concept of the transmission to naïve pigs through the consumption of PEDV-contaminated feed, the ability of feed additives to reduce this risk and evaluated survival of the virus in different feed ingredients.[1,2]

Building on this theme, working with SDSU (E. Nelson), we developed a Trans-Pacific model to simulate the movement of PEDV in contaminated feed ingredients between from China to the US.[3] Following support of the Swine Health Information Center, the project was expanded to include 11 additional viruses, including African swine fever virus. Working with SDSU (D. Diel) and KSU (M. Niederwerder), we studied the effect of certain feed ingredients on viral survival and introduced the concept of the "High Risk Combination".[4]

Over time, the model has proven to be a unique, repeatable entity and is currently in use across several research institutions. It consists of a rigorous, yet flexible experimental design that can take into account variability across feed ingredients, pathogens, transport times between countries and environmental conditions over land and sea. In closing, the model has given us new insight into the possibility that transport of pathogens in feed may be a frequent event and which ingredients and viruses may pose greater risk for transboundary spread. Its current application focuses on the evaluation of mitigation strategies and the development of storage periods in an effort to reduce this novel risk factor. Data from these studies now provide a basis for the platform known as "Responsible Imports", a science-based plan to safely introduce essential ingredients from countries of high risk; a concept that will protect both global trade and animal health, in the face of the current African swine fever crisis.

References

1. Dee S, Clement T, Schelkopf A, et al. An evaluation of contaminated complete feed as a vehicle for porcine epidemic diarrhea virus infection of naïve pigs following consumption via natural feeding behavior: Proof of concept. *BMC Vet Res* 2014,10:176.

2. Dee SA, Neill C, Clement T, et al. An evaluation of PEDV survival in individual feed ingredients in the presence or absence of a liquid antimicrobial. *Porcine Health Management* 2015, 1:9.

3. Dee SA, Neill C, Singrey A, et al. E. Modeling the transboundary risk of feed ingredients contaminated with porcine epidemic diarrhea virus. *BMC Vet Res* 2016, 12:51.

4. Dee SA, Bauermann F, Niederwerder MC, et al. Survival of viral pathogens in animal feed ingredients under transboundary shipping models. *PLOS ONE* 2018, 13(3): e0194509.

Industry and AASV response to foreign animal disease risk

Paul Sundberg[1]; Eliana Paladino[2]; Auguste Brihn[2]; Sol Perez[2]; Andres Perez[2]; John Deen[2]
[1]Executive Director, Swine Health Information Center, Ames, Iowa [2]University of Minnesota, St. Paul, Minnesota

Introduction

The Swine Health Information Center was funded by a National Pork Board grant in July of 2015. Its mission is to protect and enhance the health of the United States swine herd through coordinated global disease monitoring, targeted research investments that minimize the impact of future disease threats, and analysis of swine health data.

Funded by SHIC in 2017, a program for systematically monitoring swine diseases around the world was developed at the University of Minnesota using a private-public-academic partnership including collaboration with the USDA/APHIS Center for Epidemiology and Animal Health (USDA-CEAH). The aim of these reports is to have a support system for near real-time identification of hazards that will contribute to the mission of assessing risks to the industry and ultimately, early detect, identify, or prevent occurrence of events, in partnership with official agencies, and with our international network of collaborators.

Monthly reports are created based on the systematic screening of multiple official data sources, such as government and international organization websites, and soft data sources like blogs, newspapers and unstructured electronic information from around the world that then are curated to build a raw repository. Afterward, a group of experts uses a multi-criteria rubric to score each event, based on novelty, potential direct and indirect financial impacts on the US market, credibility, scale and speed of the outbreak, connectedness, and local capacity to respond average is calculated. The output of the rubric is a final single score for each event which is then published in the report.

On November 16, ASF was reported in Yibin City in the southeastern area of Sichuan province. Sichuan and Henan are the two provinces with the highest pig production in China, representing over 30% of total production. Sichuan province slaughtered 69 million of the approximately 700 million pigs produced by China in 2016.

The outbreak in Sichuan continues a southwestward trend in the epidemic. With 64 outbreaks reported in 18/34 provinces across China, the concern is that the disease may have steadily spread, and that major impact to pork suppliers may be seen in coming months.

Also on November 16th the Chinese Ministry of Agriculture confirmed the first case of ASF in a wild boar (Jilin province, northeastern China). This significantly increasing concerns around the effectiveness of control strategies implemented by authorities since last August.

On average 8,000 pork derived products are annually confiscated by Customs and Border Protection at the United States (US) ports of entry such as international airports, harbors or mail offices.[1] This study aimed at analyzing the risk of African swine fever (ASF) and classical swine fever (CSF) being introduced into the US through prohibited swine products carried by air passengers and identifying locations and time periods at higher risk where and when preventive and mitigation measures should be implemented.

Focusing on ASF, the the overall mean annual probability of ASF entry was estimated as 0.061 at 95% confidence interval (CI) [0.007, 0.216]. July and May were the months at highest risk for entry. For ASF, the origin countries of those air passengers that represented the highest risk (above 70% of the total risk) were Ghana, Cape Verde, Ethiopia, and the Russian Federation.

The study used international travel and USDA Customs and Border Protection data from January of 2010 to March of 2016. Thus, the results were based on data prior to the August 2018 notification of the beginning of the ASF outbreak in China. In addition, World Bank data on air transport in and from China shows a 13% increase from 2016 to 2018. The combination of the 2018 ASF epidemic in China and the increased travel in and from China may change the estimate of the risk of ASF entry through US airports. The Swine Health Information Center and the National Pork Board are co-funding a project with the University of Minnesota to update this risk estimate.

Pork Industry/USDA collaboration

On September 5, 2018, U.S. Department of Agriculture (USDA) and Food and Drug Administration (FDA) officials met with U.S. pork sector groups – including the American Association of Swine Veterinarians, the National Pork Board, the National Pork Producers Council and the Swine Health Information Center – to evaluate additional measures to prevent the spread to the United States of African swine fever (ASF) currently active in China and some European nations.

The industry organizations and USDA agree that prevention could be enhanced by Customs and Border Protection (CBP) and Plant Protection and Quarantine (PPQ) ensuring flights from China and Russia get enhanced passenger and cargo inspection attention. USDA has made a request to that effect to CBP. Sampling and monitoring of imported products that might pose a risk of ASF transmission is also being considered.

The USDA Center for Epidemiology and Animal Health's Risk Identification Unit (RIU) is monitoring diseases around the world, including China. Increased, regular communication through contact or reports is another action that could enhance prevention through raising awareness. This provides a vehicle for two-way communication to provide information from industry sources to RIU and for RIU to provide the most up-to-date information they have back to industry.

The Swine Health Protection Act gives USDA-APHIS regulatory authority for compliance inspections for licensed food waste feeders and searches for non-licensed waste feeding facilities. The pork industry organizations are communicating to USDA their request for increased inspections and searches.

The inspection and compliance processes for imported pork casings and other food products was another topic of discussion. While USDA does not allow the import of casings originating in ASF positive regions, the risk of US origin casings that have been sorted in China and returned to the US is being considered.

Response initiatives include a survey of the National Animal Health Laboratory Network (NAHLN) for their surge capacity should the industry need ASF testing to identify issues or shortcomings. USDA is already conducting the survey.

A project to validate a PCR test for ASF – and FMD and CSF – on oral fluids has been underway for over a year. Only whole blood and tonsil are currently USDA-validated for ASF surveillance via PCR testing. Diagnostic capabilities need to be updated to enable spleen and lymph nodes, tissues likely to be submitted during routine animal health event investigations, to be USDA-validated for ASF detection and surveillance. And response capability needs to be able to be applied at the herd level through plans for USDA accredited veterinarians being integrated into national response should federal resources become overwhelmed during an outbreak.

ASF-specific emergency response exercises are being planned, including allied industry, NAHLN and state and federal animal health officials' participation. And response and communication coordination with Canada and Mexico are additional items being discussed and have already been initiated by USDA.

Studying potential risk from imported feed ingredients

One of the major concerns is that commercial feed may become ASF-contaminated. In a statement to the Shenzhen Stock Exchange, the Tangrenshen Group reported that ASF was detected in feed samples in one of their units. The feed was manufactured by their 51 percent owned subsidiary, Bili Meiyingwei Nutrition Feedstuff. Following this statement, on November 13th, Tangrenshen Group followed up stating that test made by one of the units has actually ruled out the presence of ASF in feed. Subsequently, the operation resumed work as normal.

USDA and FDA believe there are currently many unknowns and data gaps that should be identified to help define or validate feed risk. In the absence of information regarding the predictive ability of unvalidated test results to accurately determine the potential risk associated with feed, they believe the design and implementation of a testing strategy is not feasible. At this time, it is up to the pork industry to address potential risk of foreign animal disease introduction into the US from imported feed ingredients.

Collaboration can help to address concerns about potential risk associated with feed and feed component imports. We know from research conducted by Dr. Scott Dee, et al.[2] that certain feedstuffs are able to support ASF during a simulated trans-Pacific shipment. A validated method to test bulk feed products and applying it to monitoring shipments for pathogen contamination has been funded by SHIC and is currently being researched. Also, currently being researched with results expected soon are feed pathogen mitigation options. Feed additives, component holding time and temperature before processing, Hazard Analysis and Risk-Based Preventive Controls and blockchain are being investigated with urgency.

The Swine Health Information Center (SHIC) has collaborated directly with USDA Ag Research Service on Plum Island, USDA-APHIS, National Pork Board, and other key researchers to determine the minimum infectious dose (MID) of foot and mouth disease (FMD) virus needed to infect pigs via feed ingredients using normal feeding behaviors. Simultaneously, this project will look at the effectiveness of possible mitigants being added to the feed to help neutralize the virus. To date, no study has looked at the FMD dose needed to infect a pig via feed, using normal feeding behaviors. Successful completion of the project will provide critical knowledge pertaining to the risk of introduction of FMD into the US pig production system by imported feed components as well as potential mitigation of that risk.

A meeting of stakeholders, including representatives of USDA, FDA, universities, industry organizations, producers, the feed processing industry, and SHIC, was held in June 2018. The objective of the meeting, hosted by SHIC and the National Pork Board, was to review current government policies and regulations and to make recommendations about research to help reduce the risk for pathogen transmission via feed and feed ingredients. A report from the meeting can be found on the Swine Health Information Center's website, `www.swinehealth.org`.

Risk mitigation research is testing several commercially available feed additives that may be added to feed during milling or other processes to neutralize or reduce the load of these pathogens in feed and help mitigate the potential risk of transport. The initial phase of the project consists of screening a panel of 10 mitigants against identified higher-risk combination of viruses and feed ingredients. The panel of viruses, which were selected based on their ability to survive in feed ingredients, include SVA (FMDV surrogate), BHV-1 (surrogate for PRV), PRRSV, PEDV, and ASFV.

At the KSU high biosecurity lab, researchers are working directly with ASF, CSF and PRV. They are determining the potential for survival in feed and feed ingredients under the transboundary model as well as assessing tools for mitigating the risk of virus transport in feed and feed ingredients.

KSU researchers are investigating using dust samples to monitor for swine pathogens in products and US feed mills. There is potential for the findings to lead to development of a diagnostic laboratory panel of assays where a single submitted swab of feed mill dust could be analyzed for multiple feed-based bacteria and viruses – a low-cost tool that could be used to help address feed safety.

ASF control in China will be extremely difficult, if even possible. The Chinese pork industry has had difficulties in controlling FMD and CSF and has relied heavily on the use of vaccines. As a vaccine is not available for ASF, the industry is thus reliant on heightened biosecurity, rapid diagnosis, complete isolation, and then elimination of infected pigs and contaminated materials. US plans for prevention and response have to be considered to be long-term adjustments to the biosecurity of our national herd.

Reference

1. Cristina Jurado, et. al, doi: 10.1111/tbed.12996, 2018.

2. Dee SA, Bauermann FV, Niederwerder MC, Singrey A, Clement T, de Lima M, et al. Survival of viral pathogens in animal feed ingredients under transboundary shipping models. PLoS One. 2018;13(3):e0194509.

50+ years of US swine disease eradication

J. F. Connor, DVM, MS

Disease emergence is triggered by a number of interrelated factors including human and animal demographics, dynamics of climate change, increased people mobility, trade globalization, urbanization, and land use changes. Pathogens that have historically been confined to a region are widely available for transmission. As global markets predominate, individual countries and country segments will need to eliminate diseases that would restrict their access to markets and manage factors providing sustainable food supply and safety. Closure of markets or cessation of exports would have a dramatic economic impact on a country's economic viability. For US swine industry, loss or interruption of markets with a confirmation of one or any of the OIE recognized foreign animal diseases (FMD, CSF, ASF) will have a direct and long-term economic impact. Our industry has realized growth in global markets by low cost production which is driven substantially by our ability to manage and over time eliminate economic diseases.

Our industry has eliminated many diseases including classical swine fever, foot-and-mouth disease, Aujeszky's, and vesicular exanthema from the US swine population. From individual herds the list is extensive including porcine reproductive and respiratory syndrome virus, Seneca Valley virus, transmissible gastroenteritis, porcine epidemic diarrhea virus, porcine deltacoronavirus, *Actinobacillus pleuropneumoniae, Brachyspira hyodysenteriae,* influenza A-S, *Leptospira* sp, *Bordetella*, and *Pasteurella*. Parasites that have been eliminated include *Sarcoptes scabei, Haematopinus suis, Trichuris suis, Metastrongylus elongatus, Stephanurus dentatus*, and *Oesophagostomum* spp. (Table 1) Elimination strategies intertwine a variety of procedures and protocols including biosecurity, sanitation, depopulation, partial depopulation, unidirectional flow, wean pig segregation, parity segregation, batching, medicated early weaning, modified medicated early weaning, herd closure, immunological, immunological plus therapeutics, confinement housing, and test and removal. These strategies evolve over time and depend on pathogen attributes of infectivity, virulence, pathogenicity, modes of transmission, vectors, environmental survivability, carrier states, and resistance to antimicrobials and disinfectants. Elimination generally initiates on a herd basis, expands to an area or region, then to a country, and finally to a continent. The cost of the strategies varies widely and disease elimination decisions should be predicated on prevention of reinfection and economics.

Industry elimination

Foot-and-mouth disease history

The last outbreak in United States occurred in January 1929 in a 3,271-head swine herd in California which was receiving garbage from a San Pedro, California seaport.[1] Confirmation of diagnostics was made January 14, 1929. Differential diagnosis was made by inoculation of two horses, four pigs, two calves, and four guinea pigs. A quarantine remained in effect for a three-mile radius.

Traceback included identification of a dead hog collector removing one head on January 13. Movements of hauler were traced. One other producer had fed garbage from the same source during the period. Forty laborers were assigned the work of cleaning and disinfecting the premises consisting of approximately 10 acres of pens and feed lots. The Bureau of Animal Industry and the State department purchased 14 hogs, 1 cow, 8 heifers, and 1 bull, as well as all goats, 1 cow, 1 heifer, and a few pigs in a nearby village. In addition, dogs on the infected ranch and others in the vicinity were euthanized and buried.

Eighteen dairies and another ranch of ranch of 3,000 hogs were within the three-mile radius under quarantine and observation.

Table 1: Pathogens that can be eliminated

Virus	Bacteria	Parasite
Aujeszky's disease virus (pseudorabies virus)	*Actinobacillus pleuropneumoniae*	*Ascaris suis*
Classical swine fever (hog cholera)	*Bordetella bronchiseptica*	*Metastrongylus elongatus*
Foot-and-mouth disease	*Brachyspira hyodysenteriae*	*Oesophagostomum* spp.
Influenza A virus – swine	*Leptospira* sp.	*Sarcoptes scabei* var *suis*
Porcine deltacoronavirus	*Pasteurella multocida* toxigenic	*Stephanurus dentatus*
Porcine epidemic diarrhea virus 2		*Trichuris suis*
Porcine reproductive and respiratory syndrome virus		
Seneca Valley virus		
Transmissible gastroenteritis virus		
Vesicular exanthema		

The second case was confirmed in a dairy cow in a herd of 51 head on January 30, 1929, six miles from the initial case. Guards were placed immediately on the premises and 7-foot high chicken-wire fence was placed around the site. It was noted that the erection of a fence was an important strategy to keep observers out and farm staff contained. The source of infection was most likely a traveling sack buyer.

The third herd was adjacent to Herd Two. Symptoms appeared on February 2. A fourth herd of 33 head of cows was identified February 6. The source was likely a calf buyer. A fifth herd of 67 cattle was identified February 16, and the source was the same calf buyer. All other dairies the calf buyer visited were quarantined, and no other cases occurred.

In cases 2-5, animals were appraised, slaughtered, and buried the same day. All five herds were within a radius of 5 miles wide and 15 miles long, resulting in a closed quarantine of 18 × 22 miles. Quarantine had little effect on market.

The last confirmed case in Mexico occurred in 1946-54 and in Canada in 1952. Stringent importation regulations such as the Tariff Act of 1930 on animals and animal products have effectively protected the FMD-free status of the United States.

Foot-and-mouth disease can spread when infected animals bring the virus into physical contact with susceptible animals (i.e., divided hoof animals). An outbreak can also occur when susceptible animals are held in contaminated facilities; are moved in contaminated vehicles; are fed raw or improperly cooked garbage containing infected meat or animal products; have contact with people wearing contaminated clothes or shoes or using contaminated equipment; are exposed to contaminated materials such as hay, feed, hides, or biologic products; drink contaminated water; or are inseminated by semen from an infected animal.[2]

Hayes, et al. examined the economy wide impacts of the elimination of export markets for 10 years for beef and pork because of a foreign animal disease such as FMD.[3] The loss of export markets would create a surplus of meat on the domestic market, with significant price reductions throughout the agri-food system. As a result, price reductions in competing proteins such as chicken, eggs and cheese would occur, feed-grain use would fall and employment in the US livestock sector and its affiliated industries would suffer. Second round impacts would include a reduction in the US trade balance and in rural employment. The cumulative impact on both sectors over the 10-year period is $128.23 billion, an average of $12.9 billion per year. The annual jobs impact of a $12.9 billion reduction in industry revenue is 58,066 in direct employment and 153,876 in total employment.

Schroeder, et al. examined the economic impacts of alternative FMD disease management strategies and assessed the economic impacts of alternative FMD vaccine deployments on producers, consumers, and the government.[4] The study concluded that because FMD is highly contagious, and animal health officials must be prepared with disease mitigation plans in place. In their simulations, government costs would total $11 billion if control relied exclusively on movement control of animals, biosecurity, and stamping out with no emergency vaccination program. A viable and aggressive vaccination strategy, on the other hand, would have the ability to potentially reduce government costs to $1.1 billion in the event of an FMD outbreak.

Classical swine fever (hog cholera) history

Hog cholera was the first major hog disease affecting pigs in the US. The first outbreak was reported in Ohio in 1833.[5] It was reported that there were only 10 outbreaks from 1833-1845 but occurring in 10 states; however, from 1846-1855 there were 95 outbreaks reported in 13 states. By 1887, hog cholera was reported in 35 states. Outbreaks in 1886, 1887, and 1896 each killed more than 13% of the nation's hogs, and more than 10% died during a 1913 outbreak. The USDA has estimated that from 1884-1913, annual mortality from hog cholera was 7.5% with six of those years over 10%. The last major epizootic was 1912-1913. The virus was identified by Dorset of the USDA's Bureau of Animal Industry in 1903. Successful preventative measures involving both anti-hog cholera serum and the virus were successfully tested in 1907 with the combination giving lifelong immunity in most hogs. In 1927, a swine sanitation system named for McLean County Illinois became a valuable adjunct to immunization. Even though this system was developed primarily to prevent roundworm infestation, this system reduced the fecal-borne intestinal diseases and dramatically reduced serious vaccine side effects.

In 1949, a variant of the hog cholera virus was identified clarifying why some of the new vaccines did not confer immunity. Researchers in Iowa identified that the virus remained viable in carcasses for up to 11 weeks during winter conditions. Numerous studies also identified house and stable fly transmission with prevalence of biting flies and incidence of hog cholera closely connected. An effective MLV vaccine was developed by Baker et al. in 1951.

In 1950, a resolution to eradicate hog cholera and endorsing formulation of a representative committee aimed at nationwide eradication was put forth. Committee was formed in 1951. Eradication was mandated by law in 1961 and in 1963 regulatory officials prohibited interstate movement of virulent viruses or of breeding stock or feeder pigs vaccinated with MLV vaccine. Studies were conducted in Florida, Iowa, and Georgia in the early 1960s using killed vaccine. In 1963, a rapid diagnostic test for hog cholera was developed at NADC in Ames, Iowa and contributed greatly to the eradication success. Classical swine fever was eradicated from the US on January 31, 1978. This was 99 years after the USDA began hog cholera research and 17 years after the start of the federal-state eradication program.

Recently, hog cholera cases have occurred in Japan. It's been 26 years since Japan reported its last hog cholera case in 1992 and declared the virus eradicated in 2007. Wild boars are suspected to be the source of infection.

Pseudorabies (PRV; Aujeszky's disease) history

Aujeszky's disease (pseudorabies) was first described in 1813 in cattle.[6] The terminology "pseudorabies" was first used in Switzerland in 1849 because symptoms in cattle were similar to those of rabies. Aujeszky's was confirmed as a nonbacterial in 1910 by Schmiedhofer, and in 1934 Sabin and Wright identified it as a Herpesvirus. Prior to 1960, PRV was not considered of economic importance in the US and the world. Subsequently, the virus emerged as an important disease domestically and globally. Several reasons have been postulated for the increase in severity, prevalence, and transmission including an increase in virulency, a decline in hog cholera which may have been misdiagnosed, and the increase in total confinement with continuous farrowing and flows.

The first meeting discussing eradication was held in Peoria, Illinois with a national symposium during the Livestock Conservation Institute's annual meeting. By 1978, 40 states had a basic PRV control program involving case reporting and quarantine, and in May 1979 APHIS issued the final draft of federal rules. In 1981, ISU announced development of a "subunit" vaccine that differentiated field virus from vaccine. In 1986, the first gene deleted vaccine was licensed. Some states utilized split or regional status to accelerate the program. Funding, differential vaccines, and quarantine of animal movements resulted in successful PRV eradication.

In April 2005, the National Institute for Animal Agriculture announced successful PRV eradication.

Estimates of the value of economic and welfare impacts with PRV program with an assumed parallel supply-curve shift were determined: consumers gained $336.5 million, producers gained $35.9 million, government expenditures were $197.1 million, and the benefit-cost ratio of the program was 89:1.[7]

Vesicular exanthema

In 1932, clinical disease first thought to be FMD occurred in a garbage-fed hoop in California.[8] In 1934, this was determined to be vesicular exanthema (VE). It remained confined to California for 20 years with 95% of cases associated with garbage feeding. In 1952, a transcontinental train from California left dining garbage at Cheyenne, Wyoming which was fed to swine. These swine moved through stockyards and within two months, the disease was reported in 19 states by 1953. A total of 42 states and District of Columbia were infected with most cases associated with garbage feeding. An eradication program was launched and was highly successful. The last case was detected and eliminated in 1956 in New Jersey. VE was declared eradicated in 1959 and is now considered a foreign animal disease.

Individual herd elimination

Porcine reproductive and respiratory syndrome

PRRSV elimination occurs with herd virus exposure, vaccine, herd closure, restrictive pig movements, and unilateral pig flow.[9]

Transmissible gastroenteritis virus

TGE was first reported by Doyle and Hutchings in 1946, but it likely was present well before.[10] TGE has been eliminated by exposing the entire sow herd and the replacement gilts (12 weeks) to the virus to stimulate the immune response and removal of suckling pigs. After the herd is exposed to the virus, the agent is eliminated through the use of all-in/all-out pig flow, thorough cleaning and disinfecting of the facilities, and herd closure for 12 weeks.[11]

Porcine endemic diarrhea virus

PEDV entered the US swine population in spring of 2013 with a similar strain to endemic PEDV in China.[12] During 2013-14 the virus spread quickly through the industry. This virus can be eliminated by entire herd exposure including 12 weeks of replacements, wean down of suckling piglets, intensive sanitation, and herd closure for 12 weeks. In pig flow, PEDV is eliminated through the use of all-in/all-out site flow with thorough cleaning and disinfecting of the facilities.

Porcine deltacoronavirus

PDCoV entered the US swine population shortly after PEDV. This virus can be eliminated by herd exposure, wean down of suckling piglets, intensive sanitation, and herd closure for six weeks.[13] After the herd is exposed to the virus, PDCoV is eliminated through the use of all-in/all-out site flow with thorough cleaning and disinfecting of the facilities.

Influenza A virus – swine

Influenza A virus – swine has become increasingly problematic resulting in emerging strains through drift, shift, and reassortment. IAV-S can be eliminated in six weeks through a strategic combination of homologous prefarrowing vaccination, restricted movement of piglets and nurse sows, and unidirectional pig flow.[14] Prevention of new infections in swine dense areas has proved to be unsuccessful.

Seneca Valley virus

Seneca Valley virus was unknown until identified in cell culture in 2002.[15] Significance of SVV is vesicles that must be differentiated from FMD. This virus can be eliminated by combination of herd exposure including growing gilts, closure (12 weeks), restricted suckling pig movements, and unidirectional flow of piglets.[16]

Actinobacillus pleuropneumoniae

APP is of lower prevalence today in US herds due to the recognition of the economic significance but remains very common in most pig producing countries. APP is eliminated with new APP-free stockings or depopulation of existing herds.[17] The bacteria can also be eliminated through MMEW, MEW, and then repopulation of the breeding herd. Certain strains have been eliminated by a combination of vaccination, gestation barn segregation, medication, early weaning, and unidirectional flow.

Bordetella bronchiseptica and *Pasteurella multocida* toxigenic

Numerous reports of elimination primarily in small populations have used vaccine, medication, and test and removal. Most common methods are depopulation and repopulation and/or establishment of new herds from negative sources.[18]

Brachyspira hyodysenteriae

Brachyspira hyodysenteriae can be eliminated by depopulation, strategic medication with intensive sanitation and rodent control, and unidirectional pig flow.[19]American Association of Swine Veterinarians and the National Pork Producers developed and published strategic programs that detail the procedures used for elimination of swine dysentery. Reemergence in recent years should be noted and a program for industry elimination developed.

Leptospira sp.

Entry source is generally through wildlife contaminated water sources. *Leptospira* is an uncommon occurrence in modern confinement production. Economic costs can be very high through reproductive losses and carcass contamination. Elimination can be successful through intensive breeding herd medication combined with individual mass therapy, segregated production, intensive sanitation, and unidirectional pig flow.[20]

Parasites

Sarcoptes scabei

Eradication has been widely successfully with injections or oral administration of ivermectins under the supervision of veterinarians. Key attributes of the program include simultaneous whole-herd two-time treatment at a recommended interval, segregation of offspring or two-time treatment, and early marketings to minimize cost and adhere to withdrawals. Economic cost of mange on the American swine industry was estimated to be $230 million annually.[21] Endemic herds should be identified and elimination completed followed by surveillance that would validate eradication.

Common internal parasites

Common parasites of swine include *Ascaris suis* (roundworm), *Stephanurus dentatus* (kidney worm), *Trichuris suis* (whipworm), *Oesophagostomum* spp., and *Metastrongylus elongatus* (lung worm). All of these are eliminated with modern confinement housing except *Ascaris suis*.

Summary

This paper is a partial review of the diseases and/or pathogens that have been eradicated from the US swine industry or eliminated from the individual herds. The bioexclusion of a pathogen originates with the economic recognition of the disease, availability of diagnostic tools, availability of biological and therapeutic interventions, and defined programs with producer, veterinarian, and government cooperation. Industry eradication is a culmination of intensive efforts of veterinarians, producers, diagnostic laboratories, researchers, biological and pharmaceutical companies, and state and federal authorities. Bioexclusion frequently initiates with individual herd strategies modified over time as interventions and diagnostics change, supported by veterinary and producer persistence ultimately moving to area, regional, and finally to country eradication. It is important to note the commonalities of pathogen introduction from the history of hog cholera, foot-and-mouth disease, Aujeszky's, and vesicular exanthema to today. These are transmissions through contaminated garbage feeding, animal movement into herds without quarantine and diagnostics, transport and collection point contamination, and people entry. Our introduction risks today are similar but increasingly well managed by the industry, but with globalization there are new risks that must be prudently managed. Our industry has to move to active surveillance for the pathogens that will disrupt our export market. The barriers to active surveillance can be overcome, and management of the short-term effect will pale in comparison to large area containment and eradication.

References

1. Mohler JR, Snyder, R. The 1929 Outbreak of Foot-and-Mouth Disease in Southern California. USDA Miscellaneous Publication No. 68. 1930.

2. USDA APHIS. Veterinary Services Factsheet: Foot-and-Mouth Disease. 2013. Available at **aphis.usda.gov/publications/animal_health/2013/fs_fmd_general.pdf**. Accessed 21 December 2018.

3. Hayes et al. Economy wide impacts of a foreign animal disease in the United States. Food and Agricultural Policy Research Institute at Iowa State University. 2011.

4. Schroeder et al. Economic impact of alternative FMD emergency vaccination strategies in the Midwestern United States. Kansas State University of Agricultural Economics Extension Publication. 2015.

5. Hog Cholera and Its Eradication: A Review of US Experience. USDA APHIS. 1981.

6. Lowell AA et al. Pseudorabies (Aujeszky's disease and its eradication) a review of the US experience. USDA Technical Bulletin No. 1923. 2008.

7. Miller G et al. Benefit-cost analysis of the national pseudorabies virus eradication program. *J AM Vet Med Assoc.* 1996; 208(2):208-213.

8. Smith AW, Akers TG. Vesicular exanthema of swine. *J AM Vet Med Assoc.* 1976;169(7):700-703.

9. 2003PRRS Compendium Producer Edition. National Pork Board. 2003.

10. Doyle, LP, Hutchings, LM. A transmissible gastroenteritis in pigs. *J Am Vet Med Assoc.* 108:257-259.

11. Wiseman B. TGE eradication, without depopulation. *Int'l Pigletter.* 1989; 9(2):5-6.

12. Geiger JO, Connor JF. Porcine diarrhea, diagnosis, and elimination. AASV. 2013.

13. PDCoV. Lower et al. 2016. Personal communication.

14. Connor et al. Management and challenges of dealing with swine influenza. Allen D. Leman Swine Conference. 2013; 167-168.

15. Seneca Valley Virus. Swine Health Information Center. 2015. Available at **swinehealth.org/wp-content/uploads/2016/03/Seneca-Valley-virus-fact-sheet.pdf**. Accessed 21 December 2018.

16. Johnson et al. 2017. Personal communication.

17. Lowe JF. *Actinobacillus pleuropneumoniae* eradication in a large breeding herd. ISU Swine Disease Conference for Swine Practitioners. 2009.

18. 17. Lowe JF. *Actinobacillus pleuropneumoniae* eradication in a large breeding herd. ISU Swine Disease Conference for Swine Practitioners. 2009.

19. Harris, DL. The epidemiology of swine dysentery as it relates to the eradicaiton of the disease. *Squibb International Swine Update.* 1982; (1).

20. Tesic M et al. Applying the Cost-Benefit method in making a program for the eradication of leptospirosis on a pig farm. Institute for Veterinary Medicine, Yugoslavia.

21. Arends J. Managing mange, losses to the swine industry may reach $230 million annually. *Lg Animal Vet.* 46:6-10.

What's next for disease elimination: *Mycoplasma hyopneumoniae*, PRRS, PEDV or ileitis?

Paul Yeske, DVM, MS
Swine Vet Center, St. Peter, Minnesota

Introduction

The swine industry has a good reputation for disease elimination over the last 50 years as Dr. Connor has detailed. The elimination of Swine Fever (Hog Cholera) and Pseudorabies virus (PRV) are examples that come to mind right away. These are national programs that have been completed which have made our industry more successful.

There are many ways to eliminate diseases from herds. The method used is dependent on the pathogen to be eliminated, the cost the producer is willing to spend and the risk that the producer is willing to live with.

The most reliable and expensive way is depopulation and repopulation. This method has been used when you want to be sure that the disease is eliminated and have negative stock available to repopulate with. This method was used extensively in the 1990's as herds began to expand and had many health challenges. This was an effective way to get rid of multiple pathogens at once and have the herd on a high productive track in a relatively short amount of time. This was not only successful for disease elimination but also production concerns as well, such as the need to change genetics or unbalanced parity distribution. To quote Dr. Al Leman in an AASV preconvention workshop on depopulation and repopulation, "There is such a benefit we should just plan on this every 5 years and build it into the budget for the farm". This is still the most successful and fastest method of disease elimination.

Medication and herd closure are the most common forms of disease elimination that are frequently used today. Just as with the depopulation repopulation method, multiple diseases can be eliminated at the same time such as PRRS, *Mycoplasma hyopneumoniae* and PEDV. There are many different variations of the medications and closures that have been customized to fit the needs of a specific farm based on pathogens present, the producer's risk tolerance, timeline and how much money the producer is willing to spend.

Methods and pathogens

What diseases should be next for national programs of elimination? *Mycoplasma hyopneumoniae*, PRRS, PEDV or ileitis. Other very good candidates include TGEV, PDCoV, swine dysentery and APP just to name a few. Over the next 50 years I'm sure there will be many more that will have good potential.

Mycoplasma hyopneumoniae elimination has probably been the most successful on a per herd basis in the US and has been done in some of the European countries (Switzerland and Finland); there is no reason it could not be done for the entire US population.[1] The longest track record of success has been using the herd closure of at least 240 days and various medication programs. It is similar to the programs used in Europe except there is no break in farrowing for a period of time. 165,000 sows have been done with an approximate success rate of 81% on average over 4.6 years (Figure 1).[2] Today the numbers continue to increase and would be close to 300,000 sows. *Mycoplasma hyopneumoniae* elimination has a very rapid return on investment with it taking 5 months to repay the investment.[3] There are protocols for just medication alone that have been successful but just at a lower rate of 63%.[2] Today some farms are doing a hybrid of these two programs using an extended but shorter than a 240-day (150-180 days) closure followed by injectable medication to the whole herd with good success. This is one that could be done with minimal effects on sow performance for all farrow to wean farms. Herds that present bigger challenges are farrow to feeder or farrow to finish or part finish. This could be done with some coordination and many producers are currently choosing to do so for economic reasons.

PRRS has been successfully eliminated from many herds multiple times. The technology to eliminate is very well known and we have the tools to do this whether using herd closure, depopulation and repopulation or other methods. The biggest challenge is keeping the herds from rebreaking. If the industry were able to

Figure 1: Survival analysis of *Mycoplasma hyopneumoniae* elimination by the method of elimination.

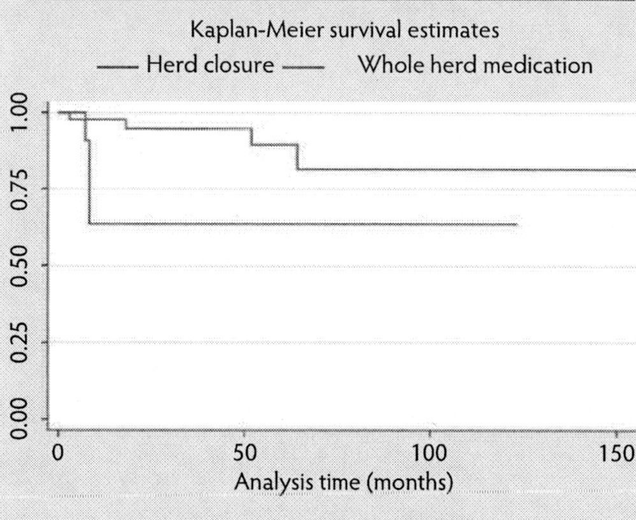

eliminate this, it would take this concern and risk away. Although the incidence is high, I still believe that it is possible to do an elimination program (Figure 2).[4] There is certainly an economic incentive to eliminate the disease due to the dramatic cost to the industry, a total of $580 million per year or $1.6 million per day each year.[5] When this is totaled up there could be a large investment made into the process. The biggest challenge is getting the industry to move forward on this with a practical end date in mind. Many projects have made progress to only slide back again, but, with a coordinated effort it should be possible to accomplish. We just need to have the will as an industry and swine veterinarians to do it.

PEDV is a disease of relative low incidence (Figure 3) and is a candidate to eliminate on a national basis.[4] The biggest challenge will be in getting producers to stop ongoing exposure to negative animals to maintain immunity in herds. One of the concerns has been that even if we were successful at elimination, what would keep 2013 from repeating itself? The recent industry changes in relation to feed additives and other products in relationship to African Swine Fever have increased the awareness and systems have started to adapt hold times on these products for specified times and temperatures. This would also help with some of the late finishing breaks once harvest plants are infected in the winter and spread back to the country. With this information and the cleanup of currently infected herds, if timed well could be done and be one less concern for the swine industry to deal with and cost for affected producers to bear.

TGEV is another disease in which the incidence is down dramatically since PEDV came into the population and today has relatively low incidence.[6] Now that the industry is at this low of a level, why wouldn't we want to eliminate TGEV and again have one less thing to deal with.

Ileitis may be a possibility since good immunity can be established with natural exposure. There is a possibility with herd closure and cleaning up the environment that it could be done. Looking at the cost of prevention and loss of performance, there is motivation to look at the possibilities.

There are many other "low incidence" diseases that can be eliminated as well. Lice and mange are practically done but still show up in some herds. Many herds have eliminated the more pathogenic strains of APP so there would be a relatively low number of herds for the 1, 5 and 7 strains.

Figure 2: National PRRS incidence/prevalence July 2018 – June 2019

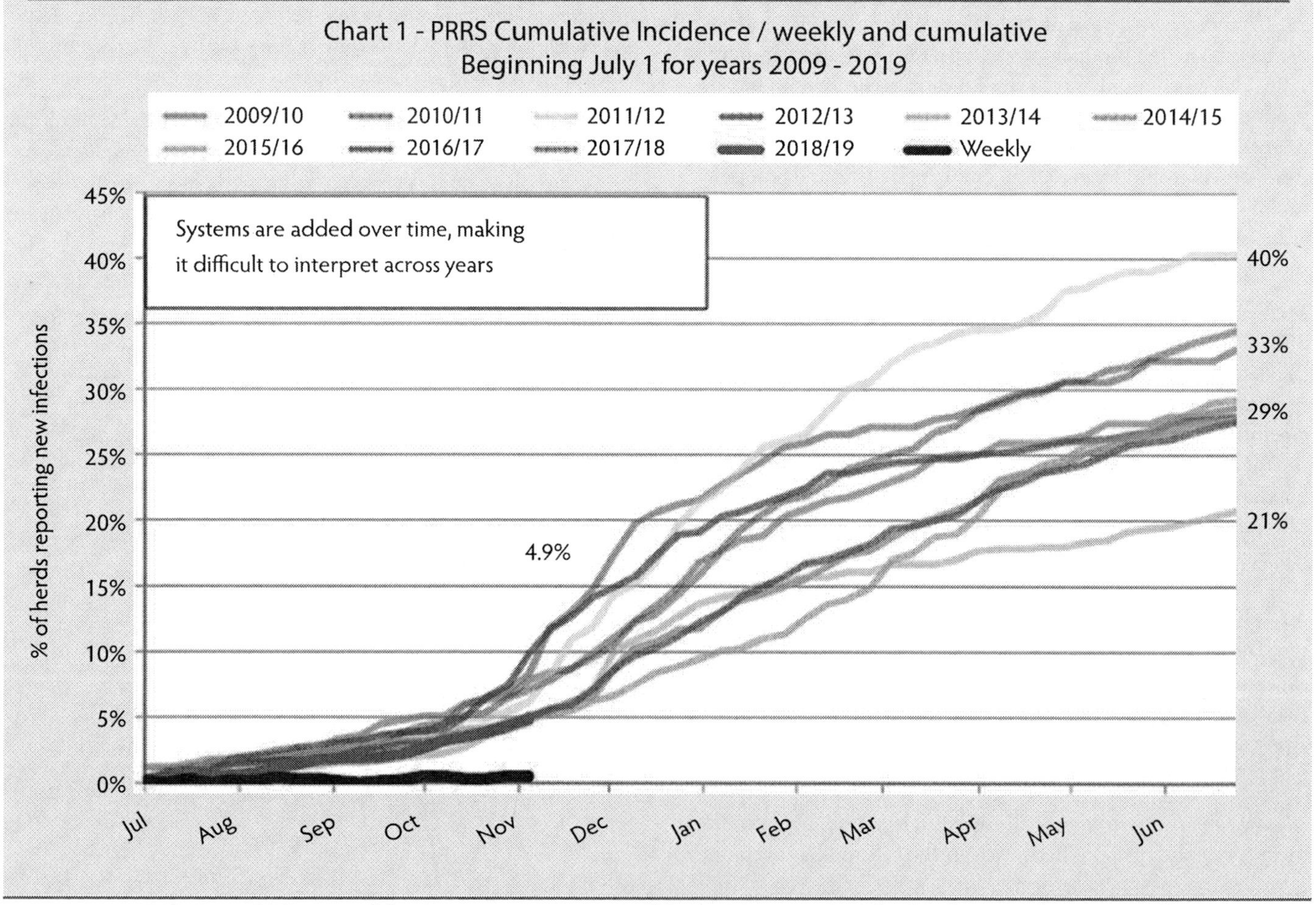

Conclusions and summary

As we continue to have better diagnostics and diagnostic methods. Just look what we have learned from processing fluids that we didn't know a couple of years ago. Understanding where the diseases are being maintained in our herds will help stop the transmission of disease. We have also learned that exposing all the gilts into a herd can reduce the impact of *Mycoplasma hyopneumoniae* in the herd by using better immunity and less shedding to offspring. We will be able to track disease in herds better, have a better understanding of the disease ecology, and can develop new plans and strategies for elimination.

The multi-site production model with farrow to wean sow farms, and ideally all in all out by site wean to finish flows has made the opportunity for disease elimination much more possible. No matter which method is used, depopulation and repopulation or herd closure, all are much more difficult if more production is on one site. As new antibiotics and vaccines come along we will only be limited by our own creativity on what can be done. The industry will need to have the courage to step outside the box and try things that we haven't done.

Removal of these diseases would not only make our industry even more competitive due to the increased productivity and lower cost of production but would also eliminate some potential non-tariff trade barriers as well. With the continued pressure to reduce the use of antibiotics in production, more disease elimination will be helpful in our efforts to have higher health animals. By eliminating many of these pathogens there should be less need for the use of antibiotics.

This will result in the most efficient and best performance, allowing the full genetic potential of animals to be expressed. This will allow us to continue to take advantage of size and scale and the resources we have available to be long term sustainable - to grow more with less - permitting our industry to remain one of the most efficient and cost competitive in the world.

References

1. Holst S, Yeske P, Pieters M. Elimination of *Mycoplasma hyopneumoniae* from breed-to-wean farms: A review of current protocols with emphasis on herd closure and medication. J Swine Health Prod. 2015;23(6):321–330.

2. Yeske P. A survival analysis of *Mycoplasma hyopneumoniae* elimination: Leman Conference 2017.

3. Linhares D. A field study on economics of *M. hyopneumoniae* elimination: 2017 James D McKean Swine Disease Conference, 2017.

4. Corzo C. Bob Morrison Swine Health Monitoring Project: November 9, 2018

5. Holtkamp D. Economic impact of PRRS infection to US swine industry: Check off funded research 1997-2016. 2016.

6. Main R. Offense from a diagnsotic prespective: 2018 Jame D McKean Swine Disease Conference, 2018.

Figure 3: National Porcine Enteric Coronavirus incidence/prevalence July 2018 – June 2019

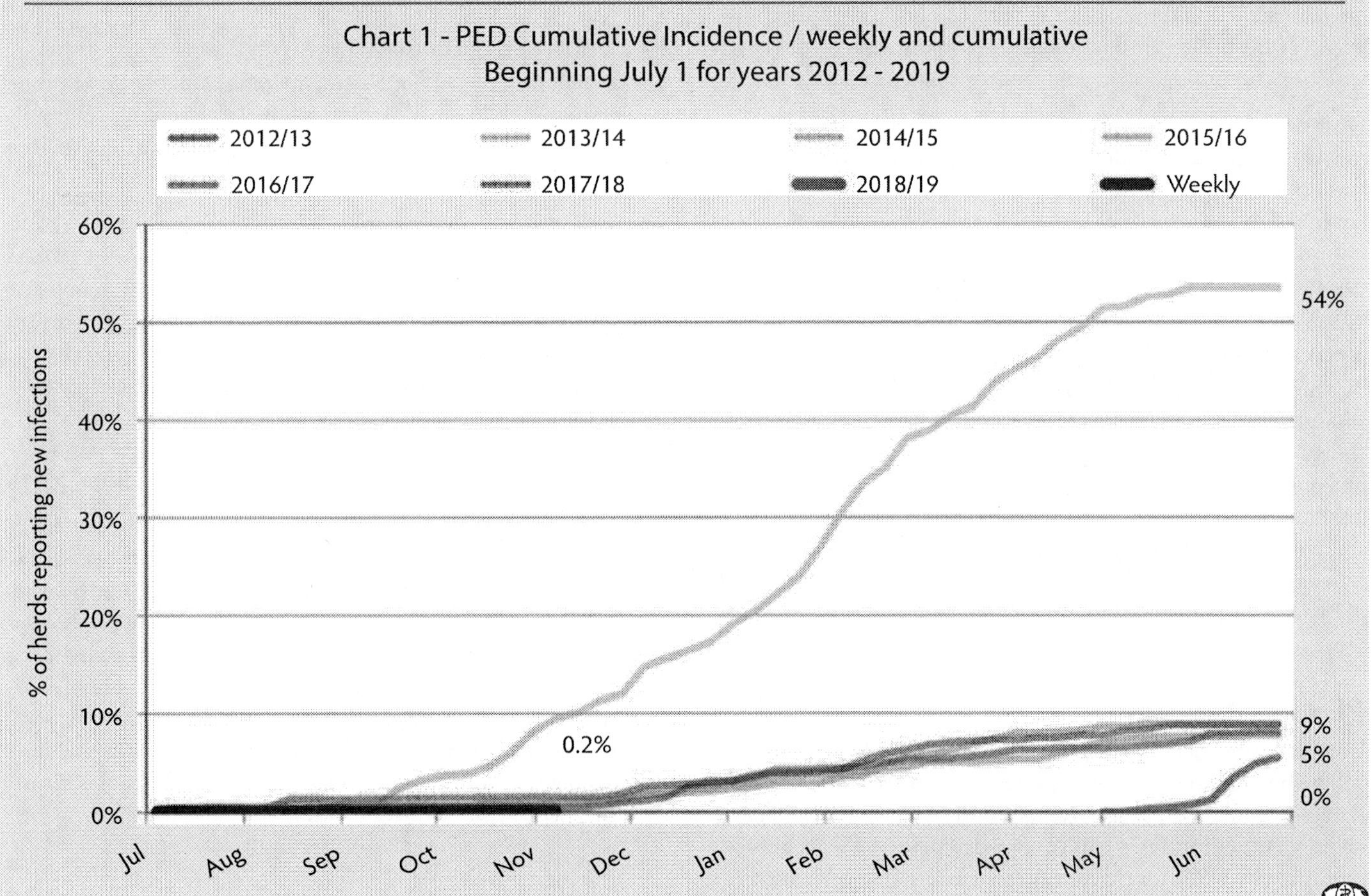

When "it" hits the fan: Will we be prepared?

Patrick Webb, DVM

Introduction

One similarity between the 2018 African swine fever outbreak in China and the 2001 foot-and-mouth disease outbreak in the United Kingdom is how each incident grabbed the attention of producers in the United States. The outbreak in the UK created a sense of urgency in the US for preparedness. Emergency planning positions were quickly set in place at both the state and federal levels and a concerted effort towards increasing foreign animal disease (FAD) preparedness began. Where are we now?

Movements, exports and money

In 2001, the pork industry harvested 98 million head[1] and moved 26.9 million head across state lines for production.[2] Exports of muscle cuts accounted for 8.1 percent of production and the total value of pork and pork products returned to producers was $16.37 per head.[3] In 2017, the industry harvested 121.3 million head[4] and moved 55.2 million head across state lines for production.[5] Exports (including variety meats) accounted for 26.6 percent of production, with an average return of $53.47 per head back to producers.[6] These numbers do not include intrastate movements within production systems or numbers of pigs moving on swine health production plans (commuter agreements), which are becoming increasingly more popular.

From an emergency response perspective, the increase in pig movements presents challenges for disease control, trace-in and trace-out activities, and animal welfare when stop movements are enacted. The growth of exports and their value, increases the industry's exposure to significant financial losses if exports and commerce are disrupted by a foreign animal disease outbreak.

Workforce

The number of state and federal individuals focused solely on emergency planning for FADs has fluctuated based on the availability of funding for those positions. In 2001, almost every state had one person hired or designated to focus on preparedness. As urgency and funding decreased, so did the percentage of people and resources focused on this topic. There has been a lot of attrition among response planners over the years resulting in the loss of historical perspective in more current conversations leading to some retreading of old ground.

Plans

Even in times where FAD's were not the "hot topic", progress was still being made on planning procedures. When I started as the Foreign Animal Disease Program Coordinator at Iowa's Department of Agriculture and Land Stewardship, the FAD Red Books published by USDA were considered the response bible used to develop state plans. These books covered disease containment, depopulation, and disposal. However they did not cover in detail what the industry would consider necessary for business continuity. Over the past several years, USDA's Veterinary Services has updated, expanded and written new plans which have been valuable to the process of developing industry strategies and plans. These efforts have been an important step forward for response planning, especially the development of the Secure Food Supply Plans.

The development of site specific response plans needs more focus from pork producers. The secure pork supply plan provides tools for producers to address business continuity when officials are ready to start permitting movements. Producers also need to have plans to address animal welfare during stop movements, depopulation and carcass disposal. To be successful producers need to engage animal health officials so they understand the parameters for plan development. Response strategies for packers and processors have been developed and reviewed on a limited basis by officials and the packing industry. More work is necessary to gain broader acceptance.

Pre-harvest traceability

The current lack of a nationally-mandated pre-harvest traceability system, that includes mandatory premises registration, complicates response activities especially when considering diseases that affect multiple species.

Specific to premises registration, unless a State has made it mandatory, officials have to rely on producers voluntarily registering their premises. One of the challenges of a voluntary system is officials are lacking an accurate count of active livestock premises, therefore increasing the time it would take to understand the scope of an outbreak in a disease control area. Premises registration in the commercial pork industry has become a standard process due, in part, to packer requirements for nationally-standardized premises identification numbers (PIN) as a condition of sale.

Related to movements, inshipments and commuter agreements require health papers and movements to be reported to officials. Intrastate movements and movements to harvest are not tracked by officials. For these movements, officials rely on the producer to be the authoritative source of that information. Traceability data needed by officials is held in disparate databases and variable formats which hinders the rapid sharing of movement information with officials if a disease outbreak were to occur. To increase the speed and efficiency of tracing movements in and out of a control area, this issue must be further addressed.

Labs, diagnostics and surveillance

In 2002, funding was provided to establish the National Animal Health Laboratory Network (NAHLN). That network has grown and is filling an important need related to FAD surveillance and response capabilities. Efforts underway to expand sample types officially approved for FAD detection in NAHLN laboratories are also positive steps to enhance preparedness. The implementation of the classical swine fever surveillance program has been invaluable. More work is needed, however, to design and implement active surveillance programs for African swine fever and foot-and-mouth disease.

Depopulation and carcass disposal

The increased size of swine operations, and thus the number of pigs located on a premises, presents a tremendous challenge for depopulation and the disposal of carcasses in the event of an FAD. The American Veterinary Medical Association has developed guidance documents for euthanasia and depopulation that can inform on potential methods. There are also a variety of guidance documents and resources for managing mortalities in emergency situations. Carrying out these activities in a timely manner, however, is complicated and/or nearly impossible as the size and number of animals increase. The pork industry still has work to do to address this important issue. Producers will need to work with officials on developing workable plans and strategies.

Vaccine

In 2001, mentioning vaccination as a tool for responding to an FAD outbreak was almost sacrilegious because it affected the ability of a country using vaccine to regain lost export markets. Over time, international markets became more favorable to the use of vaccines for FAD control. Today, vaccines for foot-and-mouth disease and classical swine fever are considered important tools that could be used in an outbreak. One of the current challenges we face is the need for major restructuring and funding of a US FAD vaccine bank, to ensure the pork industry has enough vaccine to mount an effective vaccination campaign.

Data and technology

Data and information needed by officials to mount a rapid and more effective response to an FAD outbreak exists within disparate databases held by multiple private and governmental entities. Unfortunately, the data is not standardized and not easily shared which slows a response. The swine industry is investing in data standardization and the development of a database and dashboard to allow producers to share data in a rapid, efficient and secure way with animal health officials.

Summary

Over the years, emergency preparedness has improved, however there are still gaps which must be addressed. The 2018 outbreak of ASF in China resulted in the U.S. swine industry pushing for increased preparedness at the federal, state and local levels. We must strike while the iron is hot and take advantage of the current momentum. Over time, the urgency will disappear as newer issues catch people's attention. FAD preparedness is a major priority for the National Pork Board, the National Pork Producers Council, the American Association of Swine Veterinarians and the Swine Health Information Center.

References

1. Agricultural Statistics Board. Livestock Slaughter 2001 Summary. NASS USDA. 2002; Mt An 1-2-1 (02):1.
2. Shields DA Mathews KH. Interstate Livestock Movements. ERS USDA. 2003; LDP-M-108-01:10.
3. Meyer SR. Kerns and Associates. Personal communication. 14 November 2018.
4. Agricultural Statistics Board. Livestock Slaughter 2017 Summary. NASS USDA. 2018; ISSN: 0499-0544:6.
5. Agricultural Statistics Board. Meat Animals Production, Disposition, and Income 2017 Summary. NASS USDA. 2018; ISSN: 0748-0318:12.
6. National Pork Board. 2018. Available at www.pork.org/news/u-s-pork-exports-set-new-volume-records-2017. Accessed 10 November 2018.

American Association of Swine Veterinarians
830 26th Street
Perry, IA 50220-2328
USA

CURRAN ASSOCIATES INC.
proceedings.com

ISBN 978-1-5108-8324-6
9 781510 883246